Trace Elements in Waterlogged Soils and Sediments

ADVANCES IN TRACE ELEMENTS IN THE ENVIRONMENT

Series Editor: H. Magdi Selim

Louisiana State University, Baton Rouge, USA

Trace Elements in Waterlogged Soils and Sediments
edited by Jörg Rinklebe, Anna Sophia Knox, Michael Paller

Phosphate in Soils: Interaction with Micronutrients, Radionuclides and Heavy Metals
edited by H. Magdi Selim

Permeable Reactive Barrier: Sustainable Groundwater Remediation
edited by Ravi Naidu and Volker Birke

Trace Elements in Waterlogged Soils and Sediments

Edited by
Jörg Rinklebe
Anna Sophia Knox
Michael Paller

CRC Press
Taylor & Francis Group
Boca Raton London New York

CRC Press is an imprint of the
Taylor & Francis Group, an **informa** business

CRC Press
Taylor & Francis Group
6000 Broken Sound Parkway NW, Suite 300
Boca Raton, FL 33487-2742

Printed on acid-free paper
Version Date: 20160309

International Standard Book Number-13: 978-1-4822-4051-1 (Hardback)

Library of Congress Cataloging-in-Publication Data

Names: Rinklebe, Jèorg, editor. | Knox, Anna Sophia, editor. | Paller, Michael H., 1951- editor.
Title: Trace elements in waterlogged soils and sediments / edited by Jèorg Rinklebe, Anna Sophia Knox, and Michael H. Paller.
Description: Boca Raton : Taylor & Francis, 2017. | Series: Advances in trace elements in the environment | "A CRC title." | Includes bibliographical references and index.
Identifiers: LCCN 2016010913 | ISBN 9781482240511 (hardcover : alk. paper)
Subjects: LCSH: Soils--Trace element content. | Contaminated sediments. | Sediments (Geology) | Trace elements. | Waterlogging (Soils) | Wetland ecology.
Classification: LCC S592.6.T7 T735 2017 | DDC 631.4--dc23
LC record available at http://lccn.loc.gov/2016010913

Visit the Taylor & Francis Web site at
http://www.taylorandfrancis.com

and the CRC Press Web site at
http://www.crcpress.com

Printed and bound in the United States of America by Publishers Graphics, LLC on sustainably sourced paper.

Contents

Preface

The intention of the authors of this book is to provide up-to-date knowledge on trace elements in waterlogged soils and sediments because these environments act as sinks for contaminants, especially trace elements. Water and associated sediments are precious resources that support more than half of the world's animal and plant species. Therefore, protection of water and sediment quality is highly important. Rapid industrialization and urbanization have led to the worldwide contamination of waterlogged soils and sediments with trace elements. These contaminants originate from point and nonpoint sources. Point sources of trace element contaminants are specific and identifiable, including municipal sewage treatment plants, overflow from combined sanitary and storm sewers, stormwater discharges from municipal and industrial facilities, and waste discharges from industry. Nonpoint sources include stormwater from hazardous and solid-waste sites, runoff from cropland, livestock pens, mining and manufacturing operations, and storage sites; and atmospheric deposition. Nonpoint source pollution cannot be traced to a specific spot, making effective remediation especially difficult.

The understanding of fundamental principles and phenomena that control the transfer of trace elements in the soil/sediment–plant–consumer chain can contribute to protection of the environment and human health. Compared with terrestrial environments, the geochemistry and biogeochemistry of trace elements in aquatic and semiaquatic environments is more complex and less understood. This book brings together current knowledge about trace elements in waterlogged soils and sediments. Factors controlling the dynamics and release kinetics of trace elements and the underlying biogeochemical processes are discussed. Also included are current technologies for remediating contaminated sites with trace metals and the role of bioavailability in risk assessment and regulatory decisions.

The book is composed of three sections. Section I (Chapters 1 through 10), "Understanding, Processes, and Needs," provides fundamental knowledge concerning trace element geochemistry in waterlogged soils and sediments. The crucial role of redox-driven processes for mobilization/immobilization of trace elements in waterlogged soils and sediments, which are mainly caused by different flood-dry cycles, are discussed, including the chemistry of the redox-sensitive elements iron, sulfur, and selenium. Sorption–desorption interactions of trace elements such as copper, lead, nickel, and zinc are also presented. The potentially toxic trace elements arsenic, cadmium, chromium, uranium, and zinc as well as their speciation and release kinetics in waterlogged soils receive special attention in this section of the book. Nanoparticles and rare earth elements, as emerging topics, are also included in this section.

Section II (Chapters 11 through 16), "Bioavailability," provides detailed information on the bioavailability of trace elements in the aquatic and semiaquatic ecosystems, with an emphasis on waterlogged soils and sediments. Contaminated sediments pose a potential risk to the environment and human health because they release recalcitrant chemicals that can harm organisms and enter aquatic food chains that may lead to humans. However, only the contaminant fraction that is available for biological uptake has the potential to cause human health or ecological risks. Bioavailability is the degree to which chemicals present in the soil or sediment may be absorbed or metabolized by human or ecological receptors or are available for interactions with biological systems. Bioavailability has rarely been used in risk assessment and regulatory decisions, largely because of uncertainties in our fundamental understanding in this area.

Section III (Chapters 17 and 18), "Remediation," discusses the remediation of metal-contaminated sediments. Although contaminated sediment poses difficult assessment and remediation challenges, recent advancements have improved our ability to accurately estimate the risks posed by contaminated sediments and effectively control them. Improved in situ containment and remediation methods have enhanced our ability to effectively manage sediments at reduced costs and minimize environmentally intrusive remedial actions. This section of the book discusses contaminants in marine and freshwater sediments and current technologies for remediating contaminated sediments. It also gives an overview of remediation options for metal-contaminated sediments and specifically explains passive and active capping for the in situ remediation of contaminated sediments.

The authors hope that the information presented herein will encourage both young and experienced scientists to pursue a better understanding of the complex processes controlling trace element mobility, bioavailability, and toxicity in the aquatic environment and develop better methods for the remediation of contaminated waterlogged soils and sediments. This book is intended for professionals around the world in disciplines related to contaminant bioavailability to aquatic organisms, contaminant fate and transport in contaminated sediments (e.g., rice paddies), remediation technologies for contaminated sediments, and risk assessment of aquatic and wetland ecosystems.

Jörg Rinklebe

Anna Sophia Knox

Michael H. Paller

Acknowledgments

The editors wish to sincerely thank the contributors to this book for their enormous efforts, brilliant trustworthiness, diligence, and kind cooperation in achieving our goal and making this book a reality. We are most grateful for their time in critiquing the various chapters. The editors also gratefully thank Professor H. Magdi Selim, series editor for his continuous encouragement, valuable advice, and great support. Special thanks also go to the Taylor & Francis staff, in particular, Irma Shagla Britton, Delroy Lowe, Joselyn Banks, and Ariel Crockett, for their continuous help, cooperation, and support throughout the publication of this book.

Editors

Jörg Rinklebe is a professor for soil and groundwater management at the University of Wuppertal (Germany). From 1997 to 2006, Dr. Rinklebe worked as a scientist and project leader at the department of soil sciences at the UFZ Centre for Environmental Research Leipzig-Halle, Germany. He studied ecology for one year at the University of Edinburgh, United Kingdom (1992–1993). He studied agriculture, specializing in soil science and plant nutrition, at the Martin Luther University of Halle-Wittenberg, Germany, and earned his PhD in soil science at the same university. Currently his research is mainly focused on wetland soils, sediments, waters, plants, and the related pollution (trace elements and nutrients) and biogeochemical issues. He also has a certain expertise in remediation of soils and soil microbiology. Professor Rinklebe is internationally recognized, particularly for his research in the area of the redox-chemistry of trace elements in flooded soils. He has published many scientific papers in international and national journals as well as numerous book chapters. He serves on serval editorial boards (*Geoderma, Water, Air, and Soil Pollution, Ecotoxicology, Archive of Agronomy and Soil Science)*. He is also a reviewer for many international journals. He has co-organized several special symposia at various international conferences, such as Biogeochemistry of Trace Elements (10th, 11th, 12th, and 13th ICOBTE) and International Conference on Heavy Metals in the Environment (15th, 16th, and 17th ICHMET). He has been an invited speaker at many international conferences.

CO-EDITORS

Anna Sophia Knox is a fellow scientist at the Savannah River National Laboratory (SRNL) in Aiken, South Carolina, where she conducts research on the remediation of contaminated sediments and soils and development of new materials for the stabilization of contaminants. She earned a PhD (1993) in agronomy and soil science and was certified as a professional soil scientist by the Soil Science Society of America on December 20, 1999. Dr. Knox has 25 years of responsible experience in environmental science, with emphasis on the biogeochemistry and geochemistry of metals and radionuclides in natural and contaminated soils/sediments; the transformation, transport, and bioavailability of contaminants; and the remediation of contaminated soils/sediments.

She has published more than 95 scientific papers, book chapters, and patents; has presented numerous papers at scientific meetings and international conferences; and has organized several sessions and special symposia on contaminants in soils and sediments for international conferences/meetings such as the International Conference on Remediation of Contaminated Sediments, International Conference of Biogeochemistry of Trace Elements (10th, 11th, 12th, and 13th ICOBTE), International Conference on Heavy Metals in the Environment (15th, 16th, and 17th ICHMET), and others.

Dr. Knox is nationally and internationally recognized as one of the leading authorities in the area of active capping research. She has received several awards, including a DOE Women of Excellence in Science and Engineering award and several Key Contributor Awards because of crucial contributions to the strategic goals of the Savannah River National Laboratory. Her participation in professional societies includes serving on the editorial boards for both the *International Society of Environmental Forensic Journal* and *Journal of Archives of Agronomy and Soil Science.*

Michael H. Paller is a senior fellow scientist at the Savannah River National Laboratory in Aiken, South Carolina. He earned a PhD in zoology at Southern Illinois University and has over 35 years of experience studying aquatic ecosystems. Areas of particular interest include fish and macroinvertebrate ecology, environmental impact assessment, aquatic toxicology, ecological risk assessment, remediation of contaminated sediments, the fate and transfer of radionuclides through aquatic ecosystems, and the application of passive samplers to measure environmental contamination. Recent activities include the development of ecological risk assessment protocols for the Savannah River Site, a 780-km^2 U.S. Department of Energy facility in South Carolina, and a multiyear effort to develop ecological reference models for coastal plain streams on Department of Defense installations in the southeastern United States. Dr. Paller has served on the editorial board of the American Fisheries Society, has served on several river basin committees, and has helped to organize special symposia on the fate and transport of metals in sediments. He is the author of over 140 refereed publications and technical reports and one book. Dr. Paller also serves as a faculty member at Georgia Regents University in Augusta, Georgia, where he teaches biology and environmental science.

Contributors

Keaton M. Belli
School of Earth and Atmospheric Sciences
Georgia Institute of Technology
Atlanta, Georgia

Nanthi Bolan
Global Centre for Environmental Remediation
University of Newcastle
New South Wales, Australia

Richard Bush
Southern Cross GeoScience
Southern Cross University
Lismore, New South Wales, Australia

Girish Choppala
Southern Cross GeoScience
Southern Cross University
Lismore, New South Wales, Australia

Martine Le Coz-Bouhnik
Géosciences Rennes
University of Rennes 1
Rennes, France

Mélanie Davranche
Géosciences Rennes
University of Rennes 1
Rennes, France

R. D. DeLaune
College of the Coast and Environment, Wetland Biogeochemistry
Department of Oceanography and Coastal Sciences
Louisiana State University
Baton Rouge, Louisiana

Aline Dia
Géosciences Rennes
University of Rennes 1
Rennes, France

R. P. Gambrell
School of the Coast and Environment
Department of Oceanography and Coastal Sciences
Louisiana State University
Baton Rouge, Louisiana

Brandy N. Gartman
Pacific Northwest National Laboratory
Richland, Washington

and

ViZn Energy Systems, Inc.
Columbia Falls, Montana

Kenan Gedik
Faculty of Fisheries
Recep Tayyip Erdogan University
Rize, Turkey

Gérard Grau
Géosciences Rennes
University of Rennes 1
Rennes, France

Saengdao Khaokaew
Soil Science Department
Kasetsart University
Bangkok, Thailand

Anna Sophia Knox
Savannah River National Laboratory
Aiken, South Carolina

Frederik Van Koetsem
Laboratory of Analytical Chemistry and Applied Ecochemistry
Department of Applied Analytical and Physical Chemistry
Ghent University
Ghent, Belgium

M. Kongchum
H. Rouse Caffey Rice Research Station
Louisiana State University Agricultural Center
Rayne, Louisiana

Anitha Kunhikrishnan
Chemical Safety Division
Department of Agro-Food Safety
National Academy of Agricultural Science
Jeollabuk-do, Republic of Korea

Gijs Du Laing
Laboratory of Analytical Chemistry and Applied Ecochemistry
Department of Applied Analytical and Physical Chemistry
Ghent University
Ghent, Belgium

Gautier Landrot
Environmental Engineering Department
Kasetsart University
Bangkok, Thailand

Rémi Marsac
Institute of Technology
Institute for Nuclear Waste Disposal
Karlsruhe, Germany

Maibam Dhanaraj Meitei
Department of Plant Sciences
University of Hyderabad
Hyderabad, India

Woranan Nakbanpote
Department of Biology
Faculty of Science
Mahasarakham University
Mahasarakham, Thailand

Yong Sik Ok
Korea Biochar Research Center and Department of Biological Environment
Kangwon National University
Chuncheon, Republic of Korea

Jim Olsta
Olsta Consulting LLC
Bartlett, Illinois

Michael H. Paller
Savannah River National Laboratory
Aiken, South Carolina

Natthawoot Panitlertumpai
Department of Biology
Faculty of Science
Mahasarakham University
Mahasarakham, Thailand

Mathieu Pédrot
Géosciences Rennes
University of Rennes 1
Rennes, France

Olivier Pourret
Institut Polytechnique LaSalle Beauvais
Beauvais, France

Majeti Narasimha Vara Prasad
Department of Plant Sciences
University of Hyderabad
Hyderabad, India

Nikolla P. Qafoku
Geosciences Group, Earth System Sciences Division
Pacific Northwest National Laboratory
Richland, Washington

and

ViZn Energy Systems, Inc.
Columbia Falls, Montana

Jörg Rinklebe
Institute for Soil Engineering, Water- and Waste-Management
University of Wuppertal
Wuppertal, Germany

Rahul Sahajpal
Geosciences Group, Earth System Sciences Division
Pacific Northwest National Laboratory
Richland, Washington

Abin Sebastian
Department of Plant Sciences
University of Hyderabad
Hyderabad, India

H. Magdi Selim
School of Plant, Environmental, and Soil Sciences
Louisiana State University
Baton Rouge, Louisiana

Balaji Seshadri
Global Centre for Environmental Remediation
University of Newcastle
New South Wales, Australia

Sabry M. Shaheen
University of Kafrelsheikh
Faculty of Agriculture
Department of Soil and Water Sciences
Kafr El-Sheikh, Egypt

Shiv Shankar
Global Centre for Environmental Remediation
University of Newcastle
New South Wales, Australia

Donald L. Sparks
Plant and Soil Science Department
University of Delaware
Newark, Delaware

Filip M. G. Tack
Ghent University
Department of Analytical Chemistry and Applied Ecochemistry
Gent, Belgium

Martial Taillefert
School of Earth and Atmospheric Sciences
Georgia Institute of Technology
Atlanta, Georgia

Ramya Thangarajan
Centre for Environmental Risk Assessment and Remediation
University of South Australia
Mawson Lakes, South Australia, Australia

Christos D. Tsadilas
Hellenic Agricultural Organization
General Directorship of Agricultural Research
Institute of Industrial and Forage Crops
Larissa, Greece

Section I

Understanding, Processes, and Needs

1 Processes Related to Release Dynamics of Trace Elements in Flooded Soils at Various Scales

Jörg Rinklebe

CONTENTS

1.1 INTRODUCTION

Wetlands, especially floodplains, offer a variety of ecosystem functions. One of the most important functions is their ability to act as water regulators by buffering the extremes associated with the discharge from rivers (de Groot et al., 2002; Golladay and Battle, 2002; Rupp et al., 2010). Floodplains also fulfil important retention functions in relation to the cycling and treatment of nutrients and contaminants that are transported through the river system (Costanza et al., 1997; Venterink et al., 2003; Rinklebe et al., 2007). Therefore, the contamination of floodplain soils by potentially toxic trace elements is of serious environmental concern, as many floodplain soils worldwide are polluted with trace elements (e.g., Rinklebe et al., 2007; Shaheen and Rinklebe, 2014; Shaheen et al., 2014a). In selected floodplain areas, hot spots emerge as a consequence of contaminated sediment deposition (Martin, 2000; Middelkoop, 2000; Krüger et al., 2005; Benson, 2006; Baborowski et al., 2007). Floodplain soils are often enriched with trace elements that originate from geogenic and anthropogenic sources (e.g., industrial discharge of waste into rivers or diffuse agricultural input from catchment areas) (Frohne et al., 2011, 2015). An example is the Elbe River, Germany, in which soils are contaminated mainly by wastes that originate from different industrial activities and that were discharged into the river during the previous centuries (e.g., Overesch et al., 2007; Rinklebe et al., 2007). Toxic trace elements such as arsenic (As) and chromium (Cr) are transported by the river water either in dissolved form or via suspended material, accumulate in floodplain soils during flooding, and exhibit low flow velocities (Du Laing et al., 2009c; Rennert and Rinklebe, 2009; Frohne et al., 2014). Over the decades, the Elbe River and its tributaries, which carry high concentrations of trace elements, have become one of the worst polluted, major river systems in Europe (e.g., Zimmer et al., 2011; Rinklebe and Shaheen, 2014; Shaheen and Rinklebe, 2014).

Elevated levels of trace elements in riverine soils may increase the solubility and leaching of these metals, resulting in adverse impacts on the agricultural environment (Hobbelen et al., 2004; Yang et al., 2013). The release of many trace elements poses a concern for floodplain ecosystems and could also affect the quality of both surface water and groundwater. These elements can be released from the soil solid phase into the soil solution, particularly under different flood-dry cycles, and can be transferred through the food chain, thus posing a hazard to both environmental and human health (Overesch et al., 2007; Rinklebe and Shaheen, 2014; Shaheen and Rinklebe, 2014). Therefore, the dynamics and release kinetics of trace elements, including their controlling factors in floodplain soils, are highly relevant, because they affect both scientific and practical issues pertaining to protection of groundwater and plants, sustainable management of soils, and our understanding of environmental pathways that have been affected by harmful substances (Overesch et al., 2007; Wennrich et al., 2012; Rinklebe and Shaheen, 2014; Shaheen and Rinklebe, 2014). Simultaneously, it is scientifically challenging to elucidate the underlying biogeochemical processes and drivers that control the dynamics and release kinetics of trace elements in flooded soils; this is due to the high natural complexity of floodplain ecosystems.

In general, the mobilization of trace elements is largely determined by various factors, such as the total content of trace elements, adsorption and/or desorption on soil components (clay minerals, organic matter, and pedogenic oxides), sulfur (S) chemistry, pH, and redox processes (Du Laing et al., 2009c; Rinklebe and Du Laing, 2011). Redox potential (E_H) is a particularly important factor that affects the dynamics of trace elements in flooded soils (Borch et al., 2010; Husson, 2013; Frohne et al., 2011, 2015). Fluctuating redox conditions that occur in soils during flood-dry cycles can affect the dynamics of trace elements either directly or indirectly through related changes in pH, dissolved organic carbon (DOC), and the chemistry of Fe, Mn, and S (Frohne et al., 2011, 2014; Shaheen et al., 2014b,c). Temporal inundation and a subsequent decrease of E_H can lead to a reduction of insoluble Fe and Mn(hydr)oxides, resulting in soluble Mn^{2+} and Fe^{2+} (Reddy and DeLaune, 2008; Shaheen et al., 2014a,b). Trace elements that are associated with these (hydr)oxides can then be released during the solid phase at low E_H. In addition to the reduction of Fe and Mn(hydr)oxides and the concurrent release of associated As, the reduction of As(V) to As(III) can lead to enhanced As mobilization under reducing conditions (Masscheleyn et al., 1991; Fox and Doner, 2003; Mitsunobu et al., 2006). An immobilization of trace elements can occur due to co-precipitation with or adsorption to Fe and Mn (hydr)oxides at high E_H (Frohne et al., 2011, 2014; Shaheen et al., 2014a,b,c). In addition, metal immobilization can result from the presence of largely insoluble sulfides under anoxic conditions (Du Laing et al., 2009b,c). Although the dynamics of trace elements in soils is sometimes studied, redox reactions have not yet received the same attention as pH does, although redox potential is regarded as a master variable for metal(loid) fate. Studies that specifically address the elucidation of redox-induced mobilization and the immobilization processes of metal(loid)s in flooded soils are still relatively scarce.

The processes and dynamics mentioned earlier are largely determined by the scale of investigation. Figure 1.1 shows that the process complexity increases during scaling-up while representativeness and realism increase simultaneously. In contrast, when the scale of investigation decreases, the resolution, precision, and repeatability of experiments increase (Figure 1.1). The scale of investigation, therefore, largely determines the results and should thus be considered before planning the respective study. In any case, proper capture of the processes in floodplain soils requires a combination of various methods at different scales. Biogeochemical microcosms, a groundwater lysimeter, and soil hydrological field measurement facilities have proved to be excellent tools for studying the processes that control the release kinetics of trace elements in flooded soils (Rupp et al., 2010). Several studies have been conducted at the laboratory scale by using an automatic biogeochemical microcosm system that allows the computer-controlled adjustment of E_H with the aim of studying the effect of E_H mechanistically (Frohne et al., 2011, 2014, 2015; Shaheen et al., 2014a; Rinklebe et al., 2016a,b,c). At the lysimeter scale, a few studies have been presented by Rupp et al. (2010) and Shaheen et al. (2014b,c) which have addressed the release kinetics of trace elements under different flood-dry cycles. As per these studies, the water level was controlled in an undisturbed soil monolith by specially developed techniques (Meißner et al., 2010). The release dynamics of trace elements and the factors controlling these dynamic processes, such as E_H, pH, DOC, Fe, Mn, and S, were measured in parallel to the adjustment of the water level in the soil monolith.

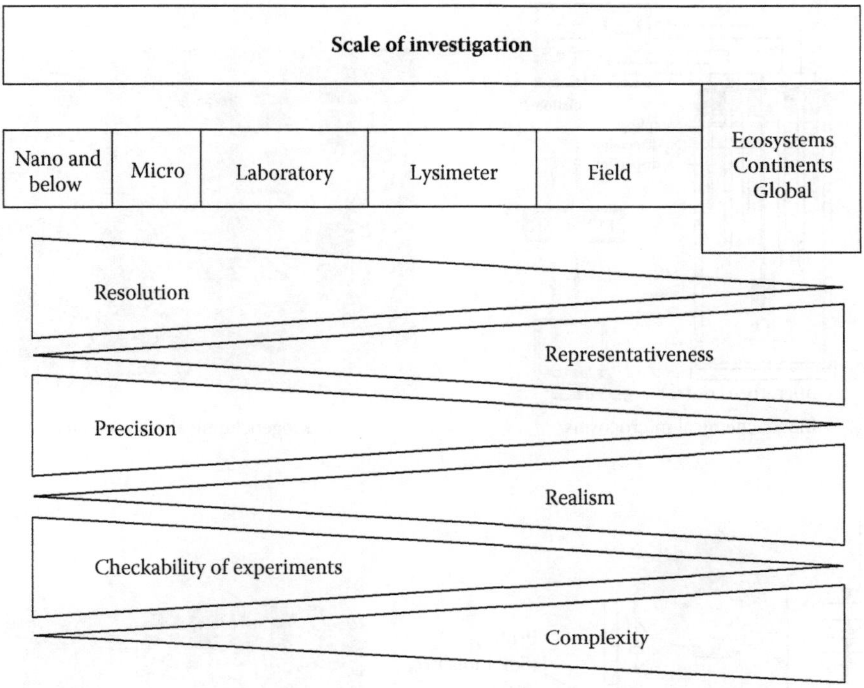

FIGURE 1.1 Scale of investigation related to processes of release dynamics of trace elements in flooded soils.

At the field scale, Rinklebe (2004), Rinklebe et al. (2005), and Schulz-Zunkel (2013) have monitored the release of trace elements and hydro-dynamics in floodplain soils of the Elbe River. Those studies were designed to monitor the release dynamics of trace elements, including factors controlling these dynamic processes, and to verify the orders of magnitude of matter fluxes and biogeochemical processes detected at the laboratory and lysimeter scale. This chapter provides one example of each of the following scales: (1) the laboratory scale using a highly sophisticated automatic biogeochemical microcosm system, (2) the lysimeter scale using undisturbed soil monoliths, and (3) the field scale using soil hydrological field monitoring stations. The focus of this chapter is to compare the results obtained at these different scales; the techniques are shown in Figure 1.2, and the methodology is described in detail in the cited references. Floodplain soils collected at the Elbe River are used as an example of flooded soils that are loaded with high levels of trace elements. Results at the three different scales—biogeochemical microcosms, groundwater lysimeters, and soil hydrological field measurements—are presented and discussed.

1.2 RESULTS AT DIFFERENT SCALES

1.2.1 Biogeochemical Microcosm Scale

Figure 1.3 shows the results gained by an experiment that uses a highly sophisticated, automated biogeochemical microcosm setup. This advanced system simulates the inundation of soil at a microcosm scale in the laboratory. It is described in detail by Yu et al. (2007), Yu and Rinklebe (2011), Frohne et al. (2011), and Yu and Rinklebe (2013). An important advantage of this system is that redox conditions are reproducible and well defined, and they can be changed rapidly. Moreover, predefined redox windows can be set, which results in mechanistic studies that elucidate the release kinetics of trace elements and governing factors.

As shown in Figure 1.3, Cr concentrations reflect E_H adjustments in the microcosms: The highest Cr concentrations in soil solution temporally coincided with the maximum E_H, and the lowest Cr

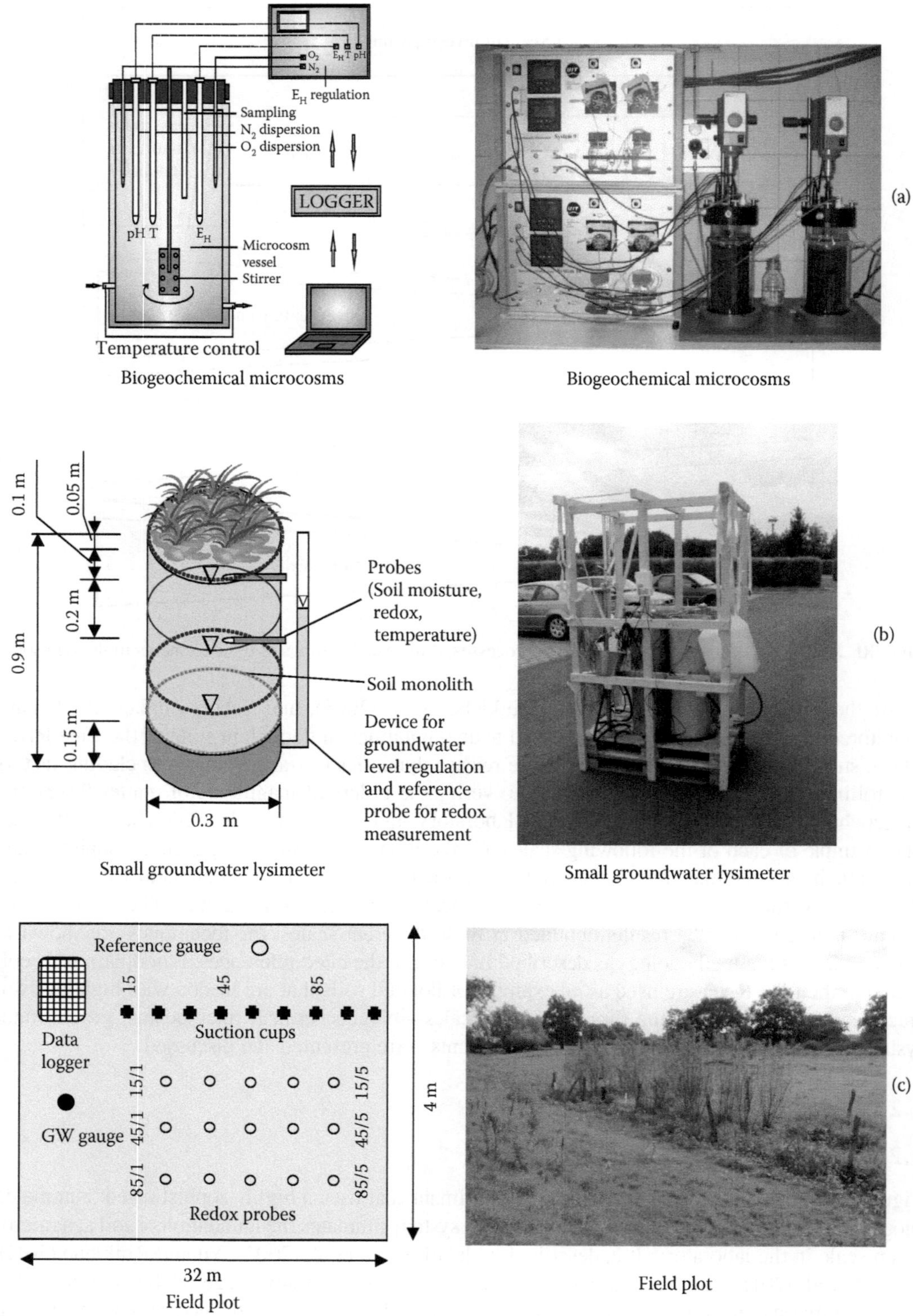

FIGURE 1.2 **(See color insert.)** Experimental setups. (a) Biogeochemical microcosm, (b) groundwater lysimeter, and (c) field plot. (Reprinted from *Ecol Eng*., 36, Rupp, H., Rinklebe, J., Bolze, S., and Meissner, R., A scale depended approach to study pollution control processes in wetland soils using three different techniques, 1439–1447, Copyright 2010, with permission from Elsevier.)

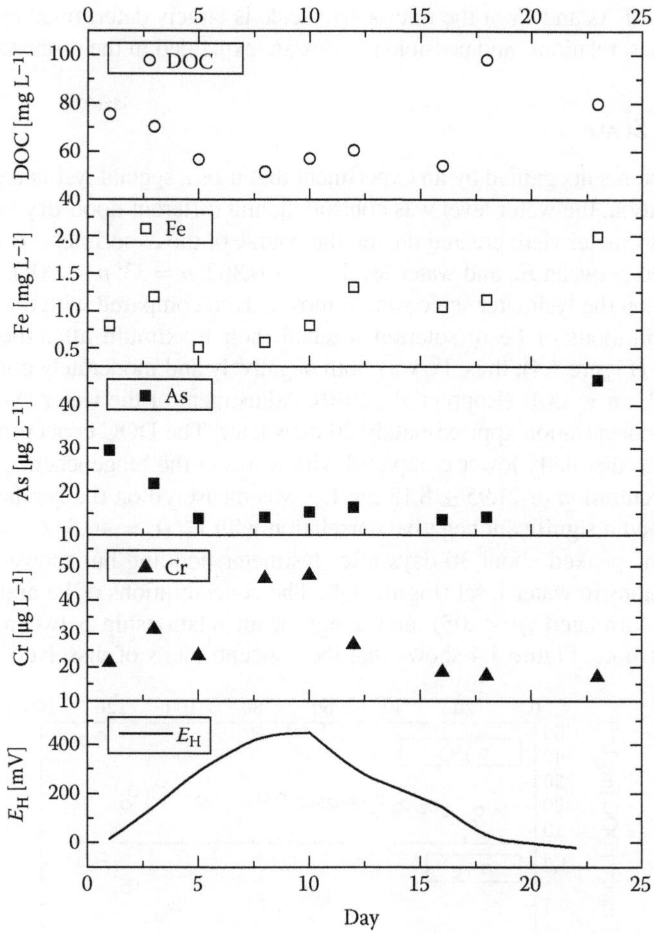

FIGURE 1.3 Time-dependent course of the redox potential (E_H) and the concentrations of Cr, As, Fe, and DOC at the microcosm scale. (Reprinted from *Ecol Eng*., 36, Rupp, H., Rinklebe, J., Bolze, S., and Meissner, R., A scale depended approach to study pollution control processes in wetland soils using three different techniques, 1439–1447, Copyright 2010, with permission from Elsevier.)

concentrations coincided with the lowest E_H (Rupp et al. 2010). Chromium was significantly correlated with E_H (Spearman's Rank correlation coefficient [r_s] of 0.85; $n = 9$; $p < .01$), because an increase in E_H was associated with an increase in dissolved Cr (Rupp et al., 2010). Figure 1.3 also shows that the highest As concentrations of 29.5 and 45.8 µg L^{-1} were measured at the beginning and end of the experiment, respectively, when the E_H was the lowest (16 and –30 mV). However, this relationship was not reflected in a significant correlation coefficient. The concentration of Fe in solution was relatively high and ranged between 0.51 and 1.99 mg L^{-1} (Rupp et al., 2010). The decrease in E_H that started at day 10 of the experiment resulted in an increase in Fe concentration. The maximum concentration of Fe was reached at day 23 when E_H was low. The inverse relationship between Fe and E_H was not statistically significant. Arsenic behaved similarly to Fe and DOC (Figure 1.2). Changes in E_H were also reflected in the temporal course of DOC concentrations in soil solution. The DOC ranged between 51.68 mg L^{-1} and 98.15 mg L^{-1}, with the highest DOC concentrations at low E_H values and the lowest DOC concentrations at higher E_H values between approximately 300 and 400 mV. The DOC concentrations were significantly negatively correlated with the E_H ($r_s = -0.817$; $n = 9$; $p < .01$) and significantly positively correlated with the As ($r_s = 0.70$, $n = 9$, and $p < .05$); however, the relationship between DOC and Cr was not statistically significant (Rupp et al., 2010). Summarizing, these results demonstrate that the release kinetics of As and Cr at the microcosm scale is largely determined by E_H, Fe, and DOC. Underlying processes, relations, and possible reasons are explained in the discussion.

1.2.2 Lysimeter Scale

Figure 1.4 shows the results gained by an experiment that uses a special lysimeter setup (Rupp et al., 2010). In this condition, the water level was controlled, and different flood-dry periods in the undisturbed soil of the lysimeter were created during the course of the experiment. A significant negative correlation occurred between E_H and water level ($r_s = -0.862$; $n = 33$; $p < .001$). The concentrations of Fe in pore water on the lysimeter scale were almost halved compared with the microcosm experiments. The concentrations of Fe in solution reached their maximum after the water level hit its peak at the surface (Figure 1.4); thus, Fe was both negatively and moderately correlated with the E_H ($r_s = -0.643$; $n = 33$; $p < .001$) (Rupp et al., 2010). Adjustment of the water level at day 83 had an impact on the Fe concentration approximately 20 days later. The DOC concentrations in the lysimeter pore water were distinctly lower compared with those in the biogeochemical microcosms. An average DOC concentration of 21.95 ± 8.15 mg L^{-1} was measured on the lysimeter scale; the DOC concentration showed a significant negative correlation with E_H ($r_s = -0.542$; $n = 33$; $p < .05$). The DOC concentrations peaked about 30 days after lysimeter flooding but showed no clear response to preceding variations in water level (Figure 1.4). The concentrations of Fe and As in soil solution were significantly correlated ($p < .05$), and a significant relationship between DOC and As was also found. Furthermore, Figure 1.4 shows that the concentrations of dissolved As in the lysimeter

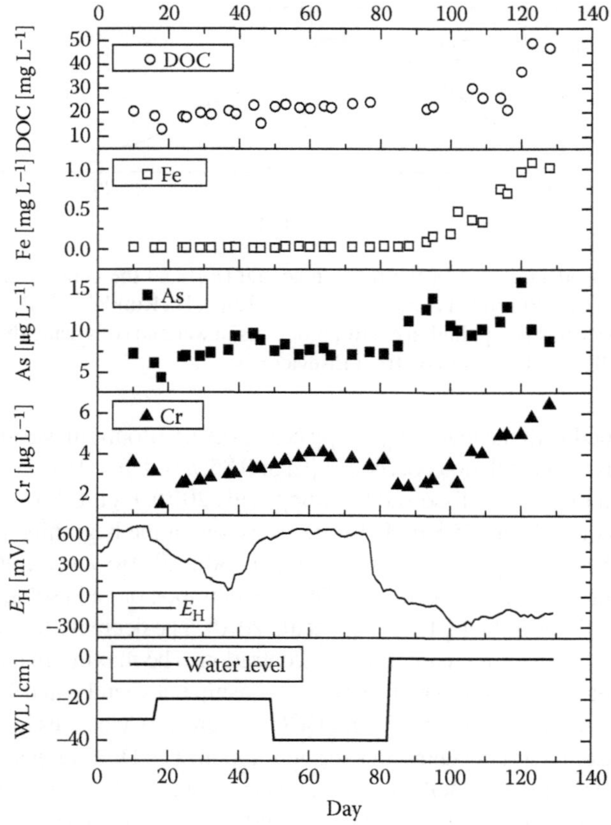

FIGURE 1.4 Time-dependent course of the redox potential (E_H) and the concentrations of Cr, As, Fe, and DOC at the lysimeter scale (May 2009–October 2009). (Reprinted from *Ecol Eng.*, 36, Rupp, H., Rinklebe, J., Bolze, S., and Meissner, R., A scale depended approach to study pollution control processes in wetland soils using three different techniques, 1439–1447, Copyright 2010, with permission from Elsevier.)

experiment were slightly lower than those in the biogeochemical microcosm experiment. The flooding of the lysimeter at day 80 was accompanied by a rapid decline in E_H. Concentrations of dissolved As reached maximal concentrations about 3 to 5 weeks later. The concentrations of dissolved As were significantly negatively correlated with E_H ($r_s = -0.668$; $n = 33$; $p < .001$) (Rupp et al., 2010), and one can state that changes in E_H that were caused by the soil water regime affected As concentrations within 1 week.

The Cr concentrations measured in the lysimeter experiment were about ten times lower than those in the biogeochemical microcosm experiment; however, the trend in Cr concentrations (Figure 1.4) followed a distinct temporal pattern. The concentrations of Cr in soil solution were high when E_H was low, particularly after day 80. Thus, the release of Cr into soil solution at the lysimeter scale revealed a different pattern than that observed at the microcosm and field scales.

1.2.3 Field Scale

Temporal changes in the water level of the Elbe River, in situ E_H measurements, and associated concentrations of Cr, As, Fe, and DOC in soil solution are shown in Figure 1.5 (results by Rupp et al., 2010). It is clear that changes in the water level of the Elbe River were reflected in the changes in E_H: When the soil was dry, the E_H was high; when the soil was flooded, the E_H was low (Figure 1.5). The E_H was significantly negatively correlated with the Elbe River gauge ($r_s = -0.367$; $n = 735$; $p < .001$). The E_H varied in its temporal course from oxidizing (644 mV) to moderately reducing conditions (–88 mV) due to the hydrologic conditions imposed by the Elbe River (Rupp et al., 2010).

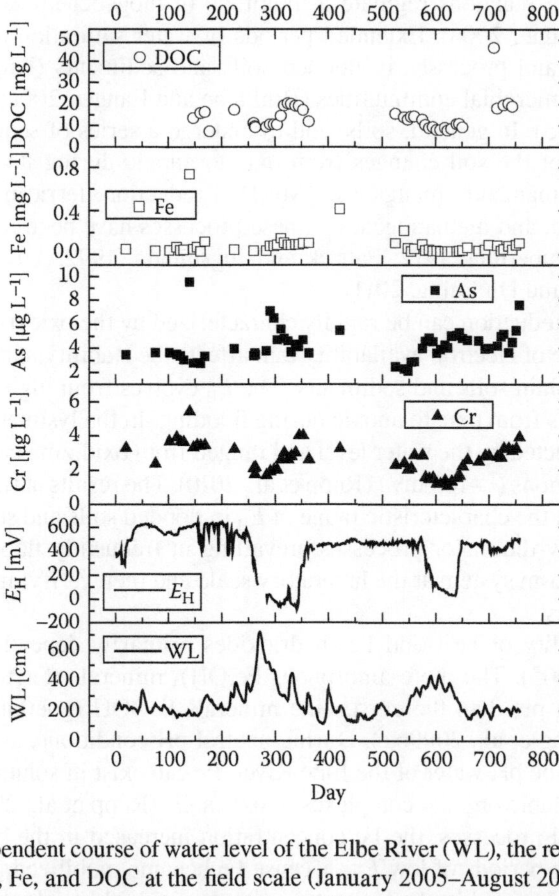

FIGURE 1.5 Time-dependent course of water level of the Elbe River (WL), the redox potential (E_H), and the concentrations of Cr, As, Fe, and DOC at the field scale (January 2005–August 2007). (Reprinted from *Ecol Eng.*, 36, Rupp, H., Rinklebe, J., Bolze, S., and Meissner, R., A scale depended approach to study pollution control processes in wetland soils using three different techniques, 1439–1447, Copyright 2010, with permission from Elsevier.)

The Cr concentrations were comparable to those measured at the lysimeter scale. However, the concentrations of dissolved Cr showed a statistically significant relationship with E_H ($r_s = 0.581$; $n = 56$; $p < .001$) (Rupp et al., 2010). The lowest Cr concentration was reached when the E_H was at the lowest level and vice versa. Patterns between concentrations of dissolved Cr and the E_H at the microcosm scale were basically similar to those at the field scale. This allows us to conclude that the biogeochemical microcosm apparatus is a powerful and useful tool that realistically mimics field conditions.

Figure 1.5 also shows that the inundation in the years 2006 and 2007 (days 260–334 and 588–609, respectively) led to an increase in dissolved As concentrations. These concentrations reached their maximum 4 weeks after the first inundation started and were accompanied by a decrease in E_H. The second inundation was also associated with an increase in As concentrations; however, a delay of 4 weeks occurred. Nevertheless, in general, one can state that a decrease in E_H resulted in an increase in the concentrations of As in soil solution, and concentrations of dissolved As were significantly negatively correlated with E_H ($r_s = -0.384$; $n = 56$; $p < .01$) (Rupp et al., 2010). The Fe concentrations were comparatively constant during the monitoring period ranging from 0.02 to 0.74 mg L^{-1}. Only the flood event of April 2006 (days 260–334) was reflected in a change in DOC concentration, with a time lag of approximately 60 days (Rupp et al., 2010) (Figure 1.5).

1.3 DISCUSSION

Oxidation and reduction reactions regulate a multitude of biogeochemical processes in flooded soils (Reddy and DeLaune, 2008). Extended periods of water saturation often result in changes in chemical properties and processes in flooded soils and sediments (Rinklebe et al., 2005) as well as changes in soil microbial communities (Rinklebe and Langer, 2006; Langer and Rinklebe, 2009; Moche et al., 2015). In general, soils tend to undergo a series of sequential redox reactions when the redox status of the soil changes from oxic to anoxic during flooding. Major reactions include denitrification, manganic manganese [Mn(IV)] reduction, ferric iron [Fe(III)] reduction, sulfate (SO_4^{2-}) reduction, and methanogenesis (these processes have been reviewed by Patrick and DeLaune, 1972; Ponnamperuna, 1972; Patrick and Jugsujinda, 1992; Yu et al., 2007; Du Laing et al., 2009c; Rinklebe and Du Laing, 2011).

The intensity of soil reduction can be rapidly characterized by the oxidation–reduction potential (E_H), which is a measure of electron availability that affects the stability and availability of various trace elements in floodplain soils and sediments. The E_H evolves from high to low when the redox status of the soil changes from oxic to anoxic during flooding. In the lysimeter and field-scale studies, E_H was strongly affected by the water level and ranged from oxidizing conditions (>400 mV) to strongly reducing conditions (<−100 mV) (Rupp et al., 2010). The results at the three different scales presented here represent the characteristic range of E_H in flooded soils and sediments. Accordingly, we were able to identify the major processes prevailing in frequently flooded soils by using the biogeochemical microcosm system at the laboratory scale and then verifying these processes at the lysimeter and field scales.

In general, the stability of Fe^{2+} and Fe (hydr)oxides primarily depends on a combination of E_H and pH (Takeno, 2005). The more amorphous $Fe(OH)_3$ minerals (ferrihydrite) are reduced at a higher E_H for a given pH than the crystalline minerals FeOOH (goethite) or Fe_2O_3 (hematite) (Takeno, 2005; Du Laing et al., 2009b,c). During neutral pH conditions, as was the case with the neutral or slightly alkaline pH water of the Elbe River, Fe can exist in solution at low E_H (probably mainly as Fe^{2+}) or as soluble organic complexes in oxic soils (Rupp et al., 2010). Corresponding to these basic geochemical processes, the Fe concentration increased in the biogeochemical microcosm experiment during periods of low E_H, because Fe became mobilized, most likely as Fe^{2+}. In general, the same was observed in the lysimeter as in the field, although other factors might have had additional impacts (Rupp et al., 2010).

Arsenic has a high affinity for Fe and, therefore, the temporal dynamics of As in soil solution generally behaves similar to Fe. Under oxic conditions, Fe will precipitate as Fe (hydr)oxides, and trace metals can co-precipitate or adsorb to these oxides (Du Laing et al., 2009a,c). The results at the lysimeter scale showed an increase in affinity between Fe, As, and Cr with an increase in E_H (Rupp et al., 2010).

Soil oxidation–reduction reactions play a crucial role in determining the solubility and mobilization of As. Under natural conditions, arsenate [As(V)] and arsenite [As(III)] are the most abundant forms of As (Smith et al., 1998; Mandal and Suzuki, 2002). In soils and water systems, As(V) is dominant under aerobic conditions and As(III) is dominant under anaerobic conditions (Reddy and DeLaune, 2008). However, due to the slow kinetics of As redox reactions, both As(V) and As(III) are often found in the soil environment regardless of redox conditions (Masscheleyn et al., 1991; Fox and Doner, 2003; Mitsunobu et al., 2006). In general, the reduction of soil conditions ($E_H <$ 0 mV) greatly enhances the solubility of As, and the majority of soluble As is present as As(III) (Reddy and DeLaune, 2008). Accordingly, the studies conducted at the three scales of biogeochemical microcosm, lysimeter, and field presented here revealed that the concentrations of dissolved As are high when E_H is low and low when E_H is high. Thus, the As concentration in solution largely depends on the E_H, which was shown at the three scales under study.

Dissolved organic carbon (DOC) is known to be the most mobile soil organic fraction involved in the cotransport of trace elements through physical and chemical binding, and it has a significant impact on trace element bioavailability (Zsolnay, 1996). Therefore, DOC in soils plays an important role in the transport as well as in the biogeochemistry of pollutants and nutrients in soils (Kalbitz et al., 2000). Dissolved organic ligands, such as low- to medium-molecular-weight carboxylic acids, amino acids, and fulvic acids (FAs), can form soluble metal complexes (Du Laing et al., 2009c). Enhanced mobilization of trace elements as dissolved organic complexes was observed for Cr and As (Kalbitz and Wennrich, 1998). Significant correlations between the concentrations of As and E_H in the microcosm and lysimeter studies confirmed these literature statements (Rupp et al., 2010). In the microcosm and lysimeter experiments, DOC increased with a decrease in E_H. The release dynamics of As and DOC showed similar patterns; therefore, one can conclude that there are interactions between As and DOC. However, these interactions are complex, and the role of DOC as the main driver of microbial reductive dissolution is insufficient to explain the variation in DOC–As relationships. Other processes that may also influence the mobilization of As include the complexation of As by dissolved humic substances, and competitive sorption and electron shuttling reactions that are mediated by humics (Mladenov et al., 2015). Mladenov et al. (2015) proposed that associations between dissolved As, Fe, and FA concentration and the fluorescence properties of isolated FA suggest that aromatic, terrestrially derived FAs promote As–Fe–FA complexation reactions that may enhance As mobilization. Similar processes might have also happened in our field study. However, it was impossible to prove this, because several other factors might have also impacted the release dynamics of As, and interferences might have occurred. For instance, the flood events in 2006 and 2007 differed considerably in their supply of humic substances due to differences in the inundation period and flood-water level (7.18 m versus 4.36 m, c.f. Figure 1.4). Flood events can differ considerably due to their genesis and hydrologic boundary conditions. Consequently, their impact on the dynamics of DOC and their associated trace elements also differ (Rupp et al., 2010).

The mobilization and immobilization of Cr in flooded soils is affected by its speciation. Chromium occurs mainly in the oxidation states +3 and +6 in the environment, depending on E_H and pH conditions (Choppala et al., 2013). More details about the biogeochemistry and bioavailability of chromium are presented in Chapters 8 and 15 of this book (Gartman and Qafoku, 2016; Choppala et al., 2016). At low E_H, relatively immobile Cr(III) species such as $Cr(OH)^{2+}$ and Cr_2O_3 (s) dominate due to the reduction of Cr(VI) in the presence of reductants such as organic matter and/or Fe(II). During this reduction process, intermediate Cr(V) is produced, which decays into Cr(III) within a few days. This process is favored by acidic conditions and occurs within minutes (Richard and Bourg, 1991; Graham and Bouwer, 2010; Lee et al., 2013). The soil in our studies was slightly acidic and contained high amounts of organic matter and Fe, which can promote Cr reduction. Therefore, Cr(III) can be expected to be present under reducing conditions.

The E_H-dependent effect on Cr concentrations was observed at the microcosm and field scales. Chromium was mobilized at high E_H values, and reduced conditions resulted in the immobilization of Cr. One reason might be the reduction of Cr(VI) to Cr(III) (Klaas et al., 2007). The lysimeter study showed an abnormal release of Cr into soil solution, especially during the last phase of the simulated flooding, where high Cr concentrations were measured at low E_H values (Rupp et al., 2010). This effect was likely caused by retarded or slowed reactions and a time lag of approximately 30 days. Nevertheless, results at the field scale also showed a few instances of comparatively high and dissolved Cr concentrations at low E_H (Figure 1.4), which might be attributed to the feed of particulate matter (e.g., colloids of humic substances, clay, and silt) during flood events.

Shaheen et al. (2014) have also used lysimeter trials to assess the impact of different flood-dry cycles on the temporal dynamics of pore water concentrations of As, Cr, Mo, and V in a contaminated floodplain soil. These were found to be affected by changes in E_H and pH and in the dynamics of DOC, Fe, Mn, and SO_4^{2-} in contaminated floodplain soil collected at the Elbe River. An important finding was that interactions between the trace elements and carriers were stronger during long flood-dry cycles than during short cycles. They concluded that the dynamics of As, Cr, Mo, and V is determined by the period during which the soils are exposed to flooding, because drivers of element mobilization need time to provoke reactions in soils under changing conditions.

1.4 CONCLUSIONS

Experiments in controlled biogeochemical microcosms and groundwater lysimeters have been shown to yield generally similar results regarding the functional interrelationships of the investigated parameters. The results of field experiments, however, are frequently characterized by interference from various factors. Nevertheless, the basic functional relationship between E_H and the concentrations of trace elements in pore water is illustrated by the example of As and Cr (Rupp et al., 2010). Dissolved As tended to be mobilized during flooding due to a decrease in E_H. Chromium showed an unexpected response, with peak concentrations at the highest E_H values. However, the patterns between concentrations of dissolved Cr and E_H at the microcosm scale were basically similar to those observed at the field scale. This leads us to conclude that the automated biogeochemical microcosm apparatus is a powerful and useful tool for realistically mimicking field conditions to a certain extent. However, tools used at each of the three scales have specific advantages. Biogeochemical microcosm and lysimeter setups enable targeted variation in site factors (e.g., E_H), which results in the identification, elucidation, and quantification of the prevailing processes. Therefore, these methods are predominately used in basic research. Lysimeter and soil hydrological field measurement facilities enable one to monitor and verify natural processes observed at the microcosm scale and permit the investigation of scale-related effects. The combination of studies at different scales is optimal to provide a picture approximating real conditions in nature. It also offers a range of scientific opportunities for a comprehensive understanding of the processes determining the dynamics of trace element release in floodplain soils. In conclusion, the multiscale approach, in terms of both space (microcosms, lysimeters, field) and time (ranging from the 23-day microcosm scale to the 760-day field scale), effectively identified and verified the major biogeochemical processes in floodplain soils that explain the release dynamics of trace elements. This integrated methodology has great potential for future scientific and applied studies at many wetland sites worldwide. In the future, the integration of studies using synchrotron facilities, as described in Chapters 5, 8, and 10 of this book (Gartman and Qafoku, 2016; Khaokaew et al., 2016; Qafoku and Sahajpal, 2016), is highly promising within this context of dynamic processes. Regarding scale, soil column experiments are the bridge between microcosms and lysimeters (Rennert et al., 2010). Column experiments basically allow to simulate conditions such as in lysimeters; however, they need less space and equipment in comparison to undisturbed lysimeters. Therefore, they are also a useful tool that is worth integrating into the scale concept. Much future scientific effort is also required to transfer the knowledge of identified processes to the landscape scale. Although many studies have been conducted to determine the distribution of trace elements on the scales of field plots, landscapes, ecosystems, continents, and globally, there are still many questions that remain unanswered. In addition, distributional studies are static and do not address dynamic processes that are described here. The transfer of the understanding of dynamic processes to these larger scales is a challenging and huge scientific task for the coming decades.

REFERENCES

Baborowski, M., Büttner, O., Morgenstern, P., Krüger, F., Lobe, I., Rupp, H., Von Tümpling, W., 2007. Spatial and temporal variability of sediment deposition on artificial-lawn traps in a floodplain of the River Elbe. *Environ Pollut.* 148: 770–778.

Benson, N.U., 2006. Lead, nickel, vanadium, cobalt, copper and manganese distributions in intensively cultivated floodplain Ultisol of Cross River, Nigeria. *Int J Soil Sci.* 1: 140–145.

Borch, T., Kretzschmar, R., Kappler, A., Van Cappellen, P., Ginder-Vogel, M., Voegelin, A., Campbell, K., 2010. Biogeochemical redox processes and their impact on contaminant dynamics. *Environ Sci Technol.* 44: 15–23.

Choppala, G, Bolan, N, Lamb, D, Kunhikrishnan, A., 2013. Comparative sorption and mobility of Cr(III) and Cr(VI) species in a range of soils: Implications to bioavailability. *Water Air Soil Pollut.* 224. DOI: 10.1007/s11270-11013-11699-11276.

Choppala, G, Kunhikrishnan, A., Seshadri, B., Bolan, S., Bush, R., Bolan, N., 2016. Reduction induced immobilization of chromium and its bioavailability in soils and sediments. In: Rinklebe et al. (eds), *Trace Elements in Waterlogged Soils and Sediments*. Boca Raton, FL: CRC Press.

Costanza, R., d'Arge, R., de Groot, R., Farber, S., Grasso, M., Hannon, B., Limburg, K., Naeem, S., O'Neil, R.V., Paruelo, J., Raskin, R.G., Sutton, P., van den Belt, M., 1997. The value of the world's ecosystem services and natural capital. *Nature* 387: 253.

de Groot, R.S., Wilson, M.A., Boumans, R.M.J., 2002. A typology for the classification, description and valuation of ecosystem functions, goods and services. *Ecol Econ.* 41: 393.

Du Laing, G., Chapagain, S.K., Dewispelaere, M., Meers, E., Kazama, F., Tack, F.M.G., Rinklebe, J., Verloo, M.G., 2009a. Presence and mobility of arsenic in estuarine wetland soils of the Scheldt estuary (Belgium). *J Environ Monit.* 11: 873–881.

Du Laing, G., Meers, E., Dewispelaere, M., Vandecasteele, B., Rinklebe, J., Tack, F.M.G., Verloo, M.G., 2009b. Heavy metal mobility in intertidal sediments of the Scheldt estuary: Field monitoring. *Sci Total Environ.* 407: 2919–2930.

Du Laing, G., Rinklebe, J., Vandecasteele, B., Meers, E., Tack, F.M.G., 2009c. Trace metal behaviour in estuarine and riverine floodplain soils and sediments: A review. *Sci Total Environ.* 407: 3972–3985.

Fox, P.M., Doner, H.E., 2003. Accumulation, release, and solubility of arsenic, molybdenum, and vanadium in wetland sediments. *J Environ Qual.* 32: 2428–2435.

Frohne, T., Diaz-Bone, R.A., Du Laing, G., Rinklebe, J., 2015. Impact of systematic change of redox potential on the leaching of Ba, Cr, Sr, and V from a riverine soil into water. *J Soils Sed.* 15: 623–633.

Frohne, T., Rinklebe, J., Diaz-Bone, R.A., 2014. Contamination of floodplain soils along the Wupper River, Germany, with As, Co, Cu, Ni, Sb, and Zn and the impact of pre-definite redox variations on the mobility of these elements. *Soil Sediment Contam.* 23: 779–799.

Frohne, T., Rinklebe, J., Diaz-Bone, R.A., Du Laing, G., 2011. Controlled variation of redox conditions in a floodplain soil: Impact on metal mobilization and biomethylation of arsenic and antimony. *Geoderma* 160: 414–424.

Gartman, B.N., Qafoku, N.P., 2016. Uranium interaction with soil minerals in the presence of co-contaminants: Case study-subsurface sediments at or below the water table. In: Rinklebe et al. (eds.), *Trace Elements in Waterlogged Soils and Sediments.* Boca Raton, FL: CRC Press.

Golladay, S.W., Battle, J., 2002. Effects of flooding and drought on water quality in Gulf coastal plain streams in Georgia. *J Environ Qual.* 31: 1266.

Graham, A.M., Bouwer, E.J., 2010. Rates of hexavalent chromium reduction in anoxic estuarine sediments: pH effects and the role of acid volatile sulfides. *Environ Sci Technol.* 44: 136–142.

Hobbelen, P.H.F., Koolhaas, J.E., van Gestel, C.A.M., 2004. Risk assessment of heavy metal pollution for detritivores in floodplain soils in the Biesbosch, The Netherlands, taking bioavailability into account. *Environ Pollut.* 129: 409–419.

Husson, O. 2013. Redox potential (Eh) and pH as drivers of soil/plant/microorganism systems: A transdisciplinary overview pointing to integrative opportunities for agronomy. *Plant Soil.* 362: 389–417.

Kalbitz, K., Solinger, S., Park, J.H., Michalzik, B., Matzner, E., 2000. Controls on the dynamics of dissolved organic matter in soils: A review. *Soil Sci.* 165: 277–304.

Kalbitz, K., Wennrich, R., 1998. Mobilization of heavy metals and arsenic in polluted wetland soils and its dependence on dissolved organic matter. *Sci Total Environ.* 209: 27–39.

Khaokaew, S., Landrot, G., Sparks, D.L., 2016. Speciation and release kinetics of cadmium and zinc in contaminated paddy soils. In: Rinklebe et al. (eds.), *Trace Elements in Waterlogged Soils and Sediments.* Boca Raton, FL: CRC Press.

Krüger, F., Meissner, R., Gröngröft, A., Grunewald, K., 2005. Flood induced heavy metal and arsenic contamination of Elbe River floodplain soils. *Acta Hydrochim Hydrobiol.* 33: 455.

Langer, U., Rinklebe, J. 2009. Lipid biomarkers for assessment of microbial communities in floodplain soils of the Elbe River (Germany). *Wetlands* 29: 353–362.

Lee, G., Park, J., Harvey, O.R., 2013. Reduction of chromium(VI) mediated by zero-valent magnesium under neutral pH conditions. *Water Res.* 47: 1136–1146

Mandal, B.K., Suzuki, K.T., 2002. Arsenic round the world: A review. *Talanta* 58: 201–235.

Martin, C.W., 2000. Heavy metal trends in floodplain sediments and valley fill, River Lahn, Germany. *Catena* 39: 53.

Masscheleyn, P.H., Delaune, R.D., Patrick, W.H., 1991. Effect of redox potential and Ph on arsenic speciation and solubility in a contaminated soil. *Environ Sci Technol.* 25: 1414–1419.

Meißner, R., Prasad, M.N.V., Du Laing G., Rinklebe, J., 2010. Lysimeters application for measuring the water and solute fluxes with high precision. *Curr Sci.* 99, 5: 601–607.

Middelkoop, H., 2000. Heavy-metal pollution of the river Rhine and Meuse floodplains in the Netherlands. *Netherlands J Geosci.* 79: 411–428.

Mitsunobu, S., Harada, T., Takahashi, Y., 2006. Comparison of antimony behavior with that of arsenic under various soil redox conditions. *Environ Sci Technol.* 40: 7270–7276.

Mladenov, N., Zheng, Y., Bailey, S., Bilinski, T.M., McKnight, D.M., Nemergut, D., Radloff, K.A., Rahman, M.M., Ahmed K.M. 2015. Dissolved organic matter quality in a shallow aquifer of Bangladesh: Implications for arsenic mobility. *Environ Sci Technol.* 49: 10815–10824.

Moche, M., Gutknecht, J., Schulz, E., Langer, U., Rinklebe, J., 2015. Monthly dynamics of microbial community structure and their controlling factors in three floodplain soils. *Soil Biol Biochem.* 90: 169–178.

Overesch, M., Rinklebe, J., Broll, G.; Neue, H., 2007. Metals and arsenic in soils and corresponding vegetation at Central Elbe river floodplains (Germany). *Environ Pollut.* 145: 800–812.

Patrick, J.W.H., DeLaune, R.D., 1972. Characterization of the oxidized and reduced zones in flooded soil. *Soil Sci Soc Am Proc.* 36: 573–576.

Patrick, J.W.H., Jugsujinda, A., 1992. Sequential reduction and oxidation of inorganic nitrogen, manganese, and iron in flooded soil. *Soil Sci Soc Am J.* 56: 1071–1073.

Ponnamperuna, F.N., 1972. The chemistry of submerged soils. *Adv Agronomy* 24: 29–96.

Qafoku, N.P., Sahajpal, R., 2016. Subsoil contaminant Cr fate and transport: The complex reality of the Hanford subsurface. In: Rinklebe et al. (eds.), *Trace Elements in Waterlogged Soils and Sediments.* Boca Raton, FL: CRC Press.

Reddy, K.R., DeLaune, R.D., 2008. *Biogeochemistry of Wetlands: Science and Applications.* Boca Raton, FL: CRC Press.

Rennert, T., Meißner, S., Rinklebe, J., Totsche, K.U., 2010. Dissolved inorganic contaminants in a floodplain soil: Comparison of in-situ soil solutions and laboratory methods. *Water Air Soil Pollut.* 209: 489–500.

Richard, F.C, Bourg, A.C.M., 1991. Aqueous geochemistry of chromium—A review. *Water Res.* 25: 807–816.

Rinklebe, J., 2004. Differenzierung von Auenböden der Mittleren Elbe und Quantifizierung des Einflusses von deren Bodenkennwerten auf die mikrobielle Biomasse und die Bodenenzymaktivitäten von β-Glucosidase, Protease und alkalischer Phosphatase. PhD Thesis. With English Summary. Agricultural Faculty of the Martin Luther University of Halle-Wittenberg, Germany, 113 pp. and appendix.

Rinklebe, J., Antic-Mladenovic, S., Frohne, T., Stärk, H.-J., Tomić, Z., Licina, V., 2016a. Nickel in a serpentine-enriched Fluvisol: Redox affected dynamics and binding forms. *Geoderma* 263: 203–214.

Rinklebe, J., Du Laing, G., 2011. Factors controlling the dynamics of trace metals in frequently flooded soils. In: Selim, H. M. (ed.), *Dynamics and Bioavailability of Heavy Metals in the Root Zone.* Boca Raton, FL: CRC Press, Taylor & Francis, pp. 245–270.

Rinklebe, J., Franke, C., Neue H.U., 2007. Aggregation of floodplain soils as an instrument for predicting concentrations of nutrients and pollutants. *Geoderma* 141: 210–223.
Rinklebe, J., Langer, U., 2006. Microbial diversity in three floodplain soils at the Elbe River (Germany). *Soil Biol Biochem.* 38: 2144–2151.
Rinklebe, J., Shaheen, S.M., 2014. Assessing the mobilization of cadmium, lead, and nickel using a seven-step sequential extraction technique in contaminated floodplain soil profiles along the central Elbe River, Germany. *Water Air Soil Pollut.* 225: 2039. DOI 10.1007/s11270-014-2039-1.
Rinklebe, J., Shaheen, S.M., Frohne, T., 2016b. Amendment of biochar reduces the release of toxic elements under dynamic redox conditions in a contaminated floodplain soil. *Chemosphere.* 142: 41–47.
Rinklebe, J., Shaheen, S.M., Yu, K., 2016c. Release of As, Ba, Cd, Cu, Pb, and Sr under pre-definite redox conditions in different rice paddy soils originating from the U.S.A. and Asia. *Geoderma.* In press. http://dx.doi.org/10.1016/j.geoderma.2015.10.011
Rinklebe, J., Stubbe, A., Staerk, H.-J., Wennrich, R., Neue, H.-U., 2005. Factors controlling the dynamics of As, Cd, Zn, Pb in alluvial soils of the Elbe river (Germany). In: Lyon, W.G., Hong, J., Reddy, R.K. (eds.), *Proceedings of Environmental Science and Technology*, vol. 2. New Orleans: American Science Press, pp. 265–270.
Rupp, H., Rinklebe, J., Bolze, S., Meissner, R., 2010. A scale depended approach to study pollution control processes in wetland soils using three different techniques. *Ecol Eng.* 36: 1439–1447.
Schulz-Zunkel, C., Krueger, F., Rupp, H., Meissner, R., Gruber, B., Gerisch, M., Bork, H., 2013. Spatial and seasonal distribution of trace metals in floodplain soils. A case study with the Middle Elbe River, Germany. *Geoderma* 211–212: 128–137.
Schulz-Zunkel, C., Rinklebe, J., Bork, H.-R., 2015. Trace element release patterns from three floodplain soils under simulated oxidized-reduced cycles. *Ecol Eng.* 83: 485–495.
Shaheen, S.M., Rinklebe, J., 2014. Geochemical fractions of chromium, copper, and zinc and their vertical distribution in floodplain soil profiles along the Central Elbe River, Germany. *Geoderma* 228–229: 142–159.
Shaheen, S.M., Rinklebe, J., Frohne, T., White, J., DeLaune, R., 2014a. Biogeochemical factors governing Co, Ni, Se, and V dynamics in periodically flooded Egyptian north nile delta rice soils. *Soil Sci Soc Am J.* 78: 1065–1078.
Shaheen, S.M., Rinklebe, J., Rupp, H., Meissner, R., 2014b. Lysimeter trials to assess the impact of different flood–dry-cycles on the dynamics of pore water concentrations of As, Cr, Mo and V in a contaminated floodplain soil. *Geoderma.* 228–229: 5–13.
Shaheen, S.M., Rinklebe, J., Rupp, H., Meissner, R., 2014c. Temporal dynamics of pore water concentrations of Cd, Co, Cu, Ni, and Zn and their controlling factors in a contaminated floodplain soil assessed by undisturbed groundwater lysimeters. *Environ Pollut.* 191: 223–231.
Smith, E., Naidu, R.M., Olston, A.M., 1998. Arsenic in the soil environment: A review. *Adv Agronomy* 64: 149–195.
Takeno, N., 2005. National Institute of Advanced Industrial Science and Technology. Atlas of Eh-pH diagrams. Geological Survey of Japan Open File Report No. 409: 11–183.
Venterink, H.O., Wiegman, F., Van der Lee, G.E.M., Vermaata, J.E., 2003. Role of active floodplains for nutrient retention in the River Rhine. *J Environ Qual.* 32: 1430.
Wennrich, R., Daus, B., Müller, K., Stärk, H-J., Brüggemann, L., Morgenstern, P., 2012. Behaviour of metalloids and metals from highly polluted soil samples when mobilized by water evaluation of static versus dynamic leaching. *Environ Pollut.* 165: 59–66.
Yang, S., Zhou, D., Yu, H., Wei, R., Pan, B., 2013. Distribution and speciation of metals (Cu, Zn, Cd, and Pb) in agricultural and non-agricultural soils near a stream upriver from the Pearl River, China. *Environ Pollut.* 177: 64–70.
Yu, K.W., Bohme, F., Rinklebe, J., Neue, H.U., DeLaune, R.D., 2007. Major biogeochemical processes in soils—A microcosm incubation from reducing to oxidizing conditions. *Soil Sci Soc Am J.* 71: 1406–1417.
Yu, K., Rinklebe, J., 2013. Soil redox potential and pH controllers. In: DeLaune, R.D., Reddy, K.R., Richardson, C.J., Megonigal, J.P. (eds.), *Methods in Biogeochemistry of Wetlands.* Soil Science Society of America, pp. 107–116.
Yu, K.W., Rinklebe, J., 2011. Advancement in soil microcosm apparatus for biogeochemical research. *Ecol Eng.* 37: 2071–2075.
Zimmer, D., Kiersch, K., Baum, C., Meissner, R., Muller, R., Jand, G., Leinweber, P., 2011. Scale-dependent variability of As and heavy metals in a River Elbe floodplain. *CLEAN – Soil, Air, Water* 39: 328–337.
Zsolnay, A., 1996. Dissolved humus in soil waters. In: Piccolo, A. (ed.), *Humic Substances in Terrestrial Ecosystems.* Amsterdam: Elsevier, pp. 171–223.

2 Physicochemical Factors Controlling Stability of Toxic Heavy Metals and Metalloids in Wetland Soils and Sediments

Kenan Gedik, R. D. DeLaune,
M. Kongchum, and R. P. Gambrell

CONTENTS

Trace and toxic metals, commonly called *heavy metals,* in wetland soils and sediments can be of concern because of their potential toxicity. Numerous sources of heavy metals are entering wetland environments. Wetland soils and sediments have unique properties that can influence metal distribution reactivity, mobility, and toxicity. This chapter focuses on the physiochemical properties of the wetland soil that govern the transformations of heavy metals in wetland soils and sediments.

2.1 INTRODUCTION

Trace and toxic metals exist in several chemical forms in surface waters, wetland soils, and sediments (Table 2.1). The forms of metals that are most readily available to aquatic and benthic organisms and plants include those dissolved in soil solution, surfaces, and interstitial waters, and those bound to the solid phase by cation-exchange processes. Figure 2.1 shows some general chemical forms of metals in wetland soils and sediments. In a soil or sediment–water system, pH and redox conditions are the two most important factors that determine whether metals are bound to the solid phase or released to the mobile or biological available phase (Gambrell, 2013). Metals fixed within the crystalline lattice of primary and secondary minerals by isomorphic substitution are unavailable to plants or animals. There are a number of chemical forms between these availability extremes that are potentially available. Some of the processes immobilizing metals in potentially available forms include: (a) formation of metal oxides, hydroxides, and carbonates of low solubility; (b) adsorption to colloidal hydrous oxides of iron and manganese in aerobic, neutral, or alkaline pH environments; (c) precipitation as highly insoluble sulfides under strongly reducing conditions; and (d) complex formation with insoluble humic materials (Gambrell et al., 1980). Metals may undergo transformations between these forms, affecting their mobility and availability due to changes, such as pH, redox potential, and salinity, in the physicochemical properties of the system (Gambrell et al., 1983).

TABLE 2.1
General Chemical Forms of Trace and Toxic Metals in Sediments

Readily Available	Potentially Available	Unavailable
Dissolved	• Exchangeable	Fixed within the crystalline lattice structure of clay minerals
Exchangeable	• Precipitated, i.e., $Me(OH)_2$ • Complexed with organic matter • Co-precipitated with hydrous oxides • Precipitated as sulfides	

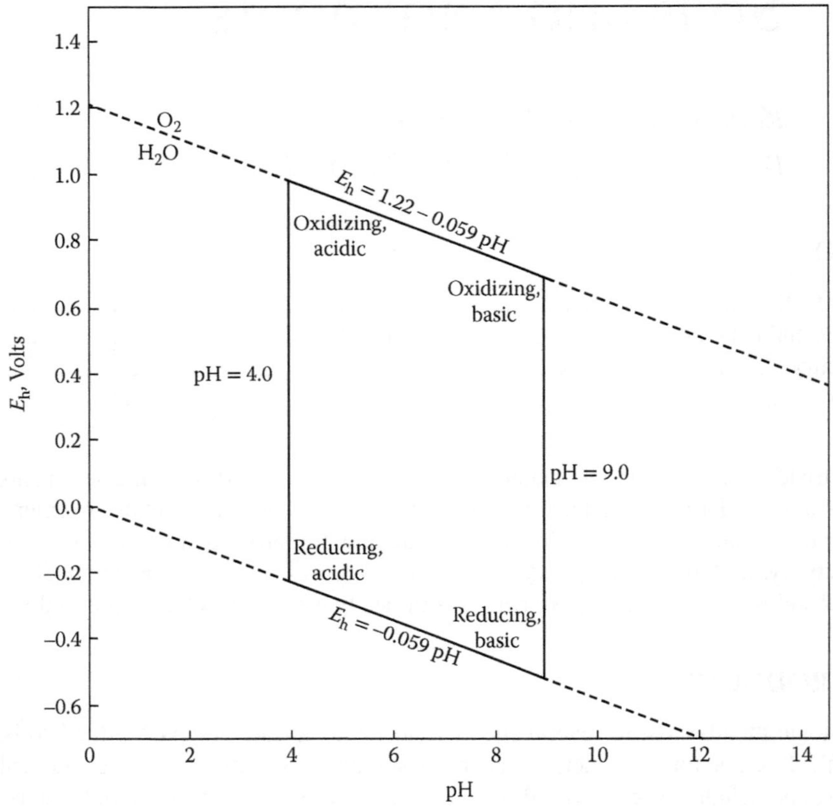

FIGURE 2.1 Framework of E_h–pH diagrams. The parallelogram outlines the usual limits of E_h and pH found in near-surface environments. (Modified from Krauskoft, K.B., *Introduction to Geochemistry*, McGraw-Hill, New York, 1967.)

2.2 REDOX AND pH CONDITIONS: EFFECTS ON METAL TRANSFORMATION

Metal cycling in wetlands is very dependent on pH. Soil redox conditions (or E_h status) also govern the oxidation and reduction of some trace metals that are found in wetland soils and sediments. The range of these two controlling parameters can be viewed as acid-alkaline and oxidized-reduced, respectively. The pH of a soil is a measure of the acidity/alkalinity of the sediment. The redox potential or E_h of a soil characterizes the intensity of oxidation or reduction. The oxidation status of the sediment has a large effect on the chemical form of many toxic heavy metals. Shaheen et al. (2014) showed that the combined effect of these two parameters has a very large effect on the solubility of most toxic heavy metals. The ranges of the two parameters usually encountered in nature are shown in Figure 2.1. The acidity of a soil has a large influence on the solubility of toxic, heavy metal ions. Changes in oxidation–reduction (redox) potential can result from natural (e.g., flooding) and anthropogenic (e.g., dredging) alterations of soil and sediment conditions. Changes in metal and metalloid speciation generally follow these changes in sediment E_h–pH conditions (Yu et al., 2007). Alterations in sediment E_h–pH conditions are often driven by microbially mediated reduction of alternate electron acceptors, including nitrate, sulfate, and oxidized forms of iron and manganese. After the depletion of oxygen, the reduction of these various alternate electron acceptors occurs in a well-defined sequence that can affect the solubility and speciation of trace metals present in the sediment (Figure 2.2) (DeLaune et al., 2004).

Concentrations of metals in sediment porewater, in turn, are controlled by sorption and precipitation reactions that are dependent on sediment properties, metal speciation, porewater composition, and fundamental physicochemical conditions such as pH and oxidation–reduction (redox) potential, or E_h. Changes in sediment E_h–pH conditions can cause changes in metal speciation and solubility, which can, subsequently, result in a flux from bottom sediments to overlying water. Changes in sediment E_h–pH conditions have also been shown to produce wide differences in metal and metalloid speciation and in solubility under controlled conditions (Frohne et al., 2011). Sediment E_h status is a complex, difficult-to-measure parameter. It is possible to classically define E_h in terms of a single chemical system (e.g., using the Nernst equation), whereas sediment porewater is a complex mixture of compounds undergoing redox reactions in various stages of nonequilibrium.

Measurements of redox potential in these systems, therefore, represent "mixed" potentials with contributions from several redox couples that may or may not be fully responsive to the E_h electrodes. The measurement of E_h is commonly performed by using platinum electrodes (Yu et al., 2007). Oxides of Fe, and perhaps Mn and Al, effectively adsorb or occlude most toxic metals (Figure 2.3). These oxides exist in mineral soils in large quantities. When soils are reduced, the metals that are bound to Fe and Mn oxides may be, to some degree, entirely transformed into readily available forms due to dissolution of the Fe and Mn oxides or partially transformed into other potentially available forms. During flooding and drainage cycles of wetlands, the formation of iron oxyhydroxides is important in retaining metals in surface soils (Gambrell, 1994; Pentrakova et al., 2013). Metals that are bound to oxides, hydroxides, and carbonates are effectively immobilized

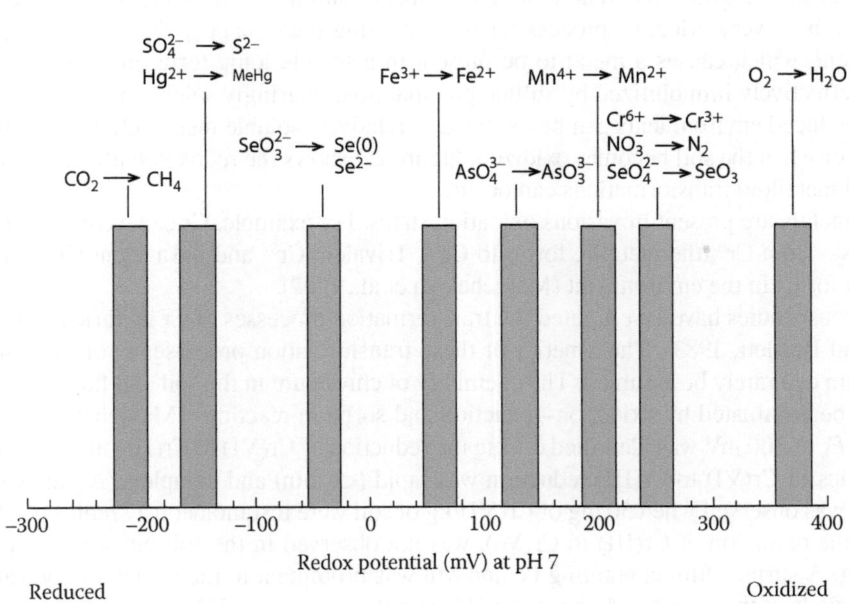

FIGURE 2.2 Soil redox conditions influencing metal transformation in wetland soils and sediments. (Modified from Reddy, K.R. and DeLaune, R.D., *Biogeochemistry of Wetlands: Science and Application*, CRC Press, Taylor & Francis Group, Boca Raton, FL, 800 pp., 2008.)

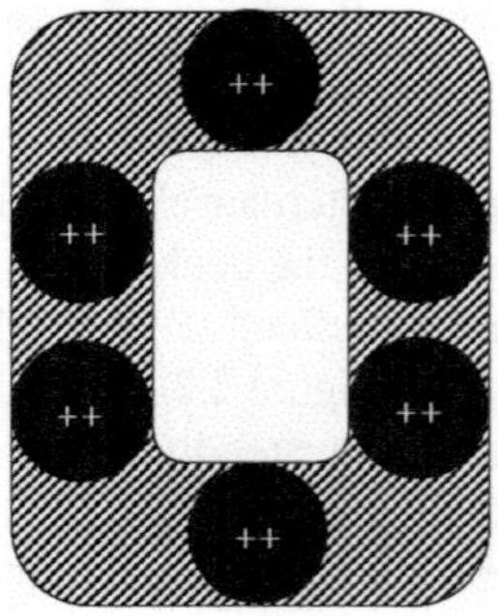

FIGURE 2.3 Sediment particle with a coating of amorphous iron oxyhydroxide containing co-precipitated trace and toxic metals.

at near-neutral to somewhat alkaline pH conditions. If, however, the pH becomes moderately to strongly acidic, as can sometimes occur when reduced soils become oxidized, these metals can be transformed into readily available forms (Figure 2.4).

Metals that are complexed with insoluble, large-molecular-weight, and naturally occurring humic compounds are effectively immobilized. There is some evidence that these metals are less effectively immobilized if reduced soils are oxidized (Figure 2.5). Oxidation of wetland soils can result in a significant release of heavy metals that are tied up with humic compounds (Gambrell and Patrick, 1988, 1994). Low pH and redox potentials in sediment–water systems tend to favor the formation of soluble species of many metals; whereas in oxidized, nonacidic systems, slightly soluble or insoluble forms tend to predominate (Figure 2.6). However, pH, and particularly E_h in combination, may regulate other processes such as sulfide formation, which influences the solubility of metals. In wetlands with high sulfate inputs and reduced sediment environments, the formation and accumulation of metal sulfides occur. The solubility of divalent metal sulfides in these systems is extremely low (Figure 2.7). Where large amounts of sulfide are present, sulfide precipitation is believed to be a very effective process for immobilizing trace metals. Thus, a strongly reducing environment, which causes a metal to be present in a soluble ionic form, may also contribute to its being effectively immobilized by sulfide precipitation. Sparingly soluble metal sulfides that are stable in reduced environments can be oxidized to relatively soluble metal sulfates in aerobic environments or when the soil becomes oxidized. Figure 2.2 shows the redox potential at which various metal and metalloid transformations can occur.

Trace metals are present in various oxidation states. For example, Cr can exist in several oxidation states—from Cr^0 (the metallic form) to Cr^{6+}. Trivalent Cr^{3+} and hexavalent Cr^{6+} are the most important forms in the environment (Masscheleyn et al., 1992).

Numerous studies have investigated the transformation processes of Cr in various environments (James and Bartlett, 1983). The kinetics of these transformation processes is of interest, because equilibrium can rarely be assumed. The chemistry of chromium in the soil and floodwater has been shown to be dominated by oxidation–reduction and sorption reactions (Masscheleyn et al., 1992). A critical E_h of 300 mV was identified during the reduction of Cr(VI) to Cr(III) in soils (Figure 2.8). The kinetics of Cr(VI) to Cr(III) reduction was rapid (<1 min) and complete. A high capacity for reduction was observed when 50 mg of Cr(VI)/kg of soil were instantaneously reduced. The reverse reaction, the oxidation of Cr(III) to Cr(VI), was not observed in the soil but was observed in the floodwater. A surface film containing Fe and Mn was prominent in the wetland floodwater during certain periods of the year. Oxidation of Cr (Masscheleyn et al., 1992) has been shown to be dependent on the presence of amorphous manganese oxides. The presence of these compounds in the surface film was attributed to the oxidation of Cr in floodwater.

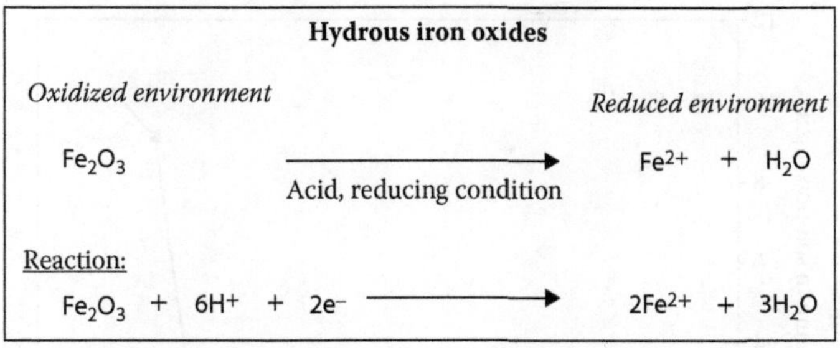

FIGURE 2.4 Hydrous iron oxide reaction in wetland soils and sediments.

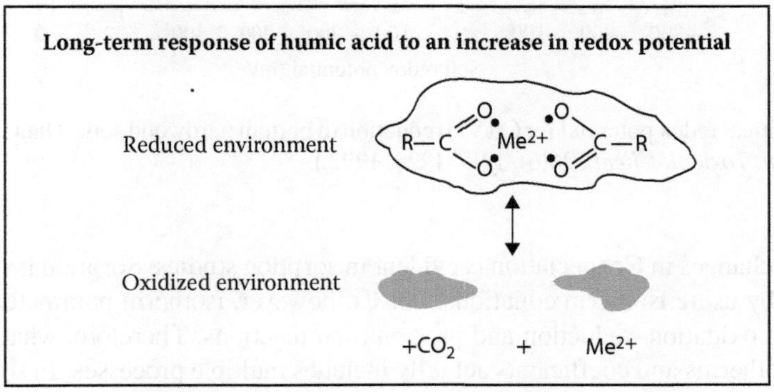

FIGURE 2.5 Long-term response of humic acid to an increase in redox potential.

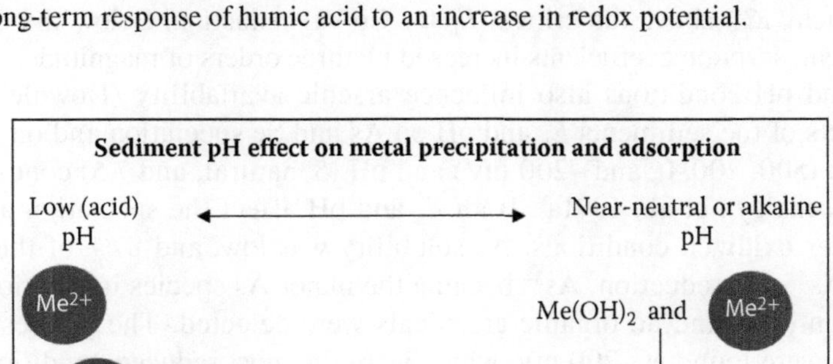

FIGURE 2.6 Sediment pH effect on metal precipitation and adsorption.

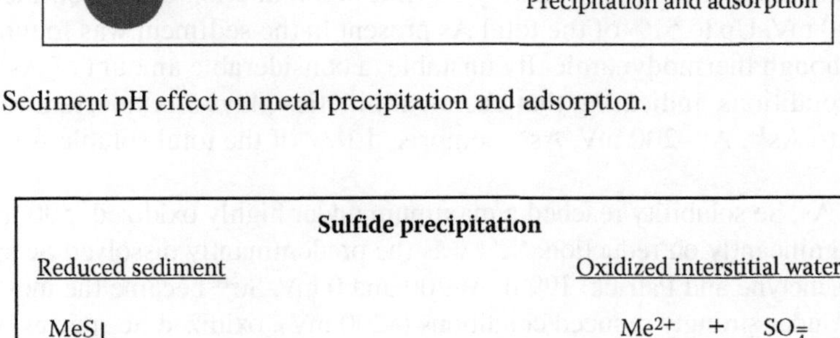

FIGURE 2.7 Role of sulfide in metal precipitation and release in wetland soils and sediments.

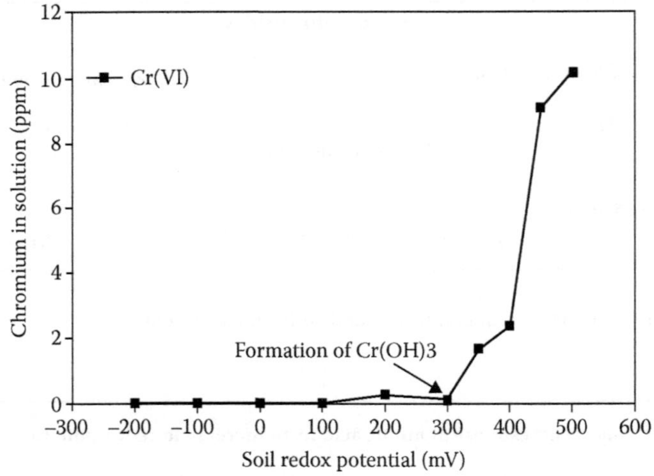

FIGURE 2.8 Critical redox potential for Cr(VI) reduction in bottom hardwood soil. (Data from Masscheleyn, P.H. et al., *Environ. Toxicol. Chem.*, 26(6), 1217–1226, 1992.)

The effect of changes in Cr speciation is evident in sorption studies. Sorption is often mathematically described by using isotherm equations. For Cr, however, isotherm parameters lump not only sorption but also oxidation–reduction and precipitation reactions. Therefore, what is referred to as Cr "sorption" isotherms and coefficients actually includes multiple processes. In sorption isotherms conducted under controlled E_h conditions, sorption was weak (as indicated by low sorption coefficients, Freundlich "K") above the critical E_h for Cr(VI) reduction. Below the critical E_h, when Cr(III) was present, sorption coefficients increased by three orders of magnitude.

Soil redox and pH conditions also influence arsenic availability (Dowdle and Oremland, 1998). The effects of the sediments E_h and pH on As and Se speciation and on solubility under controlled redox (500, 200, 0, and –200 mV) and pH (5, natural, and 7.5) conditions have been quantified (Masscheleyn et al., 1991a). Both E_h and pH affect the speciation and solubility of As and Se. Under oxidized conditions, As solubility was low, and 87% of the As in solution was present as As^{5+}. On reduction, As^{3+} became the major As species in solution, As solubility increased substantially, and no organic arsenicals were detected. The greatest As concentrations in solution were found at –200 mV, which were the most reduced conditions studied. The total As in solution increased approximately 25 times on reduction of the sediment suspension from 500 to –200 mV. Up to 51% of the total As present in the sediment was found to be soluble at –200 mV. Although thermodynamically unstable, a considerable amount of As^{5+} was present under reduced conditions, indicating that chemical kinetics plays an important role in the conversion of As $^{5+}$ to As^{3+}. At –200 mV, As^{5+} comprised 18% of the total soluble As (Masscheleyn et al., 1991b).

In contrast to As, Se solubility reached a maximum under highly oxidized (500 mV) conditions and decreased significantly on reduction. Se^{6+} was the predominantly dissolved Se species present at 500 mV (Masschelyne and Patrick, 1993). At 200 and 0 mV, Se^{4+} became the most stable oxidation state of Se. Under strongly reduced conditions (–200 mV), oxidized Se species were no longer detectable and Se solubility was controlled by the formation of elemental Se and/or metal selenides. Biomethylation of Se was important under oxidized and moderately reduced conditions (500, 200, and 0 mV). More alkaline conditions (pH 7.5) resulted in greater dissolved As and Se concentrations in solution (increased up to ten and six times, respectively) compared with the more acidic equilibrations (Masscheleyn et al., 1991b,c). Figure 2.9 shows the effect of redox potential on As and Se solubility in sediments.

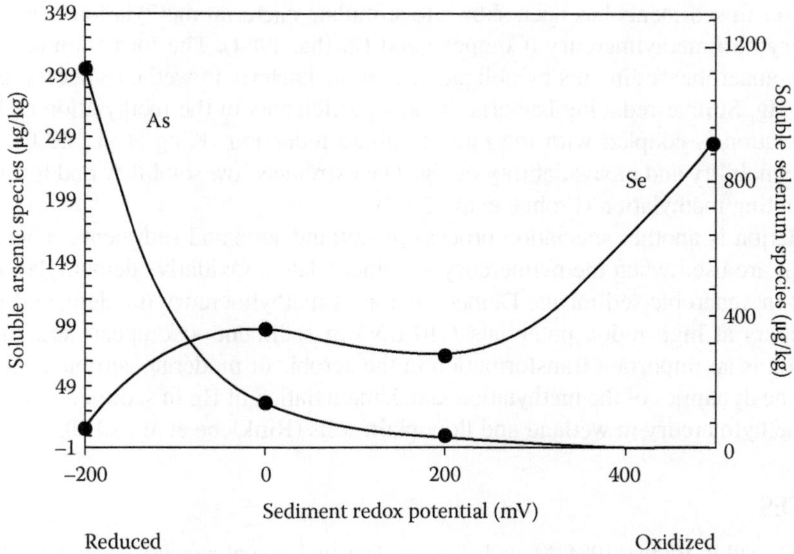

FIGURE 2.9 Influence of redox potential on solubility of As and Se in a sediment suspension at pH 7.0.

2.3 MERCURY

Methylation and demethylation of mercury in wetland soils and sediment are influenced by factors such as redox potential, pH, sulfate concentration, and microbial activity. Methylation in sediments is attributed primarily to anaerobic sulfate-reducing bacteria such as Desulfovibrio (Compeau and Bartha, 1984). DeLaune et al. (2004) showed that methylation of added Hg in sediments was greater under reduced conditions as compared with oxidized conditions (Figure 2.10). High salinity (sulfate) levels inhibited methylation, and methylation increased when sulfate levels were low.

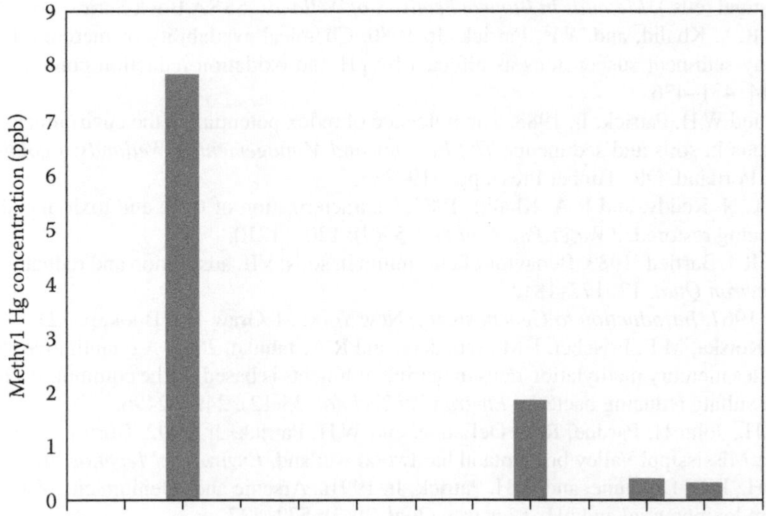

FIGURE 2.10 Effect of sediment redox conditions on changes in methylmercury concentration in Louisiana River sediment. Each point represents an average of two replications. (Modified from DeLaune, R.D. et al., *J. Environ. Sci. Health Part A*, 39(8), 1925–1935, 2004.)

Organic matter in sediments has been shown to stimulate bacterial methylation that converts inorganic mercury into methylmercury (Compeau and Bartha, 1984). The formation of soluble methylmercury in anaerobic sediments by obligate anaerobic bacteria in wetlands serves as a source of bioavailable Hg. Sulfate-reducing bacteria are key participants in the methylation of Hg. The rate of Hg methylation is coupled with the rate of sulfate reduction (King et al., 2001). Methylation enhances the mobility and bioavailability of Hg. The extremely low solubility and formation of H_2S help in preventing methylation (Frohne et al., 2012).

Demethylation is another speciation process in wetland soils and sediments. There are several pathways that are used when methylmercury is demethylated. Oxidative demethylation is a dominant process in anaerobic sediments. Demethylaiton of methylmercury (to elemental mercury and methane) occurs at high redox potentials (110 mV) in sediments (Compeau and Bartha, 1984). Demethylation is an important transformation in the aerobic or moderate aerobic zone in soils and sediments. The dynamics of the methylation and demethylation of Hg in sediments is a key factor in the flux of methylmercury in wetland and floodplain soils (Rinklebe et al., 2010).

REFERENCES

Compeau, G.C., and R. Bartha. 1984. Methylation and demethylation of mercury under controlled redox, pH and salinity conditions. *Appl Environ Microbiol.* 48: 1203–1207.

DeLaune, R.D., A. Jugsujinda, I. Devai, and W.H. Patrick, Jr. 2004. Relationship of sediment redox conditions to methylmercury in surface sediment of Louisiana Lakes. *J Environ Sci Health Part A* 39(8): 1925–1935.

Dowdle, P.R., and R.S. Oremland. 1998. Microbial oxidation of elemental selenium in soul slurries and bacterial cultures. *Environ Sci Technol.* 32: 3749–3755. doi:10.1021/es970940s.

Frohne, T., J. Rinklebe, R.A. Diaz-Bone, and G. Du Laing. 2011. Controlled variation of redox conditions in a floodplain soil: Impact on metal mobilization and biomethylation of arsenic and antimony. *Geoderma* 160: 414–424. doi:10.1016/j-geoderma.2010.10.012.

Frohne, T., J. Rinklebe, U. Langer, G. Du Laing, S. Mothes, and R. Wennrich. 2012. Biogeochemical factors affecting mercury methylation rate in two contaminated floodplain soils. *Biogeosciences* 9: 493–507. doi:10.5194/bg-9-493-2012.

Gambrell, R.P. 1994. Trace and toxic metals in wetlands: A review. *J Environ Qual.* 23(5): 883–891.

Gambrell, R.P. 2013. Total metal analyses and extractions. In: DeLaune, R.D., K.R. Reddy, C.J. Richardson, and J.P. Megonigal (eds.), *Methods in Biogeochemistry of Wetlands*. SSSA Book Series, no. 10. pp. 775–799.

Gambrell, R.P., R.A. Khalid, and W.H. Patrick, Jr. 1980. Chemical availability of mercury, lead and zinc in Mobile Bay sediment suspensions as affected by pH and oxidation-reduction conditions. *Environ Sci Technol.* 14: 431–436.

Gambrell, R.P., and W.H. Patrick, Jr. 1988. The influence of redox potential on the environmental chemistry of contaminants in soils and sediments. *The Ecology and Management of Wetlands*: *Ecology of Wetlands*, volume 1. Portland, OR: Timber Press, pp. 319–333.

Gambrell, R.P., C.N. Reddy, and R.A. Khalid. 1983. Characterization of trace and toxic metals in sediments of a lake being restored. *J Water Poll Con Fed.* 55(9): 1201–1210.

James, B.R. and R.J. Bartlett. 1983. Behavior of chromium in soils: VII, adsorption and reduction of hexavalent form. *J Environ Qual.* 12: 177–181.

Krauskoft, K.B. 1967. *Introduction to Geochemistry*. New York: McGraw-Hill Book, p. 721.

King, J.K., J.E. Kotska, M.E. Frischer, F.M. Saunders, and R.A. Jahnke. 2001. A quantitative relationship that demonstrates mercury methylation rates in marine sediments is based on the community composition and activity of sulfate reducing bacteria. *Environ Sci Technol.* 35(12): 2491–2496.

Masscheleyn, P.H., John H. Pardue, R.D. DeLaune, and W.H. Patrick, Jr. 1992. Chromium redox chemistry in a Lower Mississippi Valley bottomland hardwood wetland. *Environ Sci Technol.* 26(6): 1217–1226.

Masscheleyn, P.H., R.D. DeLaune, and W.H. Patrick, Jr. 1991a. Arsenic and selenium chemistry as affected by sediment redox potential and pH. *J Environ Qual.* 20(3): 522–527.

Masscheleyn, P.H., R.D. DeLaune, and W.H. Patrick, Jr. 1991b. Biogeochemical behavior of selenium in anoxic soils and sediments: An equilibrium thermodynamics approach. *J Environ Sci Health Part A* 26(4): 555–573.

Masscheleyn, P.H., R.D. DeLaune, and W.H. Patrick, Jr. 1991c. Effect of redox potential and pH on arsenic speciation and solubility in a contaminated soil. *Environ Sci Technol.* 25(8): 1414–1419.

Masscheleyn, P.H., and W.H. Patrick, Jr. 1993. Biogeochemical processes affecting selenium cycling in wetlands. *Environ Toxicol Chem.* 12: 2235–2243.

Pentrakova, L., K. Su., M. Pentrak, and J.W. Stucki. 2013. A review of microbial redox interactions with structural Fe in clay minerals. *Clay Miner.* 48: 543–560.

Reddy, K.R., and R.D. DeLaune. 2008. *Biogeochemistry of Wetlands: Science and Application*. Boca Raton, FL: CRC Press, Taylor & Francis Group, p. 800.

Rinklebe, J., A. During, M. Overesch, G. Du Laing, R. Wennrich, H.J. Stark, and S. Mothes. 2010. Dynamics of mercury fluxes and their controlling factors in large Hg-polluted floodplain areas. *Environ Pollut.* 158: 308–318. doi:10.1016/j.envpol.2009.07.001.

Shaheen, S.M., J. Rinklebe, T. Frohne, J.R. White, and R.D. DeLaune. 2014. Biogeochemical factors governing cobalt, nickel, selenium, and vanadium dynamics in periodically flooded Egyptian North Nile delta rice soils. *Soil Sci Soc Am J.* doi:10.2136/sssaj2013.10.0441.

Yu, K., F. Böhme, J. Rinklebe, H.U. Neue, and R.D. DeLaune. 2007. Major biogeochemical processes in soils—A microcosm incubation from reducing to oxidizing conditions. *Soil Sci Soc Am J.* 71: 1406–1417. doi:10.2136/ssaj2006.0155.

3 Redox Reactions of Heavy Metal(loid)s in Soils and Sediments in Relation to Bioavailability and Remediation

Anitha Kunhikrishnan, Balaji Seshadri, Girish Choppala, Shiv Shankar, Ramya Thangarajan, and Nanthi Bolan

CONTENTS

3.1 INTRODUCTION

Soils and sediments represent the major sink for heavy metal(loid)s that are released into the biosphere through both geogenic (i.e., weathering or pedogenic) and anthropogenic (i.e., human activities) processes. The mobility and bioavailability of heavy metal(loid)s in soils and sediments are affected by both physicochemical and redox reactions (Figure 3.1) (Alexander 2000; Adriano 2001). Redox reactions play a key role in the behavior and fate of toxic heavy metal(loid)s, especially arsenic (As), chromium (Cr), mercury (Hg), and selenium (Se), in soils and sediments by influencing their speciation (Gadd 2010). For example, As contamination of surface- and groundwater, mediated through redox reactions of geogenic As, became a major human health issue at several points around the globe (Mahimairaja et al. 2005). Thus, a greater understanding of redox reactions will help in monitoring the environmental fate of the heavy metal(loid)s, and will also aid in developing *in situ* bioremediation technologies that are environmentally compatible. It is unlikely that the natural phenomena (e.g., natural attenuation) are optimal for the removal of toxic heavy metal(loid)s

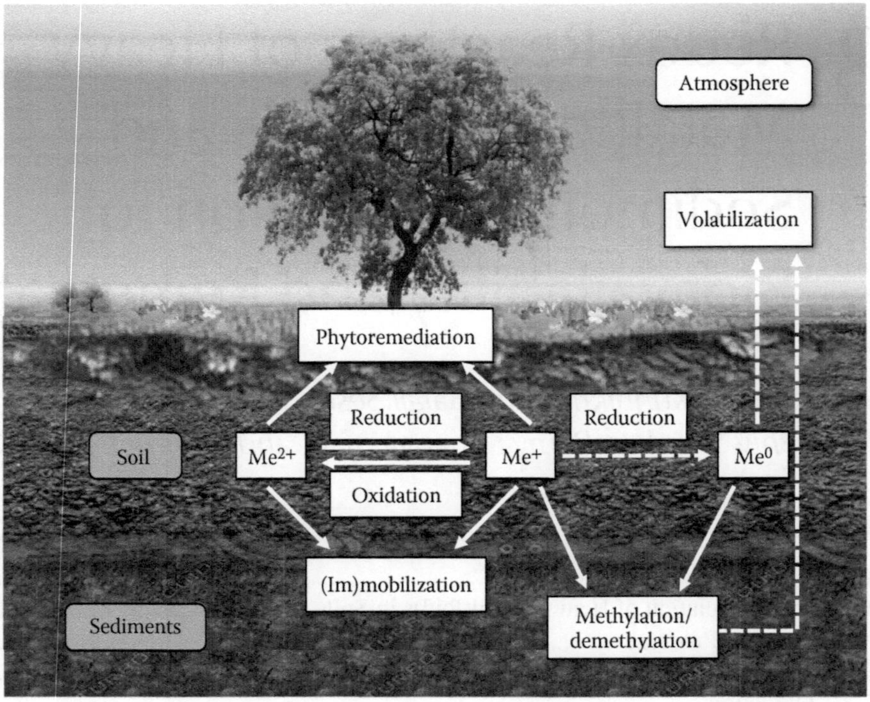

FIGURE 3.1 (**See color insert.**) Dynamics of heavy metal(loid) redox transformation in soils and sediments.

from contaminated sites. Redox reactions can readily be managed and enhanced for the efficient removal of contaminants, provided the biochemistry of these processes is understood.

In this chapter, we first briefly introduce the sources and speciation of four major heavy metal(loid)s that are subject to redox reactions in soils and sediments. These include As, Cr, Hg, and Se. We then describe two case studies of redox reactions in the context of the practical implications they have for remediation and ecotoxicology.

3.2 SOURCES AND SPECIATION OF HEAVY METAL(LOID)S IN SOILS AND SEDIMENTS

All heavy metal(loid)s originate from the crust of the earth and appear in soils due to geogenic (natural weathering of igneous and sedimentary rocks) and anthropogenic (mining) activities. Arsenic is among the most researched heavy metalloids in terms of human health because of its carcinogenicity and transformation in the soil environment; this is evident through its interaction with other elements, including iron (Fe) and aluminum (Al). Apart from the natural weathering of rocks, significant anthropogenic sources of As include fossil fuel combustion, leaching from mining wastes and landfills, mineral processing, and metal(loid) production. Application of a range of agricultural byproducts (e.g., poultry manure) also contributes large quantities of As to the land (Christen 2001). Although the anthropogenic As source is becoming increasingly important, the recent episode of extensive As contamination of groundwater in Bangladesh and West Bengal is of geological origin (Mahimairaja et al. 2005).

In soils, As is present as arsenite [As(III)], arsenate [As(V)], and organic As (monomethyl arsenic acid and dimethyl arsenic acid or cacodylic acid) (Sadiq 1997; Smith et al. 1998; Mahimairaja et al. 2005). Arsenic species are adsorbed onto iron (Fe), manganese (Mn), and aluminum (Al) compounds (Smith et al. 1998). In aquatic systems, As is predominantly bound to sediments and As concentration in suspended solids and sediments is many times higher than that in water (Mahimairaja et al. 2005). Arsenic oxidation from As(III) to As(V) is a natural process that helps in alleviating toxicity in aquatic environments, because As(V) is adsorbed onto the sediments and becomes relatively immobilized (Aposhian et al. 2003; Rubinos et al. 2011).

Chromium is another potentially toxic metal that reaches the soil environment via industrial waste disposal from coal-fired power plants, electroplating activities, leather tanning, timber treatment, pulp production, and mineral ore and petroleum refining (Bolan et al. 2003; Choppala et al. 2012). Chromium exists in hexavalent [Cr(VI)] and trivalent [Cr(III)] forms. It has been found that Cr(VI) is toxic and highly soluble in water, whereas Cr(III) is less toxic, insoluble in water, and hence less mobile in soils (Barnhart 1997; Kosolapov et al. 2004). In soils and sediments, Cr exists mainly as Cr(III) unless oxidizing agents such as manganous oxide [Mn(IV)] are present (Gong and Donahoe 1997).

The burning of fossil fuels and gold recovery in mining are the major sources of Hg (Pacyna et al. 2001; de Lacerda 2003). Similar to As and Cr, Hg also occurs in two ionic states: mercurous [Hg(I)] and mercuric [Hg(II)] salts, with the latter being much more common in the environment than the former (Schroeder and Munthe 1998). Mercury also forms organometallic compounds, and elemental Hg gives rise to a vapor that is only slightly soluble in water (Boening 2000). In an aquatic environment, Hg is dominated by methyl mercury and Hg(II)–ligand pairs in water and Hg(II) in sediments (Weiner et al. 2003).

Selenium is used in xerography, as a semiconductor in photocells, and in the manufacture of batteries, glass, electronic equipment, anti-dandruff products, veterinary therapeutic agents, feed additives, and fertilizers. Elemental Se exists in a zero-valence state and is often associated with sulfur in compounds such as selenium sulfide (SeS) and polysulfides. Selenium occurs in four oxidation states: selenate [Se(VI); SeO_4^{2-}], selenite [Se(IV); SeO_3^{2-}], elemental selenium [Se(0); Se^0], and selenide [Se(-II); Se^{2-}]. Selenate and Se(IV) are common ions in soil and sediments. Under reducing conditions, selenides and Se(0) are the common Se species in the presence of organic matter. Reduced Se compounds also include volatile methylated species such as dimethyl selenide [DMSe, $Se(CH_3)_2$], dimethyl diselenide [DMDSe, $Se_2(CH_3)_2$] and dimethyl selenone [$(CH_3)_2$, SeO_2], and sulfur-containing amino acids, including selenomethionine, selenocysteine, and selenocystine.

3.3 REDOX REACTION PROCESSES

Heavy metal(loid)s, including As, Cr, Hg, and Se, are the most commonly subjected metal(loid)s to microbial oxidation/reduction reactions (Table 3.1). Redox reactions influence the speciation and mobility of heavy metal(loid)s. For example, metals are generally less soluble in their higher oxidation state, whereas the solubility and mobility of metalloids depend on both the oxidation state and the ionic form (Ross 1994). The oxidation/reduction reactions for various metal(loid)s and the optimum redox values for these reactions are given in Table 3.2. Redox reactions of heavy metal(loid)s in soils and sediments are induced by both biotic and abiotic processes, and they are grouped into three categories: reduction, oxidation, and methylation/demethylation.

3.3.1 Reduction

Under reduced conditions, As(III) dominates in soils, but elemental arsenic [As(0)] and arsine (H_2As) may also be present. Arsenite is more toxic and mobile than As(V). The distribution and mobilization of As species in the soil and sediments is controlled by both redox and adsorption reactions (Adriano et al. 2004; Mahimairaja et al. 2005). The reduction and methylation reactions of As in sediments are generally mediated by the bacterial degradation of organic matter coupled with the reduction and use of sulfate as the terminal electron acceptor (Adriano et al. 2004). Ferrous iron[Fe(II)] can also serve as an electron acceptor in the bacterial oxidization of organic matter, resulting in the decomposition of ferric [Fe(III)] oxides and hydroxides.

The reduction of Cr(VI) to Cr(III) is mediated through both biotic and abiotic processes (Choppala et al. 2012). Chromate can be reduced to Cr(III) in environments where a ready source

TABLE 3.1
Selected References on the Redox Reactions of Arsenic, Chromium, Mercury, and Selenium in Soil and Aquatic Environments

Trace Element	Medium	Observation	Reference
Arsenic(V)	Soil	Dissimilatory As(V)–reducing bacterium, *Bacillus selenatarsenatis* increased the removal of As from contaminated soils.	Yamamura et al. (2008)
	Soil	Both biotic and abiotic (S^{2-}, Fe^{2+}, and $H_2(g)$) factors were responsible for the reduction of As(V).	Jones et al. (2000)
	Soil	Reduction mediated by H_2S was dominant in reducing conditions.	Rochette (2000)
	Soil	Attachment of *Shewanella putrefaciens* cells to oxide mineral surfaces promoted As(V) desorption, thereby facilitating its reduction.	Huang et al. (2011)
	Sediments	Dissimilatory As(V) reduction by extremely halophilic, anaerobic archaea	Kulp et al. (2006)
	Sediments	Anaerobic metal(loid)-reducing bacteria formed As(III) in sediments from the Ganges delta.	Islam et al. (2004)
Arsenic(III)	Soil	Fe(III) and Mn(III) oxides oxidized As(III) to As(V) through an electron-transfer reaction.	Mahimairaja et al. (2005)
	Water	Sorption onto magnetite oxidized As(III) to As(V) under oxic conditions at neutral pH.	Ona-Nguema et al. (2010)
	Sediments	*Acinetobacter junni* and *Marinobacter* sp. isolated in the contaminated area contained the *aox* genes and were able to oxidize As(III) to As(V).	Chang et al. (2011)
	Sediments	As(III) was oxidized to As(V) by Mn minerals present in the oxidized sediment.	Stollenwerk et al. (2007)
Chromium(VI)	Soil	Application of Fe(II) under flow conditions increased reduction of Cr(VI).	Franco et al. (2009)
	Soil	Root exudates of *Typha latifolia* and *Carex lurida* increased sulfide species, which facilitated Cr(VI) reduction in sediment pore water.	Zazo et al. (2008)
	Soil	Organic amendments increased DOC, which reduced Cr(VI) to Cr(III) in soils.	Bolan et al. (2003a)
	Soil	Addition of glucose promoted both biotic and abiotic Cr(VI) reduction in soils.	Leita et al. (2011)
	Sediments	Presence of plants enhanced Cr(VI) reduction in wetland sediments through the release of root exudates and evapotranspiration.	Zazo et al. (2008)
	Sediments	Actinomycetes (SM-11, SM-20) isolated from the marine sediments showed considerable Cr reduction activity even at high Cr concentrations.	Jain et al. (2012)
	Sediments	Halophilic Cr (VI)-resistant bacterial strain TA-04 was isolated from polluted marine sediments, and Cr reduction was significant at high NaCl concentrations.	Focardi et al. (2012)
Chromium(III)	Soil	Cr(III) was oxidized by atmospheric oxygen at a high temperature to Cr(VI) in tannery sludge-contaminated sites.	Apte et al. (2006)
	Soil	Hydrous Mn(IV) oxides reacted with Cr(III) hydroxides and influenced the rate of Cr(III) oxidation.	Landrot et al. (2009)

TABLE 3.1 ***(Continued)***
Selected References on the Redox Reactions of Arsenic, Chromium, Mercury, and Selenium in Soil and Aquatic Environments

Trace Element	Medium	Observation	Reference
	Soil	In the alkaline soil with moderate organic C at the ore processing site, most of the Cr(VI) remained dissolved after H_2O_2 had decayed, indicating the mobilization of Cr(VI). As the oxidation of organic C promotes disintegration of soil structure, it increases the access of solution to Cr(VI) mineral phases.	Rock et al. (2001)
	Soil	The Mn oxide salts, birnessite and todorokite had the same capacity to oxidize Cr(III) to Cr(VI) in soils.	Kim et al. (2002)
	Sediments and water	Mn oxides oxidized Cr(III) to Cr(VI).	Marafatto et al. (2013)
	Sediments	Rates of aeration-driven Cr(VI) reoccurrence [from the abiotic oxidation of Cr(III)] increased with pH and positively correlated with dissolved Mn decline at pH ≥ 7.	Wadhawan et al. (2013)
Mercury(II)	Water	Carboxylic groups in the humic acids reduced Hg(II) to Hg(0).	Allard and Arsenie (1991)
	Water	Magnetite reduced Hg(II) to Hg(0). The reaction rates increased with an increase in magnetite surface area and solution pH, and they decreased with an increase in chloride concentration.	Wiatrowski et al. (2009)
	Sediments and water	Hg(II) can be effectively reduced to Hg(0) in the presence of as little as 0.2 mg/L reduced humic acids, whereas the production of Hg(0) is inhibited by complexation as HA concentration increases.	Gu et al. (2011)
	Water	Hg(II) reaction with biogenic magnetite kinetically favored the pathway for Hg(II) reduction by dissimilatory Fe-reducing bacteria.	Yee et al. (2010)
	Water	Hg(0) production coupled to a reduction by Fe(II) under dark anoxic conditions increased with an increase in pH and aqueous Fe(II) concentrations.	Amirbahman et al. (2013)
	Sediments	In subsurface sediment incubations, Hg may inhibit denitrification and that inhibition may be alleviated when Hg-resistant denitrifying *Bradyrhizobium* spp. detoxify Hg by its reduction to Hg(0).	Wang et al. (2013)
Selenium(VI)	Soil	In suboxic conditions, green rust [Fe(II,III)] reduced Se(VI) to Se(0).	Myneni et al. (1997)
	Water	Microorganisms reduced Se(VI) to Se(IV), and that reduction increased in the presence of lactate.	Maiers et al. (1988)
	Soil	Organic amendments and low oxygen level in the soil reduced Se(VI) to Se(IV).	Guo et al. (1999)
	Soil and water	*Moraxea bovis* and bacterial consortia reduced Se(VI) to Se(IV) and Se(0).	Biswas et al. (2011)
	Sediments	*Ferrimonas futtsuensis* sp. nov. and *Ferrimonaskyonanensis* sp. nov., Se(VI)-reducing bacteria, reduced Se(VI) to Se(0) and used lactate, pyruvate, yeast extract, tryptone, and Casamino acids as electron donors and carbon sources.	Nakagawa et al. (2006)

(Continued)

TABLE 3.1 ***(Continued)***
Selected References on the Redox Reactions of Arsenic, Chromium, Mercury, and Selenium in Soil and Aquatic Environments

Trace Element	Medium	Observation	Reference
	Water	Zerovalent Fe nanoparticles efficiently reduced Se(VI) to Se(-II).	Olegario et al. (2010)
Selenium(VI) and (IV)	Water	Rice straw, a good source of carbon and energy, helped several bacteria in reducing Se(VI) and Se(IV) to Se(0).	Frankenberger et al. (2005)
	Sediments and water	Molasses, an organic-carbon source, was used by several bacteria to reduce Se(VI) and Se(IV) to Se(0).	Zhang et al. (2008)
	Sediments	Abiotic reduction of Se(IV) to Se(0) occurred in the presence of ferrous iron.	Chen et al. (2009)
	Sediments	*Desulfurispirillum indicum* sp. nov., Se(VI)-and Se(IV)-reducing bacteria, which was isolated from estuarine sediments, used lactate, pyruvate, and acetate as electron donors and carbon sources for reduction.	Rauschenbach et al. (2011)
	Sediments	*Duganella* sp. and *Agrobacterium* sp. were capable of reducing Se(IV) to Se(0) under aerobic conditions.	Bajaj et al. (2012)
Se(0)	Sediments	*Bacillus selenitireducens* and some other bacteria were capable of reducing Se(0) to Se(-II) in anoxic sediments.	Herbel et al. (2003)

TABLE 3.2
Oxidation–Reduction Reactions of Selected Elements and the Optimum Redox Potentials

Trace Element	Transformation	Reaction	E_o (mV)
Arsenic(0)	$As(0) + 3H_2O \rightarrow H_3AsO_3 + 3H^+ - 3e^-$	Oxidation	250
Arsenic(V)	$H_3AsO_4 + 2H^+ - 2e^- \rightarrow H_3AsO_3 + H_2O$	Reduction	560
Cadmium(0)	$Cd \rightarrow Cd^{2+} + 2e^-$	Oxidation	402
Cadmium(II)	$Cd^{2+} + 2e^- \rightarrow Cd$	Reduction	–400
Chromium(III)	$2Cr^{3+} + 3H_20 + 2MnO_4^- \rightarrow 2Cr_2O_7^{2-} + 6H^+ + 2MnO_2$	Oxidation	350
Chromium(VI)	$Cr_2O_7^{2-} + 14H^+ + 6e^- \rightarrow 2Cr^{3+} + 7H_2O$	Reduction	1360
Iron(II)	$Fe^{2+} + 2e^- \rightarrow Fe(s)$	Oxidation	–440
Iron(III)	$Fe^{3+} + e^- \rightarrow Fe^{2+}$	Reduction	770
Manganese(II)	$Mn^{2+} + 2e^- \rightarrow Mn$	Reduction	–1180
Manganese(IV)	$Mn^{4+} + 2e^- \rightarrow Mn^{2+}$	Reduction	1210
Manganese(VII)	$MnO_4^- + 8H^+ + 5e^- \rightarrow Mn^{2+} + 4H_2O$	Reduction	1510
Manganese(VII)	$MnO_4^- + + e^- \rightarrow MnO_4^{2-}$	Reduction	564
Mercury	$Hg_2^{2+} + 2e^- \rightarrow 2Hg$	Reduction	790
Nitrogen(V)	$2NO_3^- + 4H^+ + 2e^- \rightarrow 2NO_2 + 2H_2O$	Reduction	803
Selenium(0)	$Se(0)/H_2Se$	Reduction	–730
Selenium(VI)	SeO_4^{2-}/SeO_3^{2-}	Reduction	440
Selenium(VI)	$SeO_3^{2-}/Se(0)$	Reduction	180
Sulfur(II)	$S^{2-} + 2H^+ \rightarrow H_2S$	Reduction	–220
Sulfur(VI)	$SO_4^{2-} + H_2O + 2\,e^- \rightarrow SO_3^{2-} + 2\,OH^-$	Reduction	–520

Source: Bolan, N.S. et al., *Rev. Environ. Contam. Toxicol.*, 225, 1–56, 2013a.

of electrons [Fe(II)] is available. Suitable conditions for microbial Cr(VI) reduction occur when organic matter is present and can, thus, act as an electron donor, and Cr(VI) reduction is enhanced under acid rather than alkaline conditions (Hsu et al. 2009; Chen et al. 2010).

In living systems, Se tends to be reduced rather than oxidized, and reduction occurs under both aerobic and anaerobic conditions. Dissimilatory Se(IV) reduction to Se(0) is the major biological transformation with regard to the remediation of Se oxyanions in anoxic sediments (Lens et al. 2006). Selenite is either readily reduced to the elemental state by chemical reductants such as sulfide or hydroxylamine or biochemically reduced to elemental state by systems such as glutathione reductase. Hence, the precipitation of Se in its elemental form, which has been associated with bacterial dissimilatory Se(VI) reduction, has great environmental significance (Oremland et al. 1989, 2004).

Microorganisms play a major role in reducing reactive Hg(II) to nonreactive Hg(0), and this process may be subjected to volatilization losses. Bacteria are more important than eukaryotic phytoplankton in the reduction of Hg(II). Mercury-resistant bacteria can transform ionic mercury [Hg(II)] into metallic mercury [Hg(0)] by enzymatic reduction (Von Canstein et al. 2002). It is known that Hg(II) is reduced to Hg(0) by mercuric reductase, mercury resistance operon, and genetic system encoding transporters and regulators. The dissimilatory metal(loid)-reducing bacterium *Shewanella oneidensis* has been shown to reduce Hg(II) to Hg(0), and this process requires the presence of electron donors (Wiatrowski et al. 2006).

3.3.2 Oxidation

Arsenic in soils and sediments can be oxidized to As(V) by bacteria (Table 3.1) (Battaglia-Brunet et al. 2002; Bachate et al. 2012). Since As(V) is strongly retained by inorganic soil components, microbial oxidation results in the immobilization of As. Under well-drained conditions, As would present as $H_2AsO_4^-$ in acidic soils and as $HAsO_4^{2-}$ in alkaline soils.

The oxidation of Cr(III) to Cr(VI) is primarily mediated abiotically through oxidizing agents such as Mn(IV), and to a lesser extent by Fe(III). Although Cr(III) is strongly retained on soil particles, Cr(VI) is very weakly adsorbed in soils that are net negatively charged and is readily available for plant uptake and leaching to groundwater (James and Bartlett 1983; Leita et al. 2011). The oxidation of Cr(III) to Cr(VI) can also enhance the mobilization and bioavailability of Cr.

Redox potential and pH are key factors in the biogeochemistry of Se (Masscheleyn et al. 1990). A study conducted on them indicated that the oxidation of Se(-II,0) to Se(IV) was rapid and above 200 mV, and Se(IV) was slowly oxidized to Se(VI). Selenate was the predominant dissolved species present, and it constituted from 95% at a higher pH (8.9, 9) to 75% at a lower pH (7.5, 6.5) of the total soluble Se at 450 mV. Losi and Frankenberger (1998) observed that Se^0 oxidation in soils is largely biotic in nature, occurs at relatively slow rates, and yields both Se(VI) and Se(IV). They observed this in both heterotrophic and autotrophic oxidation.

Zawislanski and Zavarin (1996) observed that the oxidation rates correlated positively with temperature, with slow changes at 15°C and faster changes at 35°C. Their results did not vary significantly with regard to the different moisture conditions. Dowdle and Oremland (1998) studied the microbial oxidation of elemental Se in soil slurries. Hydrologic changes, where water tables are lowered, may result in drying out soils and reexposing them to atmospheric O_2. Under such conditions, Se(0) may be reoxidized to mobile Se(IV) and Se(VI). Zawislanski and Zavarin found a constant production rate of Se(IV) from Se(0); however, Se(VI) was produced in only small quantities and accounted for less than 5% of the Se(IV) production. This is because of the adsorption of Se(IV) onto soil particles, thereby limiting their availability to microbes. The oxidation and transformation of Hg(0) usually occur in the atmosphere because of the high volatility of Hg(0); soil emissions of Hg occur as a result of solar radiation and the resultant rise in temperature (Kikuchi et al. 2013).

3.3.3 Methylation/Demethylation

Methylation has been proposed as a biological mechanism for detoxification and the removal of toxic heavy metal(loid)s by converting them to methyl derivatives that are subsequently removed by volatilization or extraction with solvents (Frankenberger and Losi 1995). Methylation has been shown to be the major process by which As, Hg, and Se are volatilized from soils and sediments, and also by which poisonous methyl gas is released (Adriano et al. 2004). Volatilization occurs through the microbial conversion of metal(loid)s into their respective metallic, hydride, or methylated forms. These forms have low boiling points and/or high vapor pressure and are, therefore, susceptible for volatilization (Table 3.3).

Microorganisms in soils and sediments act as biologically active methylators (Frankenberger and Arshad 2001). Organic matter provides the methyl-donor source for methylation in soils and sediments. Methylation of Hg is controlled by low-molecular-weight fractions of fulvic acid in soils (Ravichandran 2004). Biomethylation is effective in forming volatile compounds of As, such as alkylarsines, which could easily be lost to the atmosphere (Lehr 2003; Yin et al. 2011).

Selenium biomethylation is of interest, because it represents a potential mechanism for removing Se from contaminated environments. Fungi are more active in the methylation of Se in soils, although some Se-methylating bacterial isolates have also been identified (Adriano et al. 2004). Dimethyl selenide can be demethylated in anoxic sediments as well as anaerobically by an obligate methylotroph that is similar to *Methanococcides methylutens* in pure culture. An anaerobic demethylation reaction may result in the formation of toxic and reactive hydrogen selenide (H_2Se) from less toxic DMSe. Although H_2Se undergoes rapid chemical oxidation under oxic conditions, it can exist for long periods in an aerobic environment (Tarze et al. 2007).

TABLE 3.3
Boiling Points and Vapor Pressure of Elemental, Hydride, and Methyl Species of Arsenic, Mercury, and Selenium

Trace Element	Species	Boiling Point (°C)	Vapor Pressure (kPa)
Arsenic	Elemental – As(0)	603	0 (approx.)
	Hydride – AsH	–55	1461.2 at 21.1°C
	Methylated –		
	• Monomethyl arsine – $As(CH_3)$	1.19	231.9 at 25°C
	• Dimethyl arsine – $As(CH_3)_2$	36	68.26 at 25°C
	• Trimethyl arsine – $As(CH_3)_3$	53.8	42.92 at 25°C
Chromium	Elemental – Cr(0)	2672	0.99 at 1857°C
Mercury	Elemental – Hg(0)	356.58	2×10^{-7} at –38.72°C
	Hydride – HgH_2	–34	–
	Methylated –		
	• Monomethyl mercury $Hg(CH_3)$	–	–
	• Dimethyl mercury $Hg(CH_3)_2$	93.5	–
Selenium	Elemental – Se(0)	688	6.95×10^{-4} at 221°C
	Hydride – SeH_4	–42	–
	Methylated –		
	• Dimethyl selenide – $Se(CH_3)_2$	49.8	41.1 at 25°C
	• Dimethyl selenite – $Se_2(CH_3)_2$	–39.042	826.85 at 25°C
	• Dimethyl selenone – $Se(CH_3)_2O_2$	153	–

Source: Bolan, N.S. et al., *Rev. Environ. Contam. Toxicol.*, 225, 1–56, 2013a.

3.4 FACTORS AFFECTING REDOX REACTION PROCESSES

Redox reactions of As, Cr, Hg, and Se in soil and sediments are affected by physicochemical characteristics of the media (e.g., pH and microbial activity), heavy metal(loid) characteristics (e.g., concentration and speciation), and environmental factors (e.g., moisture content and temperature). The redox reactions can also be manipulated through the addition of organic and inorganic amendments (Park et al. 2011; Bolan et al. 2013a).

3.4.1 Solute Factors

The rate of a redox reaction of heavy metal(loid)s depends on the bioavailability of the metal(loid) concerned, as measured by its concentration and speciation. For example, the complexation of Hg with dissolved organic carbon (DOC) decreases methylation (Miskimmin et al. 1992; Ravichandran 2004). The release of hydrogen sulfide (HgS) during sulfate reduction inhibits methylation, whereas the precipitation of sulfur (S) with Fe increases the Hg that is available for methylation (Choi and Bartha 1994). Similarly, the complexation of Hg(II) with sulfide has been shown to strongly affect the availability of Hg for methylation by microbes (Benoit et al. 1999; Drexel et al. 2002).

The reduction of Se by bacteria has often been shown to be affected by the initial Se concentration. For example, Lortie et al. (1992) observed that the Se reduction rate by *Pseudomonas stutzeri* increased with an increase in Se(VI) and Se(IV) concentrations up to 19 mM. At anSe concentration higher than 19 mM, Se(IV) reduction decreased; however, Se(VI) reduction remained constant, which might be attributed to the higher toxicity of the Se(IV) than Se(VI). Negatively charged Se oxyanions form ternary Se-cation-organic matter complexes, resulting in the immobilization of Se.

When the organic matter contents are high in shallow subsurface environments, reducing conditions in the solid phase decrease the dissolved As concentration in pore water, as affected by a relatively high amount of sulfide. Arsenic is sorbed with the formation of Fe sulfides. Under oxidizing conditions, surface waters are *undersaturated* with As(V) mineral, and as a result, secondary As(V) dissolves and As concentration increases in the water.

Choppala et al. (2013) observed that the addition of Fe(III) oxide to Cr(VI)-contaminated soils resulted in a decrease in the rate of reduction of Cr(VI), as measured by their half-life values. This phenomenon may be due to the increased retention of Cr(VI) by Fe(III) oxide, thereby decreasing the bioavailability of Cr(VI) for microorganisms.

3.4.2 Soil Factors

Soil pH affects redox reaction processes through its effects on the microorganisms, supply of protons, and adsorption and speciation of metal(loid)s. For example, protons are required for reducing Cr(VI) to Cr(III). It has often been observed that Cr(VI) reduction, being a proton consumption (or hydroxyl release) reaction, increases as soil pH decreases (Eary and Rai 1991; Choppala et al. 2012).

Mercury methylation decreased with a decrease in the pH of sediment, and methylation was not detected at a pH value less than 5.0, which may be related to the unavailability of inorganic Hg (Ramial et al. 1985). Fulladosa et al. (2004) noticed that As(V) toxicity as measured by EC_{50} value decreased as pH became basic. The optimum pH for methylation of Se was 6.5 (Frankenberger and Arshad 2001). Dissolved Se(VI) constituted 95% of the total soluble Se at pH 9 and decreased to 75% at pH 6.5. Bolan et al. (2003) and Choppala et al. (2012) have shown that the addition of organic manure increased the rate of reduction of Cr(VI), indicating the importance of C supply as an electron donor to initiate the reduction process.

Rhizosphere influences the redox reaction of heavy metal(loid)s through its effect on microbial activity, pH, and the release of organic compounds. For example, *in vitro* Se volatilization, using samples taken from a constructed wetland contaminated with Se(IV), showed that microbial cultures prepared from rhizosphere soils had higher rates of volatilization than cultures prepared from bulk soil (Azaizeh et al. 1997). Similarly, the rate of Hg methylation is found to be higher in the rhizosphere than in the non-rhizosphere bulk soil (Sun et al. 2011). Achá et al. (2005) showed that Hg methylation was influenced by rhizospheres of *Polygonum densiflorum* and *Eichhornia crassipes*. Sulfate-reducing bacteria activity and Hg methylation potentials were higher in the rhizosphere of *P. densiflorum* compared with that of *E. crassipes,* which was attributed to the higher C and N concentrations in the former plant species.

Plant roots enhance the reduction of metal(loid)s, such as As and Cr, externally by releasing root exudates or internally through endogenous metal(loid) reductase enzyme activity in the root, mainly through an increase in microbial activity (Dhankher et al. 2006; Xu et al. 2007). The rhizosphere environment can reduce Cr(VI) to Cr(III), because the pH-dependent redox reaction is related to Fe^{2+}, organic matter, and S contents in the rhizosphere (Zeng et al. 2008). Chen et al. (2000) reported that Cr(VI) reduction in a fresh wheat rhizosphere was induced by a decrease of pH. Microbial metabolism in the rhizosphere also caused the reduction of Cr(VI) to Cr(III). Low-molecular-weight organic acids, such as formic and acetic acids in the rhizosphere, can contribute to both Cr(VI) reduction and Cr(III) chelation (Bluskov et al. 2005).

3.4.3 Environmental Factors

Soil moisture influences redox reactions by controlling the activity of microorganisms and also redox conditions of the microenvironment (Alexander 2000). An increase in both moisture content and the amount of available C tends to increase the net loss of methyl Hg (Schlüter 2000; Oiffer and Siciliano 2009). Air-drying soil inhibits methylation of Se, whereas saturation with water causes anaerobiosis, thus decreasing the transfer of volatile Se from soil to air (Calderone et al. 1990). Alternate wetting and drying enhances the release of volatile Se compounds, which is attributed to the release of nutrients through organic matter mineralization (Hechun et al. 1996).

When the soil moisture content is high, the diffusion of O_2 is limited and a local anaerobic environment is created. In aerobic soils, the predominant As species is As(V) in soil pore water whereas As(III) comprises up to 80% of the total As in anaerobic soils. Chemical-reducing conditions increase As(III) in anaerobic soils. The chemical conversion of As(V) to As(III) may reduce microbial activity and production of monomethylarsonic acid (Haswell et al. 1985).

Temperature influences redox reactions of heavy metal(loid)s, mainly by controlling microbial activity and functions (Alexander 2000). Schwesig and Matzner (2001) and Heyes et al. (2006) observed that both the production and the volatilization loss of methyl Hg were directly proportional to temperature. Kocman and Horvat (2010) noticed a strong positive correlation between the soil surface temperature and Hg emission flux. They suggested that this thermally controlled emission of Hg from soils depended on the equilibrium of Hg(0) between the soil matrix and the soil gas. As suggested by Schlüter (2000), because of an increase in thermal motion, the vapor pressure of highly volatile Hg(0) is increased, and the sorption by soil is decreased.

Temperature is one of the most important environmental factors that affects the rate of Se volatilization (Frankenberger and Karlson 1994a). For every 10°C increase in the temperature, the vapor pressure of volatile Se is increased three- to four-fold (Karlson et al. 1994). Duncan and Frankenberger (2000) observed that the optimum temperature for Se volatilization was 35°C, with the rate of Se volatilization increasing as the temperature increased from 12 to 35°C. At 40°C, the rate of Se volatilization was slightly less than 35°C; however, it was greater than what it was at 30°C.

Camargo et al. (2003) found that Cr(VI) reduction by a Cr-resistant bacteria (*Bacillus sp.*) increased with an increase in soil temperature, with maximum reduction occurring at 30°C. Bacterial growth and Cr(VI) reduction by the strain *Amphibacillus* sp. KSUCr3 were studied at various temperatures (25–45°C) by Ibrahim et al. (2011). Chromate reduction was increased with an increase in temperature up to 40°C, which appeared to be the optimal temperature for growth of the strain KSUCr3. The effect of temperature on the leaching loss of As was attributed to the microbially induced transformation of As. Kim (2010) studied the effect of temperature (11 ± 1°C and 28 ± 1°C) on the adsorption of As(V) and noticed that the highest adsorption was observed at 28°C, implying that the adsorption process was endothermic.

3.4.4 Soil Amendments

The redox reaction of heavy metal(loid)s in soils and sediments is affected by organic and inorganic amendments (Table 3.4) (Park et al. 2011). Most redox reactions require an energy source, which is often organic but can also be inorganic.

It has often been shown that the addition of organic matter-rich soil amendments enhances the reduction of metal(loid)s such as Cr and Se (Figure 3.2) (Frankenberger and Karlson 1994b; Park et al. 2011). For example, several studies showed that the addition of cattle manure enhanced the reduction of Cr(VI) to Cr(III) (Losi et al. 1994; Cifuentes et al. 1996; Higgins et al. 1998). Various reasons could be given for the increase in the reduction of Cr(VI) in the presence of the organic manure composts. These include the supply of C and protons; the stimulation of microorganisms that are considered the major factors enhancing the reduction of Cr(VI) to Cr(III) (Losi et al. 1994). For example, Choppala et al. (2012) observed a decrease in Cr(VI) toxicity in soils treated with black carbon (BC); this was attributed to the supply of electrons for the reduction of toxic Cr(VI) to nontoxic Cr(III) species.

The easily oxidizable organic C fractions, such as DOC, provide the energy source for the soil microorganisms that are involved in the reduction of metal(loid)s (e.g., Cr(VI)) and non-metal(loid)s (e.g., NO_3^-) (Paul and Beauchamp 1989; Jardine et al. 1999; Vera et al. 2001; Bolan et al. 2011). Although manure addition induces the remediation of Cr-contaminated soils by reducing mobile toxic Cr(VI) to nontoxic and less mobile Cr(III), it is likely that redox reactions of the aromatic As and Hg compounds could occur, which, in turn, may result in the production of more toxic, inorganic species (Cullen and Reimer 1989; Kumagai and Sumi 2007).

The addition of Fe(II) decreased net Hg methylation in sediments and sulfide [S(-II)] concentration. The reduction in net Hg methylation can be attributed to the decreased concentration of uncharged, bioavailable Hg, which is positively related to S(-II) (Mehrotra and Sedlak 2005). The effect of clay addition on the Hg methylation depends on surface coatings. However, clays prevent Hg methylation through adsorption or sometimes promote demethylation by microorganisms. Humic substances facilitate Hg methylation, whereas humic coatings on clay stimulate demethylation in freshwater sediments (Jackson 1989; Zhang and Hsu-Kim 2010).

Most of the As forms As sulfides, such as realgar (AsS), orpiment (As_2S_3), and arsenopyrite (FeAsS). These have low solubility and mobility when S is abundant, as a result of biosolid amendments in the soil. Organic amendments such as biosolids and manure significantly reduce the potential environmental risks of As contamination under highly anoxic conditions (Carbonell-Barrachina et al. 1999). Similarly, Yadav et al. (2009) showed that the addition of dairy sludge and biofertilizer reduced the bioavailability of As and Cr, and promoted plant growth. The addition of manure increased the loss of As from contaminated soil, as affected by microbial methylation process thereby increasing As volatility. The rates of As loss were closely related to the microbial respiration because of nutrient supplementation to microbes. Bioaugmentation by As methylating fungi increased H_2As evolution rates in field-contaminated soils (Edvantoro et al. 2004).

3.5 CASE STUDIES RELATED TO BIOAVAILABILITY AND REMEDIATION

Redox reactions play a major role in the bioavailablity and remediation of heavy metal(loid)-contaminated soil, sediment, and water. From toxicological or environmental viewpoints, these reactions are important for three reasons. They may alter (a) the toxicity, (b) the water solubility,

TABLE 3.4
Selected References on the Remediation of Arsenic, Chromium, Mercury, and Selenium Toxicity by Organic Amendments

Element	Amendment	Plant Used	Observation	Reference
Arsenic	Municipal solid waste, biosolids compost	*Pteris vittata*	Increased soil water-soluble As and reduced As(V) to As(III), but decreased leaching in the presence of fern.	Cao et al. (2003)
	Biosolids compost	*Daucus carota* L. and *Lactuca sativa* L.	Biosolids compost reduced plant As uptake by 79%–86%, which might be due to the adsorption of As by biosolids organic matter.	Cao and Ma (2004)
	Green waste compost and biochar	*Miscanthus* species	Green waste compost substantially increased the plant yield; however, it increased the water-soluble and surface-adsorbed fractions of As.	Hartley et al. (2009)
	Dairy sludge	*Jatropha curcas*	Dairy sludge decreased DTPA-extractable As in the soil.	Yadav et al. (2009)
Chromium	Cow manure	*Festuca arundinacea*	Reduced Cr in roots and no change in Cr concentration in shoots.	Banks et al. (2006)
	Biosolids compost	*Brassica juncea*	Reduced Cr concentration in plant tissue.	Bolan et al. (2003)
	Cattle compost and straw	*Lactuca sativa*	Cr content in aerial biomass decreased with the addition of amendments, which may be due to the decreased association of Cr with carbonates and amorphous oxides and the increased association with humic substances.	Rendina et al. (2006)
	Hog manure and cattle dung compost	*Triticum vulgare*	Hog manure decreased soil available Cr(VI), which is attributed to its low C/N ratio and thus increased microbial reduction of Cr(VI).	Lee et al. (2006)
	Bark of *Pinus radiata*	*Helianthus annuus*	Reduced availability of Cr for plant uptake.	Bolan and Thiagarajan (2001)
	Biosolids compost	*Sesbania punicea*	Decreased Cr in plant extracts.	Branzini and Zubillaga (2010)
	Dairy sludge	*Jatropha curcas*	Dairy sludge decreased DTPA-extractable Cr in the soil.	Yadav et al. (2009)
	Farm yard manure	*Spinacea oleracea*	Increased root and shoot growth by decreasing Cr(VI) toxicity.	Singh et al. (2007)
Mercury	Humic acid	*Lactuca sativa*	Humic acids decreased the amount of Hg in soil and the translocation of Hg into plants.	Wang et al. (1997)
	Green waste compost	*Vulpia myuros* L.	Compost addition showed a negative relationship with soluble Hg and Hg tissue concentration, which may be due to the adsorption by compost.	Heeraman et al. (2001)
	Reactivated carbon	–	Powder reactivated carbon (PAC) increased stabilization/solidification of Hg in solid wastes, and pretreatment of the PAC with carbon disulfide (CS_2) increased adsorption efficiency.	Zhang and Bishop (2001)

TABLE 3.4 *(Continued)*
Selected References on the Remediation of Arsenic, Chromium, Mercury, and Selenium Toxicity by Organic Amendments

Element	Amendment	Plant Used	Observation	Reference
	Fulvic acid	–	Presence of fulvic acid increased the adsorption of Hg on goethite, which might be due to the strong affinity between sulfur groups within the fulvic acid and Hg.	Bäckström et al. (2003)
	Humic acid	*Brassica juncea*	Mercury translocation to aerial tissues of plants was restricted in the presence of humic acid.	Moreno et al. (2005a)
	Thiosulfates	*Brassica juncea*	Thiosulfates increased Hg accumulation in the plant, and Hg could be removed by phytoextraction from contaminated soils.	Moreno et al. (2005b)
Selenium	Insoluble (casein) and soluble (casamino acids) organic materials	–	Organic amendments enhanced Se removal by providing an energy source and methyl donor to the methylating microorganisms, which increased Se volatilization from the soil.	Zhang and Frankenberger (1999)
	Orange peel, cattle manure, gluten, and casein	–	The addition of organic amendments promoted the volatilization of Se; gluten was more effective, and it increased volatilization 1.2- to 3.2-fold compared with the control.	Calderone et al. (1990)
	Compost manure and gluten	–	Reduction of Se(VI) to Se(IV) increased in the presence of organic amendments under low oxygen concentration, thereby retarding Se mobility.	Guo et al. (1999)
	Press mud and poultry manure	*Triticum aestivum* L. and *Brassica napus*	Application of amendments reduced Se accumulation by enhancing volatilization, thereby reducing the transfer of Se from soil to plants.	Dhillon et al. (2010)
	Poultry manure, sugar cane press mud, and farmyard manure	*Triticum aestivum* L. and *Brassica napus*	Addition of organic amendments decreased Se accumulation and increased grain quality; however, the extent of reduction depended on the type of organic amendment applied.	Sharma et al. (2011)

and/or (c) the mobility of the element (Alexander 2000). An increase in solubility and mobility can be exploited to form bioremediate insoluble forms of elements in the soil, because the bio-transformed product is released from the solid phase into the solution phase. Conversely, a decrease in element solubility can be used to remove the element from surface- or groundwater through precipitation. In some cases, gaseous metal(loid) products can be removed through volatilization.

Metal(loid) reduction has the potential to be helpful for both intrinsic and engineered bioremediation of contaminated environments. The reduction of Se, Cr, and possibly other metal(loid)s can result in the conversion of soluble metal(loid) species into insoluble forms that can readily be removed from contaminated waters or waste streams (Crowley and Dungan 2002). The reduction of Hg can volatilize Hg from surface water and oceans (Lovley 1995; Moreno et al. 2005a). Arsenic can be

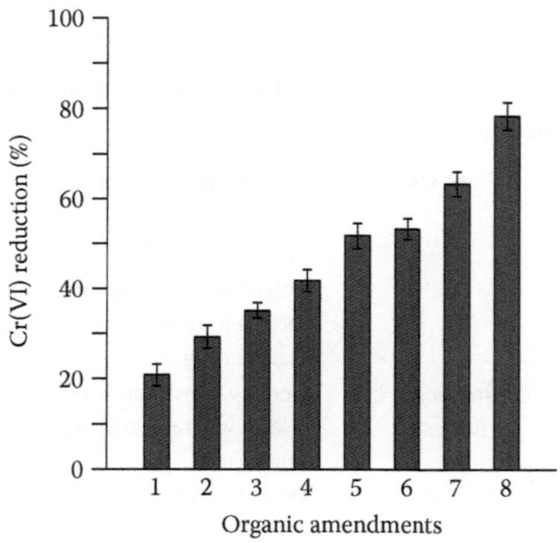

FIGURE 3.2 Effect of organic amendments on the reduction of chromate [Cr(VI)] in soil (1, soil; 2, soil + horse manure; 3, soil + farmyard manure; 4, soil + fish manure; 5, soil + spent manure; 6, soil + piggery manure; 7, soil + poultry manure; 8, soil + biosolids compost). (From Bolan, N.S. et al., *J. Environ. Qual.*, 32, 120–128, 2003.)

reduced to As(0), which is, subsequently, precipitated as As_2S_3 as a result of microbial sulfate reduction. Because As(III) is more soluble than As(V), the latter can be reduced by using bacteria in the soil and can be subsequently leached. Conversely, As(III) is oxidized to As(V) by using microbes; it is subsequently precipitated by using ferric ions (Williams and Silver 1984). *Desulfotomaculum auripigmentum* reduces both As(V) to As(III) and sulfate to H_2S, and this results in As_2S_3 precipitation (Newman et al. 1997).

Because of the lower solubility of Cr(III) compared with Cr(VI), the reduction reaction will eventually result in the immobilization of Cr, thereby diminishing the mobility and transport. The reduction of Cr(VI) to Cr(III) and the subsequent hydroxide precipitation of Cr(III) ion is the most common method that is used for treating Cr(VI)-contaminated industrial effluents (Blowes et al. 1997; James 2001). Similarly, Choppala et al. (2012) have noticed that the reduction of Cr(VI) to Cr(III) in variable-charge soils is likely to result in the adsorption of Cr(III). This happens due to an increase in pH-induced negative charges and the precipitation of $Cr(OH)_3$, resulting from the reduction-induced release of OH^- ions.

Biological immobilization of Se(VI) by a reduction to Se(0) is a practical approach that is adopted for remediation. Anaerobic bacteria can be grown in the contaminated medium with a C source, such as acetate as an electron donor and Se(VI) as an electron acceptor. The extent of Se(VI) reduction depends on the availability of the C source, and the reduced Se in the elemental form, which is insoluble, can be physically separated from contaminated water. Similarly, a methylation reaction can be used to form gaseous metal(loid) species, which can easily be removed through volatilization (Thompson-Eagle and Frankenberger 1992; Bañuelos and Lin 2007).

3.5.1 Soil Amendments to Manage Redox Reactions of Chromium

Chromium toxicity in soils can be mitigated by a reduction of Cr(VI) to Cr(III) that is influenced by the presence of free Cr(VI) species in soil solution, and the supply of protons and electrons. Choppala et al. (2012, 2013) examined the effect of organic C sources [BC derived from common weed, *Solanum elaeagnifolium* and chicken manure biochar (CMB)] on the reduction, microbial respiration, and phytoavailability of Cr(VI) in acidic and alkaline-contaminated soils. The organic C sources were applied at a rate of 5%. Chromium reduction was examined by incubating the soils and by using $K_2Cr_2O_7$ at 500 mg kg^{-1} and respiration and phytoavailability (using sunflower plants) studies at various levels of 0–500 mg Cr kg^{-1}.

The results indicated that the reduction was higher in acidic than in alkaline soils (Figure 3.3a,b). An increase in soil pH decreased Cr(VI) reduction, as the reduction of Cr(VI) to Cr(III) is a proton-consuming (or hydroxyl releasing) reaction. The rate of Cr(VI) reduction increased with the addition of organic amendments, and there was a significant difference in the effect of amendments on the rate of reduction. The rate of reduction is as follows: BC > CMB > control soil (Figure 3.3a,b). It has been reported that several organic C sources, including brown seaweed (Park et al. 2004), biosolids composts, farm yard manure, poultry manure, vermicompost (Bolan et al. 2003; Sunitha et al. 2014), cow dung, bermuda grass, yeast extract (Cifuentes et al. 1996), phenols (Elovitz and Fish 1995), and organic acids (Deng and Stone 1996), are able to reduce Cr(VI) to Cr(III) with subsequent immobilization of Cr(III).

Black carbon showed the highest rate of Cr(VI) reduction compared with biochar, which was attributed to the differences between DOC and functional groups (phenolic, hydroxyl, carbonyl, and amides) that provide electrons for the reduction of Cr(VI). Also, the high pH of manure biochar (8.8) may prevent the dissociation and oxidation of phenolic and hydroxyl groups, thereby limiting the supply of protons for Cr(VI) reduction. Hsu et al. (2009) evaluated the Cr(VI) sorption kinetics at pH levels 3–7 and examined the reaction mechanism of Cr(VI) with BC derived from burning rice straw. The results showed that Cr(VI) was sorbed and subsequently reduced to Cr(III), which was bound to the carbonyl/carboxyl groups on the BC surface through surface complexation and precipitation. There was a negative relationship between the half-life of Cr(VI) reduction and the concentration of DOC in soils that were amended with organic C sources, indicating that the rate of reduction increased when there was an increase in the concentration of DOC. Easily oxidizable C and DOC have been shown to correlate with Cr(VI) reduction, and only certain components of DOC act as electron donors when this reduction occurs (Bolan et al. 2003).

Soil respiration, as measured by the amount of CO_2 released, significantly decreased when Cr(VI) levels increased in acidic and alkaline soils, with the effect being more pronounced in the alkaline soil. Since acidic soils reduce Cr(VI) faster than alkaline soils (Figure 3.4a,b), there is less microbial toxicity of Cr(VI) in the former soil, as indicated by higher respiration.

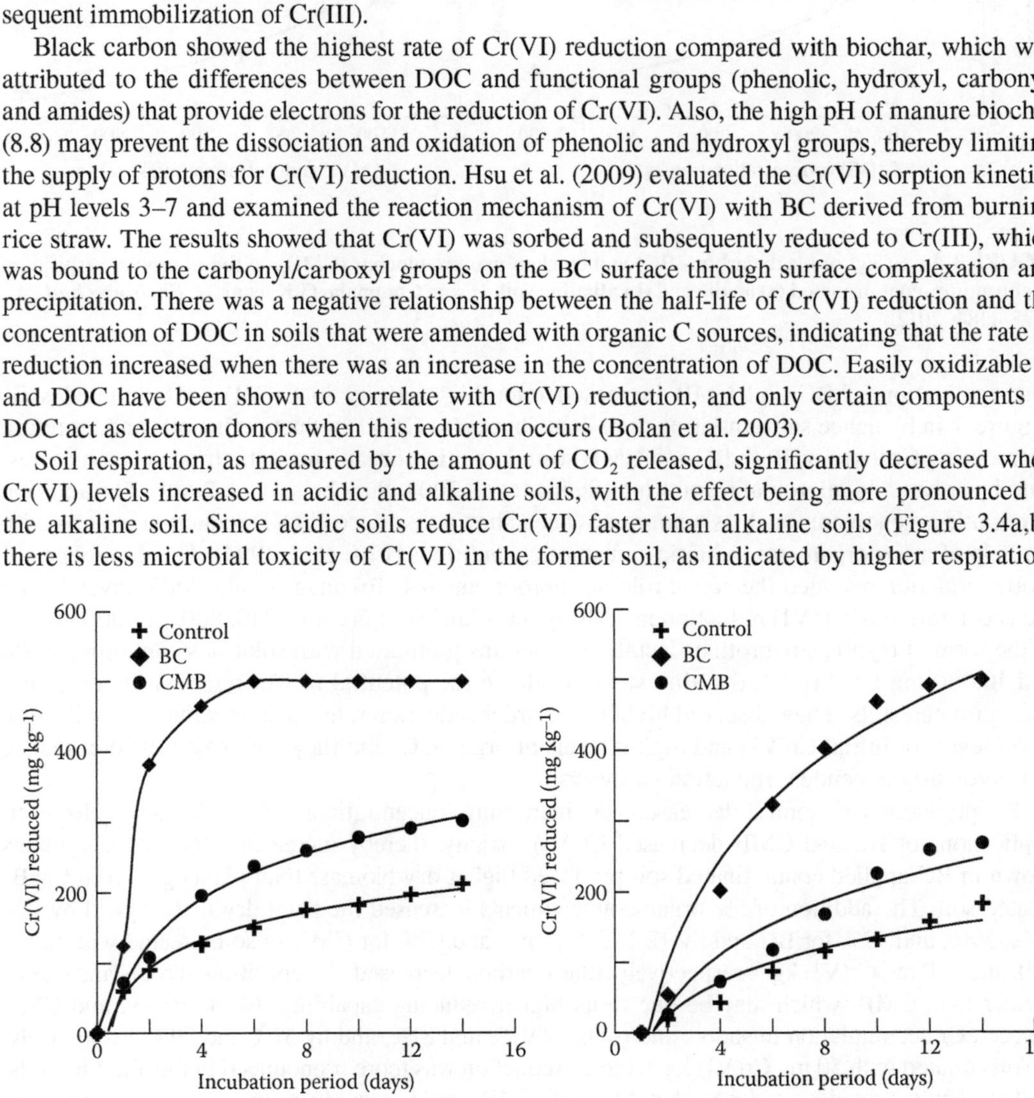

FIGURE 3.3 Effect of black carbon (BC) and chicken manure biochar (CMB) on the reduction of Cr(VI) in chromium-contaminated (a) acidic and (b) alkaline soils. (From Choppala, G. et al., *J. Hazard. Mater.* 261, 718–724, 2013.)

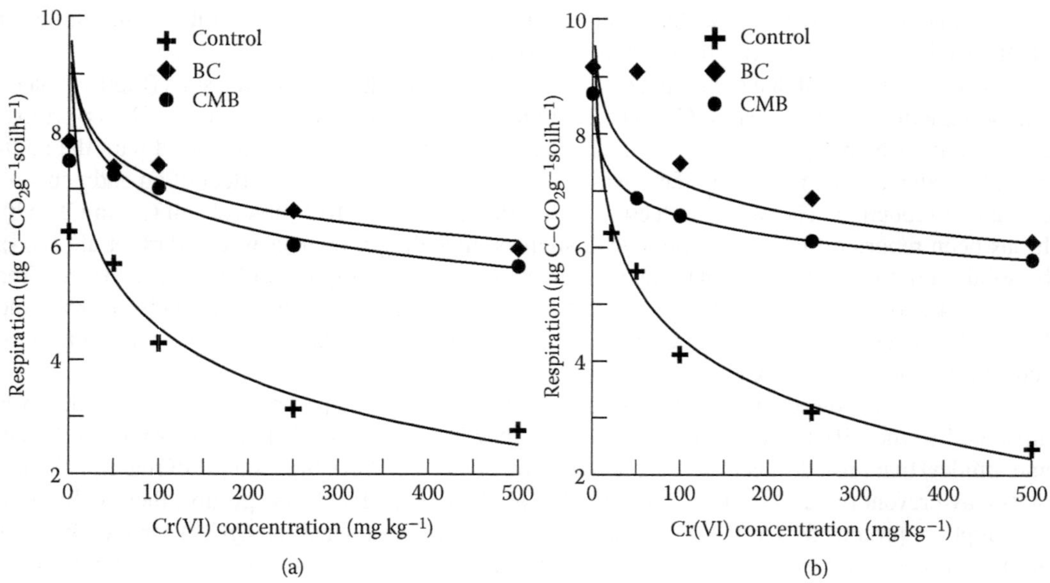

FIGURE 3.4 Effect of black carbon (BC) and chicken manure biochar (CMB) on the microbial respiration in chromium-contaminated (a) acidic and (b) alkaline soil. (From Choppala, G.K. et al., *J. Environ. Qual.*, 41, 1175–1184, 2012.)

The application of BC and CMB increased microbial activity in Cr(VI)-contaminated soils (Figure 3.4a,b), hence substantiating their contribution to Cr(VI) reduction in soils and therefore overcoming Cr toxicity. Both BC and CMB stimulate microorganisms in their porous structures, thereby enhancing microbial respiration (Steiner et al. 2004; Steinbeiss et al. 2009). The addition of organic amendments such as manure has been shown to induce Cr(VI) reduction in soils with the help of microorganisms in biotic and abiotic conditions (Losi et al. 1994). The dominance of biotic reduction revealed the active role of microorganisms. Tokunaga et al. (2003) investigated the acceleration of Cr(VI) reduction in soils by the addition of organic C (0, 800, or 4000 mg L^{-1} in the form of tryptic soy broth or lactate) in columns pretreated with solutions containing 1000 and 10,000 mg L^{-1} Cr(VI); the aim was to evaluate the potential *in situ* remediation of highly contaminated soils. They observed higher first-order reduction rate constant values for soils with lower levels of initial Cr(VI) and higher levels of organic C, and they confirmed the dominance of microbially dependent reduction pathways.

The production of biomass decreased with increasing concentrations of Cr(VI) in soils. However, applications of BC and CMB decreased Cr(VI) toxicity, thereby increasing dry biomass. Plants grown in BC-applied contaminated soil produced higher dry biomass than plants grown in CMB-treated soil. The addition of 5% organic amendments increased the plant dry matter yield by 7%, 38%, 59%, and 72% for BC and by 12%, 35%, 51%, and 67% for CMB in soils treated with 0, 50, 100, and 250 mg Cr(VI) kg^{-1}, respectively. Black carbon decreased Cr deposition in roots and shoots greater than CMB, which may be due to its higher reducing capability. Black carbon and CMB reduced Cr accumulation in shoots and roots by 95% and 81%, and by 57% and 29%, respectively, in soils treated with 50 mg Cr(VI) kg^{-1}; the Cr reduction was more pronounced in the alkaline soils.

This study provides evidence that BC and CMB could strongly mitigate Cr contamination, because they are highly reactive with many functional groups and are able to donate electrons to reduce Cr(VI) in soils. The reduction capacity of Cr(VI) was higher in acidic soil, and the addition of BC and CMB significantly increased reduction in alkaline soil. The application of BC and CMB increased microbial respiration and decreased phytotoxicity. Dissolved organic matter enhanced the reduction of toxic Cr(VI) to less toxic and relatively immobile Cr(III); the consequent increase in pH due to H^+ consumption in Cr(VI) reduction likely resulted in the immobilization of Cr(III) through adsorption/precipitation reactions.

3.5.2 Rhizoreduction of Chromate and Arsenate

The rhizosphere has been shown to influence contaminant interactions in soils (Cofield et al. 2008; Gaskin and Bentham 2010). The fate and bioavailability of heavy metal(loid)s in the rhizosphere can be different from bulk soil (Fitz et al. 2003). Bolan et al. (2013b) examined the reduction of As(V) and Cr(VI) in rhizosphere soils obtained from a range of Australian native plants. These two metal(loid)s were selected in their study because they differ in reduction-induced mobility and bioavailability. The reduction of As(V) to As(III) enhances its mobility and bioavailability, whereas the reduction of Cr(VI) to Cr(III) has an opposite effect.

Both maximum reduction and rate of reduction were higher for Cr(VI) than for As(V), indicating that Cr(VI) reduction occurs more readily in soils in the presence of electron donors such as organic matter (Choppala et al. 2013). However, As(V) reduction is found to be slow even at an E_h of –200 mV (Ascar et al. 2008). Although there was no significant difference in maximum reduction between the rhizosphere and non-rhizosphere soils, the rate of reduction was higher in the former soil, indicating that the rhizosphere soils caused up to 2.41- and 5.07-fold increases in the rate of As(V) and Cr(VI) reduction, respectively (Table 3.5).

The difference in the rate of As(V) and Cr(VI) reduction between rhizosphere and non-rhizosphere soils can be attributed to a number of reasons, which include increased microbial activity, DOC and organic acid production, and decreased pH and E_h in the rhizosphere soil (Hinsinger et al. 2009). It has often been shown that rhizosphere soil tends to have higher microbial activity due to the presence of higher levels of C and nutrients (Bowen and Rovira 1992; Hinsinger et al. 2009). In this case study, Bolan et al. (2013b) obtained a positive relationship between the difference in DOC between rhizosphere and non-rhizosphere soils (ΔDOC) and Δ basal respiration, indicating that DOC provides a source of C for microorganisms (Figure 3.5a). Similarly, there was a significant relationship between Δ basal respiration and Δ rate of reduction, indicating the role of increased microbial activity in rhizosphere soil in metal(loid) reduction (Figure 3.5b). Similarly, Gonzaga et al. (2006) have attributed the difference in the extent of As(V) reduction between a hyper-accumulator (*Pterisvittata*) and a non-accumulator (*Nephrolepis exaltata*) to the difference in the amount of DOC released to the rhizosphere. The onset of reducing conditions in the subsurface as induced by DOC not only dissolved As-rich Fe oxy-hydroxides but also released As to groundwater (Nickson et al. 2000).

Low-molecular-weight organic acids in root exudates influence metal(loid) reduction by providing a C substrate for microorganisms, decreasing E_h, and enhancing the mineral-induced catalytic reduction of metal(loid)s (Masscheleyn et al. 1990; Adriano 2001). Zhong and Yang (2012) noticed that the addition of malic acid to two Fe oxide-rich soils (Ultisol and Oxisol) accelerated the catalytic reduction of Cr(VI). Kantar et al. (2008) showed that the addition of organic ligands such as galacturonic, glucuronic, and alginic acids achieved *in situ* stabilization of Cr(VI) through increased reduction to Cr(III) followed by its subsequent retention by soil particles. Zhang et al. (2005) have shown that low-molecular-weight organic acids such as citrate, malate, and oxalate enhanced the release of As(V) and As(III), thereby influencing their redox reactions.

Soil pH is one of the important properties that controls the reduction of metal(loid)s. The difference in pH between rhizosphere and non-rhizosphere soil depends on the amount of organic acids/anions released, as influenced by cation-anion uptake balance (Tang and Rengel 2003; Hinsinger et al. 2009). Plant roots influence redox conditions of soil by releasing root exudates containing C and carbon dioxide through respiration (Cheng et al. 2010). For example, Zazo et al. (2008) obtained an E_h of –350 mV in a vegetated microcosm, which corresponds to the sulfate reduction zone, thereby facilitating the sulfate-induced reduction of Cr(VI). In the absence of free O_2

TABLE 3.5
Parameters of the Equation Describing the Rate of Reduction of As(V) and Cr(VI) in Rhizosphere and Non-Rhizosphere Soils

		As			Cr		
Plant Species	**Root Region**	**Y_m (mg As kg^{-1})**	***r***	**Rhizosphere Effect[a]**	**Y_m (mg Cr kg^{-1})**	***r***	**Rhizosphere Effect[a]**
Acacia pubescens	Rhizosphere	187.4 ± 13.2a	0.230 ± 0.012a		376.2 ± 18.7a	0.567 ± 0.011a	
	Non-rhizosphere	167.3 ± 7.81a	0.187 ± 0.007b	1.24	353.4 ± 12.2a	0.231 ±0.034b	2.45
Eucalyptus camaldulensis	Rhizosphere	201.5 ± 5.87a	0.198 ± 0.021a		345.7 ± 8.76a	0.456 ± 0.023a	
	Non-rhizosphere	196.5 ± 6.25a	0.172 ± 0.012a	1.15	356.7 ± 10.3a	0.178 ± 0.032b	2.56
Enchylaena tomentosa	Rhizosphere	179.3 ± 11.2a	0.201 ± 0.022a		381.2 ± 11.2a	0.478 ± 0.041a	
	Non-rhizosphere	186.3 ± 13.5a	0.176 ± 0.018a	1.14	367.3 ± 12.3a	0.123 ± 0.052b	3.89
Templetonia retusa	Rhizosphere	175.4 ± 9.21a	0.236 ± 0.021a		345.6 ± 18.1a	0.543 ± 0.037a	
	Non-rhizosphere	154.7 ± 8.45a	0.098 ± 0.013b	2.41	326.2 ± 15.4a	0.143 ± 0.012b	3.80
Dichantheum sericeum	Rhizosphere	167.9 ± 7.65a	0.231 ± 0.014a		378.3 ± 8.81a	0.623 ± 0.067a	
	Non-rhizosphere	171.3 ± 5.26a	0.123 ± 0.008b	1.88	376.4 ± 9.32a	0.123 ± 0.012b	5.07
Austrodanthonia richardsonii	Rhizosphere	167.7 ± 6.25a	0.198 ± 0.016a		345.6 ± 14.3a	0.534 ± 0.036a	
	Non-rhizosphere	157.6 ± 4.51a	0.123 ± 0.012b	1.61	356.2 ± 102a	0.201 ± 0.023b	2.66

Source: Bolan, N.S. et al., *Plant Soil,* 367, 615–625, 2013b.

[a] Rhizosphere effect = rhizosphere/non-rhizosphere.

Note: $Y = Y_m (1 - \text{Exp}^{-rx})$, where Y = amount of As(V) or Cr(VI) reduction (mg kg^{-1}); Y_m = maximum amount of As(V) or Cr(VI) reduction (mg kg^{-1}); r = rate constant; and x = incubation period (days). Values are mean ± standard deviation of triplicate, and the different letters within a column indicate a significant difference between rhizosphere and non-rhizosphere soils for each plant species at $p < .05$ according to Duncan's multiple-range tests.

as an electron acceptor under reduced conditions, the oxidized form of other elements, including heavy metal(loid)s, serves as an electron acceptor for energy generation by microorganisms, thereby resulting in the reduction of metal(loid)s (Harris and Arnold 1995).

Depending on the nature of metal(loid)s present in the soil, the rhizosphere-induced reduction has implications for their bioavailability with regard to both higher plants and microorganisms, and the remediation of contaminated soils. In the case of Cr, the reduction of Cr(VI) to Cr(III) decreases its bioavailability, because Cr(III) is more strongly retained and therefore becomes less mobile. Furthermore, the increase in pH resulting from the reduction reaction enhances the immobilization of Cr(III) through adsorption (via pH-induced increase in surface negative charge) and precipitation reactions. However, in the case of As, the reduction of As(V) to As(III) increases its bioavailability, because As(III) is less strongly retained and therefore becomes more mobile. The mobility and

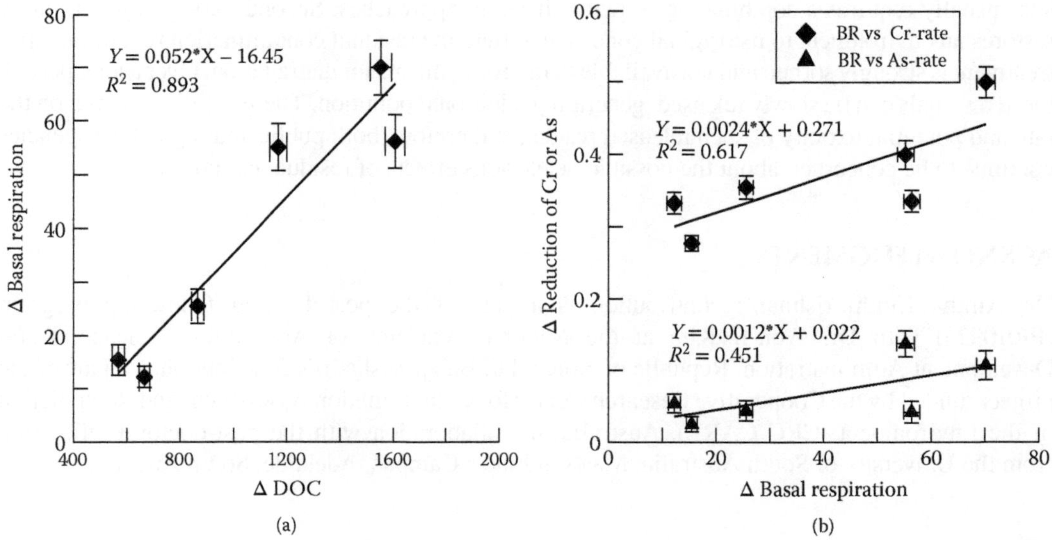

FIGURE 3.5 Relationships (a) between the difference in DOC between rhizosphere and non-rhizosphere soils (ΔDOC) and Δ basal respiration, and (b) between Δ basal respiration and ΔAs and Cr reduction. (From Bolan, *N.S.* et al., *Plant Soil*, 367, 615–625, 2013b.)

bioavailability of As(III) is likely to be exacerbated by decreased As adsorption due to an increase in surface-negative charge resulting from an increase in pH due to As(V) reduction.

3.6 CONCLUSIONS AND FUTURE RESEARCH NEEDS

Redox reactions can influence the solubility and subsequent mobility of heavy metal(loid)s, especially As, Cr, Hg, and Se, in soils and sediments by altering their speciation and redox state. Because many of the environments that receive heavy metal(loid)-containing wastes can be characterized as anoxic, for example, subsurface-saturated soils or organic-rich marsh sediments, redox reactions of metal(loid)s play a vital role in their mobilization and bioavailability. Thus, redox reactions can be readily managed and enhanced for efficient removal of contaminants, thereby enabling the development of *in situ* bioremediation technologies.

Desorption and remobilization of metal(loid)s, such as Cr and As from sediments, are controlled by pH, E_h, and metal(loid) concentration in the sediment interstitial water, as well as by contents in total Fe, Mn, and mineral hydrous oxides. Physical disturbances of the sediments by storm or flooding may move the underlying sediments to oxidizing environments where the sulfides undergo oxidation, thus resulting in the release of large quantities of metal(loid)s into the water. Similarly, depending on the nature of metal(loid)s present in the soil, the rhizosphere-induced redox reactions have implications with regard to their bioavailability to higher plants and microorganisms, and also for the remediation of contaminated soils.

Most bioremediation technologies are based on redox reaction processes that are designed to remove metal(loid)s mainly from aquatic systems. The viability and metabolic activity of microorganisms are the major limiting factors in terms of the efficiency of bio-transforming metal(loid)s in soils. Therefore, it is important to manipulate these redox reactions by controlling the factors affecting them and also by using appropriate soil amendments.

Two important issues need to be addressed when applying redox reactions to manage heavy metal(loid)s contamination. First, the implementation of bioremediation methods should be done with caution, because many sites contain multiple metal(loid)s, organic compounds, and organisms that affect the output of bioremediation approaches. Therefore, the remediation of contaminated sites usually requires a combination of many different approaches. Second, bioremediation rarely restores an environment to its original condition. Often, the residual contamination remaining after treatment is strongly sorbed and not available to microorganisms for degradation. Over a long period, these residuals can be slowly released, generating additional pollution. There is little research on the fate and potential toxicity of such released residuals; therefore, both public and regulatory agencies continue to be concerned about the possible deleterious effects of residual contamination.

ACKNOWLEDGMENTS

Dr. Anitha Kunhikrishnan's contribution is a part of the post-doctoral fellowship program (PJ010923) with Dr. Won-Il Kim at the National Academy of Agricultural Science, Rural Development Administration, Republic of Korea. Dr. Balaji Seshadri's contribution is a part of the project funded by the Cooperative Research Center for Contamination Assessment and Remediation of the Environment (CRC CARE), Australia, in collaboration with the post-doctoral fellowship from the University of South Australia, Mawson Lakes Campus, Adelaide, South Australia.

REFERENCES

Acha, D., Iñiguez, V., Roulet, M., Guimarães, J.R.D., Luna, R., Alanoca, L., and Sanchez, S. 2005. Sulfate-reducing bacteria in floating macrophyte rhizospheres from an Amazonian floodplain lake in Bolivia and their association with Hg methylation. *Appl Environ Microbiol.* 71:7531–7535.

Adriano, D.C. 2001. *Trace Elements in Terrestrial Environments: Biogeochemistry, Bioavailability and Risks of Metals*. 2nd edn. New York: Springer.

Adriano, D.C., Wenzel, W.W., Vangronsveld, J., and Bolan, N.S. 2004. Role of assisted natural remediation in environmental cleanup. *Geoderma* 122:121–142.

Alexander, M. 2000. Aging, bioavailability, and overestimation of risk from environmental pollutants. *Environ Sci Technol.* 34:4259–4265.

Allard, B., and Arsenie, I. 1991. Abiotic reduction of mercury by humic substances in aquatic system—An important process for the mercury cycle. *Water Air Soil Pollut.* 56:457–464.

Amirbahman, A., Kent, D.B., Curtis, G.P., and Marvin-DiPasquale, M.C. 2013. Kinetics of homogeneous and surface-catalyzed mercury(II) reduction by iron(II). *Environ Sci Technol.* 47(13):7204–7213.

Aposhian, H.V., Zakharyan, R.A., Avram, M.D., Kopplin, M.J., and Wollenberg, M.L. 2003. Oxidation and detoxification of trivalent arsenic species. *Toxicol Appl Pharmacol.* 193:1–8.

Apte, A.D., Tare, V., and Bose, P. 2006. Extent of oxidation of Cr(III) to Cr(VI) under various conditions pertaining to natural environment. *J Hazard Mater.* 128:164–174.

Ascar, L., Ahumada, I., and Richter, P. 2008. Influence of redox potential (Eh) on the availability of arsenic species in soils and soils amended with biosolid. *Chemosphere* 72:1548–1552.

Azaizeh, H.A., Gowthaman, S., and Terry, N. 1997. Microbial selenium volatilization in rhizosphere and bulk soil from a constructed wetland. *J Environ Qual.* 26:666–672.

Bachate, S.P., Khapare, R.M., and Kodam, K.M. 2012. Oxidation of arsenite by two β-proteobacteria isolated from soil. *Appl Microbiol Biotechnol.* 93:2135–2145.

Bäckström, M., Dario, M., Karlsson, S., and Allard, B. 2003. Effects of a fulvic acid on the adsorption of mercury and cadmium on goethite. *Sci Total Environ.* 304:257–268.

Bajaj, M., Schmidt, S., and Winter, J. 2012. Formation of Se (O) nanoparticles by *Dugenella sp.* and *Agrobacterium sp.* isolated from Se-laden soil of north-east Punjab, India. *Microbial Cell Factories* 11(1):64.

Banks, M., Schwab, A., and Henderson, C. 2006. Leaching and reduction of chromium in soil as affected by soil organic content and plants. *Chemosphere* 62:255–264.

Bañuelos, G.S. and Li, Z.Q. 2007. Acceleration of selenium volatilization in seleniferous agricultural drainage sediments amended with methionine and casein. *Environ Pollut.* 150:306–312.

Barnhart, J. 1997. Chromium chemistry and implications for environmental fate and toxicity. *Soil Sediment Contam.* 6:561–568.

Battaglia-Brunet, F., Dictor, M.C., Garrido, F., Crouzet, C., Morin, D., Dekeyser, K., Clarens, M., and Baranger, P. 2002. An arsenic (III)-oxidizing bacterial population: Selection, characterization, and performance in reactors. *J Appl Microbiol.* 93:656–667.

Benoit, J.M., Gilmour, C.C., Mason, R.P., and Heyes, A. 1999. Sulfide controls on mercury speciation and bioavailability to methylating bacteria in sediment and pore waters. *Environ Sci Technol.* 33:951–957.

Biswas, K.C., Barton, L.L., Tsui, W.L., Shuman, K., Gillespie, J., and Eze, C.S. 2011. A novel method for the measurement of elemental selenium produced by bacterial reduction of selenite. *J Microbiol Methods* 86:140–144.

Blowes, D.W., Ptacek, C.J., and Jambor. J.L. 1997. In-situ remediation of Cr (VI)-contaminated groundwater using permeable reactive walls: Laboratory studies. *Environ Sci Technol.* 31:3348–3357.

Bluskov, S., Arocena, J., Omotoso, O., and Young, J. 2005. Uptake, distribution, and speciation of chromium in *Brassica juncea*. *Int J Phytoremediation* 7:153–165.

Boening, D.W. 2000. Ecological effects, transport, and fate of mercury: A general review. *Chemosphere* 40:1335–1351.

Bolan, N.S., Adriano, D.C., Kunhikrishnan, A., James, T., McDowell, R., and Senesi, N. 2011. Dissolved organic matter: Biogeochemistry, dynamics, and environmental significance in soils. *Adv Agron.* 110:1–75.

Bolan, N.S., Adriano, D.C., Natesan, R., and Koo, B.J. 2003. Effects of organic amendments on the reduction and phytoavailability of chromate in mineral soil. *J Environ Qual.* 32:120–128.

Bolan, N.S., Choppala, G., Kunhikrishnan, A., Park, J.H., and Naidu, R. 2013a. Microbial transformation of trace elements in soils in relation to bioavailability and remediation. *Rev Environ Contam Toxicol.* 225:1–56.

Bolan, N.S., Kunhikrishnan, A., and Gibbs, J. 2013b. Rhizoreduction of chromate and arsenate in Australian native vegetation. *Plant Soil* 367:615–625.

Bolan, N.S. and Thiagarajan, S. 2001. Retention and plant availability of chromium in soils as affected by lime and organic matter amendments. *Aust J Soil Res*. 39:1091–1104.

Bowen, G.D., and Rovira, A.D. 1992. The rhizosphere: the hidden half of the hidden half. In: Waisel, Y., Eshel, A., Kafkafi, U. (eds.). *Plant Roots: The Hidden Half.* New York: Marcel Decker, pp. 641–669.

Branzini, A., and Zubillaga, M. 2010. Assessing phytotoxicity of heavy metals in remediated soil. *Int J Phytoremediation* 12:335–342.

Calderone, S., Frankenberger, W., Parker, D., and Karlson, U. 1990. Influence of temperature and organic amendments on the mobilization of selenium in sediments. *Soil Biol Biochem.* 22:615–620.

Camargo, F.A., Okeke, B.C., Bento, F.M., and Frankenberger, W.T. 2003. In vitro reduction of hexavalent chromium by a cellfree extract of *Bacillus* sp. ES 29 stimulated by Cu^{2+}. *Appl Microbiol Biotechnol.* 62:569–573.

Cao, X. and Ma, L.Q. 2004. Effects of compost and phosphate on plant arsenic accumulation from soils near pressure-treated wood. *Environ Pollut.* 132:435–442.

Cao, X., Ma, L.Q., and Shiralipour, A. 2003. Effects of compost and phosphate amendments on arsenic mobility in soils and arsenic uptake by the hyperaccumulator, *Pteris vittata* L. *Environ Pollut.* 126:157–167.

Carbonell-Barrachina, A.A., Jugsujinda, A., Sirisukhodom, S., Anurakpongsatorn, P., Burló, F., DeLaune, R.D., and Patrick Jr., W.H. 1999. The influence of redox chemistry and pH on chemically active forms of arsenic in sewagesludge-amended soil. *Environ Int.* 25:613–618.

Chang, J.S., Lee, J.H., and Kim, I.S. 2011. Bacterial *aox* genotype from arsenic contaminated mine to adjacent coastal sediment: Evidences for potential biogeochemical arsenic oxidation. *J Hazard Mater.* 193:233–242.

Chen, C.P., Juang, K.W., Lin, T.H., and Lee, D.Y. 2010. Assessing the phytotoxicity of chromium in Cr(VI)-spiked soils by Cr speciation using XANES and resin extractable Cr(III) and Cr(VI). *Plant Soil* 334:299–309.

Chen, N.C., Kanazawa, S., and Horiguchi, T. 2000. Chromium(VI) reduction in wheat rhizosphere. *Pedosphere* 10:31–36.

Chen, Y.W., Truong, H.Y.T., and Belzile, N. 2009. Abiotic formation of elemental selenium and role of iron oxide surfaces. *Chemosphere.* 74(8):1079–1084.

Cheng, L., Zhu, J., Chen, G., Zheng, X., Oh, N.H., Rufty, T.W., Richter, D.D., and Hu, S. 2010. Atmospheric CO_2 enrichment facilitates cation release from soil. *Ecol Lett.* 13:284–291.

Choi, S.C., and Bartha, R. 1994. Cobalamin-mediated mercury methylation by *Desulfovibrio desulfuricans* LS. *Appl Environ Microbiol.* 59(1):290–295.

Choppala, G., Bolan, N.S., and Seshadri, B. 2013. Chemodynamics of chromium reduction in contaminated soils: Implications for bioremediation. *J Hazard Mater*. 261: 718–724.

Choppala, G.K., Bolan, N.S., Megharaj, M., Chen, Z., and Naidu, R. 2012. The influence of biochar and black carbon on reduction and bioavailability of chromate in soils. *J Environ Qual.* 41:1175–1184.

Christen, K. 2001. Chickens, manure, and arsenic. *Environ Sci Technol.* 35:184–185.

Cifuentes, F., Lindemann, W., and Barton, L. 1996. Chromium sorption and reduction in soil with implications to bioremediation. *Soil Sci.* 161:233.

Cofield, N., Banks, M.K., and Schwab, A.P. 2008. Lability of polycyclic aromatic hydrocarbons in the rhizosphere. *Chemosphere* 70:1644–1652.

Crowley, D.E., and Dungan, R.S. 2002. *Metals: Microbial Processes Affecting Metals. Encyclopedia of Environmental Microbiology*. New York: Wiley and Sons, Inc., pp. 1878–1893.

Cullen, W.R., and Reimer, K.J. 1989. Arsenic speciation in the environment. *Chem Rev.* 89:713–729.

de Lacerda, L. 2003. Updating global Hg emissions from small-scale gold mining and assessing its environmental impacts. *Environ Geol.* 43:308–314.

Deng, B. and Stone, A.T. 1996. Surface-catalyzed chromium(VI) reduction. Reactivity comparisons of different organic reductants and different oxide surfaces. *Environ Sci Technol.* 30:2484–2494.

Dhankher, O.P., Rosen, B.P., McKinney, E.C., and Meagher, R.B. 2006. Hyperaccumulation of arsenic in the shoots of Arabidopsis silenced for arsenate reductase (ACR2). *Proc Natl Acad Sci USA* 103:5413–5418.

Dhillon, K., Dhillon, S., and Dogra, R. 2010. Selenium accumulation by forage and grain crops and volatilization from seleniferous soils amended with different organic materials. *Chemosphere* 78:548–556.

Dowdle, P.R. and Oremland, R.M. 1998. Microbial oxidation of elemental selenium in soils and bacterial cultures. *Environ Sci Technol.* 32:3749–3755.

Drexel, R.T., Haitzer, M., Ryan, J.N., Aiken, G.R., and Nagy, K.L. 2002. Mercury(II) sorption to two Florida Everglades peats: Evidence for strong and weak binding and competition by dissolved organic matter released from the peat. *Environ. Sci Technol.* 36:4058–4064.

Duncan, R.S., and Frankenberger, W.T.F. Jr. 1998. Reduction of selenite to elemental selenium by *Enterobacter cloacae* SL D1a-1. *J Environ Qual.* 27:1301–1306.

Eary, L.E., and Rai, D. 1991. Chromate reduction by subsurface soils under acidic conditions. *Soil Sci Soc Am J.* 55:676.

Edvantoro, B.B., Naidu, R., Megharaj, M., Merrington, G., and Singleton, I. 2004. Microbial formation of volatile arsenic in cattle dip site soils contaminated with arsenic and DDT. *Appl Soil Ecol.* 25:207–217.

Elovitz, M.S., and Fish, W. 1995. Redox interaction of Cr(VI) and substituted phenols: Products and mechanism. *Environ Sci Technol.* 29:1933–1943.

Focardi, S., Pepi, M., Landi, G., Gasperini, S., Ruta, M., Di Biasio, P., and Focardi, S.E. 2012. Hexavalent chromium reduction by whole cells and cell free extract of the moderate halophilic bacterial strain *Halomonas* sp. TA-04. *Int Biodeterioration Biodegrad.* 66(1):63–70.

Franco, D.V., Da Silva, L.M., and Jardim, W.F. 2009. Chemical reduction of hexavalent chromium present in contaminated soil using a packed-bed column reactor. *CLEAN* 37:858–865.

Frankenberger Jr., W.T., and Arshad, M. 2001. Bioremediation of selenium-contaminated sediments and water. *Biofactors* 14:241–254.

Frankenberger, W.T., Arshad, M., Siddique, T., Han, S.K., Okeke, B.C., and Zhang, Y. 2005. Bacterial diversity in selenium reduction of agricultural drainage water amended with rice straw. *J Environ Qual.* 34:217–226.

Frankenberger Jr., W.T., and Karlson, U. 1994a. Soil management factors affecting volatilization of selenium from dewatered sediments. *Geomicrobiol J.* 12:265–278.

Frankenberger Jr., W.T., and Karlson, U. 1994b. Microbial volatilization of selenium from soils and sediments. In: Frankenberger Jr., W.T., and Benson, S. (eds.). *Selenium in the Environment*. New York: Marcel Dekker, pp. 369–387.

Frankenberger, W.T., and Losi, M.E. 1995. Application of bioremediation in the cleanup of heavy elements and metalloids. In: Skipper, H.D. and Turco, R.F. (eds.). *Bioremediation: Science and Applications*. Soil Science Special Publication No. 43, Madison, WI: Soil Science Society of America Inc., pp. 173–210.

Fulladosa, E., Murat, J.C., Martinez, M., and Villaescusal, I. 2004. Effect of pH on arsenate andarsenite toxicity to luminescent bacteria (*Vibrio fischeri*). *Arch Environ Contam Toxicol.* 46:176–182.

Gadd, G.M. 2010. Metals, minerals and microbes: Geomicrobiology and bioremediation. *Microbiol.* 156:609–643.

Gaskin, S.E., and Bentham, R.H. 2010. Rhizoremediation of hydrocarbon contaminated soil using Australian native grasses. *Sci Total Environ.* 408:3683–3688.

Gong, C., and Donahoe, R.J. 1997. An experimental study of heavy metal attenuation and mobility in sandy loam soils. *Appl Geochem.* 12:243–254.

Gonzaga, M.I.S., Santos, J.A.G., and Ma, L.Q. 2006. Arsenic chemistry in the rhizosphere of *Pteris vittata* L. and *Nephrolepis exaltata* L. *Environ Pollut.* 143:254–260.

Gu, B., Bian, Y., Miller, C.L., Dong, W., Jiang, X., and Liang, L. 2011. Mercury reduction and complexation by natural organic matter in anoxic environments. *Proc Natl Acad Sci.* 108(4):1479–1483.

Guo, L., Frankenberger Jr., W.T., and Jury, W.A. 1999. Evaluation of simultaneous reduction and transport of selenium in saturated soil columns. *Water Resour Res.* 35:663–669.

Harris, R.F., and Arnold, S.M. 1995. Redox and energy aspects of soil bioremediation. In: *Bioremediation: Science and Applications*. Madison, WI: Soil Science Society of America, pp. 55–85.

Hartley, W., Dickinson, N.M., Riby, P., and Lepp, N.W. 2009. Arsenic mobility in brownfield soils amended with green waste compost or biochar and planted with *Miscanthus*. *Environ Pollut.* 157:2654–2662.

Haswell, S.J., O'Neill, P., and Bancroft, K.C. 1985. Arsenic speciation in soil-pore waters from mineralized and unmineralized areas of south-west England. *Talanta* 32:69–72.

Hechun, P., Guangshen, L., Zhiyun, Y., and Yetang, H. 1996. Acceleration of selenate reduction by alternative drying and wetting of soils. *Chin J Geochem.* 15:278–284.

Heeraman, D., Claassen, V., and Zasoski, R. 2001. Interaction of lime, organic matter and fertilizer on growth and uptake of arsenic and mercury by Zorro fescue (*Vulpia myuros* L.). *Plant Soil* 234:215–231.

Herbel, M.J., Blum, J.S., Oremland, R.S., and Borglin, S.E. 2003. Reduction of elemental selenium to selenide: Experiments with anoxic sediments and bacteria that respire Se-oxyanions. *Geomicrobiol J.* 20:587–602.

Heyes, A., Mason, R.P., Kim, E.H., and Sunderland, E. 2006. Mercury methylation in estuaries: Insights from using measuring rates using stable mercury isotopes. *Mat Chem.* 102:134–147.

Higgins, T.E., Halloran, A., Dobbins, M., and Pittignano, A. 1998. In situ reduction of hexavalent chromium in alkaline soils enriched with chromite ore processing residue. *J Air Waste Manage Assoc.* 48:1100–1106.

Hinsinger, P., Bengough, G., Vetterlein, D, and Young, I.M. 2009. Rhizosphere: Biophysics, biochemistry and ecological relevance. *Plant Soil* 321:117–152.

Hsu, N.H., Wang, S.L., Lin, Y.C., Sheng, G.D., and Lee, J.F. 2009. Reduction of Cr(VI) by crop-residue-derived black carbon. *Environ Sci Technol.* 43:8801–8806.

Huang, J.H., Voegelin, A., Pombo, S.A., Lazzaro, A., Zeyer, J., and Kretzschmar, R. 2011. Influence of arsenate adsorption to ferrihydrite, goethite, and boehmite on the kinetics of arsenate reduction by *Shewanella putrefaciens* strain CN-32. *Environ Sci Technol.* 44:6202–6208.

Ibrahim, A.S.S., El-Tayeb, M.A., Elbadawi, Y.B., and Al-Salamah, A.A. 2011. Isolation and characterization of novel potent Cr(VI). reducing alkaliphilic *Amphibacillus* sp. KSUCr3 from hypersaline soda lakes. *Electron J Biotechnol.* 4:1–14.

Islam, F.S., Gault, A.G., Boothman, C., Polya, D.A., Charnock, J.M., Chatterjee, D., and Lloyd, J.R. 2004. Role of metal-reducing bacteria in arsenic release from Bengal delta sediments. *Nature* 430(6995):68–71.

Jackson, T.A. 1989. The influence of clay minerals, oxides, and humic matter on the methylation and demethylation of mercury by micro-organisms in freshwater sediments. *Appl Organomet Chem.* 3:1–30.

Jain, P., Amatullah, A., Rajib, S.A., and Reza, H.M. 2012. Antibiotic resistance and chromium reduction pattern among actinomycetes. *Am J Biochem Biotechnol.* 8(2):111–117.

James, B.R. 2001. Remediation-by-reduction strategies for chromate-contaminated soils. *Environ Geochem Health* 23:175–179.

James, B.R., and Barlett, R.J. 1983. Behaviour of chromium in soils: VII. Adsorption and reduction of hexavalent forms. *J Environ Qual.* 12(2):177–181.

Jardine, P., Fendorf, S., Mayes, M., Larsen, I., Brooks, S., and Bailey, W. 1999. Fate and transport of hexavalent chromium in undisturbed heterogeneous soil. *Environ Sci Technol.* 33:2939–2944.

Jones, C., Anderson, H., McDermott K., and Inskeep, T. 2000. Rates of microbially mediated arsenate reduction and solubilization. *Soil Sci Soc Am J.* 64:600.

Kantar, C., Cetin, Z., and Demiray, H. 2008. *In situ* stabilization of chromium(VI) in polluted soils using organic ligands: The role of galacturonic, glucoronic and alginic acid. *J Hazard Mater.* 159:287–293.

Karlson, U., Frankenberger, W.T. Jr., and Spencer, W.F. 1994. Physico-chemicalproperties of dimethyl selenide. *J Chem Eng Data* 39:608–610.

Kikuchi, T., Ikemoto, H., Takahashi, K., Hasome, H., and Ueda, H. 2013. Parameterizing soil emission and atmospheric oxidation-reduction in a model of the global biogeochemical cycle of mercury. *Environ Sci Technol.* 47:12266–12274.

Kim, J.G.D., Chusuei, J.B., and Deng, C.C. 2002. Oxidation of chromium(III) to (VI) by manganese oxides. *Soil Sci Soc Am J.* 66:306–315.

Kim, M.J. 2010. Effects of pH, adsorbate/adsorbent ratio, temperature and ionic strength on the adsorption of arsenate onto soil. *Geochem Explor Environ.* 10:407–412.

Kocman, D., and Horvat, M. 2010. A laboratory based experimental study of mercury emission from contaminated soils in the River Idrijca catchment. *Atmos Chem Phys.* 10:1417–1426.

Kosolapov, D., Kuschk, P., Vainshtein, M., Vatsourina, A., Wiessner, A., Kästner, M., and Müller, R. 2004. Microbial processes of heavy metal removal from carbon-deficient effluents in constructed wetlands. *Eng Life Sci.* 4:403–411.

Kulp, T.R., Hoeft, S.E., Miller, L.G., Saltikov, C., Murphy, J.N., Han, S., and Oremland, R.S. 2006. Dissimilatory arsenate and sulfate reduction in sediments of two hypersaline, arsenic-rich soda lakes: Mono and Searles Lakes, California. *Appl Environ Microbiol.* 72(10):6514–6526.

Kumagai, Y., and Sumi, D. 2007. Arsenic: Signal transduction, transcription factor, and biotransformation involved in cellular response and toxicity. *Ann Rev Pharmacol Toxicol.* 47:243–262.

Landrot, G., Ginder-Vogel, M., and Sparks, D.L. 2009. Kinetics of chromium(III) oxidation by manganese(IV) oxides using quick scanning X-ray absorption fine structure spectroscopy (Q-XAFS). *Environ Sci Technol.* 44:143–149.

Lee, D.Y., Shih, Y.N., Zheng, H.C., Chen, C.P., Juang, K.W., Lee, J.F., and Tsui, L. 2006. Using the selective ion exchange resin extraction and XANES methods to evaluate the effect of compost amendments on soil chromium(VI) phytotoxicity. *Plant Soil* 281:87–96.

Lehr, C.R. 2003. Microbial methylation and volatilization of arsenic. PhD thesis. Department of Chemistry, The University of British Columbia, Canada.

Leita, L., Margon, A., Sinicco, T., and Mondini, C. 2011. Glucose promotes the reduction of hexavalent chromium in soil. *Geoderma,* 164:122–127.

Lens, P., Van Hullebusch, E., and Astratinei, V. 2006. Bioconversion of selenate in methanogenic anaerobic granular sludge. *J Environ Qual.* 35:1873–1883.

Lortie, L., Gould, W., Rajan, S., McCready, R., and Cheng, K.J. 1992. Reduction of selenate and selenite to elemental selenium by a *Pseudomonas stutzeri* isolate. *Appl Environ Microbiol.* 58:4042–4044.

Losi, M., Amrhein, C., and Frankenberger, Jr. W. 1994. Factors affecting chemical and biological reduction of hexavalent chromium in soil. *Environ Toxicol Chem.* 13:1727–1735.

Losi, M.E., and Frankenberger, W.T. 1998. Reduction of selenium oxyanions by *Enterobacter cloacae* strain SLDIa-I. In: Frankenberger, W.T. and Engberg, R.A. (eds.). *Environmental Chemistry of Selenium*, vol. 64. New York: Marcel Dekker, pp. 515–544.

Lovley, D.R. 1995. Bioremediation of organic and metal contaminants with dissimilatory metal reduction. *J Ind Microbiol.* 14:85–93.

Mahimairaja, S., Bolan, N.S., Adriano, D., and Robinson, B. 2005. Arsenic contamination and its risk management in complex environmental settings. *Adv Agron.* 86:1–82.

Maiers, D., Wichlacz, P., Thompson, D., and Bruhn, D. 1988. Selenate reduction by bacteria from a selenium-rich environment. *Appl Environ Microbiol.* 54:2591–2593.

Marafatto, F.F., Petrini, R., Pinzino, C., Pezzetta, E., Slejko, F., and Lutman, A. 2013. Redox reactions and the influence of natural Mn oxides on Cr oxidation in a contaminated site in northern Italy: Evidence from Cr stable-isotopes and EPR spectroscopy. In: *E3S Web of Conferences,* vol. 1. p. 33007, EDP Sciences.

Masscheleyn, P.H., Delaune, R.D., and Patrick Jr., W.H. 1990. Transformations of selenium as affected by sediment oxidation-reduction potential and pH. *Environ Sci Technol.* 24:91–96.

Mehrotra, A.S. and Sedlak, D.L. 2005. Decrease in net mercury methylation rates following iron amendment to anoxic wetland sediment slurries. *Environ Sci Technol.* 39:2564–2570.

Miskimmin, B.M., Rudd, J.W.M., and Kelly, C.A. 1992. Influences of DOC, pH, and microbial respiration rates of mercury methylation and demethylation in lake water. *Can J Fish Aquat Sci.* 49:17–22.

Moreno, F.N., Anderson, C.W.N., Stewart, R.B., Robinson, B.H., Ghomshei, M., and Meech, J.A. 2005a. Induced plant uptake and transport of mercury in the presence of sulphur-containing ligands and humic acid. *New Phytol.* 166:445–454.

Moreno, F.N., Anderson, C.W.N., Stewart, R.B., Robinson, B.H., Nomura, R., Ghomshei, M., and Meech, J.A. 2005b. Effect of thioligands on plant-Hg accumulation and volatilisation from mercury-contaminated mine tailings. *Plant Soil.* 275:233–246.

Myneni, S., Tokunaga, T.K., and Brown Jr., G.E. 1997. Abiotic selenium redox transformations in the presence of FeII, III oxides. *Science* 278:1106–1109.

Nakagawa, T., Iino, T., Suzuki, K.I., andHarayama, S. 2006. *Ferrimonas futtsuensis* sp. nov. and *Ferrimonas kyonanensis* sp. nov., selenate-reducing bacteria belonging to the Gammaproteobacteria isolated from Tokyo Bay. *Int J Syst Evol Microbiol.* 56(11): 2639–2645.

Newman, D.K., Beveridge, T.J., and Morel, F.M.M. 1997. Precipitation of arsenic trisulfide by *Desulfotomaculum auripigmentum. Appl Environ Microbiol.* 63:2022–2028.

Nickson, R.T., McArthur, J.M., Ravenscroft, P., Burgess, W.G., and Ahmed, K.M. 2000. Mechanisms of arsenic release to groundwater, Bangladesh and West Bengal. *Appl Geochem.* 15:403–413.

Oiffer, L. and Siciliano, S.D. 2009. Methyl mercury production and loss in Arctic soil. *Sci Total Environ.* 407:1691–1700.

Olegario, J.T., Yee, N., Miller, M., Sczepaniak, J., and Manning, B. 2010. Reduction of Se(VI) to Se(-II) by zerovalent iron nanoparticle suspensions. *J Nanoparticle Res.* 12(6): 2057–2068.

Ona-Nguema, G., Morin, G., Wang, Y., Foster, A. L., Juillot, F., Calas, G., and Brown Jr, G.E. 2010. XANES evidence for rapid arsenic (III) oxidation at magnetite and ferrihydrite surfaces by dissolved O_2 via Fe^{2+}-mediated reactions. *Environ Sci Technol.* 44(14):5416–5422.

Oremland, R.S., Herbel, M.J., Blum, J.S., Langley, S., Beveridge, T.J., Ajayan, P.M., Sutto, T., Ellis, A.V., and Curran, S. 2004. Structural and spectral features of selenium nanospheres produced by Se-respiring bacteria. *Appl Environ Microbiol.* 70:52–60.
Oremland, R.S., Hollibaugh, J.T., Maest, A.S., Presser, T.S., Miller, L.G., and Culbertson, C.W. 1989. Selenate reduction to elemental selenium by anaerobic bacteria in sediments and culture: Biogeochemical significance of a novel, sulfate-independent respiration. *Appl Environ Microbiol.* 55:2333–2343.
Pacyna, E., Pacyna, J., and Pirrone, N. 2001. European emissions of atmospheric mercury from anthropogenic sources in 1995. *Atmos Environ.* 35:2987–2996.
Park, D., Yun, Y.S., and Park, J.M. 2004. Reduction of hexavalent chromium with the brown seaweed *Ecklonia* biomass. *Environ Sci Technol.* 38:4860–4864.
Park, J.H., Lamb, D., Paneerselvam, P., Choppala, G., Bolan, N.S., and Chung, J.W. 2011. Role of organic amendments on enhanced bioremediation of heavy metal(loid) contaminated soils. *J Hazard Mater.* 185:549–574.
Paul, J. and Beauchamp, E. 1989. Effect of carbon constituents in manure on denitrification in soil. *Can J Soil Sci.* 69:49–61.
Ramial, P., John, W.M.R., Furutam, A., and Xun, L. 1985. The effect of pH on methyl mercury production and decomposition in lake sediments. *Can J Fish Aquat Sci.* 42:685–692.
Rauschenbach, I., Narasingarao, P., and Häggblom, M.M. 2011. *Desulfurispirillum indicum* sp. nov., a selenate-and selenite-respiring bacterium isolated from an estuarine canal. *Int J Syst Evol Microbiol.* 61(3):654–658.
Ravichandran, M. 2004. Interactions between mercury and dissolved organic matter—A review. *Chemosphere* 55:319–331.
Rendina, A., Barros, M., and de lorio, A. 2006. Phytoavailability and solid-phase distribution of chromium in a soil amended with organic matter. *Bull Environ Contam Toxicol.* 76:1031–1037.
Rochette, E.A., Bostick, B.C., Li, G.C., and Fendorf, S. 2000. Kinetics of arsenate reduction by dissolved sulfide. *Environ Sci Technol.* 34:4714–4720.
Rock, M.L., James, B.R., and Helz, G.R. 2001. Hydrogen peroxide effects on chromium oxidation state and solubility in four diverse, chromium-enriched soils. *Environ Sci Technol.* 35:4054–4059.
Ross, S.M. 1994. Retention, transformation and mobility of toxic metals in soils. In: Ross, S.M. (ed.). *Toxic Metals in Soil–Plant Systems*. New York: Wiley, pp. 63–152.
Rubinos, D.A., Iglesias, L., Díaz-Fierros, F., and Barral, M.T. 2011. Interacting effect of pH, phosphate and time on the release of arsenic from polluted river sediments Anllóns River, Spain. *Aquat Geochem.* 17:281–306.
Sadiq, M. 1997. Arsenic chemistry in soils: An overview of thermodynamic predictions and field observations. *Water Air Soil Pollut.* 93:117–136.
Schlüter, K. 2000. Review: Evaporation of mercury from soils. An integration and synthesis of current knowledge. *Environ Geol.* 39:249–271.
Schroeder, W.H., and Munthe, J. 1998. Atmospheric mercury - an overview. *Atmos Environ.* 32:809–822.
Schwesig, D., and Matzner, E. 2001. Dynamics of mercury and methylmercury in forest floor and runoff of a forested watershed in Central Europe. *Biogeochemistry* 53:181–200.
Sharma, S., Bansal, A., Dogra, R., Dhillon, S.K., and Dhillon, K.S. 2011. Effect of organic amendments on uptake of selenium and biochemical grain composition of wheat and rape grown on seleniferous soils in northwestern India. *J Plant Nutr Soil Sci.* 174:269–275.
Singh, G., Brar, M., and Malhi, S. 2007. Decontamination of chromium by farm yard manure application in spinach grown in two texturally different Cr-contaminated soils. *J Plant Nutr.* 30:289–308.
Smith, E., Naidu, R., and Alston, A.M. 1998. Arsenic in the soil environment: A review. *Adv Agron.* 66:149–195.
Steinbeiss, S., Gleixner, G., and Antonietti, M. 2009. Effect of biochar amendment on soil carbon balance and soil microbial activity. *Soil Biol Biochem.* 41:1301–1310.
Steiner, C., Teixeira, W.J., Lehmann, J., and Zech, W. 2004. Microbial response to charcoal amendments of highly weathered soils and Amazonian dark earths in Central Amazonia preliminary results. In: Glaser, E. and Woods, W.I. (eds.). *Amazonian Dark Earths: Explorations in Space and Time*. Heidelberg: Springer Verlag, pp. 195–212.
Stollenwerk, K.G., Breit, G.N., Welch, A.H., Yount, J.C., Whitney, J., Foster, A.L., and Ahmed, N. 2007. Arsenic attenuation by oxidized aquifer sediments in Bangladesh. *Sci Total Environ.* 379(2):133–150.
Sun, X., Wang, Q., Ma, H., Wang, Z., Yang, S., Zhao, C., and Xu, L. 2011. Effects of plant rhizosphere on mercury methylation in sediments. *J Soils Sediments.* 11:1062–1069.
Sunitha, R., Mahimairaja, S., Bharani, A., and Gayathri, P. 2014. Enhanced phytoremediation technology for chromium contaminated soils using biological amendments. *Int J Sci Technol.* 33:153–162.

Tang, C. and Rengel, Z. 2003. Role of plant cation/anion uptake ratio in soil acidification. In: Rengel, Z. (ed.). *Handbook of Soil Acidity*. New York: Marcel Dekker, pp. 57–81.

Tarze, A., Dauplais, M., Grigoras, I., Lazard, M., Ha-Duong, N.T., Barbier, F., Blanquet, S., and Plateau, P. 2007. Extracellular production of hydrogen selenide accounts for thiol-assisted toxicity of selenite against *Saccharomyces cerevisiae*. *J Biol Chem.* 282:8759–8767.

Thompson-Eagle, E.T., and Franakenberger Jr., W.T. 1992. Bioremediation of soils contaminated with selenium. In: Lal, R., Stewart, B.A. (eds.). *Advances in Soil Science*. New York: Springer-Verlag, pp. 261–310.

Tokunaga, T.K., Wan, J., Hazen, T.C., Schwartz, E., Firestone, M.K., and Sutton, S.R. 2003. Distribution of chromium contamination and microbial activity in soil aggregates. *J Environ Qual.* 32:541–549.

Vera, S.M., Werth, C.J., and Sanford, R.A. 2001. Evaluation of different polymeric organic materials for creating conditions that favor reductive processes in groundwater. *Bioremed J.* 5:169–181.

Von Canstein, H., Kelly, S., Li, Y., and Wagner-Döbler, I. 2002. Species diversity improves the efficiency of mercury-reducing biofilms under changing environmental conditions. *Appl Environ Microbiol.* 68:2829–2837.

Wadhawan, A.R., Stone, A.T., and Bouwer, E.J. 2013. Biogeochemical controls on hexavalent chromium formation in estuarine sediments. *Environ Sci Technol.* 47(15):8220–8228.

Wang, D., Qing, C., Guo, T., and Guo, Y. 1997. Effects of humic acid on transport and transformation of mercury in soil-plant systems. *Water Air Soil Pollut*. 95:35–43.

Wang, Y., Wiatrowski, H.A., John, R., Lin, C.C., Young, L.Y., Kerkhof, L.J., and Barkay, T. 2013. Impact of mercury on denitrification and denitrifying microbial communities in nitrate enrichments of subsurface sediments. *Biodegrad.* 24(1):33–46.

Weiner, J.G., Gilmour, C.C., and Krabbenhoft, D.P. 2003. Mercury strategy for the bay-delta ecosystem: Aunifying framework for science, adaptive management, and ecological restoration. Report to the California Bay Delta authority, Sacramento, CA.

Wiatrowski, H.A., Ward, P.M., and Barkay, T. 2006. Novel reduction of mercury (II) by mercury-sensitive dissimilatory metal reducing bacteria. *Environ Sci Technol.* 40:6690–6696.

Williams, J.W., and Silver, S. 1984. Bacterial resistance and detoxification of heavy metals. *Enzyme Microb Technol.* 12:530–537.

Xu, X., McGrath, S., and Zhao, F. 2007. Rapid reduction of arsenate in the medium mediated by plant roots. *New Phytol.* 176:590–599.

Yadav, S.K., Juwarkar, A.A., Kumar, G.P., Thawale, P.R., Singh, S.K., and Chakrabarti, T. 2009. Bioaccumulation and phyto-translocation of arsenic, chromium and zinc by *Jatropha curcas* L.: Impact of dairy sludge and biofertilizer. *Bioresour Technol.* 100:4616–4622.

Yamamura, S., Watanabe, M., Kanzaki, M., Soda, S., and Ike, M.2008. Removal of arsenic from contaminated soils by microbial reduction of arsenate and quinone. *Environ Sci Technol.* 42:6154–6159.

Yee, N., Barkay, T., Parikh, M., Lin, C., Wiatrowski, H., and Das, S. 2010. Biotic/abiotic pathways of Hg(II) reduction by dissimilatory iron reducing bacteria. *Geol Soc Am Abs Prog*. 42(1):178.

Yin, X.X., Chen, J., Qin, J., Sun, G.X., Rosen, B.P., and Zhu, Y.G. 2011. Biotransformation and volatilization of arsenic by three photosynthetic cyanobacteria. *Plant Physiol.* 156:1631–1638.

Zawislanski, P.T., and Zavarin, M. 1996. Nature and rates of selenium transformations in soils: A laboratory study. *Soil Sci Soc Am J.* 60:791–800.

Zazo, J.A., Paul, J.S., and Jaffe, P.R. 2008. Influence of plants on the reduction of hexavalent chromium in wetland sediments. *Environ Pollut.* 156:29–35.

Zeng, F., Chen, S., Miao, Y., Wu, F., and Zhang, G. 2008. Changes of organic acid exudation and rhizosphere pH in rice plant under chromium stress. *Environ Pollut.* 155:284–289.

Zhang, J., and Bishop, P.L. 2002. Stabilization/solidification (S/S) of mercury-containing wastes using reactivated carbon and Portland cement. *J Hazard Mater.* 92:199–212.

Zhang, S., Li, W., Shan, X., Lu, A., and Zhou, P. 2005. Effects of low molecular weight organic anions on the release of arsenite and arsenate from a contaminated soil. *Water Air Soil Pollut.* 167(1–4):111–122.

Zhang, T. and Hsu-Kim, H. 2010. Photolytic degradation of methylmercury enhanced by binding to natural organic ligands. *Nat Geosci.* 3:473–476.

Zhang, Y., Okeke, B.C., and Frankenberger Jr., W.T. 2008. Bacterial reduction of selenate to elemental selenium utilizing molasses as a carbon source. *Biores Technol.* 99(5):1267–1273.

Zhang, Y.Q., and Frankenberger, W.T. 1999. Effects of soil moisture, depth, and organic amendments on selenium volatilization. *J Environ Qual.* 28:1321–1326.

Zhong, L., and Yang, J. 2012. Reduction of Cr(VI) by malic acid in aqueous Fe-rich soil suspension. *Chemosphere* 86:973–978.

4 Sorption–Desorption of Trace Elements in Soils

Influence of Kinetics

H. Magdi Selim

CONTENTS

Knowledge of interactions between trace elements in the soil–water environment is essential in assessing their bioavailability and potential toxicity in soils and wetlands. Trace elements include several heavy metals such as zinc, copper, arsenic, cadmium, and vanadium, among others. Several heavy metals such as Zn and copper are essential micronutrients that are required in the growth of both plants and animals. Micronutrients are often applied in the form of fertilizers or as supplements in animal feed. Heavy metals are extensively used as fungicides and as bactericides in numerous pharmaceuticals. The bioavailability of trace elements in the soil–water environment is dependent on an array of soil properties, including soil pH, organic matter content, amount and type of dominant clay, and carbonates, among others. In addition, the counterions present in the soil system greatly influence the fate of trace metals in soils. In fact, several studies suggested varied interactions of heavy metals with phosphates in soils as discussed in a later section.

Adsorption isotherms, or more accurately sorption isotherms, are convenient ways of graphically representing the amount of an adsorbed compound, or adsorbate, in relation to its concentration in the equilibrium solution or adsorbent. In other words, an adsorption isotherm is a relationship between the concentration of a solute on the surface of an adsorbent and the concentration of the solute in the liquid with which it is in contact at a constant temperature. Freundlich and Langmuir sorption isotherms are extensively used to describe sorption isotherms for a wide range of chemicals. Knowledge of sorption isotherms and adsorption phenomena is essential for understanding heavy solute retention and transport in soils and geological media. It is crucial for assessing the environmental risk of contamination and/or pollution provoked by these elements. Studies on solute adsorption in soils are often conducted as a one-component system, where the ions or molecules are treated individually, or they can be conducted as a multicomponent system, where the ions are subjected to competition among themselves.

For several decades, it has been observed that sorption and desorption of various chemicals with matrix surface are kinetic- or time-dependent. Numerous studies on the kinetic behavior of solutes in soils are available in the literature. Recent reviews on kinetics are available (Sparks and Suarez

1991, Sparks 2003, Carrillo-Gonzalez et al. 2006). The extent of kinetics varied extensively among the different solute species and soils considered. Generally, trace elements and heavy metal species exhibit strong sorption as well as extensive kinetic behavior during sorption as well as during release or desorption. In contrast, weak sorption and less extensive kinetic behavior are often observed for organic chemicals in soils and porous media. According to Aharoni and Sparks (1991) and Sparks (2003), a number of transport and chemical reaction processes affect the rate of soil chemical reactions. The slowest of these will limit the rate of a particular reaction. The actual chemical reaction at the surface, for example, adsorption, is usually very rapid and not rate limiting. Transport processes include: (1) transport in the solution phase, which is rapid, and which can be eliminated by rapid mixing in the laboratory; (2) transport across a liquid film at the particle/liquid interface (film diffusion); (3) transport in liquid-filled macropores (>2 nm), all of which are nonactivated diffusion processes and occur in mobile regions; (4) diffusion of a sorbate along pore wall surfaces (surface diffusion); (5) diffusion of sorbate occluded in micropores (<2 nm) (pore diffusion); and (6) diffusion processes in the bulk of the solid, all of which are activated diffusion processes. Pore and surface diffusion can be referred to as interparticle diffusion, whereas diffusion in the solid is intraparticle diffusion.

The form of chemical retention reactions in soils and geological porous media must be clearly identified if predictions of their potential mobility, toxicity, and impact on the environment are sought. In general, chemical retention processes with matrix surfaces have been quantified by scientists by using a number of empirically based approaches. One approach represents an equilibrium type in which sorption reactions are assumed to be fast or instantaneous in nature. Under such conditions, apparent equilibrium may be observed in a relatively short reaction time (minutes or hours). Freundlich and Langmuir models are perhaps the most commonly used equilibrium models for the description of fertilizer chemicals, especially phosphorus, heavy metals, and pesticides. These equilibrium models include the linear and Freundlich (nonlinear) and the one- and two-site Langmuir type.

Soils and other geochemical systems are quite complex, and various sorption reactions are likely to occur. Such reactions are perhaps a series of consecutive and/or concurrent reactions or of the simultaneous type. Amacher (1991) and Selim and Amacher (1997) dealt with several types of reactions that occur in soils and the time ranges required to attain equilibrium by these reactions. The ion association, multivalent ion hydrolysis, and mineral crystallization reactions are homogeneous, because they occur within a single phase. The first two of these occur in the liquid phase, whereas the last occurs in the solid phase. The other reaction types are heterogeneous, because they involve the transfer of chemical species across the interfaces between phases. Ion association reactions refer to ion pairing, complexation (inner- and outer sphere), and chelation-type reactions in solution. Gas–water reactions refer to the exchange of gases across the air–liquid interface. Ion-exchange reactions refer to electrostatic ion replacement reactions on charged solid surfaces. Sorption reactions refer to simple physical adsorption, surface complexation (inner- and outer-sphere), and surface precipitation reactions. Mineral–solution reactions refer to precipitation/dissolution reactions involving discrete mineral phases and co-precipitation reactions by which trace constituents can become incorporated into the structure of discrete mineral phases.

4.1 MODELING OF KINETIC SORPTION

Kinetic models represent slow reactions in which the amount of solute sorption or transformation is a function of contact time. The most commonly encountered model is the first-order kinetic reversible reaction that is used for describing time-dependent adsorption/desorption in soils. Others include linear, irreversible and nonlinear, reversible kinetic models. Recently, a combination of equilibrium- and kinetic-type (two-site) models, and consecutive and concurrent multireaction-type models has been proposed.

4.1.1 First-Order and Freundlich Kinetics

The first-order kinetic approach is perhaps one of the earliest single form of reactions used to describe the sorption versus time for several dissolved chemicals in soils. This may be written as follows:

$$\frac{\partial S}{\partial t} = k_f\left(\frac{\theta}{\rho}\right)C - k_b\, S \tag{4.1}$$

where the parameters k_f and k_b represent the forward and backward rates of reactions (h^{-1}) for the retention mechanism, respectively. The first-order reaction was first incorporated into the classical convection–dispersion equation by Lapidus and Amundson (1952) to describe solute retention during transport under steady-state water flow conditions. Integration of Equation 4.1 subject to the initial conditions of $C = C_i$ and $S = 0$ at $t = 0$, for several C_i values, yields a system of linear sorption isotherms. That is, for any reaction time, t, a linear relationship between S and C is obtained.

There are numerous examples in the literature on the kinetics of pesticides and other organic sorption in various soils. Examples are shown in Figure 4.1 for imidacloprid that illustrate experimental observations in which sorption over time appears linear (Jeong and Selim, 2010).

It was argued that such apparent linear behavior is not surprising for a number of reasons, including the concentration range of the solute in solution, and adsorption optima are not attained. In addition, Selim (2011, 2012) suggested that linear behavior is also due to the uniform or homogeneous nature of the sorbing matrix. It is safe to consider organic matter as the dominant sorbent for imidaclorpid in a Vacherie soil with high-organic matter. Other examples of apparent linear kinetics for other solutes are not abundant, with the exception of cations of low affinity such as Ca, K, and Na (Gaston and Selim 1990a, b).

Kinetic sorption that exhibits nonlinear or curve-linear retention behavior is commonly observed for several reactive chemicals, as depicted by the nonlinear isotherms for arsenic shown in Figure 4.2 (Zhang and Selim 2005). A second example is illustrated in Figure 4.3 for the curve-linear and kinetic behavior of Cu in a McLaren soil. To describe such nonlinear behavior, the single reaction given in Equation 4.1 is commonly extended to include nonlinear kinetics such that (Selim 1992)

$$\frac{\partial S}{\partial t} = k_f\left(\frac{\theta}{\rho}\right)C^b - k_b\, S \tag{4.2}$$

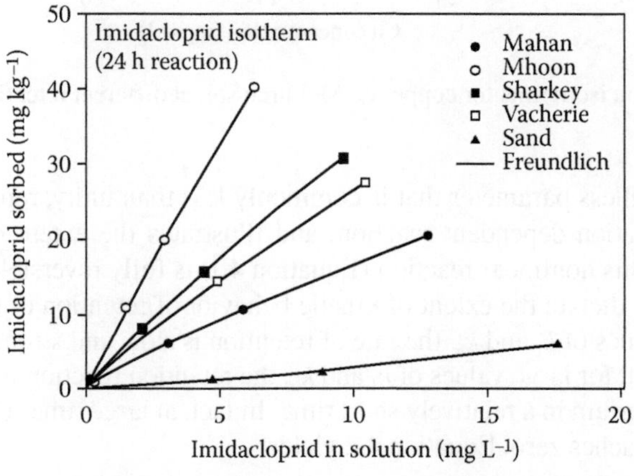

FIGURE 4.1 Adsorption isotherms for imidacloprid for five different retention soils. The solid curves are based on the Freundlich equation.

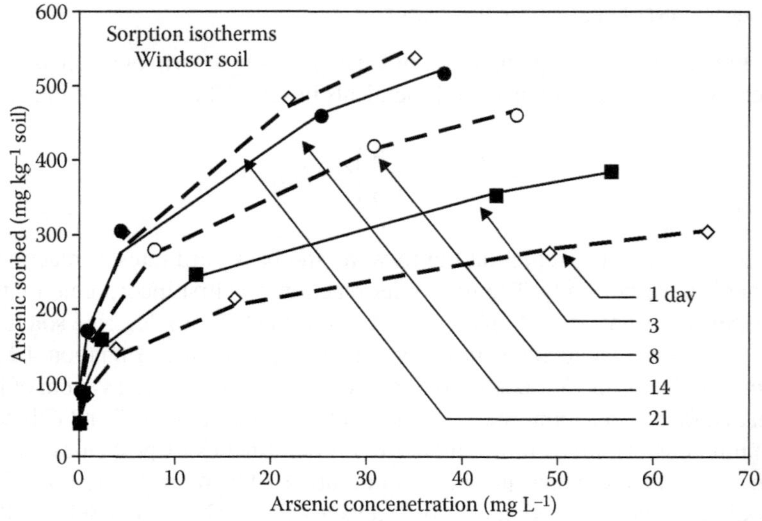

FIGURE 4.2 Adsorption isotherms for arsenic on Windsor soil at different retention times.

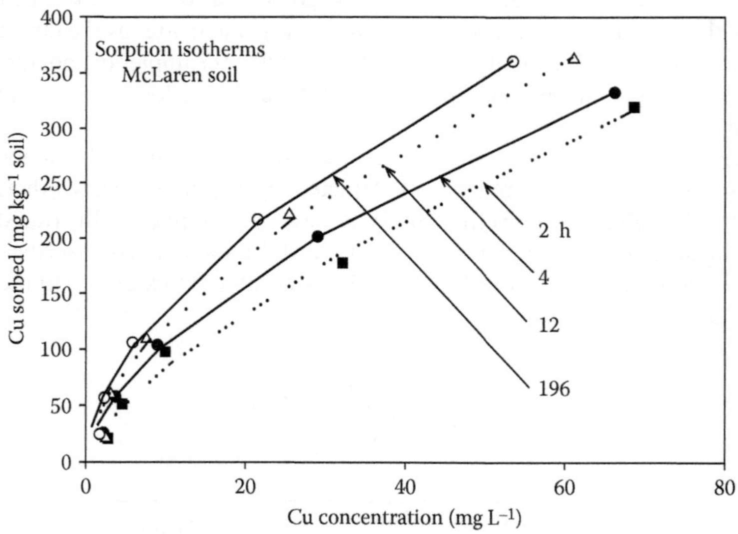

FIGURE 4.3 Adsorption isotherms for copper on McLaren soil at different retention times.

where b is a dimensionless parameter that is commonly less than unity, represents the order of the nonlinear or concentration-dependent reaction, and illustrates the extent of heterogeneity of the retention processes. This nonlinear reaction (Equation 4.1) is fully reversible when the magnitudes of the rate coefficients dictate the extent of kinetic behavior of retention of the solute from the soil solution. For small values of k_f and k_b, the rate of retention is slow and strong kinetic dependence is anticipated. In contrast, for large values of k_f and k_b, the retention reaction is a rapid one and should approach quasi-equilibrium in a relatively short time. In fact, at large times (i.e., as $t \to \infty$), when the rate of retention approaches zero, Equation 4.2 yields

$$S = K_f C^b \quad \text{where } K_f = \left(\frac{\theta k_f}{\rho k_b}\right) \tag{4.3}$$

Equation 4.3 is analogous to the Freundlich equilibrium equation, where K_f is the solute partitioning coefficient (cm^3/g). Therefore, one may regard the parameter K_f as the ratio of the rate coefficients for sorption (forward reaction) to that for desorption or release (backward reaction).

The parameter b is a measure of the extent of the heterogeneity of sorption sites of the soil matrix. In other words, sorption sites have different affinities for heavy metal retention by matrix surfaces, where sorption by the highest energy sites takes place preferentially at the lowest solution concentrations.

For the simple case where $b = 1$, we have the linear form

$$S = K_d C \quad \text{where } K_d = \left(\frac{\theta k_f}{\rho k_b}\right) \tag{4.4}$$

where the parameter K_d is the solute distribution coefficient (cm^3/g) and of similar form to the Freundlich parameter K_f. There are numerous examples of cations and heavy metal retention, which were described successfully by use of the linear or the Freundlich equation (Buchter et al. 1989). The lack of nonlinear or concentration-dependent behavior of sorption patterns as indicated by the linear case of Equation 4.1 is indicative of the lack of heterogeneity of sorption-site energies. For this special case, sorption-site energies for linear sorption processes of heavy metals may be best regarded as relatively homogenous. A partial list of equilibrium and kinetic retention models for solutes is presented in Table 4.1.

TABLE 4.1
Selected Equilibrium and Kinetic-Type Models for Heavy Metal Retention in Soils

Model	Formulation[a]
Equilibrium Type	
Linear	$S = K_d C$
Freundlich	$S = K_f C^b$
General Freundlich	$S / S_{max} = [\omega C / (1 + \omega C)]^\beta$
Rothmund–Kornfeld ion exchange	$S_i / S_T = K_{RK} (C_i / C_T)^n$
Langmuir	$S / S_{max} = \omega C / [1 + \omega C]$
General Freundlich and Langmuir	$S / S_{max} = (\omega C)^\beta / [1 + (\omega C)^\beta]$
Langmuir with sigmoidicity	$S / S_{max} = \omega C / [1 + \omega C + \sigma / C]$
Kinetic Type	
First order	$\partial S / \partial t = k_f (\theta / \rho) C - k_b S$
*n*th order	$\partial S / \partial t = k_f (\theta / \rho) C^n - k_b S$
Irreversible (sink/source)	$\partial S / \partial t = k_s (\theta / \rho)(C - C_p)$
Second-order irreversible	$\partial S / \partial t = k_s (\theta / \rho) C (S_{max} - S)$
Langmuir kinetic	$\partial S / \partial t = k_f (\theta / \rho) C (S_{max} - S) - k_b S$
Elovich	$\partial S / \partial t = A \exp(-BS)$
Power	$\partial S / \partial t = K (\theta / \rho) C^n S^m$
Mass transfer	$\partial S / \partial t = K (\theta / \rho)(C - C^*)$

[a] A, B, b, C^*, C_p, K, K_d, K_{RK}, k_b, k_f, k_s, n, m, S_{max}, ω, β, and σ are adjustable model parameters; ρ is bulk density; θ is volumetric soil water content; C_T is total solute concentration; and S_T is total amount sorbed among all competing species.

4.2 DESORPTION AND HYSTERESIS

Desorption of sorbed solutes from matrix surfaces is the process of detachment or release of ions or molecules to the bulk solution or the liquid phase. Knowledge-associated mechanisms are significant in understanding release or desorption behavior and in providing the necessary tools for predictions and risk assessment. As with water infiltration and subsequent redistribution in the soil profile, solute release often continues for extended periods when compared with the duration for adsorption. Most applications of chemicals on soils and accidental spills occur over a relatively short time (hours or days). It is obvious that adsorption is dominant during such applications. This adsorption is commonly followed by extended periods of release or desorption ranging from months to years or decades.

Figure 4.4 shows adsorption and desorption isotherms with emphasis on desorption subsequent to adsorption. The major advantage here is that two sets of separate isotherms are not required. One should note that the solid curve in the family of desorption isotherms represents that of adsorption after 504 h, whereas all dashed curves represent successive desorption isotherms. These isotherms clearly illustrate the kinetic behavior of solute sorption and that short-term isotherms (h or d) do not necessarily provide an accurate description of their affinity to a specific type of matrix surface or a given soil. The extent of hysteresis is further illustrated in Figure 4.5 by the results for Zn desorption versus time for soils with distinctly different sorption affinity for Zn. In Webster soil, desorption or release of Zn appears to be slow, which is indicative of strong sorption. In contrast, rapid release was observed in Windsor soil. The respective isotherms for the two soils are shown in a later figure and clearly illustrate extensive hysteresis for Webster sorption that is consistent with strong kinetic behavior and possible irreversible reactions.

The hysteresis phenomenon has been reported in numerous studies published in colloids and colloidal chemistry literature for several decades. In fact, the term *hysteresis* is not restricted to solute sorption isotherms; it is used in other disciplines such as soil physics and hydrology. In water-unsaturated porous media, hysteresis was observed in soil-moisture content and applied suction. Discrepancies between wetting and drying curves and subsequent scanning curves give rise to the term *hysteresis* (Figures 4.4 and 4.5).

Reasons for the observed hysteresis have been discussed, and various explanations have been advanced in the literature. Most of them center on irreversible reactions, change of phase, and

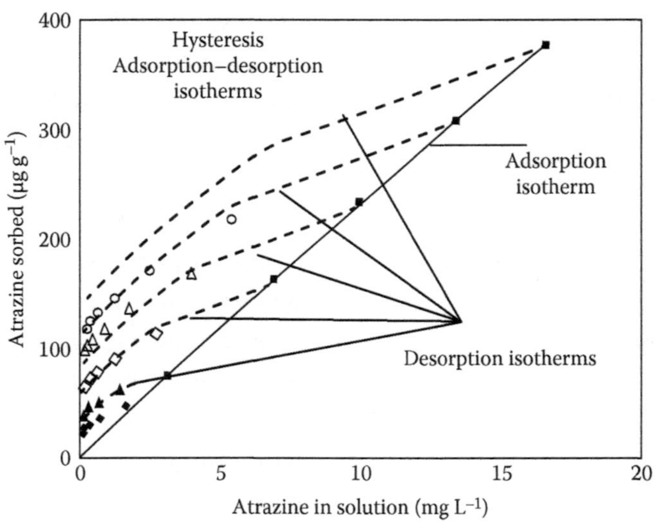

FIGURE 4.4 Traditional desorption isotherms of atrazine by sugarcane mulch residue. Solid line is the adsorption isotherm for 504-h reaction. Dashed curves are prediction using (a) multireaction model (upper), and (b) Freundlich model (lower).

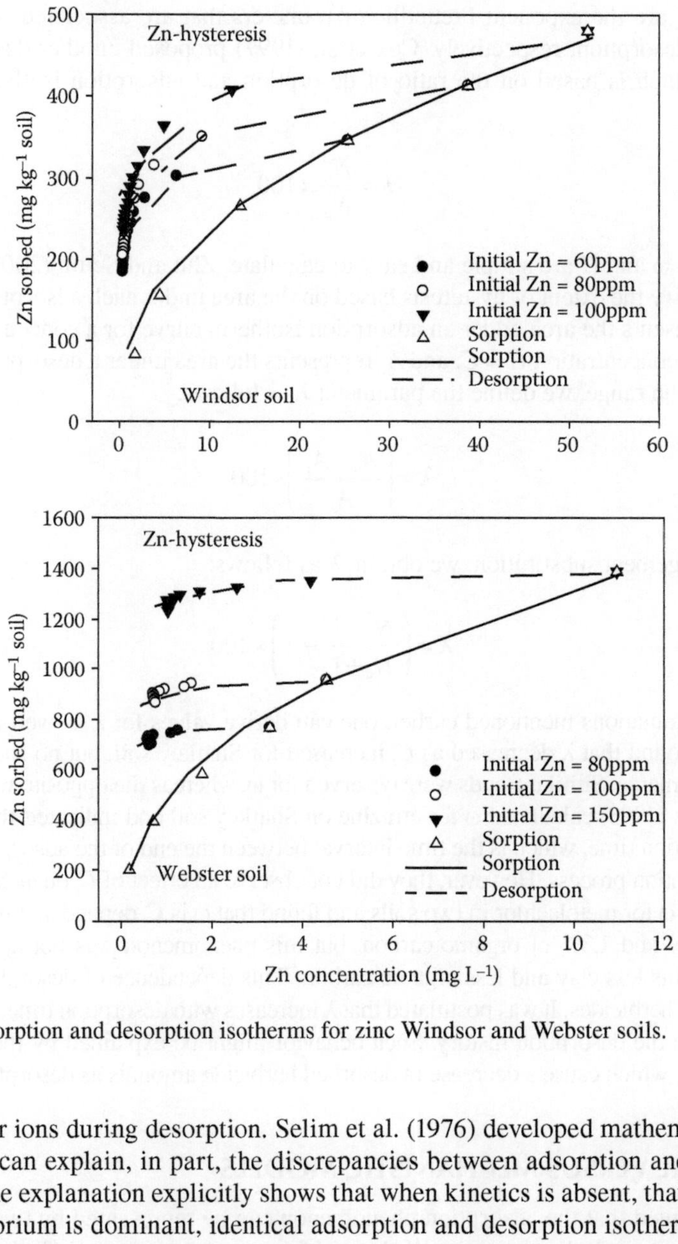

FIGURE 4.5 Adsorption and desorption isotherms for zinc Windsor and Webster soils.

formation of other ions during desorption. Selim et al. (1976) developed mathematical proof that sorption kinetics can explain, in part, the discrepancies between adsorption and desorption isotherms. In fact, the explanation explicitly shows that when kinetics is absent, that is, when instantaneous or equilibrium is dominant, identical adsorption and desorption isotherms are obtained. In other words, both isotherms coalesce and nonsingularity or hysteresis is not observed.

4.2.1 Empirical Hysteresis Coefficient

Several efforts have been made to quantify hysteresis based on adsorption and desorption parameters that are associated with the Freundlich equation. Ma et al. (1993) defined *hysteresis* based on the difference between adsorption and desorption isotherms as a direct way to quantify the discrepancy between sorption and desorption. They derived the following equation as a hysteresis parameter based on the maximum difference between an adsorption and a desorption isotherm:

$$\omega = \left(\frac{N_a}{N_d} - 1 \right) \times 100 \tag{4.5}$$

where N_a and N_d are the exponent Freundlich parameters that are associated with the respective adsorption and desorption, respectively. Cox et al. (1997) proposed another desorption hysteresis coefficient H, which is based on the ratio of desorption and adsorption isotherm parameters as follows:

$$H = \frac{N_a}{N_d} \times 100 \tag{4.6}$$

Both coefficients ω and H are simple and easy to calculate. Zhu and Selim (2000) derived another formula to quantify the extent of hysteresis based on the area under each adsorption and desorption curve. If A_a represents the area under an adsorption isotherm curve for a concentration from $C = 0$ to some solution concentration $C = C$, and A_d represents the area under a desorption isotherm at the same concentration range, we define the parameter λ as follows:

$$\lambda = \left(\frac{A_d - A_a}{A_a} \right) \times 100 \tag{4.7}$$

On further arrangement substitution, we obtain λ as follows:

$$\lambda = \left(\frac{N_a + 1}{N_d + 1} - 1 \right) \times 100 \tag{4.8}$$

Based on the formulations mentioned earlier, one can derive values for λ as well as for ω and Selim and Zhu (2005) found that λ decreased as C_i increased for Sharkey soil, but no such relationship was observed for commerce. Similar trends were observed for ω, whereas the opposite was observed for the H value. Ma et al. (1993) calculated ω for atrazine on Sharkey soil and indicated that ω increased linearly with incubation time, which is the time interval between the end of the adsorption and the beginning of the desorption process. However, they did not observe an effect of C_i on ω. Seybold and Mersie (1996) calculated ω for metolachlor in two soils and found that ω is C_i dependent on Cullen soil, which contains 31% clay and 1.3% of organic carbon, but this phenomenon was not apparent in Emporia soil, which contains less clay and less organic carbon. This dependence of desorption on C_i has been reported for other herbicides. It was postulated that λ increases with desorption time, which is indicative of dependency on the desorption history. Such behavior might be explained by the existence of irreversible reactions, which cause a decrease in desorbed herbicide amounts as desorption time increases.

4.3 EMPIRICAL VERSUS MECHANISTIC MODELS

Fontes (2012) argued that the adsorption phenomenon can be represented by two main conceptual models: (1) the empirical, the ones initially derived from experiments, and (2) the semiempirical or mechanistic, the ones that are based on reaction mechanisms. The main difference between these two models is the lack of an electrostatic term in the empirical models, whereas its presence is mandatory in mechanistic models. Mechanistic-type models, also known as chemical models, are expected to provide "a close representation of the real adsorption phenomenon in the soil system." Nevertheless, due to the complexity of chemical models, empirical models are usually utilized in most solute studies with soils and geological media. Moreover, empirical models have been widely used in soil science and environmental studies related to metals, anion adsorption, and pesticide retention in soils. A listing of such models is presented in Table 4.1. These models do not consider the electrostatic influence of the electrically charged surfaces in the solution, as well as the influence of changes in surface charges due to the composition of soil solution. In the empirical model, the model form is chosen a posteriori from the observed adsorption data. To enable a satisfying fitting of the experimental data, the mathematical form and the number of parameters are chosen to be as simple as possible (Bradl 2004).

4.4 EFFECT OF SOIL PROPERTIES

Buchter et al. (1989) studied the retention of 15 elements by 11 soils from 10 soil orders to determine the effects of element and soil properties on the magnitude of the Freundlich parameters K_f and *b*. They also explored the correlation of the Freundlich parameters with selected soil properties and found that pH, cation-exchange capacity (CEC), and iron/aluminum oxide contents were the most important factors for correlation with the partitioning coefficients. The names, taxonomic classification, and selected properties of the 11 soils used in their study were measured and estimated values for K_f and *b* for selected heavy metals were obtained. A wide range of K_f values, from 0.0419 to 4.32×10^7 ml g^{-1}, were obtained, which illustrates the extent of affinity of heavy metals among various soil types. However, such a wide range of values was not obtained for the exponent parameter *b*. The magnitude of K_f and *b* was related to both soil and element properties. Strongly retained elements such as Cu, Hg, Pb, and V had the highest K_f values. The transition metal cations Co and Ni had similar K_f and *b* values as did the group IIB elements Zn and Cd. Oxyanion species tended to have lower *b* values than did cation species. Soil pH and CEC were significantly correlated to log K_f values for cation species. High pH and high CEC soils retained greater quantities of the cation species than did low pH and low CEC soils. A significant negative correlation between soil pH and the Freundlich parameter *b* was observed for cation species, whereas a significant positive correlation between soil pH and *b* for Cr(VI) was found. Greater quantities of anion species were retained by soils with high amounts of amorphous iron oxides, aluminum oxides, and amorphous material than were retained by soils with low amounts of these minerals. Several anion species were not retained by high pH soils. Despite the facts that element retention by soils is the result of many interacting processes and that many factors influence retention, significant relationships among retention parameters and soil and element properties exist even among soils with greatly different characteristics. Buchter et al. (1989) made the following conclusions:

1. pH is the most important soil property that affects K_f and *b*.
2. Cation-exchange capacity influences K_f for cation species.
3. The amounts of amorphous iron oxides, aluminum oxides, and amorphous material in soils influence both cation and anion retention parameters.
4. Except for Cu and Hg, transition metal (Co and Ni) and group IIB cations (Zn and Cd) have similar K_f and *b* values for a given soil.
5. Significant relationships between soil properties and retention parameters exist even in a group of soils with greatly different characteristics.

The relationships between soil properties and retention parameters (e.g., Figure 4.6) can be used to estimate retention parameters when retention data for a particular element and soil type are lacking, but soil property data are available. For example, the retention characteristics of Co, Ni, Zn, and Cd are sufficiently similar such that these elements can be grouped together and an estimated *b* value for any one of them could be estimated from soil pH data by using the regression equation for curve *A* shown in Figure 4.6. For many purposes, such an estimate would be useful, at least as a first approximation, in describing the retention characteristics of soil.

$$\text{Curve A:} \quad b = 1.24 - 0.0831 \text{ pH} \quad (r = 0.83) \tag{4.9}$$

$$\text{Curve B:} \quad b = -0.0846 + 0.116 \text{ pH} \quad (r = 0.98) \tag{4.10}$$

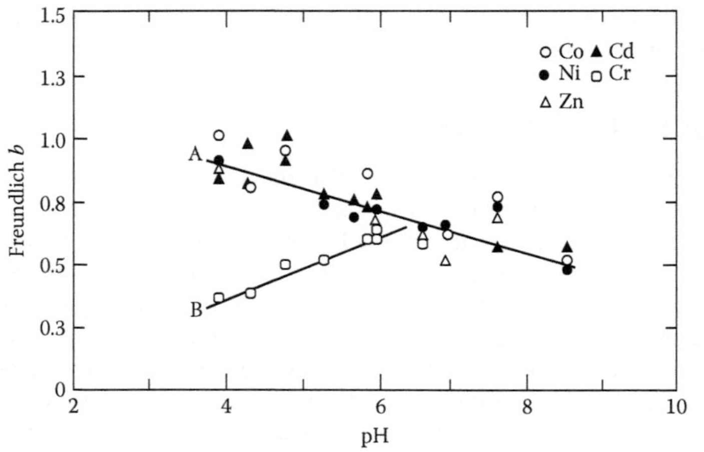

FIGURE 4.6 Correlation between soil pH and Freundlich parameter *b*. Curve *A* is a regression line for Co, Ni, Zn, and Cd ($b = 1.24 - 0.0831$ pH, $r = 0.83^{**}$). Curve *B* is for Cr(VI) ($b = -0.0846 + 0.116$ pH, $r = 0.98^{**}$). (After Buchter, B. et al., *Soil Sci.*, 148, 370–379, 1989.)

Sauvé et al. (2000) compiled data from more than 70 studies of various origins in an effort to correlate the distribution coefficient K_d with soil properties for five heavy metals: cadmium, copper, lead, nickel, and zinc. Specifically, the relationships between the reported K_d values were explored relative to variations in soil solution pH, soil organic carbon (SOC), and total metal retained by the soil. Sauvé et al. (2000) proposed two models to predict K_d values for several heavy metals based on chemical properties of the soil. These models were developed based on the regression analysis of extensive K_d values published in the literature for a wide range of soils. The proposed models are as follows:

$$\text{Model I:} \quad \log(K_d) = A + B\,(\text{pH}) \tag{4.11}$$

$$\text{Model II:} \quad \log(K_d) = A + B(\text{pH}) + C(\text{SOC}) \tag{4.12}$$

where *A*, *B*, and *C* are fitting parameters. Sauvé et al. (2000) proposed a third model that incorporated retained heavy metals by the soil prior to K_d measurements. With the exception of contaminated sites, however, amounts of heavy metals retained are often extremely small.

Recently, experimental studies were carried out to quantify the retention of 5 heavy elements by 10 soils from 10 soil orders to determine the effects of element and soil properties on the magnitude of the Freundlich parameters K_f and *b*. The adsorption was measured after days 1 and 7 of adsorption, and desorption was measured after adsorption (based on the successive dilution method). The heavy metals investigated were cadmium, copper, lead, nickel, and zinc. These five metals are the same as those considered by Sauvé et al. (2000) in their regression study. The names, taxonomic classifications, and selected properties of the 10 soils used in this study are listed in Tables 4.2 and 4.3. The *Ap* horizons of all soils were used in this retention study. The only exception is the sandy-candor subsurface sample that was sampled at a depth of 90 cm. The physicochemical properties of the 10 soils were also quantified. A comparison of K_d parameter values for all 10 soils and 5 heavy metals is provided in Table 4.4 for days 1 and 7 of sorption. Freundlich model K_f and *b* parameter values for all 10 soils and 5 heavy metals are provided in Table 4.5 for days 1 and 7 of sorption. These results illustrate the sorption affinity of the various soils for each of the heavy metals.

TABLE 4.2
Selected Soil Properties of the 10 Soils Used in the Study

Soil Series	State	Texture	pH	%C	CEC	Sand %	Silt %	Clay %	Carbonates (%)
Arapahoe	NC	FSL	5.02	10.22	30.10	64.7	25.2	10.1	
Candor: Surface	NC	LCoS	4.39	2.24	5.30	84.1	6.6	9.3	
Candor: Subsurface	NC	CoS	4.05	0.56	1.70	91.0	4.1	4.9	
Olney	CO	FSL	8.12	1.14	10.10	71.1	11.6	17.3	3.32
Lincoln	OK	VFSL	7.54	1.27	3.00	59.5	29.7	10.8	3.6
Nada	TX	FSL	6.61	0.76	6.30	56.8	33.6	9.6	
Morey	TX	L	7.74	1.05	27.80	29.4	46.5	24.1	
Crowley	LA	SiL	5.22	1.16	16.50	8.3	77.3	14.4	
Sharkey	LA	SiC	5.49	2.76	39.70	6.4	48.6	45.0	
Houston	TX	SiC	7.78	4.26	47.70	9.6	41.2	49.2	26.0

TABLE 4.3
Taxonomic Classification of the 10 Soils Used in the Study

Soil Series	State	Taxonomic Classification
Arapahoe	NC	Coarse-loamy, mixed, semiactive, nonacid, thermic *Typic Humaquepts*
Candor: Surface	NC	Sandy, kaolinitic, thermic *Grossarenic Kandiudults*
Candor: Subsurface	NC	Sandy, kaolinitic, thermic *Grossarenic Kandiudults*
Olney	CO	Fine-loamy, mixed, superactive, mesic *Ustic Haplargids*
Lincoln	OK	Sandy, mixed, thermic *Typic Ustifluvents*
Nada	TX	Fine-loamy, siliceous, active, hyperthermic *Albaquic Hapludalfs*
Morey	TX	Fine-silty, siliceous, superactive, hyperthermic *Oxyaquic Argiudolls*
Crowley	LA	Fine, smectitic, thermic *Typic Albaqualfs*
Sharkey	LA	Fine-silty, mixed, active, thermic *Typic Glossaqualfs*
Houston	TX	Fine, smectitic, thermic *Udic Haplusterts*

TABLE 4.4
Linear Model K_d Parameter Values (g mL^{-1}) for 10 Soils and 5 Heavy Metals

Element	Arapahoe	Candor Surface	Candor Subsurface	Olney	Lincoln	Nada	Morey	Crowley	Sharkey	Houston
					K_d - 1 day					
Cu	1142.28	6.23	2.52	3110.95	161.45	25.31	754.59	42.63	272.98	11,748.66
Zn	16.51	1.66	3.21	86.30	10.65	7.30	38.45	8.83	32.32	415.76
Cd	144.68	6.82	5.21	42.34	13.09	9.69	56.67	26.68	87.84	380.02
Ni	32.47	1.98	2.10	43.98	10.96	8.93	47.92	11.91	39.99	147.05
Pb	4035.40	19.92	5.46	34,996.46	67.83	41.65	20,696.37	109.84	1344.13	ND
					K_d - 7 days					
Cu	1776.43	11.49	5.59	4680.96	3406.10	32.13	5233.00	53.11	420.37	13,124.76
Zn	17.91	2.26	2.94	384.38	19.83	8.00	47.32	9.26	33.60	817.07
Cd	156.42	7.02	5.51	60.85	15.57	10.62	59.16	27.47	87.62	664.69
Ni	38.68	1.95	2.19	111.64	34.28	9.19	44.51	12.23	42.70	281.76
Pb	5497.26	26.47	7.67	57,251.48	4329.19	53.63	85,573.42	121.94	2311.93	ND

Note: ND, not detected.

TABLE 4.5
Freundlich Model K_f (g mL^{-1}) and b Parameter Values for 10 Soils and 5 Heavy Metals

Element	Arapahoe	Candor Surface	Candor Subsurface	Olney	Lincoln	Nada	Morey	Crowley	Sharkey	Houston
					***K_f* - 1 day**					
Cu	1004.91	64.51	16.15	5027.78	512.40	235.45	833.78	213.88	501.73	66,205.54
Zn	150.51	0.00	0.00	552.83	181.27	60.29	323.29	37.21	91.07	988.81
Cd	313.18	13.62	6.30	367.21	157.04	92.94	267.61	93.42	220.88	738.24
Ni	78.51	–	0.00	169.79	64.04	16.65	159.24	17.72	65.92	373.10
Pb	3792.51	93.99	7.09	5169.46	1679.58	1360.44	4224.09	1280.45	2825.04	C S
					***K_f* - 7 days**					
Cu	1305.73	103.12	72.88	2,209,492.85	3993.56	350.52	4842.30	283.91	630.16	29,606,459.71
Zn	22.37	0.51	1.53	847.56	243.53	50.98	387.64	45.00	96.24	1258.36
Cd	323.09	16.68	7.94	514.97	0.39	93.58	301.51	103.19	224.12	935.35
Ni	89.39	–	0.00	261.10	134.28	28.04	130.48	23.51	70.94	467.14
Pb	4705.95	174.11	33.53	6889.88	3863.83	1497.59	5296.20	1452.22	3275.69	C S
					***b* - 1 day**					
Cu	0.54	0.40	0.54	1.32	0.29	0.30	0.41	0.41	0.46	1.16
Zn	0.43	2.85	2.50	0.31	0.32	0.51	0.37	0.66	0.70	0.33
Cd	0.65	0.84	0.96	0.35	1.00	0.47	0.49	0.65	0.65	0.50
Ni	0.73	–	2.74	0.56	0.56	0.85	0.60	0.90	0.84	0.54
Pb	0.62	0.65	0.95	0.16	0.18	0.18	0.08	0.28	0.27	C S
					***b* - 7 days**					
Cu	0.58	0.39	0.34	3.96	1.10	0.22	0.96	0.35	0.44	3.48
Zn	0.83	1.32	1.14	0.37	0.33	0.56	0.34	0.62	0.69	0.28
Cd	0.66	0.80	0.92	0.30	0.33	0.48	0.46	0.63	0.65	0.54
Ni	0.74	–	2.83	0.62	0.57	0.73	0.65	0.83	0.84	0.64
Pb	0.69	0.56	0.70	0.24	0.36	0.18	0.12	0.26	0.29	C S

Copper isotherms for different soils are shown in Figures 4.7 and 4.8, and they indicate a wide range of copper sorption among the different soils. The results for Houston and Olney indicate the extremely high sorption of copper by both soils as manifested by the low concentration in the soil solution after 24-hour sorption (Figure 4.7). This high sorption is due to the high clay content of Houston clay as well as due to the presence of carbonates. The dashed and solid curves represent linear and Freundlich model simulations. As shown in Figure 4.8, Arapaho soil exhibits extensive copper sorption. In contrast, Crowley and Nada Cu isotherms indicate the lowest copper sorption. Solid dashed curves are simulations using the Freundlich model that best describe the isotherms for all 10 soils. Best-fit parameter values for the linear and Freundlich models along with their r^2 values are given in Tables 4.4 and 4.5.

Cadmium isotherms for all soils are shown in Figures 4.9 and 4.10. Houston, Arapahoe, and Sharkey exhibited the highest cadmium sorption, which was indicative of strong copper affinity for soils with high clay content and organic matter. The lowest sorption for copper is shown in Figure 4.10 for Lincoln, Nada, and Candor soils. Solid and dashed curves shown in Figures 4.9 and 4.10 are simulations using the Freundlich model that best describe the isotherms for all 10 soils. Parameter values for K_d of the linear model and for K_f and b of the Freundlich model along

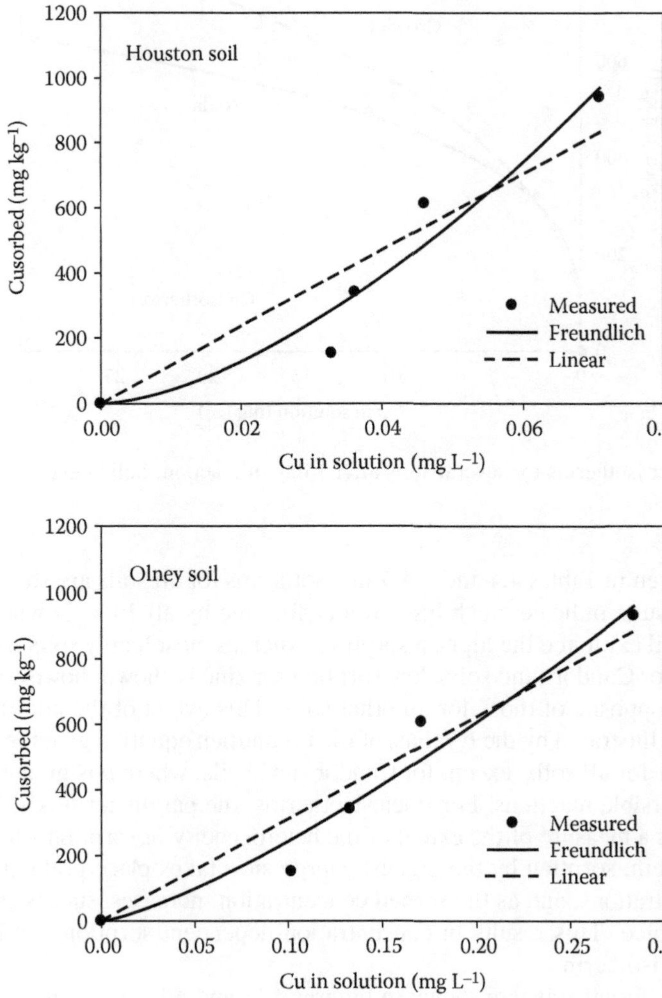

FIGURE 4.7 Copper isotherms for Houston and Olney soils after 1 day of reaction. Solid and dashed curves are simulations using the linear and Freundlich models.

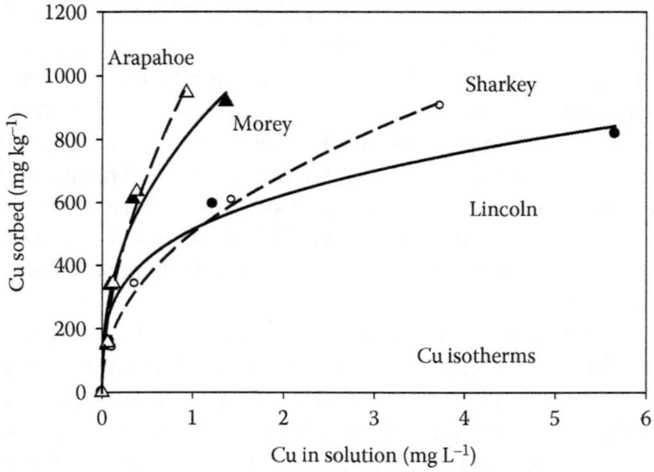

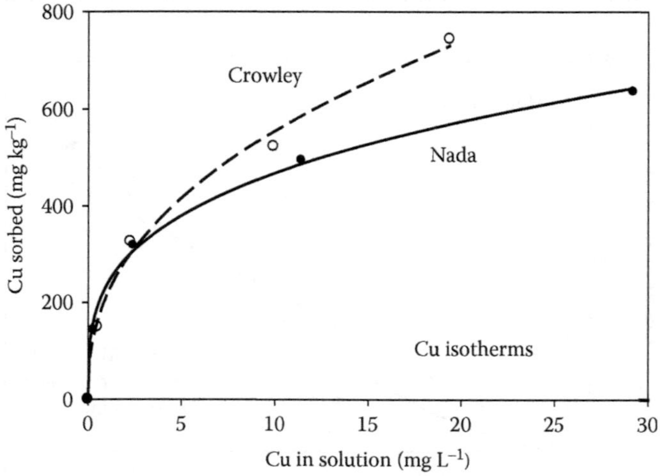

FIGURE 4.8 Copper isotherms for several soils after 1 day of reaction. Solid curves are simulations using the Freundlich models.

with their r^2 are given in Tables 4.4 and 4.5. Zinc isotherms for all soils are shown in Figures 4.11 and 4.12. These results indicate much less affinity for zinc by all 10 soils when compared with copper. Houston soil exhibited the highest sorption, whereas most loamy soils indicated moderate sorption for zinc. For Candor sand soils, low sorption for zinc is shown; however, the shape of the isotherms was the opposite of those for all other soils. This extent of the nonlinearity and shape of the isotherms is illustrated by the b values of the Freundlich equation given in Table 4.5. Values for b are less than 1 for all soils, except for Candor sand soils, where b is greater than 1, which is indicative of irreversible reactions. For linear isotherms, the parameter b is 1. This parameter b is often regarded as a measure of the extent of the heterogeneity of sorption sites in the soil . In a heterogeneous system, sorption by the highest energy sites takes place preferentially at the lowest solution concentrations, and as the sorbed concentration increases, successively lower energy sites become occupied. This results in concentration-dependent sorption equilibrium behavior, that is, a nonlinear isotherm.

Nickel isotherms for all soils are shown in Figures 4.13 and 4.14. These results indicate similar affinities for nickel by all 10 soils when compared with zinc. Houston soil exhibited the highest sorption, whereas most loamy soils indicated moderate sorption affinities. Among all soils, Candor

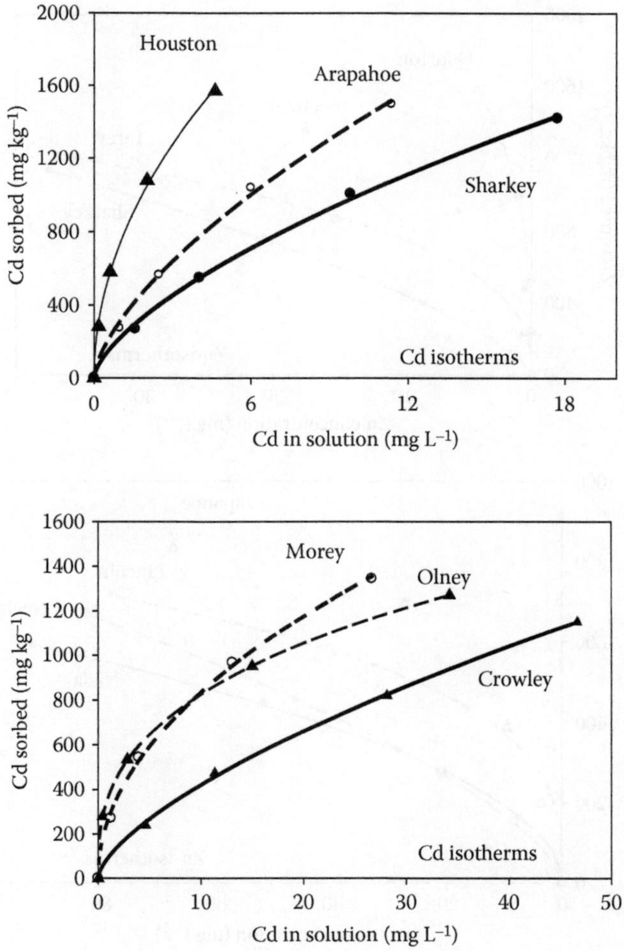

FIGURE 4.9 Cadmium isotherms for several soils after 1 day of reaction. Solid and dashed curves are simulations using the Freundlich model.

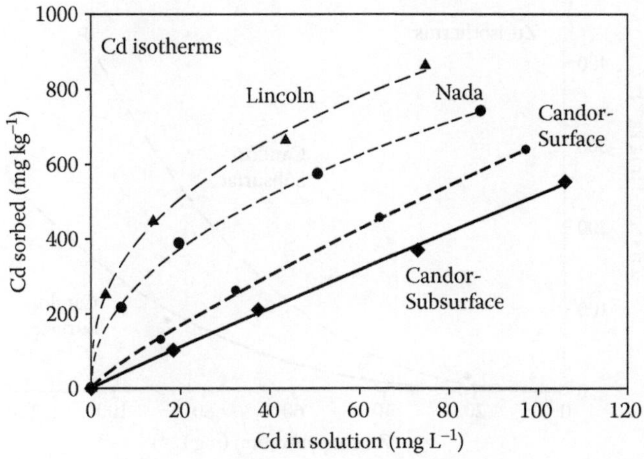

FIGURE 4.10 Cadmium isotherms for four soils after 1 day of reaction. Solid and dashed curves are simulations using the Freundlich model.

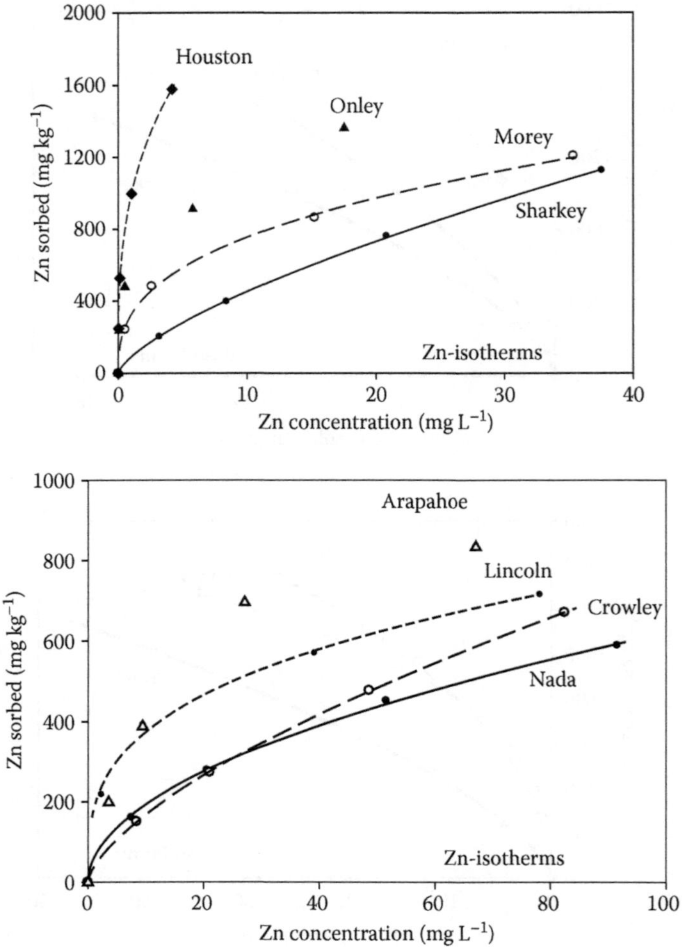

FIGURE 4.11 Zinc isotherms for several soils after 1 day of reaction. Solid and dashed curves are simulations using the Freundlich model.

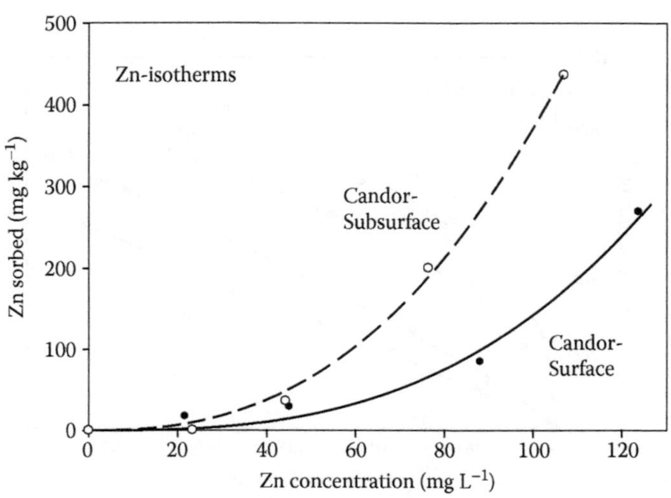

FIGURE 4.12 Zinc isotherms for Candor soil after 1 day of reaction. Curves are simulations using the Freundlich models.

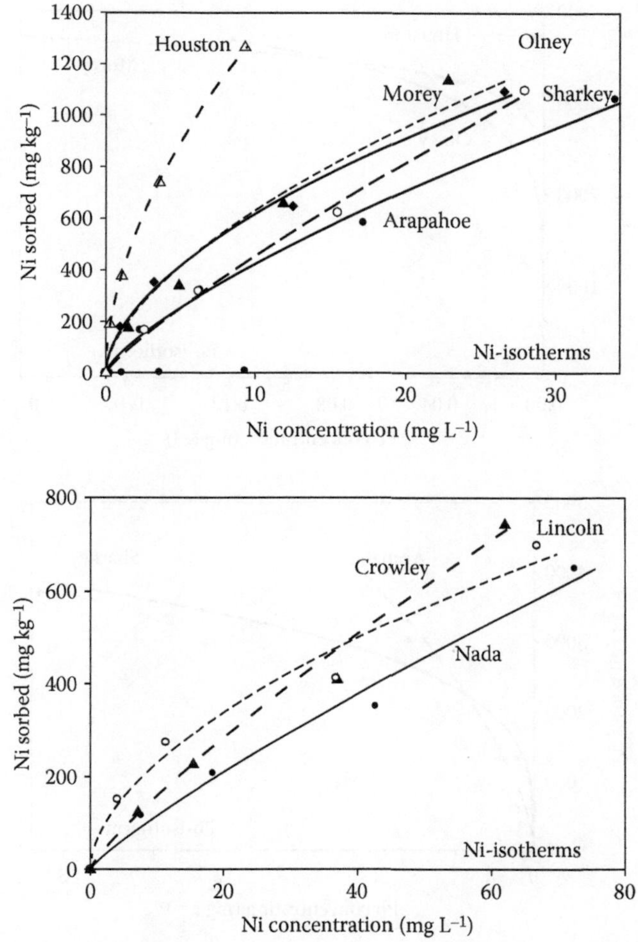

FIGURE 4.13 Nickel isotherms for several soils after 1 day of reaction. Solid and dashed curves are simulations using the Freundlich model.

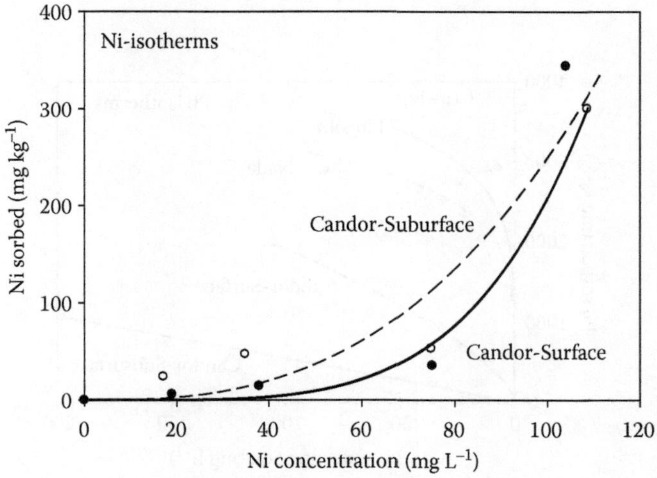

FIGURE 4.14 Nickel isotherms for Candor surface and subsurface soils after 1 day of reaction. Solid and dashed curves are simulations using the Freundlich model.

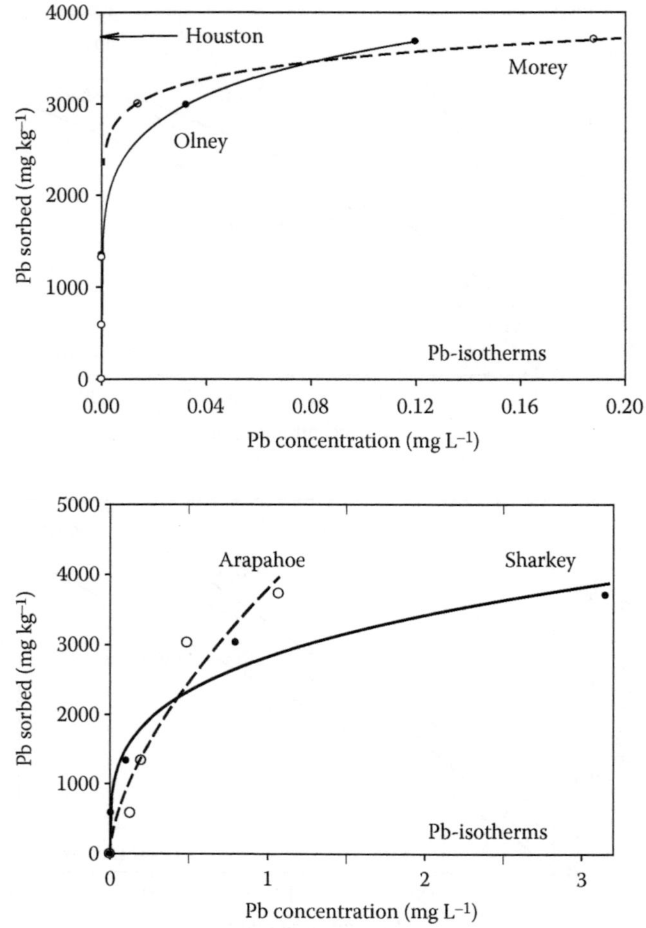

FIGURE 4.15 Lead isotherms for several soils after 1 day of reaction. Curves are simulations using the Freundlich models.

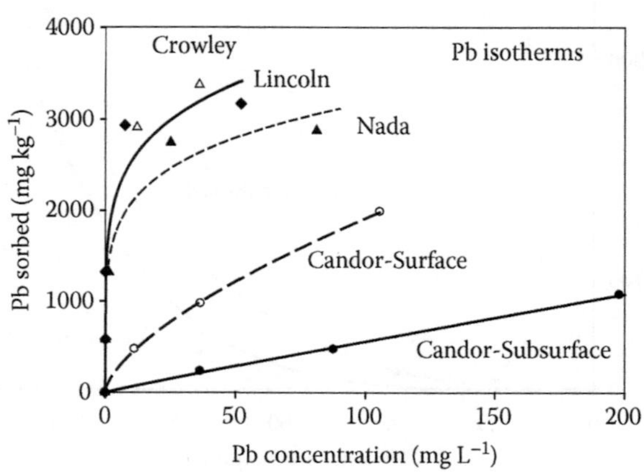

FIGURE 4.16 Lead isotherms for five different soils after 1 day of reaction. Curves are simulations using the Freundlich models.

sand soils exhibited the lowest affinity for nickel. Parameter values for K_d of the linear model and for K_f and b of the Freundlich models along with their r^2 are given in Tables 4.4 and 4.5. Lead isotherms for all soils are shown in Figures 4.15 and 4.16. These results indicate a wide range of affinities for lead in the 10 soils. Houston soil exhibited the highest sorption, where the solution of lead in the soil was below detection for the entire range of input concentration. The detection limit for Pb using ICP-AES was 28 μg L^{-1} (ppb). The other nine soils exhibited different degrees of affinities for lead, as shown in Figures 4.15 and 4.16. Soils with high organic matter content such as Arapahoe as well as those with high clay content such as Sharkey exhibited high sorption. It should also be emphasized that Olney soil with high pH and carbonate content exhibited high sorption for lead. Parameter values for K_d of the linear model and for K_f and b of the Freundlich model along with their r^2 are given in Tables 4.4 and 4.5.

4.4.1 Predictions

Here, Sauvé et al. (2000) used two models to predict the K_d values given in Equations 4.3 and 4.4. These models were tested for the prediction of measured K_d values from the 5 heavy metals and the 10 soils discussed earlier. As discussed earlier, for model I, K_d predictions are based only on soil pH. In model II, K_d predictions are based on two variables: pH and percentage of SOC. A comparison of measured K_d values from this study for all 10 soils by using models I and II is shown in Figures 4.17 through 4.21. For copper, predictions were overestimated at a low K_d range. For both cadmium and zinc, the predictions did not illustrate any pattern and were highly inadequate. Somewhat improved predictions were obtained for nickel and lead when model I was used. Overall, both models yielded inadequate predictions for all heavy metals used in this study. In contrast, when the Buchter (1989) model was used to predict the Freundlich parameter b, good overall predictions were obtained (see Figure 4.22). For all heavy metals, extremely good trends were observed. The best prediction was obtained for Zn. Based on this investigation, several findings can be made as follows:

1. Adsorption of all 5 heavy metals was nonlinear.
2. For all 10 soils used in this study, adsorption of heavy metals follows the order Pb > Cu > Cd > Zn > Ni. In the presence of carbonates, adsorption of heavy metals follows the order Pb > Cu > Zn > Cd > Ni.

For all 10 soils used in this study, models of Sauvé et al. (2000) provided less than adequate predictions of measured K_d values for all five heavy metals. In contrast, good overall predictions were obtained for the Freundlich parameter b when the Buchter (1989) model was used.

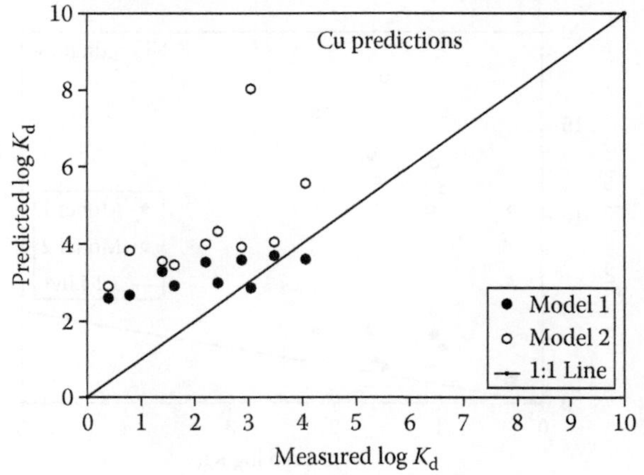

FIGURE 4.17 Measured and calculated K_d values for Cu for all soils. Calculated K_d were obtained using models 1 and 2 of Sauvé et al. (2000).

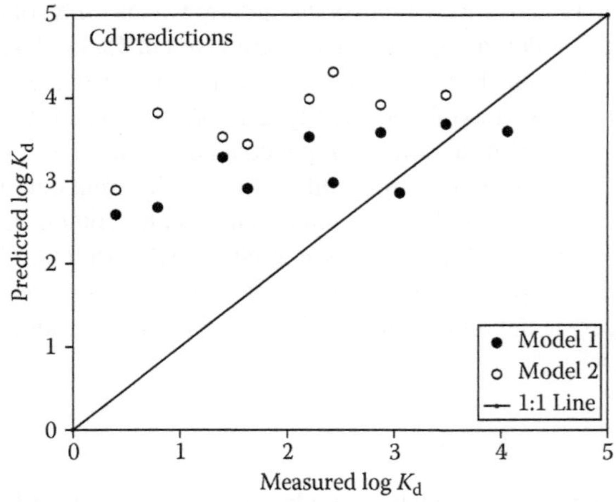

FIGURE 4.18 Measured and calculated K_d values for Cd for all soils. Calculated K_d were obtained using models 1 and 2 of Sauvé et al. (2000).

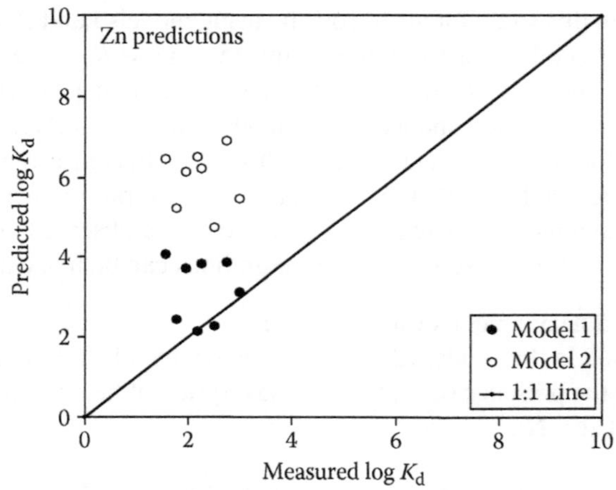

FIGURE 4.19 Measured and calculated K_d values for Zn for all soils. Calculated K_d were obtained using models 1 and 2 of Sauvé et al. (2000).

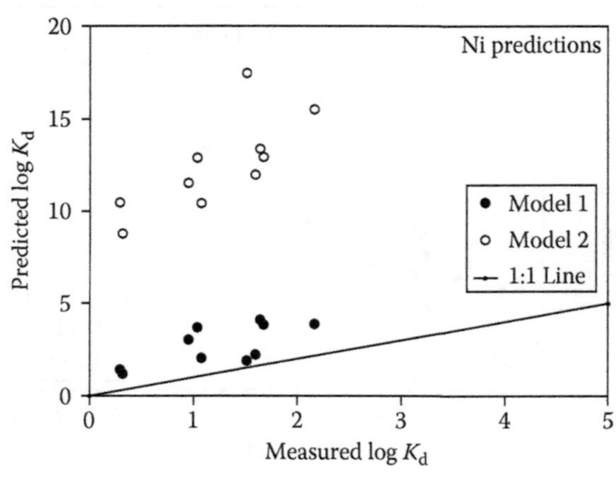

FIGURE 4.20 Measured and calculated K_d values for Ni for all soils. Calculated K_d were obtained using models 1 and 2 of Sauvé et al. (2000).

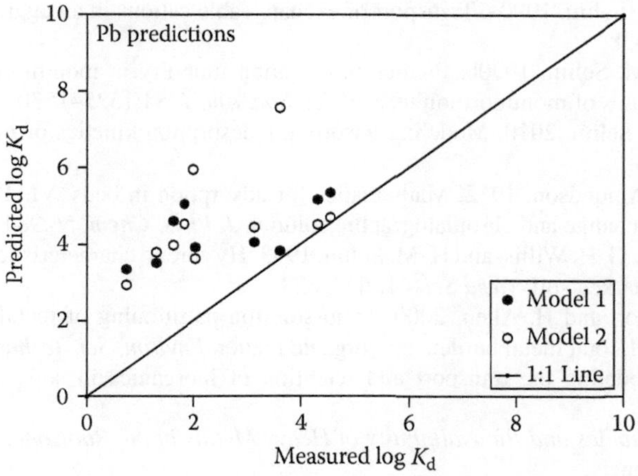

FIGURE 4.21 Measured and calculated K_d values for Pb for all soils. Calculated K_d were obtained using models 1 and 2 of Sauvé et al. (2000).

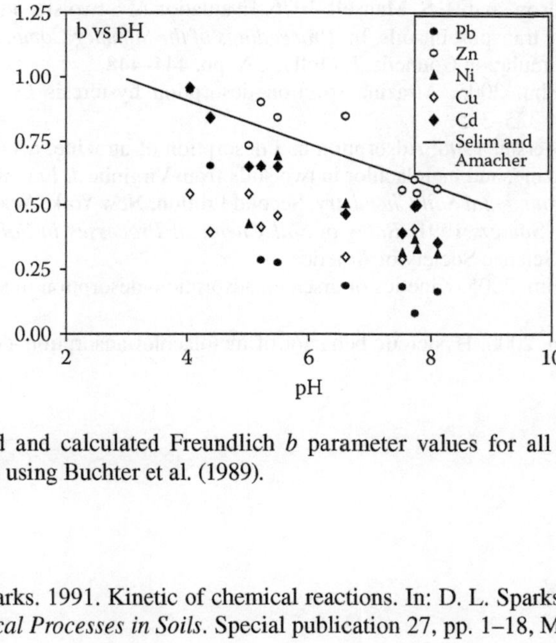

FIGURE 4.22 Measured and calculated Freundlich *b* parameter values for all heavy metals and soils. Calculated *b* was obtained using Buchter et al. (1989).

REFERENCES

Aharoni, C., and D. L. Sparks. 1991. Kinetic of chemical reactions. In: D. L. Sparks and D. L. Suarez (eds.), *Rates of Soil Chemical Processes in Soils*. Special publication 27, pp. 1–18, Madison, WI: Soil Science Society of America.

Amacher, M. C. 1991. Methods of obtaining and analyzing kinetic data. In: D. L. Sparks and D. L. Suarez (eds.), *Rates of Soil Chemical Processes in Soils*. Special publication 27, pp. 19–59, Madison, WI: Soil Science Society of America.

Bradl, H. B. 2004. Adsorption of heavy metal ions on soils and soil constituents. *J. Colloid. Interf. Sci.* 277: 1–18.

Buchter, B., B. Davidoff, M. C. Amacher, C. Hinz, I. K. Iskandar, and H. M. Selim. 1989. Correlation of Freundlich K_d and *n* retention parameters with soils and elements. *Soil Sci.* 148:370–379.

Carrillo-Gonzalez, R., J. Simunek, S. Sauve, and D. Adriano. 2006. Mechanisms and pathways of trace element mobility in soils. *Adv. Agron.* 91:111–178.

Cox, L., W. C. Koskinen, and P. Y. Yen. 1997. Sorption–desorption of imidacloprid and its metabolites in soils. *J. Agric. Food Chem.* 45:1468–1472.

Fontes, M. 2012. Behavior of heavy metals in soils: Individual and multiple competitive adsorption. In: H. M. Selim (ed.), *Competitive Sorption and Transport of Trace Elements in Soils and Geological Media*, pp. 77–117, Boca Raton, FL: CRC Press.

Gaston, L. A., and H. M. Selim. 1990a. Transport of exchangeable cations in an aggregated clay soil. *Soil Sci. Soc. Am. J.* 54:31–38.

Gaston, L. A., and H. M. Selim. 1990b. Prediction of cation mobility in montmorillonitic media based on exchange selectivities of montmorillonite. *Soil Sci. Soc. Am. J.* 54:1525–1530.

Jeong, C. Y., and H. M. Selim. 2010. Modeling adsorption–desorption kinetics of imidacloprid in soils. *Soil Sci.* 175:214–222.

Lapidus, L., and N. L. Amundson. 1952. Mathematics for adsorption in beds. VI. The effect of longitudinal diffusion in ion exchange and chromatographic column. *J. Phys. Chem.* 56:984–988.

Ma, L., L. M. Southwick, G. H. Willis, and H. M. Selim. 1993. Hysteretic characteristics of atrazine adsorption–desorption by a Sharkey soil. *Weed Sci.* 41: 627–633.

Sauvé, S., W. Hendershot, and H. Allen. 2000. Solid-solution partitioning of metals in contaminated soils: Dependence on pH, total metal burden, and organic matter. *Environ. Sci. Technol.* 34:1125–1131.

Selim, H. M. 1992. Modeling the transport and retention of inorganics in soils. *Adv. Agron* 47:331–384 (Academic Press).

Selim, H. M. 2011. *Dynamics and Bioavailability of Heavy Metals in the Rootzone*, 315 p., Boca Raton, FL: CRC/Taylor & Francis.

Selim, H. M. 2012. *Competitive Sorption and Transport of Trace Elements in Soils and Geological Media*, 425 p., Boca Raton, FL: CRC/Taylor & Francis.

Selim, H. M., and M. C. Amacher. 1997. *Reactivity and Transport of Heavy Metals in Soils*, 240 p., Boca Raton, FL: CRC.

Selim, H. M., J. M. Davidson, and R. S. Mansell. 1976. Evaluation of a two site adsorption–desorption model for describing solute transport in soils. In: *Proceedings of the Summer Computer Simulation Conference*, Washington, DC, Simulation Councils, La Jolla, CA, pp. 444–448.

Selim, H. M., and H. Zhu. 2005. Atrazine sorption–desorption hysteresis by sugarcane mulch residue. *J. Environ. Qual.* 34:325–335.

Seybold, C. A., and W. Mersie. 1996. Adsorption and desorption of atrazine, deethylatrazine, deisopropylatrazine, hydroxyatrazine, and metolachlor in two soils from Virginia. *J. Environ. Qual.* 25:1179–1185.

Sparks, D. L. 2003. *Environmental Soil Chemistry*, Second Edition, New York: Academic Press.

Sparks, D. L., and D. L. Suarez. 1991. *Rates of Soil Chemical Processes in Soils*. Special Publication 27, Madison, WI: Soil Science Society of America.

Zhang, H., and H. M. Selim. 2005. Kinetics of arsenate adsorption–desorption in soils. *Environ. Sci. Technol.* 39:6101–6108.

Zhu, H., and H. M. Selim. 2000. Hysteretic behavior of metolachlor adsorption–desorption in soils. *Soil Sci.* 165:632–643.

5 Speciation and Release Kinetics of Cadmium and Zinc in Paddy Soils

Saengdao Khaokaew, Gautier Landrot, and Donald L. Sparks

CONTENTS

5.1 INTRODUCTION

Rice is the main food staple of more than 50% of the world population (Wallace, 1998). Its production will have to significantly increase in the future to satisfy the needs of the growing population worldwide, especially in developing countries. This represents a challenge, as the development of economies and industries often implies higher anthropogenic inputs to the environment, including agricultural fields, such as those where rice is cultivated. The latter, referred to as *paddy soils*, will be defined and described in the first part of this chapter.

Reports of paddy soils located in different regions of the world and contaminated by heavy metals that have been released into the environment as a result of human activities have been increasing, especially over the past decade and in Asian countries. Indeed, paddy soil contamination by heavy metals, such as cadmium (Cd) and zinc (Zn), due to the proximity of the agricultural soils to urban or industrialized areas, has been reported in China (Römkens et al., 2009; Søvik et al., 2011; Lin et al., 2012; Huang et al., 2013; Zhao et al., 2014), India (Singh et al., 2011), Iran (Jalali and Hemati, 2013), Japan (Kikuchi et al., 2007), and Vietnam (Nguyen Ngoc et al., 2009). In particular, paddy soil contamination by Cd and Zn contained in waste products that are generated by nearby mining and tailing activities has been reported in China (Yang et al., 2008; Williams et al., 2009; Lei et al., 2010), Korea (Lee et al., 2001; Lee, 2006), Macedonia (Rogan et al., 2009), and Thailand (Simmons et al., 2005). The sources of Cd or Zn pollution can also be derived from the applications to the paddy soils of fertilizers, soil amendments, and manures that contain significant amounts of trace elements, as reported in China (Li et al., 2009a, b), India (Reddy et al., 2013), and Malaysia (Jamil et al., 2011). The presence of heavy metals, such as Cd and Zn, in paddy soils poses a serious environmental threat, as the contaminants may be transferred to crops, and hence the human food chain. It is well documented that some heavy metals present in the human body at certain levels can cause many health problems, including some cancers that may lead to death. For example, Cd present in highly contaminated paddy soils can be transferred to rice grain (Simmons et al., 2005), which can notably cause hypertension, diabetes, and renal dysfunctions to those consuming it on a daily basis (Swaddiwudhipong et al., 2012). Therefore, to produce rice that is safe for consumption, paddy soils containing high concentrations of Cd should be remediated to decrease the metal content to safe levels. Typical concentrations of Cd and Zn and other aspects related to the presence of these two metals in impacted or noncontaminated paddy soils will be discussed in the second part of this chapter.

To choose the best strategy to remediate paddy fields that are contaminated by Cd or Zn, one must primarily understand the chemical forms and behaviors of the metals in soils. The degree of urgency of cleaning up a paddy soil containing a metal at a high concentration partly depends on the time required by the contaminant to be released into solution. Therefore, one must accurately understand the release kinetics of Cd and Zn in paddy soils and the factors controlling them, and also develop efficient methodologies to measure these rates. These aspects will be discussed in the third part of this chapter. Cadmium and zinc speciation, which will be focused on in the fourth part of this chapter, is also one of the most important aspects to be taken into account when choosing a remediation method to clean up a paddy soil. In addition, it directly controls the desorption kinetics of Cd and Zn in paddy fields. Lastly, a summary and a discussion on future studies will be provided at the end of this chapter.

5.2 DESCRIPTION OF PADDY SOIL

Paddy soil refers to a type of wetland that is either mainly used or potentially used to grow rice. The word *paddy* has several meanings. It is derived from the Malaysian word *padi*, which means *a rice plant, an unhusked rice grain,* or *a rice field.* Paddy soils in the world, especially in Asia, are mainly cultivated by following lowland practices, which consist of alternatively flooding and draining the agricultural soil. In contrast, upland practices do not involve using these flooding and draining cycles. Therefore, *lowland* and *upland* do not refer to a specific altitude at which paddy soils are cultivated. These terms rather refer to two different farming practices. Lowland rice cultivation is often chosen by farmers over other kinds of cropping practices due to its high yields, low maintenance of paddy soils, and resistance of the rice plants to harsh climatic conditions (Kyuma, 2004). The cyclical process of submerged and drained conditions alters the chemical, physical, and biological properties of the paddy soil. Under flooding conditions, the amounts of dissolved oxygen in soil solution, gas exchanges between the soil and the atmosphere, and decomposition of soil organic matter are limited (Ponnamperuma, 1972). Consequently, the soil is mostly anoxic under flooding conditions. In a flooded soil, E_h usually decreases with an increase in soil depth, flooding period, and the amount of water. Typical layers, E_h values, and redox reactions occurring at various soil depths in paddy soils under flooding conditions are shown in Figure 5.1. The E_h values associated with the transformation of the redox couples that are shown in Figure 5.1 are reported in DeLaune and Reddy (2005).

The soil, however, becomes more oxic as the field shifts to drained conditions. Soil redox potential (E_h) then dramatically changes, and also soil pH changes (Kyuma, 2004). The extent of E_h change depends on several factors, which are essentially the types and amounts of electron donors (e.g., soil organic matter) and acceptors (e.g., oxide minerals) in the system. These modifications can greatly impact the chemical properties of sorbents that are present in paddy soils, such as organic matter, clays, (oxy)hydroxides, carbonates, and sulfide minerals, in which nutrients and contaminants can be retained (Bingham et al., 1976; Bostick et al., 2001). In addition, the modifications in soil chemical properties, such as E_h and pH, occurring during agricultural cycle shifts may affect Cd and Zn speciation and release kinetics in paddy soils. General information on Cd and Zn occurrences in soils, including paddy fields, are provided in the next section.

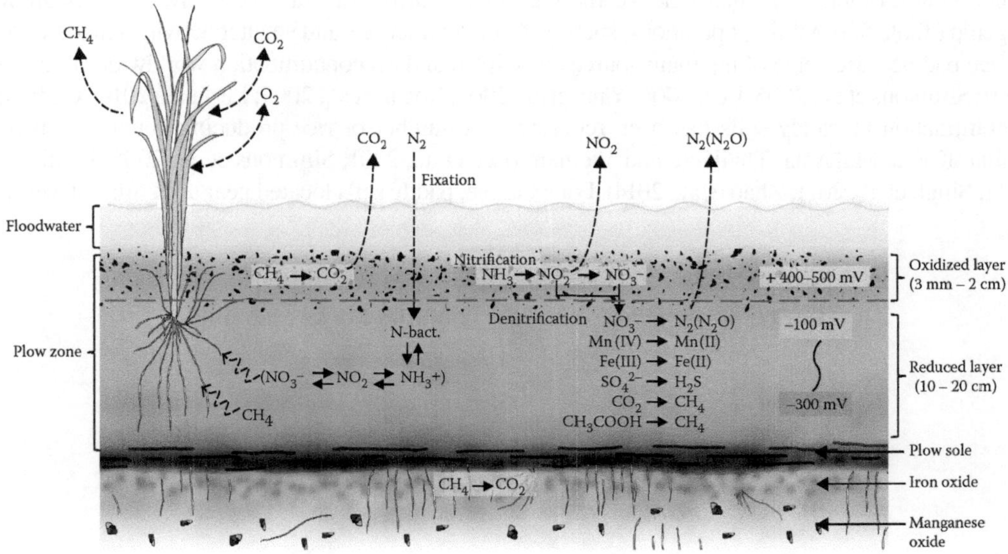

FIGURE 5.1 **(See color insert.)** Typical E_h values and redox reactions occurring at various soil depths in a paddy soil under flooded conditions. (From Wakatsuki, T. In: Kyuma, K. (ed.), *New Soil Science*, Asakura Publishing, Co., Tokyo, Japan, 1997. Modified with permission.)

5.3 CADMIUM AND ZINC IN PADDY SOIL

Cadmium and zinc are two transition metals that have some chemical properties that are similar to each other (Table 5.1), such as electronic configurations, valence states, and affinities for S, N, and O donor ligands (Kummerová et al., 2010). Cadmium is mostly present in soils in the structure of mineral phases, such as greenockite (CdS), otavite ($CdCO_3$), and monteponite (CdO). These minerals are, in soils, often associated with Zn ores, such as franklinite ($ZnFe_2O_4$), smithsonite ($ZnCO_3$), willmite ($ZnSiO_4$), sphalerite (ZnS), zinkosite ($ZnSO_4$), and hopeite ($Zn_3(PO_4)_2$) (McLaughlin and Singh, 1999; Kabata-Pendias, 2011). Zinc ores usually contain between 0.1 and 5% Cd (Nan et al., 2002). The levels of Cd and Zn in the crust of the earth are 0.1 and 70 mg/kg, respectively. The world average levels of Cd and Zn in surface soils are 0.41 and 70 mg/kg, respectively (Kabata-Pendias, 2011).

Despite geochemical similarities between Cd and Zn, the latter is an important nutrient to plants, animals, and humans, whereas the former has no biological function in plants and can be very toxic to humans, even at very low levels. Humans can be exposed to Cd, notably via consumption of rice grains that are grown in contaminated paddy soils (Simmons et al., 2005). The sources of pollution can be

TABLE 5.1
Selected Chemical Properties of Cadmium and Zinc

Element	Symbol	Ground-State Electronic Configuration	Oxidation State, Ionic Charge	Radius in Metallic Substances (Å)	Density (g > cm³)	First Ionization Energy (kJ > mol)
Cadmium	Cd	[Kr] $4d^{10}$ $5s^2$	(II), 2+	1.48	8.6	868
Zinc	Zn	[Ar] $3d^{10}$ $4s^2$	(II), 2+	1.37	7.1	906

Source: Brown, T.E. et al. *Chemistry: The Central Science* (13th Edition), Prentice Hall, Upper Saddle River, NJ, 2014. With permission.

either natural (Table 5.2), due to the weathering of Cd-bearing minerals (Alloway, 1995), or anthropogenic (Table 5.3). Mining operations, such as Zn ore extractions and smelter activities that are close to rice paddies, are some of the main sources of soil Cd and Zn contamination worldwide (Lee et al., 2001; Simmons et al., 2005; Lee, 2006; Yang et al., 2008; Rogan et al., 2009; Lei et al., 2010). Cadmium contamination in paddy soils has been reported in a number of rice-producing countries, such as China, Korea, Malaysia, Thailand, and Vietnam (Lee et al., 2001; Simmons et al., 2005; Jamil et al., 2011; Singh et al., 2011; Zhao et al., 2014). For example, paddy soils located near a Zn mine at Mae Sot

TABLE 5.2
Concentrations of Cd and Zn Found in Rocks and Soils

	Concentration (mg/kg)	
Type of Rock	**Cd**	**Zn**
Igneous rock		
Ultramafic	0.12	58
Mafic	0.13	100
Granitic	0.09	52
Sedimentary rock		
Limestone	0.03	20
Sandstone	0.05	30
Shales/clays	0.22	120

Source: Rose, A.W. et al. *Geochemistry in Mineral Exploration* (2nd Edition), Academic Press, London, UK, 1979. With permission.

TABLE 5.3
Cadmium and Zinc Levels in Various Environments as Well as Maximum Concentrations Set by CODEX

	Element (mg/kg)	
Considered Values	**Cd**	**Zn**
Average background content in crust	0.1	70
Average ranges on crust of the earth	0.1–0.2	52–80
World soil average background content	0.41	70
Total concentration ranges in world surface soils	0.01–2.7	3.5–770
Average mean of total concentration in world surface soils	0.53	64
Total concentration ranges in phosphate rocks	0.01–0.1	4–345
Total concentration ranges in phosphate fertilizers	7–170	50–1450
World total concentration ranges in sewage sludge	2–1500	550–49,000
Ranges in dry, digested sewage biosolids in the United States	1–3410	101–49,000
Total concentration ranges in pig manure	0.25–43	680–2287
Average total concentration in fly ash	1.3	221
Critical concentration in plant tissue for insensitive species	5–10	150–200
Critical concentration in plant tissue for 10% yield loss	150–200	100–500
Content in rice grains	<0.3	1–41
Maximum limit in rice grain by CODEX	<0.4	–

Source: Kabata-Pendias, A., *Trace Elements in Soils and Plants* (4th Edition), CRC Press, Boca Raton, FL, 2010; Mortvedt, J.J., *Fertilizer Research* 43, 55–61, 1996. With permission.

District, Tak Province, Thailand, contained more than 200 and 3000 mg/kg of Cd and Zn, respectively (Simmons et al., 2005). Rice grown in these contaminated paddy soils had Cd levels in their grains that were higher than the world standard maximum limit (Simmons et al., 2005; Sriprachote et al., 2012). In China and Korea, paddy soils impacted by lead (Pb) and Zn mines contained Cd and Zn in a range of 9–11 and 753–875 mg/kg, respectively (Lee et al., 2001; Lei et al., 2010). Applications in paddy soils of phosphate fertilizers, sewage sludge, and pig manure that contain Cd and Zn also have become a major environmental problem in paddy soils (Li et al., 2009a, b; Jamil et al., 2011; Reddy et al., 2013). Indeed, since intensive animal farming has developed quickly in recent years, supplements and growth stimulants, which often contain traces of heavy metals, including Cd and Zn, have been added to commercial feedstuffs to increase productivity (Nicholson et al., 1999). Liu et al. (2005) reported that pig manure is the main animal manure used in agricultural fields in China, and it can have Cd and Zn levels up to 43 and 2287 mg/kg, respectively. Average contents of Cd and Zn found in phosphate rocks, phosphate fertilizers, sewage sludge, and pig manures are shown in Table 5.3.

The maximum standard limits of Cd and Zn in agricultural soils, which can differ from country to country, often refer to the maximum total metal concentration in the agricultural field. For example, the maximum standard limit of total Cd concentration in agricultural soils in Thailand is 37 mg/kg (Paijitprapapon et al., 2006), whereas the European maximum permissible Cd in sludge-amended soils is 3 mg/kg (Simmons et al., 2014). The maximum Zn concentration allowed in agricultural soils can be higher than 400 mg/kg, as this metal can act as a plant nutrient at lower concentrations. Studies that have investigated Cd and Zn levels in contaminated paddy fields measured the total metal amounts in the soils, by using total digestion methods such as aqua regia or nitric–perchloric acids (Khaokaew et al., 2011; Sriprachote et al., 2012) and/or their extractable metal fractions, by using, for instance, 0.1 M HCl, 0.1 M $CaCl_2$, 1 M $BaCl_2$, and diethylene triamine pentaacetic acid (DTPA) or ethylenediaminetetraacetic acid (EDTA) (Lindsay and Norvell, 1978; Fulda et al., 2013). However, measurement of the total amounts of metals and even their exchangeable fractions does not provide information on how rapidly the metals can be released into solution due to changing soil conditions, such as those that may occur when using lowland farming practices, and on how stable the associations are between Cd and Zn and soil components. To better understand these important aspects, which must be clearly constrained to elaborate efficient remediation strategies that are used to clean up the contaminated soils, release kinetic desorption experiments should be performed. A description of these experiments is provided in the next section.

5.4 RELEASE KINETICS OF CADMIUM AND ZINC IN PADDY SOIL

Sorption–desorption processes control the bioavailability and fate of heavy metals in the environment. A number of studies have focused on metal sorption, whereas little is known about metal desorption. To improve remediation strategies and risk assessments, and to better predict the mobility of contaminants, it is critical to understand time-dependent metal desorption behaviors from the soil (Strawn and Sparks, 2000). To date, only a few studies have investigated Cd and Zn desorption in soils, especially wetlands, such as paddy soils. The methodologies employed in these former investigations, factors controlling Cd and Zn releases, and aspects on this topic that still require further understanding will be described in this section.

5.4.1 Experimental Approaches to Study Release Kinetics of Cd and Zn in Paddy Soils

Traditional batch technique and stirred-flow methods have been widely used to assess metal sorption and desorption kinetics in soils (Yin et al., 1997; Strawn and Sparks, 2000; Sukreeyapongse et al., 2002; McNear, 2006; Shi, 2006). Detailed information on these techniques is provided in Sparks (1996). During batch desorption experiments, reverse reactions are not controlled and the released species are not removed. Consequently, re-adsorption or secondary precipitation may occur (Sparks, 2003). Moreover, an increase in metal concentrations in solution may inhibit further release of adsorbate in the desorption process (Shi, 2006). These problems are limited with the stirred-flow technique, which enables continuous removal of products from the reaction chamber. Several studies have employed a stirred-flow method to investigate at the macroscopic scale Cd and Zn desorption behaviors from paddy soils (Khaokaew et al., 2011, 2012). A typical stirred-flow apparatus is shown as a diagram and pictures in Figures 5.2 and 5.3, respectively.

To study the release kinetics of Cd and Zn in paddy soil by using the stirred-flow method, a small amount of soil sample (sorbent) is placed in the chamber and retained there by using a prefilter and a

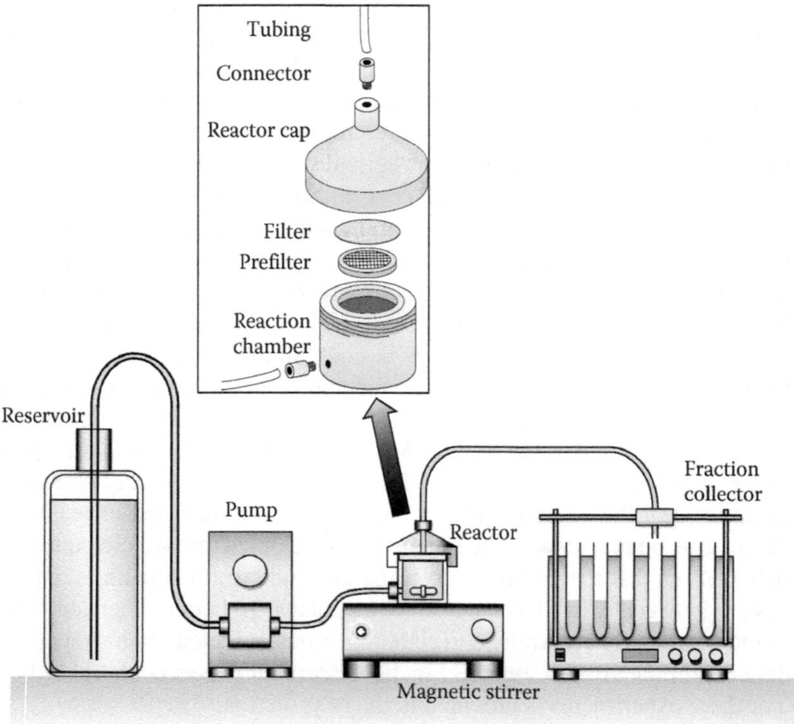

FIGURE 5.2 Sketch of a standard stirred-flow apparatus. Box: sketch of a reactor.

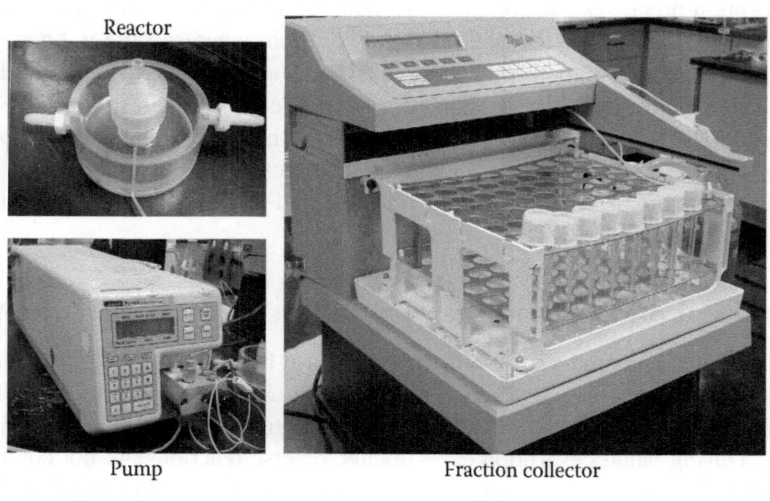

FIGURE 5.3 Pictures of a reaction chamber, pump, and fraction collector used in a stirred-flow setup.

membrane filter that are fitted just below the effluent port (Figure 5.2). Bar-Tal et al. (1990) and Seyfried et al. (1989) suggested that suitable solution-to-solid weight ratios in the reactor should be between 5:1 and 10:1. This ratio should not be too small for studies lasting for more than a few minutes, as clogging of soil particles on the filter can occur in the stirred-flow chamber. This can happen especially when the sample contains a fair amount of clay minerals and organic matter, which is often the case with paddy soils. Therefore, for paddy soils that contain high amounts of clays and organic matter, the ratio can be increased from 16:1 to 40:1 (Khaokaew et al., 2011, 2012). The nature of the desorbing agent employed to release Cd and Zn from paddy soils depends on the purpose of the study. For example, one can employ low-molecular-weight organic acids (LMWOAs) such as citric acid, oxalic acid, or malic acid to study the effects of root exudates on metal availability (Khaokaew, 2010). Alternatively, one can use, for example, EDTA, DTPA, Melich 3, or $CaCl_2$, to study the amounts of exchangeable metals that can be removed from the paddy soil by rice (Zhang et al., 2010).

Although the period of a desorption experiment that is used to study the release of Cd and Zn in paddy soils can be chosen randomly, it should last at least a few minutes. Indeed, a study showed that a majority of the amounts of Cd and Zn desorbed from the paddy soils were released during the first 10 minutes (Khaokaew et al., 2011, 2012).

For a study that employed an 8-mL reaction chamber, a flow rate of about 0.4–1 mL/min was used (Khaokaew, 2010). Increasing this flow rate was not suitable for the setup employed, as the pressure in the reaction chamber increased and threatened to break the filter paper. Decreasing this flow rate was not desirable, as not enough liquid of the effluent fraction was collected in each tube at the outlet of the stirred-flow setup. With the flow rate employed, about 1–2 mL was collected in each tube every 2 min, which was suitable for subsequent elemental analysis. One can calculate the amount of Cd or Zn desorbed at a given period of the experiment using Equation 5.1, which is modified from Schnabel and Fitting (1988):

$$q(ti) = \frac{\Sigma[Ci(ti) \times Q \times \Delta t] + (\Sigma[C(ti) \times Q \times \Delta t]) / V]}{m} \quad (5.1)$$

where ti = total time period between the beginning of the experiment and end of a sample fraction collection (min); $q(ti)$ = total amount of sorptive (Cd or Zn) desorbed during ti (mg/l); Q = flow rate (l/min); Δt = time period of a sample fraction collection (min); C = averaged concentration of sorptive (Cd or Zn) in a collected sample fraction (mg/l) and the following one (for the last fraction, C is equal to the concentration in this fraction only); V = volume of solution in chamber (l); and m = mass of paddy soil in chamber (mg).

A typical desorption plot features on the abscissa the period of the experiment, and on the ordinate the value q features, which are calculated using Equation 5.1. Alternatively, one can plot on the ordinate the fraction of metal cumulative release (in %), which is expressed as the amount of metal cumulative release (in mg/L) over the total amount of metal initially present in the reaction chamber (in mg/L) (Figure 5.4).

The stirred-flow apparatus shown in Figures 5.2 and 5.3 is suitable to measure the release kinetics of metals in various soil systems. However, this experimental setup presents one drawback when it specifically comes to studying metal desorption in soil samples in which lowland conditions are mimicked. This will be discussed in the next section, and in other aspects related to incubation experiments.

5.4.2 Incubation Experiments to Mimic Lowland Rice Culture Conditions

Paddy soils are often cultivated after using lowland farming practices, which consist of cyclically flooding and draining the agricultural field during the period of rice growth (Figure 5.5). Since these processes can dramatically alter soil chemical properties, such as E_h (Figure 5.1), they must be taken into account when experimentally determining Cd and Zn speciation and release kinetics in paddy

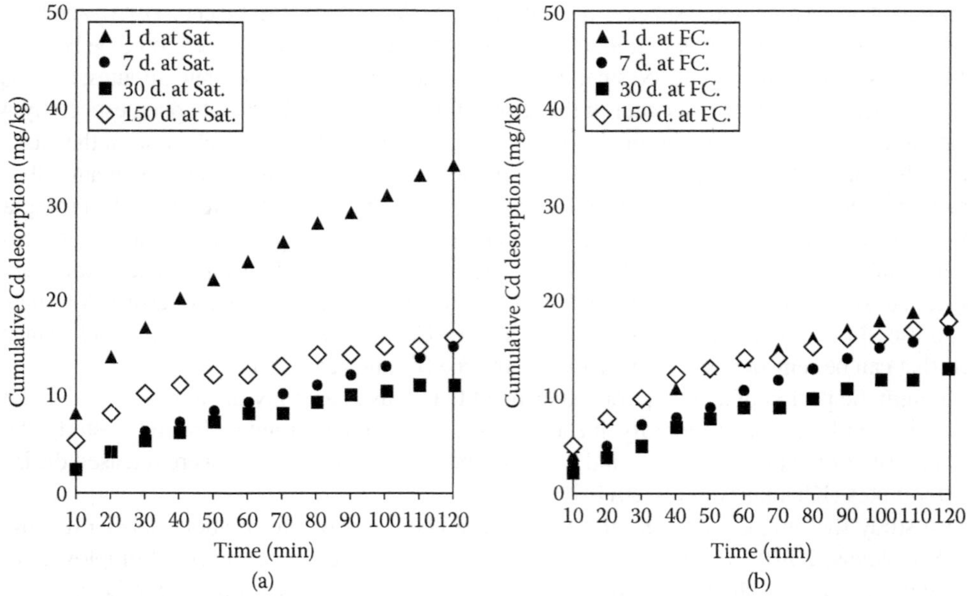

FIGURE 5.4 Example of Cd desorption graph plotted as cumulative release: Cd desorption in a paddy soil (mg/kg) by DTPA-TEA-Ca, after 1, 7, 30, and 150 days of flooding and draining to (a) saturation (Sat.), and (b) field capacity (FC). (Reprinted with permission from Khaokaew, S. et al., *Environ. Sci. Technol.*, 45, 4249–4255, 2011. Copyright 2011 American Chemical Society.)

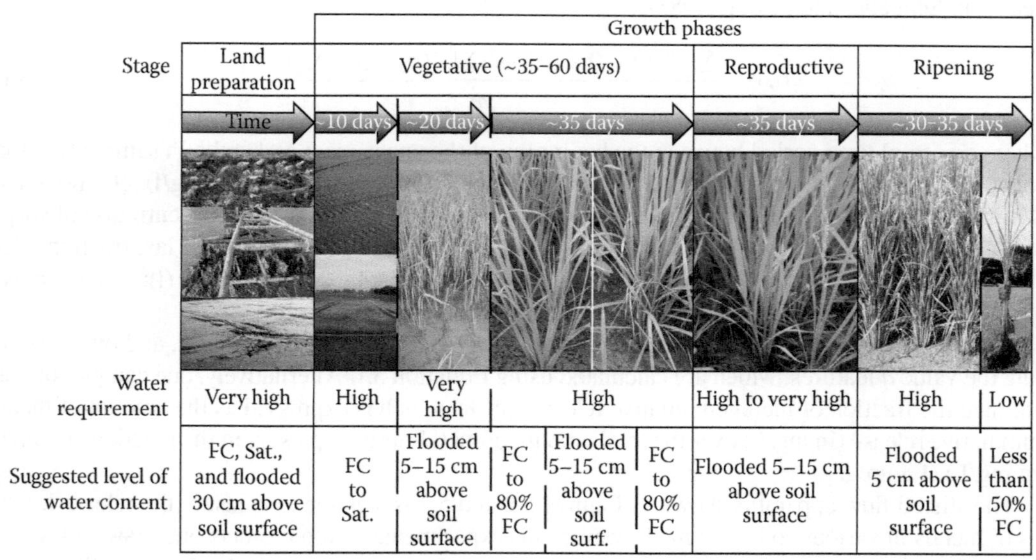

FIGURE 5.5 **(See color insert.)** Examples of flooding and draining cycles used in lowland rice cultivation, with FC: soil water content at field capacity, and Sat.: soil water content at saturation point.

soils under lowland conditions. Incubation experiments, which involve alternatively flooding and draining the soil sample, are often employed to mimic soil conditions under lowland rice practices. The period of flooding used to incubate the soil sample should be similar to the periods employed in paddy fields. This should not last longer than 150 days, as this represents the longest period that farmers usually employ to submerge their paddy fields. In addition, this period should last long enough for the system to reach a near-steady state. Typically, when the field shifts to flooding conditions, soil chemical properties dramatically change during the first 2 weeks and reach equilibrium within

30 days (Kyuma, 2004). Therefore, the flooding period used in incubation experiments should last for at least 2 weeks.

During flooding periods, farmers often keep the level of water at about 5–25 cm above the soil surface, which corresponds to saturation point (Sat.) condition. However, in paddy fields that are well irrigated, some farmers can keep the level of water >30 cm above the soil surface. This practice is, in theory, not recommended, as it can negatively impact rice growth by notably increasing iron solubility, which can decrease the uptake of other nutrients by rice plants. To avoid this problem, it is recommended that farmers keep the level of water at 5–15 cm above the soil surface (Ponnamperuma, 1972; Kyuma, 2004). The draining period used in incubation experiments is equal to the time required to drain the soil, initially under flooding conditions, to a moisture content that is specifically chosen by the farmer. This soil moisture content, which must be lower than the one corresponding to Sat., can be, for instance, field capacity (FC), or a value lower than the one corresponding to FC. FC is approximately the amount of water that is held in a soil after it has been fully wetted and all gravitational water has been drained away. An example of an incubation experiment that employed two soil moisture contents, Sat. and FC, can be found in Khaokaew (2010).

Several studies have measured the release kinetics of Cd or Zn at various redox conditions (de Livera et al., 2011; Khaokaew et al., 2011; Fulda et al., 2013). Most of these studies used a batch experimental approach, such as the one employed by de Livera et al. (2011). The inherent drawbacks of the batch methods were discussed in Section 5.2.1. The positive aspect of the batch approach used by de Livera et al. (2011) was that the setup enabled to keep the system in reducing conditions by flowing N_2 gas inside the reaction cell. The latter was sealed and submerged in a vessel full of water. A water heater was also placed in this vessel to keep the system at a constant temperature. These features, enabling the control of E_h and temperature in the reactor, have, to our knowledge, never been adapted to a stirred-flow setup to study the desorption behaviors of Cd and Zn in paddy soils under anoxic conditions. Soil samples under anoxic conditions have been usually transferred from a glove box under an N_2 environment to the stirred-flow reaction chamber before promptly starting the desorption experiment. Therefore, with this type of stirred-flow setup where E_h cannot be controlled, one must assume that the hypothetical increase in the redox potential of the sample in the reaction chamber occurs slower than the time scale of the desorption reactions studied. Therefore, future studies that would investigate Cd and Zn release kinetics in paddy soils should idealistically use an experimental approach that would combine a stirred-flow method with features of setups that are similar to the one employed by de Livera et al. (2011), or soil microcosm systems that allow control of E_h, pH, and temperature throughout the experiment (Yu et al., 2007).

5.4.3 Factors Controlling Release Kinetics of Cd and Zn in Paddy Soils

Release kinetics of Cd and Zn in paddy soils are often characterized by two periods: an initial rapid release during the first 30 min, followed by a slower release (Khaokaew et al., 2011, 2012). The former is attributed to metals that are weakly sorbed to soil components or other fractions that can be rapidly desorbed via, for example, ion-exchange mechanisms, whereas the latter could be attributed to intra- and interparticle diffusion processes, sites of differing reactivity, and surface precipitation phenomena (Sparks, 1989). For example, in alkaline soil, the rapid release can be due to the fast exchange between Cd^{2+} and Ca^{2+} on the surfaces of soil components, such as $CaCO_3$ (McBride, 1980). The rapid release of Cd and Zn is also controlled by the speciation of these two metals. For example, more Zn is released when the metal is bound to organic matter than inorganic minerals (Candelaria and Chang, 1997).

Cadmium and Zn can be associated to various soil components in paddy soils. For example, Cd was mainly associated to humic acid, kaolinite, sulfide, and ferrihydrite in an acid Thai paddy soil (Khaokaew, 2010). These associations may affect the solubility of Cd and Zn. Under oxic conditions, CdS has been considered the most important species controlling Cd solubility in paddy soils (Bingham et al., 1976; Chaney, 2010). However, the formation of CdS is limited by the amounts of sulfur in the system, especially under multimetal environments (Fulda et al., 2013). Since the release kinetics of Cd and Zn in paddy soils is greatly influenced by metal speciation, more information will be provided on the chemical forms of these metals in paddy soils in Section 5.3.

Cadmium and Zn desorption behaviors are also controlled by soil properties, especially E_h, pH, and the types and amounts of sorbents, ions, and LMWOAs (Krishnamurti et al., 1997; Khaokaew, 2010). One sorbent that can either enhance or inhibit metal mobility via different mechanisms is humic acid, which is a major component of organic matter (Schnitzer, 1986). For example, it increases the amounts of metals bound to oxide minerals via the formation of metal-ternary surface complexes (Floroiu et al., 2001). The association between metals and humic acids is highly pH dependent, and it decreases with a decrease in pH (Pehlivan and Arslan, 2006). In addition, the capacity of (oxy)hydroxide minerals in retaining metals in soils can be enhanced by adding organic ligands (Burnett et al., 2006). This increases the amounts of ternary surface complexes on the surfaces of (oxy)hydroxide minerals (Alcacio et al., 2001).

Zinc is one of the most soluble and mobile trace metal cations under acidic, oxidizing conditions (McBride, 1994). However, under alkaline conditions, Zn is rather immobilized. For example, Khaokaew et al. (2012) reported that Zn was mostly associated with a layered-double hydroxide (LDH) phase in an alkaline Thai paddy soil, which greatly limited Zn solubility. The capacities of soil components to retain Cd and Zn usually increase with an increase in soil pH (Naidu et al., 1997; Roberts et al., 2002; Jacquat et al., 2009). This is partly due to the amounts of negative charges on the surface of some soil components, for example, clay and oxide minerals or organic matter, which increase with an increase in pH (Sparks, 2003). Moreover, an increase in soil pH can result in precipitation, chemisorption, or stronger adsorption of metals, especially on the surfaces of carbonate and (oxy)hydroxide minerals (Papadopoulos and Rowell, 1989; McBride, 1994). Consequently, cadmium and zinc are less mobile in alkaline soils than acidic soils (Figures 5.6 and 5.7) (Khaokaew et al., 2011, 2012). Approximately 85% of total Cd can be then released from soils at pH 2.8 (Tyler, 1978). Similarly, the solubility of Zn can be high at low-pH values, and it can decrease markedly with an increase in soil pH (Farrah and Pickering, 1977; Papadopoulos and Rowell, 1989).

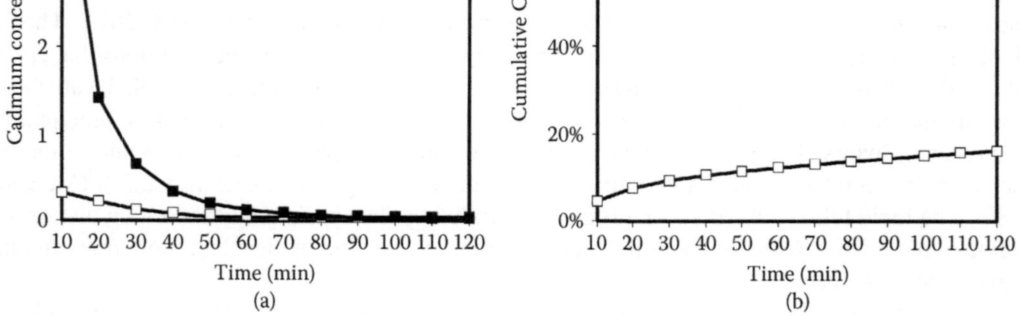

FIGURE 5.6 (a) Cadmium release kinetics and (b) cumulative Cd desorption (in %) by DTPA-TEA-Ca in an alkaline, air-dried paddy soil and acidified, air-dried paddy soil. (From Khaokaew, S. *Speciation and Release Kinetics of Cadmium and Zinc in Paddy Soils*, Plant and Soil Department, University of Delaware, Newark, DE, 2010. With permission.)

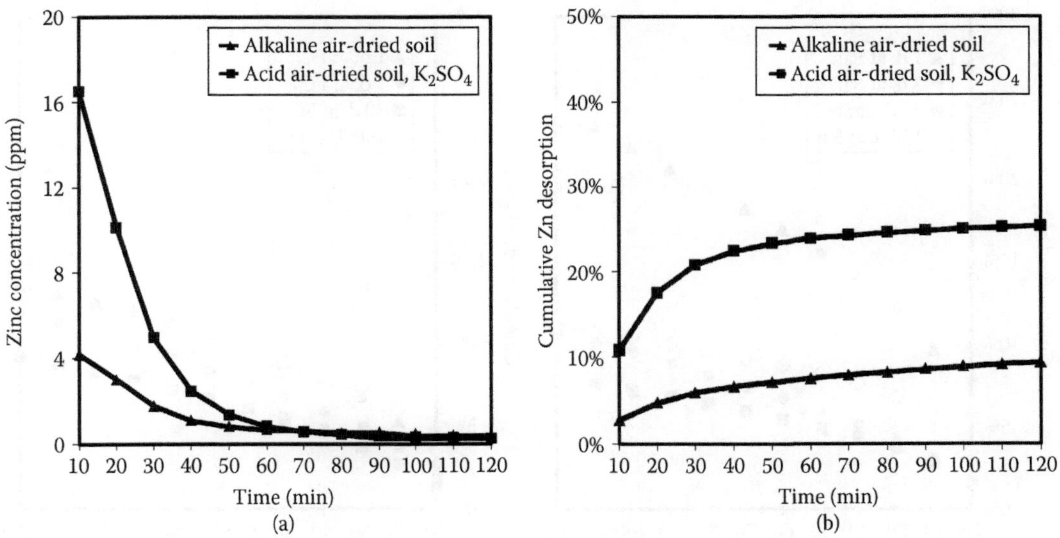

FIGURE 5.7 (a) Zinc release kinetics and (b) cumulative Zn desorption (in %) by DTPA-TEA-Ca in an alkaline, air-dried paddy soil and acidified, air-dried paddy soil. (From Khaokaew, S., *Speciation and Release Kinetics of Cadmium and Zinc in Paddy Soils*, Plant and Soil Department, University of Delaware, Newark, DE, 2010. With permission.)

Soil moisture contents have lower effects on Cd release in alkaline soil compared with acidic soil containing high amounts of sulfur. Indeed, desorption experiments showed that a small amount of Cd was released into alkaline soil that was subjected to various flooding periods and draining conditions (Khaokaew et al., 2011). The cumulative Cd amounts released for 2 h were only about 10% in most treatments. Small variations in the amounts of Cd released from the samples that were drained to Sat. or FC were observed when the flooding period was varied (Figure 5.8). Varying draining conditions had even lesser effects on Cd desorption in alkaline soil. These results supported those reported by van Gestel and van Diepen (1997), which showed that Cd sorption to soil components was not significantly affected by soil moisture content. However, after the pH and ion contents in the alkaline soil were modified by using nitric acid and K_2SO_4, the cumulative desorption of Cd increased dramatically. In the acid soil spiked with K_2SO_4, results showed that cumulative Cd desorption reached almost 90% for 30 days under flooding and draining to Sat. conditions, and >50% for other treatments. Under the two types of draining conditions studied and for the first 10 min of Cd desorption, the amounts of Cd released followed this order: 30 days > 1 day > 150 days of flooding. The amounts of released Cd were higher when soils were drained at Sat. compared with at FC, especially at 30 days of flooding (Khaokaew, 2010).

In heterogeneous paddy soil systems, rice roots can influence Cd and Zn release, by secreting LMWOAs. The types and amounts of LMWOAs depend on the nature of rice cultivar and soil chemical parameters, such as soil pH. Depending on their types and levels in solution, LMWOAs can enhance, decrease, or have no effect on Cd and Zn solubility in soils (White, 1981; Jones, 1998; Greger and Landberg, 2008; Fang et al., 2008).

LMWOAs can affect Cd solubility by acidifying soil pH or forming metal–organic complexes in solution (Chen et al., 2001). The nature of metals and LMWOAs control the extent of metal–organic complexation and thus metal solubility (Mench and Martin, 1991; Chen et al., 2001).

The solubilities of Cd and Zn around the rhizosphere generally increase with an increase in concentrations of LMWOAs (Cieśliński et al., 1998; Nigam et al., 2001; Chiang et al., 2006). The major LMWOAs secreted from rice roots are oxalic, malic, maleic, and citric acids (Liu et al., 2007; Khaokaew, 2010). Hoffland et al. (2006) studied seven lowland rice cultivars and found

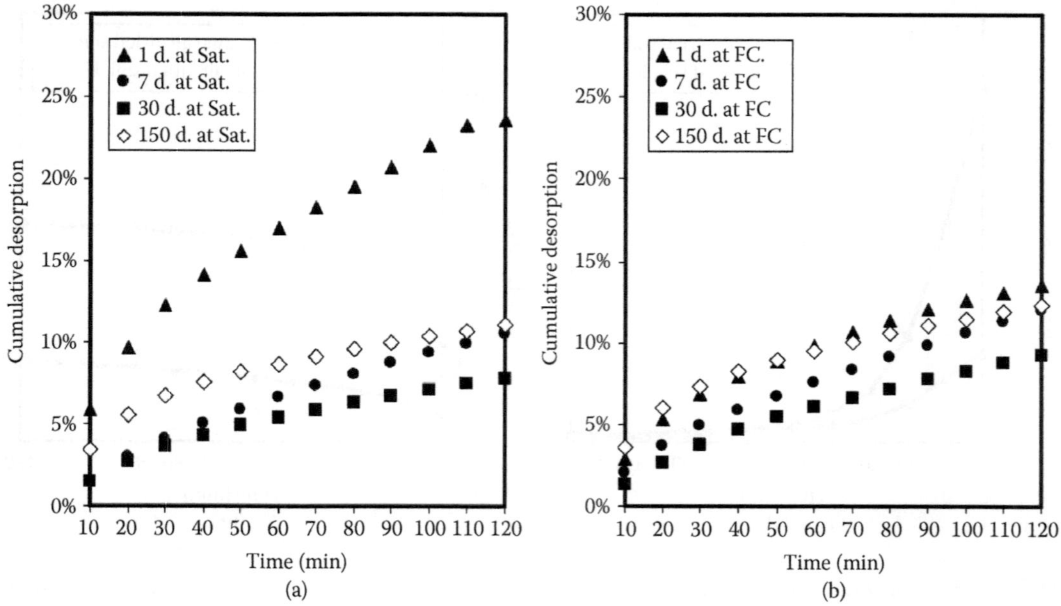

FIGURE 5.8 Cumulative Cd desorption (%) in a paddy soil by DTPA-TEA-Ca, after 1, 7, 30, and 150 days of flooding and draining to (a) saturation (Sat.) and (b) field capacity (FC). (From Khaokaew, S., *Speciation and Release Kinetics of Cadmium and Zinc in Paddy Soils*, Plant and Soil Department, University of Delaware, Newark, DE, 2010. With permission.)

that citrate was the most effective root exudate in mobilizing Zn in paddy soils, although oxalate was the most abundant in solution. Similarly, the abundance of organic acids secreted from two Thai rice cultivars grown in high Cd concentration media followed this order: oxalic acid > citric acid > maleic acid (Khaokaew, 2010). The ability of these three organic acids to extract Cd and Zn from Thai-contaminated paddy soils was citric acid > maleic acid > oxalic acid. Khaokaew (2010) also reported that soil pH and metal speciation are important parameters in controlling the ability of LMWOAs to complex with Cd and Zn and, hence, affect the solubility of the two metals. LMWOAs can affect Cd and Zn solubilities, especially in acidic soils where metal species can be easily dissolved. Examples of release kinetics of Cd and Zn in paddy soils, using LMWOAs secreted by rice plants as the desorbing agent, are shown in Figures 5.9 and 5.10, respectively. Krishnamurti et al. (1997) demonstrated that various types of LMWOAs could affect the rate of Cd release and increase Cd solubility in different soils via the formations of soluble Cd–LMWOA complexes. Wu et al. (2003) conducted a pot experiment to study the effects of LMWOAs on Cd and Zn desorption in a paddy soil. They found that the addition of 3 mmol/kg EDTA or LMWOAs to the soil increased the total concentrations of Cd and Zn in soil solution, following this order: EDTA > malic acid > citric acid > oxalic acid for Cd, and EDTA > malic acid > citric acid ≈ oxalic acid for Zn. Lastly, Jiang et al. (2011) conducted batch experiments to assess the effects of pH and LMWOAs on Cd desorption in Chinese paddy soils. They found that the presence of LMWOAs inhibited Cd desorption at low concentrations (≤0.1 mmol/L) but promoted Cd desorption at higher concentrations (≥0.5 mmol/L for citric acid and ≥1 mmol/L for malic and oxalic acids).

To summarize, there are a number of factors that can affect the release kinetics of Cd and Zn in paddy soils. They consist of not only various physicochemical properties of the paddy soil at a given period during the lowland farming cycle but also the nature of the rice plant cultivated. As previously mentioned, the chemical forms of the metals and their associations with soil components are one of the parameters that can drastically affect metal desorption behavior. The methodologies

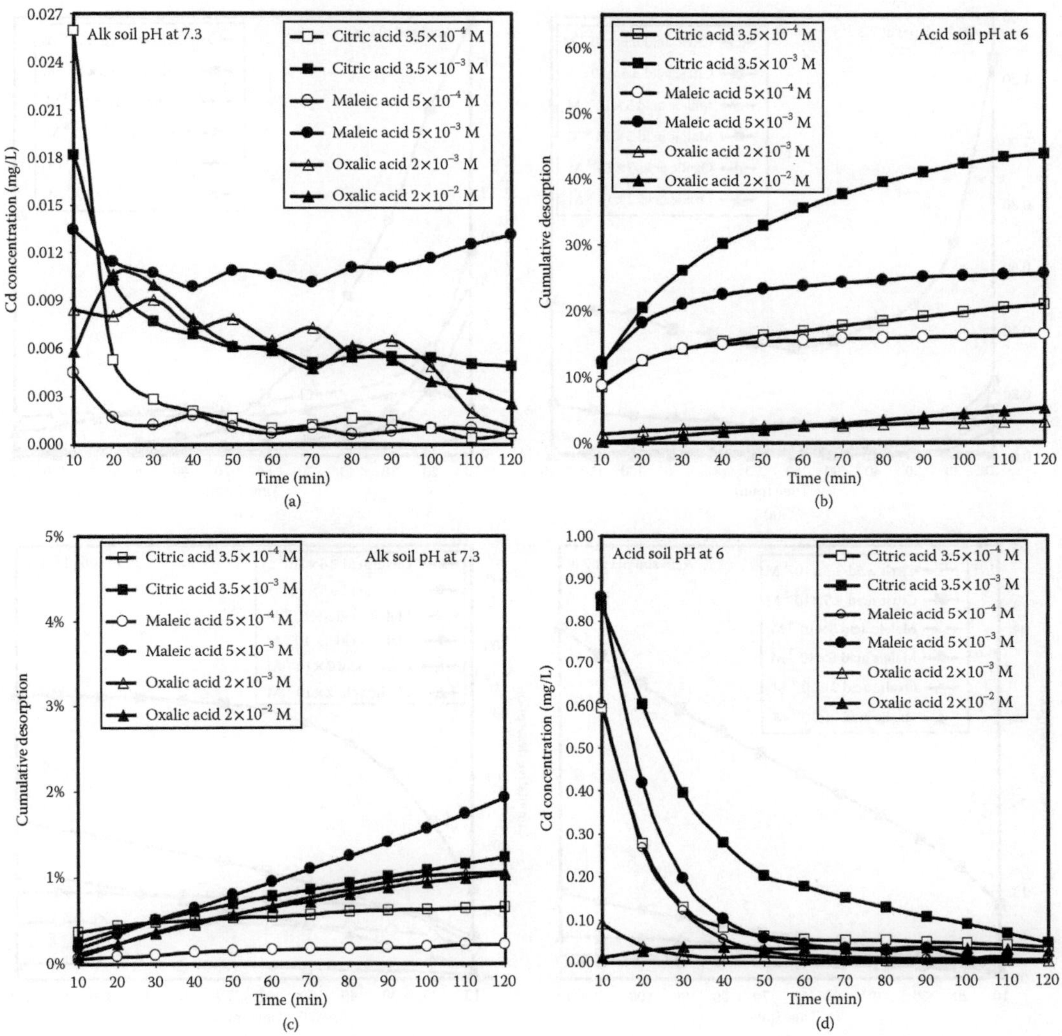

FIGURE 5.9 (a) Cd release kinetics and (b) cumulative Cd desorption (in %) in an alkaline paddy soil by oxalic acid, citric acid, and maleic acid at 0.35 or 3.5 mM citric acid, 0.5 or 5 mM maleic acid, and 2 or 20 mM oxalic acid. (c) Cd release kinetics and (d) cumulative Cd desorption (in %) in an acidified paddy soil by oxalic acid, citric acid, and maleic acid at 0.35 or 3.5 mM citric acid, 0.5 or 5 mM maleic acid, and 2 or 20 mM oxalic acid. (From Khaokaew, S., *Speciation and Release Kinetics of Cadmium and Zinc in Paddy Soils*, Plant and Soil Department, University of Delaware, Newark, DE, 2010. With permission.)

employed to determine Cd and Zn speciation, as well as the factors that may control the chemical forms of the two metals, will be discussed in the next section.

5.5 SPECIATION OF CADMIUM AND ZINC IN PADDY SOILS

5.5.1 Methodologies to Study Cd and Zn Speciation in Paddy Soils

To elucidate Cd and Zn associations with soil components in paddy fields, sequential extraction methods have been one of the most employed techniques. The information obtained from them can be useful in predicting metal availability, mobility, and toxicity. Some examples of sequential extraction procedures that have been employed to elucidate Cd and Zn distributions in paddy soils

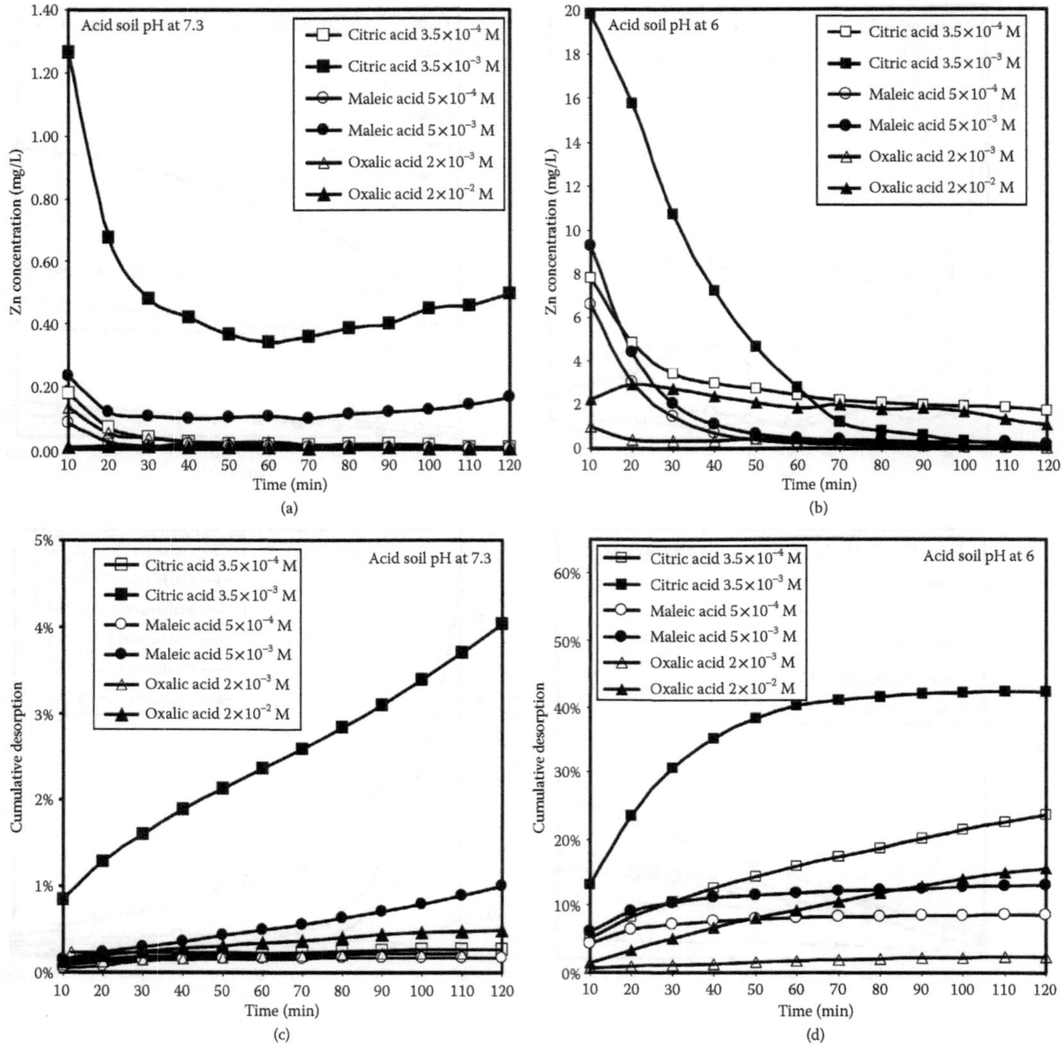

FIGURE 5.10 (a) Zn release kinetics and (b) cumulative Zn desorption (in %) in an alkaline paddy soil by oxalic acid, citric acid, and maleic acid at 0.35 or 3.5 mM citric acid, 0.5 or 5 mM maleic acid, and 2 or 20 mM oxalic acid. (c) Zn release kinetics and (d) cumulative Zn desorption (in %) in an acidified paddy soil by oxalic acid, citric acid, and maleic acid at 0.35 or 3.5 mM citric acid, 0.5 or 5 mM maleic acid, and 2 or 20 mM oxalic acid. (From Khaokaew, S., *Speciation and Release Kinetics of Cadmium and Zinc in Paddy Soils*, Plant and Soil Department, University of Delaware, Newark, DE, 2010. With permission.)

can be found in Zeien and Brümmer (1989), Salbu and Krekling (1998), Kashem et al. (2007), Khanmirzaei et al. (2013), and Sriprachote et al. (2012). However, sequential extraction techniques can result in incomplete dissolution of the target fraction, incomplete removal of dissolved species due to resorption, removal of a nontarget species, and change in the valence of redox-sensitive target elements (Sparks, 2003). Over the past few decades, synchrotron-based x-ray absorption fine structure (XAFS) spectroscopy has become an alternative technique to sequential extraction methods for determining metal speciation in soils. It enables determination of the local environments at the molecular scale and the electronic structures of Cd and Zn in paddy soils (Khaokaew et al., 2011; Fulda et al., 2013). Manceau et al. (2005) was among the first groups to employ XAFS to elucidate Zn speciation in paddy soil, whereas Khaokaew et al. (2011, 2012) have used this technique to determine Cd and Zn speciation under various redox regimes. Descriptions of XAFS principles, analyses, and data processing are beyond the scope of this chapter and can be found elsewhere (Calvin, 2013). Cadmium speciation in paddy soils is determined by using XAFS at the Cd K-edge (i.e., 26.71 keV). Although XAFS data may be also collected at the Cd L-edges, this type of analysis is often not suitable for paddy soil samples, as they may contain high potassium content, given that the energy differences between Cd L-edges (i.e., L-I = 4.02 keV, L-II = 3.73 keV, L-III = 3.54 keV) and K K-edges (i.e., 3.61 keV) are small. Despite the fact that many synchrotron facilities exist around the world, only a few of them allow Cd XAFS analyses of soil samples, due to the high Cd K-edge that involves the collection of spectra at a 30 keV energy, which is often above the workable energy range of hard x-ray beamlines. For example, XAFS analyses of soil samples at the Cd K-edge are possible at beamline 10-ID, Advanced Photon Source (APS), Chicago, IL, USA, or beamline 7-3 and 4-1 at the Stanford Synchrotron Radiation Light source (SSRL), Stanford University, CA, USA. The minimum metal concentration in the soil sample that is suitable for XAFS analyses depends on many factors, including the nature of the beamline and synchrotron (Calvin, 2013). For example, a soil sample containing 20 mg/kg Cd has been analyzed by extended x-ray absorption fine structure (EXAFS) (Fulda et al., 2013) at the SuperXAS beamline (X10DA), Swiss Light Source, Paul Scherrer Institute, Switzerland. Cadmium K-edge EXAFS spectra that correspond to soil samples featuring [Cd] of at least 60 mg/kg were collected at beamline 10-ID (Khaokaew et al., 2011). Zinc K-edge EXAFS spectra having limited amounts of noise below 8 1/Å and corresponding to soil samples containing [Cd] of about 100 mg/kg or higher have been collected at beamlines XAS (Angströmquelle Karlsruche, ANKA, Karlruhe, Germany) and SNBL (European Synchrotron Radiation Facility, ESRF, Grenoble, France), as shown in Figure 5.11. The Zn K-edge has been mainly used to analyze paddy soils by XAFS since its corresponding energy (i.e., 9.65 keV) falls within the workable energy range of many hard x-ray beamlines, such as SNBL (European Synchrotron Radiation Facility), XAS (Angströmquelle Karlsruche), X-11A (National Synchrotron Light Source, Brookhaven, Upton, New York, USA), and BL8 (Synchrotron Light Research Institute, Nakhon Ratchasima, Thailand).

In addition, one can locally measure, at the microscopic scale and in two dimensions, the distribution in the paddy soil sample of Cd and Zn along with other elements, such as Ca, Cu, Fe, K, Zn, and Mn, by using μ-x-ray fluorescence (μ-XRF) spectroscopy, at microprobe beamlines

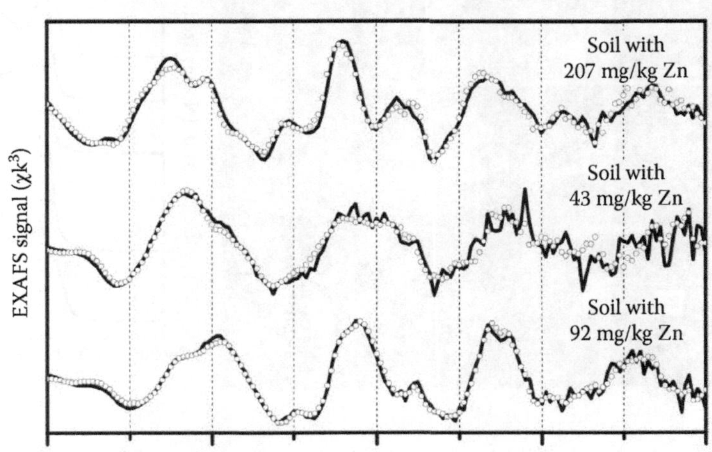

FIGURE 5.11 EXAFS spectra of three limestone soils containing specific Zn concentrations (Reprinted from *Geochim. Cosmochim. Acta*, 73, Jacquat, O. et al., Changes in Zn speciation during soil formation from Zn-rich limestones, 5534–5571, Copyright 2009, with permission from Elsevier.)

(e.g., 13-ID-C, APS, USA, and 10.3.2, Advanced Light Source [ALS], USA, for Cd and Zn as the primary target metal, respectively). Examples of μ-XRF elemental maps corresponding to a Cd- and Zn-contaminated paddy soil sample are shown in Figure 5.12. These maps allow for a determination of whether any correlations exist between the distributions of the target metal and other elements. This can be notably helpful in determining the nature of the associations between the target metal and soil components in the sample. For example, if Cd was mainly distributed at the microscopic scale in regions of the sample that mainly contained Fe based on μ-XRF maps, cadmium would be perhaps principally associated with iron oxides at these locations. In addition, mineralogical analyses of soil particles observed in the μ-XRF maps can be done, usually in an area coarser than 4 μm^2, by μ-x-ray diffraction (μ-XRD) spectroscopy, which is a complementary technique that is often offered at microprobe beamlines. The regions of interests observed in the μ-XRF maps, where the target element is locally more concentrated than the rest of the probed regions, can be analyzed by the μ-XAFS technique, which is also available at many microprobe beamlines, such as 13-ID-C, APS, or 10.3.2, ALS. These μ-XAFS analyses probe the soil sample in a microscopic or submicroscopic area, depending on the nature of the beamline. This allows a determination of the molecular environments of the target element locally in a soil sample, which

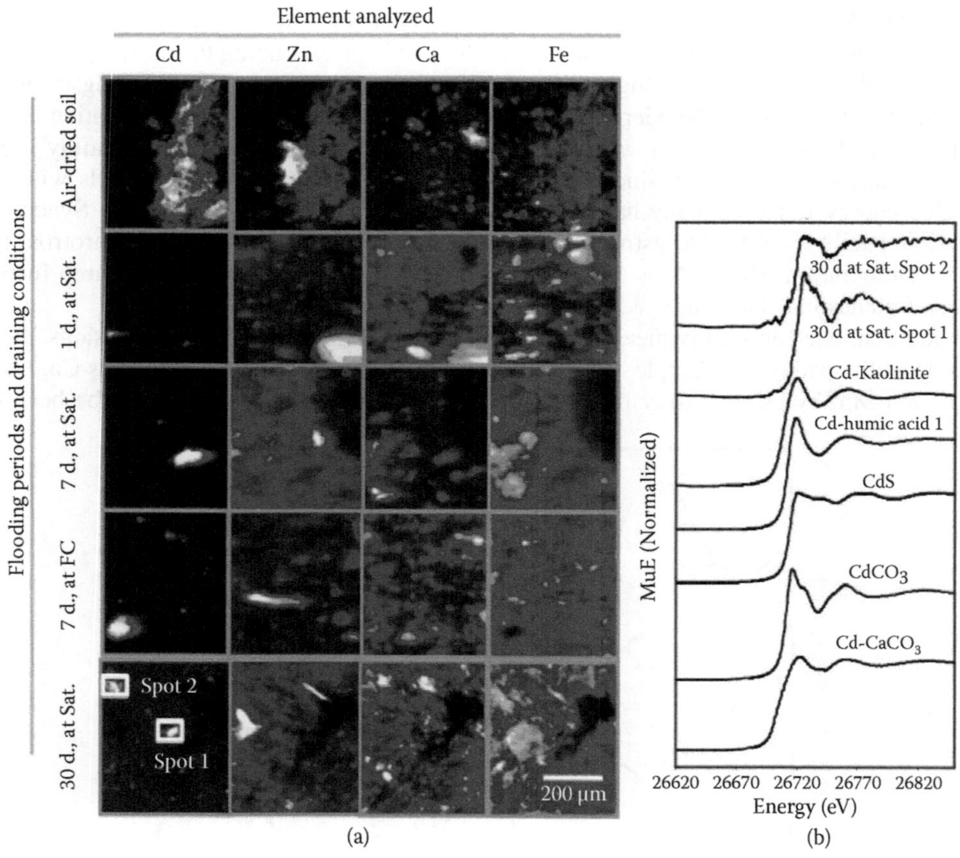

FIGURE 5.12 (See color insert.) (a) Elemental distributions of cadmium (Cd), zinc (Zn), calcium (Ca), and iron (Fe) in an air-dried paddy soil flooded for different flooding periods and draining conditions by using μ-XRF mapping; (b) μ-XANES spectra, at the Cd K-edge, of Cd-$CdCO_3$, $CdCO_3$, CdS, Cd-humic acid 1, and Cd-kaolinite standards, and μ-XANES spectra taken at spots 1 and 2. (Reprinted with permission from Khaokaew, S. et al., *Environ. Sci. Technol.*, 45, 4249–4255, 2011. Copyright 2011, the American Chemical Society.)

differs from the data obtained with bulk-XAFS, as the latter technique entirely probes the soil sample. Therefore, bulk-XAFS and μ-XAFS/μ-XRF/μ-XRD are two complementary methods and are useful in constraining Cd and Zn speciation at multiple scales. In addition, although some paddy soils have Cd or Zn levels that are too low for EXAFS analyses, one might still be able to locally determine metal speciation in these samples, by using μ-XAFS combined with μ-XRF. This would be indeed possible if Cd or Zn was discretely distributed in the soil, in a way that *hot spots* would be observed in the μ-XRF metal elemental map corresponding to the sample, such as those featured in the Cd and Zn elemental maps shown in Figure 5.12, which could be analyzed by μ-XAFS to determine metal speciation at these locations. Conversely, if a paddy soil contained Cd or Zn at a concentration that is too low for bulk-EXAFS analyses, and the metal amount in the soil was homogeneously distributed in a way that Cd or Zn hot spots would be not distinctively observed in the μ-XRF map or not concentrated enough for μ-XAFS analyses, one would at least still be able to use sequential extraction methods to determine metal speciation in the sample at the macroscopic scale.

Lastly, basic physicochemical characterizations of the soil samples (e.g., mineralogy constrained by powder x-ray diffraction analyses, elemental composition, soil organic matter content, cation-exchange capacity, point of zero charge, E_h, pH, soil texture, soil particle distribution), whose standard methodologies can be found, for instance, in Sparks (1996), are always useful in complementing results obtained from macroscopy (e.g., sequential extraction methods) and/or spectroscopy (e.g., bulk-XAFS, μ-XRF, μ-XAFS, μ-XRD) to accurately constrain the speciation of metals present in paddy soils.

5.5.2 Factors Controlling Cd and Zn Speciation in Paddy Soil

Soil pH, redox and moisture regimes, competitive cations, anions, residence time, metal concentration, and type and amount of sorbents can affect Cd and Zn speciation in paddy soils (Xue and Sigg, 1998; Elbanaa and Selim, 2010; Khaokaew et al., 2011, 2012). Soil pH is one of the most crucial factors, and is used in controlling the speciation of Cd and Zn that are sorbed to various soil components in paddy fields (Farrah and Pickering, 1977; Roberts et al., 2002; Jacquat et al., 2009). Naidu et al. (1997) observed that soluble $CdOH_2$ species can form at a pH above 8, and Cd precipitation to $Cd(OH)_2$ is likely to occur at a pH above 10. In addition, in alkaline conditions, Cd can substitute for Ca in carbonate mineral phases, which could be due to similarities in ionic radius between Cd (i.e., 1.01 Å) and Ca (i.e., 1.1 Å) (McBride, 1980). Lastly, Farrah and Pickering (1977) found that chemisorption of Cd on soil components in alkaline conditions was more significant than that of Zn. Instead, the latter metal forms sparingly soluble compounds such as $ZnCO_3$, $2Zn(OH)_2$, and $Zn(OH)_2$ at a pH > 7.7.

Redox potential also plays a crucial role in controlling Cd and Zn speciation in soil. Under reducing conditions, a number of studies have suggested that Cd, a chalcophile metal, could be effectively immobilized in soil via metal sulfide precipitation (Benjamin and Leckie, 1982) or sorption to Fe oxides (Lee, 2006). Similarly, although Zn only has one stable oxidation state in the environment (i.e., Zn[II]), variations in soil redox conditions can affect Zn partitioning (Bostick et al., 2001). Under acidic, oxidizing conditions, Zn is one of the most soluble and mobile trace metal cations (McBride, 1994). Gao (2007) reported that E_h and pH caused variations in the bioavailability of Zn, which was immobilized under aerobic conditions. However, Mandal et al. (2000) reported that Zn uptake in rice was higher under a pre-flooding treatment compared with the control where water was not formerly added to the soil. Cadmium can sorb to several soil components in flooded paddy fields, including carbonates, kaolinite, ferrihydrite, humic acid, and sulfide (Khaokaew et al., 2011). Precipitation may be the predominant reaction in the presence in solution of significant amounts of anions, such as S^{2-} and OH^- (Naidu et al., 1997). Moreover, in oxidizing environments and alkaline soils, Cd is likely to precipitate to minerals such as octavite ($CdCO_3$) as well as to CdO and $Cd(OH)_2$ (McBride, 1994). Cadmium and Zn speciation has been determined by XAFS in various sorbents (Sparks, 2003). For example, when Cd sorbs to montmorillonite, one of the most important metal sorbents in paddy soils, the metal sorption process determined by XAFS is mainly an outer-sphere complex mechanism and surface precipitation to a mineral phase with a Cd5(OH)8(NO_3)2(H_2O)-like structure on the clay surface in acidic and alkaline paddy soils, respectively (Takamatsu et al., 2006). Macroscopic results from a number of studies have suggested that CdS is one of the most important species controlling Cd uptake by rice in paddy soils under anoxic conditions during flooding periods. However, based on XAFS results, the anoxic condition is not the only requirement for CdS to form in the system. The amounts of sulfide and competitive metals also play a crucial rule in the formation of CdS (Khaokaew, 2010; Fulda et al., 2013). Fulda et al. (2013) showed that the extent of Cd-sulfide precipitation decreased with a decrease in sulfate and an increase in Cu contents, even when sulfate amounts present in the system exceeded those of Cd. Examples of speciation of Cd in an alkaline paddy soil and of Zn in an acidified paddy soil at various flooding periods and draining conditions are shown in Figures 5.13 and 5.14, respectively.

One can see in Figure 5.13 that Cd association with kaolinite can be found in Thai paddy fields, especially in acidic soil. Kaolinite is one of the most common phyllosilicate clays found in soils,

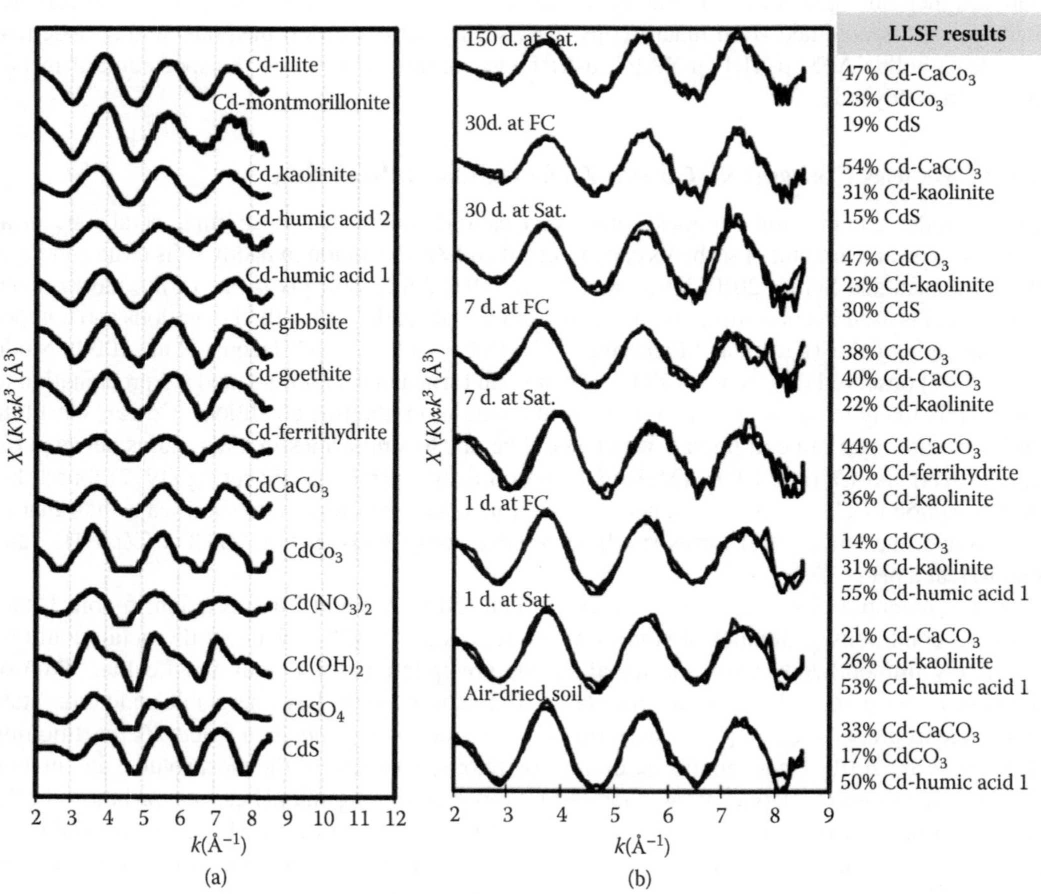

FIGURE 5.13 Linear least-squares fitting results for Cd bulk-XAFS spectra of an alkali, air-dried Thai paddy soil under various flooding periods and draining conditions. Solid lines represent the k^3-weighted χ-spectra, and the dotted lines indicate the best fits obtained by using linear least-squares fitting. (Reprinted with permission from Khaokaew, S. et al., *Environ. Sci. Technol.*, 45, 4249–4255, 2011. Copyright 2011, the American Chemical Society.)

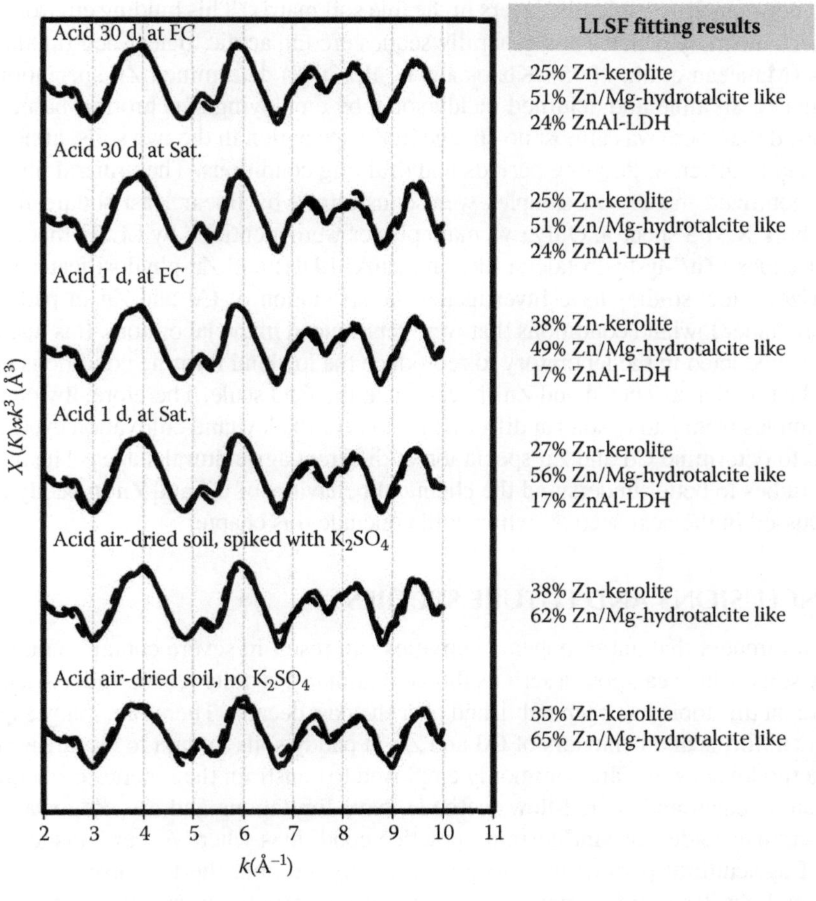

FIGURE 5.14 Linear least-squares fitting results for Zn bulk-XAFS spectra of acidified paddy soil samples, air-dried or flooded for different periods (d = days), and drained to saturation (Sat.) or field capacity (FC). The solid lines represent the k^3-weighted spectra, and the dotted lines represent the best fits obtained by using linear least-squares fitting. (Reprinted with permission from Khaokaew, S. et al., *Environ. Sci. Technol.*, 46, 3957–3963. Copyright 2012, the American Chemical Society.)

and it has been used as a metal scavenger for soil remediation applications (Gale and Wendt, 2003; Bhattacharyya and Gupta, 2007). Farrah and Pickering (1977) found that the capacities of clays to sorb metals followed these orders: Pb > Cd > Zn for kaolinite; Pb > Cu > Zn > Cd for illite; and Pb > Cu > Cd ~ Zn for montmorillonite. The mobility of Cd can decrease when the metal is associated with stable kaolinite particles, or it can increase when it is transported via colloids (Vasconcelos et al., 2008). At pH 6, more than 75% of Cd can sorb to kaolinite in an outer-sphere complex mechanism (Gräfe et al., 2007). At a pH higher than 6, Cd can form inner-sphere adsorption complexes on the edge sites of kaolinite (Vasconcelos et al., 2008). At a pH lower than pH 6, Cd sorption to kaolinite can occur via two primary mechanisms: adsorption via an ion-exchange reaction at permanent negatively charged sites, and association with aluminol groups that are present on the edges of the clay (Angove et al., 1998). For the first time, Manceau et al. (2005) determined Zn speciation in paddy soil by using the XAFS technique. They studied the natural speciation of Zn (42 mg/kg) in the argillic horizon of an Ultisol (pH = 5.6) from a paddy soil in northern Taiwan, and they found that Zn (r = 0.74 Å) could fill vacant sites in the gibbsitic layer of natural lithiophorite, in a similar manner as lithium (r = 0.74 Å). Results from x-ray diffraction, powder, and polarized EXAFS spectroscopy analyses showed that Zn was predominantly bound to hydroxy–Al interlayers that were sandwiched between 2:1 vermiculite layers in the fine soil matrix. This binding environment may be the main mechanism by which Zn is generally sequestered in acidic to near-neutral aluminum-rich clayey soils (Manceau et al., 2005). Khaokaew et al. (2012) determined Zn speciation in Cd–Zn co-contaminated alkaline and acidified paddy soils, by employing synchrotron-based techniques. Results showed that there was almost no change in Zn speciation in the two soils, although the soils were subjected to different flooding periods and draining conditions. The mineral phases in which Zn was immobilized in the soil samples were constrained by linear least square fitting (LLSF) analyses of bulk XAFS spectra. Only two main phases were identified by LLSF, that is, Zn-layered double hydroxides (Zn/Mg-hydrotalcite like and ZnAl-LDH) and Zn-phyllosilicates (Zn-kerolite). To summarize, a few studies have investigated the speciation of Cd and Zn in paddy soils. The samples were under lowland conditions that were reproduced in the laboratory. It is unclear whether experiments conducted in the laboratory to reproduce the lowland farming conditions can take into account all factors that affect Cd and Zn speciation at the field scale. Therefore, it would be needed to collect samples from paddy soils at different periods of the lowland cultivation cycle and analyze them in situ to determine Cd and Zn speciation at different agricultural stages. The need for more field-scale studies to better understand the chemical behaviors of Cd and Zn in paddy soils will be further discussed in the next section, which will conclude this chapter.

5.6 CONCLUSIONS AND FUTURE STUDIES

The global awareness that anthropogenic activities can result in severe contaminations of Cd and Zn in paddy soils, which can pose a serious threat to humans, is quite recent. Indeed, a major part of the literature on this topic has been published over the past decade. Therefore, many aspects related to the chemical forms and behaviors of Cd and Zn in paddy soils remain to be further understood. First, the methodologies that are commonly employed to constrain the kinetics of metal desorption in soils, such as batch and stirred-flow methods, have limitations and are not always adapted to paddy soil samples under lowland farming practice conditions where E_h can drastically vary over the different agricultural periods used to grow rice. In addition, the few investigations that have studied Cd and Zn desorption kinetics in paddy soils measured these rates from samples under lowland conditions that were reproduced in the laboratory, such as flooding and draining conditions (Elbanaa and Selim, 2010; Khaokaew, 2010; Fulda et al., 2013). The systems studied were also simple and ideal, as the water used for submerging the samples consisted of deionized water with a background electrolyte, to artificially reproduce the composition of water flooding in real agricultural fields. Moreover, specific desorbing agents were used to desorb Cd and Zn from soil components. To our knowledge, no studies have conducted experiments at the field scale to determine Cd and Zn desorption in paddy soils. It would be important to determine whether results obtained from desorption experiments carried out in the laboratory could be correlated to the field scale, where many environmental factors, such as the soil microbiota, may influence Cd and Zn release and cannot be fully taken into account at the lab scale.

An accurate determination of the speciation of Cd and Zn and their chemical behaviors in paddy soils is needed to effectively choose the best strategies to remediate paddy soils. For example, a phytoremediation approach was recently employed at the field scale to remediate Cd- and Zn-contaminated Thai paddy soils (Khaokaew and Landrot, 2015). The methodology employed in this remediation effort was chosen after constraining the speciation and release kinetics of the two metals based on spectroscopic and macroscopic analyses of the paddy soils (Khaokaew et al., 2011, 2012). More studies demonstrating how to efficiently apply the knowledge on Cd and Zn speciation and release kinetics to solve the contamination problems of paddy soils at the field scale would be needed. Lastly, methodologies can be used to extract from metal desorption experiments the kinetic parameters that are associated with the reaction studied, such as the rate constant k (Sparks, 1989). These kinetics parameters could be used in reactive transport modeling to accurately predict the fates of Cd and Zn in paddy soils. Studies on this topic are also needed.

ACKNOWLEDGMENTS

The authors would like to thank Emeritus Professor Toshiyuki Wakatsuki, Shimane University, Japan, for his advice and guidance on Figure 5.1, and Chaiyaporn Promphan, also known as the "Thai Millionaire Farmer," for his contribution toward Figure 5.5.

REFERENCES

Alcacio, T.E., Hesterberg, D., Chou, J.W., Martin, J.D., Beauchemin, S., Sayers, D.E., 2001. Molecular scale characteristics of Cu(II) bonding in goethite-humate complexes. *Geochimica et Cosmochimica Acta* 65, 1355–1366.

Alloway, B.J., 1995. *Heavy Metals in Soils*. London, UK: Chapman and Hall.

Angove, M.J., Johnson, B.B., Wells, J.D., 1998. The influence of temperature on the adsorption of cadmium(II) and cobalt(II) on kaolinite. *Journal of Colloid and Interface Science* 204, 93–103.

Bar-Tal, A., Sparks, D.L., Pesek, J.D., Feigenbaum, S., 1990. Analyses of adsorption kinetics using a stirred-flow chamber: I. Theory and critical tests. *Soil Science Society of America Journal* 54, 1273–1278.

Benjamin, M.M., Leckie, J.O., 1982. Effects of complexation by Cl^-, SO_4^{2-}, and S_2O_3 on adsorption behavior of Cd on oxide surfaces. *Environmental Science and Technology* 16, 162–170.

Bhattacharyya, K.G., Gupta, S.S., 2007. Influence of acid activation of kaolinite and montmorillonite on adsorptive removal of Cd(II) from water. *Industrial and Engineering Chemistry Research* 46, 3734–3742.

Bingham, F.T., Page, A.L., Mahler, R.J., Ganje, T.J., 1976. Cadmium availability to rice in sludge-amended soil under "flood" and "nonflood" culture. *Soil Science Society of America Journal* 40, 715–719.

Bostick, B.C., Hansel, C.M., La Force, M.J., Fendorf, S., 2001. Seasonal fluctuations in zinc speciation within a contaminated wetland. *Environmental Science and Technology* 35, 3823–3829.

Brown, T.E., LeMay, E.H., Bursten, B.E., Murphy, C., Woodward, P., Stoltzfus, M.E., 2014. *Chemistry: The Central Science* (13th Edition). Upper Saddle River, NJ: Prentice Hall.

Burnett, P.-G.G., Daughney, C.J., Peak, D., 2006. Cd adsorption onto *Anoxybacillus flavithermus*: Surface complexation modeling and spectroscopic investigations. *Geochimica et Cosmochimica Acta* 70, 5253–5269.

Calvin, S., 2013. *XAFS for Everyone*. Boca Raton, FL: CRC Press.

Candelaria, L.M., Chang, A.C., 1997. Cadmium activities, solution speciation and solid phase distribution in cadmium nitrate and sewage sludge treated soil systems. *Soil Science* 162, 722–732.

Chaney, R.L., 2010. Cadmium and zinc. In: Hooda, P.S. (ed.). *Trace Elements in Soils*. Chichester, UK: John Wiley & Sons, Ltd.

Chen, M.-C., Wang, M.-K., Chiu, C.-Y., Huang, P.-M., King, H.-B., 2001. Determination of low molecular weight dicarboxylic acids and organic functional groups in rhizosphere and bulk soils of Tsuga and Yushania in a temperate rain forest. *Plant and Soil* 231, 37–44.

Chiang, P.N., Wang, M.K., Chiu, C.Y., Chou, S.Y., 2006. Effects of cadmium amendments on low-molecular-weight organic acid exudates in rhizosphere soils of tobacco and sunflower. *Environmental Toxicology* 21, 479–488.

Cieśliński, G., Van Rees, K.C.J., Szmigielska, A.M., Krishnamurti, G.S.R., Huang, P.M., 1998. Low-molecular-weight organic acids in rhizosphere soils of durum wheat and their effect on cadmium bioaccumulation. *Plant and Soil* 203, 109–117.

DeLaune, R.D., Reddy, K.R., 2005. Redox potential. In: Hillel, D. (ed.). *Encyclopedia of Soils in the Environment*. Oxford, UK: Academic Press, pp. 366–371.

de Livera, J., McLaughlin, M.J., Hettiarachchi, G.M., Kirby, J.K., Beak, D.G., 2011. Cadmium solubility in paddy soils: Effects of soil oxidation, metal sulfides and competitive ions. *Science of the Total Environment* 409, 1489–1497.

Elbanaa, T.A., Selim, H.M., 2010. Cadmium transport in alkaline and acidic soils: Miscible displacement experiments. *Soil Science Society of America Journal* 74, 1956–1966.

Farrah, H., Pickering, W.F., 1977. The sorption of lead and cadmium species by clay minerals. *Australian Journal of Chemistry* 30, 1417–1422.

Floroiu, R.M., Davis, A.P., Torrents, A., 2001. Cadmium adsorption on aluminum oxide in the presence of polyacrylic acid. *Environmental Science and Technology* 35, 348–353.

Fulda, B., Voegelin, A., Kretzschmar, R., 2013. Redox-controlled changes in cadmium solubility and solid-phase speciation in a paddy soil as affected by reducible sulfate and copper. *Environmental Science and Technology* 47, 12775–12783.

Gale, T.K., Wendt, J.O.L., 2003. Mechanisms and models describing sodium and lead scavenging by a kaolinite aerosol at high temperatures. *Aerosol Science and Technology* 37, 865–876.

Gao, X., 2007. *Bioavailability of Zinc to Aerobic Rice*. Wageningen, the Netherlands: Wageningen University.

Gräfe, M., Singh, B., Balasubramanian, M., 2007. Surface speciation of Cd(II) and Pb(II) on kaolinite by XAFS spectroscopy. *Journal of Colloid and Interface Science* 315, 21–32.

Hoffland, E., Wei, C., Wissuwa, M., 2006. Organic anion exudation by lowland rice (*Oryza Sativa* L.) at zinc and phosphorus deficiency. *Plant and Soil* 283, 155–162.

Huang, J.-H., Wang, S.-L., Lin, J.-H., Chen, Y.-M., Wang, M.-K., 2013. Dynamics of cadmium concentration in contaminated rice paddy soils with submerging time. *Paddy and Water Environment* 11, 483–491.

Jacquat, O., Voegelin, A., Villard, A., Juillot, F., Kretzschmar, R., 2009. Changes in Zn speciation during soil formation from Zn-rich limestones. *Geochimica et Cosmochimica Acta* 73, 5534–5571.

Jalali, M., Hemati, N., 2013. Chemical fractionation of seven heavy metals (Cd, Cu, Fe, Mn, Ni, Pb, and Zn) in selected paddy soils of Iran. *Paddy and Water Environment* 11, 299–309.

Jamil, H., Theng, L.P., Jusoh, K., Razali, A.M., Ali, F.B., Ismail, B.S., 2011. Speciation of heavy metals in paddy soils from selected areas in Kedah and Penang, Malaysia. *African Journal of Biotechnology* 10, 13505–13513.

Jiang, H., Li, T., Han, X., Yang, X., He, Z., 2011. Effects of pH and low molecular weight organic acids on competitive adsorption and desorption of cadmium and lead in paddy soils. *Environmental Monitoring and Assessment* 184, 6325–6335.

Kabata-Pendias, A., 2011. *Trace Elements in Soils and Plants* (4th Edition). Boca Raton, FL: CRC Press.

Kashem, A., Singh, B.R., Kawai, S., 2007. Mobility and distribution of cadmium, nickel and zinc in contaminated soil profiles from Bangladesh. *Nutrient Cycling in Agroecosystems* 77, 187–198.

Khanmirzaei, A., Bazargan, K., Amir Moezzi, A., Richards, B.K., Shahbazi, K., 2013. Single and sequential extraction of cadmium in some highly calcareous soils of southwestern Iran. *Journal of Soil Science and Plant Nutrition* 13, 153–164.

Khaokaew, S., 2010. *Speciation and Release Kinetics of Cadmium and Zinc in Paddy Soils*. Plant and Soil Department, University of Delaware, Newark, DE.

Khaokaew, S., Chaney, R.L., Landrot, G., Ginder-Vogel, M., Sparks, D.L., 2011. Speciation and release kinetics of cadmium in an alkaline paddy soil under various flooding periods and draining conditions. *Environmental Science and Technology* 45, 4249–4255.

Khaokaew, S., Landrot, G., 2015. A field-scale study of cadmium phytoremediation in a contaminated agricultural soil at Mae Sot District, Tak Province, Thailand: (1) Determination of Cd-hyperaccumulating plants. *Chemosphere* 138, 883–887.

Khaokaew, S., Landrot, G., Chaney, R., Pandya, K., Sparks, D.L., 2012. Speciation and release kinetics of zinc in contaminated paddy soils. *Environmental Science and Technology* 46, 3957–3963.

Kikuchi, T., Okazaki, M., Toyota, K., Motobayashi, T., Kato, M., 2007. The input–output balance of cadmium in a paddy field of Tokyo. *Chemosphere* 67, 920–927.

Krishnamurti, G.S.R., Cieslinski, G., Huang, P.M., Van Rees, K.C.J., 1997. Kinetics of cadmium release from soils as influenced by organic acids: Implication in cadmium availability. *Journal of Environmental Quality* 26, 271–277.

Kummerová, M., Zezulka, Š., Kráľová, K., Masarovičová, E., 2010. Effect of zinc and cadmium on physiological and production characteristics in *Matricaria recutita*. *Biologia Plantarum* 54, 308–314.

Kyuma, K., 2004. *Paddy Soil Science*. Japan: Kyoto University Press.

Lee, C.G., Chon, H.-T., Jung, M.C., 2001. Heavy metal contamination in the vicinity of the Daduk Au–Ag–Pb–Zn mine in Korea. *Applied Geochemistry* 16, 1377–1386.

Lee, S., 2006. Geochemistry and partitioning of trace metals in paddy soils affected by metal mine tailings in Korea. *Geoderma* 135, 26–37.

Lei, M., Zhang, Y., Khan, S., Qin, P.-F., Liao, B.-H., 2010. Pollution, fractionation, and mobility of Pb, Cd, Cu, and Zn in garden and paddy soils from a Pb/Zn mining area. *Environmental Monitoring and Assessment* 168, 215–222.

Li, P., Wang, X.-X., Zhang, T.-L., Zhou, D.-M., He, Y.-Q., 2009a. Distribution and accumulation of copper and cadmium in soil–rice system as affected by soil amendments. *Water, Air, and Soil Pollution* 196, 29–40.

Li, S., Liu, R., Wang, H., Shan, H., 2009b. Accumulations of cadmium, zinc, and copper by rice plants grown on soils amended with composted pig manure. *Communications in Soil Science and Plant Analysis* 40, 1889–1905.

Lin, Y.-T., Weng, C.-H., Lee, S.-Y., 2012. Spatial distribution of heavy metals in contaminated agricultural soils exemplified by Cr, Cu, and Zn. *Journal of Environmental Engineering* 138, 299–306.

Lindsay, W.L., Norvell, W.A., 1978. Development of a DTPA soil test for zinc, iron, manganese, and copper. *Soil Science Society of America Journal* 42, 421–428.

Liu, J., Qian, M., Cai, G., Zhu, Q., Wong, M.H., 2007. Variations between rice cultivars in root secretion of organic acids and the relationship with plant cadmium uptake. *Environmental Geochemistry and Health* 29, 189–195.

Liu, R.L., Li, S.T., Wang, X.B., Wang, M., 2005. Contents of heavy metal in commercial organic fertilizers and organic wastes. *Journal of Agro-environmental Science* 24, 392–397.

Manceau, A., Tommaseo, C., Rihs, S., Geoffroy, N., Chateigner, D., Schlegel, M., Tisserand, D., Marcus, M.A., Tamura, N., Chen, Z.-S., 2005. Natural speciation of Mn, Ni, and Zn at the micrometer scale in a clayey paddy soil using X-ray fluorescence, absorption, and diffraction. *Geochimica et Cosmochimica Acta* 69, 4007–4034.

Mandal, B., Hazra, G.C., Mandal, L.N., 2000. Soil management influences on zinc desorption for rice and maize nutrition. *Soil Science Society of America Journal* 64, 1699–1705.

McBride, M.B., 1980. Chemisorption of Cd^{2+} on calcite surfaces. *Soil Science Society of America Journal* 44, 26–28.

McBride, M.B., 1994. *Environmental Chemistry of Soils*. New York City, NY: Oxford University Press.

McLaughlin, M.J., Singh, B.R., 1999. *Cadmium in Soils and Plants*. Dordrecht, the Netherlands: Kluwer Academic Publishers.

McNear, D.H., 2006. *The Plant Soil Interface: Nickel Bioavailability and the Mechanisms of Plant Hyperaccumulation*. Newark, DE: Plant and Soil Department, University of Delaware.

Mench, M., Martin, E., 1991. Mobilization of cadmium and other metals from two soils by root exudates of Zea mays L., *Nicotiana tabacum* L. and *Nicotiana rustica* L. *Plant and Soil* 132, 187–196.

Mortvedt, J.J., 1996. Heavy metal contaminants in inorganic and organic fertilizers. *Fertilizer Research* 43, 55–61.

Naidu, R., Kookana, R.S., Sumner, M.E., Harter, R.D., Tiller, K.G., 1997. Cadmium sorption and transport in variable charge soils: A review. *Journal of Environmental Quality* 26, 602–617.

Nan, Z., Li, J., Zhang, J., Cheng, G., 2002. Cadmium and zinc interactions and their transfer in soil-crop system under actual field conditions. *Science of the Total Environment* 285, 187–195.

Nguyen Ngoc, M., Dultz, S., Kasbohm, J., 2009. Simulation of retention and transport of copper, lead and zinc in a paddy soil of the Red River Delta, Vietnam. *Agriculture, Ecosystems and Environment* 129, 8–16.

Nicholson, F.A., Chambers, B.J., Williams, J.R., Unwin, R.J., 1999. Heavy metal contents of livestock feeds and animal manures in England and Wales. *Bioresource Technology* 70, 23–31.

Nigam, R., Srivastava, S., Prakash, S., Srivastava, M.M., 2001. Cadmium mobilisation and plant availability—The impact of organic acids commonly exuded from roots. *Plant and Soil* 230, 107–113.

Paijitprapapon, A., Udomtanateera, K., Udomtanateera, O., Chariyapisuthi, V., Chengsuksawat, T., 2006. Environmental investigation for causality and remediation of cadmium contaminated soils in Tak Province, Thailand. *Chinese Journal of Geochemistry* 25, 253–254.

Papadopoulos, P., Rowell, D.L., 1989. The reactions of copper and zinc with calcium carbonate surfaces. *European Journal of Soil Science* 40, 39–48.

Pehlivan, E., Arslan, G., 2006. Uptake of metal ions on humic acids. *Energy Sources, Part A: Recovery, Utilization, and Environmental Effects* 28, 1099–1112.

Ponnamperuma, F.N., 1972. The chemistry of submerged soils. *Advances in Agronomy* 24, 29–96.

Reddy, M.V., Satpathy, D., Dhiviya, K.S., 2013. Assessment of heavy metals (Cd and Pb) and micronutrients (Cu, Mn, and Zn) of paddy (*Oryza sativa* L.) field surface soil and water in a predominantly paddy-cultivated area at Puducherry (Pondicherry, India), and effects of the agricultural runoff on the elemental concentrations of a receiving rivulet. *Environmental Monitoring and Assessment* 185, 6693–6704.

Roberts, D.R., Scheinost, A.C., Sparks, D.L., 2002. Zinc speciation in a smelter-contaminated soil profile using bulk and microspectroscopic techniques. *Environmental Science and Technology* 36, 1742–1750.

Rogan, N., Serafimovski, T., Dolenec, M., Tasev, G., Dolenec, T., 2009. Heavy metal contamination of paddy soils and rice (*Oryza sativa* L.) from Kocani Field (Macedonia). *Environmental Geochemistry and Health* 31, 439–451.

Römkens, P.F., Guo, H.-Y., Chu, C.-L., Liu, T.-S., Chiang, C.-F., Koopmans, G.F., 2009. Characterization of soil heavy metal pools in paddy fields in Taiwan: Chemical extraction and solid-solution partitioning. *Journal of Soils and Sediments* 9, 216–228.

Rose, A.W., Hawkes, H.E., Webb, J.S., 1979. *Geochemistry in Mineral Exploration* (2nd Edition). London, UK: Academic Press.

Salbu, B., Krekling, T., 1998. Characterisation of radioactive particles in the environment. *Analyst* 123, 843–850.

Schnabel, R.R., Fitting, D.J., 1988. Analysis of chemical kinetics data from dilute, dispersed, well-mixed flow-through systems. *Soil Science Society of America Journal* 52, 1270–1273.

Schnitzer, M., 1986. Binding of humic substances by soil mineral colloids. In: Huang, P.M., Schnitzer, M. (eds.). *Interactions of Soil Minerals with Natural Organics and Microbes*. Madison, WI: Soil Science Society of America, pp. 78–102.

Seyfried, M.S., Sparks, D.L., Bar-Tal, A., Feigenbaum, S., 1989. Kinetics of Ca–Mg exchange on soil using a stirred-flow reaction chamber. *Soil Science Society of America Journal* 52, 1270–1273.

Shi, Z., 2006. *Kinetics of Trace Metals Sorption on and Desorption from Soils Developing Predictive Models*. Newark, DE: Department of Civil and Environmental Engineering, University of Delaware.

Simmons, R.W., Chaney, R.L., Angle, J.S., Kruatrachue, M., Klinphoklap, S., Reeves, R.D., Bellamy, P., 2014. Towards practical Cd phytoextraction with *Noccaea caerulescens*. *International Journal of Phytoremediation*. DOI:10.1080/15226514.2013.876961.

Simmons, R.W., Pongsakul, P., Saiyasitpanich, D., Klinphoklap, S., 2005. Elevated levels of cadmium and zinc in paddy soils and elevated levels of cadmium in rice grain downstream of a zinc mineralized area in Thailand: Implications for public health. *Environmental Geochemistry and Health* 27, 501–511.

Singh, J., Upadhyay, S.K., Pathak, R.K., Gupta, V., 2011. Accumulation of heavy metals in soil and paddy crop (*Oryza sativa*), irrigated with water of Ramgarh Lake, Gorakhpur, UP, India. *Toxicological and Environmental Chemistry* 93, 462–473.

Søvik, M.-L., Larssen, T., Vogt, R.D., Wibetoe, G., Feng, X., 2011. Potentially harmful elements in rice paddy fields in mercury hot spots in Guizhou, China. *Applied Geochemistry* 26, 167–173.

Sparks, D.L., 1989. *Kinetics of Soil Chemical Processes*. San Diego, CA: Academic Press, Inc.

Sparks, D.L., 1996. *Methods of Soil Analysis*. Part 3, Chemical methods. Soil Science Society of America, Madison, WI: American Society of Agronomy.

Sparks, D.L., 2003. *Environmental Soil Chemistry*. San Diego, CA: Academic Press.

Sriprachote, A., Kanyawongha, P., Ochiai, K., Matoh, T., 2012. Current situation of cadmium-polluted paddy soil, rice and soybean in the Mae Sot District, Tak Province, Thailand. *Soil Science and Plant Nutrition* 58, 349–359.

Strawn, D.G., Sparks, D.L., 2000. Effects of soil organic matter on the kinetics and mechanisms of Pb(II) sorption and desorption in soil. *Soil Science Society of America Journal* 64, 144–156.

Sukreeyapongse, O., Holm, P.E., Strobel, B.W., Panichsakpatana, S., Magid, J., Hansen, H.C., 2002. pH-dependent release of cadmium, copper, and lead from natural and sludge-amended soils. *Journal of Environmental Quality* 31, 1901–1909.

Swaddiwudhipong, W., Limpatanachote, P., Mahasakpan, P., Krintratun, S., Punta, B., Funkhiew, T., 2012. Progress in cadmium-related health effects in persons with high environmental exposure in northwestern Thailand: A five-year follow-up. *Environmental Research* 112, 194–198.

Takamatsu, R., Asakura, K., Chun, W.-J., Miyazaki, T., Nakano, M., 2006. EXAFS studies about the sorption of cadmium ions on montmorillonite. *Chemistry Letters* 35, 224–225.

Tyler, G., 1978. Leaching rates of heavy metal ions in forest soil. *Water, Air, and Soil Pollution* 9, 137–148.

van Gestel, C.A., van Diepen, A.M., 1997. The influence of soil moisture content on the bioavailability and toxicity of cadmium for *Folsomia candida* Willem (Collembola: Isotomidae). *Ecotoxicology and Environmental Safety* 36, 123–132.

Vasconcelos, I.F., Haack, E.A., Maurice, P.A., Bunker, B.A., 2008. EXAFS analysis of cadmium(II) adsorption to kaolinite. *Chemical Geology* 249, 237–249.

Wakatsuki, T., 1997. Paddy (Sawah) soils. In: Kyuma, K. (ed.). *New Soil Science*. Tokyo, Japan: Asakura Publishing, Co. (in Japanese).

Wallace, I., 1998. Delivering information to rice scientists around the world. *Information Development* 14, 198–202.

Williams, P.N., Lei, M., Sun, G., Huang, Q., Lu, Y., Deacon, C., Meharg, A.A., Zhu, Y.-G., 2009. Occurrence and partitioning of cadmium, arsenic and lead in mine impacted paddy rice: Hunan, China. *Environmental Science and Technology* 43, 637–642.

Wu, L.H., Luo, Y.M., Christie, P., Wong, M.H., 2003. Effects of EDTA and low molecular weight organic acids on soil solution properties of a heavy metal polluted soil. *Chemosphere* 50, 819–822.

Xue, H.-B., Sigg, L., 1998. Cadmium speciation and complexation by natural organic ligands in fresh water. *Analytica Chimica Acta* 363, 249–259.

Yang, Q.-W., Lan, C.-Y., Shu, W.-S., 2008. Copper and zinc in a paddy field and their potential ecological impacts affected by wastewater from a lead/zinc mine, P. R. China. *Environmental Monitoring and Assessment* 147, 65–73.

Yin, Y., Allen, H.E., Huang, C.P., Sparks, D.L., Sanders, P.F., 1997. Kinetics of mercury(II) adsorption and desorption on soil. *Environmental Science and Technology* 31, 469–503.

Yu, K., Böhme, F., Rinklebe, J., Neue, H.-U., DeLaune, R.D., 2007. Major biogeochemical processes in soils—A microcosm incubation from reducing to oxidizing conditions. *Soil Science Society of America Journal* 71, 1406–1417.

Zeien, H., Brümmer, G.W., 1989. Chemische extraction zur bestimmung von schwermetallbindungs-formen in Böden. Mitteilgn. *Dtsch Bodenkundl Gesellsch* 59, 505–510 (in German).

Zhang, M., Liu, Z., Wang, H., 2010. Use of single extraction method to predict bioavailability of heavy metals in polluted soils to rice. *Communications in Soil Science and Plant Analysis* 41, 820–831.

Zhao, X., Jiang, T., Du, B., 2014. Effect of organic matter and calcium carbonate on behaviors of cadmium adsorption–desorption on/from purple paddy soils. *Chemosphere* 99, 41–48.

6 Analysis and Fate of Metal-Based Engineered Nanoparticles in Aquatic Environments, Wetlands, and Floodplain Soils

Frederik Van Koetsem, Jörg Rinklebe, and Gijs Du Laing

CONTENTS

6.1 INTRODUCTION

Nanomaterials (NMs) are generally defined as particulate objects having a size between 1 and 100 nm in at least one dimension. These materials include nanofilms or nanoplates (one dimension on the nanoscale), nanofibers or nanotubes (two dimensions on the nanoscale), and nanoparticles (three dimensions on the nanoscale) (Handy et al., 2008; Batley et al., 2013).

Nanoparticles occur in aquatic, terrestrial, and atmospheric environments, from both natural sources and human activities, and can by definition be considered as ultrafine particles (airborne) or a subset of colloidal particles (1 nm–1 μm in aquatic or terrestrial systems). They can exist in different forms (e.g., as single, aggregated, or agglomerated particles) and have various morphologies (e.g., spherical, tubular, or irregular shapes), coatings, and surface functionalities (Nowack and Bucheli, 2007; Klaine et al., 2008). Depending on their origin, nanoparticles can be categorized further into natural (i.e., geogenic, biogenic, or atmospheric processes), anthropogenic (i.e., unintentionally produced), and engineered nanoparticles (ENPs) (i.e., intentionally manufactured), and be divided according to their chemical composition into organic and inorganic nanoparticles (Figure 6.1).

Due to their small size, composition, structure, and surface characteristics, ENPs often exhibit unique physicochemical properties and reactivities that are not present at a larger scale. Such novel properties have led to a rapid increase in potential applications in various areas of the economy, including textiles, electronics, optics, cosmetics, medical devices, food packaging, catalysts, fuel cells, (waste)water treatment, and environmental remediation (Handy et al., 2008; Navarro et al., 2008). Since the beginning of the twenty-first century, the annual production of ENPs has increased exponentially from about 400 tons to an expected 58,000 tons during 2011–2015. Accordingly, the global investment in the nanomaterial industry has grown from \$10 billion in 2005 to an estimated \$1 trillion by 2015 (Navarro et al., 2008; Sharma, 2009a).

Metal-based ENPs are among the most widely used nanomaterials to date (Mudunkotuwa and Grassian, 2011; Weinberg et al., 2011). Au and Ag ENPs have found uses in biomedical applications for cancer diagnosis and therapy, as catalysts, or in the pharmaceutical sector (Ju-Nam and Lead, 2008; Sharma et al., 2009b). Ag ENPs have been demonstrated to possess antimicrobial properties, which have led to their use in many applications, including textiles, food packaging, refrigerator coatings, and cosmetics, and they are nowadays considered the largest and fastest growing class of ENPs in product applications (Bradford et al., 2009; Fabrega et al., 2011; Ferreira da Silva et al., 2011; Sharma et al., 2014). Nanoscale zero-valent iron has been used to degrade organic and inorganic contaminants such as halogenated hydrocarbons, pesticides, and heavy metals in groundwater and soil, whereas TiO_2 is the most commonly used photocatalyst for the removal of nitrates or nonbiodegradable organic compounds during (waste)water treatment (Theron et al., 2008; Sharma et al., 2009a; Grieger et al., 2010; Sánchez et al., 2011). Other commonly used ENPs include metal oxides (e.g., CeO_2, CuO, Fe_2O_3, SnO_2, and ZnO) and chalcogenides (e.g., CdS, CdSe,

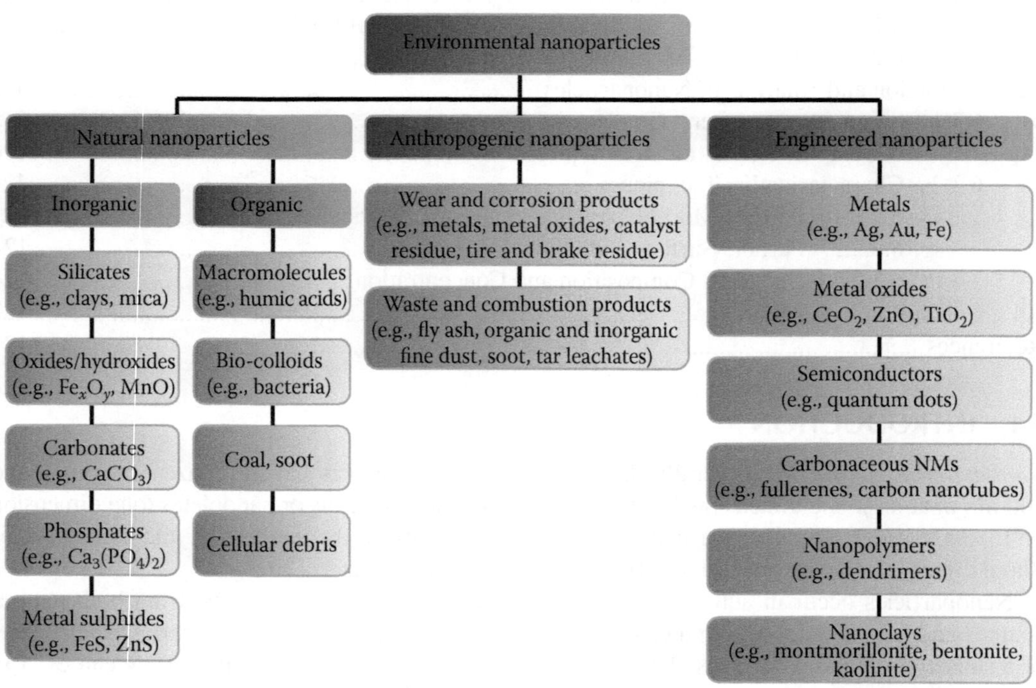

FIGURE 6.1 Concise schematic overview of environmental nanoparticles categorized according to origin and composition. (After Christian, P. et al., *Ecotoxicology*, 17, 326–343, 2008.)

CdTe, and ZnS) (Ju-Nam and Lead, 2008; Lin et al., 2010; Bhatt and Tripathi, 2011; Mudunkotuwa and Grassian, 2011).

The increasing production and widespread use of ENPs will inevitably result in their discharge into the environment, where their novel characteristics, such as high-specific surface area and abundant reactive sites on the particle surface, as well as their high mobility and bioavailability due to their small size, could lead to unexpected health and/or environmental hazards (Navarro et al., 2008). It is, therefore, important to gain extensive knowledge on the environmental behavior and fate of ENPs on their release into the environment, in order to assess their potential risks (Klaine et al., 2008). The next few sections provide an overview on the fate and behavior of ENPs in the aquatic environment, and the most commonly used analytical methodologies that can be applied to assess the environmental occurrence, behavior, and fate of ENPs.

6.2 ENVIRONMENTAL BEHAVIOR AND FATE OF ENGINEERED NANOPARTICLES

6.2.1 Potential Pathways into the Environment

There is a scientific consensus that the manufacturing, transportation, use, and disposal of ENPs will result in their entry into the atmosphere, water bodies, sediment, soil, and biota (Lowry et al., 2010; Nowack et al., 2012). Although ENPs are not yet regulated, they are already included in lists of emerging contaminants, and similar to many pollutants, water bodies, and soils, they can receive ENPs from point sources, such as production facilities, landfills, and (waste)water treatment plants, or from nonpoint sources, including wear from materials containing ENPs. In addition, deliberate introduction of ENPs into the environment also occurs, for example, on their application in groundwater or soil remediation, or the use of ENP-containing agrochemicals (Scown et al., 2010; Ferreira da Silva et al., 2011; Gottschalk and Nowack, 2011). Due to the diverse input routes, aquatic systems are highly susceptible to contamination with ENPs. However, very little data on the actual emissions of ENPs from products and their releases into the environment are available (Nowack et al., 2012). Besides quantification of the emissions, it is also important to investigate in which form ENPs are released (e.g., as single nanoparticles, aggregates, or agglomerates, or embedded in a matrix) (Gottschalk and Nowack, 2011). For example, Benn and Westerhoff (2008) have demonstrated the leaching of both colloidal (100–200 nm) and ionic Ag from socks on repeated washing. Benn et al. (2010) reported the emission of Ag ENPs from several consumer products, including a t-shirt, shampoo, and toothpaste. Kaegi et al. (2008a) provided evidence for the release of TiO_2 ENPs into the aquatic environment, due to natural weathering of painted house façades. In addition, Limbach et al. (2008) illustrated that a small yet significant fraction (6 wt%) of CeO_2 ENPs fed to a model wastewater treatment plant could be retrieved in the effluent at concentrations from 2 to 5 mg/L.

6.2.2 Processes and Factors Affecting Behavior and Fate in Aquatic Systems

On their release into the environment, ENPs can undergo a number of potential physical, chemical, and even biological transformation processes that govern their behavior and fate, including aggregation, sedimentation, dissolution, interactions with colloidal components, sorption to particulates and other solid surfaces, biological degradation, abiotic degradation (e.g., photolysis), and oxidation and/or reduction. These transformation processes depend on both the characteristics of the nanoparticles (e.g., size, chemical composition, and surface properties) and the prevailing conditions of the receiving medium (e.g., solution pH, ionic strength (IS), natural organic matter [NOM] content, and clay content). Aggregation and dissolution are considered the two most important contributors to the environmental impacts of ENPs in aquatic systems (Hotze et al., 2010; Lowry et al., 2010; Scown et al., 2010; Mudunkotuwa and Grassian, 2011; Nowack et al., 2012; Batley et al., 2013). Potential entry routes as well as the occurring transformation processes of ENPs in the environment are shown in Figure 6.2.

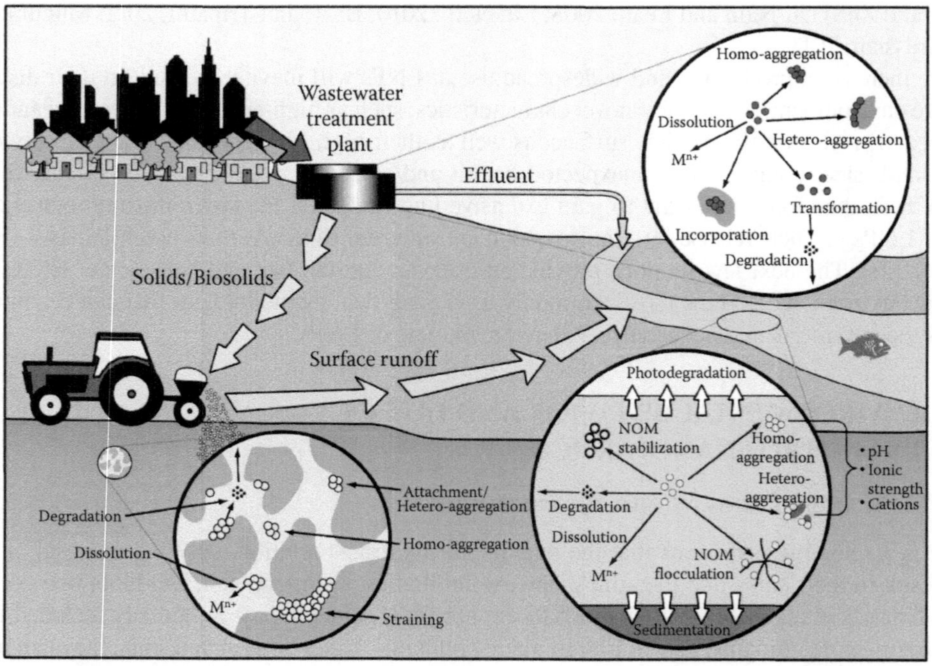

FIGURE 6.2 Potential entry routes and transformation processes of ENPs in the environment. (Reprinted from Batley, G.E. et al., *Acc. Chem. Res.*, 46, 854–862, 2013. With permission.)

6.2.2.1 Aggregation–Sedimentation

Stability in suspension is a key factor that controls the transport and ultimate fate of ENPs in aquatic systems. Aggregation and subsequent sedimentation restrict their mobility and bioavailability, whereas well-dispersed ENPs can be transported widely, thereby increasing the risk for potential harmful interactions with organisms. Aggregation also limits the reactivity of ENPs by reducing the overall specific surface area and interfacial free energy (Boxall et al., 2007; Lin et al., 2010).

Aggregation occurs when particles collide due to physical processes, and thermodynamic interactions enable particle–particle attachment to take place. These particle–particle collisions originate from three fundamental mechanisms: (1) Brownian diffusion, leading to perikinetic aggregation; (2) particles moving at different velocities in a shear flow, resulting in orthokinetic aggregation; and (3) particles of different sizes or densities undergoing differential settling. Considering particles smaller than 100 nm and a fluid at rest, the long-range forces between individual particles, and hence the collisions between particles, are controlled by Brownian motion, and mechanisms (2) and (3) can be neglected. Contact between particles can result in attachment or repulsion, which is controlled by short-range thermodynamic interactions. Such interactions can be understood in the context of Derjaguin–Landau–Verwey–Overbeek (DLVO) theory (Handy et al., 2008b; Hotze et al., 2010).

According to classical DLVO theory, attachment between spherical particles is determined by the sum of attractive Van der Waals forces (V_{VdW}) and repulsive electrostatic double-layer forces (V_{EDL}) (Mudunkotuwa and Grassian, 2011). The total interaction energy (V_T) is therefore

$$V_{\text{T}} = V_{\text{VdW}} + V_{\text{EDL}} \tag{6.1}$$

Attractive Van der Waals forces are given by

$$V_{\text{VdW}} = -\frac{A_{\text{H}}}{6}\left[\frac{2a_1a_2}{R^2-(a_1+a_2)^2}+\frac{2a_1a_2}{R^2-(a_1-a_2)^2}+\ln\frac{R^2-(a_1+a_2)^2}{R^2-(a_1-a_2)^2}\right] \tag{6.2}$$

where V_{VdW} [J] is the Van der Waals attraction, A_H [J] is the overall Hamaker constant (taking particle composition into account), a_1 and a_2 [m] are the particle radii, and R [m] is the distance between the centers of adjacent particles.

Repulsive electrostatic forces, which depend on the particle radius (a) and the thickness of the electrical double layer ($1/\kappa$), are given by Equations 6.3 and 6.4, in the case of $\kappa a > 5$ or $\kappa a < 5$, respectively:

$$V_{EDL} = 4\pi\varepsilon\psi_0^2 \frac{a_1 a_2}{a_1 + a_2} \ln\left[1 + \exp(-\kappa x)\right] \tag{6.3}$$

$$V_{EDL} = 4\pi\varepsilon Y_1 Y_2 a_1 a_2 \left(\frac{k_B T}{e}\right)^2 \frac{\exp(-\kappa x)}{a_1 + a_2 + x} \tag{6.4}$$

with

$$Y_i = \frac{8 \tan h(e\psi_0 / 4k_B T)}{1 + \sqrt{\left[1 - \left((2\kappa a_i + 1)/(\kappa a_i + 1)^2\right) \tan h^2\left(e\psi_0 / 4k_B T\right)\right]}} \tag{6.5}$$

where V_{EDL} [J] is the electrostatic double-layer repulsion, ε [–] is the relative permittivity of the fluid, a_1 and a_2 [m] are the radii of the particles, ψ_0 [V] is the surface potential, k_B [J/K] is the Boltzmann constant, T [K] is the absolute temperature, e [C] is the elementary charge of an electron, x [m] is the interparticle distance, and κ [1/m] is the inverse Debeye length.

When the attractive forces are larger than the repulsive forces, aggregation will occur. Summation of these forces also demonstrates that particles can have a net attraction in a primary or secondary minimum. The primary minimum implies irreversible aggregation of particles, whereas the secondary minimum indicates reversible aggregation (Hotze et al., 2010). The Stern–Gouy–Chapman electrical double-layer model as well as particle–particle interaction energy profiles are shown in Figure 6.3.

ENPs are most often coated with surface ligands, such as polymers, polyelectrolytes, or NOM, resulting in (electro)steric stabilization of the particles in suspension, which arises from elastic and osmotic contributions. Therefore, extended DLVO (XDLVO) theory has been developed to account for these additional repulsion forces. Both DLVO and XDLVO theories stem from colloid chemistry, and although these theories have been shown to be able to semiquantitatively describe aggregation and deposition of ENPs under laboratory conditions, prediction of the aggregation behavior of ENPs in more complex matrices has been proven to be a lot more challenging. Nevertheless, DLVO and XDLVO theories are important modeling tools that are used to gain insight into the processes and influences of diverse parameters affecting aggregation of ENPs (Hotze et al., 2010; Petosa et al., 2010; Mudunkotuwa and Grassian, 2011). For instance, studies conducted by Elzey and Grassian (2010a), Li et al. (2010), von der Kammer et al. (2010), Bian et al. (2011), and Stebounova et al. (2011) have successfully applied (X)DLVO theories to investigate the aggregation and sedimentation behavior of Ag, TiO_2, and ZnO ENPs as a function of aquatic parameters, including pH, IS, and NOM content.

Two types of aggregation of ENPs can occur in the environment: homo-aggregation and hetero-aggregation. The former refers to the attachment of similar particles, whereas the latter addresses the attachment of dissimilar particles (Hotze et al., 2010). Natural colloids (e.g., clays, iron, and manganese hydrous oxides, dissolved organic matter, fibrillary colloids (exopolymers), and micro-organisms) are ubiquitous in the aquatic environment, and their concentrations (mg/L) are typically several orders of magnitude higher than the expected environmental concentrations of ENPs (μg/L or less). Hence, hetero-aggregation is likely to be the dominant process controlling the fate of most ENPs, and the attachment of ENPs to natural colloids followed by sedimentation is considered the most important removal process of ENPs from natural water bodies. Still, most studies have been focused on examining homo-aggregation or are performed by using artificial suspension media instead of natural aquatic matrices (Lin et al., 2010; Quik et al., 2010, 2012; Batley et al., 2013).

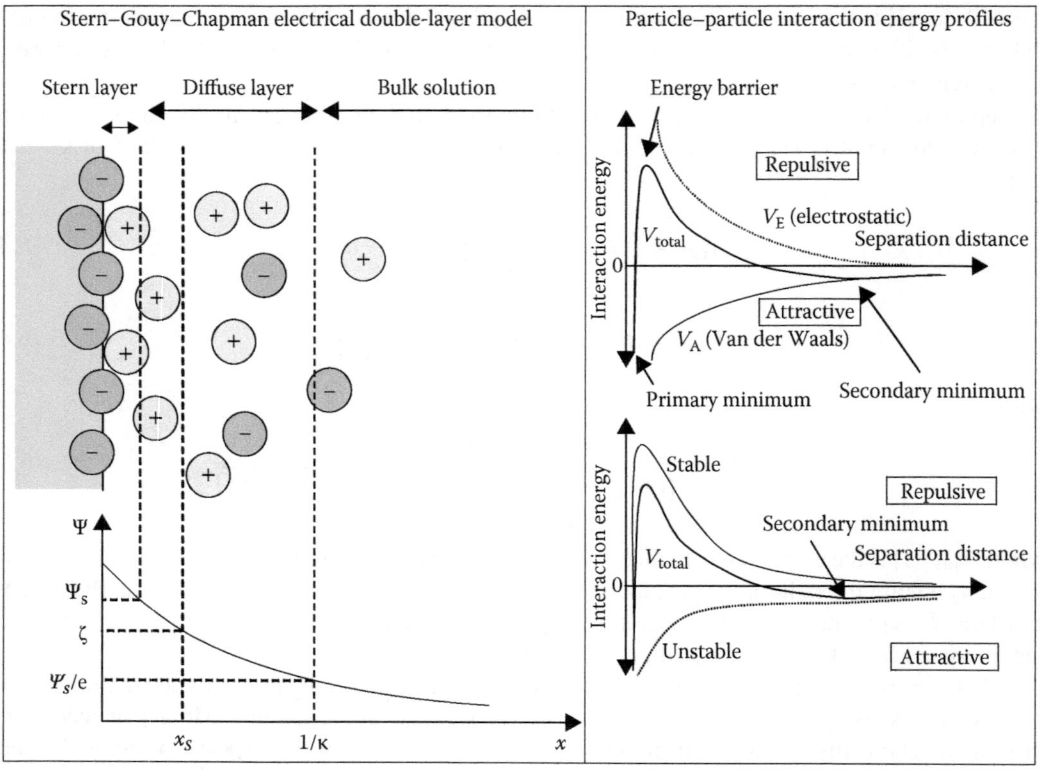

FIGURE 6.3 Schematic diagram showing (left panel) the electrical double layer (EDL) on the surface of a particle, with the different potentials to be considered [i.e., the electrostatic potential (ψ), the potential at the Stern layer (ψ_s), the zeta potential (ζ), and the potential at the diffuse layer (ψ_s/e)], the Debeye length ($1/\kappa$) at which the potential has fallen to a value of $1/e$ of the Stern potential, and the shear plane (X_s) where ions and molecules are mobile. An increased ionic strength will effectively compress the EDL, thereby influencing the Debeye length (e.g., from a couple of nm to nearly 1 μm in the case of seawater or ultrapure water, respectively). The right panel shows a simplified graph summarizing the DLVO interaction energies and the resulting sum function. The upper graph shows a situation where repulsive electrostatic forces are working against attractive Van der Waals forces, and an activation energy is required to achieve particle–particle attachment in either the secondary or primary minimum. The bottom graph shows three possible situations: a fully stabilized system, a system having a secondary and primary minimum, and a fully destabilized system. (With kind permission from Springer Science+Business Media: Ecotoxicology, The ecotoxicology and chemistry of manufactured nanoparticles, 17, 2008, 287–314, Handy, R.D. et al.)

IS is a measure for the concentration of dissolved salts in solution, and it is an important parameter in regard to colloidal stability. An increase in IS results in compression of the EDL, which leads to a decrease in the repulsion forces and potentially to aggregation of ENPs. The ionic concentration at which the repulsive energy barrier is completely screened and rapid aggregation occurs is called the *critical coagulation concentration (CCC)* (Hotze et al., 2010). In addition, according to the Schulze–Hardy rule, the valence of the counterions also has an important impact on colloidal stability (Lin et al., 2010). ENPs will, therefore, have a higher tendency to aggregate and settle out of solution in seawater or brackish water, than in freshwater, as was demonstrated in a study by Keller et al. (2010). French et al. (2009) have also shown that aggregation of TiO_2 ENPs is stimulated by increasing IS and cation valence, whereas Li et al. (2010) identified CCC values of 30, 40, and 2 mM for bare 82 ± 1.3 nm Ag ENPs in the case of $NaNO_3$, NaCl, and $CaCl_2$, respectively.

Solution pH also plays a crucial role in the aggregation of ENPs, by affecting the surface charge of particles or rather their zeta (ζ) potential, which is the potential difference between the dispersion medium and the stationary layer of fluid attached to the particle (see also Figure 6.3) and which actually determines the interparticle forces (Handy et al., 2008b; Lin et al., 2010). Colloids and ENPs having a ζ-potential between −30 and +30 mV are generally considered unstable in suspension and tend to aggregate. When the pH of an aquatic system is at the point of zero charge (PZC) or isoelectric point (IEP), the colloidal system exhibits minimum stability (Lin et al., 2010). A pH below the PZC results in a positively charged particle surface, and the ζ-potential will increase further with a decrease in pH; however, a pH above the PZC results in a negatively charged particle surface, and a further increase in the pH will lead to a decrease in ζ-potential. A high-absolute ζ-potential, therefore, implies the stability of ENPs in dispersion. Because the PZC differs between different ENPs, their stability in suspension also varies at a certain solution pH (Lin et al., 2010). Among numerous studies, Ghosh et al. (2008), for instance, examined the colloidal behavior of Al_2O_3 ENPs as a function of pH, showing that the surface charge of the particles decreased with an increase in pH, and reached a PZC at pH 7.9. The effect of pH and NOM on the ζ-potential of CeO_2 ENPs that are dispersed in deionized water and algae medium is shown in Figure 6.4.

As mentioned earlier, ENPs are often coated with synthetic surface ligands to improve their colloidal stability and to overcome rapid aggregation and sedimentation. Limbach et al. (2008) have demonstrated that different surface coatings have a major effect on the stability of CeO_2 ENPs in solutions of varying NaCl concentration. Similar results were obtained in a study by Phenrat et al. (2007), who compared the sedimentation of coated Fe ENPs with bare Fe ENPs, and in a study by El Badawy et al. (2010), who examined the impact of capping agents and environmental conditions (e.g., pH, IS, and background electrolytes) on the surface charge and aggregation potential of Ag ENPs. In addition to stabilization by synthetic ligands, ENPs can also be coated and stabilized by NOM on their release into the environment. NOM has been shown to sorb onto various metal oxide ENPs (e.g., TiO_2, Al_2O_3, and ZnO), resulting in a decrease in particle ζ-potential and suggesting a higher stability in suspension (Scown et al., 2010). For instance, over a 12-day period, up to 40% of CeO_2 ENPs initially spiked to artificial fresh water have been shown to remain stable in suspension in the presence of NOM (Quik et al., 2010). Furthermore, von der Kammer et al. (2010) found that about 70%–90% of initially added TiO_2 ENPs remained in suspension after 15 h of settling in the presence of NOM. In contrast, ENPs can also be bridged by NOM, resulting in destabilization, aggregation, and settling. Whether attachment results in either stabilization or destabilization depends on various factors, such as type and concentration of NOM, as well as on solution chemistry (Hotze et al., 2010). Baalousha et al. (2008), for example, demonstrated that the commencement of aggregation of Fe_2O_3 ENPs shifted from a pH of 5–6 to a pH of 4–5 on addition of Suwannee River humic acid (SRHA), and that increasingly larger aggregates were formed with an increase in SRHA concentration (0–25 mg/L) and pH. In a different study, it was also shown that divalent

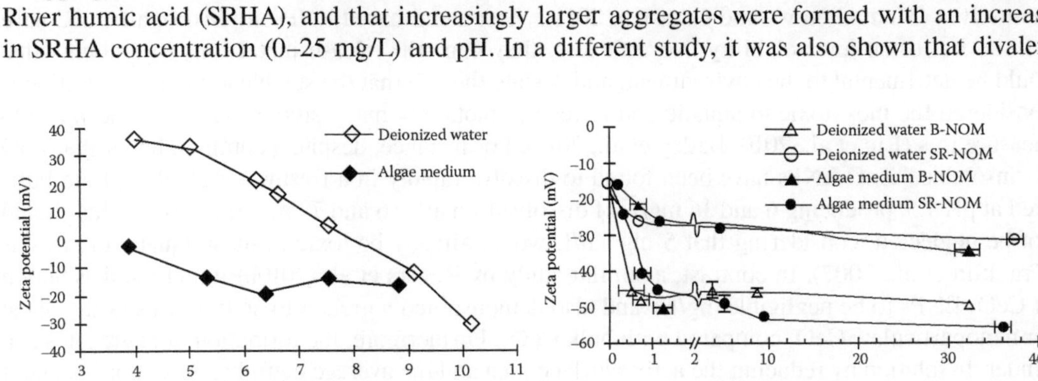

FIGURE 6.4 Effect of pH (left graph), and of various concentrations of Bilhain (B) and Suwannee River (SR) NOM (right graph), on the ζ-potential of CeO_2 ENPs dispersed in deionized water and algae medium. (Reprinted from *Chemosphere* 81, Quik, J.T.K. et al., Effect of natural organic matter on cerium dioxide nanoparticles settling in model fresh water, 711–715, Copyright 2010 with permission from Elsevier.)

cations (0.04–0.06 M Ca^{2+}) could induce aggregation of NOM-coated metal oxide (ZnO, NiO, TiO_2, Fe_2O_3, and SiO_2) ENPs (Zhang et al., 2009).

Contaminants in natural aquatic systems are mostly bound to particle surfaces or form complexes with, for instance, NOM. They can be sorbed, co-precipitated, or trapped on aggregation of ENPs, and, thus, ENPs could play an important role in the solid–liquid partitioning of pollutants. Interactions between contaminants and ENPs depend on characteristics, such as particle size, structure and composition, and solution chemistry (Christian et al., 2008). Stabilization of ENPs in the environment induces a higher mobility, thereby not only enhancing the risks directly associated with the ENPs but also increasing the potential exposure to associated contaminants, as ENPs could potentially act as cotransporters of other pollutants (Scown et al., 2010).

6.2.2.2 Dissolution

Essentially, the dissolution of ENPs can be considered the transformation of a chemical compound in its nanoparticulate form to individual ions or molecules that are soluble in water (Quik et al., 2011; Nowack et al., 2012). It is a heterogeneous process that takes place on the solid–liquid phase boundaries in two steps: (1) a reaction at the solid–liquid interface and (2) a transfer of the dissolved matter away from the reaction site (Dokoumetzidis et al., 2008). According to Liu et al. (2009), a modified Kelvin equation (Equation 6.6) (also known as the Ostwald–Freundlich equation) has commonly been used to describe the thermodynamic size-dependent dissolution of nanoparticles:

$$\frac{S}{S_0} = \exp\left(\frac{2\gamma V}{RTr}\right) \tag{6.6}$$

where S [mol/kg] is the solubility of a spherical nanoparticle with radius r [m], S_0 [mol/kg] is the solubility of the bulk material, γ [J/m^2] is the surface free energy, V [m^3/mol] is the molecular volume, R [J/mol/K] is the gas constant, and T [K] is the absolute temperature.

Following Equation 6.6, the solubility increases exponentially with a decrease in particle size, and it is clear that surface energy and particle size strongly influence nanoparticle dissolution. However, due to a lack of detailed knowledge on the surface energy, its size dependency, and its dependency on the details of the surface structure, deviations from predictions that are rendered by Equation 6.6 can be expected (Mudunkotuwa and Grassian, 2011).

The dissolution behavior of ENPs is important in terms of understanding their overall stability in the aquatic environment, and especially their potential (eco)toxicity, as the latter has generally been attributed to originate from the release of ionic species rather than being induced by the ENPs themselves (Mudunkotuwa and Grassian, 2011). Most metal-based ENPs are hydrophilic and have a finite but often low solubility. Nevertheless, they can still release metal ions in amounts that could be detrimental to the environment, and despite the fact that the soluble ionic metal fraction is considered the most toxic to aquatic and terrestrial biota, not many studies have been designed to measure this (Lin et al., 2010; Batley et al., 2013). For instance, despite a common belief that ZnO is "insoluble," ZnO ENPs have been found to dissolve rapidly in a freshwater algal medium buffered at pH 7.5, producing 6 and 16 mg/L of dissolved Zn after 6 and 72 h, respectively. This should not be neglected, considering that 5 mg Zn/L would already be toxic to most aquatic organisms (Franklin et al., 2007). In contrast, a similar study by Rogers et al. (2010b) found the dissolution of CeO_2 ENPs to be negligible (ng/L), and also demonstrated a greater toxicity to freshwater algae for nanoparticulate CeO_2 compared with bulk CeO_2. Furthermore, the formation of aggregates can hinder dissolution by reducing the active surface area and the average equilibrium solubility, and it can lead to kinetic hindrance of the diffusion process (Lin et al., 2010).

In addition to nanoparticle properties, including particle size, morphology, and composition, dissolution of ENPs is also greatly affected by environmental parameters, such as pH, temperature, and NOM content (Lin et al., 2010). The dissolution of Cu ENPs in low pH aqueous suspension was examined by Elzey and Grassian (2010b), and they postulated that copper oxide ($CuO{:}Cu_2O$) layers present on the surface of the nanoparticles likely facilitated nanoparticle dissolution. Baalousha et al. (2008) have reported that the dissolution of Fe_2O_3 ENPs was negligible at pH values above 4, whereas 35% of total iron from the Fe_2O_3 ENPs was found in the dissolved phase (<1 kDa) at low pH. Furthermore, Bian et al. (2011) found that humic acid can increase the release of Zn^{2+} ions from ZnO ENPs at a pH greater than 9, whereas no effect of the presence of humic acid on dissolution was seen at a pH less than 6. According to Liu and Hurt (2010), the ionic release from Ag ENPs is a cooperative oxidation process requiring both protons and dissolved oxygen. The ionic release was found to increase with temperature (0–37°C), and decrease with an increase in pH, or the addition of NOM. In addition, sea salts appear to have only a minor effect on the release of dissolved silver. Kittler et al. (2010) demonstrated that in addition to temperature, surface functionalization of ENPs could also affect nanoparticle dissolution. Considerable differences in the release of Ag^+ ions were observed when examining the rate and degree of dissolution of citrate-stabilized and polyvinylpyrrolidone-stabilized Ag ENPs in ultrapure water over a time scale of several days. In all studies, none of the Ag ENPs dissolved completely; however, the nanoparticles released up to 90 wt% of silver in some cases. Ag ENPs have been shown to be more toxic than their micrometer-sized counterparts, mainly due to the larger release of dissolved silver (Kittler et al., 2010; Batley et al., 2013).

6.2.2.3 Additional Transformation Processes

Once released into the environment, ENPs can undergo additional biotic or abiotic alterations, such as biological transformations, disaggregation, photolysis, hydrolysis, and redox transformations, which can alter the size, shape, surface chemistry, and surface coating of nanoparticles and affect their concentration, persistence, and potential toxicity in the environment (Lowry and Casman, 2009).

Biological uptake and biodegradation are considered important cleansing routes for environmental contaminants, and they can therefore also be of significance for the fate of ENPs. The uptake of ENPs by aquatic organisms generally occurs through normal feeding behavior and/or through water filtration via gills. After their uptake, ENPs could pass through the digestive tract and then be either translocated to other body parts or excreted (Lin et al., 2010). For instance, it has been reported that exposure to 80 nm Cu ENPs resulted in higher Cu content in the gills of zebrafish, and this result could not solely be attributed to dissolved Cu. However, the translocation of Ag ENPs to the brain, heart, yolk, and blood of zebrafish embryos has also been observed (Lin et al., 2010). In addition, microorganisms are capable of oxidizing or reducing ENPs or their coating, for instance, by electron transport in pili of bacteria or through direct reduction by membrane-bound cytochromes, or could induce localized dissolution. However, detailed information regarding such biological transformation processes is scarce, and additional research is required (Lowry and Casman, 2009).

Abiotic redox transformations are also vital processes that affect the environmental fate of ENPs. Redox reactions could not only result in an ionic release in solution but also occur at the surface layer of ENPs, leading to alterations in their crystalline nature (Auffan et al., 2009). Fe ENPs, for instance, have been used for *in situ* remediation and are designed to supply electrons for the reductive dehalogenation of chlorinated solvents or for the reductive sequestration of heavy metals (Lowry and Casman, 2009). It has been reported that Fe ENPs could be oxidized over time (i.e., aging) to form iron oxides, such as magnetite, maghemite, hematite, and goethite, and that the surface oxidation affects the redox activity, aggregation, sedimentation, and toxicity of Fe ENPs (Lin et al., 2010). CdSe quantum dots (QDs) have also been shown to release substantial amounts of Cd^{2+} ions, which was attributed to selenide oxidation (Batley et al., 2013). In addition, redox reactions involving metal-based ENPs could also result in the generation of reactive oxygen species (ROS), leading to oxidative stress toward cellular organisms. Strong cytotoxic effects due to oxidative stress have been demonstrated for some bacteria, humans, and rodent cells (Auffan et al., 2009).

Metal-based nanoparticles and their ionic dissolution products readily react with ionic species (e.g., Cl^-, HS^-, HCO_3^-, $H_2PO_4^-$, S^{2-}, CO_3^{2-}, and HPO_4^{2-}) that are present in aquatic environments, which could result in the formation of rather insoluble solid compounds that might precipitate out of solution. The solubility product constants (K_{sp}) of a selection of metal salts are tabulated in

Table 6.1. It is, for instance, recognized that Ag ENPs are not stable in aquatic systems under most prevailing environmental conditions, but they will dissolve or transform into other species, of which AgCl and Ag_2S seem to be the most relevant inorganic forms (Levard et al., 2012; Thalmann et al., 2014). E_h–pH conditions and solution composition will dictate the speciation of metals in natural waters. Low Cl/Ag ratios typically found in freshwater systems are expected to result in $AgCl_{(s)}$ precipitation, whereas high Cl/Ag ratios occurring in seawater are predicted to lead to the formation of soluble $AgCl_{(aq)}$, $AgCl^{2-}$, $AgCl_3^{2-}$, and $AgCl_4^{3-}$ species under aerobic conditions. However, under anaerobic conditions and in sulfide-rich environments (e.g., anaerobic (waste) water, wetlands, and anoxic sediments), $Ag_2S_{(s)}$ precipitation is expected to predominate, as it is thermodynamically favored compared with AgCl formation (Levard et al., 2012; Thalmann et al., 2014). Sequestration of metals through precipitation of metal sulfides in aqueous and subsurface systems as well as their environmental implications have been well documented (Kim et al., 2010; Lewis, 2010). Sulfidation is, therefore, also considered an important transformation process that affects the environmental behavior, fate, transport, and toxicity of metal-containing ENPs. According to Liu et al. (2011), the sulfidation of Ag ENPs can take place via two reaction mechanisms, depending on the amount of sulfides present in solution. At high sulfide concentrations, nanosilver oxysulfidation occurs by a direct particle–fluid heterogeneous reaction at the particle surface; however, at low

TABLE 6.1
Solubility Product Constants (K_{sp}) at 25°C of a Selection of Inorganic Compounds

Compound	Formula	Reaction	K_{sp} at 25°C
Aluminum hydroxide	$Al(OH)_3$	$AlOH_3 \rightleftharpoons Al^{3+} + 3OH^-$	4.6×10^{-33}
Aluminum phosphate	$AlPO_4$	$AlPO_4 \rightleftharpoons Al^{3+} + PO_4^{3-}$	9.8×10^{-21}
Cadmium hydroxide	$Cd(OH)_2$	$Cd(OH)_2 \rightleftharpoons Cd^{2+} + 2OH^-$	7.2×10^{-15}
Cadmium phosphate	$Cd_3(PO_4)_2$	$Cd_3(PO_4)_2 \rightleftharpoons 3Cd^{2+} + 2PO_4^{3-}$	2.5×10^{-33}
Cadmium sulfide	CdS	$CdS \rightleftharpoons Cd^{2+} + S^{2-}$	1.4×10^{-29}
Copper(I) chloride	CuCl	$CuCl \rightleftharpoons Cu^+ + Cl^-$	1.7×10^{-7}
Copper(I) sulfide	Cu_2S	$Cu_2S \rightleftharpoons 2Cu^+ + S^{2-}$	2.3×10^{-48}
Copper(II) hydroxide	$CuOH_2$	$CuOH_2 \rightleftharpoons Cu^{2+} + 2OH^-$	1.6×10^{-19}
Copper(II) phosphate	$Cu_3(PO_4)_2$	$Cu_3(PO_4)_2 \rightleftharpoons 3Cu^{2+} + 2PO_4^{3-}$	1.4×10^{-37}
Copper(II) sulfide	CuS	$CuS \rightleftharpoons Cu^{2+} + S^{2-}$	1.3×10^{-36}
Iron(II) carbonate	$FeCO_3$	$FeCO_3 \rightleftharpoons Fe^{2+} + CO_3^{2-}$	3.1×10^{-11}
Iron(II) hydroxide	$Fe(OH)_2$	$Fe(OH)_2 \rightleftharpoons Fe^{2+} + 2OH^-$	4.9×10^{-17}
Iron(II) sulfide	FeS	$FeS \rightleftharpoons Fe^{2+} + S^{2-}$	1.6×10^{-19}
Iron(III) hydroxide	$Fe(OH)_3$	$Fe(OH)_3 \rightleftharpoons Fe^{3+} + 3OH^-$	2.8×10^{-39}
Iron(III) phosphate	$FePO_4$	$FePO_4 \rightleftharpoons Fe^{3+} + PO_4^{3-}$	9.9×10^{-16}
Silver carbonate	Ag_2CO_3	$Ag_2CO_3 \rightleftharpoons 2Ag^+ + CO_3^{2-}$	8.5×10^{-12}
Silver chloride	AgCl	$AgCl \rightleftharpoons Ag^+ + Cl^-$	1.8×10^{-10}
Silver phosphate	Ag_3PO_4	$Ag_3PO_4 \rightleftharpoons 3Ag^+ + PO_4^{3-}$	8.9×10^{-17}
Silver sulfide	Ag_2S	$Ag_2S \rightleftharpoons 2Ag^+ + S^{2-}$	5.9×10^{-51}
Tin(II) hydroxide	$Sn(OH)_2$	$Sn(OH)_2 \rightleftharpoons Sn^{2+} + 2OH^-$	5.5×10^{-27}
Tin(II) sulfide	SnS	$SnS \rightleftharpoons Sn^{2+} + S^{2-}$	3.3×10^{-28}
Zinc carbonate	$ZnCO_3$	$ZnCO_3 \rightleftharpoons Zn^{2+} + CO_3^{2-}$	1.5×10^{-10}
Zinc hydroxide	$Zn(OH)_2$	$Zn(OH)_2 \rightleftharpoons Zn^{2+} + 2OH^-$	3.0×10^{-17}
Zinc sulfide	ZnS	$ZnS \rightleftharpoons Zn^{2+} + S^{2-}$	2.9×10^{-25}

Sources: Reger, D.L. et al., *Chemistry: Principles and Practice*, 3rd edition, Brooks/Cole, Cengage Learning, Inc., Belmont, CA, pp. A.17–A.18, 2010; Levard, C. et al., *Environ. Sci. Technol.*, 46, 6900–6914, 2012. With permission.

sulfide concentrations, sulfidation proceeds indirectly via oxidative dissolution of Ag to Ag^+ ions followed by precipitation of Ag_2S particles. Metal sulfides are generally considered very stable compounds (see also Table 6.1) (Lewis, 2010). For instance, the dissolution of Ag_2S on reoxidation has been reported to be negligible over a period of months (Thalmann et al., 2014). Moreover, even partial sulfidation of Ag ENPs to Ag/Ag_2S core/shell particles has been shown to drastically reduce nanosilver toxicity toward aquatic and terrestrial eukaryotic organisms, such as *Danio rerio* (zebrafish), *Fundulus heteroclitus* (killifish), *Caenorhabditis elegans* (nematode worm), and *Lemna minuta* (least duckweed), as a result of a decreased release of Ag^+ ions (Levard et al., 2013). However, findings by Ma et al. (2014) suggest that sulfidation of CuO ENPs may enhance their apparent solubility, and hence increase their bioavailability and eco-toxicity, which have been attributed to toxic Cu^{2+} ions. The observed increased solubility was caused by the oxidative dissolution of Cu_xS_y clusters. In addition, Ma et al. (2013) have demonstrated that the solubility of partially sulfidized ZnO ENPs is controlled by the ZnO core and is not quenched by the formation of the ZnS shell, contrary to what was observed with partially sulfidized Ag ENPs. Sulfidation of ZnO ENPs also led to a decrease in surface charge and to aggregation, suggesting that sulfidation alters their behavior, fate, and toxicity in the environment.

Disaggregation of ENPs is as substantial as aggregation processes are (Christian et al., 2008). Hetero-aggregation of ENPs with NOM or biocolloids (e.g., bacteria) could not only prevent aggregation (see Section 6.2.2.1) but also promote disaggregation (Hotze et al., 2010). The presence of SRHA has been shown to induce (partial) disaggregation of Ag ENPs aggregates (Fabrega et al., 2011) and Fe_2O_3 ENPs aggregates (Baalousha, 2009). CeO_2 ENPs have been reported to sorb onto the surface of bacteria (15 mg CeO2/m2) due to electrostatic attraction, whereas disaggregation of TiO_2 ENPs in the presence of microorganisms and enhanced stabilization of Fe ENPs in suspension on attachment with bacteria have also been mentioned (Hotze et al., 2010).

Certain ENPs are photoreactive (e.g., TiO_2, ZnO, CdS, CdSe, and WO_3), which has also led to their use in environmental remediation. For instance, TiO_2 ENPs have been frequently used as photocatalysts in (waste)water treatment (e.g., for photodegradation of organic pollutants) due to their high-catalytic activity resulting from a high surface-to-volume ratio, low toxicity, chemical stability, low cost, and the large availability of raw materials (Qu et al., 2013). After their release into the environment, photochemical transformations could occur when incident light reaches photoreactive ENPs, thereby, for instance, inducing excitation and generation of free radicals or leading to structural degradation at the surface of the particles or disaggregation (Hotze et al., 2010; Nowack et al., 2012). The extent to which such processes could take place depends on, for example, the wavelength of the incident light, the capacity of the light to penetrate the outer layers of the particles (e.g., aggregation or particle surface ligands might decrease the light penetration efficiency), and the capacity of the photosensitive portion of the ENPs to be excited or photodegraded (Nowack et al., 2012). Photochemical alterations are fast, with the rate-determining step being the mass transfer from the surface of the particles to the surrounding medium, and they could alter the interactions of ENPs with environmental components, such as binding to NOM (Nowack et al., 2012). Yin et al. (2014), for instance, have studied the effect of natural and synthetic sunlight on the reduction of ionic Au^{3+} complexes by NOM in river water, and they have reported the photo-induced formation of elemental Au nanoparticles from ionic Au in the presence of SRHA. Furthermore, it was shown that SRHA served not only as a reducing agent but also as a coating agent, stabilizing the formed Au nanoparticles in dispersion. Li et al. (2014) examined photochemical transformations of Ag ENPs in the presence of perfluorocarboxylic acids (PCFAs) under ultraviolet (UV) irradiation and observed a decrease in dissolution, aggregation, ROS generation, and photo-induced toxicity, suggesting that on their release into surface waters, the lower amounts of Ag^+ ions and ROS produced from Ag ENPs could be less hazardous to aquatic organisms under natural solar radiation.

6.2.3 Fate in Terrestrial Systems

Although extensive information on the behavior and fate of ENPs in terrestrial environments is still lacking, sediments and soils are generally considered to serve as key receptors for ENPs. The physicochemical processes and factors mentioned and described in the earlier sections are also likely to play a major role in the alterations, transport, and environmental effects of ENPs in soil and sediment environments (Boxall et al., 2007; Lin et al., 2010; Ferreira da Silva et al., 2011; Peralta-Videa et al., 2011; Batley et al., 2013).

6.2.3.1 Dissolution–Aggregation–Deposition–Partitioning

The solid phase present in sediment and soil systems provides a large, reactive sink for ENPs. The dissolution of ENPs in sediments and soils could be promoted by the large available surface area and exchange capacity for cations and anions, by acting as a sink for dissolution products, and by providing protons, thereby enhancing dissolution of compounds with a pH-dependent solubility (Batley and McLaughlin, 2010).

Aggregation and deposition are two closely related phenomena that affect the fate and transport of ENPs. *Aggregation* describes the interaction between two mobile objects, whereas attachment of a mobile particle onto an immobile phase (i.e., a collector) is referred to as *deposition* (Navarro et al., 2008; Hotze et al., 2010). Due to the higher colloidal load and IS of soil pore waters compared with natural water bodies, an even greater aggregation of ENPs is expected in soils and sediments (Batley and McLaughlin, 2010).

Studies on the partitioning of ENPs between liquid and solid soil phases are scant; however, hydrophilic ENPs having a net positive surface charge can be expected to bind strongly to the predominantly negatively charged surfaces of soil minerals and soil organic matter and, thus, be retained in soils and sediments. ENPs could, however, also stay mobile (or be remobilized) depending on particle properties (e.g., surface charge and surface coating) and the (changing) environmental conditions (Batley and McLaughlin, 2010; Peralta-Videa et al., 2011).

Among only a handful of studies investigating the fate of ENPs in real soil systems, Cornelis et al. (2011a,b) examined the retention and dissolution behavior of citrate-stabilized CeO_2 ENPs and polyvinylpyrrolidone-stabilized Ag ENPs in different soil solutions. They found homo-aggregation to be unlikely in the studied soil suspensions and hetero-aggregation to be dominating. Both negatively charged Ag and CeO_2 ENPs were found to primarily bind at positively charged sites on clay particles. In addition, almost no ionic Ag or Ce was detected, suggesting that either dissolution of the particles was negligible or the dissolved species partitioned strongly toward the solid phase. In addition, Fang et al. (2009) reported that TiO_2 ENPs could remain suspended in real soil solutions even after settling for 10 days, and that the suspended TiO_2 ENPs content after 24 h of settling was positively correlated with NOM and clay content of the soils; however, negative correlations were obtained with IS, pH, and ζ-potential.

6.2.3.2 Transport–Mobility

The extent of movement of ENPs away from their source, as well as their potential to be removed from (waste)water by filtration, is determined by their transport in environmental matrices (e.g., in porous media) (Lowry et al., 2010). Knowledge on the transport behavior of ENPs in natural sediment and soil systems is essential not only for assessing their potential impact on biota (e.g., on plant roots and benthic organisms) but also for the further development of environmental applications such as groundwater remediation, which requires ENPs to be transported through the soil or sediment (Lin et al., 2010; Peralta-Videa et al., 2011). Generally, columns packed with a porous matrix (e.g., glass beads, silica sand, or model soils) are used to study the mobility of ENPs, and particle filtration theory can be used to aid in understanding the transport behavior of ENPs in sediments and soils (Lin et al., 2010). A schematic diagram of the transport of ENPs in porous media is shown in Figure 6.5.

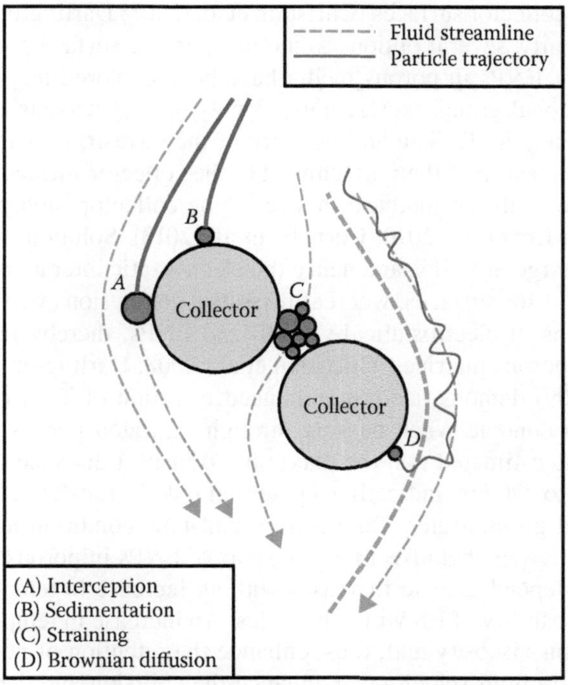

FIGURE 6.5 Schematic illustration of basic particle filtration mechanisms affecting the transport of ENPs in porous media. (After Yao, K.-M. et al., *Environ. Sci. Technol.*, 5, 1105–1112, 1971.)

The transport or filtration of ENPs in porous media such as soils, sediments, or aquifers largely depends on the rate of collision and the attachment of ENPs to stationary soil or sediment grain surfaces (i.e., the collector). Attachment is mainly a function of electrostatic interactions between the ENPs and the collector, whereas collisions occur as a result of three processes: Brownian diffusion, interception, and gravitational sedimentation (Christian et al., 2008). For small particles, Brownian diffusion is considered the dominant transport mechanism, whereas interception and sedimentation mainly regulate the transport of larger particles and aggregates. In addition, particles can also be retained by pore size exclusion when their size is larger than the pore size through which the fluid flows (i.e., straining) (Christian et al., 2008; Hotze et al., 2010; Lin et al., 2010).

The transport of ENPs in porous media depends on particle properties (e.g., size, morphology, and surface properties), solution conditions (e.g., pH, IS, NOM content, temperature, and fluid velocity), and collector characteristics (e.g., porosity and grain size) (Christian et al., 2008; Navarro et al., 2008; Darlington et al., 2009; Lin et al., 2010; Peralta-Videa et al., 2011). Particle transport is theoretically predicted to increase with a decrease in particle size; however, since ENPs are prone to aggregation, straining of aggregates can greatly inhibit nanoparticle transport (Christian et al., 2008; Lin et al., 2010). A correlation between aggregate size and the transportability of Al ENPs through soils has been demonstrated (Darlington et al., 2009), whereas Cornelis et al. (2013) also reported the irreversible attachment of Ag ENPs to mobile soil colloids in natural soil columns, leading to straining and favorable deposition. Surface characteristics (e.g., surface chemistry, charge, and coating) also play a key role. ENPs that possess a hydrophilic surface will generally be easily dispersed and transported, whereas hydrophobic ENPs tend to aggregate and settle out of solution, followed by deposition onto a collector surface (Lin et al., 2010). As mentioned earlier, environmental sediment and soil particles are usually negatively charged; ENPs possessing a negative surface charge, therefore, tend to remain more mobile in the terrestrial matrix as a result of the stronger electrostatic repulsion between the ENPs and sediment or soil particles, whereas positively charged ENPs are readily attracted electrostatically to the collector surfaces (Christian et al., 2008; Darlington et al., 2009; Cornelis et al., 2013). Consequently, several options to modify particle surface properties to enhance or restrict the transport of ENPs in porous media have been explored (e.g., surface modification with hydrophilic functional groups, surfactants, or polymers) (Christian et al., 2008; Lin et al., 2010; Peralta-Videa et al., 2011). Size and porosity of the terrestrial matrix can also affect the mobility of ENPs by determining their attachment to the collector surfaces. Generally, nanoparticle transport increases with an increase in size of the collectors and porosity of the matrix (Christian et al., 2008; Lin et al., 2010; Cornelis et al., 2013). Solution pH determines the dissolution and surface charge of ENPs, and hence the electrostatic interactions between ENPs and between ENPs and collector surfaces, whereas dissolved counterions will screen the long-range electrostatic interactions of electrostatically stabilized ENPs, thereby reducing their stability and transport through porous matrices (Christian et al., 2008; Darlington et al., 2009; Lin et al., 2010). Fang et al. (2009) demonstrated an enhanced retention of TiO_2 ENPs as a function of increasing IS and clay content, when passing through saturated porous columns packed with various soils. They also estimated that the maximum transport distance of TiO_2 ENPs in some soils ranged from 41.3 to 370 cm, indicating a potential risk of transfer of ENPs from soil media to deep soil layers and groundwater. Temperature and flow condition (e.g., flow velocity) are additional solution properties that affect the transport of ENPs in porous media. Brownian diffusion is temperature dependent, and increases with an increase in temperature, leading to an enhanced collision probability of Brownian particles. An increase in temperature can also result in a decrease in solution viscosity and, thus, enhance the retention of non-Brownian particles. Moreover, an increase in fluid velocity can enhance the collision and deposition of ENPs, and hence their transport (Christian et al., 2008; Lin et al., 2010).

It has been recognized that sediments can serve as sinks for a variety of organic (e.g., herbicides and pesticides) and inorganic (e.g., heavy metals) contaminants that have a high tendency to sorb onto colloidal particles (e.g., due to their high-specific surface). Understanding the transport behavior and mobility of ENPs is, therefore, also important to assess the interactions of ENPs with existing contaminants, as the association of ENPs with pollutants could potentially lead to particle-facilitated contaminant transport (see also Section 6.2.2.1) (Peralta-Videa et al., 2011). Plathe et al. (2010, 2013), for example, studied the associations between potentially toxic trace metals (e.g., Pb, As, Cu, and Zn) and metal oxide nanoparticles in riverbed sediment, and their transport and distribution in riverine systems. Furthermore, TiO_2 ENPs have been shown to act as a Cu carrier and to potentially facilitate the transport of Cu through the soil (Fang et al., 2011).

6.3 EXTRACTION AND ANALYSIS OF NANOPARTICLES

Assessment of the occurrence, fate, and behavior of ENPs in the aquatic environment is a prerequisite to be able to understand and evaluate their potential environmental impact and risks. Therefore, a suitable analytical toolbox that can provide both qualitative and quantitative information is essential. Currently, no single method or protocol to fully characterize nanoparticles in complex environmental media (e.g., natural surface waters, aquatic sediments, and soils) exists, but rather a combination of (complicated) analytical techniques is required (Tiede et al., 2009a; Paterson et al., 2011; Weinberg et al., 2011).

Key parameters that should be considered during characterization analyses of nanoparticles in environmental matrices include: average particle size and particle-size distribution, particle shape and structure, aggregation state, surface charge, surface chemistry and surface area, dissolution potential, particle number, and particle chemical composition and concentration (Tiede et al., 2009a; Stone et al., 2010; Farré et al., 2011; Paterson et al., 2011).

The next sections provide a concise overview of a selection of analytical methodologies that are commonly used to physically or chemically characterize nanoparticles when conducting environmental fate studies.

6.3.1 Sample Pretreatment–(Pre)fractionation

Depending on the analytical method(s) that will be used to analyze the sample, sample preparation and/or digestion is often necessary. The main purpose of sample pretreatment is to extract the nanoparticles from the environmental matrix or to reduce the complexity or the volume of the original sample prior to analysis. Sample pretreatment can, however, impact the dispersion state of nanoparticles, and thus the obtained results could differ when compared with the results obtained from direct *in situ* characterization of the nanoparticles. It is, therefore, recommended to use analytical techniques that do not require (extensive) sample preparation, or to minimize sample perturbation from the point of sampling till analysis (Tiede et al., 2008; Simonet and Valcárcel, 2009; Weinberg et al., 2011). The most common traditional techniques for a coarse prefractionation of samples are based on settling, centrifugation, or filtration methodologies. Advantages of these analytical tools include simplicity, low cost, and short time consumption (Hassellöv et al., 2008; Tiede et al., 2008).

Settling or centrifugation can only remove particles from a solution if their settling velocities overcome their Brownian motion. Settling velocities are a function of parameters, such as particle volume and shape, and the density difference between particles and water. As a result, settling or centrifugation is more effective in removing more dense mineral particles than, for instance, organic particles (Hassellöv et al., 2008; Ferreira da Silva et al., 2011). Centrifugation allows working with a broad range of particle sizes (10–1000 nm), causes only minimum sample perturbation, and requires little sample preparation (i.e., it only requires that particles are in suspension) (Handy et al., 2008). The technique could, however, induce aggregation, as settling particles can scavenge other smaller particles as a result of differential settling velocities. Although centrifugation is mainly considered a prefractionation method, it can also be used as a general nanoparticle fractionation technique, especially when applied as ultracentrifugation (UC). For instance, Kaegi et al. (2008b) demonstrated a stepwise particle-size fractionation approach during drinking water processing, where sedimentation was followed by differential centrifugation, to achieve stepwise size cutoff stages of 9000, 750, 180, and 12 nm. Bai et al. (2010) applied ultracentrifugation in the separation and concomitant purification of Au, Ag, and CdSe nanoparticles. The authors also successfully separated Au nanowires from their spherical counterparts, thereby showing that nanoparticles can not only be separated by size but also by morphology.

Traditional membrane filtration (0.2–1 μm) and microfiltration (>0.1 μm) are considered common prefractionation techniques, mainly due to their simplicity of operation, low cost, field portability, ability to treat high volumes, and the fact that they only cause low-sample perturbation. Conventional, dead-end filtration is, however, also prone to artifacts, such as nanoparticle deposition, concentration polarization, surface interactions, and filter-cake formation (Buffle et al., 1992; Morrison and Benoit, 2001). These artifacts become greater with a decrease in pore size of the filter membrane (ultra- and nanofiltration).

In order to reduce concentration polarization and pore clogging, cross-flow or tangential flow filtration (CFF or TFF), whereby the sample is recirculated on top of the membrane, has been developed (Hassellöv et al., 2008). Liu and Lead (2006) have applied CFF in the separation of colloids and particles, and they have validated its efficacy against atomic force microscopy (AFM). Cross-flow ultrafiltration (CFUF) has been evaluated for the size fractionation of freshwater colloids and particles (1 nm to 1 μm) against AFM and scanning electron microscopy (SEM) by Doucet et al. (2004). They concluded that CFUF is not fully quantitative and that fractionation is not always based solely on size.

Ultrafiltration (UF) (1–30 nm) is a size fractionation method that is applicable to process large sample volumes and that is capable of producing large quantities of isolated nanomaterials (Hassellöv et al., 2008). When the pore size of the filter membrane is smaller than 1 nm, the filtration method is typically defined as nanofiltration (NF). Although fractionation by filtration is limited to only two fractions (i.e., larger and smaller than the membrane pore size), multistage membrane filtration (MMF) can allow for a crude size fractionation. This method is, however, extremely labor intensive and time consuming (Hassellöv et al., 2008; Fedotov et al., 2011).

Dialysis is an ultra- or nanofiltration technique in which the driving force of operation relies on the diffusion of solutes across a membrane, arising from concentration gradients and osmotic pressure, rather than driven by a pressure gradient. This mild fractionation technique can be used to separate "truly dissolved" components (i.e., ions and small molecules) from their (nano)particulate counterparts. Dialysis has been applied to examine nanoparticle dissolution and nanoparticle-solute sorption behavior (Franklin et al., 2007; Kittler et al., 2010), whereas Liu et al. (2010) investigated the ion release kinetics from Ag ENPs in aqueous solutions by application of centrifugal ultrafiltration.

6.3.1.1 Fractionation Techniques

The analysis of nanoparticles in environmental samples generally requires extraction or separation from the matrix. Therefore, in order to meet detector sensitivity limitations, high preconcentration factors are required. Chromatographic techniques that have been widely used to separate nanoparticles include: high-performance liquid chromatography (HPLC), size-exclusion chromatography (SEC), capillary electrophoresis (CE), hydrodynamic chromatography (HDC), and field flow fractionation (FFF). These techniques are characterized as being sensitive and nondestructive, and they allow further characterization of the sample when combined with other techniques (Weinberg et al., 2011). In Table 6.2, a summary is presented of these separation techniques, as well as of the previously mentioned (pre)fractionation methods.

HPLC is a pressure-driven fractionation technique that separates components in a mixture based on the affinity of the analytes for a stationary phase. The interactions between analyte and stationary and mobile phases are based on dipole–dipole forces, electrostatic interactions, and dispersion forces. As a result, compounds differing in polarity or chemical structure are easier to separate. HPLC can be operated in two modes: (1) normal phase (NP-HPLC), utilizing a polar stationary phase (e.g., silica) and a nonpolar mobile phase (e.g., chloroform); and (2) reversed phase (RP-HPLC), where a nonpolar stationary phase (e.g., C18 surface-modified silica) and a (moderately) polar mobile phase (aqueous) is used. Although usually applied in the separation of solutes rather than as a particle-size fractionation method, Helfrich et al. (2006) did achieve size separation of Au nanoparticles by coupling RP-HPLC to inductively coupled plasma mass spectrometry (ICP-MS). It was shown that gold colloids that were 5–20 nm in diameter could be separated by means of a standard C18 column, owing to a size-dependent retention behavior, comparable to the separation mechanism that occurs during SEC. In a different study, HPLC–ICP-MS also proved to be a valuable tool in monitoring and characterizing the formation of citrate-stabilized Au nanoparticles (Helfrich and Bettmer, 2011).

Size-exclusion chromatography is one of the most commonly used and well-known size fractionation techniques for submicron particles, and it is frequently combined with detection methods, including voltammetry, multiangle (laser) light scattering (MA(L)LS), dynamic light scattering (DLS), and ICP-MS, to further characterize the particles (Pérez et al., 2009). The sample is brought onto an SEC column, which is packed with porous beads (i.e., the stationary phase), and as the particles pass through the column, they are retained in the pores according to their size and shape, whereby the largest particles are eluted first (Lespes and Gigault, 2011). Size-exclusion chromatography is often referred to as gel filtration or gel permeation chromatography (GFC or GPC), when an aqueous or an organic solvent is used as the mobile phase, respectively. To ensure an efficient separation, the pore size distribution of the stationary phase has to be in accordance with the size of the particles that are to be separated. As a result, several columns are needed to cover a broad size range. Separation efficiency also depends on column length and mobile phase flow rate (Lespes and Gigault, 2011; Weinberg et al., 2011). Other disadvantages of SEC include possible interactions of the analyte with the stationary phase, leading to irreversible sorption, risk of sample loss, and clogging of the pores when working with complex environmental samples (Lead and Wilkinson, 2006). Addition of surfactants to the mobile phase has proved to tackle such clogging issues. The anionic surfactant sodium dodecyl sulfate (SDS) has been shown to improve the separation of

TABLE 6.2
Summary of a Selection of Methods Applied for (Pre)fractionation of Nanoparticles in Environmental Samples

Technique		Approximate Size Range	Fractionation according to	Benefits or Advantages	Drawbacks or Limitations
Membrane filtration	Microfiltration	100 nm to 1 μm	Size	Simplicity, low cost, high volume and speed, and little sample preparation	Filtration artifacts, clogging, membrane interactions, and poor size resolution
	Ultrafiltration	1–30 nm	Size		
	Nanofiltration	0.5–1 nm	Size		
	Dialysis	0.5–100 nm	Size		
Centrifugation	Conventional	10 nm to 1 μm	Size, density	Low cost, high volume, little sample preparation, and perturbation	Low size resolution, can induce aggregation
	UC	100 Da to 10 GDa	Size, density		
Chromatography	HPLC	3 nm to 1 μm	Size, affinity for stationary phase	Allows combination with detection methods; equipment is generally available	Mobile and stationary phase interactions, size range limited by column, and poor size resolution
	SEC	0.5–100 nm	Size, shape	Good separation efficiency, simple	Unwanted solvent and column interactions, limited size separation range
	HDC	5 nm to 1.2 μm	Size	Robust, simple, large separation range, and independent of particle density	Poor selectivity, poor peak resolution
Field flow fractionation	Fl-FFF	1 nm to 50 μm	Size, density	High resolution, applicable to a large size range, less interactions compared with SEC or HDC, most selective over 50 nm, analysis of subfractions with defined particle size is possible, and low sample perturbation	Requires preconcentration of sample and standards for calibration, prone to errors due to particle shape or density, and requires coupling to a detector that can measure nanoparticle number or mass concentration (e.g., ICP-MS, MALLS, and UV–vis)
	Sd-FFF	50 nm to 50 μm	Size, density		
	Th-FFF	30 nm to 10 μm	Size, chemical composition		
	El-FFF	10 nm to 2.5 μm	Size, surface charge		
Capillary electrophoresis	CE	3 nm to 1 μm	Size, shape, and surface charge	Sensitivity, direct data on charge, faster than EM, low sample and reagent consumption	Mobile phase interactions, high concentration is necessary, and interpretation of migration times can be cumbersome

Sources: Handy, R.D. et al., *Ecotoxicology*, 17, 287–314, 2008; Hassellöv, M. et al., *Ecotoxicology*, 17, 344–361, 2008; Tiede, K. et al., *Food Addit. Contam.*, 25, 795–821, 2008; Simonet, B.M. and Valcárcel, M., *Anal. Bioanal. Chem.*, 393, 17–21, 2009; Ferreira da Silva, B. et al., *Trends Anal. Chem.*, 30, 528–540, 2011; Lespes, G. and Gigault, J., *Analyt. Chim. Acta*, 692, 26–41, 2011. With permission.

5.3 nm and 38.3 nm citrate-stabilized Au nanoparticles, by electrostatic repulsion of the negatively charged particles from the stationary phase (Wei and Liu, 1999). Size-exclusion chromatography has also been used for the characterization of CdSe QDs (Krueger et al., 2005).

Another valuable size-based fractionation tool for nanoparticles is CE, which separates compounds in solution according to their electrophoretic mobility, based on their size to charge ratio. The fractionation takes place in the interior of fine capillaries that are filled with an electrolyte; this happens on application of a high voltage. Parameters of the electrolyte, such as pH or IS, are important in optimizing nanoparticle fractionation (Fedotov et al., 2011). According to Surugau and Urban (2009), CE represents a major advancement in the size separation of nanoparticles. The technique is less prone to surface effects, in comparison to SEC, is considered faster than electron microscopy (EM), and only requires a low amount of sample and reagents during operation. In addition, useful information regarding the ζ-potential of the nanoparticles can also be obtained, as the ζ-potential is derived from the electrophoretic mobility (Weinberg et al., 2011). Data interpretation can be quite complex due to the fact that CE does not solely rely on particle size. Additional drawbacks include the need to work with high concentrations, inevitable mobile phase interactions, and the lack of sensitivity of online detectors when working with low sample volumes (Tiede et al., 2008; Fedotov et al., 2011; Weinberg et al., 2011). A variety of examples of successful application of CE in particle separation and characterization are available. Separation of Ag nanoparticles (17–50 nm) (Liu et al., 2005), Au (5–60 nm) and Au/Ag core/shell nanoparticles (24–90 nm) (Liu et al., 2007), and CdSe QDs (Oszwałdowski et al., 2011) has been reported in the literature. Surugau and Urban (2009) published an extensive overview on CE techniques used for nanoparticles, and they concluded that a consistent correlation exists between electrophoretic mobility and nanoparticle size.

Hydrodynamic chromatography is a chromatographic technique that separates nanoparticles according to their hydrodynamic radius. Unlike with SEC, an HDC column is packed with nonporous beads creating interstitial voids, which can be considered capillaries, in which particles are separated by flow velocity and velocity gradient across the particle. Within these capillaries, a parabolic flow profile exists, creating regions exhibiting slow or fast flow rates (i.e., maximum at the center of the capillary, minimum near the wall). The migration rate of the particles is influenced by particle size, size of the packing material, and flow rate and composition of the aqueous mobile phase. Since the maximum distance of a particle to the column wall is limited by its radius, smaller particles tend to remain near the capillary walls, whereas larger particles are subjected to higher flow velocity regions near the centerline. Consequently, larger particles are eluted faster from the column than smaller particles are, similar to SEC (Tiede et al., 2008; Fedotov et al., 2011; Lespes and Gigault, 2011). Depending on column length, HDC allows separation of particles in the range of 5–1200 nm, independent of their density, making it a robust method for the analysis of nanoparticles in environmental samples (Tiede et al., 2008; Weinberg et al., 2011). Additional advantages include reduced solid-phase interactions due to the fact that the technique utilizes a nonporous stationary phase, and the potential combination with other detection methods, with the most common being a UV–vis detector (Williams et al., 2002). However, problems of analyte sorption onto packing material, as well as poor selectivity and resolution, have been reported (Fedotov et al., 2011). A hyphenated HDC–ICP-MS method that analyzes metal-based nanoparticles in environmental samples was developed and optimized by Tiede et al. (2009b); this method includes a rapid analysis time (<10 min), an extended size range (5–300 nm), and limited sample preparation requirements. In a later study, the method was applied to detect and characterize Ag nanoparticles in activated sewage sludge, showing the technique's potential to study the fate and behavior of ENPs in complex environmental samples (Tiede et al., 2010). Figure 6.6 shows a typical-size calibration curve obtained from the HDC method by using Au nanoparticles as sizing standards, as well as the detection and characterization of Ag nanoparticles in sewage sludge supernatant.

Another chromatography-like fractionation technique that has received much attention over the years and showed high promise in the size separation of ENPs in complex matrices is FFF. Unlike other chromatographic methods, FFF does not utilize a stationary phase. Particle fractionation solely relies on physical phenomena occurring in an open channel. Particles are separated by application of an external field (i.e., no solid-phase interactions), which controls the particle transport velocity depending on particle diffusion coefficients, by positioning them in different laminar flow vectors in a very thin channel (Giddings et al., 1976; Giddings, 1993; Handy et al., 2008; Hassellöv et al., 2008; Tiede et al., 2008; Fedotov et al., 2011). The different FFF subtechniques are usually categorized according to the nature of the applied external field. The most commonly used FFF subtechniques are tabulated in Table 6.3. These techniques are also the only ones that are currently commercially available (Lespes and Gigault, 2011). Figure 6.7 shows a schematic overview of the working mechanism of two types of hydraulic flow FFF (Fl-FFF).

Based on the applied external field and mode of operation, FFF allows the separation and characterization of particles in the range of 1–200 μm (Lespes and Gigault, 2011; Fedotov et al., 2011). In terms of fractionation range, selectivity, and resolution, FFF is considered one of the most versatile methods. It provides high resolution, reduced sample complexity, direct related information between time and size, versatility in mobile phase composition, and the possibility of analysis on obtained subfractions with a defined particle size (Dubascoux et al., 2008). The most notable drawbacks include that errors may arise due to particle shape and density, the necessity of sizing standards to calibrate the apparatus, and the requirement to be combined with some kind of a detector. The most commonly used detectors include UV–vis, light scattering, and ICP-MS (von der Kammer et al., 2005; Dubascoux et al., 2010;

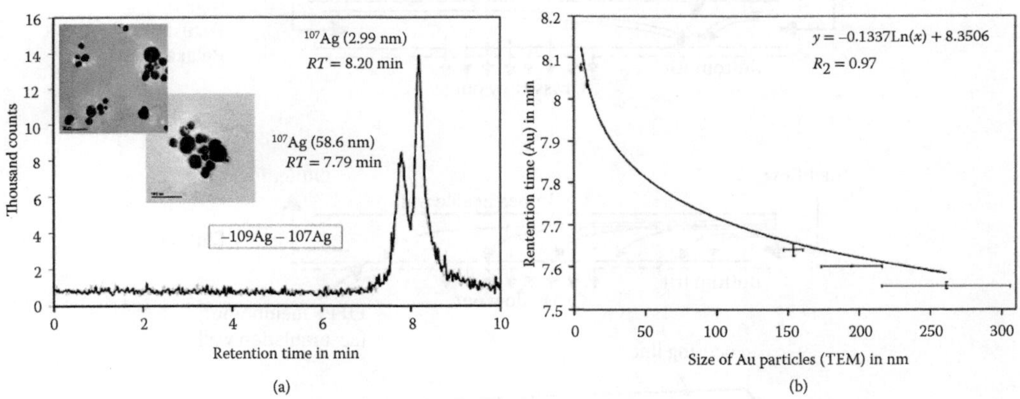

FIGURE 6.6 (a) HDC-ICP-MS chromatogram and TEM images obtained from the supernatant of sewage sludge samples spiked with Ag nanoparticles (10 mg Ag/L, 2 g MLSS/L), and (b) the corresponding calibration curve using Au nanoparticles as size calibration standards (*y*-axis error bars indicate standard deviation of retention times, whereas *x*-axis error bars represent standard deviation of Au particle size as determined by TEM). (Tiede, K. et al., A robust size-characterisation methodology for studying nanoparticle behaviour in "real" environmental samples, using hydrodynamic chromatography coupled to ICP-MS. *J. Anal. At. Spectrom.* 2009b, 24, 964–972. Reproduced by permission of the Royal Society of Chemistry.)

TABLE 6.3
Overview of the Most Commonly Applied FFF Subtechniques

FFF Subtechnique	Acronym	Applied External Field
Hydraulic Flow FFF	Fl-FFF	Two crossed flow streams, superimposed on the same channel; two types of Fl-FFF: symmetrical and asymmetrical (AsFl-FFF or AF4)
Sedimentation FFF	Sd-FFF	Centrifugation (40,000×*g*)-induced force field
Thermal FFF	Th-FFF	Temperature gradient is applied across the channel
Electrical FFF	El-FFF	Electrical potential is applied across the channel

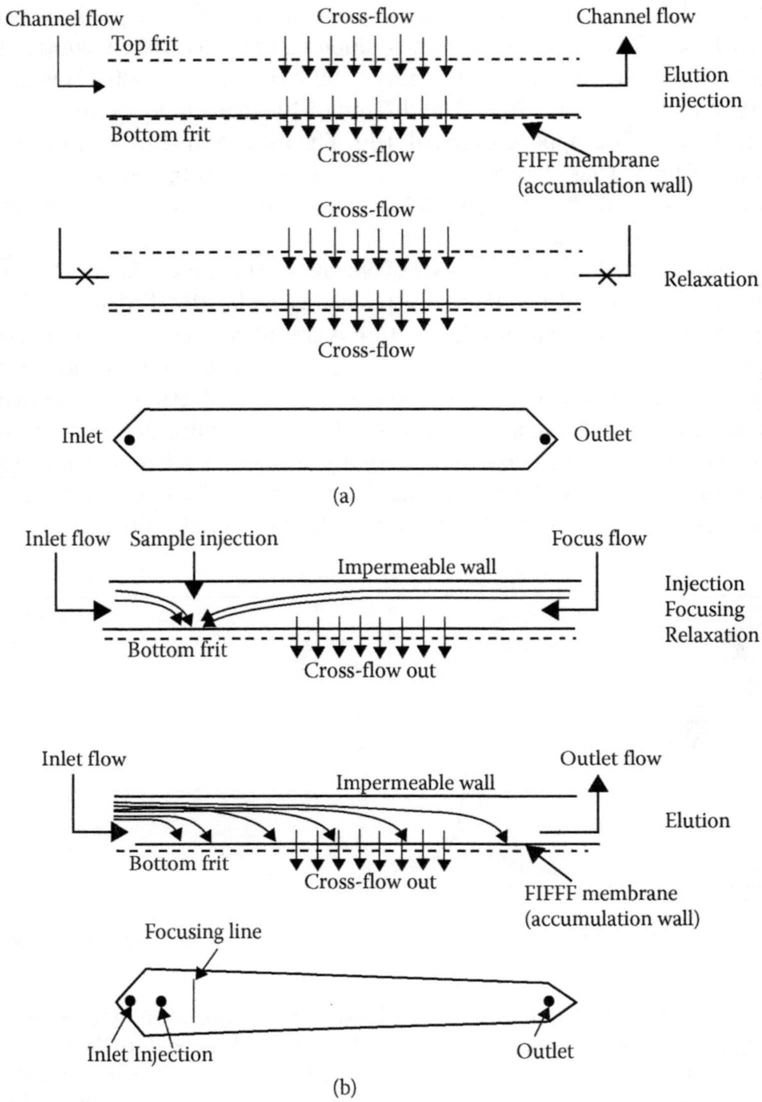

FIGURE 6.7 Schematic representation of (a) a symmetrical Fl-FFF flow diagram and channel geometry, and (b) a typical asymmetrical Fl-FFF channel. (Reprinted from *J. Chromatogr.* A 1218, Baalousha, M.A. et al., Flow field-flow fractionation for the analysis and characterization of natural colloids and manufactured nanoparticles in environmental systems: A critical review, 4078–4103, Copyright 2011, with permission from Elsevier.)

Fedotov et al., 2011). Numerous studies applying FFF in the fractionation and characterization of ENPs have been published in the literature. For instance, Gray et al. (2012) performed a comparison study of AsFl-FFF versus HDC coupled to ICP-MS, for the analysis of Au nanoparticles in aqueous samples. They concluded that AsFl-FFF is capable of separating mixtures of 5, 20, 50, and 100 nm with a significantly greater resolution than HDC. However, recoveries tended to be generally lower in the case of AsFl-FFF, and HDC also provided an additional benefit for being able to distinguish between dissolved Au and nanoparticulate Au. Ag nanoparticles have also been successfully detected and characterized by AsFl-FFF and Fl-FFF (Bolea et al., 2011; Poda et al., 2011), whereas Schmidt et al. (2011) established an analytical platform consisting of AsFl-FFF–MALLS–DLS–ICP-MS, and used it for separation and quantitative analysis of Au nanoparticles in aqueous suspensions.

6.3.2 Characterization Techniques

6.3.2.1 Particle Morphology, Size, and Size Distribution

Particle-size distribution and shape are important nanoparticle characteristics. Not only are nanoparticles, per definition, classified by particle size, but also size and shape play fundamental roles in controlling properties of nanoparticles, such as chemical reactivity, transport behavior, accessibility, diffusivity, and the ability to permeate cell membranes (Brar and Verma, 2011).

An established and traditional technique that is used to determine the size distribution of nanoparticles that are suspended in a liquid is DLS, also referred to as photon correlation spectroscopy (PCS) or quasi-elastic light scattering (QELS). DLS utilizes a laser beam to probe the random movement (i.e., Brownian motion) of nanoparticles in suspension, causing the laser light to scatter at different intensities. These time-dependent intensity fluctuations are mathematically related to the diffusion coefficient and, consequently, to the particle hydrodynamic radius according to the Stokes–Einstein relationship (Equation 6.7) (Filella et al., 1997; Hassellöv et al., 2008; Brar and Verma, 2011):

$$D = \frac{k_B T}{6\,\pi\,\eta\,r_h} \tag{6.7}$$

where D is the diffusion coefficient (m^2/s), k_B is Boltzmann's constant (J/K), T is the absolute temperature (K), η is the solution's dynamic viscosity (Pa/s), and r_h is the hydrodynamic radius of a spherical particle (m).

Dynamic light scattering is a relatively rapid and easy-to-perform, nondestructive technique that can be combined with other analytical tools, such as SEC, MALLS, and ICP-MS. However, DLS is not suitable to determine very broad size distributions, and due to the fact that it is an intensity-based methodology, the determined particle size results are biased toward large particles (Ledin et al., 1994; Handy et al., 2008). In addition, DLS is a nonelement specific analysis tool and cannot be used to analyze complex environmental samples. However, in combination with a fractionation tool such as FFF, it can allow a more thorough characterization of complex matrices (Weinberg et al., 2011; Isaacson and Bouchard, 2010). Römer et al. (2011) applied DLS for studying the aggregation behavior of Ag nanoparticles in exposure media used for aquatic toxicity tests, and they compared it with AsFl-FFF.

Nanoparticle tracking analysis (NTA) is a relatively novel technique that not only allows for size measurements of nanoparticles in suspensions but can also provide information regarding particle count and concentration (Malloy, 2011). The principle of operation is closely related to DLS; however, DLS measures the changes in scattering intensities caused by Brownian motion of the particles, whereas NTA measures the observed particle diffusion directly using video microscopy. Nanoparticle tracking analysis can therefore overcome the intensity-biased results potentially occurring during DLS (Gallego-Urrea et al., 2011; Malloy, 2011). The method, however, cannot provide information on elemental composition, and it suffers from detection issues for particles smaller than 50 nm (Domingos et al., 2009; Paterson et al., 2011).

Microscopic techniques based on optical (i.e., confocal) and electron or scanning probe approaches can be applied to visualize and characterize nanoparticles based on size, size distribution, aggregation state, shape, structure, and surface topology (Mavrocordatos et al., 2004; Tiede et al., 2008). The typical dimensions of nanoparticles are below the diffraction limit of visible light, and hence below the detection limit of conventional optical microscopy. Near-field scanning optical microscopy (NSOM) can, however, obtain a spatial resolution of 50–100 nm by using a subwavelength diameter aperture, and it can potentially be used for optical imaging of nanoparticle aggregates (Tiede et al., 2008). Confocal laser scanning microscopy (CLSM) can be used to characterize colloids, as well as thick and fluorescent specimens with resolutions of up to 200 nm, and has been applied in the analysis of more complex systems when combined with fluorescence correlation spectroscopy (FCS) (Lead et al., 2000; Prasad et al., 2007). The most preferred analytical tools for visualization of ENPs are electron and scanning probe microscopy, including scanning and transmission electron microscopy (SEM and TEM), and AFM, due to their high resolving power (i.e., subnanometer range) (Tiede et al., 2008). Scanning electron microscopy utilizes a low-energy beam of electrons (1–30 keV) to scan across the surface of a sample. A microscopic image of the sample interface is then created by detection of the scattered electrons. Focused ion beam SEM (FIB-SEM) makes use of an ion beam to cut into the sample material and allows for 3D imaging of a solid specimen (Dudkiewicz et al., 2011). The principle of TEM is based on interactions of a high-energy electron beam (80–300 keV) with a very thin sample (<200 nm). Electrons that are transmitted through the sample are focused onto a detector, generating a microscopic image (Dudkiewicz et al., 2011). Both SEM and TEM are quite time consuming and expensive, and they often require extensive sample preparation [e.g., dehydration, (cryo)fixation, embedding, and staining], which, in turn, can lead to unwanted errors when characterizing ENPs in aqueous media (e.g., induced aggregation). Some of these artifacts can, however, be tackled through the use of relatively new developments in EM, such as environmental SEM and TEM (ESEM and ETEM), and WetSEM™ and WetSTEM, which allow imaging of hydrated or even fully liquid samples (Tiede et al., 2008, 2009c; Dudkiewicz et al., 2011). Atomic force microscopy is a high-resolution type of scanning probe microscopy that utilizes an oscillating cantilever with a silicon or silicon nitride tip to scan the surface of a specimen and that produces a topographical image of the surface of the sample with atomic resolution (<1 nm) (Tiede et al., 2008). The technique is capable of analyzing liquid samples, although artifacts might occur due to sample smearing or because of attachment of the sample to the probe (Hassellöv et al., 2008; Tiede et al., 2008). Figure 6.8 shows microscopic images of ZnO and TiO_2 ENPs obtained from SEM, AFM, and TEM measurements. A brief summary of selected techniques applied in nanoparticle size or shape analysis is presented in Table 6.4.

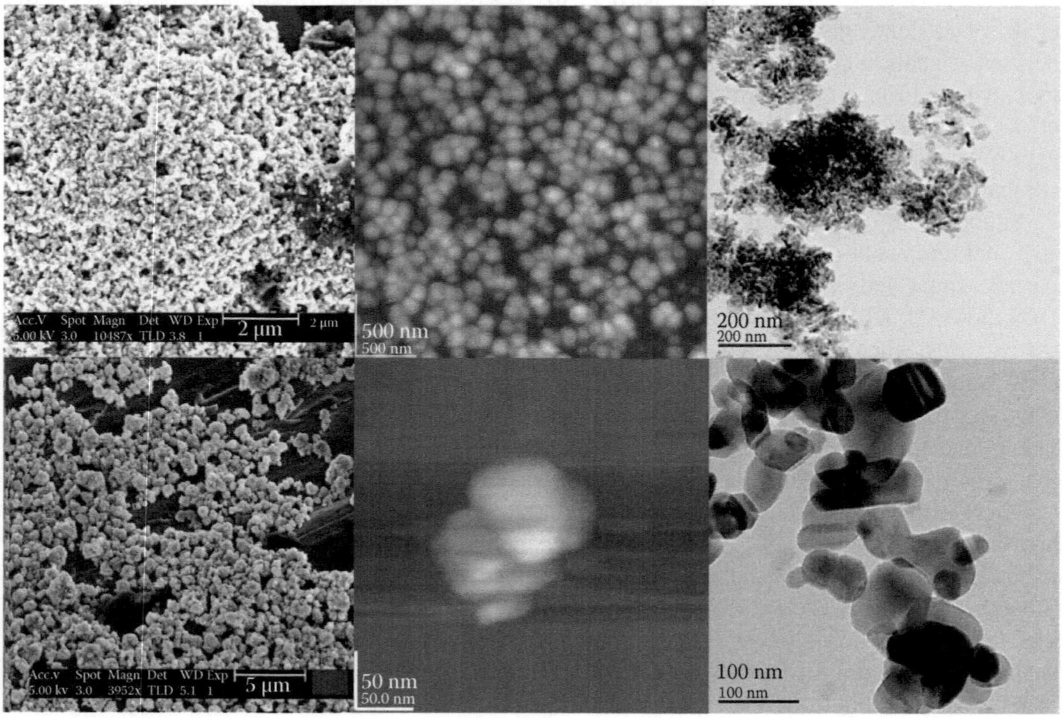

FIGURE 6.8 **(See color insert.)** ZnO (1st row) and TiO_2 (second row) nanoparticles suspended in distilled water, allowed to dry, and imaged in order from left to right by SEM, AFM, and TEM, respectively. (Reprinted from Tiede, K. et al., *Food Addit. Contam.*, 25, 795–821, 2008. With permission from Taylor & Francis Ltd.)

TABLE 6.4
Summary of a Selection of Characterization Techniques Applied in Determination of the Particle Size, Size Distribution, and Morphology of Nanoparticles in Environmental Samples

Technique	Information Obtained	Approximate Size Range	Benefits or Advantages	Drawbacks or Limitations
DLS (PCS, QELS)	Particle size, size distribution	3 nm to 1 μm	Rapid, simple, readily available, nondestructive, and aggregation process follow-up	Not suitable for complex heterogeneous samples, biased toward large particles
NTA	Particle size, size distribution, and particle count	30 nm to 1 μm	Overcomes intensity-biased results seen in DLS, low sample preparation	Nonelement specific, detection issues for particles <50 nm
NSOM	Particle visualization, size, and optical properties	50 nm to 1 μm	Optical imaging, can provide information on size, topography, and optical properties	Spatial resolution, sensitivity, and requires thin samples
SEM	Particle visualization, size, shape, structure, and aggregation	10 nm to 1 μm	High resolution, suitable for elemental analysis by complementary techniques (e.g., EDS/EDX)	Requires high vacuum, extensive sample preparation, prone to artifacts, and expensive
ESEM	Particle visualization, size, and shape	40 nm to 1 μm	No sample preparation or charging effects, able to handle moist samples	Loss in resolution, contrasting, and imaging not possible under atmospheric pressure
WetSEM™	Particle visualization, size, and shape	50 nm to 1 μm	Imaging under fully liquid conditions	Loss in resolution, sensitive membrane
TEM/HR-TEM	Particle visualization, size, shape, structure, and aggregation	1 nm to 1 μm	Very high resolution, suitable for elemental analysis by complementary techniques (e.g., EDS/EDX, EELS)	Requires high vacuum, extensive sample preparation, prone to artifacts, and very expensive
AFM	Particle visualization, size, shape, structure, topography, and electrical and mechanical properties	40 nm to 1 μm	Dry, moist samples, or liquid samples under ambient conditions, sub-nm resolution, 3D surface plots	Poor lateral dimension accuracy, artifacts due to particle smearing, and attachment to cantilever tip

Sources: Hassellöv, M. et al., *Ecotoxicology*, 17, 344–361, 2008; Tiede, K. et al., *Food Addit. Contam.*, 25, 795–821, 2008; Tiede, K. et al., *J. Chromatogr. A*, 1216, 503–509, 2009a; Dudkiewicz, A. et al., *Trends Anal. Chem.* 30, 28–43, 2011; Ferreira da Silva, B. et al., *Trends Anal. Chem.*, 30, 528–540, 2011; Gallego-Urrea, J.A. et al., *Trends Anal. Chem.* 30, 473–483, 2011; Malloy, A., *Mater. Today*, 14, 170–173, 2011; Paterson, G. et al., *Anal. Methods* 3, 1461–1467, 2011. With permission.

6.3.2.2 Particle Surface and Optical Properties

Nanoparticles are characterized by a large specific surface area (SSA), hence the particle–liquid interface plays a key role in the properties and behavior of nanoparticulate matter. The Brunauer–Emmet–Teller (BET) method is the most commonly applied technique to determine the SSA of solids. Under ultrahigh vacuum conditions, the SSA is determined by measuring the adsorption of N_2 gas on the surface and in the micropores of the nanoparticles. The SSA of ENPs can also be calculated from TEM measurements; however, this approach is not very straightforward (Handy et al., 2008; Hassellöv et al., 2008).

One of the most important nanoparticle properties is particle surface charge, since it determines their stability in dispersion and their tendency to aggregate, as well as their possible chemical or biological reactivity at the particle surface. As no easy method to measure the actual net surface charge is available, the surface potential of nanoparticles is assessed by determination of the ζ-potential (see also Section 6.2.2.1) (Tiede et al., 2008; Hassellöv et al., 2008). Electrophoretic light scattering (ELS), also called laser Doppler velocimetry, is a well-established technique to determine the electrophoretic mobility by electrophoresis. Using Smoluchowski's theories, the electrophoretic mobility can then be converted to ζ-potential (Hassellöv et al., 2008; Fedotov et al., 2011).

Information on optical properties of nanoparticles, such as reflection, absorption, and transmission characteristics, can be obtained from UV–vis spectroscopy. The technique can also be applied to confirm the presence of metal-based ENPs in suspension, monitor nanoparticles' aggregation processes, and give an indication on particle concentration according to Lambert–Beer's law. UV–vis is a rapid, simple, and inexpensive technique that does not require extensive sample preparation (Hassellöv et al., 2008; Stebounova et al., 2011). Fluorescence is a property used in many fields, including medical imaging, immunoassays, and photonics. The fluorescent properties of ENPs can be assessed by fluorescence spectroscopy or even via NSOM (Kim and Song, 2007; Hassellöv et al., 2008).

6.3.2.3 Chemical Composition and Concentration

Laser-induced breakdown detection (LIBD) is a quite novel, extremely sensitive technique that is suitable for the determination of the bulk concentration of nanoparticles in suspension, as well as for their number-weighted mean diameter. A pulsed laser beam is used to generate selective dielectric breakdowns on the particles in suspension. The measured breakdown probability depends on total particle count and size. Although considered a highly promising tool, LIBD cannot discriminate between different types of particles, and it requires particle-specific calibration (Hassellöv et al., 2008; Tiede et al., 2008). Kaegi et al. (2008b) have, for instance, demonstrated the abilities of LIBD by characterizing colloidal particles in drinking water from a lake source.

X-ray spectroscopic techniques, such as energy-dispersive x-ray spectroscopy (EDS or EDX), x-ray absorption spectroscopy (XAS), and x-ray diffraction (XRD), are useful tools in the determination of the chemical composition of single nanoparticles. In EDS, an x-ray spectrum is obtained after exposing a solid sample to a high-energy beam of electrons or x-rays. Energy-dispersive x-ray spectroscopy is generally combined with SEM and TEM for elemental analysis of a sample (Mavrocordatos et al., 2004; Hassellöv et al., 2008). The principle of XAS is based on the fact that each element has unique electron binding energies, resulting in x-ray absorption edges occurring at unique energies. Besides knowledge on elemental composition, XAS also provides information on the oxidation state and structure of nanoparticles (Schulze and Bertsch, 1995; Tiede et al., 2008). X-ray diffraction makes use of the fact that each crystalline solid generates a unique, characteristic x-ray scattering pattern on being irradiated with an x-ray beam. The method is usually applied to determine the crystal structure of mineral particles, but information on elemental composition can also be obtained (Hassellöv et al., 2008). Another technique to determine the chemical and structural composition of single particles is electron energy loss spectrometry (EELS), in which a sample is bombarded with electrons and the resulting loss in energy is determined. The changes in the kinetic energy of electrons, caused by inelastic scattering processes when passing through the sample, are element specific. Electron energy loss spectrometry is often combined with TEM (Hassellöv et al., 2008).

Inductively coupled plasma optical emission spectrometry and mass spectrometry (ICP-OES and ICP-MS) are the most conventional and widely used methods to perform elemental analysis and to determine the elemental concentration of bulk metals or metal-based nanoparticles. In ICP-OES, a liquid sample is introduced into a nebulizer, by means of a peristaltic pump, and a fine aerosol is formed by using Ar-gas. A spray chamber is used to eliminate large aerosol droplets before the fine aerosol is introduced into the plasma torch, where excited atoms and ions that emit electromagnetic radiation that is specific to a particular element are generated. The intensity of emission is proportional to the elemental concentration. Sample introduction and ionization in ICP-MS is similar to that in ICP-OES, but ICP-MS utilizes a mass spectrometer (e.g., a quadrupole) to detect analytes based on their mass-to-charge ratio. Inductively coupled plasma mass spectrometry is more sensitive than ICP-OES (ca. 3 orders of magnitude), making it the preferred option when maximum detection power is required (Scheffer et al., 2008). Both techniques generally require extensive sample preparation, including acid digestion, making them destructive methods (Ferreira da Silva et al., 2011). However, Allabashi et al. (2009) have demonstrated the potential of ICP-MS to analyze Au ENPs directly, without prior dissolving, as illustrated in Figure 6.9. In addition, recent developments have allowed ICP-MS to gain information on elemental concentration, as well as on nanoparticle size and count when operated in single particle mode (i.e., SP-ICP-MS). In SP-ICP-MS, extremely short measurement times are used in order to ensure detection of the ion pulses resulting from the ionization of individual nanoparticles introduced into the plasma (one ion cloud per particle). The number of observed pulses is related to the particle concentration by the nebulization efficiency and the total number of nanoparticles in the sample, whereas the pulse intensity is related to the size of the nanoparticles (Degueldre et al., 2006; Mitrano et al., 2012; Tuoriniemi et al., 2012). Inductively coupled plasma mass spectrometry can also be combined with fractionation techniques

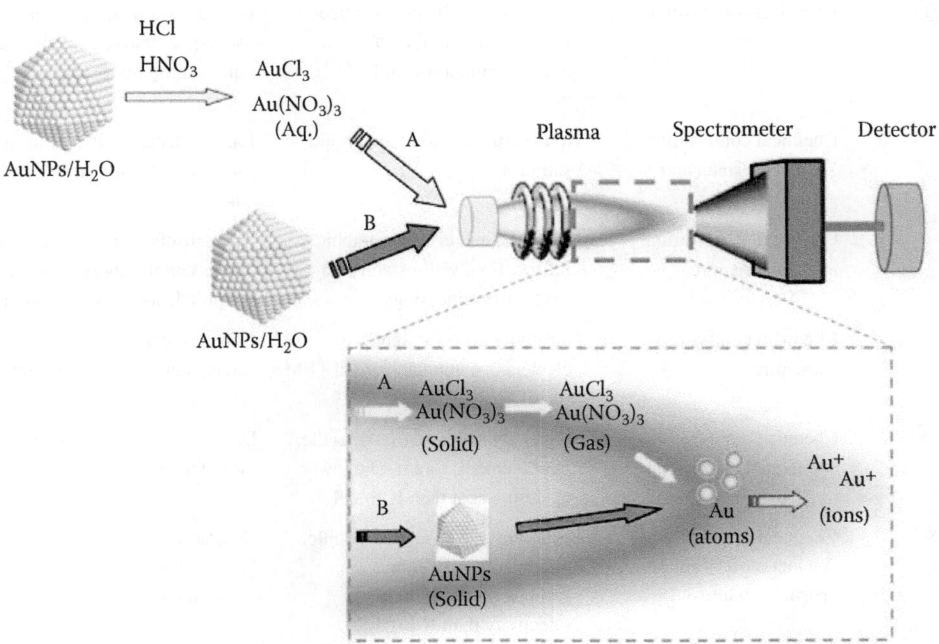

FIGURE 6.9 Schematic overview of processes involving ICP-MS analysis of Au nanoparticles (A) with and (B) without previous dissolving. (With kind permission from Springer Science+Business Media: *J. Nanopart. Res.* ICP-MS: A powerful technique for quantitative determination of gold nanoparticles without previous dissolving, 11, 2009, 2003–2011, Allabashi, R. et al.)

(e.g., FFF, HDC, HPLC, and SEC) to create very powerful hyphenated analytical tools that can be applied in nanoparticle environmental fate studies (Stolpe et al., 2005; Tiede et al., 2008; Fedotov et al., 2011; Lespes and Gigault, 2011; Weinberg et al., 2011). Table 6.5 provides a concise summary of techniques used for measuring the surface and optical properties, chemical composition, and concentration of nanoparticles.

TABLE 6.5
Summary of a Selection of Characterization Techniques Applied in Determination of the Particle Surface and Optical Properties, Chemical Composition, and Concentration of Nanoparticles in Environmental Samples

Technique	Information Obtained	Benefits or Advantages	Drawbacks or Limitations
BET	Specific surface area, porosity	Well-known technique in material science, high precision, and direct measurement of SSA	Artifacts due to sample preparation, requires ultrahigh vacuum, and solid sample
ELS	ζ-Potential	Rapid, simple, and minimal sample perturbation	Interference from complex samples, prone to contamination
UV–vis	Optical properties, particle concentration, and aggregation state	Cheap, rapid, simple, robust, nondestructive, equipment is readily available, and little sample preparation	Signal depends on concentration and extinction coefficient, interference from sample matrix
LIBD	Particle count, size	Sensitivity, accuracy, little sample preparation, nondestructive, and information on particle concentration and size	Not element specific, external size calibration required, and research-grade instrumentation
EDS (EDX)	Chemical composition	Sensitivity (for heavier elements), rapid, high spatial resolution, and combination with EM possible	Biased toward heavier elements, semiquantitative, and measurement uncertainty up to 20%
XAS	Chemical composition, oxidation state, and structure	Nondestructive, solid and liquid samples	Lack of accuracy in identifying scattering from nearest neighboring atoms
XRD	Chemical composition, crystal structure	Elemental and crystallographic information, well-established method in mineralogy	Low sensitivity and spatial resolution, large sample amount (1–3 wt%) required, and only solid samples
EELS	Chemical composition, structure	Sensitivity (also for lighter elements), combination with TEM	Complex, expensive, time consuming, and requires thin samples
ICP-OES	Chemical composition, elemental concentration	Sensitivity (mg to μg/L), specific, multielement, robust, and lower operating cost than ICP-MS	Destructive, possible spectral interferences
ICP-MS	Chemical composition, elemental concentration, particle count, and size	Sensitivity (μg to ng/L), specific, multielement, isotopic composition, and higher selectivity than ICP-OES	Destructive, possible isobaric interferences, and expensive equipment

Sources: Handy, R.D. et al., *Ecotoxicology*, 17, 287–314, 2008; Hassellöv, M. et al., *Ecotoxicology*, 17, 344–361, 2008; Tiede, K. et al., *Food Addit. Contam.*, 25, 795–821, 2008; Tiede, K. et al., *J. Chromatogr. A*, 1216, 503–509, 2009a; Ferreira da Silva, B. et al., *Trends Anal. Chem.*, 30, 528–540, 2011; Lespes, G. and Gigault, J., *Analyt. Chim. Acta*, 692, 26–41, 2011. With permission.

6.4 CONCLUSION

The rapid growth of nanotechnology over the previous decade was governed by an increase in a variety of potential applications involving engineered nanomaterials, such as metal-based ENPs. However, the increasing production, distribution, use, and disposal of ENPs will, undoubtedly, result in enhanced emissions into the environment, raising concerns on their potential impact on human and environmental health. The aquatic environment is considered to be particularly at risk for exposure to ENPs, as it acts as a sink for most environmental contaminants.

On their release in the aquatic environment, ENPs can undergo various physical, chemical, or biological alterations that depend on the characteristics of both the ENPs and the receiving environmental matrix. Ultimately, ENPs are expected to accumulate in sediments and soils, due to an increased probability of hetero-aggregation, subsequent sedimentation, and/or deposition in natural aquatic systems.

Existing theories stemming from colloid science have aided in understanding the physicochemical behavior of ENPs in synthetic media under controlled laboratory conditions; however, a prediction of their behavior in more complex matrices has proved to be a lot more challenging. In addition, it is still not exactly clear how, in what form, and at which concentrations ENPs will be released into the aquatic environment. Therefore, despite recent progress in understanding the transformation, transport, and fate of ENPs in model aquatic systems, there is still a lack of information regarding the physicochemical behavior and fate of ENPs, especially in real complex environmental media (e.g., surface waters and real sediment and soil solutions).

Until now, the lack of studies on the physicochemical fate of ENPs in aquatic systems has, in part, been due to a lack of effective and reliable analytical methods to routinely detect, identify, and quantify ENPs in aqueous mixtures. Characterization of nanoparticles in environmental samples is challenging. For instance, the heterogeneous nature of environmental media, as well as the expected low environmental concentrations of ENPs, might impede the assessment of nano-sized materials in complex environmental matrices. Recently, great progress in the development of analytical tools for the analysis of ENPs in environmental samples has been made (e.g., hyphenated techniques). However, numerous challenges remain, including how to minimize sampling and handling artifacts and being able to distinguish between naturally occurring and anthropogenically generated nanoparticles. Currently, the most reliable way to perform environmental analysis of ENPs involves a combination of analytical methods, including isolation of ENPs by filtration, centrifugation, and size-exclusion techniques, followed by bulk characterization via ICP-OES or ICP-MS, as well as characterization of individual nanoparticulate objects by means of microscopy-based techniques (e.g., SEM and TEM), which are often combined with x-ray spectrometry (e.g., EDS and XRD).

REFERENCES

Allabashi, R., Stach, W., de la Escosura-Muñiz, A., Liste-Calleja, L., and Merkoçi, A., 2009. ICP-MS: A powerful technique for quantitative determination of gold nanoparticles without previous dissolving. *J. Nanopart. Res.* 11, 2003–2011.

Auffan, M., Rose, J., Bottero, J.-Y., Lowry, G.V., Jolivet, J.-P., and Wiesner, M.R., 2009. Towards a definition of inorganic nanoparticles from an environmental, health and safety perspective. *Nat. Nanotechnol.* 4, 634–641.

Baalousha, M.A., 2009. Aggregation and disaggregation of iron oxide nanoparticles: Influence of particle concentration, pH and natural organic matter. *Sci. Total Environ.* 407, 2093–2101.

Baalousha, M.A., Manciulea, A., Cumberland, S., Kendall, K., and Lead, J.R., 2008. Aggregation and surface properties of iron oxide nanoparticles: Influence of pH and natural organic matter. *Environ. Toxicol. Chem.* 27, 1875–1882.

Baalousha, M.A., Stolpe, B., and Lead, J.R., 2011. Flow field-flow fractionation for the analysis and characterization of natural colloids and manufactured nanoparticles in environmental systems: A critical review. *J. Chromatogr. A* 1218, 4078–4103.

Bai, L., Ma, X., Junfeng, L., Sun, X., Zhao, D., and Evans, D.G., 2010. Rapid separation and purification of nanoparticles in organic density gradients. *J. Am. Chem. Soc.* 132, 2333–2337.

Batley, G.E., Kirby, J.K., and McLaughlin, M.J., 2013. Fate and risks of nanomaterials in aquatic and terrestrial environments. *Acc. Chem. Res.* 46, 854–862.

Batley, G.E., and McLaughlin, M.J., 2010. Fate of Manufactured Nanomaterials in the Australian Environment. CSIRO Niche Manufacturing Flagship Report, Lucas Heights, NSW, Australia, pp. 85. Available online (16/05/2014) at: http://www.environment.gov.au/node/21212

Benn, T., Cavanagh, B., Hristovski, K., Posner, J.D., and Westerhoff, P., 2010. The release of nanosilver form consumer products used in the home. *J. Environ. Qual.* 39, 1875–1882.

Benn, T., and Westerhoff, P., 2008. Nanoparticle silver released into water from commercially available sock fabrics. *Environ. Sci. Technol.* 42, 4133–4139.

Bhatt, I., and Tripathi, B.N., 2011. Interaction of engineered nanoparticles with various components of the environment and possible strategies for their risk assessment. *Chemosphere* 82, 308–317.

Bian, S.-W., Mudunkotuwa, I.A., Rupasinghe, T., and Grassian, V.H., 2011. Aggregation and dissolution of 4 nm ZnO nanoparticles in aqueous environments: Influence of pH, ionic strength, size, and adsorption of humic acid. *Langmuir* 27, 6059–6068.

Bolea, E., Jiménez-Lamana, J., Laborda, F., and Castillo, J.R., 2011. Size characterization and quantification of silver nanoparticles by asymmetric flow field-flow fractionation coupled with inductively coupled plasma mass spectrometry. *Anal. Bioanal. Chem.* 401, 2723–2732.

Boxall, A.B.A., Tiede, K., and Chaudry, Q., 2007. Engineered nanoparticles in soils and water: How do they behave and could they pose a risk to human health? *Nanomedicine* 2, 919–927.

Bradford, A., Handy, R.H., Readman, J.W., Atfield, A., and Mühling, M., 2009. Impact of silver nanoparticle contamination on genetic diversity of natural bacterial assemblages in estuarine sediments. *Environ. Sci. technol.* 43, 4530–4536.

Brar, S.K., and Verma, M., 2011. Measurement of nanoparticles by light-scattering techniques. *Trends Anal. Chem.* 30, 4–17.

Buffle, J., Perret, J., and Newman, J., 1992. The use of filtration and ultrafiltration for size fractionation of aquatic particles, colloids and macromolecules. In: Buffle, J. and van Leeuwen, H.P. (eds.), *Environmental Particles I*. Chelsea, MI: Lewis Publishers, pp. 171–230.

Christian, P., von der Kammer, F., Baalousha, M., and Hofmann, T., 2008. Nanoparticles: Structure, properties, preparation and behaviour in environmental media. *Ecotoxicology* 17, 326–343.

Cornelis, G., Brooke, R., McLaughlin, M.J., Kirby, J.K., Beak, D., and Chittleborough, D., 2011a. Solubility and batch retention of CeO_2 nanoparticles in soils. *Environ. Sci. Technol.* 45, 2777–2782.

Cornelis, G., Doolette, C., Thomas, M., McLaughlin, M.J., Kirby, J.K., Beak, D.G., and Chittleborough, D., 2011b. Retention and dissolution of silver nanoparticles in natural soils. *Soil Sci. Soc. Am. J.* 76, 891–902.

Cornelis, G., Pang, L., Doolette, C., Kirby, J.K., and McLaughlin, M.J., 2013. Transport of silver nanoparticles in saturated columns of natural soils. *Sci. Total Environ.* 463–464, 120–130.

Darlington, T.K., Neigh, A.M., Spencer, M.T., Nguyen, O.T., and Oldenburg, S.J., 2009. Nanoparticle characteristics affecting environmental fate and transport through soil. *Environ. Toxicol. Chem.* 28, 1191–1199.

Degueldre, C., Favarger, P.-Y., and Wold, S., 2006. Gold colloid analysis by inductively coupled plasma-mass spectrometry in a single particle mode. *Anal. Chim. Acta* 555, 263–268.

Dokoumetzidis, A., Papadopoulou, V., Valsami, G., and Macheras, P., 2008. Development of a reaction-limited model of dissolution: Application to official dissolution test experiments. *Int. J. Pharm.* 355, 114–125.

Domingos, R.F., Baalousha, M.A., Ju-Nam, Y., Reid, M.M., Tufenkji, N., Lead, J.R., Leppard, G.G., and Wilkinson, K.J., 2009. Characterizing manufactured nanoparticles in the environment: Multimethod determination of particle sizes. *Environ. Sci. Technol.* 43, 7277–7284.

Doucet, F.J., Maguire, L., and Lead, J.R., 2004. Size fractionation of aquatic colloids and particles by cross-flow filtration: Analysis by scanning electron and atomic force microscopy. *Anal. Chim. Acta* 522, 59–71.

Dubascoux, S., Le Hécho, I., Hassellöv, M., von der Kammer, F., Potin Gautier, M., and Lespes, G., 2010. Field-flow fractionation and inductively coupled plasma mass spectrometry coupling: History, development and applications. *J. Anal. At. Spectrom.* 25, 613–623.

Dubascoux, S., von der Kammer, F., Le Hécho, I., Potin Gautier, M., and Lespes, G., 2008. Optimisation of asymmetrical flow field flow fractionation for environmental nanoparticles separation. *J. Chromatogr. A* 1206, 160–165.

Dudkiewicz, A., Tiede, K., Loeschner, K., Helene, L., Jensen, S., Jensen, E., Wierzbicki, R., Boxall, A.B.A., and Molhave, K., 2011. Characterization of nanomaterials in food by electron microscopy. *Trends Anal. Chem.* 30, 28–43.

El Badawy, A.M., Luxton, T.P., Silva, G.R., Scheckel, K.G., Suidan, M.T., and Tolaymat, T.M., 2010. Impact of environmental conditions (pH, ionic strength, and electrolyte type) on the surface charge and aggregation of silver nanoparticles suspensions. *Environ. Sci. Technol.* 44, 1260–1266.

Elzey, S., and Grassian, V.H., 2010a. Agglomeration, isolation and dissolution of commercially manufactured silver nanoparticles in aqueous environments. *J. Nanopart. Res.* 12, 1945–1958.

Elzey, S., and Grassian, V.H., 2010b. Nanoparticle dissolution from particle perspective: Insights from particle sizing measurements. *Langmuir* 26, 12505–12508.

Fabrega, J., Luoma, S.N., Tyler, C.R., Galloway, T.S., and Lead, J.R., 2011. Silver nanoparticles: Behaviour and effects in the aquatic environment. *Environ. Int.* 37, 517–531.

Fang, J., Shan, X.-Q., Wen, B., Lin, J.-M., and Owens, G., 2009. Stability of Titania nanoparticles in soil suspensions and transport in saturated homogeneous soil columns. *Environ. Pollut.* 157, 1101–1109.

Fang, J., Shan, X.-Q., Wen, B., Lin, J.-M., Owens, G., and Zhou, S.-R., 2011. Transport of copper as affected by Titania nanoparticles in soil columns. *Environ. Pollut.* 159, 1248–1256.

Farré, M., Sanchís, J., and Barceló, D., 2011. Analysis and assessment of the occurrence, the fate and the behavior of nanomaterials in the environment. *Trends Anal. Chem.* 30, 517–527.

Fedotov, P.S., Vanifatova, N.G., Shkinev, V.M., and Spivakov, B.Y., 2011. Fractionation and characterization of nano- and microparticles in liquid media. *Anal. Bioanl. Chem.* 400, 1787–1804.

Ferreira da Silva, B., Pérez, S., Gardinalli, P., Singhal, R.K., Mozeto, A.A., and Barceló, D., 2011. Analytical chemistry of metallic nanoparticles in natural environments. *Trends Anal. Chem.* 30, 528–540.

Filella, M., Zhang, J., Newman, M.E., and Buffle, J., 1997. Analytical applications of photon correlation spectroscopy for size distribution measurements of natural colloidal suspensions: Capabilities and limitations. *Colloids Surf. A* 120, 27–46.

Franklin, N.M., Rogers, N.J., Apte, S.C., Batley, G.E., Gadd, G.E., and Casey, P.S., 2007. Comparative toxicity of nanoparticulate ZnO, bulk ZnO, and $ZnCl_2$ to a freshwater microalga (*Pseudokirchneriella subcapitata*): The importance of particle solubility. *Environ. Sci. Technol.* 41, 8484–8490.

French, R.A., Jacobsen, A.R., Kim, B., Isley, S.L., Penn, R.L., and Baveye, P.C., 2009. Influence of ionic strength, pH, and cation valence on aggregation kinetics of titanium dioxide nanoparticles. *Environ. Sci. Technol.* 43, 1354–1359.

Gallego-Urrea, J.A., Tuoriniemi, J., and Hassellöv, M., 2011. Applications of particle-tracking analysis to the determination of size distributions and concentrations of nanoparticles in environmental, biological and food samples. *Trends Anal. Chem.* 30, 473–483.

Ghosh, S., Mashayekhi, H., Pan, B., Bhowmik, P., and Xing, B., 2008. Colloidal behavior of aluminum oxide nanoparticles as affected by pH and natural organic matter. *Langmuir* 24, 12385–12391.

Giddings, J.C., 1993. Field-flow fractionation: Analysis of macromolecular, colloidal, and particulate materials. *Science* 260, 1456–1465.

Giddings, J.C., Yang, F.J., and Myers, M.N., 1976. Flow field-flow fractionation—Versatile new separation method. *Science* 193, 1244–1245.

Gottschalk, F., and Nowack, B., 2011. The release of engineered nanomaterials to the environment. *J. Environ. Monit.* 13, 1145–1155.

Gray, E.P., Bruton, T.A., Higgins, C.P., Halden, R.U., Westerhoff, P., and Ranville, J.F., 2012. Analysis of gold nanoparticle mixtures: A comparison of hydrodynamic chromatography (HDC) and asymmetrical flow field-flow fractionation (AF4) coupled to ICP-MS. *J. Anal. At. Spectrom.* 27, 1532–1539.

Grieger, K.D., Fjordbøge, A., Hartmann, N.B., Eriksson, E., Bjerg, P.L., and Baun, A., 2010. Environmental benefits and risks of zero-valent iron nanoparticles (nZVI) for *in situ* remediation: Risk mitigation or trade-off? *J. Contam. Hydrol.* 118, 165–183.

Handy, R.D., von der Kammer, F., Lead, J.R., Hassellöv, M., Owen, R., and Crane, M., 2008. The ecotoxicology and chemistry of manufactured nanoparticles. *Ecotoxicology* 17, 287–314.

Hassellöv, M., Readman, J.W., Ranville, J.F., and Tiede, K., 2008. Nanoparticle analysis and characterization methodologies in environmental risk assessment of engineered nanoparticles. *Ecotoxicology* 17, 344–361.

Helfrich, A., and Bettmer, J., 2011. Analysis of gold nanoparticles using ICP-MS-based hyphenated and complementary ESI-MS techniques. *Int. J. Mass Spectrom.* 307, 92–98.

Helfrich, A., Brüchert, W., and Bettmer, J., 2006. Size characterisation of Au nanoparticles by ICP-MS coupling techniques. *J. Anal. At. Spectrom.* 21, 431–434.

Hotze, E.M., Phenrat, T., and Lowry, G.V., 2010. Nanoparticle aggregation: Challenges to understanding transport and reactivity in the environment. *J. Environ. Qual.* 39, 1909–1924.

Isaacson, C.W., and Bouchard, D., 2010. Asymmetric flow field flow fractionation of aqueous C60 nanoparticles with size determination by dynamic light scattering and quantification by liquid chromatography atmospheric pressure photo-ionization mass spectrometry. *J. Chromatogr. A* 1217, 1506–1512.

Ju-Nam, Y., and Lead, J.R., 2008. Manufactured nanoparticles: An overview of their chemistry, interactions and potential environmental implications. *Sci. Total Environ.* 400, 396–414.

Kaegi, R., Ulrich, A., Sinnet, B., Vonbank, R., Wichser, A., Zuleeg, S., Simmler, H., Brunner, S., Vonmont, H., Burkhardt, M., and Boller, M., 2008a. Synthetic TiO_2 nanoparticle emission from exterior facades into the aquatic environment. *Environ. Pollut.* 156, 233–239.

Kaegi, R., Wagner, T., Hetzer, B., Sinnet, B., Tzvetkov, G., and Boller, M., 2008b. Size, number and chemical composition of nanosized particles in drinking water determined by analytical microscopy and LIBD. *Water Res.* 42, 2778–2786.

Keller, A.A., Wang, G., Zhou, D., Lenihan, H.S., Cherr, G., Cardinale, B.J., Miller, R., and Ji, Z., 2010. Stability and aggregation of metal oxide nanoparticles in natural aqueous matrices. *Environ. Sci. Technol.* 44, 1962–1967.

Kim, B., Park, C.-S., Murayama, M., and Hochella Jr., M.F., 2010. Discovery and characterization of silver sulfide nanoparticles in final sewage sludge products. *Environ. Sci. Technol.* 44, 7509–7514.

Kim, J., and Song, K.-B., 2007. Recent progress of nano-technology with NSOM. *Micron* 38, 409–426.

Kittler, S., Greulich, C., Diendorf, J., Köller, M., and Epple, M., 2010. Toxicity of silver nanoparticles increases during storage because of slow dissolution under release of silver ions. *Chem. Mater.* 22, 4548–4554.

Klaine, S.J., Alvarez, P.J.J., Batley, G.E., Fernandes, T.F., Handy, R.D., Lyon, D.Y., Mahendra, S., McLaughlin, M.J., and Lead, J.R., 2008. Nanomaterials in the environment: Behavior, fate, bioavailability, and effects. *Environ. Toxicol. Chem.* 27, 1825–1851.

Krueger, K.M., Al-Somali, A.M., Falkner, J.C., and Colvin, V.L., 2005. Characterization of nanocrystalline CdSe by size exclusion chromatography. *Anal. Chem.* 77, 3511–3515.

Lead, J.R., and Wilkinson, K.J., 2006. Aquatic colloids and nanoparticles: Current knowledge and future trends. *Environ. Chem.* 3, 159–171.

Lead, J.R., Wilkinson, K.J., Starchev, K., Canonica, S., and Buffle, J., 2000. Determination of diffusion coefficients of humic substances by fluorescence correlation spectroscopy: Role of solution conditions. *Environ. Sci. Technol.* 34, 1365–1369.

Ledin, A., Karlsson, S., Düker, A., and Allard, B., 1994. Measurements *in situ* of concentration and size distribution of colloidal matter in deep groundwaters by photon correlation spectroscopy. *Water Res.* 28, 1539–1545.

Lespes, G., and Gigault, J., 2011. Hyphenated analytical techniques for multidimensional characterisation of submicron particles: A review. *Analyt. Chim. Acta* 692, 26–41.

Levard, C., Hotze, E.M., Colman, B.P., Dale, A.L., Truong, L., Yang, X.Y., Bone, A.J., Brown Jr., G.E., Tanguay, R.L., Di Giulio, R.T., Bernhardt, E.S., Meyer, J.N., Wiesner, M.R., and Lowry, G.V., 2013. Sulfidation of silver nanoparticles: Natural antidote to their toxicity. *Environ. Sci. Technol.* 47, 13440–13448.

Levard, C., Hotze, E.M., Lowry, G.V., and Brown Jr., G.E., 2012. Environmental transformations of silver nanoparticles: Impact on stability and toxicity. *Environ. Sci. Technol.* 46, 6900–6914.

Lewis, A.E., 2010. Review of metal sulphide precipitation. *Hydrometallurgy* 104, 222–234.

Li, X., Lenhart, J.J., and Walker, H.W., 2010. Dissolution-accompanied aggregation kinetics of silver nanoparticles. *Langmuir* 26, 16690–16698.

Li, Y., Niu, J., Shang, E., and Crittenden, J., 2014. Physicochemical transformations and photoinduced toxicity reduction of silver nanoparticles in the presence of perfluorocarboxylic acids under UV irradiation. *Environ. Sci. Technol.* 48, 4946–4953.

Limbach, L.K., Bereiter, R., Müller, E., Krebs, R., Gälli, R., and Stark, W.J., 2008. Removal of oxide nanoparticles in a model wastewater treatment plant: Influence of agglomeration and surfactants on clearing efficiency. *Environ. Sci. Technol.* 42, 5828–5833.

Lin, D., Tian, X., Wu, F., and Xing, B., 2010. Fate and transport of engineered nanomaterials in the environment. *J. Environ. Qual.* 39, 1896–1908.

Liu, F.-K., Ko, F.-H., Huang, P.-W., Wu, C.-H., and Chu, T.-C., 2005. Studying the size/shape separation and optical properties of silver nanoparticles by capillary electrophoresis. *J. Chromatogr. A* 1062, 139–145.

Liu, J., Aruguete, D.M., Murayama, M., and Hochella Jr., M.F., 2009. Influence of size and aggregation on the reactivity of an environmentally and industrially relevant nanomaterial (PbS). *Environ. Sci. Technol.* 43, 8178–8183.

Liu, J., and Hurt, R.H., 2010. Ion release kinetics and particle persistence in aqueous nano-silver colloids. *Environ. Sci. Technol.* 44, 2169–2175.

Liu, J., Pennell, K.G., and Hurt, R.H., 2011. Kinetics and mechanisms of nanosilver oxysulfidation. *Environ. Sci. Technol.* 45, 7345–7353.

Liu, R., and Lead, J.R., 2006. Partial validation of cross flow ultrafiltration by atomic force microscopy. *Anal. Chem.* 78, 8105–8112.

Liu, R., Lead, J.R., and Baker, A., 2007. Fluorescence characterization of cross flow ultrafiltration derived freshwater colloidal and dissolved organic matter. *Chemosphere* 68, 1304–1311.

Lowry, G.V., and Casman, E.A., 2009. Nanomaterial transport, transformation, and fate in the environment. A risk based perspective on research needs. In: Linkov, I., and Steevens, J. (eds.), *Nanomaterials: Risks and Benefits*. GX Dordrecht, the Netherlands: Springer Science + Business media B.V., pp. 125–137.

Lowry, G.V., Hotze, E.M., Bernhardt, E.S., Dionysiou, D.D., Pedersen, J.A., Wiesner, M.R., and Xing, B., 2010. Environmental occurrences, behavior, fate, and ecological effects of nanomaterials: An introduction to the special series. *J. Environ. Qual.* 39, 1867–1874.

Ma, R., Levard, C., Michel, F.M., Brown Jr., G.E., and Lowry, G.V., 2013. Sulfidation mechanisms for zinc oxide nanoparticles and the effect of sulfidation on their solubility. *Environ. Sci. Technol.* 47, 2527–2534.

Ma, R., Stegemeier, J., Levard, C., Dale, J.G., Noack, C.W., Yang, T., Brown Jr., G.E., and Lowry, G.V., 2014. Sulfidation of copper oxide nanoparticles and properties of resulting copper sulfide. *Environ. Sci. Nano* 1, 347–357.

Malloy, A., 2011. Count, size and visualize nanoparticles. *Mater. Today* 14, 170–173.

Mavrocordatos, D., Pronk, W., and Boller, M., 2004. Analysis of environmental particles by atomic force microscopy, scanning and transmission electron microscopy. *Water Sci. Technol.* 50, 9–18.

Mitrano, D.M., Barber, A., Bednar, A., Westerhoff, P., Higgins, C.P., and Ranville, J.F., 2012. Silver nanoparticle characterization using single particle ICP-MS (SP-ICP-MS) and asymmetrical flow field flow fractionation ICP-MS (AF4-ICP-MS). *J. Anal. At. Spectrom.* 27, 1131–1142.

Morrison, M.A. and Benoit, G., 2001. Filtration artifacts caused by overloading membrane filters. *Environ. Sci. Technol.* 35, 3774–3779.

Mudunkotuwa, I.A. and Grassian, V.H., 2011. The devil is in the details (or the surface): Impact of surface structure and surface energetics on understanding the behavior of nanomaterials in the environment. *J. Environ. Monit.* 13, 1135–1144.

Navarro, E., Baun, A., Behra, R., Hartmann, N.B., Filser, J., Miao, A.-J., Quigg, A., Santschi, P.H., and Sigg, L., 2008. Environmental behavior and ecotoxicity of engineered nanoparticles to algae, plants, and fungi. *Ecotoxicology* 17, 372–386.

Nowack, B., and Bucheli, T.D., 2007. Occurrence, behavior and effects of nanoparticles in the environment. *Environ. Pollut.* 150, 5–22.

Nowack, B., Ranville, J.F., Diamond, S., Gallego-Urrea, J.A., Metcalfe, C., Rose, J., Horne, N., Koelmans, A.A., and Klaine, S.J., 2012. Potential scenarios for nanomaterial release and subsequent alteration in the environment. *Environ. Toxicol. Chem.* 31, 50–59.

Oszwałdowski, S., Zawistowska-Gibuła, K., and Roberts, K.P., 2011. Capillary electrophoretic separation of nanoparticles. *Anal. Bioanal. Chem.* 399, 2831–2842.

Paterson, G., Macken, A., and Thomas, K.V., 2011. The need for standardized methods and environmental monitoring programs for anthropogenic nanoparticles. *Anal. Methods* 3, 1461–1467.

Peralta-Videa, J.R., Zhao, L., Lopez-Moreno, M.L., de la Rosa, G., Hong, J., and Gardea-Torresdey, J.L., 2011. Nanomaterials and the environment: A review for the biennium 2008-2010. *J. Hazard. Mater.* 186, 1–15.

Pérez, S., Farré, M., and Barceló, D., 2011. Analysis, behavior and ecotoxicity of carbon-based nanomaterials in the aquatic environment. *Trends Anal. Chem.* 28, 820–832.

Petosa, A.R., Jaisi, D.P., Quevedo, I.R., Elimelech, M., and Tufenkji, N., 2010. Aggregation and deposition of engineered nanomaterials in aquatic environments: Role of physicochemical interactions. *Environ. Sci. Technol.* 44, 6532–6549.

Phenrat, T., Saleh, N., Sirk, K., Tilton, R.D., and Lowry, G.V., 2007. Aggregation and sedimentation of aqueous nanoscale zero valent iron dispersions. *Environ. Sci. Technol.* 41, 284–290.

Plathe, K.L., von der Kammer, F., Hassellöv, M., Moore, J.N., Murayama, M., Hofmann, T., and Hochella Jr., M.F., 2010. Using FlFFF and aTEM to determine trace metal-nanoparticle associations in riverbed sediment. *Environ. Chem.* 7, 82–93.

Plathe, K.L., von der Kammer, F., Hassellöv, M., Moore, J.N., Murayama, M., Hofmann, T., and Hochella Jr., M.F., The role of nanominerals and mineral nanoparticles in the transport of toxic trace metals: Field-flow fractionation and analytical TEM analyses after nanoparticle isolation and density separation. *Geochim. Cosmochim. Acta* 102, 213–225.

Poda, A.R., Bednar, A.J., Kennedy, A.J., Harmon, A., Hull, M., Mitrano, D.M., Ranville, J.F., and Steevens, J., 2011. Characterization of silver nanoparticles using flow-field flow fractionation interfaced to inductively coupled plasma mass spectrometry. *J. Chromatogr. A* 1218, 4219–4225.

Prasad, V., Semwogerere, D., and Weeks, E.R., 2007. Confocal microscopy of colloids. *J. Phys. Condens. Matter.* 19, 1–25.

Qu, X., Alvarez, P.J.J., and Li, Q., 2013. Applications of nanotechnology in water and wastewater treatment. *Water Res.* 47, 3931–3946.

Quik, J.T.K., Lynch, I., Van Hoecke, K., Miermans, C.J.H., De Schamphelaere, K.A.C., Janssen, C.R., Dawson, K.A., Cohen Stuart, M.A., and Van De Meent, D., 2010. Effect of natural organic matter on cerium dioxide nanoparticles settling in model fresh water. *Chemosphere* 81, 711–715.

Quik, J.T.K., Stuart, M.C., Wouterse, M., Peijnenburg, W., Hendriks, A.J., and Van De Meent, D., 2012. Natural colloids are the dominant factor in the sedimentation of nanoparticles. *Environ. Toxicol. Chem.* 31, 1019–1022.

Quik, J.T.K., Vonk, J.A., Hansen, S.F., Baun, A., and Van De Meent, D., 2011. How to assess exposure of aquatic organisms to manufactured nanoparticles? *Environ. Int.* 37, 1068–1077.

Reger, D.L., Goode, S.R., and Ball, D.W., 2010. Appendix F: Solubility product, acid, and base constants. In: *Chemistry: Principles and Practice.* 3rd edition. Belmont, CA: Brooks/Cole, Cengage Learning, Inc., pp. A.17–A.18.

Rogers, N.J., Franklin, N.M., Apte, S.C., Spadaro, D., Angel, B., Batley, G.E., Lead, J.R., and Baalousha, M.A., 2010. Behaviour and toxicity to algae of nanoparticulate CeO_2 in freshwater. *Environ. Chem.* 7, 50–60.

Römer, I., White, T.A., Baalousha, M.A., Chipman, K., Viant, M.R., and Lead, J.R., 2011. Aggregation and dispersion of silver nanoparticles in exposure media for aquatic toxicity tests. *J. Chromatogr. A* 1218, 4226–4233.

Sánchez, A., Recillas, S., Font, X., Casals, E., González, E., and Puntes, V., 2011. Ecotoxicity of, and remediation with, engineered inorganic nanoparticles in the environment. *Trends Anal. Chem.* 30, 507–516.

Scheffer, A., Engelhard, C., Sperling, M., and Buscher, W., 2008. ICP-MS as a new tool for determination of gold nanoparticles in bioanalytical applications. *Anal. Bioanal. Chem.* 390, 249–252.

Schmidt, B., Loeschner, K., Hadrup, N., Mortensen, A., Sloth, J.J., Koch, C.B., and Larsen, E.H., 2011. Quantitative characterization of gold nanoparticles by field-flow fractionation coupled online with light scattering detection and inductively coupled plasma mass spectrometry. *Anal. Chem.* 83, 2461–2468.

Schulze, D.G., and Bertsch, P.M., 1995. Synchrotron X-ray techniques in soil, plant, and environmental research. *Adv. Argon.* 55, 1–66.

Scown, T.M., van Aerle, R., and Tyler, C.R., 2010. Review: Do engineered nanoparticles pose a significant threat to the aquatic environment? *Crit. Rev. Toxicol.* 40, 653–670.

Sharma, V.K., 2009a. Aggregation and toxicity of titanium dioxide nanoparticles in aquatic environment—A review. *J. Environ. Sci. Health A* 44, 1485–1495.

Sharma, V.K., Siskova, K.M., Zboril, R., and Gardea-Torresdey, J.L., 2014. Organic-coated silver nanoparticles in biological and environmental conditions: Fate, stability and toxicity. *Adv. Colloid Interface Sci.* 204, 15–34.

Sharma, V.K., Yngard, R.A., and Lin, Y., 2009b. Silver nanoparticles: Green synthesis and their antimicrobial activities. *Adv. Colloid Interface Sci.* 145, 83–96.

Simonet, B.M. and Valcárcel, M., 2009. Monitoring nanoparticles in the environment. *Anal. Bioanal. Chem.* 393, 17–21.

Stebounova, L.V., Guio, E., and Grassian, V.H., 2011. Silver nanoparticles in simulated biological media: A study of aggregation, sedimentation, and dissolution. *J. Nanopart. Res.* 13, 233–244.

Stolpe, B., Hassellöv, M., Andersson, K., and Turner, D.R., 2005. High resolution ICPMS as an on-line detector for flow field-flow fractionation; multi-element determination of colloidal size distributions in a natural water sample. *Analyt. Chim. Acta* 535, 109–121.

Stone, V., Nowack, B., Baun, A., van den Brink, N., von der Kammer, F., Dusinska, M., Handy, R., Hankin, S., Hassellöv, M., Joner, E., and Fernandes, T.F., 2010. Nanomaterials for environmental studies: Classification, reference material issues, and strategies for physico-chemical characterisation. *Sci. Total Environ.* 408, 1745–1754.

Surugau, N., and Urban, P.L., 2009. Electrophoretic methods for separation of nanoparticles. *J. Sep. Sci.* 32, 1889–1906.

Thalmann, B., Voegelin, A., Sinnet, B., Morgenroth, E., and Kaegi, R., 2014. Sulfidation kinetics of silver nanoparticles reacted with metal sulfides. *Environ. Sci. Technol.* 48, 4885–4892.

Theron, J., Walker, J.A., and Cloete, T.E., 2008. Nanotechnology and water treatment: Applications and emerging opportunities. *Crit. Rev. Microbiol.* 34, 43–69.

Tiede, K., Boxall, A.B.A., Tear, S.P., Lewis, J., David, H., and Hassellöv, M., 2008. Detection and characterization of engineered nanoparticles in food and the environment. *Food Addit. Contam.* 25, 795–821.

Tiede, K., Boxall, A.B.A., Tiede, D., Tear, S.P., David, H., and Lewis, J., 2009b. A robust size-characterisation methodology for studying nanoparticle behaviour in "real" environmental samples, using hydrodynamic chromatography coupled to ICP-MS. *J. Anal. At. Spectrom.* 24, 964–972.

Tiede, K., Boxall, A.B.A., Wang, X., Gore, D., Tiede, D., Baxter, M., David, H., Tear, S.P., and Lewis, J., 2010. Application of hydrodynamic chromatography-ICP-MS to investigate the fate of silver nanoparticles in activated sludge. *J. Anal. At. Spectrom.* 25, 1149–1154.

Tiede, K., Hassellöv, M., Breitbarth, E., Chaudhry, Q., and Boxall, A.B.A., 2009a. Considerations for environmental fate and ecotoxicity testing to support environmental risk assessments for engineered nanoparticles. *J. Chromatogr. A* 1216, 503–509.

Tiede, K., Tear, S.P., David, H., and Boxall, A.B.A., 2009c. Imaging of engineered nanoparticles and their aggregates under fully liquid conditions in environmental matrices. *Water Res.* 43, 3335–3343.

Tuoriniemi, J., Cornelis, G., and Hassellöv, M., 2012. Size determination and detection capabilities of single-particle ICPMS for environmental analysis of silver nanoparticles. *Anal. Chem.* 84, 3965–3972.

von der Kammer, F., Baborowski, M., and Friese, K., 2005. Field-flow fractionation coupled to multi-angle laser light scattering detectors: Applicability and analytical benefits for the analysis of environmental colloids. *Analyt. Chim. Acta* 552, 166–174.

von der Kammer, F., Ottofuelling, S., and Hofmann, T., 2010. Assessment of the physico-chemical behavior of titanium dioxide nanoparticles in aquatic environments using multi-dimensional parameter testing. *Environ. Pollut.* 158, 3472–3481.

Wei, G.-T., and Liu, F.-K., 1999. Separation of nanometer gold particles by size exclusion chromatography. *J. Chromatogr. A* 836, 253–260.

Weinberg, H., Galyean, A., and Leopold, M., 2011. Evaluating engineered nanoparticles in natural waters. *Trends Anal. Chem.* 30, 72–83.

Williams, A., Varela, E., Meehan, E., and Tribe, K., 2002. Characterisation of nanoparticulate systems by hydrodynamic chromatography. *Int. J. Pharm.* 242, 295–299.

Yao, K.-M., Habibian, M.T., and O'Melia, C.R., 1971. Water and waste water filtration: Concepts and applications. *Environ. Sci. Technol.* 5, 1105–1112.

Yin, Y., Yu, S., Liu, J., and Jiang, G., 2014. Thermal and photoinduced reduction of ionic Au(III) to elemental Au nanoparticles by dissolved organic matter in water: Possible source of naturally occurring Au nanoparticles. *Environ. Sci. Technol.* 48, 2671–2679.

Zhang, Y., Chen, Y., Westerhoff, P., and Crittenden, J., 2009. Impact of natural organic matter and divalent cations on the stability of aqueous nanoparticles. *Water Res.* 43, 4249–4257.

7 Rare Earth Elements in Wetlands

Mélanie Davranche, Gérard Grau, Aline Dia, Martine Le Coz-Bouhnik, Rémi Marsac, Mathieu Pédrot, and Olivier Pourret

CONTENTS

7.1 INTRODUCTION

In the periodic table, rare earth elements (REEs) represent a group of 15 elements, specifically the lanthanides (Table 7.1). They are often referred to as REEs, share common physiochemical properties, and, therefore, often occur together as elemental constituents of their host minerals.

The term *rare* is a carryover of the metallurgical processes that are needed to isolate the individual metal species that are complex and low productive. As a result, lanthanide metals or metal oxides are difficult to obtain and are, thus, considered rare (Sonich-Mullin et al., 2013). REEs form a coherent series of elements whose chemical properties display small but systematic changes with an increase in atomic number. This chemical coherence is due to the gradual filling of their 4f electron shell. Because outer electrons ($n = 5, 6$) shield this inner shell, there are only minor differences in the chemical reactivity along the series (e.g., de Baar et al., 1991; McLennan, 1994). On the basis of their atomic number, REEs are segregated into light REE (LREE) and heavy REE (HREE) categories with a division between Eu and Gd. Some authors distinguish middle REE (MREE) from Sm and Tb (e.g., Hannigan and Sholokovitch, 2001; Tang and Johannesson, 2003). This "weight" distinction is

TABLE 7.1
Yttrium and Lanthanide (REE) Symbols, Atomic Weight, and Ionic Radius for Coordination Number 6

	Symbol	Atomic Number (g mol^{-1})	Ionic Radius (Å)
Yttrium	Y	88.9	0.9
Lanthanum	La	138.9	1.03
Cerium	Ce	140.1	1.01
Praseodymium	Pr	140.9	0.99
Neodynium	Nd	144.2	0.98
Promethium	Pr		
Smarium	Sm	150.4	0.96
Europium	Eu	152	0.95
Gadolimium	Ga	157.2	0.94
Terbium	Tb	158.9	0.92
Dysprosium	Dy	162.5	0.91
Holmium	Ho	164.9	0.9
Erbium	Er	167.3	0.89
Terbium	Tb	168.9	0.88
Ytterbium	Yb	173	0.87
Lutetium	Lu	175	0.86

Source: Shannon, R.D., *Acta Crystallogr.*, A32, 751–767, 1976.

used to simplify the description and quantification of the inter-element relationship, typically ratios of normalized concentrations. Similarly, the anomalies of certain REEs due to the redox behavior of Ce and Eu and the large anthropogenic emission of Gd are used to interpret geochemical processes.

In aquatic systems, with regards to their slight solubility, REE concentrations are low compared with their concentration in rocks. Numerous studies suggest that the solution and interface chemistry are the major controlling factors of REE concentration in aquatic systems (Goldstein and Jacobsen, 1988; Elderfield et al., 1990; Sholkovitz, 1995). In solution, an important chemical property is that REEs can form strong complexes with a number of different ligands. In general, REE^{3+} ions prefer the donor atoms in the following order: O>N>S. The resulting chemical species tend to form mainly ionic bonds with REEs within their unoccupied lower high-energy orbitals (Weber, 2008).

For the sake of convenience, REE distribution in natural materials and water is usually illustrated by normalized REE patterns. Two geological reservoirs are used for the normalization: the upper continental crust and shales. REE abundance is therefore normalized to the Post-Archean Australian Shale (PAAS) (McLennan, 1989), North American shale composite (NASC) (Gromet et al., 1984), or upper continental crust (UCC) (Taylor and McLennan, 1985) (Table 7.2, Figure 7.1). A dominant systematic change is observed in REE chemical properties, such as solution complexation, which is caused by the decrease in the ionic radius with an increase in atomic number. This change results in a systematic REE fractionation in the pattern called the "REE contraction." A normalized REE pattern allows for the recognition of an anomalous concentration for an individual REE as a positive or negative anomaly in an otherwise smooth pattern. This type of anomaly can occur in response to the redox behavior of Ce and Eu, which can exist as a tetravalent or divalent state. Processes that convert Ce^{3+} into Ce^{4+} include biologically mediated oxidation (Moffett, 1990, 1994a,b) and abiotic oxidation on the surfaces of Mn oxides (Koeppenkastrop and De Carlo, 1992;

TABLE 7.2
Rare Earth Element Abundance in the Various References Used for REE Pattern Normalization

	UCC		NASC		PAAS	
	ppm	μmol L⁻¹	ppm	μmol L⁻¹	ppm	μmol L⁻¹
La	30	216.0	32	230.4	38.2	275.0
Ce	64	456.8	73	521.0	79.6	568.1
Pr	7.1	50.4	7.9	56.1	8.83	62.7
Nd	26	180.3	33	228.8	33.9	235.0
Sm	4.5	29.9	5.7	37.9	5.55	36.9
Eu	0.88	5.8	1.24	8.2	1.08	7.1
Gd	3.8	24.2	5.2	33.1	4.66	29.6
Tb	0.64	4.0	0.85	5.3	0.774	4.9
Dy	3.5	21.5	5.8	35.7	4.68	28.8
Ho	0.8	4.9	1.04	6.3	0.991	6.0
Er	2.3	13.8	3.4	20.3	2.85	17.0
Tm	0.33	2.0	0.5	3.0	0.405	2.4
Yb	2.2	12.7	3.1	17.9	2.82	16.3
Lu	0.32	1.8	0.48	2.7	0.433	2.5

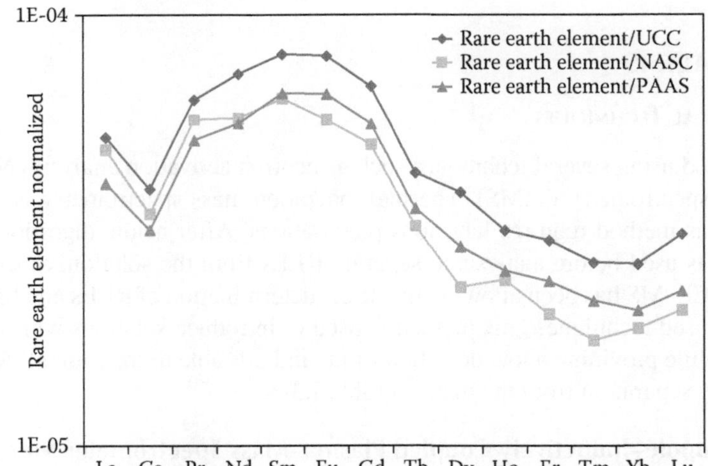

FIGURE 7.1 REE patterns normalized by using the UCC, NASC, and PAAS references. In X, a negative Ce anomaly is shown.

Sholkovitz et al., 1994; Ohta and Kawabe, 2001). The reduction of Eu generally occurs at high temperatures and pressures, such as in hydrothermal fluids (e.g., Michard et al., 1983; German et al., 1991; Klinkhammer et al., 1994). The anomalous behaviors of Ce and Eu are quantified via Ce and Eu anomalies, such as the Ce anomaly = $3Ce_N / (2La_N + 2Pr_N)$ or $2Ce_N / (La_N + Pr_N)$ and the Eu anomaly = $2Eu_N / (Sm_N + Gd_N)$, with N equal to the normalized abundance (Figure 7.1).

The REE pattern, therefore, results from the combination of several processes that are able to induce their fractionation. These processes are themselves controlled by several physicochemical mechanisms and parameters. In between, three processes can be distinguished into the following categories: (i) precipitation/dissolution, (ii) sorption onto colloids and particles, and (iii) complexation in solution with organic and inorganic ligands. The REE pattern, therefore, corresponds to the REE pattern for the mineral sources that is modified by the sorption/complexation constants of the REE with ligands, colloids, and particles. The result is evident in highly diverse REE patterns that can be measured by a degree of depletion or enrichment relative to heavy REE (La/Yb or Sm/Yb ratios) or by whether or not anomalies are present.

The coherent physicochemistry of REEs allows their abundances and fractionation in rock to be used as a tracer and fingerprint for cosmochemical, geodynamic, and petrogenetic processes (e.g., Henderson, 1984; Taylor and McLennan, 1985). The decrease in the quantification limit for REE analytical techniques such as inductively coupled plasma–mass spectrometry (ICP–MS) also offers the opportunity for REE patterns to be used as a tracer for processes occurring in hydrosystems (ocean, surface, and groundwater) (e.g., Goldberg et al., 1963; Elderfield and Greaves, 1982; de Baar et al., 1983; Byrne and Kim, 1990; German et al., 1991; Bau et al., 1996; Duncan and Shaw, 2004; Sholkovitz, 1993; Bau and Dulski, 1996; Dupré et al., 1996; Elbaz-Poulichet and Dupuy, 1999; Shiller et al., 2002; Fee et al., 1992; Möller and Bau, 1993; De Carlo and Green, 2002; Gammons et al., 2003; Johannesson and Lyons, 1995, 1996, 1997; Viers et al.,1997; Aubert et al., 2001; Janssen and Verweij, 2003). More specifically, REE fractionation patterns and abundances have been used to investigate the processes occurring in wetlands, such as the hydrology of the system, the mineral phases activated during water saturation, the trace element sources, and the fine sorption processes occurring on wetland colloidal organic matter (OM) (Dia et al., 2000; Grybos et al., 2007, 2009; Pourret et al., 2007, 2010; Marsac et al., 2011; Davranche et al., 2011, 2015).

7.2 REEs ANALYSIS

7.2.1 Analytical Techniques

REEs are analyzed using several techniques such as neutron activation analysis (NAA) or thermal ionization mass spectrometry (TIMS). Thermal ionization mass spectrometry is a highly precise technique, but this method requires laborious preparations. After acidic digestion, liquid column chromatography is used before analysis to separate REEs from the solution. Since the 1990s, the development of ICP–MS has been allowing the direct determination of REEs and has, thus, become the most widely used technique. This method is used to introduce solutions with weak concentrations of a REE while providing a low detection limit, and it is able to measure all REEs simultaneously without any separation from the matrix (Table 7.3).

7.2.1.1 Quadrupole–Inductively Coupled Plasma–Mass Spectrometer Measurements

Ions produced in high-temperature plasma are identified on the basis of the mass-to-charge ratio, *m*/*z*, which is characteristic of a given isotope. Each REE has at least one isotope that is free from isobaric overlap, and the sensitivity is relatively uniform from 139La to 175Lu.

The major analytical problem encountered with quadrupole–inductively coupled plasma–mass spectrometer (Q–ICP–MS) is the level of oxide formation in the plasma (Longerich et al., 1987; Jarvis et al., 1992) (Table 7.3). Each REE forms a continuous group from 139 to 175 *m*/*z*, and the formation of a light REE and barium oxide can produce significant middle REE interferences. In many cases, the LREE concentrations are higher than those for heavy REEs, and the potential for interferences is increased. Refractory oxide ions are influenced by the plasma operating parameters. The oxide production level is close to 1%–2%. The operating conditions are presented in Table 7.3. Mathematical corrections are required in order to suppress these spectroscopic interferences; these corrections are calculated as a function of the oxide level (Aries et al., 2000; Raut et al., 2005a,b). Previous studies have shown the importance of these corrections for accurate

REE quantification. The choice of the measured isotopes and interferences that can be applied are summarized in Table 7.3.

Calibrations can usually be performed by using synthetic, multielemental solutions that are prepared in 2% HNO_3. The quality of the blanks is fundamental, because the level of concentrations for this technique is very low. Instrument drift is monitored and corrected by spiking each sample with an internal standard (In, Re, Rh, etc.) or by introducing it online by using a peristaltic pump.

All of the experimental solutions used must be prepared with ultrapure, analytical-grade solution. Polyethylene and Teflon® vessels must have been previously decontaminated, and a clean room is recommended in order to obtain lower values for the blanks. The accurate and precise determination of REE element concentration requires very low detection limits (DL). For the whole group of REEs, the DL are typically between 0.01 and 0.1 ng mL^{-1}. These values are presented in Table 7.4.

ICP–MS analysis usually includes three consecutive replicate measurements, producing a repeatability error less than 2%. To control the quality of REE element measurements, certificated reference material is necessary. The SLRS-4 followed by SLRS-5 standards (National Research Council Canada [CNRC]) that are distributed without certificated values of REE can be used to analyze fresh water and soil solutions. Several studies have published REE concentrations (Yeghicheyan et al., 2001, 2013; Lawrence et al., 2006; Heimburger et al., 2013). For peat and organic material, no references are available, and therefore geological standards (USGS or NIST) and plant references (NIST1515, 1573a, etc.) are used (Ferrat et al., 2012).

TABLE 7.3

Preferred Isotopes for REE Analysis, Isotopic Abundance in %, and Oxide and Hydroxide Interferences in Analyte and Instrument Operating Conditions

REE Element	Isotope Mass	Abundance (%)	Oxide and Hydroxide Interferences
La	139	99.91	
Ce	140	88.48	
Pr	141	100	
Nd	146	17.19	
Sm	147–149	15	^{130}BaOH
Eu	151–153	47.8–52.2	^{135}BaO, ^{137}BaO, ^{136}BaOH,
Gd	157–158	15.65–24.84	^{142}CeO, ^{142}NdO, ^{141}PrO, ^{141}PrOH
Tb	159	100	^{143}NdO
Dy	163	24.9	^{147}SmO, ^{146}NdOH,
Ho	165	100	^{149}SmO
Er	166	33.6	^{151}EuO
Tm	169	100	
Yb	172–174	21.9–31.8	^{157}GdOH
Lu	175	97.41	^{159}TbO
Plasma Conditions			
RF power			1,450–1,550W
Carrier gas			15 L/min
Auxiliary argon flow			1 L/min
Nebulizer argon flow			0.9–1.0 L/min
CeO^+/Ce^+			1%
Ce^{2+}/Ce^+			<2%

TABLE 7.4
Detection Limits in ng L^{-1} (ppt), Set Values Obtained on HP 7700 × ICP–MS, Calculated from the Asssociation Française de Normalisation (AFNOR) Standards, and Based on the Blank Measurement

Element	Isotope	Detection Limit (ppt)
La	139	0.15
Ce	140	0.14
Pr	141	0.06
Nd	146	0.16
Sm	147	0.23
Eu	153	0.10
Gd	157	0.22
Gd	158	0.13
Tb	159	0.04
Dy	163	0.17
Ho	165	0.05
Er	166	0.15
Tm	169	0.04
Yb	174	0.17

7.2.1.2 High-Resolution or Sector-Field Inductively Coupled Plasma–Mass Spectrometer Measurements

High-resolution or sector-field inductively coupled plasma–mass spectrometry (HR-ICP–MS or SF-ICP–MS) is more sensitive than Q–ICP–MS. This technique is used to analyze REE concentration with high precision in aqueous samples. Typically, HR-ICP–MS is able to eliminate or reduce interferences due to mass overlap; it also presents resolving powers for mass separation up to 10,000 and operates at preset resolution settings for low, medium, or high resolution.

To increase the precision of REE measurements, a sample can be enriched with a pre-concentrated multispike REE solution, and the matrix can be removed by using ion chromatography (Bakers et al., 2002; Rousseau et al., 2013). The level of the detection limit is very low, close to a pg L^{-1} (ppq) level.

7.2.2 REE Measurements in Wetland Soils, Sediments, and Solutions

7.2.2.1 Sample Pretreatment

In water soil samples, the levels of REE concentrations are close to the μg L^{-1} to ng L^{-1} level (ppb to ppt). These solutions are rich in OM and Fe. Water samples must be collected and filtered by using 0.22-μm or 0.45-μm filters (cellulose acetate membrane filters) that have previously been decontaminated. The water is sampled in acid-washed polyethylene bottles and is immediately acidified after filtration with ultrapure HNO_3 up to 2% to prevent iron oxide or hydroxide precipitation. Acidified samples are stored at 4°C. The level of dissolved organic carbon (DOC) must be controlled, as a high concentration affects the ICP–MS analysis. For a high DOC concentration, the sample must be treated in order to eliminate the OM, after an acidic digestion with a mix of HNO_3 and H_2O_2 at 95°C on a hotplate or in a microwave furnace, and thus the sample is entirely solubilized.

In wetland soil and sediment samples, the levels of REE concentrations are close to the mg kg^{-1} to μg kg^{-1} (ppm to ppb) level. The organic-rich soil sample is dried at 20°C and then sieved at 2 mm. Dried soil samples (approximately 0.1–0.2 g) are first digested in an HNO_3 and HF mixture in Teflon® containers using hotplate or microwave autoclave protocols. Then, two acidic digestions

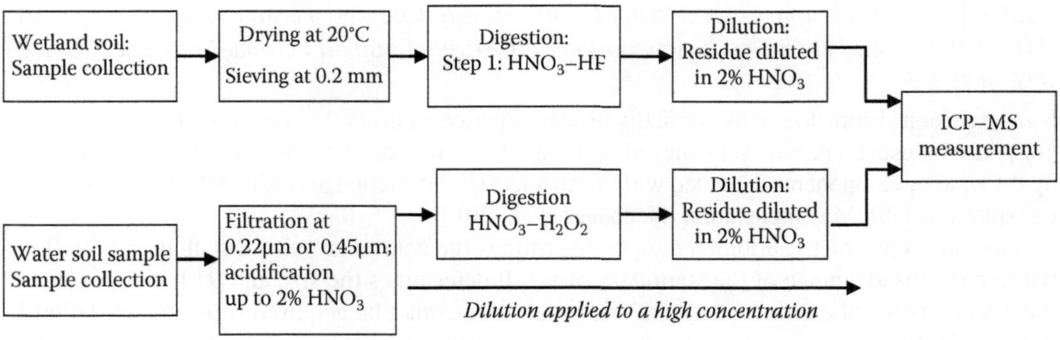

FIGURE 7.2 Pretreatment protocol for REE analysis in wetland soils and water soil samples.

are needed to remove the excess of HF, and sometimes, H_2O_2 treatment is necessary. The dried residue is dissolved to a clear nitric acid solution (2%) for analysis. Finally, the sample is diluted by a factor ranging from 500 to 2,000. The pretreatment protocols are shown in Figure 7.2.

7.2.2.2 Separation and Speciation of REE

REE element concentrations in wetland solutions are strongly associated with dissolved and colloidal OM (e.g., Dia et al., 2000; Grybos et al., 2007; Pédrot et al., 2008). The reactive fraction of this soluble OM mainly comprises humic substances (fulvic acid and humic acid [HA]), which are known to strongly bind metals and influence the sorption of trace metals onto mineral surfaces (e.g., Avena and Koopal, 1999). It is, therefore, interesting to study and to precisely identify the distribution of REE in the different organic and inorganic fractions. Several technologies can be used to identify this distribution, such as capillary electrophoresis, field-flow fractionation, and size exclusion coupled to ICP–MS or ultrafiltration systems.

7.2.2.2.1 Ultrafiltration Systems

Ultrafiltration systems are used to study the control of organic colloids on metal partitioning in water samples. To separate the colloidal bound elements from the non-colloidal elements, ultrafiltration experiments can be performed using centrifugal tubes at different molecular cutoff sizes (Amicon Ultra Millipore®, Vivaspin Sartorius®, Macrosep Pall®, etc.) that are equipped with permeable membranes of a decreasing pore size ranging from 30-20-15-10-5-3 or 2 kDa, with 1 Da = 1 g mol^{-1}. Each centrifugal filter device must be washed prior to use, in order to remove the glycerin protecting the membranes. Blank tests must be performed to determine possible contaminations (DOC and REE elements). The centrifugation speed is approximately 3,000–4,000×*g*; this is a function of the choice of ultrafiltration cells. The temperature must be controlled during centrifugation, and the length of time should be determined from the experiment.

According to the different cutoffs, REE–colloid complexes are retained by the ultrafiltration membrane whereas free ions and smaller complexes pass into the ultrafiltrate. The degree of metal–colloid complexation is usually determined from the metal concentration in the ultrafiltrate, relative to the original solution (e.g., Pourret et al., 2007; Pédrot et al., 2008; Vasuykovas et al., 2012).

7.2.2.2.2 Capillary Electrophoresis–Inductively Coupled Plasma–Mass Spectrometry

Capillary electrophoresis (CE) combined with ICP–MS (CE–ICP–MS) is used as a speciation tool in order to investigate the complexation of HA with trivalent REE. Capillary electrophoresis separates the elements with a short separation time; ICP–MS has an excellent elemental selectivity with a high sensitivity. The main advantage of this speciation method is the simultaneous detection of metal that is complexed or not by HA in only one analytical run. CE–ICP–MS studies are able to identify free and complexed HA–REE species and to qualify the ligand effect (Sonke and Salters, 2005; Stern et al., 2007; Kautenburger et al., 2014).

7.2.2.2.3 Size-Exclusion Chromatography–Inductively Coupled Plasma–Mass Spectrometry (SEC–ICP–MS) and Field-Flow Fractionation–Inductively Coupled Plasma–Mass Spectrometry (FFF–ICP–MS)

Size-exclusion chromatography is usually used to separate natural OM over a column with a stationary phase by using a porous gel material. The high molecular mass of the OM elutes first, followed by the smaller components. Coupled with ultraviolet (UV) detection and ICP–MS, SEC can be used to explore metal/OM complexation (Neubauer et al., 2013).

The field-flow fractionation technique determines the continuous size distribution of colloids without the disadvantage of the stationary phase. It determines the size and REE composition of the distinct types of colloids, information that usually cannot be acquired from standard ultrafiltration. Field-flow fractionation–inductively coupled plasma–mass spectrometry (FFF–ICP–MS) is used to characterize REE binding to colloids (Stolpe et al., 2013; Neubauer et al., 2013).

7.3 REE IN WATERLOGGED SOIL AND SEDIMENTS

Waterlogged soils and sediments are probably one of the most striking investigated surface environments involving the use of REE as efficient tracers of processes and/or matter sources.

Wetland and paddy soils are temporally or permanently flooded, which involves the development of anoxic conditions and subsequent soil reduction. Both OM adsorption and Fe(III) oxyhydroxide reduction accompanying wetland soil flooding strongly control the mobility, transfer, and fate of REE in wetlands (e.g., Grybos et al., 2007, 2009). Therefore, specific REE patterns that are displayed within these environments are related not only to the sources but also to the prevailing reduction processes and associated OM and Fe dynamics.

Mihajlovic et al. (2014), who studied REE distribution in wetland soil profiles (Eutric Fluvisols, at the Wupper River, Germany), found very small differences between the total REE concentration and distribution between the different horizons, as shown in Figure 7.3. Total REE concentrations and indicators of REE pattern fractionation (Eu/Eu*, Ce/Ce*, La/Yb ratio) do not evolve significantly. They explained these minute differences by the wetland soil flooding and the subsequent homogenization processes.

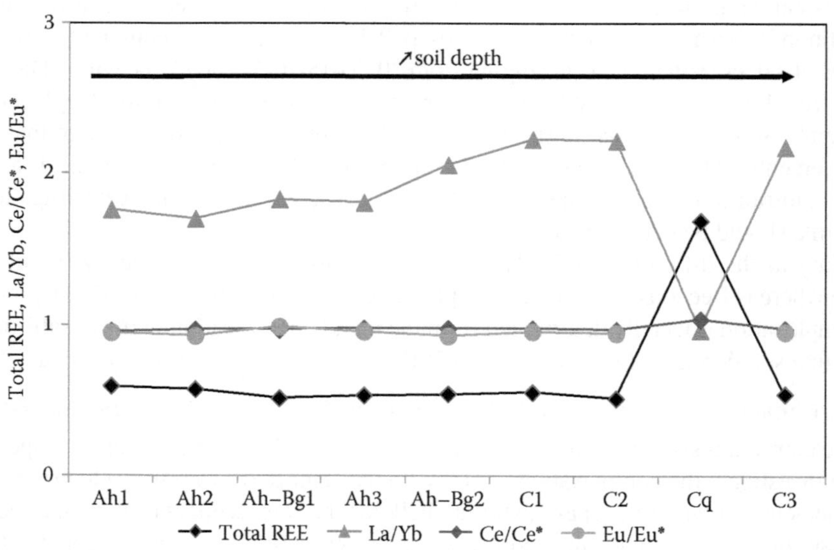

FIGURE 7.3 Evolution of the total REE concentration, La_{UCC}/Yb_{UCC}, Eu/Eu*, and Ce/Ce*, relative to the different soil horizons (Ah1, Ah2, Ah3, Ah–Bg, C1, C2, Cq, C, WRB classification) in Eutric Fluvisols, at the Wupper River, Germany. (From Mihajlovic, J. et al., *Geoderma*, 228–229, 160–172, 2014.)

Sequential extractions that are performed on soil wetlands generally show, regardless of the protocol, that REEs dominate in the residual fraction, followed by the reducible and the oxidizable fraction, with the exchangeable fraction being very low (Leybourne and Johannesson, 2008; Pédrot et al., 2008; Davranche et al., 2011; Mihajlovic et al., 2014).

7.3.1 REE Signature of the Solid Fraction

The REE patterns displayed by wetland soils or floodplain sediments could provide a large amount of information. The REE patterns of floodplain sediments taken from the Kaveri River basin (southern India) were discussed in terms of REE behavior during sedimentary processes and as provenance tracers (Singh and Rajamani, 2001). As reported elsewhere (McLennan, 1989; Morey and Setterholm, 1997; Vital and Stattegger, 2000), this study clearly pointed out that fluvial sorting processes affect REE distribution in the sediments. This type of physical sorting by fluvial processes could result in the accumulation of various minerals that are possibly enriched in REEs. As a consequence, an REE-mediated provenance assessment study should be undertaken with sediments characterized by similar granulometric grades, because they would more closely reflect the source area (Singh and Rajamani, 2001). However, it should also be considered that a large amount of REEs could be also present as surface coatings on grains in addition to those occurring within heavy minerals and clay minerals.

Paddy fields are usually distributed in flood plains along rivers and associated tributaries and valley floor plains that are incised by small rivers (Egashira et al., 1997, 2004), such as those occurring in the Mekong River. REE content and associated patterns were studied, as well as the particle-size distribution and clay mineralogical composition, to estimate both the origin and fertility potentiality of the soil materials (Figure 7.4) (Egashira et al., 1997; Singh and Rajamani, 2000). The various REE patterns displayed in the paddy field samples along the Mekong River were divided into two groups, establishing that the material origin was controlled by local composite materials and sediment carried by the Mekong River. This division was (i) confirmed through a mineralogical survey establishing differences in the origin and genesis of the soils, notably between the upper and lower areas, (ii) compared with indicators of soil potentiality such as the exchangeable Ca or organic C content. However, conversely to samples recovered along the Mekong River (Egashira et al., 1997), the unambiguous use of REE patterns was not capable of assessing the t soil origins in floodplains of the Brahmaputra, Meghna, and Ganges rivers in Bangladesh (Egashira et al., 2004).

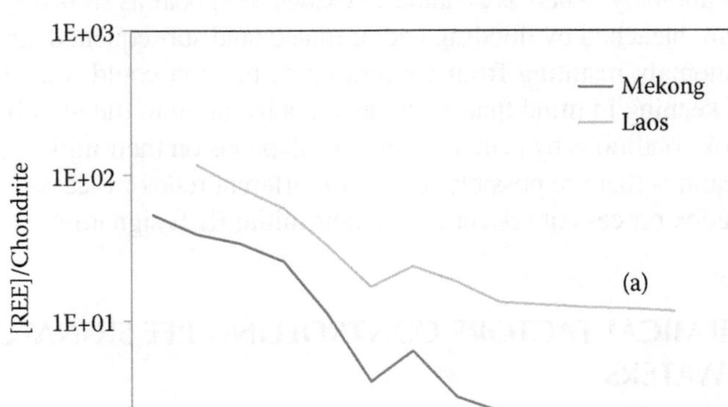

FIGURE 7.4 Chondrite-normalized REE patterns of (a) a soil recovered in a lowland of the Mekong River floodplain in Laos; and (b) a sediment collected in the floodplain of the Kaveri River basin in southern India. ((a) From Egashira, K. et al., *Geoderma*, 78, 237–249, 1997. (b) From Singh, P. and Rajamani, V., *Geochim. Cosmochim. Acta* 65, 3093–3108, 2001.)

It is of prime importance to remember that REE composition in soils has mostly been considered as being inherited from the parent rock (Taylor and McLennan, 1985), given the low solubility and relative immobility of REEs in the upper crust of the earth, whereas the environmental parameters and processes involved in soil formation as well as the anthropogenic inputs cannot be neglected (e.g., manure, phosphate fertilizers, or waste effluents) (Tsumura and Yamasaki, 1993; Yuan et al., 2001; Protano and Riccobono, 2002).

7.3.2 REE Fingerprinting of the Paleoredoximorphic Features in Waterlogged Soils

Wetlands and paddy soils are at the heart of alternating redox processes induced by waterlogging that are associated with the occurrence of large amounts of OM that are responsible for leachings, Fe-oxide dissolution or precipitation as oxide coatings, and concretions or amorphous organo-mineral colloidal phases.

Since both REE are strongly associated with Fe and Mn oxides and their fractionation is related to the drainage conditions, they are particularly interesting with regards to tracing the redox conditions, especially when considering Ce and Eu. However, all Eu fractionation must be precluded in the soils since the required reducing conditions that result in Eu reduction are far from the conditions possibly encountered within the soils (Bonnot-Courtois, 1981; Henderson, 1984; Panahi et al., 2000). By contrast, if the Eu anomaly cannot trace any redox processes occurring within the soils, the oxidation of Ce(III) to Ce(IV) can take place in soils, as previously established through field observations and experiments (Takahashi et al., 2000). This process results in the precipitation of the so-called cerianite (CeO2) coupled with the reduction of Mn(IV) to Mn(III) on the surface of the Mn oxides (Ran and Liu, 1992; Bau, 1999; Ohta and Kawabe, 2001). Oxidative conditions allow for new Fe and Mn oxide precipitation, preferentially incorporating Ce over REE and leading to the development of a positive Ce anomaly. Conversely, when reducing conditions prevail, notably during podzolization, Fe oxides are reduced by OM; Fe can be transferred within the organic complex and then possibly precipitated as ferrihydrate (Schwertmann and Fisher, 1973; Buurman and Jongmans, 2005; Sauer et al., 2007; Laveuf and Cornu, 2009). The newly precipitated Fe-rich phases might not only be enriched in HREE with regards to LREE, in response to the observed preferential transfer of HREE as organic complexes, but also be enriched in MREE when the newly formed phases remain amorphous (Aubert et al., 2001). The dissolution of the Fe and Mn oxides, which release all REEs except Ce bound to cerianite, the dissolution of which is more dependent on pH than E_h, results in the persistence of the Ce anomaly; whereas Fe and Mn oxides disappear as shown by Koppi et al. (1996) in clayey areas that are bleached by flooding and drainage (and subsequent degradation) in northern Australia. The Ce anomaly resulting from cerianite precipitation could trace the redox processes (Koppi et al., 1996), keeping in mind that the relative contribution to the mobilization of REEs that is made through redox conditions by primary minerals depends on their initial proportion in the different pedological features that are possibly related to different redox processes, their relative mobilization during the redox process of concern, and their initial REE signatures (Laveuf et al., 2012).

7.4 BIOGEOCHEMICAL FACTORS CONTROLLING REE SIGNATURE IN WETLAND WATERS

One of the critical questions concerning REE behavior in wetlands is whether REE concentrations are controlled by the sources or the physicochemistry. Besides the specific characteristics of a specific site, it seems reasonable to state that the REE dynamics in the water–soil system depends on: (i) the fractionation characteristics of the host rock/sediment, (ii) the weathering process that might improve the dissolution of a mineral that is either depleted or enriched in REEs, (iii) the water physicochemical characteristics (pH, E_h, organic and inorganic ligands, colloids/particles), and (iv) the water hydrodynamics. A systematic evolution in the REE pattern is therefore observed

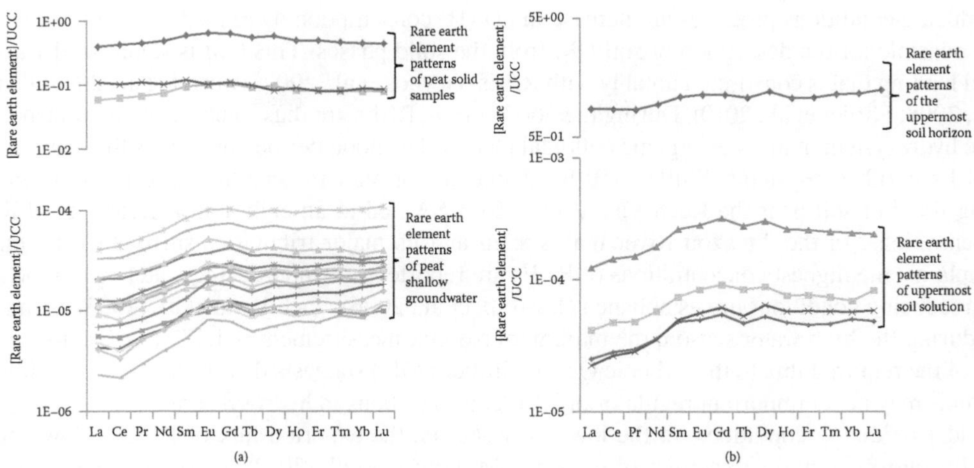

FIGURE 7.5 (a) REE patterns for a solid peat sample and peat solutions from the Contentin marshland (France). (b) REE patterns for the uppermost soil horizon and its soil solution at various sampling dates from the riparian Le Home wetland (Brittany, France). The peat and soil REE patterns are flat, whereas the peat and wetland soil solution patterns exhibit an LREE depletion that is significant of an REE fractionation during their solubilization. ((a) From Auterives, C., Ph.D. thesis, University of Rennes I, France, Mémoires du CAREN 17, ISBN 2-914375-46-8, p. 261, 2007. (b) Modified from Gruau, G. et al., *Wat. Res.*, 38, 3576–3586, 2004.)

between the wetland soil/sediments and the soil solution or shallow groundwater. This evolution indicates that soil cannot be the one single factor controlling REE dynamics in wetland solutions (Figure 7.5). Several physicochemical parameters, either combined or not, have been advanced to explain this discrepancy.

7.4.1 Seasonal Flooding and Redox Conditions

In wetlands, redox conditions and their alternations are the main factors accounting for REE solubilization and mobilization in the environment. Many authors have therefore considered Mn and Fe oxides reducing dissolution/precipitation as a major parameter controlling trace metal mobility in wetland soils (Charlatchka and Cambier, 2000; Chuan et al., 1996; Davranche et al., 2003; Francis and Dodge, 1990; Green et al., 2003; Quantin et al., 2001, 2002). However, in the case of REEs, Fe and Mn oxides seem to just act indirectly. The gradual establishment of reducing conditions in wetlands results in the release of metals, either redox sensitive or not, such as Fe or Mn; Pb and Cd; as well as a large concentration of dissolved/colloidal OM (fraction < 0.45 or $0.2\ \mu m$). Several studies report positive correlations between Mn(II) and Fe(II) concentrations in wetland soil solutions and dissolved OM (DOC) (Hagedorn et al., 2000; Olivié-Lauquet et al., 2001; Gruau et al., 2004). Grybos et al. (2009) demonstrated that OM is mainly released as humic substances that are desorbed from soil minerals in response to the rise in pH that is caused by the reduction reactions (H^+ consumption). They also reported that the colloidal fraction of this OM strongly bound a large range of metals, including REEs. They experimentally demonstrated that REE speciation was entirely dominated by their binding with the dissolved/colloidal organic fraction of the soil solution. This strong binding of REEs by OM has been confirmed by several field, experimental, and modeling studies (Tang and Johannesson, 2003, 2010; Sonke and Salters, 2006; Pourret et al., 2007a,b; Stern et al., 2007; Pédrot et al., 2008; Kerr et al., 2008; Marsac et al., 2011). Using REE patterns as a tracer for REE sources in wetland soils, Davranche et al. (2011) demonstrated that soil OM was the main source of REEs and trace metals during wetland soil reduction. Iron was mainly present in the soil as amorphous Fe(III) nanoparticles, which are poor in REEs and trace elements and are embedded within the OM. Therefore, in permanently or temporarily flooded wetlands, the establishment of reducing conditions produces an increase in pH (H^+ consumption by reductive reactions), which is responsible for the desorption of soil OM from the solid phases. This OM is solubilized with its metal loading that is constituted notably with REEs (Grybos et al., 2007; Pourret et al., 2007; Pédrot et al., 2008; Shiller et al., 2010). During the flood period, REEs are thus solubilized and transported to the hydrosystem, mainly as organic colloidal phases. The flood period constitutes the major input of REEs into hydrosystems. Shiller (2010) calculated that reducing conditions resulting from the spring flood of soil near the Loch Vale (Colorado, USA) led to an eight-fold increase in all REE concentrations. In the Amazon River mainstream and its major tributaries surrounded by many floodplains, the highest concentrations of REEs are reported in winter during water saturation and when reducing conditions are established (Barroux et al., 2006). Tachikawa et al. (2003) estimated that during the high water season, the maximum Nd flux measurement is 1,277 t yr^{-1}, constituting 30% of the required flux to the Atlantic Ocean. Shiller (2010) suggests that the seasonal flooding of wetlands may be an important regulator of REE concentrations in hydrosystems.

Under oxidizing conditions, in the low water season, the exported flux of REEs is low and is mainly controlled by the dynamics of the soil OM. Pourret et al. (2010) showed that REE speciation is controlled by colloidal OM that is present in wetland soil solution even under oxidizing conditions. In the same way, Dia et al. (2000) did not observe any significant evolution in the REE patterns in shallow groundwater from the Naizin wetland (Brittany, France) between periods of oxidized and reduced conditions. Figure 7.5 shows that E_h decreases and the Fe concentration increases in solution subsequent to the reductive solubilization of Fe(III). The establishment of the moderately reducing condition caused an increase in the dissolved REE concentration but without any drastic modification of the REE pattern (Figure 7.6). The speciation of REEs in the shallow groundwater was therefore not significantly modified between the oxidized and reduced periods. Thus, REEs are bound to the dissolved/colloidal OM present in the soil solution and shallow groundwater under oxidizing and reducing conditions.

7.4.2 Colloidal Control

In wetland soil solution or shallow groundwater surrounded by wetlands, REEs are closely associated with colloids. Studies performed on surface water that drains wetlands, wetland solution, or shallow wetland groundwater from various type of boreal, tropical, Mediterranean, or temperate wetlands used an ultrafiltration analysis and various pore sizes to demonstrate that REEs are mainly concentrated in the high-molecular-weight fraction, namely bound to the colloid phases (Viers et al., 1997; Dia et al., 2000; Tang and Johannesson, 2003; Pourret et al., 2007a; Pédrot et al., 2008; Cidu et al., 2013; Vasyukova et al., 2012; Neubauer et al., 2013). If colloids are regarded as the main transfer and binding phases of REE in wetland solutions and waters, there is no real consensus on the nature of the REE carrier phases in the colloids themselves. The Fe and OM phases are the major components of the colloids that are encountered and formed within wetlands or waterlogged soils. The term *wetland* covers a large diversity of areas that are subject to various hydrodynamic and climatic conditions that drastically influence the nature of the colloids released in solution. Andersson et al. (2006) demonstrated that in subarctic boreal rivers, draining organic-rich soils, two different REE colloid phases can be distinguished. During the sole spring flood, subsequent to soil saturation, small organic-rich colloids (~3 nm) are released. By contrast, large Fe-rich colloids (~12 nm) are formed during the winter and spring floods. REEs are bound to both C-rich and F-rich colloids. However, the amount of released REE is higher during the spring flood when organic-rich colloids are present. Moreover, the subsequent LREE-enriched REE pattern suggests that REEs are released with the organic-rich colloids that are found in the litter of the organic-rich topsoil. In a temperate climate, in swamp water, speciation modeling and the voltammetric titrations indicate that dissolved REEs in the Great Dismal Swamp water are controlled by REE complexation with natural OM (Johannesson et al., 2004). Neubauer et al. (2013) applied flow field-flow fractionation (FlowFFF) to water sampled in a small stream draining an unpolluted wetland (Tanner Moor) in Upper Austria to

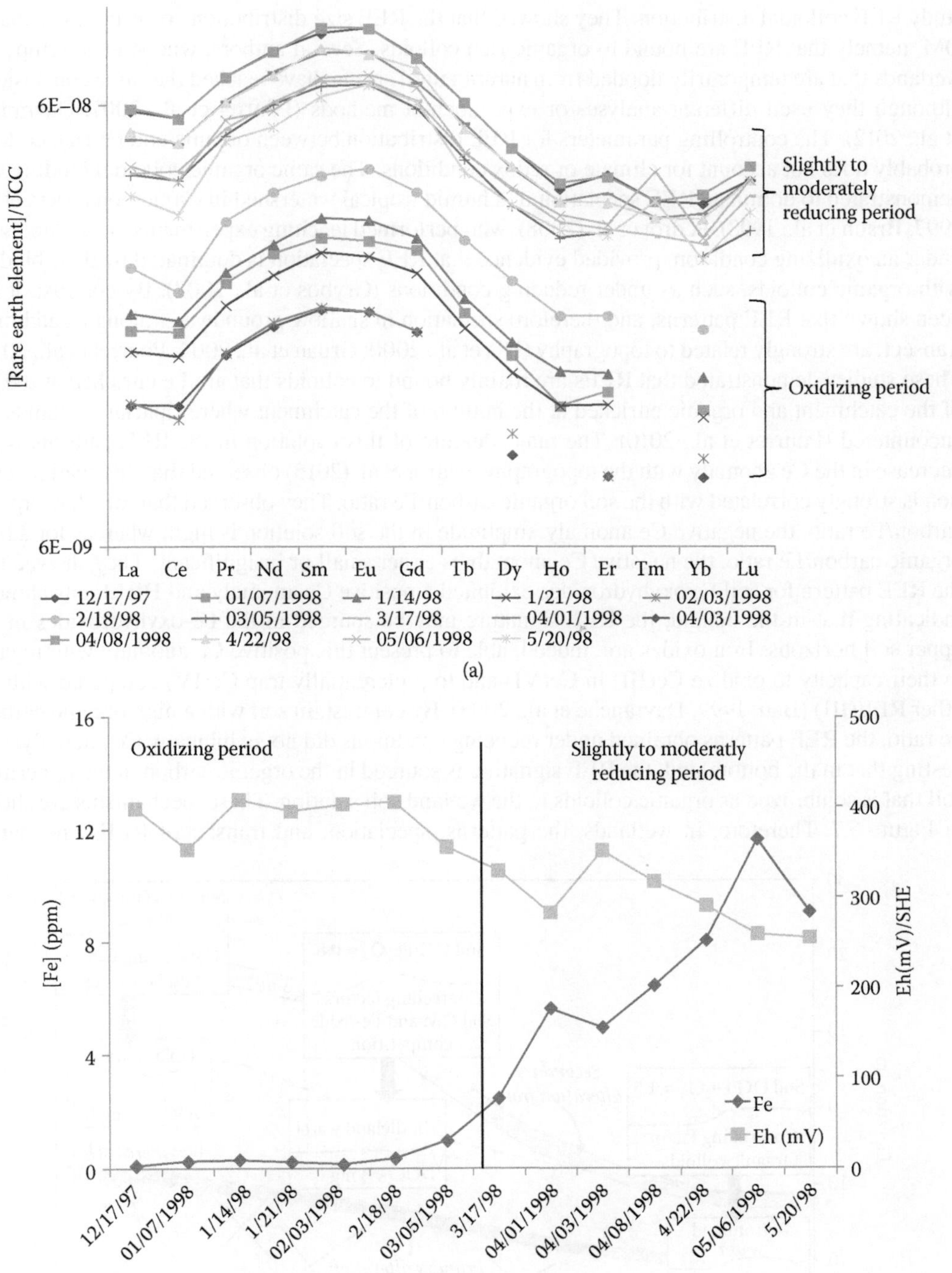

FIGURE 7.6 (a) Evolution of the REE patterns relative to time in shallow groundwater from the Naizin wetland (Brittany, France). (b) Evolution of the E_h and Fe concentration, indicating the establishment of moderately reducing conditions and the reductive dissolution of soil Fe oxides. Although the redox conditions became moderately reduced, the REE pattern was not significantly modified, suggesting no drastic damage in the REE speciation. ((b) From Dia, A. et al., *Geochim. Cosmochim. Acta*, 64, 4131–4151, 2000.)

study REE colloidal distribution. They showed that the REE size distribution corresponds to that of OM, namely that REE are bound to organic-rich colloids. Several authors, who studied temperate wetlands that are temporarily flooded from autumn until spring, have reached the same conclusions, although they used different analyses or experimental methods (Pourret et al., 2007; Davranche et al., 2012). The controlling parameters for REE distribution between organic and Fe-rich colloids probably does not account for climate or redox conditions. The same organic colloidal binding was demonstrated to dominate REE speciation in a humid tropical watershed in Cameroon (Viers et al., 1997; Braun et al., 1998). Pédrot et al. (2008), who performed leaching experiments on wetland soils under an oxidizing condition, provided evidence that REE speciation is dominated by their binding with organic colloids, such as under reducing conditions (Grybos et al., 2007). By contrast, it has been shown that REE patterns, and therefore speciation in shallow groundwater along a catchment transect, are strongly related to topography (Dia et al., 2000; Gruau et al., 2004; Pourret et al., 2010). These studies demonstrated that REEs are mainly bound to colloids that are Fe enriched at the top of the catchment and organic enriched at the bottom of the catchment where riparian wetlands are encountered (Pourret et al., 2010). The major feature of this evolution in the REE patterns is the decrease in the Ce anomaly with the topography. Pédrot et al. (2015) observed that this spatial variation is strongly correlated with the soil organic carbon/Fe ratio. They observed that for a low organic carbon/Fe ratio, the negative Ce anomaly amplitude in the soil solution is high; whereas for a high organic carbon/Fe ratio, the negative Ce anomaly is either small or insignificant. They showed that the REE pattern for soil Fe oxyhydroxides exhibited a positive Ce anomaly and HREE enrichment, indicating that in the upland, the REE signature may be sourced in the Fe-oxyhydroxides in the upper soil horizons. Iron oxides are, indeed, able to present this positive Ce anomaly with regards to their capacity to oxidize Ce(III) in Ce(VI) and to preferentially trap Ce(IV) compared with the other REE(III) (Bau, 1999; Davranche et al., 2005). By contrast, in soil with a high organic carbon/Fe ratio, the REE patterns obtained under reducing conditions did not exhibit any Ce anomaly, suggesting that in the bottomland, the REE signature is sourced in the organic carbon in the uppermost soil that is solubilized as organic colloids in the wetland soil solution. These mechanisms are shown in Figure 7.7. Therefore, in wetlands, the patterns, speciation, and transfer of REEs are mainly

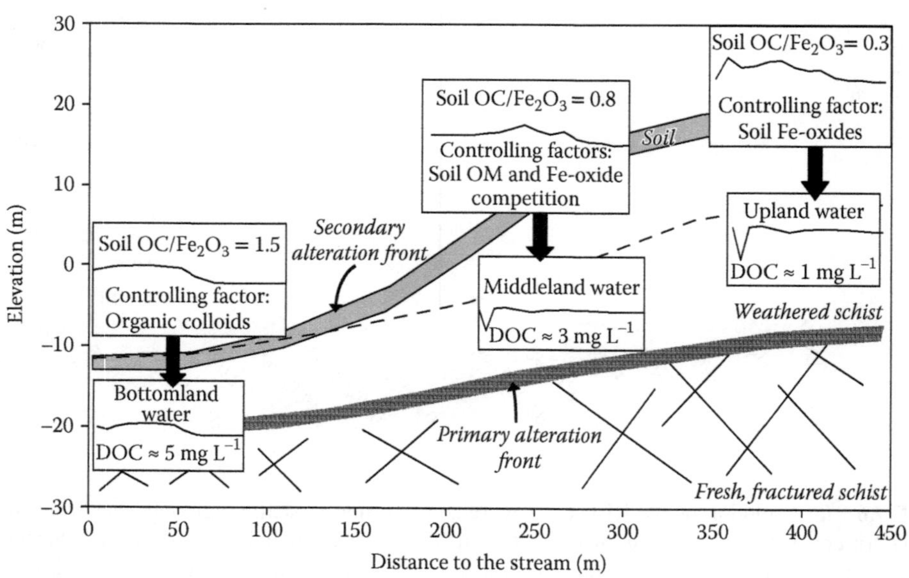

FIGURE 7.7 Sketch summarizing the processes responsible for the topography-related REE pattern in a theoretical catchment. At the top of the catchment, a negative Ce anomaly in the shallow groundwater is sourced in the presence of Fe oxide in the uppermost soil although the REE pattern is controlled by the solubilization of organic colloids in the bottomland occupied by riparian wetlands.

controlled by colloids. The composition of these colloids (organic or Fe rich) is dependent on the wetland soil composition, and on the mechanisms that themselves control the formation and transfer of colloids in solution, such as, for instance, hydrodynamic conditions.

7.4.3 Impact of Biological Parameters

Biological parameters could potentially influence REE distribution in wetland waters through both direct and indirect mechanisms. Several authors have demonstrated the ability of the bacterial cell surface to bind REEs (Takahashi et al., 2005, 2010; Ngwenya et al. 2009, 2010). The resulting REE pattern exhibits a tetrad effect and a prominent enrichment in HREEs (from Er to Lu). Based on extended x-Ray absorption fine structure (EXAFS) evidence and modeling calculations, this shape was further attributed to the binding of REEs as inner-sphere complexes with carboxylic and multiple phosphate sites occurring on the surface cells (Takahashi et al., 2010; Ngwenya et al., 2010; Martinez et al., 2014). Simpson et al. (2007) suggested that bacterially derived biomass could constitute more than 50% of the total soil organic carbon in aerobic soils. However, in wetlands, with regards to the temporary or permanent saturation of the soil, the bacterial activity is low; the nitrate respiratory or Mn and Fe reduction is less energetic compared with the O2 respiratory in soil where aerobic conditions prevail. A direct consequence is that organic molecules are transformed into humic substances rather than degraded. The proportion of bacterial biomass as compared with humic ligands is, therefore, potentially lower than 50%. Moreover, bacteria and cell residues have to also compete with the soil organic ligand present in high amounts for REE binding. Therefore, it is unlikely that REE binding by bacterial cells could account significantly for the REE pattern and distribution. Another mechanism that can be inferred by REEs behavior in wetlands is the mechanism used by plants. The absorption of REEs by plants is low (Lima e Cunha et al., 2012). REE concentration varied from 1 to 500 ppm depending on the plant species, organs, and soil concentrations. However, the internal processes of plants can fractionate the REE (Ding et al., 2006; Lima e Cunha et al., 2006). For example, Ding et al. (2006) observed MREE enrichment in the roots and MREE and HREE enrichment in the leaves. These results suggest that the REE patterns observed in the wetland solution could potentially be inherited from their fractionation in wetland plants. However, further studies need to be carried out in order to confirm or dispel this hypothesis.

Therefore, the major biological mechanism that seems to significantly account for the behavior and distribution of REEs in wetlands is the indirect bioreduction of Mn(IV) and Fe(III), which is mediated by the bacteria consortium occurring in wetlands. The saturation of wetland soil with water promotes the use of Mn and Fe oxides as the e-acceptor by the bacteria for their growth. This reduction involves the concomitant dissolution of Mn and Fe oxides and the increase in pH, which are responsible for the release of REEs into solution.

7.4.4 Organic Matter Control

In wetland soil solution or shallow groundwater, REEs are mainly associated with organic colloids. However, two types of REE patterns are generally observed: an MREE downward concavity (Figure 7.8a) and an HREE enrichment (Figure 7.8b). Pourret et al. (2007b) experimentally demonstrated that the distribution coefficient (Kd) between REE and HA increases for MREE at a pH less than 7 and at high REE concentrations with respect to HA (Figure 7.8a). Between pH values from 6 to 9 and with a low REE concentration with regards to HA, Sonke and Salters (2006) observed a regular increase from La to Lu, that is, a "lanthanide contraction" (Figure 7.8b). Marsac et al. (2010, 2011) demonstrated that this discrepancy between both kinds of REE patterns is explained by the combined effect of the metal loading and the surface heterogeneity of HA. At low loading, REEs are complexed to strong but less abundant HA sites, namely phenolic and/or multidendate sites. The resulting REE pattern exhibits a lanthanide contraction. By contrast, at high metal loading, REEs are complexed to weak but more abundant HA sites, which are indicated by the carboxylic group. The resulting REE pattern exhibits an MREE downward concavity.

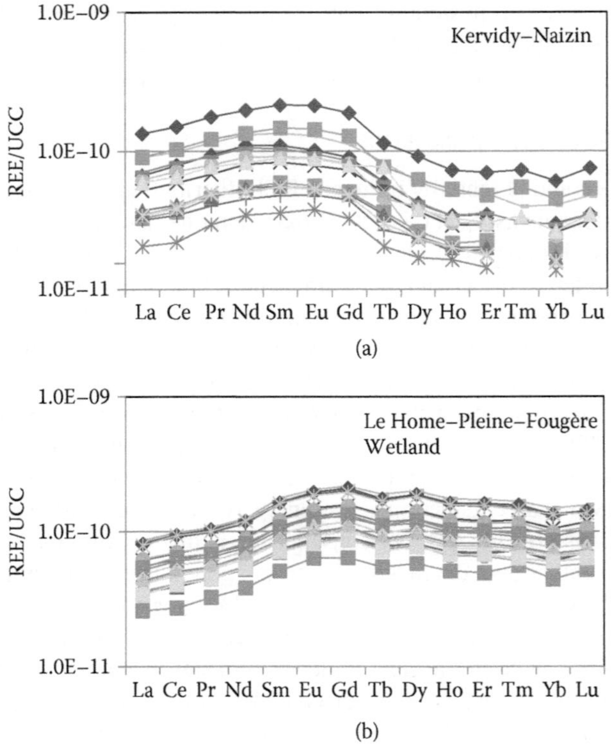

FIGURE 7.8 REE patterns in the dissolved fraction (<0.2 μm) of the organic-rich shallow groundwater from the (a) Kervidy–Naizin wetland (France) (DOC ≈ 15 ppm) and (b) Le Home–Pleine–Fougère wetland (France) (DOC ≈ 27 ppm). Both REE patterns are different, although the modeling calculations using Model VI and ultrafiltration analysis showed that REEs are bound to colloidal organic matter at around 90%. (From Gruau, G. et al., *Wat. Res.*, 38, 3576–3586, 2004; Pourret, O. et al., *Geochim. Cosmochim. Acta*, 71, 2718–2735, 2007a; Pourret, O. et al., *Chem. Geol.*, 243, 128–141, 2007b.)

Humic acids can be regarded as a group of discrete sites. It is, thus, possible to compare the binding properties of HA with those of organic ligand models. The REE pattern corresponding to the binding of REE with acetate (carboxylic group) exhibits an MREE downward concavity, whereas catechol and NTA (phenolic and chelate group) exhibit a lanthanide contraction effect corresponding to REE–HA patterns obtained, respectively, at both high and low metal loadings (Figures 7.8 and 7.9). Figure 7.9d shows the evolution of the log (K Lu–organic ligand/K La–organic ligand) relative to the average log (K REE–organic ligand) for the 101 organic ligands compiled by Byrne and Li (1995). This figure shows that when the ligand is stronger, HREEs are more strongly bound to the ligand compared with LREEs. This result is supported by infrared spectroscopy findings that demonstrate that HREEs, compared with LREEs, are preferentially bound to aromatic functional groups of OM such as phenolic sites (Gangloff et al., 2014).

In a second step, Marsac et al. (2011) used modeling to provide evidence that HREEs and LREEs are not complexed to HA via the same functional sites. At high loading and acidic pH, LREEs are bound to carboxylic groups and HREEs are bound to carboxylic and chelate groups. For circumneutral pH, LREEs are bound to carboxylic groups and HREEs are bound via phenolic groups. The denticity of the REE–HA complex is also dependent on the metal loading (Marsac et al., 2011, 2014). More recently, Marsac et al. (2015), based on EXAFS records, suggested that at high loading, REEs are bound to HA through bi-ligand complexes without any chelation effect in which REEs act

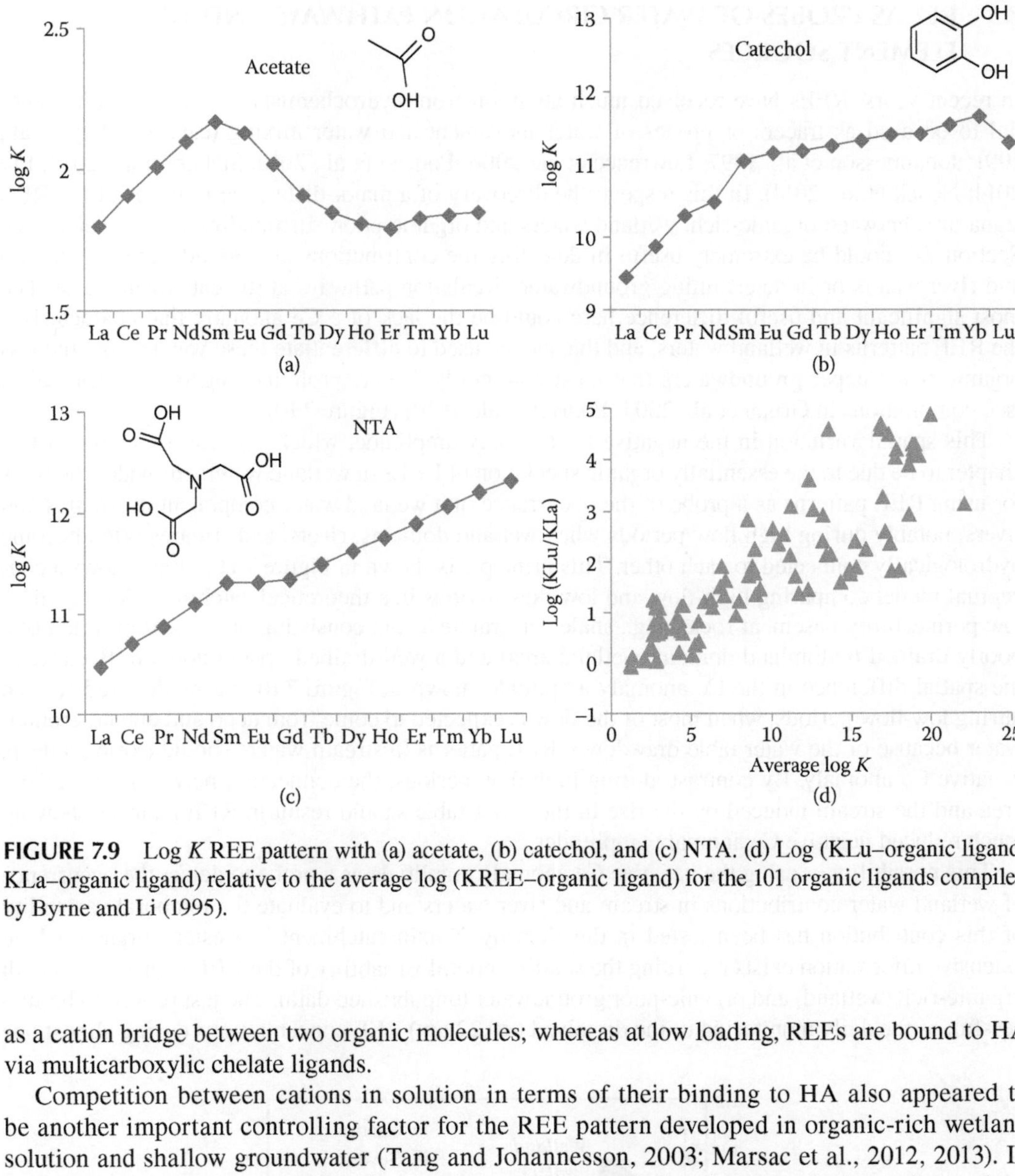

FIGURE 7.9 Log K REE pattern with (a) acetate, (b) catechol, and (c) NTA. (d) Log (KLu–organic ligand/KLa–organic ligand) relative to the average log (KREE–organic ligand) for the 101 organic ligands compiled by Byrne and Li (1995).

as a cation bridge between two organic molecules; whereas at low loading, REEs are bound to HA via multicarboxylic chelate ligands.

Competition between cations in solution in terms of their binding to HA also appeared to be another important controlling factor for the REE pattern developed in organic-rich wetland solution and shallow groundwater (Tang and Johannesson, 2003; Marsac et al., 2012, 2013). In natural water, the HA metal loading is generally imposed by other dissolved metals, such as Fe and Al, which occur in much higher concentrations than REEs. Marsac et al. (2012, 2013) demonstrated that (i) Fe^{3+} competes more strongly with HREEs than LREEs, whereas Fe species formed at higher pH values (i.e., $FeOH^{2+}$ or Fe polymer) compete equally with LREEs and HREEs; and that (ii) Al^{3+} has the same competitive effect on REE–HA binding as Fe^{3+}, but $AlOH^{2+}$ competes mainly with LREEs. Kohler et al. (2014) studied the mobilization of REEs, Al, Fe, and U in a boreal catchment. They demonstrated that OM controls their speciation in solution. However, the pH increases downstream from the catchment and involves the precipitation of Fe and Al as ferrihydrate and gibbsite, respectively. This selective removal of Al and Fe from the OM binding sites results in a higher La concentration downstream, namely a higher amount of La bound to colloidal OM present in the solution. Therefore, the pH, which controls the chemical species of the competitor present in the solution, appears to drive this competition between trivalent cations and REEs and the resulting REE pattern.

7.5 REE AS PROBES OF WATER CIRCULATION PATHWAYS AND TRACE ELEMENT SOURCES

In recent years, REEs have received much attention from hydrochemists because of their potential to be used as tracers or probes of water movement and water mixing (e.g., Smedley et al., 1991; Johannesson et al., 1997; Lawrence et al., 2006; Pourret et al., 2010; Siebert et al., 2012; Lu, 2014; Noack et al., 2014). In this respect, the discovery of a major difference in terms of the REE signatures between organic-rich, wetland waters and organic-poor, surrounding groundwaters (see Section 7.4) could be extremely useful in detecting the contributions of wetland waters to stream and river waters or in determining groundwater circulation pathways at the catchment scale. The most significant and useful difference here could be the lack of a Ce anomaly that characterizes the REE patterns in wetland waters, and that can be used to differentiate these waters from the less organic-rich, deeper groundwaters that most commonly display profound negative Ce anomalies (see compilations in Gruau et al., 2004; Pourret et al., 2010) (Figure 7.10).

This spatial variation in the negative Ce anomaly amplitude, which was earlier shown in this chapter to be due to the essentially organic speciation of REEs in wetland waters, provides the basis for using REE patterns as a probe of the occurrence of a wetland water component in streams and rivers, notably during high flow periods when wetland domains, rivers, and streams often become hydrologically connected to each other. This principle is shown in Figure 7.11, which shows a conceptual model comparing high-flow and low-flow periods in a theoretical catchment developed on low-permeability basement rocks (e.g., shales or granites), and consisting of the juxtaposition of a poorly drained bottomland domain (wetland area) and a well-drained upland domain. Because of the spatial difference in the Ce anomaly amplitude shown in Figure 7.10, the model predicts that during low-flow periods, when most of the flow is expected to come from deep and upland groundwater because of the water table drawdown, REE patterns in stream waters should exhibit a deep, negative Ce anomaly. By contrast, during high-flow periods, the connection between the wetland area and the stream induced by the rise in the water table should result in REE patterns showing much reduced negative Ce anomaly amplitudes.

This possibility of using the negative Ce anomaly amplitude as a probe to detect the occurrence of wetland water contributions in stream and river waters and to evaluate the temporal variability of this contribution has been tested in the Kervidy–Naizin catchment in western France, where extensive information exists regarding the spatiotemporal variability of the REE signatures in both organic-rich (wetland) and organic-poor groundwater (unpublished data). The test proved to be successful, as the daily monitoring of the dissolved (<0.22 μm) REE concentrations during almost one

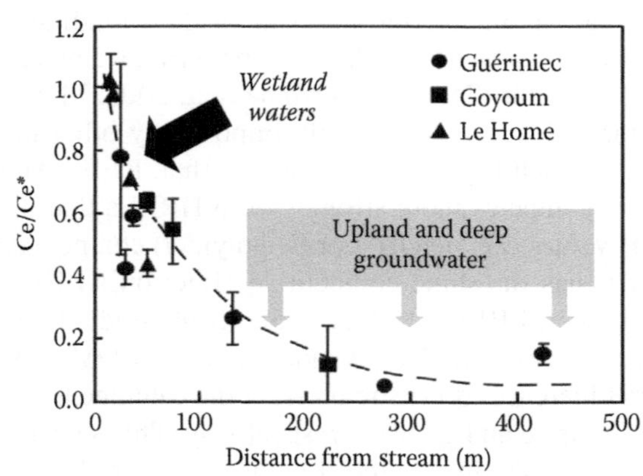

FIGURE 7.10 Strong difference in the negative Ce anomaly amplitude between wetland waters (reduced or nonreduced negative Ce anomaly) and deep/upland groundwaters (deep negative Ce anomaly) as revealed by data from three toposequences located in Europe (Gueriniec; Le Home) and Africa (Goyoum).

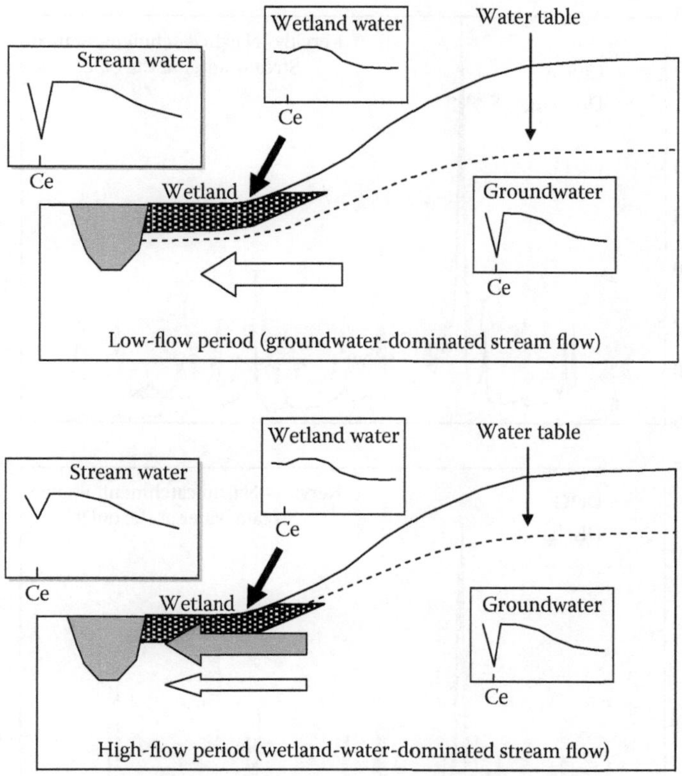

FIGURE 7.11 The manner in which the Ce anomaly amplitude in stream and river waters could be used to detect the contribution of a wetland water component and the possible temporal variability of this contribution.

entire hydrological year (1999–2000) revealed strong Ce anomaly amplitude fluctuations (Figure 7.12). Systematic reductions in the Ce anomaly amplitude were observed in phase with the stream discharge and increases in the DOC concentration (Figure 7.12), two features that are known to correspond to wetland water inputs into the stream of this catchment (Olivier-Lauquet et al., 2001; Morel et al., 2009).

This use of the Ce anomaly amplitude variation as a probe of wetland water contributions to stream and river water flows could be regarded as having limited applicability, as the example presented relies on the daily monitoring of REE concentrations, which could be seen as a severe limitation for extrapolation to other river systems. However, comparable variations (i.e., decrease in the negative Ce anomaly amplitude along with an increase in the discharge and DOC concentrations) were observed at the Loch Vale catchment outlet in Colorado and in the Kalix River in Sweden based on weekly sampling (Ingri et al., 2010; Shiller, 2010), suggesting that the method does not necessarily require a systematic daily monitoring frequency. Although possibly difficult to implement, this ability of the negative Ce anomaly to serve as a tool to identify the periods when the rivers and streams are becoming hydrologically connected to wetland soils is worth considering. For example, it could be used to show that some of the pollutants occurring at that time in the rivers and streams considered (e.g., some trace metals) have these soils as an ultimate source.

Another domain in which REE can be used as a probe concerns the identification of soil components that host trace metals that are released during wetland/floodplain soil reduction. The release of trace metals such as Pb, Cr, Ni, Zn, Cd, and As during flooding, and their subsequent reduction, is a classical feature of floodplain/wetland soils (Schulz-Zunkel and Krueger, 2009; Du Laing et al., 2009). In this type of soil, several soil phases such as Mn(IV)– and Fe(III)–oxy-hydroxides, OM, or mixed Fe–OM particles can host trace metals, and one important question

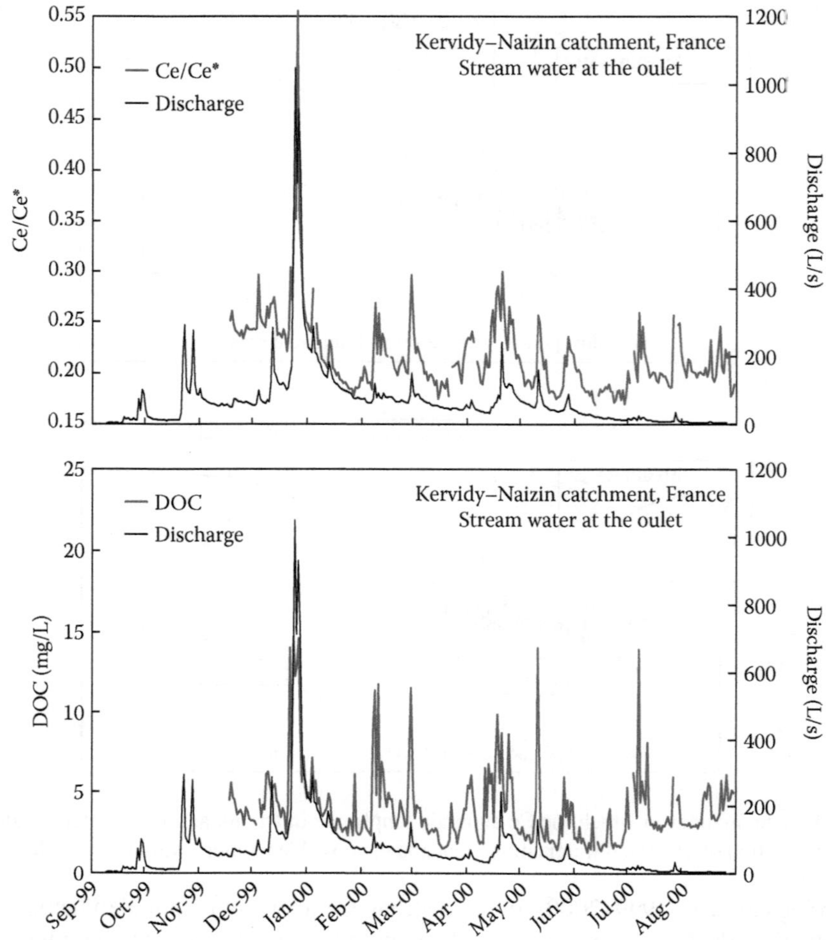

FIGURE 7.12 Synchronous Ce anomaly amplitude, discharge, and DOC concentration variations in the stream at the outlet of the Kervidy–Naizin catchment during the hydrological year 1999–2000.

is to know which of these phases is/are mobilized during the release process. Davranche et al. (2011) tackled this question by using REE as a probe of the activated phases. To do this, they used a wetland soil from the Kervidy–Naizin catchment in France for which previous field monitoring data (Dia et al., 2000) revealed trace metal release during reduction. Their work was based on (i) the difference in the REE pattern between soil OM (relative MREE enrichment) and soil Fe(III) oxhydroxides (LREE enrichment) and (ii) the virtual absence of REE in the Fe components of the mixed Fe–OM particles. By performing different incubation laboratory experiments, Davranche et al. (2011) demonstrated that the REE fraction released in solution during the reduction of the studied soil had a pattern similar to that of the soil OM fraction, suggesting, in turn, that this fraction is the main source of REE and other trace metals that are released during the reduction of this soil (Figure 7.13).

The methodology developed by Davranche et al. (2011) is very promising with regards to the potential of REE to serve as probes to identify the soil components from which trace metals are released during the reduction of wetland/floodplain soils. However, this use is clearly dependent on the occurrence of significant differences in the REE patterns between the end-members involved in the soil–water exchange process, a prerequisite condition that should be evaluated at a greater number of sites.

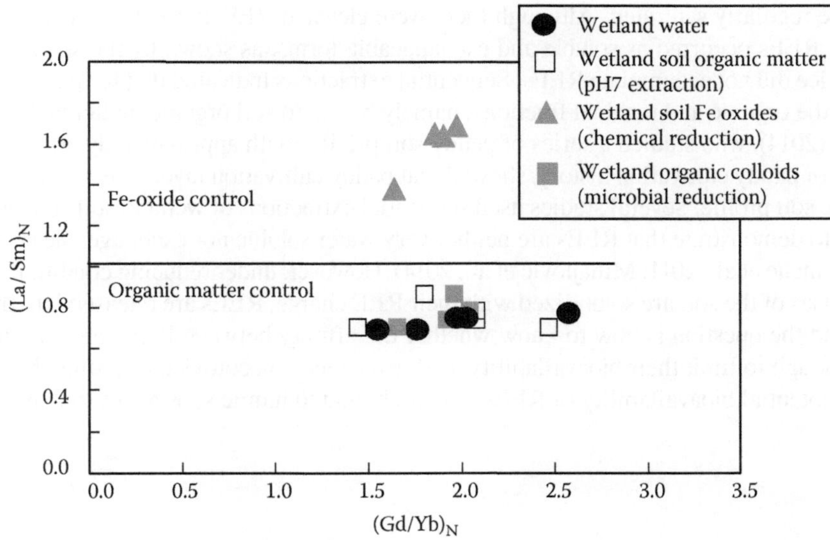

FIGURE 7.13 The manner in which the REE pattern can be used to determine the ultimate soil source of the REE and trace metals released during the reduction of wetland soils. (From Davranche, M. et al., *Chem. Geol.*, 284, 127–137, 2011.)

7.6 REE IMPACT ON HUMAN HEALTH AND ENVIRONMENTAL QUALITY

REEs have been long considered rare since the metallurgic extraction of individual elements is complicated and because of their low concentrations (ppb to ppt) in natural waters. However, over the past several decades, REE became of critical importance to many high-tech products and medical applications, and are therefore of great economic interest (U.S. GAO, 2010). The global production of REE oxides increased from 10,000 t y^{-1} in 1965 to >80,000 t y^{-1} in 2000 (Haxel et al., 2002). In 2008, the consumption of REE oxides was approximately 129,000 t (Goonan, 2011). The various applications of REE and the intense use of fertilizers in agriculture (0.1%–1% REE in natural phosphate, Otero et al., 2005) may lead to a significant release of REE into the environment (Cidu et al., 2013). Kulaksiz and Bau (2011) showed abnormally high concentrations of Gd and La in the Rhine River (France). These authors highlighted that La was extremely mobile in this environment since La contamination was present more than 400 km after the source of the contamination. Tagami and Uchida (2006) provided evidence that REEs are able to accumulate in soil and water and to bioaccumulate in the food chain. Sonich-Mullin (2013) compiled the studies concerned with the specific human health effects of elevated REE concentrations. They only found a few studies that are, for the most part, dedicated to epidemiological data mixtures of REEs rather than individual elements. These data indicate that the pulmonary toxicity of REEs in humans may be a concern. In addition, it has been shown that the larger, lighter (i.e., smaller atomic number), and less soluble REEs are primarily deposited in the liver, whereas the smaller, heavier, and more soluble REEs are similar, in terms of their ionic radius, to divalent calcium and are primarily distributed to the bones. Therefore, it appears especially important to assess the occurrence and fate of aqueous REE in the environment.

Specifically concerning wetlands, previous concerns suggest that REE could be trapped by wetland soil components, notably organic matter. Their subsequent fate would therefore be controlled by the soil components and the behavior of the organic matter, which are themselves relative to the redox conditions prevailing in the wetland. Šmuc et al. (2012) investigated the mobility of REEs in contaminated paddy soil in Macedonia. Paddy soil can be considered an anthropogenic wetland, as the soils are regularly saturated. Although there were elevated REE concentrations in the soil, low amounts of REEs occurred as soluble and exchangeable forms, as shown by the sequential extractions. The rice did not accumulate REEs. Sequential extractions indicated that REEs are mainly distributed in the organic and residual fractions, namely bound to soil organic matter and as minerals. Chen et al. (2014), who studied a series of paddy soil profiles with approximately 50, 300, 700, and 1000 years of paddy cultivation history, showed that paddy cultivation favors the accumulation of all REEs in the soil profile. Several studies used sequential extractions of wetland soils under oxidizing conditions to demonstrate that REEs are neither very water soluble nor exchangeable (Pédrot et al., 2008; Davranche et al., 2011; Mihajlovic et al., 2014). However, under reducing conditions, when the organic phases of the soil are solubilized with their REE charge, REEs are released into the wetland solution, and the question is now to know whether the affinity between REEs and organic ligands is strong enough to limit their bioavailability. Unfortunately, no ecotoxicology study is available to assess the potential bioavailability of REEs that are bound to humic substances that are sourced in wetlands.

REFERENCES

Andersson K., Dahlqvist R., Turner D., Stolpe B., Larsson T., Ingri J., Andersson P. (2006). Colloidal rare earth elements in boreal river: Changing sources and distributions during the spring flood. *Geochim. Cosmochim. Acta* 70, 3261–3274.

Aries S., Valladon M., Polvé M., Dupré B. (2000). A routine method for oxide and hydroxide interference corrections in ICP-MS chemical analysis of environmental and geological samples. *Geostand. Newsl.* 24,19–31.

Aubert D., Stille P., Probst A. (2001). REE fractionation during granite weathering and removal by waters and suspended loads: Sr and Nd isotopic evidence. *Geochim. Cosmochim. Acta* 65, 387–406.

Auterives C. (2007). Influence des flux d'eau souterraine entre une zone humide superficielle et un aquifère profond sur le fonctionnement hydrochimique des tourbières: Exemple des marais du Cotentin, Basse-Normandie. Ph.D. thesis, University of Rennes I, France. Mémoires du CAREN 17, ISBN 2-914375-46-8, p. 261.

Avena, M.J., Koopal, L.K. (1999). Kinetics of humic acid adsorption at solid-water interfaces. *Environ. Sci. Technol.* 33, 2739–2744.

Baker J., Waight T., Ulfbeck D. (2002). Rapid and highly reproducible analysis of rare earth elements by multiple collector inductively coupled plasma mass spectrometry. *Geochim. Cosmochim. Acta* 66, 3635–3646.

Barroux G., Sonke J., Boaventura G., Viers J., Godderis Y., Bonnet M.P., Sondag F., Gardoll S., Lagane C., Seyler P. (2006) Seasonal dissolved rare earth element dynamics of the Amazon River main stem, its tributaries, and the Curuaí floodplain. *Geochem. Geophys. Geosyst.* 7, 1–18.

Bau M. (1999). Scavenging of dissolved yttrium and rare earths by precipitating iron oxyhydroxide: Experimental evidence for Ce oxidation, Y-Ho fractionation, and lanthanide tetrad effect. *Geochim. Cosmochim. Acta* 63, 67–77.

Bau M., Dulski P. (1996). Comparing yttrium and rare earths in hydrothermal fluids from the Mid-Atlantic Ridge: Implications for Y and REE behaviour during nearvent mixing and for the Y/Ho ratio of Proterozoic seawater. *Chem. Geol.* 155, 77–90.

Bau M., Koschinsky A., Dulski P., Hein J.R. (1996). Comparison of the partitioning behaviours of yttrium, rare earth elements, and titanium between hydrogenetic marine ferromanganese crusts and seawater. *Geochim. Cosmochim. Acta* 60, 1709–1725.

Bonnot-Courtois C. (1981). Géochimie des terres rares dans les principaux milieux de formation et de sédimentation des argiles. Thesis of the University Paris-Sud Orsay, Orsay (France), 217 p.

Braun J.J., Viers J., Dupré B., Polvé M., Ndam J., Muller J.P. (1998). Solid/liquid REE fractionation in the lateritic system of Goyoum, East Cameroon: The implication for the present dynamics of the soil covers of the humid tropical regions. *Geochim. Cosmochim. Acta* 62, 273–299.

Buurman P., Jongmans A.G. (2005). Podzolisation and soil organic matter dynamics. *Geoderma* 125, 71–83.

Byrne R.H., Kim K.-H. (1990). Rare earth element scavenging in seawater. *Geochim. Cosmochim. Acta* 54, 2645–2656.

Byrne R.H., Li B. (1995). Comparative complexation behaviour of the rare earth elements. *Geochim. Cosmochim. Acta* 59, 4575–4589.

Charlatchka R., Cambier P. (2000). Influence of reducing conditions on solubility of trace metals in contaminated soils. *Water Air Soil Pollut.* 118, 143–167.

Chen L.M., Zhang G-L., Jin Z-D. (2014). Rare earth elements of a 1000-year paddy soil chronosequence: Implications for sediment provenances, parent material uniformity and pedological changes. *Geoderma*, in press.

Chuan M.C., Shu G.Y., Liu A. (1996). Solubility of heavy metals in a contaminated soil: Effect of redox potential and pH. *Water Air Soil Pollut.* 90, 543–556.

Cidu R., Antisari L.V., Biddau R., Buscaroli A., Carbone S., Da Pelo S., Dinelli E., Vianello G., Zannoni D. (2013). Dynamics of rare earth elements in water–soil systems: The case study of the Pineta San Vitale (Ravenna, Italy). *Geoderma* 193–194, 52–67.

Davranche M., Bollinger J.-C., Bril H. (2003). Effect of reductive conditions on metal mobility from wasteland solids: An example from the Mortagne-du-Nord site (France). *Appl. Geochem.* 18, 383–394.

Davranche M., Gruau G., Dia A., Marsac R., Pédrot M., Pourret O. (2015). Biogeochemical factors affecting rare earth element distribution in shallow wetland groundwater. *Aq. Geochem.* 21, 197–21.

Davranche M., Grybos M., Gruau G., Pédrot M., Dia A., Marsac R. (2011). Rare earth element patterns: A tool for identifying trace metal sources during wetland soil reduction. *Chem. Geol.* 284, 127–137.

Davranche M., Pourret O., Gruau G., Dia A., Le Coz-Bouhnik M. (2005). Adsorption of REE(III)-humate complexes onto MnO2: Experimental evidence for cerium anomaly and lanthanide tetrad effect suppression. *Geochim. Cosmochim. Acta* 69, 4825–4835.

de Baar H.J.W., Bacon M.P., Brewer P.G. (1983). Rare-earth distributions with a positive Ce anomaly in the western North Atlantic Ocean. *Nature* 301, 324–327.

de Baar H.J.W., Schijf J., Byrne R.H. (1991). Solution chemistry of the rare earth elements in seawater. *Eur. J. Solid State Inorg. Chem.* 28, 357–373.

De Carlo E.H., Green W.J. (2002). Rare earth elements in the water column of Lake Vanda, McMurdo Dry Valleys, Antarctica. *Geochim. Cosmochim. Acta* 66, 1323–1333.

Dia A., Gruau G., Olivié-Lauquet G., Riou C., Molénat J., Curmi P. (2000). The distribution of rare-earths in groundwater: Assessing the role of source-rock composition, redox changes and colloidal particles. *Geochim. Cosmochim. Acta* 64, 4131–4151.

Ding S.M., Liang T., Zhang C.S., Huang Z.C., Xie Y.N., Chen T.B. (2006). Fractionation mechanisms of rare earth elements (REEs) in hydroponic wheat: An application for metal accumulation by plants. *Environ. Sci. Technol.* 40, 2696–2691.

Du Laing G., Rinklebeb J., Vandecasteelec B., Meersa E., Tacka F.M.G. (2009). Trace metal behaviour in estuarine and riverine floodplain soils and sediments: A review. *Sci. Tot. Environ.* 4007, 3972–3985.

Duncan T., Shaw T.J. (2004). The mobility of rare earth elements and redox sensitive elements in the groundwater/seawater mixing zone of a shallow coastal aquifer. *Aquat. Geochem.* 9, 233–255.

Dupré B., Gaillardet J., Rousseau D., Allègre C.J. (1996). Major and trace elements of river-borne material: The Congo Basin. *Geochim. Cosmochim. Acta* 60, 1301–1321.

Egashira K., Aramaki K., Yoshimasa M., Takeda A., Yamasaki S. (2004). Rare earth elements and clay minerals of soils of the floodplains of three major rivers in Bangladesh. *Geoderma* 120, 7–15.

Egashira K., Fujii K., Yamasaki S., Virakornphanich P. (1997). Rare earth elements and clay minerals of paddy soils from the central region of the Mekong River, Laos. *Geoderma* 78, 237–249.

Elbaz-Poulichet F., Dupuy C. (1999). Behaviour of rare earth elements at the freshwater-seawater interface of two acid mine rivers: The Tinto and Odiel (Andalucia, Spain). *Appl. Geochem.* 14, 1063–1072.

Elderfield H., Greaves M.J. (1982). The rare earth elements in seawater. *Nature* 296, 214–219.

Elderfield H., Upstill-Goddard R., Sholkovitz E.R. (1990). The rare earth elements in rivers, estuaries, and coastal seas and their significance to the composition of ocean waters. *Geochim. Cosmochim. Acta* 54, 971–991.

Fee J.A., Gaudette H.E., Lyons W.B., Long D.T. (1992). Rare-earth element distribution in Lake Tyrrell groundwaters, Victoria, Australia. *Chem. Geol.* 96, 67–93.

Ferrat, M., Weiss, D.J., Strekopytov, S. (2012). A single procedure for the accurate and precise quantification of the rare earth elements, Sc, Y, Th and Pb in dust and peat for provenance tracing in climate and environmental studies. *Talanta* 93, 415–423.

Francis A.J., Dodge C.J. (1990). Anaerobic microbial remobilization of toxic metals coprecipitated with iron oxide. *Environ. Sci. Technol.* 24, 373–378.

Gammons C.H., Wood S.A., Ponas J.P., Madison J.P. (2003). Geochemistry of the rare-eath elements and uranium in the acidic Berkeley Pit Lake, Butte, Montana. *Chem. Geol.* 198, 269–288.

Gangloff S., Stille P., Pierret M.-C., Weber T., Chabaux F. (2014). Characterization and evolution of dissolved organic matter in acidic forest soil and its impact on the mobility of major and trace elements (case of the Strengbach watershed). *Geochim. Cosmochim. Acta* 130, 21–41.

German C.R., Holiday B.P., Elderfield H. (1991). Redox cycling of rare earth elements in the suboxic zone of the Black Sea. *Geochim. Cosmochim. Acta* 55, 3553–3558.

Goldberg E.D., Koide M., Schmitt R.A., Smith R.H. (1963). Rare-earth distributions in the marine environment. *J. Geophys. Res.* 68, 4209–4217.

Goldstein S.J., Jacobsen S.B. (1988). Rare earth elements in river waters. *Earth Planet. Sci. Lett.* 89, 35–47.

Goonan, T.G. (2011). Rare earth elements—End use and recyclability: U.S. Geological Survey Scientific Investigations Report 2011-5094, 15 p. Available only at http://pubs.usgs.gov/sir/2011/5094/

Green C.H., Heil D.M., Cardon G.E., Butters G.L., Kelly E.F. (2003). Solubilization of manganese and trace metals in soils affected by acid mine runoff. *J. Environ. Qual.* 32, 1323–1334.

Gromet L.P., Dymek R.F., Haskin L.A., Korotev R.L. (1984). The "North American Shale Composite": Its compilations, major and trace element characteristics. *Geochim. Cosmochim. Acta* 48, 2469–2482.

Gruau G., Dia A., Olivié-Lauquet G., Davranche M., Pinay G. (2004). Controls on the distribution of rare earth elements in shallow groundwaters. *Wat. Res.* 38, 3576–3586.

Grybos M., Davranche M., Gruau G., Petitjean P. (2007). Is trace metal release in wetland soils controlled by organic matter mobility or Fe-oxyhydroxide reduction? *J. Colloid Interface Sci.* 314, 490–501.

Grybos M., Davranche M., Gruau G., Petitjean P., Pédrot M. (2009). Increasing pH drives organic matter solubilization from wetland soils under reducing conditions. *Geoderma* 154, 13–19.

Hagedorn F., Kaiser K., Feyen H., Schleppi P. (2000). Effect of redox conditions and flow processes on the mobility of dissolved organic carbon and nitrogen in a forest soil. *J. Environ. Qual.* 29, 288–297.

Hannigan R.E., Sholkovitz E. (2001). The development of middle rare earth element enrichments in freshwaters: Weathering of phosphate minerals. *Chem. Geol.* 175, 495–508.

Haxel G.B., Hedrick J.B., Orris G.J. (2002). Rare earth elements. Critical resources for high technology. US Geological Survey, Fact Sheet 087-02. http://pubs.usgs.gov/ fs/2002/fs087-02/

Heimburger A., Tharaud M., Monna F., Losno R., Desboeufs K., Nguyen E.B. (2013). SLRS-5 elemental concentrations of thirty-three uncertified elements deduced from SLRS-5/SLRS-4 ratios. *Geostand. Geoanal. Res.* 37, 77–85.

Henderson P. (1984). *Rare Earth Element Geochemistry: Development in Geochemistry*. Vol. 2. Amsterdam: Elsevier, 510 p.

Ingri J., Widerlund A., Land M., Gustafsson O., Andersson P., Ohlander B. (2000). Temporal variations in the fractionation of the rare earth elements in a boreal river; The role of colloidal particles. *Chem. Geol.* 166, 23–45.

Janssen R.P.T., Verweij W. (2003). Geochemistry of some rare earth elements in groundwater, Vierlingsbeek, The Netherlands. *Wat. Res.* 37, 1320–1350.

Jarvis K.E., Gray A.L., Houk R.S. (1992). *Handbook of Inductively Coupled Plasma Mass Spectrometry*. Glasgow, Blackie: New York: Chapman and Hall, 380 p.

Johannesson K.H., Lyons W.B. (1995). Rare-earth element geochemistry of Colour Lake, an acidic freshwater lake on Axel Heiberg Island, Northwest Territories, Canada. *Chem. Geol.* 119, 209–223.

Johannesson K.H., Stetzenbach K.J., Hodge V.F. (1997). Rare earth elements as geochemical tracers of regional groundwater mixing. *Geochim. Cosmochim. Acta* 61, 3605–3618.

Johannesson K.H., Tang J., Daniels J.M., Bounds W.J., Burdige D.J. (2004). Rare earth element concentrations and speciation in organic-rich blackwaters of the Great Dismal Swamp, Virginia, USA. *Chem. Geol.* 209, 271–294.

Kautenburger R., Hein C., Sander J.M., Beck H.P. (2014). Influence of metal loading and humic acid functional groups on the complexation behavior of trivalent lanthanides analyzed by CE-ICP-MS. *Anal. Chim. Acta* 816, 50–59.

Kerr S.C., Shafer M.M., Overdier J., Armstrong D.E. (2008). Hydrologic and biogeochemical controls on trace element export from northern Wisconsin wetlands. *Biogeochemistry* 89, 273–294.

Klinkhammer G., German C.R., Elderfield H., Greaves M.J., Mitra A. (1994). Rare earth elements in hydrothermal fluids and plume particulates by inductively coupled plasma mass spectrometry. *Mar. Chem.* 45, 179–186.

Koeppenkastrop D., De Carlo E.H. (1992). Sorption of rare-earth elements from seawater onto synthetic mineral particles: An experimental approach. *Chem. Geol.* 95, 251–263.

Kohler S.J., Lidman F., Laudon H. (2014). Landscape types and pH control organic matter mediated mobilization of Al, Fe, U and La in boreal catchments. *Geochim. Cosmochim. Acta* 135, 190–202.

Koppi A.J., Edis R., Field D.J., Geering H.R., Klessa D.A., Cockayne D.J.H. (1996). Rare earth element trends and cerium-uranium-manganese associations in weathered rock from Koongarra, Northern Territory, Australia. *Geochim. Cosmochim. Acta* 60, 1695–1707.

Kulaksız S., Bau M. (2011). Rare earth elements in the Rhine River, Germany: First case of anthropogenic lanthanum as a dissolved microcontaminant in the hydrosphere. *Environ. Int.* 37, 973–979.

Laveuf C., Cornu S. (2009). A review of the potentiality of rare earth element to trace pedogenetic processes. *Geoderma* 154, 1–12.

Laveuf C., Cornu S., Guilherme L.R.G., Guerin A., Juillot F. (2012). The impact of redox conditions on the rare earth element signature of redoximorphic features in a soil sequence developed from limestone. *Geoderma* 170, 25–38.

Lawrence M.G., Greig A., Collerson K.D., Kamber B.S. (2006). Rare earth element and yttrium variability in South East Queensland waterways. *Aquat. Geochem.* 12, 39–72.

Leybourne M.I., Johannesson K.H. (2008). Rare earth elements (REE) and yttrium in stream waters, stream sediments, and Fe–Mn oxyhydroxides: fractionation, speciation, and controls over REE + Y patterns in the surface environment. *Geochim. Cosmochim. Acta* 72, 5962–5983.

Lima e Cunha M.C., Do Carmo M., Pereira V.P., Nardi L.V.S., Bastos Neto A.C., Vedana L.A., Formoso M.L.L. (2012). REE distribution pattern in plants and soils from Pitinga Mine—Amazon, Brazil. *Open J. Geol.* 2, 253–259.

Lima e Cunha M.C., Pereira V.P., Bastos Neto A.C.,. Nardi L.V.S., Formoso M.L.L., Menegotto L. (2006). Biogeoquímica dos Eelementos Terras Rraras Na Província Eestanífera De Pitinga (AM). *Revista Brasileira de Geociências* 39, 560–566.

Longerich H.P, Fryer B.J., Strong D.F., Kantipuly C.J.(1987). Effects of operating conditions on the determination of the rare earth elements by inductively coupled plasma-mass spectrometry (ICP-MS). *Spectrochim. Acta B Atom. Spectrosc.* 42, 75–92.

Lu H.Y. (2014). Application of water chemistry as a hydrological tracer in a volcano catchment area: A case study of the Tatun Volcano Group, North Taiwan. *J. Hydrol.* 51, 825–837.

Marsac R., Davranche M., Gruau G., Bouhnik-Le Coz M., Dia A. (2011). An improved description of the interactions between rare earth elements and humic acids by modeling: PHREEQC-Model VI coupling. *Geochim. Cosmochim. Acta* 75, 5625–5637.

Marsac R., Davranche M., Gruau G., Bouhnik-Le Coz M., Dia A. (2013). Iron competitive effect on REE binding to organic matter: implications with regards to REE patterns in waters. *Chem. Geol.* 342, 119–127.

Marsac R., Davranche M., Gruau G., Dia A. (2010). Metal loading effect on rare earth element binding to humic acid: Experimental and modelling evidence. *Geochim. Cosmochim. Acta* 74, 1749–1761.

Marsac R., Davranche M., Gruau G., Dia A., Bouhnik-Le Coz M. (2012). Aluminum competitive effect on rare earth elements binding to humic acid. *Geochim. Cosmochim. Acta* 89, 1–9.

Marsac R., Davranche M., Morin G., Takahashi Y., Gruau G., Dia A. (2015). Impact of REE loading on REE-humate binding: Sm and Yb EXAFS evidence. *Chem. Geol.* 396, 218–227.

Martinez R.E., Pourret O., Takahashi Y. (2014). Modeling of rare earth element sorption to the gram positive bacillus subtilis bacteria surface. *J Colloid interface Sci.* 413, 106–111.

McLennan S.M. (1989). Rare earth elements in sedimentary rocks: Influence of provenance and sedimentary processes. In: B.R. Lipin, G.A. McKay (eds.). *Geochemistry and Mineralogy of Rare Earth Elements*. Vol. 21. Mineralogical Society of America, Washington, DC, pp. 169–200.

McLennan S.M. (1994). Rare earth element geochemistry and the "tetrad" effect. *Geochim. Cosmochim. Acta* 58, 2025–2033.

Michard A., Albarède F., Michard G., Minster J.F., Charlou J.L. (1983). Rare-earth elements and uranium in high-temperature solutions from East Pacific Rise hydrothermal vent field (13°N). *Nature* 303, 795–797.

Mihajlovic J., Stärk H-J., Rinklebe J. (2014). Geochemical fractions of rare earth elements in two floodplain soil profiles at the Wupper River, Germany. *Geoderma* 228–229, 160–172.

Moffett J.W. (1990). Microbially mediated cerium oxidation in sea water. *Nature* 345, 421–423.

Moffett J.W. (1994a). A radiotracer study of cerium and manganese uptake onto suspended particles in Chesapeake Bay. *Geochim. Cosmochim. Acta* 58, 695–703.

Moffett J.W. (1994b). The relationship between cerium and manganese oxidation in the marine environment. *Limnol. Oceanogr.* 39, 1309–1318.

Möller P., Bau M. (1993). Rare-earth patterns with positive cerium anomaly in alkaline waters from Lake Van, Turkey. *Earth Planet. Sci. Lett.* 117, 671–676.

Morel B., Durand P., Jaffrezic A., Gruau G., Molénat J. (2009). Sources of dissolved organic carbon during stormflow in a headwater agricultural catchment. *Hydrol. Process.* 23, 2888–2901.

Morey G.B., Setterholm D.R. (1997). Rare earth elements in weathering profiles and sediments of Minnesota: Implications for provenance studies. *J. Sed. Res.* 67, 105–115.

Neubauer E., Kammer F.V.D., Hofmann T. (2013). Using FLOWFFF and HPSEC to determine trace metal colloid associations in wetland runoff. *Wat. Res.* 47, 2757–2769.

Ngwenya B.T., Magennis M., Olive M., Mosselmans J.F.W., Ellam R.M. (2010). Discrete site surface complexation constant for lanthanide adsorption to bacteria as determined by experiments and linear free energy relationship. *Environ. Sci Technol.* 44, 650–656.

Ngwenya B.T., Mosselmans J.F.W., Magennis M., Atkinson K.D., Tourney J., Olive V., Ellam R.M. (2009). Macroscopic and spectroscopic analysis of lanthanide adsorption to bacterial cells. *Geochimi. Cosmochim. Acta* 73, 3134–3147.

Noack C.W., Dzombak D.A., Karamalidis A.K. (2014). Rare earth element distributions and trends in natural waters with a focus on groundwater. *Environ. Sci. Technol.* 48, 4317–4326.

Ohta A., Kawabe I. (2001). REE(III) adsorption onto Mn dioxide (δ-MnO2) and Fe oxyhydroxide: Ce(III) oxidation by δ-MnO2. *Geochim. Cosmochim. Acta* 65, 695–703.

Olivié-Lauquet G., Allard T., Benedetti M., Muller J.-P. (1999). Chemical distribution of trivalent iron in riverine material from a tropical ecosystem: A quantitative EPR study. *Wat. Res.* 33, 2726–2734.

Olivié-Lauquet G., Gruau G., Dia A., Riou C., Jaffrezic A., Henin O. (2001). Release of trace elements in wetlands: Role of seasonal variability. *Wat. Res.* 35, 943–952.

Otero N., Vitoria L., Soler A., Canals A. (2005). Fertilizer characterization: Major, trace and rare earth elements. *Appl. Geochem.* 20, 1473–1488.

Panahi A., Young G.M., Rainbird R.H. (2000). Behavior of major and trace elements (including REE) during Paleoproterozoic pedogenesis and diagenetic alteration of an Archean granite near Ville Marie, Quebec, Canada. *Geochim. Cosmochim. Acta* 64, 2199–2220.

Pédrot M., Dia A., Davranche M., Bouhnik-Le Coz M., Henin O., Gruau G. (2008). Insights into colloid-mediated trace element release at soil/water interface. *J. Colloid Interface Sci.* 325, 187–197.

Pédrot M., Dia A., Davranche M., Gruau G. (2015). How do upper soil horizons control rare earth element patterns in shallow groundwaters? *Geoderma* 239–240, 84–96.

Pourret O., Davranche M., Gruau G., Dia A. (2007a). Organic complexation of rare earth elements in natural waters: Evaluating model calculations from ultrafiltration data. *Geochim. Cosmochim. Acta* 71, 2718–2735.

Pourret O., Davranche M., Gruau G., Dia A. (2007b). Rare earth complexation by humic acid. *Chem. Geol.* 243, 128–141.

Pourret O., Gruau G, Dia A., Davranche M., Molénat J. (2010). Colloidal control on the distribution of rare earth elements in shallow groundwaters. *Aquat. Geochem.* 16, 31–59.

Protano G., Riccobono F. (2002). High contents of rare earth elements in stream waters of a Cu-Pb-Zn mining area. *Environ. Pollut.* 117, 499–514.

Quantin C., Becquer T., Berthelin J. (2002). Mn-oxide: A major source of easily mobilisable Co and Ni under reducing conditions in New Caledonia Ferralsols. *Comptes Rendus Geosci.* 334, 273–278.

Quantin C., Becquer T., Rouiller J.H., Berthelin J. (2001). Oxide weathering and trace metal release by bacterial reduction in a New Caledonia Ferralsol. *Biogeochemistry* 53, 323–340.

Ran Y., Liu Z. (1992). Specific adsorption of trivalent La, Ce and Y by soils and ferromanganese oxides and its mechanism. *Pedosphere* 2, 13–22.

Raut N.M., Huang L.S., Aggarwal S.K., Lin K.C. (2005b). Mathematical correction for polyatomic isobaric spectral interferences in determination of lanthanides by inductively coupled plasma-mass spectrometry. *J. Chinese Chem. Soc.* 52, 589–597.

Raut N.M., Huang L.S., Lin K.C., Aggarwal S.K. (2005a). Uncertainty propagation through correction methodology for the determination of rare earth elements by quadrupole based inductively coupled plasma mass spectrometry. *Anal. Chim. Acta*, 530, 91–103.

Rousseau T.C., Sonke J.E., Chmeleff J., Candaudap F., Lacan F., Boaventura G., Jeandel C. (2013). Rare earth element analysis in natural waters by multiple isotope dilution–sector field ICP-MS. *J. Anal. Atomic Spect.* 28, 573–584.

Sauer D., Sponagel H., Sommer M., Giani L., Jahn R., Stahr K. (2007). Podzol: Soil of the year 2007. A review on its genesis, occurrence, and functions. *J. Plant Nutr. Soil Sci.* 170, 581–597.

Schulz-Zunkel C., Krueger F. (2009). Trace metal dynamics in floodplain soils of the river Elbe: A review. *J. Environ. Qual.* 38,1349–1362.

Schwertmann U., Fisher W.R. (1973). Natural amorphous ferric hydroxide. *Geoderma* 10, 237–247.

Shannon R.D. (1976). Revised effective ionic radii and systematic studies of interatomic distances in halides and chalcogenides. *Acta Crystallogr.* A32, 751–767.

Shiller A.M. (2002). Seasonality of dissolved rare earth elements in the Lower Mississippi River. *Geochem. Geophys. Geosyst.* 3, 1068.

Shiller A.M. (2010). Dissolved rare earth elements in a seasonally snow-covered, alpine/subalpine watershed, Loch Vale, Colorado. *Geochim. Cosmochim. Acta* 74, 2040–2052.

Sholkovitz E.R. (1993). The geochemistry of rare earth elements in the Amazon River estuary. *Geochim. Cosmochim. Acta* 57, 2181–2190.

Sholkovitz E.R. (1995). The aquatic chemistry of the rare earth elements in rivers and estuaries. *Aq. Geochem.* 1, 1–34.

Sholkovitz E.R., Landing W.M., Lewis B.L. (1994). Ocean particle chemistry: The fractionation of rare earth elements between suspended particles and seawater. *Geochim. Cosmochim. Acta* 58, 1567–1579.

Siebert C., Rosenthal E., Möller P., Rödiger T., Meiler M. (2012). The hydrochemical identification of groundwater flowing to the Bet She'an-Harod multiaquifer system (Lower Jordan Valley) by rare earth elements, yttrium, stable isotopes (H, O) and Tritium. *Appl. Geochem.* 27, 703–714.

Simpson A.J., Simpson M.J., Smith E., Kellher B.P. (2007). Microbially derived inputs to soil organic matter: Are current estimates too low? *Environ. Sci. Technol.* 41, 8070–8076.

Singh P., Rajamani V. (2001). REE geochemistry of recent clastic sediments from the Kaveri floodplains, southern India: Implications to source area weathering and sedimentary processes. *Geochim. Cosmochim. Acta* 65, 3093–3108.

Smedley P. (1991). The geochemistry of rare earth elements in groundwater from the Carnmenellis area, southwest England. *Geochim. Cosmochim. Acta* 55, 2267–2779.

Šmuc N.R., Dolenec T., Serafimovski T., Dolenec M., Vrhovnik P. (2012). Geochemical characteristics of rare earth elements (REEs) in the paddy soil and rice (Oryza sativa L.) system of Kočani Field, Republic of Macedonia. *Geoderma* 183–184, 1–11.

Sonich-Mullin C. (2013). Rare earth elements: A review of production, processing, recycling, and associated environmental issues. Office of Research and Development, EPA/600/R-12/572, 135 p, USA.

Sonke J.E., Salters V.J. M. (2006). Lanthanide–humic substances complexation. I. Experimental evidence for a lanthanide contraction effect. *Geochim. Cosmochim. Acta* 70, 1495–1506.

Stern J.C., Sonke J.E., Salters V.J.M. (2007). A capillary electrophoresis-ICP-MS study of rare earth elements complexation by humic acids. *Chem. Geol.* 246, 170–180.

Stolpe B., Guo L., Shiller A.M. (2013). Binding and transport of rare earth elements by organic and iron-rich nanocolloids in Alaskan rivers, as revealed by field-flow fractionation and ICP-MS. *Geochim. Cosmochim. Acta* 106, 446–462.

Tachikawa K., Athias V., Jeandel C. (2003). Neodymium budget in the modern ocean and paleo-oceanographic implications. *J. Geophys. Res.* 108, 0148–0227.

Tagami K., Uchida S. (2006). Transfer of REEs from nutrient solution to radish through fine roots and their distribution in the plant. *J. Alloys. Compd.* 408, 409–412.

Takahashi Y., Châtellier X., Hattori K.H., Kato K., Fortin D. (2005). Adsorption of rare earth elements onto bacterial cell walls and its implication for REE sorption onto natural microbial mats. *Chem. Geol.* 219, 53–67.

Takahashi Y., Shimizu H., Usui A., Kagi H., Nomura M. (2000). Direct observation of tetravalent cerium in ferromanganese nodules and crust by X-ray absorption near-edge structure (XANES). *Geochim. Cosmochim. Acta* 64, 2829–2835.

Takahashi Y., Yamamoto M., Yamamoto Y., Tanaka K. (2010). EXAFS study on the cause of enrichment of heavy REEs on bacterial cell surfaces. *Geochim. Cosmochim. Acta* 74, 5443–5462.

Tang J., Johannesson K.H. (2003). Speciation of rare earth elements in natural terrestrial waters: Assessing the role of dissolved organic matter from the modeling approach. *Geochim. Cosmochim. Acta* 67, 2321–2339.

Tang J., Johannesson K.H. (2010). Ligand extraction of rare earth elements from aquifer sediments: Implications for rare earth element complexation with organic matter in natural waters. *Geochim. Cosmochim. Acta* 74, 6690–6705.

Taylor S.R., McLennan S.M. (1985). *The Continental Crust: Its Composition and Evolution.* Oxford: Blackwell Science Publishers, Taylor and McLennan.

Tsumura A., Yamasaki S. (1993). Behaviour of uranium, thorium and lanthanides in paddy fields. *Radioisotopes* 42, 265–272.

U.S. GAO. United States Government Accountability Office. Rare earth materials in the defense supply chain, GAO-10-617R (2010). http://www.gao.gov/new.items/d10617r.pdf

Vasyukova E., Pokrovsky O., Viers J., Dupré B. (2012). New operational method of testing colloid complexation with metals in natural waters. *Appl. Geochem.* 27, 1226–1237.

Viers J., Dupré B., Polvé M., Schott J., Dandurand J.-L., Braun J.J. (1997). Chemical weathering in the drainage basin of a tropical watershed (Nsimi–Zoetele site, Cameroon): Comparison between organic poor and organic-rich waters. *Chem. Geol.* 140, 181–206.

Vital H., Stattegger K. (2000). Major and trace elements of stream sediments from the lowermost Amazon River. *Chem. Geol.* 168, 151–168.

Weber R. (2008). An experimental study of fractionation of the rare earth elements in poplar plants (populous eugenei) grown in a calcium-bearing smectite soil. Master's Thesis, Kansas State University, Manhattan. 50 p

Yeghicheyan D., Bossy C., Bouhnik Le Coz M., Douchet C., Granier G., Heimburger A., Lacan F., Lanzanova A., Rousseau T., Seidel J., Tharaud M., Candaudap F., Chmeleff J., Cloquet C., Delpoux S., Labatut M., Losno R., Pradoux C., Sivry Y., Sonke J. (2013). A compilation of silicon, rare earth element and twenty-one other trace element concentrations in the natural river water reference material SLRS-5 (NRC-CNRC). *Geos. Geoanal. Res.* 4, 449.

Yeghicheyan D., Carignan J., Valladon M., Coz M., Cornec F., Castrec-Rouelle M., Robert M., Aquilina L., Aubry E., Churlaud C., Dia A., Deberdt S., Dupré B., Freydier R., Gruau G., Hénin O., Kersabiec A., Macé J., Marin L., Morin N. (2001). A compilation of silicon and thirty one trace elements measured in the natural river water reference material SLRS-4 (NRC-CNRC). *Geos. Geoanal. Res.* 25, 2/3, 465.

Yuan D., Shan X.Q., Huai Q., Wen B., Zhu X. (2001). Uptake and distribution of rare earth elements in rice seeds cultured in fertilizer solution of rare earth elements. *Chemosphere* 43, 33–44.

8 Subsoil Contaminant Cr Fate and Transport
The Complex Reality of the Hanford Subsurface

Nikolla P. Qafoku and Rahul Sahajpal

CONTENTS

8.1 INTRODUCTION

Hexavalent chromium [Cr^{6+}] is a groundwater contaminant at numerous sites of the U.S. Department of Energy (DOE) or other sites across the United States. Globally, Cr^{6+} contamination is a major concern, and remediation of the Cr-contaminated sites remains one of the most challenging environmental issues. The Hanford site in the State of Washington (USA) was the location of the government's primary plutonium production during the World War II Manhattan Project and the Cold War. Because of its corrosion-resistant properties, Cr was used extensively at the site as sodium dichromate, which led to its release into the environment. As a result, chromate (CrO_4^{2-}) was, and in some areas still is, one of the major contaminants of concern near the Columbia River in the 100 Area at the Hanford Site (2010 Groundwater Annual Report: DOE/RL-2010-11). Oxidized aqueous Cr^{6+}, which has higher toxicity than reduced Cr^{3+}, is highly mobile under neutral and slightly alkaline conditions that are commonly present in contaminated sites and strict regulations are in place to mitigate risks. For this contaminant, which is considered highly toxic to both plants and animals (Abbasi and Soni, 1984; Paschin et al., 1983) and a human carcinogen (Ono, 1988), the aquatic water quality criterion of 11 μg L^{-1} is lower than drinking water standards (0.1 mg L^{-1}) (Dresel et al., 2008).

8.2 GENERAL OVERVIEW OF Cr(VI) CONTAMINATION AT HANFORD

A historical perspective of Cr contamination at Hanford is included in previous publications (Dresel et al., 2008; Qafoku et al., 2011). In this section, we briefly outline the mechanism and reactions that control Cr mobility, focusing on the Hanford site. The hexavalent Cr, mainly as sodium dichromate ($Na_2Cr_2O_7{\cdot}2H_2O$), was used extensively as a corrosion inhibitor in the nuclear reactor cooling water and for equipment decontamination (Peterson et al., 1996; Thornton, 1992). After passing through the reactor, cooling water was transported through large-diameter underground pipes to retention basins for thermal and radioactive cooling before release to the Columbia River. Overall, 2.0 mg/L of sodium dichromate (0.7 mg L^{-1} as Cr) was present in cooling water (Foster, 1957). Until approximately 1953, sodium dichromate solutions were made in a batch system by using 100-lb. bags of granular dichromate that were manually hoppered into large (~3600 gal.) tanks to obtain a final solution concentration of 15% $Na_2Cr_2O_7$ by weight (wt.) (Whipple, 1953). After 1953, 70% of $Na_2Cr_2O_7$ solutions by weight were delivered to the Hanford site, stored in large tanks, and diluted as required (Schroeder, 1966). These concentrated solutions were then delivered to various water treatment plants in rail cars, tanker trucks, barrels, and local pipelines as stock solutions. Additional information can be found in the 100-D Area operations and waste sites presented in the *100-D Area Technical Baseline Report* (Carpenter, 1993).

Concentrated dichromate solutions were discharged to surface or near-surface ground through spills during handling, pipeline leaks, or when discarded to cribs (Figures 8.1 and 8.2). Additional Cr was discharged to the environment from decontamination operations, likely after mixing with sulfuric acid to form chromic acid (Peterson et al., 1996). Although the exact acidity of Hanford site Cr stock solutions is not known, a 10% $Na_2Cr_2O_7$ (0.82 mol L^{-1} Cr) has a pH of 3.5, and a 70% $Na_2Cr_2O_7$ (8.96 mol L^{-1} Cr) may have a lower pH (~1.5 to 2) (Dresel et al., 2008).

In the 1990s, after the end of the production mission, increasing attention was drawn toward the chemical impacts of Cr contamination, particularly in the 100 Areas (100-B, 100-C, 100-D/DR, 100-F, 100-H, and 100-K), where the nuclear reactors were located along the Columbia River.

FIGURE 8.1 Cr-contaminated sediments in the Hanford 100 Area. (From Dresel, P.E. et al., *Geochemical Characterization of Chromate Contamination in the 100 Area Vadose Zone at the Hanford Site*, Pacific Northwest National Laboratory, PNNL-17674, 2008.)

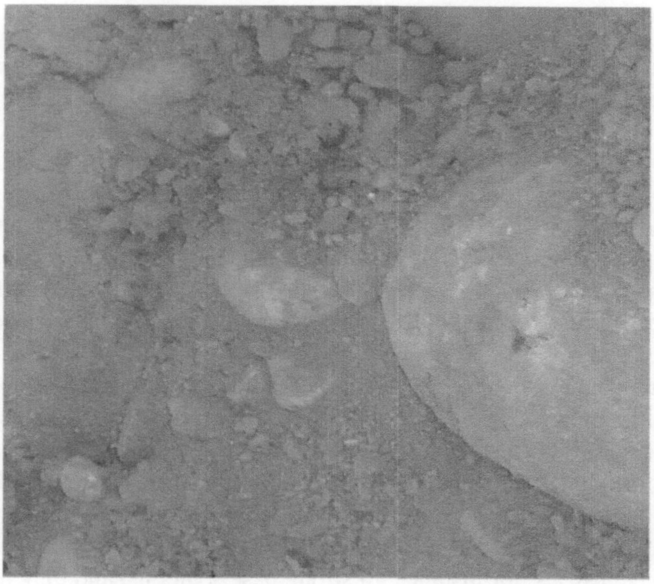

FIGURE 8.2 Rocks stained at the Hanford site. (From Dresel, P.E. et al., *Geochemical Characterization of Chromate Contamination in the 100 Area Vadose Zone at the Hanford Site*, Pacific Northwest National Laboratory, PNNL-17674, 2008.)

Potential sources of vadose zone and groundwater contamination include leaks from cooling water pipelines and retention basins, disposal of contaminated water to liquid waste cribs and trenches, and spills of sodium dichromate solids or solutions (Thornton, 1992).

Discharge of Cr-contaminated groundwater to the Columbia River has been documented through porewater sampling in the river bed and small-diameter sampling points (called aquifer tubes) along the shoreline (Hope and Peterson, 1996). Dissolved Cr in the groundwater is dominated by Cr(VI), which usually occurs as anionic chromate, CrO_4^{-2} (Thornton et al., 1995). An understanding of the nature of the vadose zone contamination is important for evaluating options for remediation and protection of groundwater and environmental receptors. Successful groundwater remediation and protection depends on the ability to understand and limit the flux of Cr(VI) to the water table from the vadose zone.

8.3 CHAPTER OBJECTIVES

The chapter summarizes the work conducted to study contaminant Cr^{6+} fate and behavior at the Hanford site under conditions imposed by different waste chemistries ranging from acidic to neutral and hyperalkaline. The objectives are to: (1) discuss different aspects of Cr interaction with minerals; (2) present evidence for similar and contrasting Cr^{6+} reactions, processes, and attenuation mechanisms operating in subsurface environments under different conditions that are imposed by different waste chemistries (i.e., acidic, neutral and highly alkaline); (3) provide information that can be used to construct conceptual Cr geochemical models (site specific or more general and suitable for many contaminated sites); and (4) present ideas on potential remedial measures that are based on reactions and processes involving aqueous Cr^{6+} and its transformation to relatively or fully insoluble phases.

8.4 CHROMIUM INTERACTIONS WITH MINERALS

Chromium may interact with the minerals via redox and adsorption reactions under a variety of conditions of contaminated soils and subsoil environments. In the next section, different aspects of these interactions will be discussed.

8.4.1 Homogeneous Cr Redox Reactions

Oxidation–reduction [electron transfer (ET)] reactions are important in both inorganic and biological processes and, regardless of the system under investigation, the ET step is integral to the overall redox process. The ET reaction has been the center of considerable experimental studies and modeling efforts in inorganic, organic, and organometallic systems over the past 50 years. Electron-transfer reactions are bimolecular reactions that can be inner-sphere, outer-sphere, and diffusion-controlled reactions. In the case of Cr, as an example, the redox reaction may occur: (1) in the aqueous phase when Fe^{2+} is present in the soil solution (homogeneous reduction); (2) on the Fe(II)-exchanged surface of a solid phase; and (3) on the surface of an Fe(II)-bearing mineral. The latter two processes are considered heterogeneous reduction, and the surface plays the role of a catalyst (Figures 8.3 and 8.4).

Chromium reduction may occur if organic or inorganic reductants are present in the system that is contaminated with oxidized Cr. For example, at the Hanford site, sediments have the potential to solubilize large quantities of Fe^{2+} that (1) consume the available O_2, creating anoxic local conditions, (2) form Fe precipitates (probably of a mixed valance), and (3) act as a reductant for Cr^{6+}.

It has been found that Fe^{2+} may be generated from different sources. Studies have shown that the dissolution of biotite and chlorite and, to a lesser extent, magnetite occurs under conditions of different wastewater chemistries, for instance, in strong base solutions, and they release Fe^{2+} into the aqueous phase, leading to the reduction of Cr^{6+} to Cr^{3+} (e.g., He et al., 2005). Other studies provided evidence for dissolving biotite particles under hyperalkaline conditions (Qafoku et al., 2004), and spectroscopic measurements demonstrated chromate reduction in sediments that were exposed to highly alkaline and saline waste simulants (Qafoku et al., 2003d).

Detailed investigations have been conducted while looking at rapid Cr^{6+} homogeneous reduction by Fe^{2+} over a range of pH values and solution composition (Eary and Rai, 1988; Pettine et al., 1998; Sedlak and Chan, 1997), and the formation of Cr(III) phases and/or solid solutions as a result of Cr^{3+} precipitation (Eary and Rai, 1989a). Studies have shown that Fe^{2+} was more effective than Mn(II) in reducing Cr^{6+} in soils at basic pHs (James, 1994). Sedlak and Chan (1997) found that Cr^{6+} reduction by Fe^{2+} followed a first-order kinetics with respect to Fe^{2+} and Cr^{6+}, in three, one-electron-transfer steps, during which Cr^{6+} is reduced to Cr^{5+}, Cr^{4+}, and, ultimately, Cr^{3+}. The half-life of Cr^{6+} reduction ranged from several minutes to several months in their experiments. They suggested that the relationships between the rate coefficient and solution conditions can be explained by considering the reactivity of each of the Fe^{2+} species that may be present at significant concentrations [Fe^{2+}, $FeOH^+$, and $Fe(OH)_2$]. At extreme alkaline pHs that were representative of some Hanford waste streams, the reaction kinetics was rapid, leading to the formation of mixed Cr(III)-Fe spinels (He and Traina, 2005).

8.4.2 Heterogeneous Cr Redox Reactions

Chromium interacts with minerals via redox and adsorption reactions under a variety of conditions found in contaminated soils and associated subsoil environments. In the next section, we will discuss the role of various minerals in either alleviating or exacerbating the Cr contamination in environments.

The role of soil minerals in the redox reactions that occur in soils and sediments is multidimensional, and their chemical properties (e.g., sorption capacity, surface charge density, elemental composition), physical properties (e.g., layered or not and their size), and mineralogical identity (e.g., Fe(II)-bearing minerals, sulfides, Mn oxides) are important determinants of the extent of their participation in such reactions. Surface reactions can often accelerate a slow homogeneous reaction, because the surface can concentrate reactants, allow for longer-lived encounter complexes, and increase the reaction driving force.

Surface-mediated redox reactions are ET processes during which the oxidant and reductant interact as adsorbate species (Sposito, 2004). Importantly, the adsorbent itself may not participate

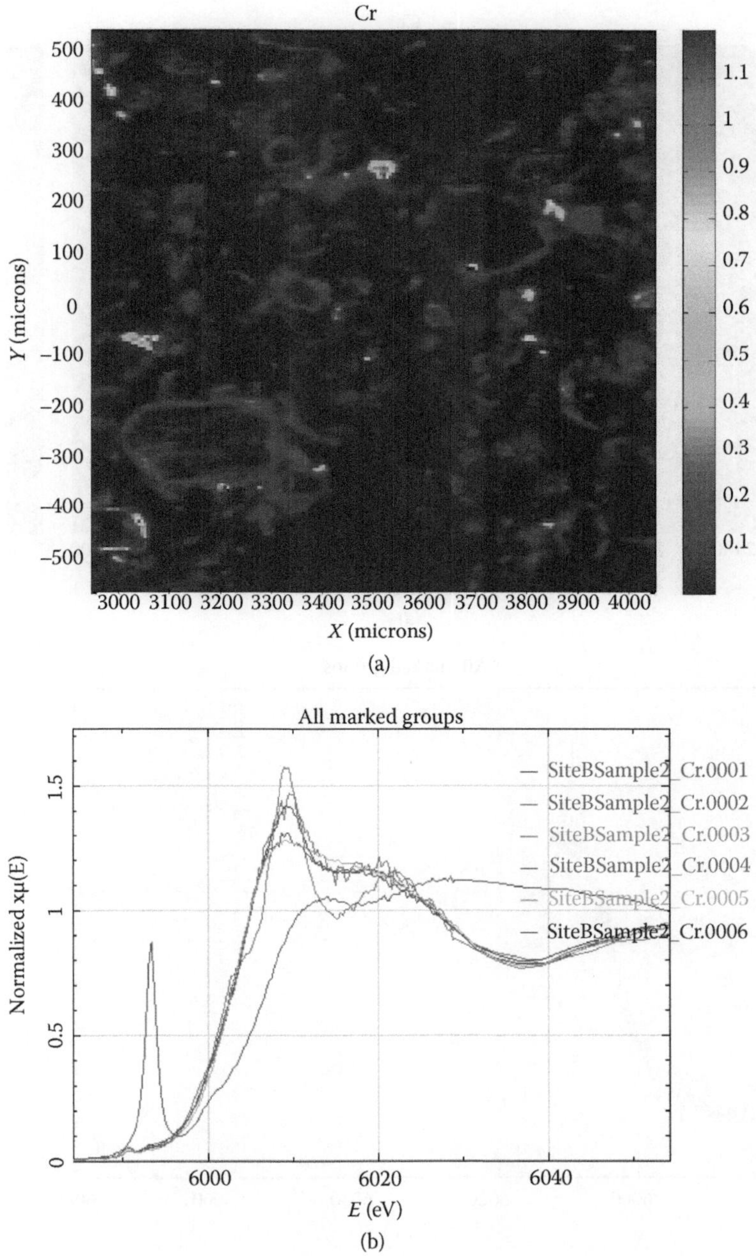

FIGURE 8.3 **(See color insert.)** (a) and (b) Chromium x-ray microprobe mapping and XANES measurements taken in high concentration Cr spots located in the sediments matrix (Hanford sediment). XANES results show varying Cr(VI) contents at different interrogated sites (indicated by the pre-edge peak intensity). (From Dresel, P.E. et al., *Geochemical Characterization of Chromate Contamination in the 100 Area Vadose Zone at the Hanford Site*, Pacific Northwest National Laboratory, PNNL-17674, 2008.)

directly in the redox reaction (Sposito, 2004), although there are reports in the literature that show that even structural elements may participate in these heterogeneous redox reactions (Williams and Scherrer, 2004). Surface redox reactions usually follow a sequence of reactions or processes, which are initiated by inner-sphere surface complexation of either the oxidant or the reductant; then, a complex forms between the adsorbed species and its counterpart reactant as a precursor to an ET

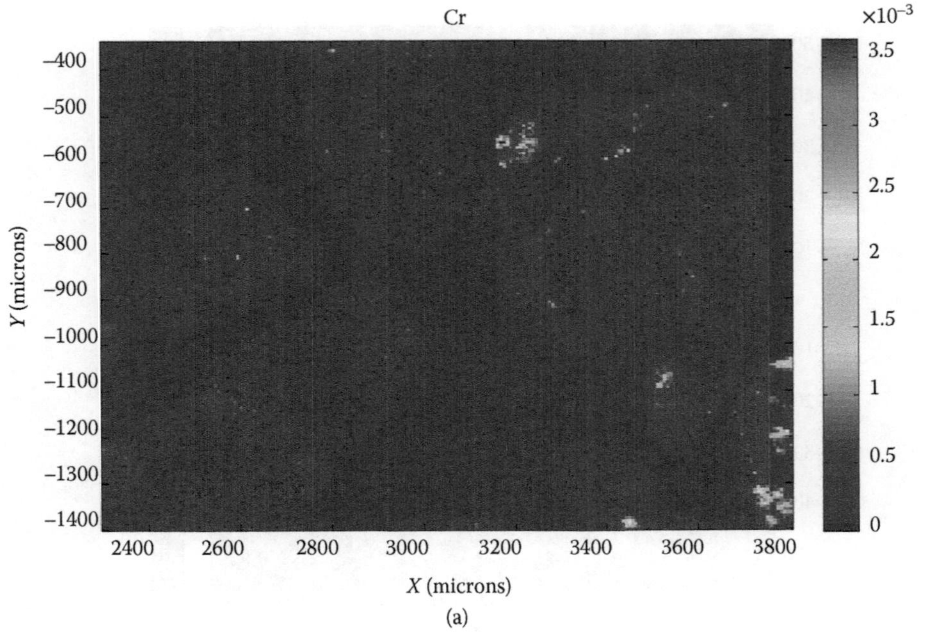

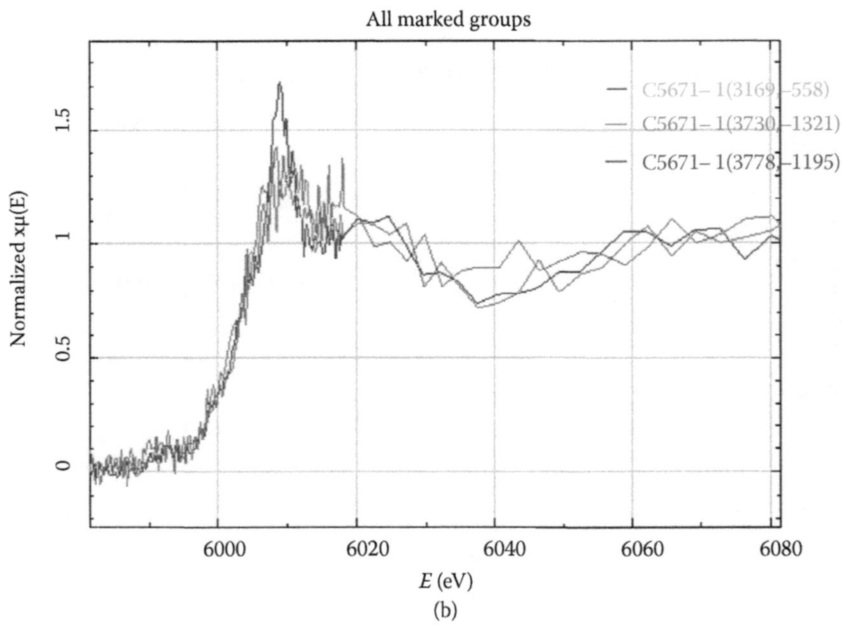

FIGURE 8.4 **(See color insert.)** Chromium x-ray microprobe mapping and XANES measurements taken in high concentration Cr hot spots. The XANES analyses show the presence of insoluble Cr(III). (From Dresel, P.E. et al., *Geochemical Characterization of Chromate Contamination in the 100 Area Vadose Zone at the Hanford Site*, Pacific Northwest National Laboratory, PNNL-17674, 2008.)

step, after which the ternary surface complex becomes destabilized by the production of newly reduced and oxidized species (Sposito, 2004).

The catalytic role of surfaces in redox reactions is clearly illustrated in the example of the Cr^{6+} reaction with sorbed Fe(II). The catalytic role of soil minerals in the redox reactions that occur in soils, sediments, and aquifers is demonstrated in recent literature (Elsner et al.,

2004a,b; Fredrickson et al., 2004; Ilton et al., 2004; Klupinski et al., 2004; Liger et al., 1999; Strathmann and Stone, 2003). It should be emphasized that processes such as advection, mass transfer, and/or diffusion-limiting conditions may affect the rate of these surface-mediated redox reactions. Other factors such as the presence of co-contaminants, ionic composition, concentration, and pH of the contacting solution that affect the rate of the surface-mediated redox reactions (Holtta et al., 1992) may also have a crucial role in determining the extent and rate of Cr(VI) reduction.

Stumm and coworkers have emphasized the importance of coupled geochemical processes and reactions with the Fe^{2+} and Fe^{3+} redox transformations (Stumm, 1992; Stumm and Morgan, 1996). As mentioned earlier, reduced Fe may be released from dissolving minerals that are exposed to high alkaline conditions, although other reactions may control the rate of release. The rate of dissolution of the minerals not only is a function of concentration of the free OH in the aqueous phase but also depends on the concentration of Al that is initially present in the waste solution; these act antagonistically, with Al content slowing the dissolution process. However, Al may be removed from the solution by homogeneous and/or heterogeneous precipitation reactions that form a series of Al-bearing secondary precipitates whose final product, rate of formation, and transient intermediates are dictated by the slow step of Si release during dissolution (Qafoku et al., 2003a,b,c, 2004; Zachara et al., 2004). This example clearly illustrates the complexity of the natural systems and how a secondary reaction (i.e., Si and Al precipitation) can control the rate of Fe^{2+} production in this system.

Many important redox reactions in soils, sediments, and waters are microbially mediated, and therefore, chemical parameters cannot be used alone to predict reaction rates (Morel and Hering, 1993). However, although biotic Cr(VI) reduction may be the principal pathway in an aerobic environment, it was demonstrated that biological pathways are not likely to contribute to Cr(VI) reduction in anaerobic systems (Fendorf et al., 2000). In these systems, Fe^{2+} will be the dominant inorganic reductant of Cr^{6+} at a pH greater than 5.5, and hydrogen sulfide will dominate this reduction process at a pH less than 5.5 (Fendorf et al., 2000). It has been found that Fe^{2+} is abundant in many suboxic and anoxic soils and sediments (Anderson et al., 1994), and Fe(II) is the dominant reductant in oligotrophic environments such as Hanford sediments.

8.4.3 Chromium Adsorption Reactions

Groundwater plume characterization at the Hanford site (conducted a few years ago) indicated the presence of an ongoing Cr source near the 100-D Area where the dichromate transfer station was once located (Petersen and Hall, 2008). The degree of chromate interaction with sediments during downward advective and diffusive transport through the thick vadose zone at Hanford has been investigated in many studies. Most of the Cr (more than 95% of total Cr) was highly mobile and only a small leaching-resistant fraction was present in the sediments under neutral and slightly basic conditions, producing a long tail of mobilization in saturated column experiments (Qafoku et al., 2009). Microscopic characterization indicated that Cr was found on grain coatings but some Cr was associated with individual "hot-spots" and in altered minerals. In addition to reduction, research has shown that aqueous Cr^{6+} may get involved in other geochemical reactions and/or processes that may affect its mobility in the vadose zone, such as adsorption to soil mineral surfaces and precipitation of Cr(VI) mineral phases with varying stabilities.

The vadose zone Cr geochemistry investigations at the Hanford site have focused on defining the controls on Cr(VI) flux to the groundwater and providing a basis for predicting the attenuation of the contaminant sources. As a part of this effort, researchers have determined the extent of oxidized and reduced Cr adsorption in these sediments. The proportion of the Cr that is readily transported (advective transport of dissolved Cr^{6+}) is a fundamental parameter for assessing the current Cr^{6+} flux to the water table. Longitudinal dispersivity and possible retardation of the chemical transport due to adsorption versus the aqueous flow have been also considered. However, the adsorption of Cr^{6+} [which usually occurs in the aqueous phase as chromate (CrO_4^{2-})] is found to be minimum in the Hanford sediments at circumneutral pH. On the other hand, physical sequestration in finer-grained particles (e.g., weathered clays) or dead-end pores may occur and contribute to diffusive transport that is indicated by significant tailing of the Cr(VI) movement and contribution to persistent flux from the source. In addition, chromate-bearing minerals [e.g., barium chromate ($BaCrO_4$) and/or solid solution such as $Ba(SO_4,CrO_4)$], and incorporation of trace levels of Cr^{6+} into other mineral phases may also contribute to the overall sorption. The scanning electron microscopy (SEM) micrographs (Figures 8.5 through 8.9) show that Cr may be associated with magnetite, ilmenite, and secondary alumino-silicates, and it may also form insoluble phases such as barium chromate.

Reduced chromium (Cr^{3+}) sorption may usually occur to phyllosilicates at the permanently charged surfaces. However, Cr^{3+} sorption is inhibited in KCl relative to NaCl solutions despite nearly identical ionic strength, pH, and activities of Cl^- and Cr^{3+}, indicating that K and Na must differentially affect Cr^{3+} sorption (Ilton et al., 2000). Permanent charge sites of phyllosilicates, and micas in particular, have a much greater affinity for K relative to Na (Sposito, 1984). In contrast, Na and K should not differentially affect the adsorption of Cr^{3+} by mica-edge (variable charge) sites. Previous studies indicate little to no specific adsorption of alkali cations on nonporous oxides at low temperatures in these variable-charge sites (e.g., Anderson and Sposito, 1991, and references therein). It appears that only a small fraction of the total energy for alkali cation adsorption in variably charged functional groups is chemical (Sahai and Sverjensky, 1997, and references therein). It is likely, therefore, that K decreases the sorption of Cr^{3+}relative to Na by restricting the access of Cr^{3+} to permanent charge sites.

Oxidized Cr^{6+} [which usually occurs in the aqueous phase as chromate (CrO_4^{2-})] sorption is affected by the changes in the background electrolyte concentration (Ilton et al., 1997, 2000). For example, an increase in NaCl concentrations from 0 to 25 mM only marginally increased sorption of Cr^{3+} but increased coupled sorption–reduction of Cr^{6+} by nearly a factor of five. It was suggested that an increase in ionic strength may lower the positive potential in the plane of sorption that is associated with edge sites, which could enhance the sorption of cations such as Cr^{3+} by variable-charge functional groups.

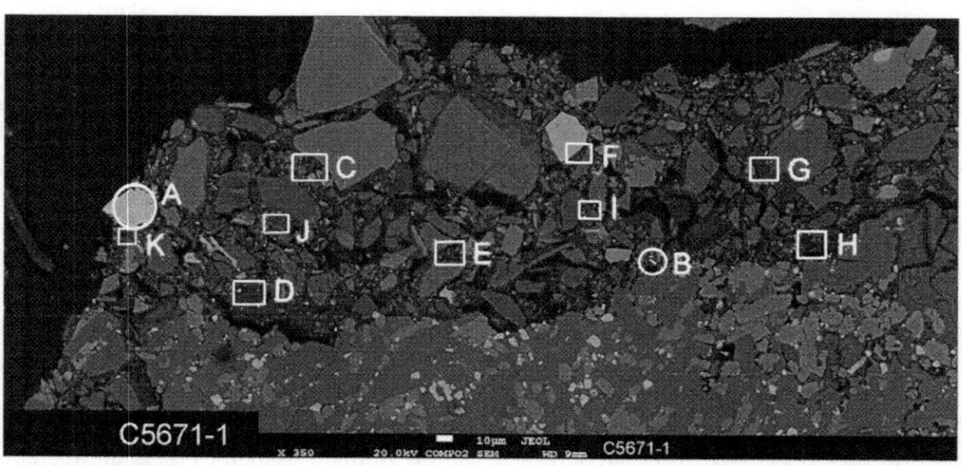

FIGURE 8.5 SEM image of Cr-contaminated sediment. Circle areas are those where Cr was observed in x-ray probe mapping. Boxed areas were measured with EDS. (From Dresel, P.E. et al., *Geochemical Characterization of Chromate Contamination in the 100 Area Vadose Zone at the Hanford Site*, PNNL-17674, 2008.)

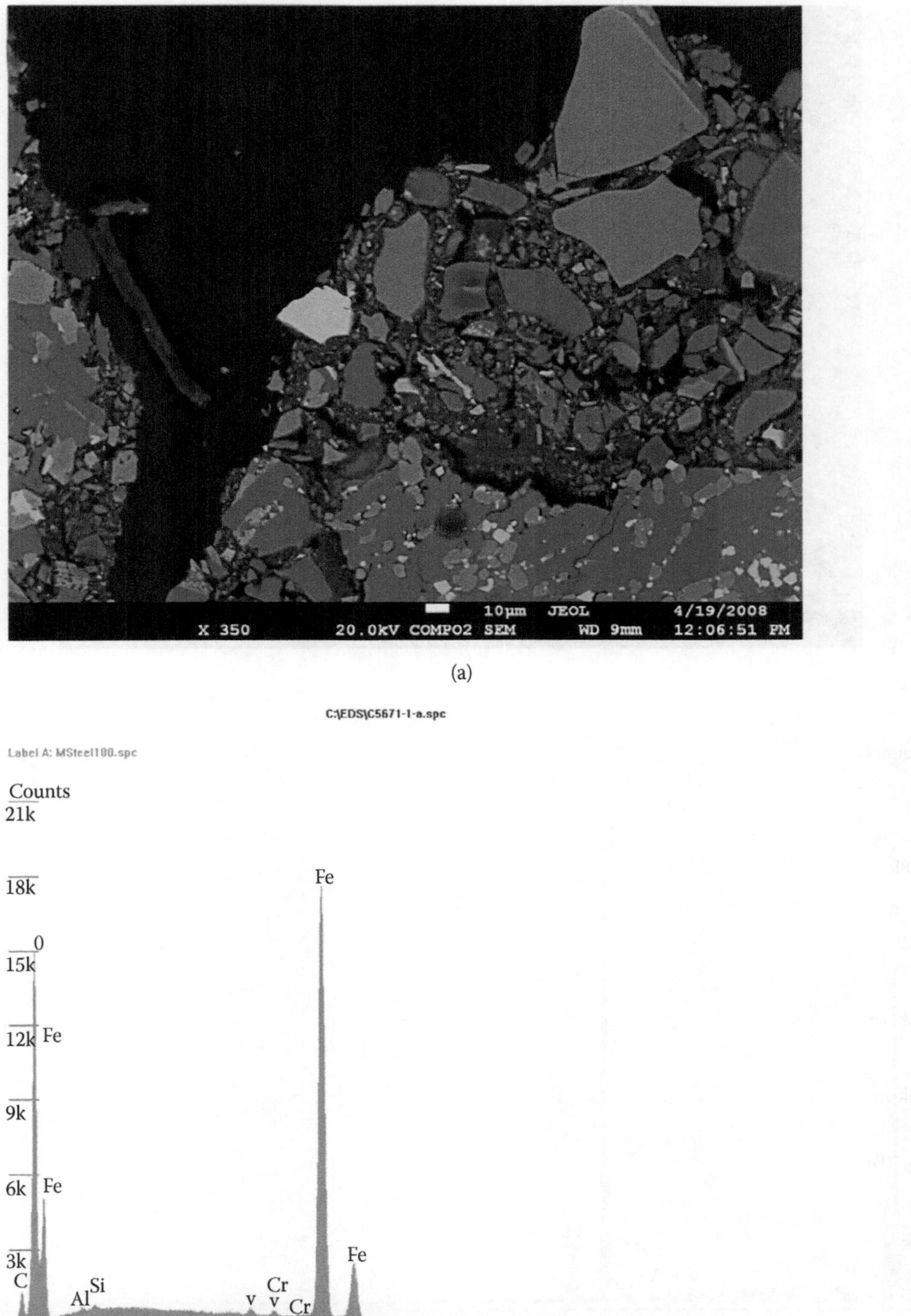

FIGURE 8.6 (a) SEM image and (b) EDS spectrum of the soil particle of area *A* in Figure 8.5 (the bright soil particle at the center of the image is most likely a particle of magnetite). (From Dresel, P.E. et al., *Geochemical Characterization of Chromate Contamination in the 100 Area Vadose Zone at the Hanford Site*, Pacific Northwest National Laboratory, PNNL-17674, 2008.)

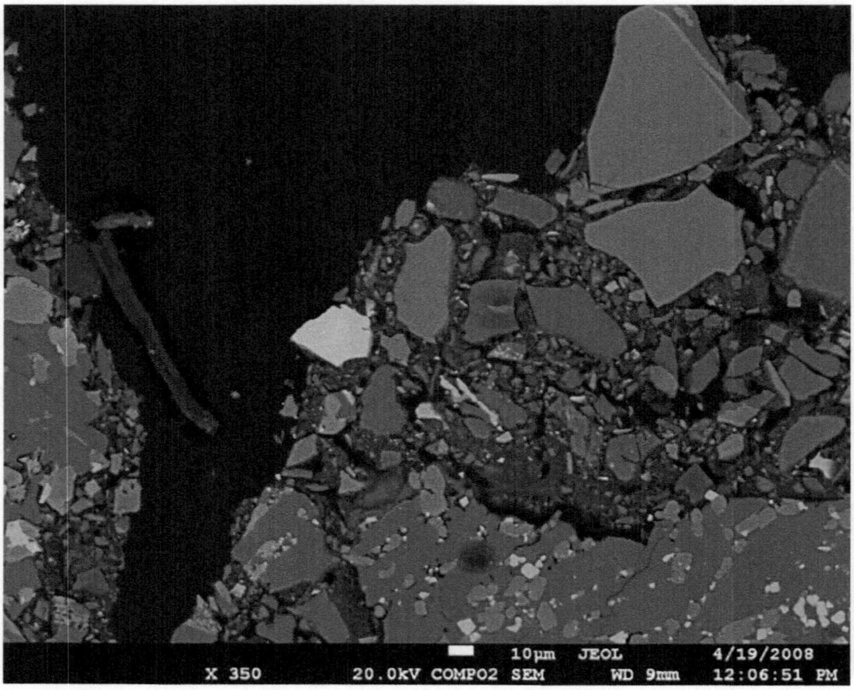

(a)

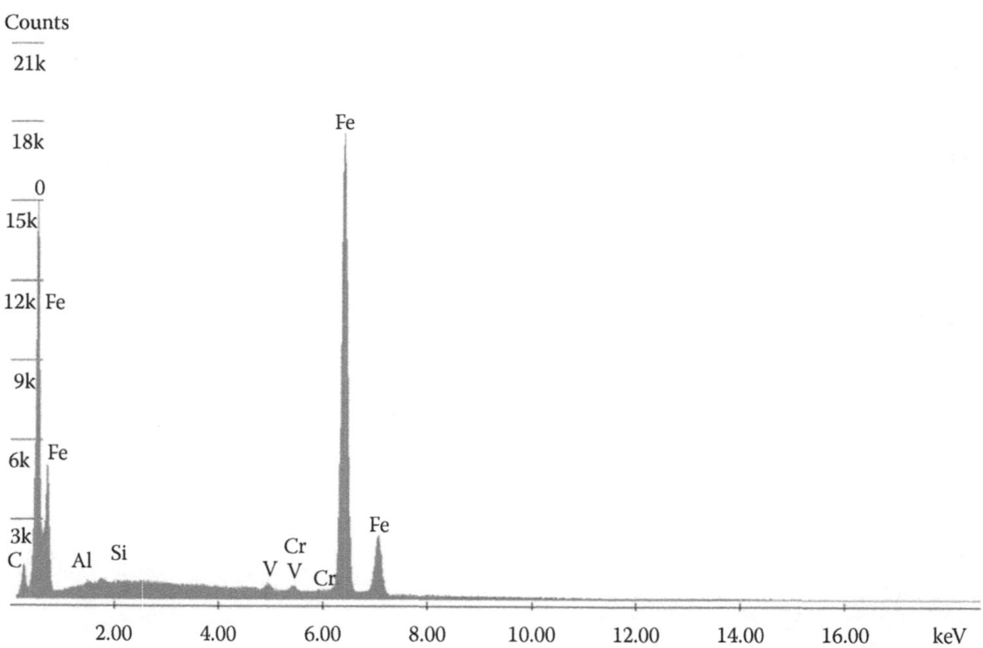

(b)

FIGURE 8.7 (a) SEM image and (b) EDS spectrum of the soil particle of area *B* in Figure 8.5 (white-colored particle at the center of the image is most likely a particle of $BaCrO_4$). (From Dresel, P.E. et al., *Geochemical Characterization of Chromate Contamination in the 100 Area Vadose Zone at the Hanford Site*, Pacific Northwest National Laboratory, PNNL-17674, 2008.)

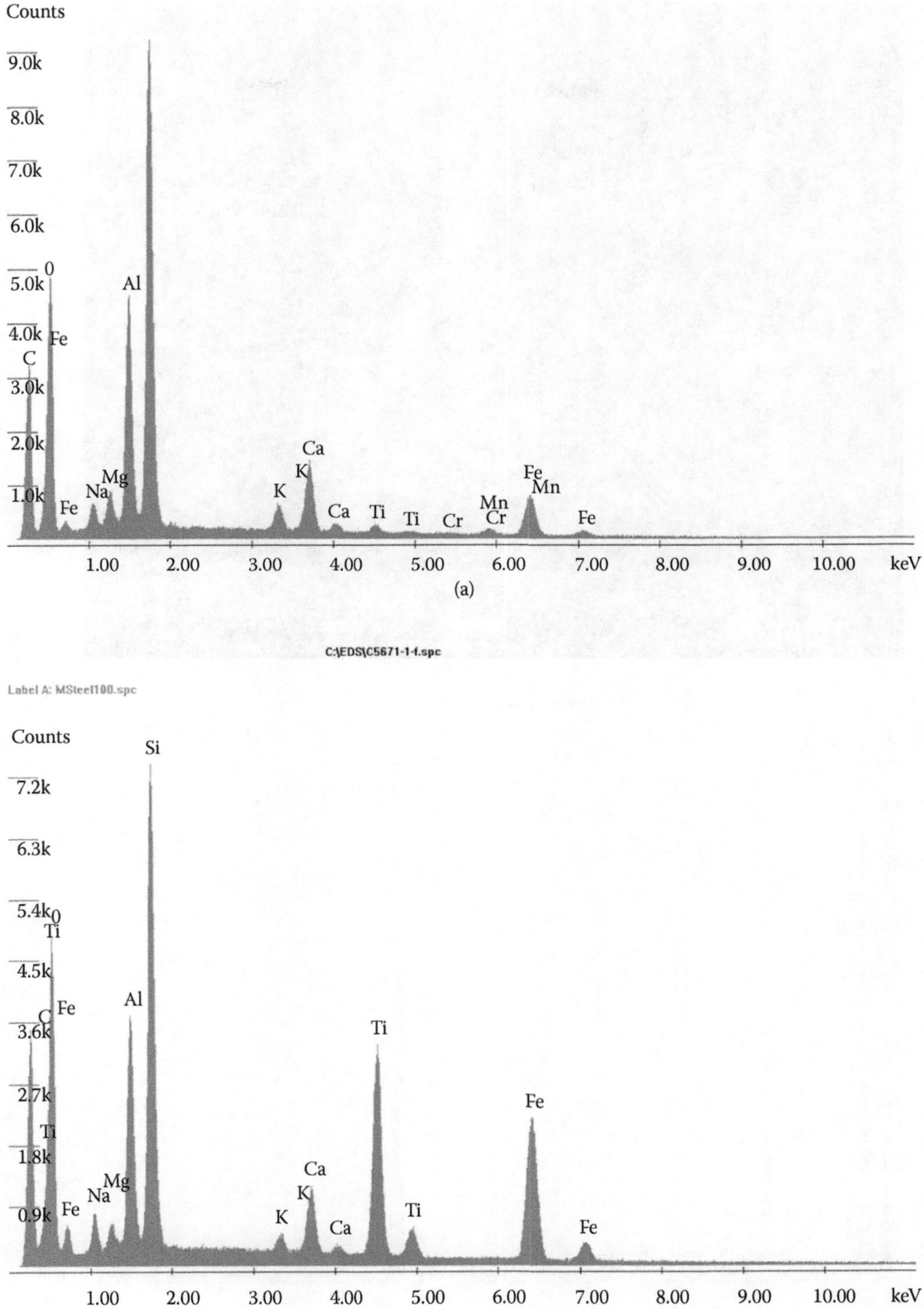

FIGURE 8.8 (a) and (b) EDS spectra collected in the boxed areas *K* and *F* in Figure 8.5. They show the presence of secondary alumino-silicate minerals and ilmenite (titanium magnetite). (From Dresel, P.E. et al., *Geochemical Characterization of Chromate Contamination in the 100 Area Vadose Zone at the Hanford Site*, Pacific Northwest National Laboratory, PNNL-17674, 2008.)

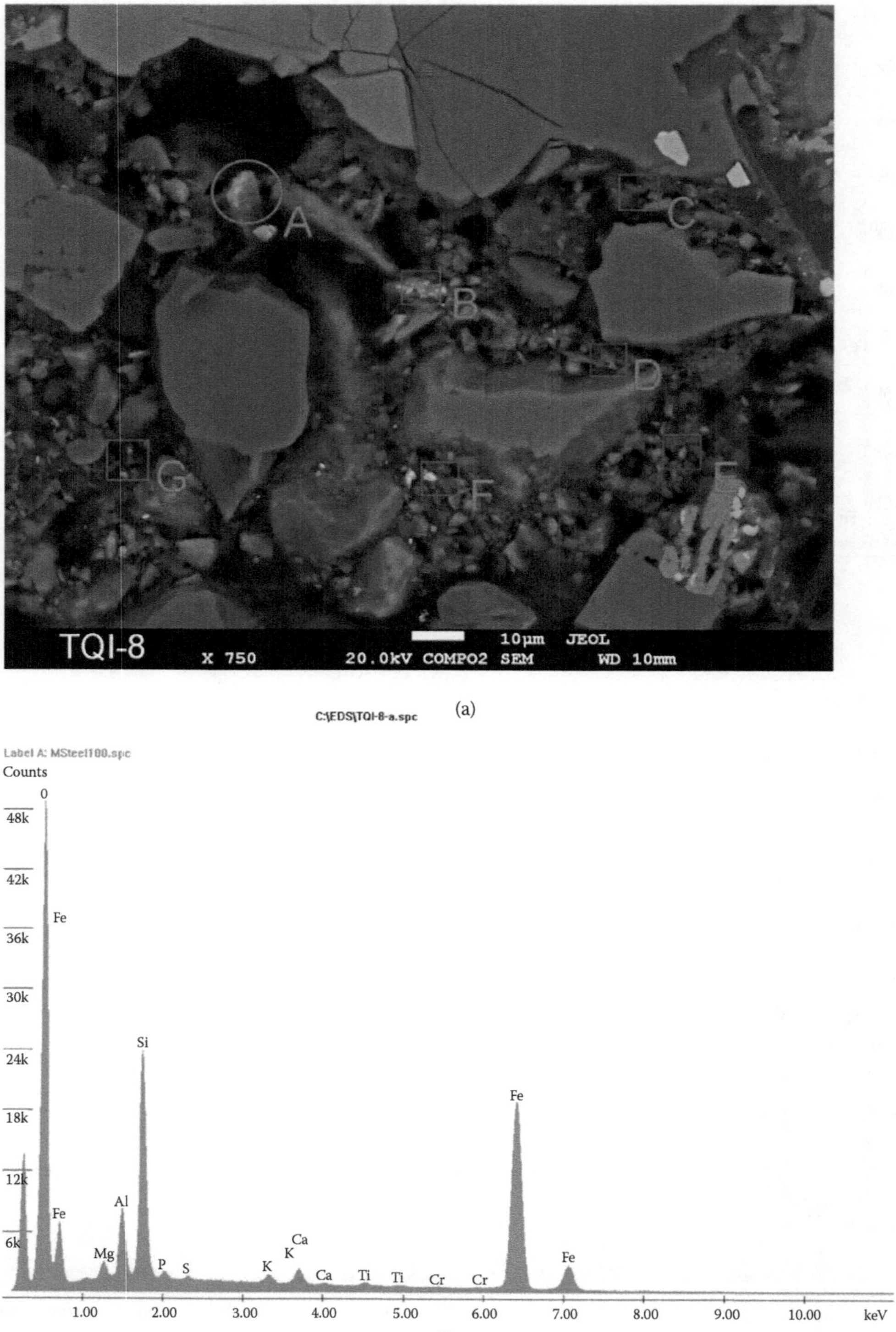

FIGURE 8.9 (a) SEM image and (b) EDS spectrum from area *A* taken in a sediment sample after leaching with Cr-free synthetic groundwater. Circles are areas where chromium was observed with an x-ray microprobe. (From Dresel, P.E. et al., *Geochemical Characterization of Chromate Contamination in the 100 Area Vadose Zone at the Hanford Site*, Pacific Northwest National Laboratory, PNNL-17674, 2008.)

8.5 CHROMIUM BEHAVIOR IN HANFORD SUBSURFACE

The role of co-disposed chemicals in reactions that occur in subsurface environments between the waste, minerals, and contaminant Cr is important. For example, it is observed that the co-disposal of acidic or hyperalkaline chemicals may be an important control on Cr(VI) mobility at some contaminated sites. The variability in contaminant Cr leaching behavior under different geochemical conditions that are imposed by chemically different waste streams makes it extremely important to develop a comprehensive understanding of Cr behavior under these conditions with the ultimate goal of site remediation.

The objective of this section is to provide data from experiments conducted with Hanford sediments to investigate the microscopic-scale Cr distribution and the Cr mobility. Continuous and stop-flow column experiments were conducted by using contaminated sediments from different contaminated areas. Several extraction and microscopic-scale techniques were also used to characterize sediment contamination and to identify possible mechanisms of chemical or physical Cr(VI) attenuation in these sediments. A fundamental understanding of Cr vadose zone geochemistry may help accelerate the cleanup by developing scientifically based remedial actions.

8.5.1 Acid Conditions

The chemical composition of the waste liquids discarded either simultaneously with the Cr^{6+}-containing liquids or existing previously at Cr-contaminated sites may be an important determinant of the Cr mobility in the vadose zone (Qafoku et al., 2010). Hexavalent Cr attenuation mechanisms may be operational under acidic conditions (Eary and Rai, 1991). It is possible that Fe(II)-bearing soil minerals may undergo dissolution when sediments are exposed to acidic waste liquids, releasing Fe(II) and ultimately reducing Cr^{6+} to Cr^{3+}. Chromate may also get heterogeneously reduced on Fe(II)-bearing mineral surfaces, such as magnetite. However, studies have demonstrated that the surface of magnetite becomes passivated, causing slowing or stopping of the Cr^{6+} reduction reaction when thicknesses of the coating layers (which in one study was composed of dominantly calcite with minor quantities of aragonite and vaterite) increase beyond 15 Å (Doyle et al., 2004). Silicate and/or Fe(III) coatings may also cause the surfaces of Fe(II)-bearing minerals to become passive, decreasing their capacity to reduce Cr^{6+} (Kendelewicz et al., 2000; Peterson et al., 1997b). However, exposure of the passive surfaces to acidic waste liquids may result in removal of the coatings, rendering reactive the surfaces of Fe(II)-bearing minerals and increasing the overall extent of Cr^{6+} reduction. For example, removal of coatings by acidic pretreatment of the sediments resulted in a reduction of Cr^{6+} to Cr^{3+} by the Fe(II)-bearing minerals (e.g., antigorite, lizardite, magnetite, and some clay minerals) (Ginder-Vogel et al., 2005).

More emphasis needs to be placed on arid environments in terms of investigating contaminant Cr^{6+} reduction (Ginder-Vogel et al., 2005) and on the coupled effect of dissolution and reduction reactions (i.e., dissolution-controlled reduction of Cr^{6+}) (Eary and Rai, 1989b; Henderson, 1994; Park et al., 2007). Very few studies have been conducted with contaminated, naturally aged, and low organic carbon sediments from arid areas. One such recent study focused on the mobility and the extent of Cr^{6+} reduction in contaminated and naturally aged (decades) sediments that were exposed to mixed-acid and dichromate waste fluids for decades (Qafoku et al., 2010). In this study, a series of columns experiments were conducted to investigate contaminant Cr mobility in these sediments, which were combined with chemical extractions and microprobe and spectroscopic inspections and interrogations to demonstrate the occurrence and to measure the extent of Cr^{6+} reduction in these natural systems.

Results from wet chemical extractions indicated the presence of a large pool of contaminant Cr that was tightly bound to sediment minerals. The effluent Cr^{6+} concentrations were low, indicating low mobility. Electron microprobe (EMPA) inspections suggested that contaminant Cr was distributed throughout the sediment matrix, and it was hard to locate areas of high concentrations. The redox state of contaminant Cr was confirmed by XPS to be mostly Cr(III), which was spatially associated with Fe, which, importantly, occurred as mixed Fe(II) (minor) and Fe(III) (dominant). These results showed that in the arid sediments of the Hanford sediments, the reduction of Cr^{6+} to Cr^{3+} has occurred, limiting the advective transport and possibly the mass transfer fluxes away from the original source of contamination. Most likely, Cr^{6+} attenuation in these sediments occurred as a result of acid removal of unreactive coatings and acid promoting the dissolution of Fe(II)-bearing soil minerals, which released Fe^{2+} into the aqueous phase. This was followed by Cr^{6+} abiotic homogeneous and/or heterogeneous reduction by aqueous, sorbed, and/or structural Fe(II), and, subsequently, formation of insoluble Cr(III) phases or [Cr(III) Fe(III)] solid solutions.

8.5.2 Neutral and Slightly Alkaline Conditions

Three mechanisms may control Cr^{6+} attenuation under oxic and neutral to slightly alkaline conditions when soils and subsurface sediments are exposed to highly concentrated Cr-containing waste fluids: (1) adsorption (partition) to soil minerals, (2) precipitation and formation of relatively insoluble Cr(VI) solid phases, and (3) reduction followed by precipitation of less mobile Cr(III) phases (Qafoku et al., 2009). In some arid sites with low precipitation, aqueous Cr^{6+} may also be sensitive to physical attenuation mechanisms (Qafoku et al., 2009).

Aqueous Cr^{6+} as chromate (CrO_4^{2-}) may be sorbed on soil mineral reactive surfaces, but studies have reported that chromate formed a relatively weak outer-sphere surface complex on Fe oxides and other minerals of subsoil media and adsorption was suppressed by other competing co-anions, mainly HCO_3^- and SO_4^{2-} (Zachara et al., 1987, 1989). Other studies have shown that at neutral and alkaline pH, CrO_4^{2-} was not significantly adsorbed and moved nearly unretarded (no adsorptive retardation) through sediments of the vadose and saturated zones (Fruchter et al., 2000; Qafoku et al., 2003c).

Studies have also investigated Cr(VI) sorption to soil minerals, soils, and sediments (Ainsworth et al., 1989; Ilton and Veblen, 1994; Jardine et al., 1999; Kent et al., 1994, 1995; Peterson et al., 1997a; Selim et al., 1989; Villalobos et al., 2001; Wu et al., 2000; Zachara et al., 1987). For example, volcanic ash soils (VAS) or Andisols are able to remove Cr(VI) from wastewater, because Andisols have high abundances of noncrystalline secondary minerals such as allophane and imogolite that exhibit variable charge characteristics and have the ability to retain cations and anions depending on soil pH and ionic strength (Babel and Opiso, 2007). Soils with high clay content and porosity and lower bulk density and that are rich in fine particles with a higher surface area and more active noncrystalline minerals have a higher sorption capacity of Cr(VI) (Babel and Opiso, 2007).

The effect of secondary mineral coating on sorption and the effect of carbonate coatings on the reduction of aqueous Cr^{6+} on the magnetite(111) surface were investigated with synchrotron-based x-ray photoemission spectroscopy (PES) and x-ray absorption near edge structure (XANES) spectroscopy, along with laboratory-based powder x-ray diffraction (XRD) and SEM (Doyle et al., 2004). This study showed that carbonate coatings on natural magnetite particles can significantly reduce or fully eliminate their ability to reduce Cr(VI) (Doyle et al., 2004).

If soils and sediments are exposed to high concentrations of Cr^{6+}, precipitation of a moderately soluble mineral (hashemite: $BaCrO_4$) and/or the formation of the lower solubility solid solutions of $BaCrO_4$-$BaSO_4$ may be promoted (Rai et al., 1989; Rai and Zachara, 1986). In addition, Cr^{6+} may co-precipitate with calcite and the concentration of Cr incorporated into calcite increases with an increase in Cr concentration in solution (Tang et al., 2007). The distribution coefficient, Kd, was the highest at low Cr solution concentration, and it decreased to a constant value with an increase in Cr solution concentration (Tang et al., 2007). XANES spectra collected as a part of this investigation confirmed that incorporated Cr is hexavalent. Consideration of possible local coordination indicated that significant distortion or disruption is required to accommodate CrO_4^{2-} in the calcite structure (Tang et al., 2007). Site-specific studies in naturally aged contaminated sediments are needed to determine the role of these solid phases in controlling Cr^{6+} mobility in highly contaminated sediments.

Studies have shown that Cr^{6+} can be immobilized under neutral to alkaline conditions after a reduction reaction that results in the formation of less mobile Cr(III) (Lee and Hering, 2005). However, there is a need to determine the extent and rates of Cr^{6+} reduction in surface and subsurface contaminated systems that are oxic and low in organic carbon content, conditions under which any substantive microbial reduction of Cr^{6+} would require major additions of NO_3^- and/or organic carbon (Oliver et al., 2003).

Specifically, studies that can measure separate and combined contributions to the overall reduction capacity of a natural medium of different reduction mechanisms and pathways, such as reduction via aqueous and sorbed Fe(II), reduced sulfur compounds, soil organic matter, and microbial processes (Anderson et al., 1994; Buerge and Hug, 1997; Buerge and Hug, 1999; Deng and Stone, 1996; Eary and Rai, 1988; Fendorf and Li, 1996; Fendorf et al., 2000; Ginder-Vogel et al., 2005; Pettine et al., 1998; Qafoku et al., 2003c; Sedlak and Chan, 1997), are definitely required. Other attenuation mechanisms, such as physical matrix potential effects holding CrO_4^{2-}-contaminated pore water against gravimetric force (Qafoku et al., 2009), should be studied as well.

A recent study of sediments from a long-term Cr-contaminated site with concentrated Cr liquids demonstrated that at least three pools of Cr^{6+} with different leaching behavior were present (Qafoku et al., 2009): (1) weakly bound to sediments and highly mobile Cr^{6+} (more than 95% of total Cr mass in all sediments); (2) a leaching resistance fraction, which contained Cr^{6+} held at physical and mineralogical remote sites that served as a longer-term continuing source of Cr contamination; and (3) a small pool consisting of reduced and immobile Cr(III) that resulted from homogeneous and/or heterogeneous redox reactions of Cr^{6+} and aqueous, sorbed, or structural Fe(II). The study found that both Cr^{6+} reduction and precipitation were not significant in the sediments under slightly alkaline and oxic conditions that were exposed to concentrated waste fluids.

Reoxidation is another pathway that requires attention. Studies have shown that oxidation of Cr(III) via an oxidative dissolution mechanism may also occur, although the formation of a secondary phase at low pH values, that is, chromic hydroxy chromate, may decrease the rate of oxidation of Cr(III) (Lee and Hering, 2005). Importantly, the same study reported that Cr^{6+} may be strongly sorbed to $Cr(OH)_3$ solids under oxidizing conditions.

8.5.3 Hyperalkaline Conditions

High temperature, alkaline, and saline radioactive waste fluids that are rich in Cr (up to 0.413 mol L^{-1}) (Jones et al., 2000) and other contaminants have leaked from the underground single-shell waste tanks at the Hanford site, WA. Recent studies have shown that Cr^{6+} reduction to less mobile trivalent Cr occurred when biotite (He et al., 2005), magnetite (He and Traina, 2005), and Hanford sediments (Qafoku et al., 2003c; Zachara et al., 2004) were brought into contact with simulated hyperalkaline waste fluids. Studies with contaminated sediments from the Hanford site have demonstrated that approximately 42% of the Cr^{6+} originally present in the waste fluids was immobilized mostly as Cr(III), although some Cr(VI) precipitates were also present (Zachara et al., 2004).

Studies have also shown that Cr^{6+} mobility was significantly decreased in the sediments that were exposed to highly alkaline and saline waste liquids (Qafoku et al., 2007a, 2003c; Zachara et al., 2004). The minerals of the sediment may undergo intense dissolution under these conditions (Qafoku et al., 2003b, 2004; Wan et al., 2004; Zachara et al., 2004), and the decrease in Cr^{6+} mobility has been attributed to a two-step reaction: (1) Cr^{6+} reduction to Cr^{3+} by aqueous (or sorbed) Fe^{2+} released from dissolving Fe(II)-bearing minerals; (2) subsequent formation of relatively insoluble Cr(III) pure phases and/or [Fe(III) Cr(III)] solid solutions. This mechanism was also supported by previous studies, which demonstrated that Cr^{6+} was reduced by aqueous or sorbed Fe^{2+} (Buerge and Hug, 1999; Eary and Rai, 1989a).

Chromate was also shown to be homogeneously reduced by Fe^{2+} in hyperalkaline solutions, and the rate of reduction increased with an increase in the OH^- concentration (He et al., 2004). The reduction of Cr^{6+} in this study was nonstoichiometric, probably due to Fe(II) precipitation and/or removal of Fe^{2+} from solution via the oxygen oxidation pathway. In a follow-up study (He and Traina, 2005), the authors found that compared with acidic and neutral pH, Cr^{6+} reduction by magnetite at high pH conditions was limited (<20% of potential reduction capacity), and the extent of reduction did not vary significantly with an increase in NaOH concentration.

Finally, in another study conducted under hyperalkaline conditions (i.e., 10 M NaOH), XRF maps combined with localized XRD analysis revealed the association of high localized Cr concentrations with several Fe(II)-bearing minerals, including antigorite $\{(Mg,Fe(II)_3[(OH)_4|Si_2O_5])\}$, lizardite $\{[Mg_xFe_{3-x}[Si_2O_5](OH)_4]\}$ [which has Mg-substituted Fe(II) in its structure], and magnetite (Ginder-Vogel et al., 2005). Importantly, the same study reported that analysis of portlandite $[Ca(OH)_2]$ particles indicated that Cr was associated with this mineral, 20% of which was in the hexavalent oxidation state (Ginder-Vogel et al., 2005). These authors concluded that (1) portlandite may catalyze surface-mediated precipitation of Cr(III) phases, (2) portlandite may act as a Cr^{6+} sorbent, and (3) portlandite may incorporate Cr^{6+} during its rapid formation under hyperalkaline conditions.

The rate at which Fe^{2+} is produced from the reaction between the Al-rich, hyperalkaline, and saline fluids and the sediment minerals, for example, biotite and chlorite, is given by (Qafoku et al., 2003c)

$$\frac{\delta Fe^{2+}}{\delta t} = \left\{k_{\text{OH-biotite}}[OH]^a + k_{\text{Na-biotite}}[Na]^b + k_{\text{NO}_3\text{-biotite}}[NO_3]^c + k_{\text{Al-biotite}}[Al]^d\right\} + \left\{k_{\text{OH-chlorite}}[OH]^e + k_{\text{Na-chlorite}}[Na]^f + k_{\text{NO}_3\text{-chlorite}}[NO_3]^g + k_{\text{Al-chlorite}}[Al]^h\right\} \quad (8.1)$$

The three-step homogeneous reaction of Cr^{6+} reduction by Fe^{2+} (Sedlak and Chan, 1997) has a theoretical removal rate of reactants (unlimited reactant concentrations) of

$$\frac{1/3\delta Fe^{2+}}{\delta t} = \frac{\delta Cr^{6+}}{\delta t} = -k_1[Cr^{6+}]^1[Fe^{2+}]^3 \quad (8.2)$$

In addition, Fe^{2+} may react with dissolved O_2 and the rate of removal is (Millero et al., 1987)

$$\frac{\delta Fe^{2+}}{\delta t} = -k_2[Fe^{2+}][O_2][OH]^2 \quad (8.3)$$

Assuming that both removal reactions occur in parallel, the total rate of Fe^{2+} removal is

$$\frac{\delta Fe^{2+}}{\delta t} = -\{3k_1[Cr^{6+}]^1[Fe^{2+}]^3 + k_2[Fe^{2+}][O_2][OH]^2\} \quad (8.4)$$

The changes in aqueous Fe^{2+} concentration along with time in the system and the steady-state Fe^{2+} concentration can be described by the following equation:

$$\frac{\delta Fe^{2+}}{\delta t} = \left\{k_{\text{OH-biotite}}[OH]^a + k_{\text{Na-biotite}}[Na]^b + k_{\text{NO}_3\text{-biotite}}[NO_3]^c + k_{\text{Al-biotite}}[Al]^d\right\} + \left\{k_{\text{OH-chlorite}}[OH]^e + k_{\text{Na-chlorite}}[Na]^f + k_{\text{NO}_3\text{-chlorite}}[NO_3]^g + k_{\text{Al-chlorite}}[Al]^h\right\} - \left\{3k_1[Cr^{6+}]^1[Fe^{2+}]^3 + k_2[O]^{0.5}[Fe^{2+}]^2[OH]^4\right\} \quad (8.5)$$

If the rate of Fe^{2+} generation (the first term of Equation 8.5) is greater than the rate of Fe^{2+} consumption by O_2 (second part of the second term), localized reducing zones may be created when Cr^{6+} is reduced to Cr^{3+}. The rate of Cr^{6+} reduction in these systems would be controlled by the rate of Fe^{2+} production.

Equation 8.5 may be used to calculate the changes in Fe^{2+} concentration along with time and to estimate its availability to enter into a redox reaction with Cr^{6+}. However, as clearly stated in previous studies (Qafoku et al., 2003c), the accurate measurement of the rate of Fe^{2+} generation and consumption in these systems is difficult, because: (1) aqueous Fe^{2+} concentration at different times during the experiment depends not only on the rate of dissolution of different soil minerals but also on the rate of diffusion of reactants to and products from the reaction sites; (2) aqueous Fe^{2+} concentration is also controlled by other reactions and processes that may simultaneously occur and effectively compete for aqueous Fe^{2+} (Fe^{2+} adsorption to soil mineral surfaces would be one of them); and (3) the accurate measurement of the Fe^{2+} aqueous concentration is problematic because of the strong interferences with relatively high concentrations of Cr and Al (Qafoku et al., 2003c). Therefore, a quantitative estimation of the Cr^{6+} reduction rate in these geochemical systems would be challenging.

Homogeneous studies indicate that Cr^{6+} reacts rapidly with Fe(II) over a range of pH values and solution composition (Eary and Rai, 1988; Pettine et al., 1998; Sedlak and Chan, 1997), by following three, one-electron-transfer steps during which Cr^{6+} is reduced to Cr^{5+}, Cr^{4+}, and, ultimately, Cr^{3+} (Sedlak and Chan, 1997). However, in natural systems, Cr^{6+} removal could be gradual rather than instantaneous and time may be required to reach the maximum possible rate of Cr(VI) reduction (Qafoku et al., 2003c).

Sediments beneath the leaking tanks at Hanford were exposed to self-boiling alkaline waste fluids, and extensive mineral alteration occurred in them (Serne et al., 2002). Studies conducted under simulated conditions that were similar to those of the sediments beneath the tank waste indicated that dissolution of existing soil minerals in the sediments was intense under simulated conditions of hyperalkalinity and hypersalinity (Ainsworth et al., 2005; Bickmore et al., 2001; Chorover et al., 2003; Deng et al., 2006; Mashal et al., 2005; Qafoku et al., 2003a,b,c, 2005, 2007b; Wan et al., 2004; Zhao et al., 2004). These studies also showed the formation of neophases with large reactive surface areas and sorption capacities, such as zeolites, feldspathoids in the groups of cancrinite and sodalite, and Fe oxides that may sorb Cr^{6+}, contributing to the overall sorption of this contamination in waste altered sediments.

ACKNOWLEDGMENTS

Pacific Northwest National Laboratory (PNNL) is operated for the DOE by Battelle Memorial Institute under Contract DE-AC06-76RLO 1830. Funding for this research effort was provided by the U.S. Department of Energy Office of Environmental Management. Some of the figures presented in this chapter were taken by Dr. Evan Dresel. The SEM images and energy dispersive spectroscopy measurements were taken by Dr. Jim McKinley and the x-ray microprobe and XANES measurements were taken by Dr. Steve Heald at Argonne National Laboratory. The research presented in this chapter was conducted in part in the Environmental Molecular Sciences Laboratory, a national scientific user facility that is sponsored by the U.S. DOE Office of Biological and Environmental Research and located at the Pacific Northwest National Laboratory in Richland, WA (USA).

REFERENCES

2010 Groundwater Annual Report: DOE/RL-2010-11, Revision 0, Hanford Site Grondwater Monitoring and Performance Report for 2009, Volumes 1 and 2. Washington, DC: U.S. Department of Energy.

Abbasi, S.A., Soni, R., 1984. Teratogenic effects of chromium(VI) in the environment as evidenced by the impact of larvae of amphibian *Rand tigrina*: Implications in the environment management of chromium. *Int. J. Environ. Stud.*, 23: 131–137.

Ainsworth, C.C. et al., 2005. Impact of highly basic solutions on sorption of Cs+ to subsurface sediments from the Hanford site, USA. *Geochim. Cosmochim. Acta*, 69(20): 4787–4800.

Ainsworth, C.C., Girvin, D.C., Zachara, J.M., Smith, S.C., 1989. Chromate adsorption on goethite: Effects of aluminum substitution. *Soil Sci. Soc. Am. J.*, 53(2): 411–418.

Anderson, L.D., Kent, D.B., Davis, J.A., 1994. Batch experiments characterizing the reduction of Cr(VI) using suboxic material from a mildly reducing sand and gravel aquifer. *Environ. Sci. Technol.*, 28: 178–185.

Anderson, S.J., Sposito, G., 1991. Cesium-adsorption method for measuring accessible structural surface charge. *Soil Sci. Soc. Am. J.*, 55: 1569–1576.

Babel, S., Opiso, E.M., 2007. Removal of Cr from synthetic wastewater by sorption into volcanic ash soil. *Int. J. Environ. Sci. Technol.*, 4(1): 99–107.

Bickmore, B.R., Nagy, K.L., Young, J.S., Drexler, J.W., 2001. Nitrate-cancrinite precipitation on quartz sand in simulated Hanford tank solutions. *Environ. Sci. Technol.*, 35: 4481–4486.

Buerge, I.J., Hug, S.J., 1997. Kinetics and pH dependence of chromium(VI) reduction by iron(II). *Environ. Sci. Technol.*, 31: 1426–1432.

Buerge, I.J., Hug, S.J., 1999. Influence of mineral surfaces on chromium(VI) reduction by iron(II). *Environ. Sci. Technol.*, 33(23): 4285–4291.

Carpenter, R.W., 1993. 100-D Area Technical Baseline Report. WHC-SD-EN-TI-181 Rev 0, Westinghouse Hanford Company, Richland, WA.

Chorover, J. et al., 2003. Linking cesium and strontium uptake to kaolinite weathering in simulated tank waste leachate. *Environ. Sci. Technol.*, 37(10): 2200–2208.

Deng, B., Stone, A.T., 1996. Surface-catalyzed chromium(VI) reduction: Reactivity comparisons of different organic reductants and different oxide surfaces. *Environ. Sci. Technol.*, 30: 2484–2494.

Deng, Y., Harsh, J.B., Flury, M., Young, J.S., Boyle, J.S., 2006. Mineral formation during simulated leaks of Hanford waste tanks. *Appl. Geochem.*, 21: 1392–1409.

Doyle, C.S., Kendelewicz, T., Brown, G.E., 2004. Inhibition of the reduction of Cr(VI) at the magnetite-water interface by calcium carbonate coatings. *Appl. Surf. Sci.*, 230(1–4): 260–271.

Dresel, P.E. et al., 2008. *Geochemical Characterization of Chromate Contamination in the 100 Area Vadose Zone at the Hanford Site*. Pacific Northwest National Laboratory, PNNL-17674.

Eary, L.E., Rai, D., 1988. Chromate removal from aqueous wastes by reduction with ferrous ion. *Environ. Sci. Technol.*, 22: 972–977.

Eary, L.E., Rai, D., 1989a. Chromate removal from aqueous wastes by reduction with ferrous ion. *Environ. Sci. Technol.*, 22: 972–977.

Eary, L.E., Rai, D., 1989b. Kinetics of chromate reduction by ferrous ions derived from hematite and biotite at 25 degree C. *Am. J. Sci.*, 289: 180–213.

Eary, L.E., Rai, D., 1991. Chromate reduction by subsurface soils under acidic conditions. *Soil Sci. Soc. Am. J.*, 55(3): 676–683.

Elsner, M., Haderlein, S.B., Kellerhais, T. et al., 2004a. Mechanisms and products of surface-mediated reductive dehalogenation of carbon tetrachloride by Fe(II) on goethite. *Environ. Sci. Technol.*, 38(7): 2058–2066.

Elsner, M., Schwarzenbach, R.P., Haderlein, S.B., 2004b. Reactivity of Fe(II)-bearing minerals toward reductive transformation of organic contaminants. *Environ. Sci. Technol.*, 38(3): 799–807.

Fendorf, S.E., Li, G., 1996. Kinetics of chromate reduction by ferrous iron. *Environ. Sci. Technol.*, 30: 1614–1617.

Fendorf, S.E., Wielinga, B.W., Hansel, C.M., 2000. Chromium transformation in natural environments: The role of biological and abiological processes in chromium (VI) reduction. *Int Geol. Rev.*, 42: 691–701.

Foster, R.F., 1957. *Recommended Limit on Addition of Dichromate to the Columbia River*. HW-49713. Richland WA: General Electric Company.

Fredrickson, J.K. et al., 2004. Reduction of TcO_4^- by sediment-associated biogenic Fe(II). *Geochim. Cosmochim. Acta*, 68(15): 3171–3187.

Fruchter, J.S. et al., 2000. Creation of a subsurface permeable treatment zone for aqueous chromate contamination using in situ redox manipulation. *Ground Water Mon. Remed.*, 20: 66–77.

Ginder-Vogel, M., Borch, T., Mayes, M.A., Jardine, P.M., Fendorf, S., 2005. Chromate reduction and retention processes within arid subsurface environments. *Environ. Sci. Technol.*, 39: 7833–7839.

He, Y.T., Bigham, J.M., Traina, S.J., 2005. Biotite dissolution and Cr(VI) reduction at elevated pH and ionic strength. *Geochim. Cosmochim. Acta*, 69(15): 3791–3800.

He, Y.T., Chen, C.C., Traina, S.J., 2004. Inhibited Cr(VI) reduction by aqueous Fe(II) under hyperalkaline conditions. *Environ. Sci. Technol.*, 38: 5535–5539.

He, Y.T., Traina, S.J., 2005. Cr(VI) reduction and immobilization by magnetite under alkaline pH conditions: The role of passivation. *Environ. Sci. Technol.*, 39(12): 4499–4504.

Henderson, T., 1994. Geochemical reduction of hexavalent chromium in the trinity sand aquifer. *Ground Water*, 32(3): 477–486.

Hope, S.J., Peterson, R.E., 1996. *Chromium in River Substrate Pore Water and Adjacent Groundwater: 100-D/DR Area, Hanford Site, WA*. BHI-00778 Rev 0. Richland, WA: Bechtel Hanford Inc..

Ilton, E.S. et al., 2004. Heterogeneous reduction of uranyl by micas: Crystal chemical and solution controls. *Geochim. Cosmochim. Acta*, 68(11): 2417–2435.

Ilton, E.S., Moses, C.O., Veblen, D.R., 2000. Using x-ray photoelectron spectroscopy to discriminate among different sorption sites on micas: With implications for heterogeneous reduction of chromate at the mica-water interface. *Geochim. Cosmochim. Acta*, 64(8): 1437–1450.

Ilton, E.S., Veblen, D.R., 1994. Chromium sorption by phlogopite and biotite in acidic solutions at 25 degree C: Insights from the x-ray photoelectron spectroscopy and electron microscopy. *Geochim. Cosmochim. Acta*, 58: 2777–2788.

Ilton, E.S., Veblen, D.R., Moses, C.O., Raeburn, S.P., 1997. The catalytic effect of sodium and lithium ions on coupled sorption–reduction of chromate at the biotite edge-fluid interface. *Geochim. Cosmochim. Acta*, 61(17): 3543–3563.

James, B.R., 1994. Hexavalent chromium solubility and reduction in alkaline soils enriched with chromite ore processing residue. *J. Environ. Qual.*, 23: 227–233.

Jardine, P.M. et al., 1999. Fate and transport of hexavalent chromium in undisturbed heterogeneous soil. *Environ. Sci. Technol.*, 33: 2939–2944.

Jones, T.E., Watrous, R.A., Maclean, G.T., 2000. *Inventory Estimates for Single-Shell Tank Leaks in S and SX Tank Farms*. RPP-6285 Rev. 0. Richland, WA: CH2M HILL Hanford Group, Inc..

Kendelewicz, T., Liu, P., Doyle, C.S., Brown, G.E., 2000. Spectroscopic study of the reaction of aqueous Cr(VI) with Fe_3O_4(111) surfaces. *Surf. Sci.*, 469: 144–163.

Kent, D.B., Davis, J.A., Anderson, C.D., Rea, B.A., 1994. Transport of chromium and selenium in the suboxic zone of a shallow aquifer: Influence of redox and adsorption reactions. *Water Resour. Res.*, 30(4): 1099–1114.

Kent, D.B., Davis, J.A., Anderson, C.D., Rea, B.A., 1995. Transport of chromium and selenium in a pristine sand and gravel aquifer: Role of adsorption processes. *Water Resour. Res.*, 31(4): 1041–1050.

Klupinski, T.P., Chin, Y.P., Traina, S.J., 2004. Abiotic degradation of pentachloronitrobenzene by Fe(II): Reaction on goethite and iron oxide nanoparticles. *Environ. Sci. Technol.*, 38(16): 4353–4360.

Lee, G.H., Hering, J.G., 2005. Oxidative dissolution of chromium(III) hydroxide at pH 9, 3, and 2 with product inhibition at pH 2. *Environ. Sci. Technol.*, 39(13): 4921–4928.

Liger, E., Charlet, L., Van Cappellen, P., 1999. Surface catalysis of uranium(VI) reduction by iron(II). *Geochim. Cosmochim. Acta*, 63(19): 2939–2955.

Mashal, K., Harsh, J.B., Flury, M., 2005. Clay mineralogical transformations over time in Hanford sediments reacted with simulated tank waste. *Soil Sci. Soc. Am. J.*, 69(2): 531–538.

Millero, F.J., Sotolongo, S., Izaguirre, M., 1987. The oxidation kinetics of Fe(II) in seawater. *Geochim. Cosmochim. Acta*, 51: 793–801.

Morel, F.M.M., Hering, J.G., 1993. *Principles and Applications of Aquatic Chemistry*. New York: John Wiley & Sons, Inc. 588 pp.

Oliver, D.S., Brockman, F.J., Bowman, R.S., Kieft, T.L., 2003. Microbial reduction of hexavalent chromium under vadose zone conditions. *J. Environ. Qual.*, 32(1): 317–324.

Ono, B.-I., 1988. Genetic approaches in the study of chromium toxicity and resistance in yeast and bacteria. In: Nriagu, J.O., Nieboer, E. (eds.), *Chromium in Natural and Human Environment*. New York: John Wiley and Sons, pp. 351–368.

Park, D.H., Lim, S.R., Yun, Y.S., Park, J.M., 2007. Reliable evidences that the removal mechanism of hexavalent chromium by natural biomaterials is adsorption-coupled reduction. *Chemosphere*, 70: 298–305.

Paschin, Y.V., Kozachenko, V.I., Sal'nikova, L.E., 1983. Differential mutagenic response at HGPRT locus in V-79 and CHO cells after treatment with chromate. *Mutat. Res.*, 122: 361–365.

Petersen, S.W., Hall, S.H., 2008. *Investigation of Hexavalent Chromium Source in the Southwest 100-D Area*. SGW-38757 Draft A. Richland, WA: Fluor Hanford.

Peterson, M.L., Brown, G.E., Parks, G.A., Stein, C.L., 1997a. Differential redox and sorption of Cr(III/VI) on natural silicate and oxide minerals: EXAFS AND XANES results. *Geochim. Cosmochim. Acta*, 61(16): 3399–3412.

Peterson, M.L., White, A.F., Brown, G.E.J., Parks, G.A., 1997b. Surface passivation of magnetite by reaction with aqueous Cr(VI): XFAS and TEM results. *Environ. Sci. Technol.*, 31: 1573–1576.

Peterson, R.E., Raidl, R.F., Denslow, C.W., 1996. *Conceptual Site Models for Groundwater Contamination at 100-BC-5, 100-KR-4, 100-HR-3, and ,100-FR 3 Operable Units*. BHI-00917 Rev 0.Richland, WA: Bechtel Hanford Inc.

Pettine, M., D'Ottone, L., Campanella, L., Millero, F.J., Passino, R., 1998. The reduction of Cr(VI) by iron (II) in aqueous solutions. *Geochim. Cosmochim. Acta*, 62(9): 1509–1519.

Qafoku, N.P. et al., 2003a. Aluminum effect on dissolution and precipitation under hyperalkaline conditions: II. Solid phase transformations. *J. Environ. Qual.*, 32(6): 2364–2372.

Qafoku, N.P. et al., 2011. *Geochemical Characterization of Chromate Contamination in the 100 Area Vadose Zone at the Hanford Site*. Richland, WA: Pacific Northwest National Laboratory.

Qafoku, N.P. et al., 2009. Pathways of aqueous chromium (VI) attenuation in a slightly alkaline oxic subsurface. *Environ. Sci. Technol.*, 43(4): 1071–1077.

Qafoku, N.P., Ainsworth, C.C., Heald, S.M., 2007a. Cr(VI) fate in mineralogically altered sediments by hyperalkaline waste fluids. *Soil Sci.*, 172(8): 598–613.

Qafoku, N.P., Ainsworth, C.C., Szecsody, J.E., Qafoku, O.S., 2003b. Aluminum effect on dissolution and precipitation under hyperalkaline conditions: I. Liquid phase transformations. *J. Environ. Qual.*, 32(6): 2354–2363.

Qafoku, N.P., Ainsworth, C.C., Szecsody, J.E., Qafoku, O.S., 2003c. Effect of coupled dissolution and redox reactions on $Cr(VI)_{aq}$ attenuation during transport in the Hanford sediments under hyperalkaline conditions. *Environ. Sci. Technol.*, 37: 3640–3646.

Qafoku, N.P., Ainsworth, C.C., Szecsody, J.E., Qafoku, O.S., 2004. Transport-controlled kinetics of dissolution and precipitation in the Hanford sediments under hyperalkaline conditions. *Geochim. Cosmochim. Acta*, 68(14): 2981–2995.

Qafoku, N.P., Evan Dresel, P., Ilton, E., McKinley, J.P., Resch, C.T., 2010. Chromium transport in an acidic waste contaminated subsurface medium: The role of reduction. *Chemosphere*, 81(11): 1492–1500.

Qafoku, N.P., Qafoku, O., Ainsworth, C.C., Dohnalkova, A., 2007b. Fe-solid phase transformations under highly basic conditions. *Appl. Geochem.*, 22(9):2054–2064. doi:10.1016/j.apgeochem.2007.04.023.

Qafoku, N.P., Zachara, J.M., Liu, C., 2005. Kinetic desorption of U(VI) during reactive transport in a contaminated Hanford sediment. *Environ. Sci. Technol.*, 39(9):3157–3165. doi:10.1021/es048462q.

Rai, D., Eary, L.E., Zachara, J.M., 1989. Environmental chemistry of chromium. *Sci. Tot. Environ.* 86: 15–23.

Rai, D., Zachara, J.M., 1986. *Geochemical Behavior of Chromium Species*. EA-4544. Richland, WA: Pacific Northwest National Laboratory.

Sahai, N., Sverjensky, D.A., 1997. Evaluation of internally consistent parameters for the triple-layer model by the systematic analysis of oxide surface titration data. *Geochim. Cosmochim. Acta*, 61(14): 2801–2826.

Schroeder, G.C., 1966. *Discharge of Sodium Dichromate Solution*. Compliance with executive order 11258. Richland, WA: Douglas United Nuclear Inc.

Sedlak, D.L., Chan, P.G., 1997. Reduction of hexavalent chromium by ferrous iron. *Geochim. Cosmochim. Acta*, 61(11): 2185–2192.

Selim, H.M., Amacher, M.C., Iskandar, I.K., 1989. Modeling the transport of chromium(VI) in soil columns. *Soil Sci. Soc. Am. J.*, 53: 996–1004.

Serne, R.J. et al., 2002. *Geologic and Geochemical Data Collected from Vadose Zone Sediments from the Slant Borehole [SX-108] in the S/SX Waste Management Area and Preliminary Interpretations*. PNNL-13757-1, Pacific Northwest National Laboratory, Richland, WA.

Sposito, G., 1984. *The Surface Chemistry of Soils*. New York: Oxford University Press.

Sposito, G., 2004. *The Surface Chemistry of Natural Particles*. New York: Oxford University Press, 242 pp.

Strathmann, T.J., Stone, A.T., 2003. Mineral suface catalysis of reactions between Fe-II and oxime carbamate pesticides. *Geochim. Cosmochim. Acta*, 67(15): 2775–2791.

Stumm, W., 1992. *Chemistry of the Solid-Water Interface. Processes at the Mineral-Water and Particle-Water Interface in Natural Systems*. New York: John Wiley & Sons, Inc..

Stumm, W., Morgan, J.J., 1996. *Aquatic Chemistry: Chemical Equilibria and Rates in Natural Waters*. New York: John Wiley & Sons, Inc..

Tang, Y.Z., Elzinga, E.J., Lee, Y.J., Reeder, R.J., 2007. Coprecipitation of chromate with calcite: Batch experiments and x-ray absorption spectroscopy. *Geochim. Cosmochim. Acta*, 71(6): 1480–1493.

Thornton, E.C., 1992. *Disposal of Hexavalent Chromium in the 100-BC Area—Implications for Environmental Remediation*. WHC-SD-EN-TI-025 Rev 0. Richland, WA: Westinghouse Hanford Co.

Thornton, E.C., Amonette, J.E., Olivier, J.A., Huang, D.L., 1995. *Speciation and Transport Characteristics of Chromium in the 100D/H Areas of the Hanford Site*. WHC-SD-EN-TI-302 Rev 0. Richland, WA: Westinghouse Hanford Company.

Villalobos, M., Trotz, M.A., Leckie, J.O., 2001. Surface complexation modeling of carbonate effects on the adsorption of Cr(VI), Pb(II) and U(VI) on goethite. *Environ. Sci. Technol.*, 35: 3849–3856.

Wan, J.M., Larsen, J.T., Tokunaga, T.K., Zheng, Z.P., 2004. pH neutralization and zonation in alkaline-saline tank waste plumes. *Environ. Sci. Technol.*, 38(5): 1321–1329.

Whipple, J.C., 1953. *Design Planning of a Dichromate System for the 100-K Areas*. HW-26913. Richland, WA.

Williams, A.G.B., Scherrer, M.M., 2004. Spectroscopic evidence for Fe(II)–Fe(III) electron transfer at the iron oxide—water interface. *Environ. Sci. Technol.*, 38: 4782–4790.

Wu, C.H., Lo, S.L., Lin, C.F., 2000. Competitive adsorption of molybdate, chromate, sulfate, selenate, and selenite on g-Al_2O_3. *Colloid Surf. A.*, 166: 251–259.
Zachara, J.M. et al., 2004. Chromium speciation and mobility in a high level nuclear waste vadose zone plume. *Geochim. Cosmochim. Acta*, 68(1): 13–30.
Zachara, J.M., Ainsworth, C.C., Cowan, C.E., Resch, C.T., 1989. Adsorption of chromate by subsurface soil horizons. *Soil Sci. Soc. Am. J.*, 53(2): 418–428.
Zachara, J.M., Girvin, D.C., Schmidt, R.L., Resch, C.T., 1987. Chromate adsorption on amorphous iron oxyhydroxide in the presence of major groundwater ions. *Environ. Sci. Technol.*, 21(6): 589–594.
Zhao, H.T., Deng, Y.J., Harsh, J.B., Flury, M., Boyle, J.S., 2004. Alteration of kaolinite to cancrinite and sodalite by simulated Hanford tank waste and its impact on cesium retention. *Clays Clay Min.*, 52(1): 1–13.

9 Biogeochemical Processes Regulating the Mobility of Uranium in Sediments

Keaton M. Belli and Martial Taillefert

CONTENTS

9.1 INTRODUCTION

The biogeochemical cycling of uranium regulates its transport in the environment, and a mechanistic understanding of its transformation is essential to identify uranium reserves for industrial uses and to predict the subsurface transport of this environmental contaminant. Since its discovery in 1789, uranium was widely used as a colorant in the glass and ceramic industry (Strahan 2001), and it has found modern use as a high-yield energy source. The demonstration of controlled uranium fission in 1942 set the stage for massive uranium mining operations to support nuclear weapons programs and, later, the nuclear energy sector (Plant et al. 1999). After a global shift from nuclear proliferation to nuclear disarmament and years of poor waste disposal practices at nuclear facilities, research interests in uranium have more recently focused on understanding its mobility in the environment as an ecological and human health hazard (Blanchard et al. 1983; Riley et al. 1992). The adverse effects of uranium exposure have been observed in microorganisms (Carvajal et al. 2012; Fortin et al. 2004, 2007; Konopka et al. 2013; Tapia-Rodriguez et al. 2012; VanEngelen et al. 2010, 2011), macrofauna (Markich et al. 2000; Tran et al. 2004), and aquaculture (Trenfield et al. 2011), and the public health effects arising from ingestion of uranium and exposure to its radioactive decay products (e.g., radium) have been documented (Blanchard et al. 1983; Hirose and Fawell 2012). As governments recognize the benefits of shifting away from fossil fuel–based energy sources and look to nuclear energy as an attractive alternative, understanding the biogeochemical cycling of uranium is imperative to ensure safe production of nuclear energy and to minimize its adverse effects on the environment.

A mechanistic understanding of uranium geochemistry has played a crucial role in identifying uranium reserves (Plant et al. 1999) and utilizing uranium as a powerful tool for dating geologic samples (Condomines et al. 2003; Edwards et al. 2003), tracing oceanic mixing processes (Moore and Shaw 2008; Santos et al. 2011), and reconstructing the redox environments of the early Earth (Tribovillard et al. 2006). The discovery of biological uranium transformations (Lovley et al. 1991) has fundamentally altered our understanding of uranium cycling in the environment and has even led to the development of biological remediation strategies that address anthropogenic nuclear waste. This area of research has greatly expanded the network of biogeochemical processes that regulate the distribution of uranium in the subsurface and has identified new caveats to the traditional view of uranium cycling that must be considered to accurately predict the fate of uranium in the environment. This chapter reviews newly discovered biogeochemical processes that regulate the transformation of uranium in saturated soils and sediments and identifies how these recent findings impact the remediation of uranium-contaminated environments.

9.2 URANIUM CHEMISTRY AND GEOCHEMICAL CONTROLS OF URANIUM MOBILITY

The actinide uranium is a hard cation, and it forms strong complexes with hard anions such as oxygen-containing ligands (e.g., OH^-, CO_3^{2-}, ΣPO_4^{3-}) (Clark et al. 1995; Cotton et al. 1999). Natural uranium primarily comprises three isotopes (^{238}U, 99.27%; ^{235}U, 0.72%; and ^{234}U, 0.01%), all of which are radioactive and undergo alpha decay with half-lives in the order of 10^5–10^9 years. Uranium exists in four oxidation states: U(III), U(IV), U(V), and U(VI); however, most attention has focused on reduced U(IV) and oxidized U(VI), as they are the most prevalent oxidation states observed in nature and are generally recognized as the dominant oxidation states that control the mobility of uranium in the environment. U(III) is unstable within the stability field of water; thus, it is not reported in natural environments (Langmuir 1978). Likewise, U(V) is commonly described as a transient intermediate during the reduction of U(VI) or oxidation of U(IV) due to its propensity for disproportionation over a wide range of conditions. Thermodynamics predicts U(V) to be the most stable oxidation state in reducing environments below pH 7 (Langmuir 1978); however, U(V) readily undergoes acid-mediated disproportionation outside a narrow pH range of around 2–2.5 (Ekstrom 1974; Mougel et al. 2010; Steele and Taylor 2007) that regenerates the two most stable oxidation states [U(VI) and U(IV)] and helps explain its elusiveness in aqueous environments. Although rare, naturally occurring U(V) minerals have been identified (Burns and Finch 1999) even though the mechanism of their formation remains poorly understood.

In oxidizing environments, uranium is present as the U(VI) uranyl ion (UO_2^{2+}), which is highly soluble over a wide range of geochemical conditions and contributes to the mobility of uranium in subsurface environments. The uranyl ion has a linear geometry that is characterized by double bonds to two axial oxygen atoms that are highly stable and remain intact during sorption to mineral surfaces and complexation with ligands. Equatorial binding sites allow for coordination of a wide variety of ligands depending on pH and solution composition that alters the physical (e.g., size, charge), thermodynamic (e.g., redox potential, solubility), and biochemical (e.g., bioavailability) properties of the uranyl ion and drastically affects how U(VI) interacts with its immediate surroundings. In pure water, the uranyl ion is mainly hydrated or hydrolyzed, and it is present as monomeric or polymeric uranyl hydroxide species depending on the concentration of U(VI), which may lead to the precipitation of schoepite [$(UO_2)_8O_2(OH)_{12}\bullet 12(H_2O)$ (s)] or its derivatives (Bargar et al. 2000; Rai et al. 1990). In most natural aquatic systems, however, uranyl can be complexed by chloride, sulfate, nitrate, and orthophosphates (ΣPO_4^{3-}) at low pH, or it can be hydrolyzed and complexed by carbonates at circumneutral pHs (Bernhard et al. 2001; Langmuir 1997) such that the "free" hydrated uranyl ion is only present in appreciable amounts in acidic environments (Figure 9.1). When carbonates and alkali earth metal cations (i.e., Ca^{2+}, Mg^{2+}, Sr^{2+}, Ba^{2+}) are present, ternary uranyl carbonate complexes represent the only significant fraction of U(VI) above circumneutral pH (Figure 9.1). Such complexation generally enhances the solubility of U(VI) compared to solutions without ligands. Certain ligands, however, decrease the solubility of U(VI), and a variety of U(VI) solids may form depending on local geochemical conditions (e.g., uranyl phosphates, vanadates, and arsenates) (Figure 9.2). Furthermore, U(VI) minerals are also commonly associated with uraninite deposits, as alteration products are generated a result of fluctuating redox conditions (for a thorough description of these processes, see Finch and Murakami, 1999). In addition, small organic ligands such as citrate, malonate, and oxalate can complex uranyl with different efficiencies (Ganesh et al. 1999; Haas and Northup 2004; Pasilis and Pemberton 2003; Robinson et al. 1998). The high number of U(VI) complexes and their variable stoichiometry make it difficult to investigate the reactivity of U(VI) with reductants, ligands, and solid phases.

Contrary to U(VI), U(IV) as the uranous ion (U^{4+}) lacks the two axial oxo ligands that are characteristic of the uranyl ion and is highly insoluble. The uranous ion undergoes hydrolysis above pH 1 and forms aqueous complexes with carbonate ligands at higher pH; however, the number of identified complexes and their equilibrium concentrations are small due to the low solubility of U(IV) (Ciavatta et al. 1983; Clark et al. 1995; Guillaumont et al. 2003; Langmuir 1978). U(IV) is present in the environment predominantly as reduced U(IV) mineral deposits such as uraninite (UO_2), coffinite ($USiO_4$), and sometimes ningyoite ($CaU(PO_4)_2•2H_2O$) (Langmuir 1978; Plant et al. 1999). The formation of

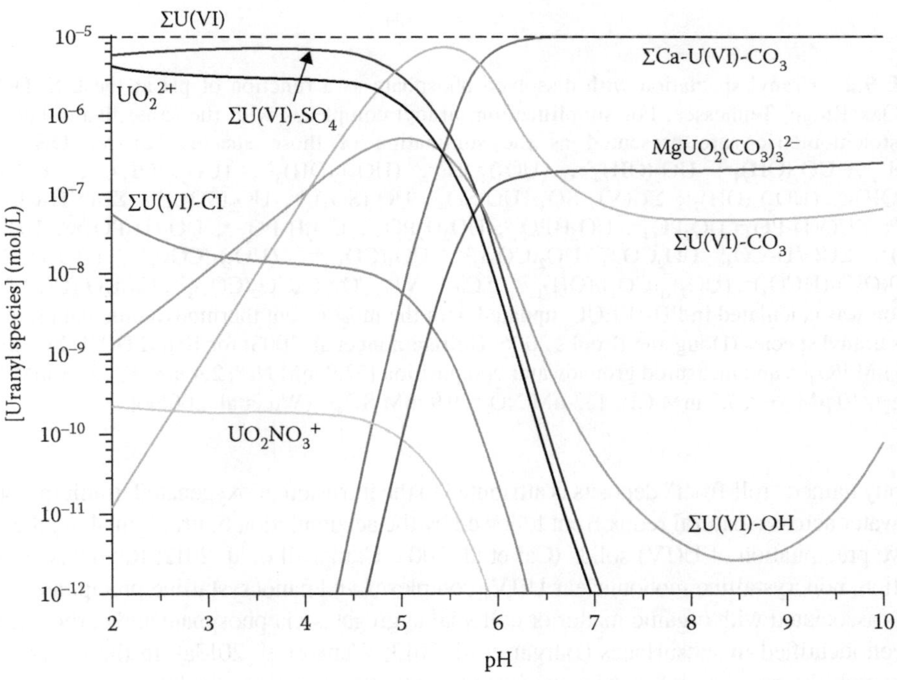

FIGURE 9.1 Uranyl speciation as a function of pH at the U.S. Department of Energy (DOE) Integrated Field Research Challenge (IFRC) site in Rifle, Colorado. For simplification, uranyl complexes with the same ligand but with different stoichiometries are presented as the summation of those species [ΣU(VI)-OH: UO_2OH^+, $UO_2(OH)_2^0$, $UO_2(OH)_3^-$, $UO_2(OH)_4^{2-}$, $(UO_2)_2OH^{3+}$, $(UO_2)_2(OH)_2^{2+}$, $(UO_2)_3(OH)_4^{2+}$, $(UO_2)_3(OH)_5^+$, $(UO_2)_3(OH)_7^-$, $(UO_2)_4(OH)_7^+$; ΣU(VI)-SO_4: $UO_2SO_4^0$, $UO_2(SO_4)_2^{2-}$, $UO_2(SO_4)_3^{4-}$; ΣU(VI)-Cl: UO_2Cl^+, $UO_2Cl_2^0$; ΣU(VI)-CO_3: $UO_2CO_3^0$, $UO_2(CO_3)_2^{2-}$, $UO_2(CO_3)_3^{4-}$, $(UO_2)_3(CO_3)_6^{6-}$, $(UO_2)_2CO_3(OH)_3^-$, $(UO_2)_3O(OH)_2(HCO_3)^+$, $(UO_2)_{11}(CO_3)_6(OH)_{12}^{2-}$; ΣCa-U(VI)-$CO_3$: $CaUO_2(CO_3)_3^{2-}$, $Ca_2UO_2(CO_3)_3^0$]. Uranyl speciation was calculated in PHREEQC, updated with the most recent thermodynamic data for solid and aqueous uranyl species (Dong and Brooks 2006; Guillaumont et al. 2003) for 10 μM U(VI) by using measured groundwater composition [3.3 mM DIC, 6.8 mM Na^+, 4.8 mM Ca^{2+}, 4.6 mM Mg^{2+}, 44 mM NH_4^+, 8.4 mM SO_4^{2-}, 2.8 mM Cl^-, 60 μM NO_3^- (Campbell et al. 2011; Zachara et al. 2013)].

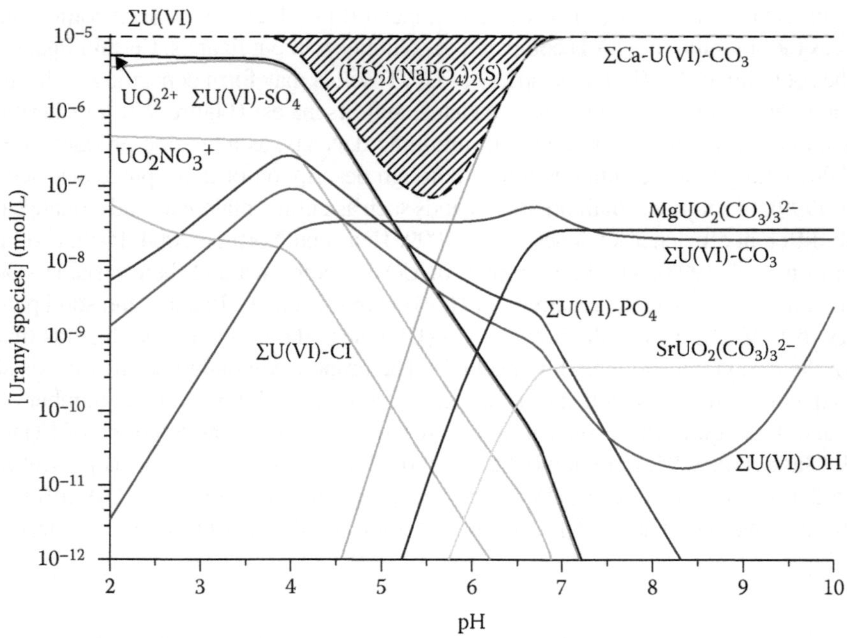

FIGURE 9.2 Uranyl speciation with dissolved phosphate as a function of pH at the U.S. DOE IFRC site in Oak Ridge, Tennessee. For simplification, uranyl complexes with the same ligand but with different stoichiometries are presented as the summation of those species [ΣU(VI)-OH: UO_2OH^+, $UO_2(OH)_2^0$, $UO_2(OH)_3^-$, $UO_2(OH)_4^{2-}$, $(UO_2)_2OH^{3+}$, $(UO_2)_2(OH)_2^{2+}$, $(UO_2)_3(OH)_4^{2+}$, $(UO_2)_3(OH)_5^+$, $(UO_2)_3(OH)_7^-$, $(UO_2)_4(OH)_7^+$; ΣU(VI)-SO_4: $UO_2SO_4^0$, $UO_2(SO_4)_2^{2-}$, $UO_2(SO_4)_3^{4-}$; ΣU(VI)-Cl: UO_2Cl^+, $UO_2Cl_2^0$; ΣU(VI)-PO_4: $UO_2PO_4^-$, $UO_2HPO_4^0$, $UO_2H_2PO_4^+$, $UO_2H_3PO_4^{2+}$, $UO_2(H_2PO_4)_2^0$, $UO_2(H_2PO_4)(H_3PO_4)^+$; ΣU(VI)-CO_3: $UO_2CO_3^0$, $UO_2(CO_3)_2^{2-}$, $UO_2(CO_3)_3^{4-}$, $(UO_2)_3(CO_3)_6^{6-}$, $(UO_2)_2CO_3(OH)_3^-$, $(UO_2)_3O(OH)_2(HCO_3)^+$, $(UO_2)_{11}(CO_3)_6(OH)_{12}^{2-}$; ΣCa-U(VI)-$CO_3$: $CaUO_2(CO_3)_3^{2-}$, $Ca_2UO_2(CO_3)_3^0$]. Uranyl speciation was calculated in PHREEQC, updated with the most recent thermodynamic data for solid and aqueous uranyl species (Dong and Brooks 2006; Guillaumont et al. 2003) for 10 μM U(VI) by using 1 mM DIC, 35 μM PO_4^{3-}, and measured groundwater composition [32.9 mM Na^+, 2.4 mM K^+, 25.1 mM Ca^{2+}, 6.8 mM Mg^{2+}, 10 μM Sr^{2+}, 7.3 mM Cl^-, 133 μM NO_3^-, 9.9 mM SO_4^{2-} (Wu et al. 2006a)].

these aptly named "roll-front" deposits is attributed to the intrusion of oxygenated uranium-containing groundwater across a natural redox front followed by the accumulation of uranium along the front via reductive precipitation of U(IV) solids (Cai et al. 2007; Campbell et al. 2012; Reynolds et al. 1982). In addition, noncrystalline mononuclear U(IV) complexes and nanocrystalline precipitates, in which U(IV) is associated with organic matter or colloidal aggregates via phosphate and carbonate ligands, have been identified in subsurfaces (Bargar et al. 2013; Wang et al. 2013a). In these environments, uranium mobility may be higher than anticipated due to the formation of soluble organic U(IV) complexes with natural and synthetic organic acids (Francis and Dodge 2008; Ganesh et al. 1997; Haas and Northup 2004; Suzuki et al. 2010) and humics (Burgos et al. 2007) or U(IV)-associated colloids that may be susceptible to advective transport (Wang et al. 2013a).

9.3 INTERACTIONS BETWEEN URANIUM AND THE MINERAL–WATER INTERFACE

Although the oxidation state of uranium provides insight into the mobility of uranium to a first degree, the strong affinity of the uranyl ion for solid-phase minerals exerts an additional control on the transport of uranium in subsurface environments. Uranyl ions can be removed by adsorption onto aluminum (Gu et al. 2003; Guo et al., 2009; Jung et al. 2012; Sylvester et al. 2000; Tao et al. 2000), manganese (Brennecka et al. 2011; Rihs et al. 2014; Wang et al. 2013b; Webb et al. 2006), and iron oxides (Bargar et al. 2000; Bruno et al. 1995; Cheng et al. 2004; Duff et al. 2002; Gabriel et al. 1998; Hsi and Langmuir 1985; Lack et al. 2002; Lenhart and Honeyman 1999; Moyes et al. 2000; Sato et al. 1997; Singh et al. 2012; Tao et al. 2000; Waite et al. 1994), with more efficient sorption onto amorphous than crystallized iron oxide phases (Hsi and Langmuir 1985; Payne et al. 1996) but less efficient sorption onto biotic than synthesized manganese oxides (Wang et al. 2013b). In natural waters, U(VI) adsorption typically increases with pH, as hydrolysis of the uranyl ion becomes progressively more important and generates negatively charged complexes that are attracted by the positively charged oxohydroxide minerals below the pH of zero point charge (pH_{zpc}). Above the pH_{zpc} of these minerals, however, the mean surface charge of the oxide phases is negative and U(VI) adsorption tends to decrease. As a result, the apex of adsorption of U(VI) onto oxohydroxide minerals, though variable depending on the exact mineral composition, tends to be maximum between pH 6 and 10 for aluminum oxides (Guo et al. 2009; Tao et al. 2000), between 4 and 9 for manganese oxides (Wang et al. 2013b), and between 7 and 10 for iron (hydr)oxides (Cheng et al. 2004; Hsi and Langmuir 1985; Tao et al. 2000). In the presence of carbonates, uranyl adsorption onto iron and manganese oxides is less efficient, because uranyl carbonate complexes are weakly sorbed (Hsi and Langmuir 1985; Lenhart and Honeyman 1999; Waite et al. 1994; Wang et al. 2013b). In fact, carbonates have been used to leach uranium from contaminated soils (Francis et al. 1999; Mason et al. 1997). This effect is exacerbated by alkali earth metal cations, which result in the formation of ternary uranyl carbonate species and further decrease uranyl-surface interactions (Nair and Merkel 2011; Stewart et al. 2010). If the pH is low enough, however, ternary complexes can form between the manganese or iron oxide, uranyl, carbonate, and sometimes calcium carbonate (Bargar et al. 2000; Foerstendorf et al. 2012; Singer et al. 2012a; Wang et al. 2013b). In addition, coprecipitation of carbonate minerals may be significant at a pH greater than 8 (Reeder et al. 2000, 2001). The presence of dissolved phosphates (ΣPO_4^{3-}) may also have an impact on the adsorption of U(VI) onto oxide minerals. At low pH, adsorption of U(VI) is enhanced in the presence of phosphate, probably by the formation of ternary complexes with phosphate-adsorbed oxides (Cheng et al. 2004; Galindo et al. 2010; Guo et al. 2009; Singh et al. 2012). At a pH greater than 7, however, the desorption of U(VI) is promoted by the presence of phosphate, likely due to the complexation of uranyl by phosphate (Cheng et al. 2004). The phosphate concentration also plays a significant role, as the coprecipitation of uranium phosphate mineral phases accompanies adsorption above a certain threshold of the phosphate concentration (Galindo et al. 2010; Singh et al. 2012). In the presence of natural organic matter, adsorption onto oxide minerals is enhanced at low pH, decreased at high pH (Lenhart and Honeyman 1999; Payne et al. 1996; Tao et al. 2000), and may be followed by coprecipitation of schoepite (Bruno et al. 1995; Duff et al. 2002) or other U(VI) minerals (Gu et al. 2003; Sato et al. 1997) if the concentration of uranyl ions in solution and/or the pH is high enough. Uranium can also be removed by adsorption onto clays (Catalano and Brown Jr. 2005; Chisholm-Brause et al. 2001; Krestou et al. 2004; McKinley et al. 1995; Prikryl et al. 2001; Sylvester et al. 2000), adsorption to (Drot and Simoni 1999; Fuller et al. 2003; Rakovan et al. 2002) or coprecipitation with (Arey et al. 1999; Krestou et al. 2004) phosphate minerals, or a combination of both depending on whether the solubility of uranium phosphate minerals is exceeded (Fuller et al. 2002; Ghafar et al. 2002; Simon et al. 2008). Finally, uranium adsorbs to cell membranes mainly via complexation by carboxyl and phosphoryl groups (Fowle et al. 2000; Haas et al. 2001; Hu et al. 1996; Kelly et al. 2002), which may act as nucleation centers for the precipitation of uranium phosphate minerals (Dunham-Cheatham et al. 2011). Biosorption is also affected by the speciation of uranium in solution, and carbonates, not surprisingly, moderate the removal of uranium by biosorption (Carvajal et al. 2012) as a result of the formation of highly negatively charged complexes at circumneutral pH that are electrostatically repulsed from the deprotonated carboxyl ($pK_a = 4.8$) and phosphoryl ($pK_a = 6.9$) groups on bacterial surfaces (Fein et al. 1997).

Because U(VI) is the most stable aqueous oxidation state of uranium, the adsorption of U(V) and U(IV) species has been vastly overlooked. Although aqueous U(V) easily disproportionates to U(VI) and U(IV), ferrous mica (Ilton et al. 2005) and magnetite (Fe_3O_4) (Ilton et al. 2010) surfaces may stabilize the pentavalent oxidation state after heterogeneous reduction of U(VI) as a polymerized U(IV)–U(V)–U(VI) surface species that would prevent contact between two U(V) atoms (Ilton et al. 2005). Ferrous micas and magnetite are quite common and represent a significant reservoir of Fe(II) in sediments, suggesting that U(V) may play a significant unrecognized role in uranium cycling. Until recently, it was commonly assumed that U(IV) adsorption on mineral surfaces was negligible and that any U(IV) would readily precipitate as either uraninite (Burgos et al. 2008; Schofield et al. 2008; Senko et al. 2007) or, more recently, mononuclear U(IV) species (Bargar et al. 2013; Bernier-Latmani et al. 2010; Boyanov et al. 2011; Fletcher et al. 2010; Kelly et al. 2008). However, U(IV) adsorption was recently observed on both rutile (TiO_2) and magnetite surfaces at low surface loading, when the availability of high affinity U(IV) binding sites was in excess (Latta et al. 2014), an indication that this process has been overlooked. Adsorbed U(IV) eventually precipitated as nano-uraninite on magnetite surfaces, but U(IV) was stable on rutile surfaces for at least a year, possibly explaining why uraninite is not always observed in biostimulated-reduced sediments (Latta et al. 2014). Although the stabilization of U(V) and U(IV) on mineral surfaces has been observed, the mechanism of formation of U(V) and U(IV) surface species is not well defined. As the solubility of U(IV) and the stability of U(V) in solution are low, aqueous-reduced uranium species are not expected under environmental conditions in the absence of organic ligands to keep U(IV) in solution (Burgos et al. 2007; Francis and Dodge 2008; Haas and Northup 2004; Suzuki et al. 2010) or carbonate ligands to inhibit disproportionation of U(V) (Docrat et al. 1999; Morris 2002; Wander et al. 2006; Wander and Shuford 2012). It is, therefore, likely that U(V) and U(IV) surface species represent stabilized products of the reduction of adsorbed U(VI) in the presence of an exogenous chemical reductant (e.g., AH_2DS) (Ilton et al. 2012; Latta et al. 2014), a reduced mineral (Ilton et al. 2005, 2010; Latta et al. 2014; Renock et al. 2013), or, in the case of U(V), a stabilized product of uraninite oxidation (Ulrich et al. 2009). Further studies on the reaction pathways of surface-bound U(V) and U(IV) are needed to understand how this process contributes to uranium cycling in the environment.

9.4 URANIUM REDOX TRANSFORMATIONS

9.4.1 Chemical Redox Processes

As demonstrated earlier, the oxidation state of uranium exerts significant influence over its mobility in the subsurface due to the large range in solubility among its reduced and oxidized forms. Basic thermodynamic calculations demonstrate the versatility of uranium as both an effective oxidant and a reductant to a number of redox-active species (Figure 9.3). Calculated Gibbs free energies of reaction (ΔG_{rxn}) for the reduction of the "free" uranyl ion to amorphous uraninite under conditions at the U.S. Department of Energy (DOE) Integrated Field Research Challenge (IFRC) site in Rifle, Colorado, as an example of a subsurface environment exposed to uranium contaminations, demonstrate the numerous possible redox transformations of uranium. The nonlinear relationship between ΔG_{rxn} and pH is a result of changes in the concentration of UO_2^{2+} with pH in the presence of ligands (Figures 9.1 and 9.2), and it highlights the challenge of determining the fate of uranium in complex, highly heterogeneous subsurface environments. Although thermodynamics provides a good approximation to the mobility of uranium in various redox environments, thermodynamic equilibrium in soils and sediments is rarely achieved due to kinetic limitations, microbial activity, and the effect of transport on microbial and chemical processes.

In anaerobic environments, sulfate-reducing bacteria (SRB) catalyze the production of sulfide, a strong reductant that plays a significant role in the cycling of metals and nonmetals, including uranium (Figure 9.3a). When iron oxides are present, their reduction by dissolved sulfides (ΣH_2S) produces Fe(II), which readily reacts with excess dissolved sulfides to form amorphous iron sulfide ($FeS_{(am)}$),

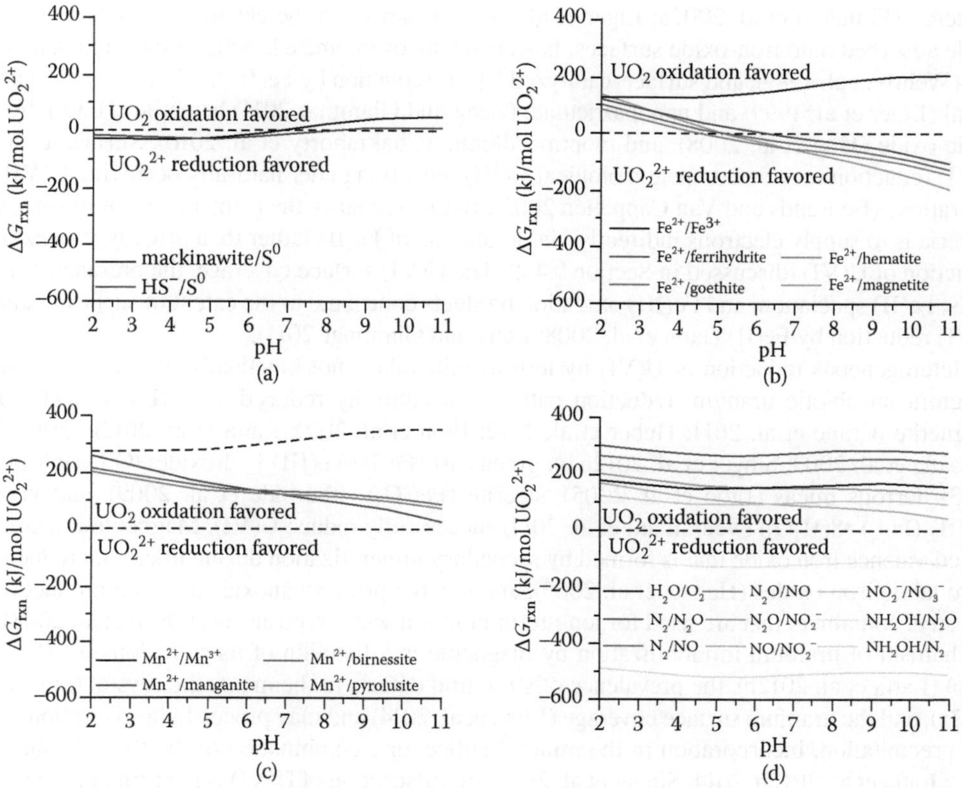

FIGURE 9.3 **(See color insert.)** Gibbs free energy of reaction (ΔG_{rxn}) as a function of pH for the reduction of UO_2^{2+} to UO_2(am) coupled to the oxidation of reduced (a) sulfur, (b) iron, (c) manganese, and (d) nitrogen in typical uranium-contaminated groundwater (only 1 and 2 e^- transfer reactions were considered). A negative ΔG_{rxn} indicates that the reduction of UO_2^{2+} is thermodynamically favorable, whereas a positive ΔG_{rxn} indicates that the oxidation of UO_2(am) is thermodynamically favorable. ΔG_{rxn} ($\Delta G_{rxn} = \Delta G_{rxn}{}^0 + RT\ \ln[Q]$, where Q is the reaction quotient) was calculated by using measured groundwater composition at the U.S. DOE IFRC site in Rifle, Colorado (Campbell et al. 2011; Zachara et al. 2013). "Free" hydrated UO_2^{2+} concentrations were calculated as a function of pH by using PHREEQC, updated with the most recent thermodynamic data for solid and aqueous uranyl species (Dong and Brooks 2006; Guillaumont et al. 2003). A concentration of 10 μM was used for HS^-, Fe^{3+}, Mn^{3+}, N_2, N_2O, NO, NO_2^-, and NH_2OH.

mackinawite (FeS), and, eventually, pyrite (FeS_2) over longer time scales (Rickard and Luther 1997). Although the kinetics of U(VI) reduction by dissolved sulfides is slow (Hua et al. 2006), the kinetics of its reduction by iron sulfides (Hua and Deng 2008; Hyun et al. 2012; Qafoku et al. 2009; Veeramani et al. 2013; Wersin et al. 1994) and metal sulfides (Wersin et al. 1994) is rapid. Reports of oxidation products of iron sulfide minerals by U(VI) vary and include elemental sulfur (S^0) (Hua and Deng 2008; Hyun et al. 2012), sulfate (SO_4^{2-}) (Veeramani et al. 2013), and polysulfides (S_x^{2-}) (Wersin et al. 1994). Regardless of the composition of the product, the reduction of U(VI) by iron sulfide minerals proceeds via adsorption of U(VI) to the iron sulfide surface, coinciding with the release of structural Fe(II) into solution and followed by electron transfer between adsorbed U(VI) and structural sulfide (Hua and Deng 2008; Hyun et al. 2012; Veeramani et al. 2013).

In anaerobic environments that are not exposed to sulfate reduction, Fe(II) is usually prevalent as the product of microbial iron reduction and provides a significant source of electrons to catalyze the reduction of U(VI). Although the homogenous reduction of U(VI) by aqueous Fe(II) is thermodynamically favorable under circumneutral to alkaline pH (Figure 9.3b) (Du et al. 2011), its significance in the reduction of U(VI) has been questioned due to reports that it may be kinetically hindered (Finneran et al. 2002a; Liger et al. 1999). Changes in the electronic structure of Fe(II), while adsorbed onto iron-oxide surfaces, however, may overcome a kinetic or thermodynamic barrier (Wehrli et al. 1989), and surface-catalyzed U(VI) reduction by Fe(II) has been observed on colloidal (Liger et al. 1999) and nanoparticulate (Zeng and Giammar 2011) hematite (Fe_2O_3), hydrous ferric oxide (Jang et al. 2008), and montmorillonite (Chakraborty et al. 2010). Surface-catalyzed U(VI) reduction may outcompete biological U(VI) reduction under naturally occurring U(VI) concentrations (Behrends and Van Cappellen 2005). In this scenario, the primary role of iron-reducing bacteria is to supply electrons indirectly via production of Fe(II) rather than directly via enzymatic reduction of U(VI) (discussed in Section 9.4.2). The U(VI) surface coverage, the presence of humic acids, Fe(II) speciation, and Fe(III) oxidation products contribute to the extent of surface-catalyzed U(VI) reduction by Fe(II) (Jang et al. 2008; Zeng and Giammar 2011).

Heterogeneous reduction of U(VI) by ferrous minerals is not kinetically inhibited and may be a significant abiotic uranium reduction pathway in naturally reduced soils (Latta et al. 2012a). Magnetite (Crane et al. 2011; Huber et al. 2012; Ilton et al. 2010; Latta et al. 2012b, 2013, 2014; Missana et al. 2003; Singer et al. 2012a,b), green rust (Fe(II)/Fe(III)-hydroxide) (O'Loughlin et al. 2003), ferrous micas (Ilton et al. 2005), siderite ($FeCO_3$) (Ithurbide et al. 2009), and vivianite ($Fe(II)_3(PO_4)_2 \cdot 8(H_2O)$) (Veeramani et al. 2011) successfully reduce U(VI). Magnetite is a common mixed-valence iron oxide that is formed by secondary mineralization during microbial reduction of more labile iron oxides (Hansel et al. 2003), and it is the primary anoxic corrosion product of carbon steel containers that are used for long-term uranium waste repositories (Duro et al. 2008). The mechanism of uranium immobilization by magnetite is a function of its stoichiometry (Fe^{2+}:Fe^{3+} ratio) (Latta et al. 2012b), the prevalence of structural defects in the magnetite crystal (Singer et al. 2012a), and the uranium surface coverage (Latta et al. 2014) and may proceed via adsorption, reductive precipitation, incorporation in the mineral lattice, or a combination of the three (Huber et al. 2012; Latta et al. 2012b, 2014; Singer et al. 2012a,b). Adsorption of U(VI) is the primary mechanism of sequestration below a magnetite Fe^{2+}:Fe^{3+} ratio of 0.4 (Latta et al. 2012b). Adsorption and reduction occur simultaneously above this threshold, primarily at surface defect sites (e.g., crystal domain boundaries and cracks) (Singer et al. 2012a) with precipitation of uraninite only at high uranium surface loadings (Latta et al. 2014). The incorporation of U(V) into the lattice of secondary iron-oxide phases or uranium precipitates on the magnetite surface has also been observed (Huber et al. 2012; Ilton et al. 2010), indicating the high sensitivity of uranium–magnetite interactions to experimental conditions. The reduction capacity of magnetite may be recharged on exposure to Fe(II), which replenishes structural Fe(II) and restores its ability to reduce U(VI) (Latta et al. 2012b). Thus, magnetite may play an important role in maintaining reducing environments and in buffering against the oxidation and mobilization of uranium during brief periods of aerobic conditions. Uraninite is commonly cited as the primary U(IV) mineral product of uranium reduction by magnetite (Latta et al. 2012b; O'Loughlin et al. 2010; Veeramani et al. 2011). However, mononuclear U(IV) species are observed with titanium-doped magnetite (Latta et al. 2013) or when phosphate is presorbed onto the magnetite surface (Veeramani et al. 2011), similar to U(VI) reduction by vivianite, demonstrating the importance of the geochemical conditions and coordination environment in the composition of the final U(IV) mineral product.

Iron oxides are also efficient oxidants of U(IV), primarily below circumneutral pH (Figure 9.3b), but the mineralogy of iron oxides affects both the thermodynamics and the extent of U(IV) oxidation (Figure 9.3b). Organic chelating agents that solubilize Fe(III) and U(IV) promote homogenous and heterogeneous reactions between Fe(III) and U(IV) and expedite electron transfer (Stewart et al. 2013). As U(IV) is oxidized by iron oxides, the accumulation of Fe(II) in solution decreases the thermodynamic driving force of the reaction (Ginder-Vogel et al. 2006) and passivates the oxide surface (Senko et al. 2005), which limits the extent of U(IV) oxidation. Furthermore, Fe(II) eventually induces the ripening and secondary mineralization of ferrihydrite ($Fe(OH)_3$) to more crystalline phases (i.e., goethite FeOOH and hematite, Fe_2O_3) (Hansel et al. 2003) that are less favorable for the oxidation of U(IV) (Figure 9.3b) (Ginder-Vogel et al. 2006). On the other hand, processes that limit the extent of Fe(II) accumulation in solution, such as adsorption or precipitation of Fe(II)-bearing minerals (e.g., siderite, mackinawite, magnetite), maintain the thermodynamic driving force of the reaction and promote the continuous oxidation of U(IV) by iron oxides (Spycher et al. 2011). Similarly, dissolved sulfides (ΣH_2S) that are produced following the development of sulfate-reducing conditions may outcompete U(IV) for oxidation by iron oxides and prevent the oxidation of uranium (Sani et al. 2004; Spycher et al. 2011).

Manganese oxides are common in soils and sediments and are powerful oxidizing agents for a variety of reduced species, including U(IV) (Figure 9.3c) (Chinni et al. 2008; Fredrickson et al. 2002; Liu et al. 2002; Wang et al. 2013c). Their prevalence is attributed to manganese-oxidizing bacteria that catalyze the otherwise kinetically inhibited abiotic oxidation of Mn(II) by dissolved oxygen (Tebo et al. 2004). In fact, in the presence of Mn(II) and dissolved oxygen, manganese-oxidizing bacteria catalyze the formation of manganese oxides and indirectly oxidize U(IV) at faster rates than with dissolved oxygen alone (Chinni et al. 2008). Microbial manganese reduction (Lin et al. 2012) and oxidation (Butterfield et al. 2013; Webb et al. 2005), however, proceed through the formation of a soluble Mn(III) intermediate that readily oxidizes U(IV) at rates that are much faster than dissolved oxygen under similar concentrations (Wang et al. 2014), suggesting that Mn(IV) oxides may not necessarily be the direct oxidant of U(IV).

Iron and manganese oxides are effective oxidants of U(IV) in laboratory experiments, whereas rates of U(IV) oxidation in subsurface environments may be limited by a lack of direct contact between these minerals and solid U(IV). These conditions provide ample opportunities for oxidation of U(IV) by soluble oxidants such as nitrogen species and dissolved oxygen, which are highly favorable oxidants of U(IV) across all pH and even at low concentrations (Figure 9.3d). Denitrification intermediates that are produced during dissimilatory nitrate reduction (i.e., nitrite, NO_2^-; nitric oxide, NO; nitrous oxide, N_2O) are effective oxidants of U(IV) (Moon et al. 2007; Senko et al. 2005), though the kinetics and mechanisms of these reactions remain to be determined. Because nitrite can also oxidize Fe(II) (Sorensen and Thorling 1991; Van Cleemput and Baert 1983), iron can serve as an electron shuttle between nitrite and U(IV) and effectively oxidize large concentrations of uranium, even at a low Fe(II) content (Senko et al. 2005). The high reactivity of Fe(II), and to some extent dissolved sulfides, with molecular oxygen compared with denitrification intermediates may not only make Fe(II) and dissolved sulfides effective redox buffers against aerobic U(IV) oxidation but also favor the oxidation of U(IV) by nitrogen species (Bi and Hayes 2014; Bi et al. 2013; Moon et al. 2007). In contrast, hydroxylamine appears to be a good reductant of U(VI) (Figure 9.3d), suggesting that ammonium-oxidizing organisms could be involved in the reduction of U(VI).

9.4.2 Biological Redox Processes

Microbial activity is largely responsible for the modern-day distribution of elements on the earth by catalyzing chemical reactions that are otherwise kinetically or thermodynamically hindered. Uranium is not immune to the activity of microbial communities, as select species of bacteria are able to utilize uranium during respiration processes, which, in some cases, affords sufficient energy for cell growth (Sanford et al. 2007; Tebo and Obraztsova 1998; Wade and DiChristina 2000) and simultaneously alters the oxidation state and, inherently, the mobility of uranium. As uranium is not an essential element and uranium-dependent biochemical processes do not exist, the ability of bacteria to utilize uranium during cellular respiration is surprising, especially given that uranium is toxic to many microorganisms (Carvajal et al. 2012; Konopka et al. 2013; Tapia-Rodriguez et al. 2012; VanEngelen et al. 2010) by compromising cell membrane integrity (Bencheikh-Latmani and Leckie 2003) and inhibiting enzymatic functions (Pible et al. 2010; VanEngelen et al. 2011). However, select species of metal-reducing bacteria (MRB) (e.g., *Shewanella* spp., *Geobacter* spp., *Anaeromyxobacter* spp.) and SRB (e.g., *Desulfotomaculum* sp., *Desulfovibrio* spp.) couple the reduction of U(VI) to U(IV) with the oxidation of organic carbon and generate energy that is sufficient for growth (Sanford et al. 2007), even if the precipitation of intercellular uranium minerals decreases cell viability (Cologgi et al. 2011). Select lithotrophic bacteria, in turn, may couple the oxidation of U(IV) to dissimilatory nitrate reduction, although not with the aim of supporting cell growth (Beller 2005; Finneran et al. 2002b; Weber et al. 2011).

The biochemical mechanism and genetic components that are responsible for the enzymatic reduction of U(VI) remain elusive. Genetic studies with pure cultures of *Shewanella oneidensis* strain MR-1, a model MRB, invoke the involvement of outer-membrane c-type cytochromes in U(VI) reduction. However, MR-1 mutants lacking genes for individual cytochromes (i.e., MtrC, OmcA, MtrC/OmcA) display moderate U(VI) reduction ability compared with wild-type cells, and a complete inability to reduce U(VI) is only observed with a mutant lacking all c-type cytochrome genes (Marshall et al. 2006). A uranium reduction-deficient mutant of *Shewanella putrefaciens* strain 200 is still able to grow on Fe(III) and Mn(IV) but lacks the ability to use nitrite as a terminal electron acceptor (TEA), suggesting that uranium and nitrite reduction pathways share molecular components (Wade and DiChristina 2000). Multiple mechanisms of enzymatic uranium reduction are supported by the observation that biogenic uraninite is formed both on the outside of cells and within the periplasmic space (Marshall et al. 2006; Senko et al. 2007). Similarly, genetic studies with *Geobacter sulfurreducens*, a model SRB, identified conductive pili as the primary mechanism of U(VI) bioreduction, in addition to secondary mechanisms involving abundant outer-membrane c-type cytochromes (Cologgi et al. 2011). As a singular terminal reductase that catalyzes the reduction of U(VI) has not been identified, it is likely that U(VI) bioreduction involves nonspecific, outer-membrane cytochromes and periplasmic reductases with sufficiently low redox potentials rather than a unique molecular system.

Significantly less is known about lithotrophic U(IV)-oxidizing bacteria that utilize U(IV) as an electron donor and couple its oxidation to dissimilatory nitrate reduction. Nitrate-dependent U(IV) oxidizers are taxonomically diverse and include members of the *Acidobacteria* and *Alpha*-, *Beta*-, *Gamma*-, and *Delta-proteobacteria* (Weber et al. 2011). C-type cytochromes have been identified as potential molecular components of U(IV) oxidation (Beller et al. 2009). Cell growth, however, has not been reported for this metabolism, likely due to the very small concentrations of U(IV) oxidized, and whether U(IV) oxidation is coupled to energy generation remains to be determined (Beller 2005; Weber et al. 2011). In any case, U(IV) oxidizers can liberate sufficient U(VI) to surpass regulatory limits, and their presence may impede remediation efforts to sequester uranium as biogenic U(IV) minerals. As U(IV) oxidation activity of genetic mutants of *Thiobacillus denitrificans* was correlated to the extent of denitrification in positive controls with thiosulfate (Beller et al. 2009) and as U(IV) oxidation was not observed in the absence of molecular hydrogen (Beller 2005), the possibility exists that U(IV) oxidation is an artifact of chemical U(IV) oxidation by denitrification intermediates (i.e., NO_2^-, NO, N_2O) (Senko et al. 2002). The lack of a kinetic rate law for U(IV) oxidation by denitrification intermediates prevents exclusion of this possibility.

The ability of bacteria to utilize uranium in respiration processes is unique in that concentrations of uranium that are high enough to support energy generation for cell growth are not common outside contaminated environments. In addition, the evolutionary advantage of developing a molecular system to reductively precipitate aqueous U(VI), which leads to deleterious intracellular precipitation of U(IV) (Cologgi et al. 2011; Senko et al. 2007), or to oxidize U(IV) solids, which increases the concentration of toxic U(VI), is perplexing, especially for bacteria that are isolated from uranium-free environments. Indeed, genetic evidence points toward uranium extremophily as an adaptive rather than an intrinsic feature by which genetic mutations that allow an organism to deal with uranium stress are preserved in the genomes of future generations (Mukherjee et al. 2012). In the case of MRB and SRB species that are isolated outside of uranium-contaminated aquifers, the ability to use U(VI) as a TEA is likely the result of nonspecific molecular pathways that are able to catalyze uranium transformations rather than a widespread evolutionary trait. Nonetheless, microorganisms play significant roles in uranium cycling and ore formation (Cai et al. 2007; Min et al. 2005), both directly by enzymatically reducing or oxidizing uranium and indirectly by altering subsurface geochemical constituents.

9.5 PRECIPITATION AND DISSOLUTION OF URANIUM-BEARING SOLIDS

9.5.1 U(VI) Precipitates in Oxidizing Environments

Uranyl hydroxides are able to precipitate in oxidizing conditions (Bargar et al. 2000; Rai et al. 1990), whereas uranium phosphate minerals are thermodynamically more stable across a broad range of pH (Figure 9.2). They dissolve easily at a pH less than 3 (Sowder et al. 2001), but they are very stable at a pH greater than 3 (Ghafar et al. 2002). Above circumneutral pH, however, the presence of carbonates eventually dissolves uranyl phosphate minerals (Langmuir 1978) (Figure 9.2); this effect has been indeed observed in laboratory experiments (Gudavalli et al. 2013a,b) and appears to be catalyzed by microorganisms (Katsenovich et al. 2012; Smeaton et al. 2008). The precipitation of uranyl phosphate prevails, however, when dissolved orthophosphate concentrations are elevated at circumneutral pH (Salome et al. 2013). Interestingly, uranium phosphate minerals are found in a wide variety of soils (Sato et al. 1997), including those in the U.S. DOE IFRC sites in Oak Ridge, Tennessee (Roh et al. 2000), and Hanford, WA (Arai et al. 2007), and the former Fernald Environmental Management Project site in Fernald, OH (Morris et al. 1996). In addition, various calcium–uranium–phosphate minerals have proved extremely stable over long periods in the presence of natural concentrations of carbonates (Raicevic et al. 2006), suggesting that phosphate precipitates may represent a significant fraction of U(VI) present in subsurface sediments. As dissolved inorganic phosphate is scarce in most soils and sediments, the presence of uranyl phosphate minerals suggests that microorganisms hydrolyzing organophosphates or polyphosphates may be able to generate enough inorganic phosphate in solution to precipitate uranyl phosphate minerals in a biomineralization process, possibly as a detoxification mechanism (Macaskie et al. 1988, 1992; Martinez et al. 2007; Renninger et al. 2004; Suzuki and Banfield 2004). Indeed, several microorganisms have been demonstrated to immobilize uranium as phosphate minerals, including *Bacillus* sp. (Macaskie et al. 1992) and *Citrobacter* sp. N14 (Macaskie et al. 2000) that are isolated from a metal-polluted soil; *Arthrobacter ilicis*, isolated from an open-pit uranium mine; *Deinococcus radiodurans*, a radiation-tolerant microorganism (Suzuki and Banfield 2004); *Rahnella* sp. Strain Y9602 and *Bacillus* sp. Strain Y9-2 (Beazley et al. 2007, 2009), two isolates from the acidic uranium-contaminated subsurface of Oak Ridge (Martinez et al. 2006, 2007); *Myxococcus xanthus* (Jroundi et al. 2009), a member of one of the most ubiquitous bacterial groups in soils that is present in uranium-contaminated sites (Petrie et al. 2003); *Cellulomonas* sp. Strain ES6, isolated from the Hanford subsurface (Sivaswamy et al. 2011); *Pelosimus* Strain UFO1, a new representative of a fermentative group of bacteria isolated from the Oak Ridge subsurface (Ray et al. 2011); *Pseudomonas aeruginas* Strain J007, isolated from acidic mines (Choudhary and Sar 2011); and *Bacillus sphaericus* Strain JG-7B and *Sphingomonas* sp. Strain S15-S1, two isolates from acidic-contaminated environments (Merroun et al. 2011).

The precipitation of uranyl phosphate compounds is generated by three different mechanisms depending on the strain used: (i) hydrolysis of exogenous organophosphate compounds by nonspecific acid phosphatases (Beazley et al. 2007, 2009; Macaskie et al. 1992, 2000), (ii) hydrolysis of endogenous organophosphate compounds that are generated during cell lysis by acid phosphatases (Jroundi et al. 2009; Merroun et al. 2011; Suzuki and Banfield 2004), and (iii) depolymerization of polyphosphates by polyphosphatases (Ray et al. 2011; Renninger et al. 2004; Sivaswamy et al. 2011; Suzuki and Banfield 2004). In addition, the uptake and precipitation of uranium inside the cells by a yet-to-be-determined mechanism that involves cellular phosphate compounds has also been observed (Choudhary and Sar 2011; Jroundi et al. 2009). The vast majority of these microorganisms appears to precipitate uranium phosphate compounds in aerobic conditions, but these findings may be biased by the fact that aerobic organisms are easier to isolate. In fact, two of these microorganisms are able to precipitate U(VI) phosphate minerals in aerobic conditions while reducing U(VI) in anaerobic conditions via electron shuttling mechanisms (Ray et al. 2011; Sivaswamy et al. 2011). So far, only two strains have been demonstrated to precipitate U(VI) phosphate minerals in anaerobic conditions by using nitrate as a TEA (Beazley et al. 2009; Shelobolina et al. 2009), and several studies using natural sediments were conducted in anaerobic conditions (Beazley et al. 2011; Geissler et al. 2009; Salome et al. 2013). In addition, the biomineralization process is mostly investigated in low pH environments, mainly because uranium-contaminated sites are usually acidic as a result of the use of nitric acid to extract uranium from ore, and most microorganisms have been isolated from acidic environments. Alkaline bacterial phosphatases, however, are able to promote the biomineralization of U(VI) phosphate minerals, though so far they have mostly been overexpressed in heterologous microorganisms (Kulkarni et al. 2013; Nilgiriwala et al. 2008). Readers who are interested in learning about the biochemistry of acid and alkaline phosphatases or polyphosphatases should read the different reviews that synthesize the knowledge on these complicated enzymes (Brown and Kornberg 2009; Keasling et al. 2000; Rigden 2008; Rossolini et al. 1998). A more recent study that investigated the eventual competition between U(VI) bioreduction and U(VI) phosphate biomineralization in anaerobic conditions in contaminated soils at both low and high pH demonstrated that U(VI) phosphate biomineralization outcompetes bioreduction in both mildly acidic and circumneutral pH conditions, partly because nitrate reduction dominated anaerobic respiration processes in these sediments (Salome et al. 2013). Although competition between these two uranium immobilization processes has to be investigated under different anaerobic respiration conditions, these findings suggest that this process may be ubiquitous in a variety of environments where nitrate reduction represents the main anaerobic respiration process.

Although the biomineralization of U(VI) phosphate minerals appears to require endogenous phosphate compounds that are produced by cell lysis or exogenous organophosphate compounds that are not naturally present in soils or sediments, phytate or inositol hexaphosphate (IP_6) is a phosphorylated inositol with six attached phosphate groups that represents the most abundant natural organophosphate compound in terrestrial environments (Turner et al. 2002a). The myo-inositol form of IP_6 is, by far, the most abundant of the nine stereoisomers that are found in the environment, perhaps because it is the sole isomer with only one axial phosphate group (Parthasarathy and Eisenberg 1991). The six phosphate groups on phytic acid carry 12 ionizable protons with pK_a's ranging from 1.1 to 12 (Costello et al. 1976). With six pK_a's below 3, phytic acid carries a highly negative charge (–6 to –9) at environmentally relevant pH. As such, phytic acid efficiently chelates multivalent cations, including Ca^{2+}, Zn^{2+}, Mg^{2+}, Mn^{2+}, and Cu^{2+} (Turner et al. 2002a; Wodzinski and Ullah 1996). Phytate also interacts strongly with soil sorbents, including metal oxides and natural organic matter, and it also displaces sorbed inorganic phosphate from iron oxides (Degroot and Golterman 1993; Johnson et al. 2012). Although IP_6 remains chemically stable at all environmentally relevant pH and is not hydrolyzed by conventional phosphohydrolases (Turner et al. 2002a), enantioselective phytase enzymes may catalyze its hydrolysis to lower inositol derivatives (IP_X, x: 2–5) and inorganic phosphate (Irving and Cosgrove 1971; Parthasarathy and Eisenberg 1991). Once hydrolysis of the initial phosphate group from IP_6 is achieved, however, hydrolysis of lower inositol derivatives (IP_X, x: 1–5) may be catalyzed by either phytase or any nonspecific acid or alkaline phosphatase (Markovarga and Gorton 1990; Meek and Nicoletti 1986; Shan et al. 1994).

Phytase enzymes are a substrate-specific subclass of phosphomonoesterases that are produced by plants, animals, and microorganisms (Oh et al. 2004) with variable optimum pH, molecular weight, structure, and metal cofactor that are required for enzyme activation (Shin et al. 2001). Based on their optimum pH, phytases may be classified broadly as either acid or alkaline phytases. Acid phytases include the subclasses purple acid phosphatases (PAPs) and histidine acid phosphatases (HAPs), which display optimum activity at acidic pH (Oh et al. 2004; Turner et al. 2002b). Alkaline phytases consist of a class of enzymes that are known as beta-propeller phytases (BPPs) (Mullaney and Ullah 2007; Yao et al. 2012) that may be widespread in prokaryotes (Cheng and Lim 2006; Lim et al. 2007). So far, calcium phytate has been used to adsorb U(VI) from contaminated soils (Nash et al. 1998; Seaman et al. 2003), but the use of phytate to biomineralize uranium phosphate minerals has only been demonstrated in a recent study using low pH uranium-contaminated soils (Salome et al. 2016). Complete hydrolysis of IP_6 could be achieved under mildly acidic conditions but not at circumneutral pH (Figure 9.4a), suggesting that natural microbial populations carry acid phytases to hydrolyze phytate. Hydrolysis of phytate

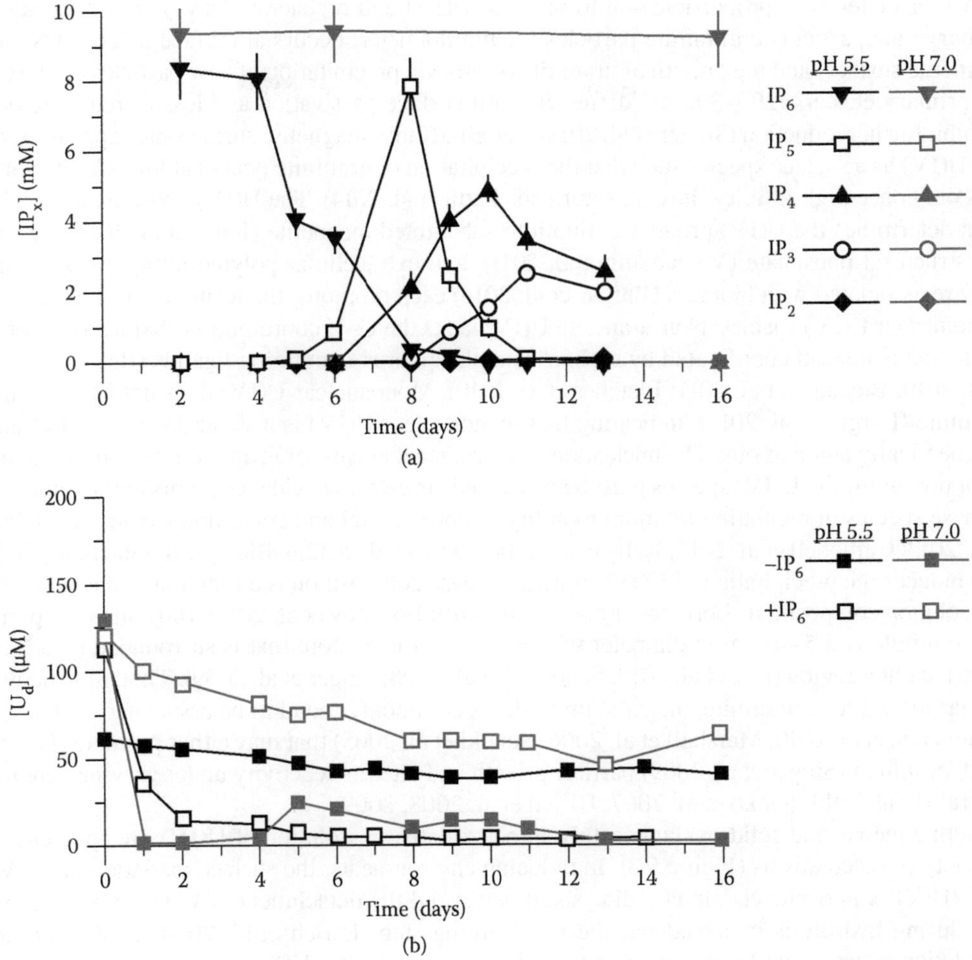

FIGURE 9.4 Evolution of (a) inositol hexaphosphate (IP_6) and its lower derivatives (IP_X = 2–5) and (b) dissolved uranium as a function of time in pH 5.5 (black) and pH 7 slurries containing 16 g/L Area 3 soils from the U.S. DOE IFRC site in Oak Ridge, Tennessee, in artificial groundwater, amended with 200 μM UO_2^{2+} with or without 10 mM phytate. Reactors buffered at pH 7 were also amended with 10 mM of dissolved inorganic carbon. Error bars include variations between triplicate reactors and analytical error from calibrations.

and its intermediate byproducts produced more than 40 mM inorganic phosphate in slurries containing 10-mM phytate initially (Salome et al. 2016), which resulted in the complete precipitation of uranium compared with controls without phytate or incubations that are conducted at pH 7 that did not hydrolyze phytate (Figure 9.4b). These findings suggest that uranium removal was mainly due to precipitation of uranium phosphate minerals and not because of adsorption or coprecipitation of uranium with phytate, as previously concluded in other investigations (Nash et al. 1998; Seaman et al. 2003).

9.5.2 U(IV) Precipitates in Reducing Environments

Although U(IV) is the main product of the chemical or microbial reduction of U(VI), different mineral products may form depending on the process involved and the chemical conditions of the aqueous environment. The reduction of U(VI) by magnetite (Ilton et al. 2010) and ferrous mica (Ilton et al. 2005) proceeds through the initial adsorption of U(VI) followed by the reduction to U(V), which may be stabilized as a U(V) surface species (adsorbed or precipitated), reduced to

U(IV), or undergo disproportionation to produce U(IV) and regenerate U(VI). The nucleation of primary nanoparticulate uraninite particles (~1 nm diameter) occurs at surface defect sites on the magnetite surface, and the growth of uraninite occurs via precipitation of new particles on the existing primary clusters (200–300 nm diameter) until surface passivation and loss of reducing power prohibit further reduction (Singer et al. 2012a). High-affinity magnetite surface sites appear to stabilize U(IV) as a surface species such that the precipitation of uraninite occurs at low-affinity surface sites only once high-affinity sites are saturated (Latta et al. 2014). The U(IV) coordination environment determines the U(IV) product as titanium-substituted magnetite (Latta et al. 2013), adsorbed and structural phosphate (Veeramani et al. 2011), and extracellular polymeric organic substances that are associated with biomass (Bargar et al. 2013) each promotes the formation of non-uraninite mononuclear U(IV) species. Non-uraninite U(IV) lacks the U–U coordination characteristic of uraninite and is instead coordinated by carbon- and phosphorus-containing ligands (Bernier-Latmani et al. 2010; Boyanov et al. 2011; Fletcher et al. 2010). Mononuclear U(IV) does not ripen or age to uraninite (Bargar et al. 2013), indicating that mononuclear U(IV) is a stable form of U(IV) and is geochemically autonomous. The nucleation kinetics, mechanisms of formation, and thermodynamics of non-uraninite U(IV) species must be addressed, as growing evidence points to the importance of these species in regulating uranium mobility in both natural and contaminated aquifers (Bargar et al. 2013; Campbell et al. 2012; Kelly et al. 2008; Latta et al. 2012a). Biological reduction of U(VI) also induces the precipitation of U(IV) minerals whose composition is a function of bacterial strain and solution composition (Bernier-Latmani et al. 2010; Boyanov et al. 2011). Biogenic nanoparticulate uraninite is 2.5–10 nm in diameter with a highly ordered core that is surrounded by a locally distorted outer region (Lee et al. 2010; Schofield et al. 2008; Singer et al. 2009). Though structurally similar to synthetic uraninite, biogenic uraninite is commonly found to be associated with organic matter (Lee et al. 2010; Marshall et al. 2006; Suzuki et al. 2005) that may either promote (Lee et al. 2010) or inhibit (Singer et al. 2009) particle growth and alter its reactivity under varying conditions (Cerrato et al. 2013; Senko et al. 2007; Ulrich et al. 2008, 2009).

Both aqueous and solid oxidants catalyze the oxidative dissolution of U(IV) precipitates under a variety of mechanisms (Figure 9.3). In reducing environments, the surface oxidation of U(IV) to U(V)/U(VI) can occur via alpha-radiolysis of water and the detachment of U(IV) oxidation products during hydrolysis is considered the rate-limiting step (Ulrich et al. 2008, 2009). Oxidative dissolution of uraninite by dissolved oxygen proceeds through a U(V) intermediate (Ulrich et al. 2009) and may be inhibited by passivation of the uraninite surface by adsorbed species or a surface precipitate resulting from either heterogeneous nucleation on the uraninite surface (Cerrato et al. 2012) or precipitation of an oxidized U(V)/U(VI) mineral phase during oxidation (Ulrich et al. 2009). Elevated concentrations of carbonates may decrease surface passivation and increase the rate and extent of oxidative dissolution by enhancing the release of adsorbed U(VI) or the dissolution of U(VI) solids (Cerrato et al. 2012; Ulrich et al. 2008, 2009). Direct contact between uraninite and a solid oxidant (i.e., iron and manganese oxides) promotes more rapid dissolution of uraninite compared with heterogeneous oxidation (Wang et al. 2013c). This mechanism, however, is likely hindered in the subsurface where the transport of aggregate minerals is strictly limited. Thus, in diffusion-controlled environments, the primary role of solid oxidants may be to sequester uranium via adsorption rather than remobilizing it via oxidation (Plathe et al. 2013). The proposed heterogeneous mechanism for uraninite oxidation by solid oxidants involves dissolution of uraninite and adsorption of U^{4+} on the oxide surface, followed by electron transfer and release of Fe(II) or Mn(II) and U(VI) (Ginder-Vogel et al. 2010; Wang et al. 2013c). Accordingly, the rate of uraninite oxidation is expected to be the fastest under conditions that promote uraninite dissolution, namely acidic or alkaline solutions and high concentrations of carbonates (Ginder-Vogel et al. 2010). Mononuclear U(IV) species are more labile than biogenically or chemically precipitated uraninite, as they readily nonoxidatively dissolve under anaerobic conditions with elevated concentrations of carbonates (Alessi et al. 2012) or oxidatively dissolve in aerobic conditions (Cerrato et al. 2013), though the mechanism of their dissolution is largely unknown. Finally, alternating reducing and oxidizing conditions may promote the coprecipitation of uranium in iron oxides, as adsorbed U(VI) becomes encompassed in the crystal lattice during successive reductive dissolution and oxidative precipitation of iron minerals (Duff et al. 2002; Huber et al. 2012; Stewart et al. 2009). Over long time scales, uranium may be incorporated deeper in the mineral and become more recalcitrant than uranium precipitates whose stability is highly sensitive to redox environments. For example, the reduction of hematite-incorporated U(VI) results in U(V), which is stabilized in the hematite lattice in uranate coordination and lacks the characteristic oxo ligands of uranyl coordination (Ilton et al. 2012).

9.6 THE EFFECT OF AQUEOUS SPECIATION ON URANIUM REACTIVITY

The aqueous speciation of uranium is highly complex, as all oxidation states of uranium form aqueous complexes with both organic and inorganic ligands that are present in subsurface environments and are strongly influenced by the pH (Langmuir 1978). Ligands alter uranium adsorption on mineral surfaces (Stewart et al. 2010; Waite et al. 1994), bioavailability to microbes (Belli et al. 2014; Brooks et al. 2003; Markich et al. 1996; Ulrich et al. 2011), reaction kinetics (Behrends and Van Cappellen 2005; Ginder-Vogel et al. 2010; Hua et al. 2006), and, ultimately, uranium mobility. The classic view of uranium geochemistry associates U(VI) with high solubility and mobility, U(V) with instability toward disproportionation, and U(IV) with low solubility and mobility. These properties, however, depend on the speciation of uranium and do not necessarily reflect the reactivity of uranium in all environments. As a result, understanding the influence of speciation on the physical, chemical, and biological processes controlling uranium mobility is crucial to successfully predict the behavior of uranium in diverse geochemical environments.

As discussed earlier, the adsorption of uranium relies, in part, on electrostatic attraction between the charged uranyl complex and the charged mineral surface, which are a function of the type of ligand, its concentration, and the pH. In addition, proper coordination between the uranyl complex and the mineral surface is affected by bulky complexing ligands (e.g., carbonates, phosphates) (Cheng et al. 2004; Hsi and Langmuir 1985; Lenhart and Honeyman 1999; Nair and Merkel 2011; Stewart et al. 2010; Waite et al. 1994; Wang et al. 2013b). Ligands also impact the redox cycling of uranium by altering the U(VI)/U(IV) redox potential and, in some cases, limit the kinetics of electron transfer to and from uranium. For example, the potential of the U(VI)/U(IV) redox couple under geochemical conditions at the Rifle site spans 70 mV due to differences in redox potential among uranyl species (Figure 9.5). The "free" hydrated uranyl ion and uranyl hydroxide species typically display the highest redox potential, whereas ternary Ca and Mg uranyl carbonate complexes display the lowest potential. Although this range of potentials is small in comparison to the range of potentials across the suite of TEAs used in microbial respiration (Figure 9.5), speciation-dependent changes in the U(VI)/U(IV) potential are significant in that they span the ferrihydrite/Fe(II) potential and may alter iron–uranium–bacteria interactions. As iron is an effective oxidant and a reductant of uranium (Figure 9.3b), changes in uranyl speciation that alter its redox potential determine the role of iron in redox transformations of uranium. This effect can be explained in both thermodynamic and mechanistic terms. Thermodynamically, carbonates enhance the effectiveness of ferrihydrite as an oxidant of uraninite by lowering the redox potential of U(VI) relative to other uranyl species (i.e., the "free" hydrated uranyl ion and uranyl hydroxides) (Figure 9.5) and shifting the equilibrium toward the formation of highly soluble uranyl carbonate complexes (Ginder-Vogel et al. 2006). Mechanistically, carbonates enhance the oxidative dissolution of uraninite by promoting the release of U(VI) from the uraninite surface and limiting the surface passivation by U(VI) precipitates (Ginder-Vogel et al. 2010; Ulrich et al. 2009). As carbonates thermodynamically promote the oxidation of U(IV) by iron oxides, they also hinder the reduction of U(VI) by Fe(II) and other reductants (Du et al. 2011). In fact, rates of abiotic reduction of U(VI) by Fe(II) (Behrends and Van Cappellen 2005) and dissolved sulfides (Hua et al. 2006) are inversely related to the concentration of carbonates. Simultaneously, the complexation of uranyl by carbonates alters its size and charge, which may increase steric hindrance and limit the kinetics of electron transfer, as confirmed

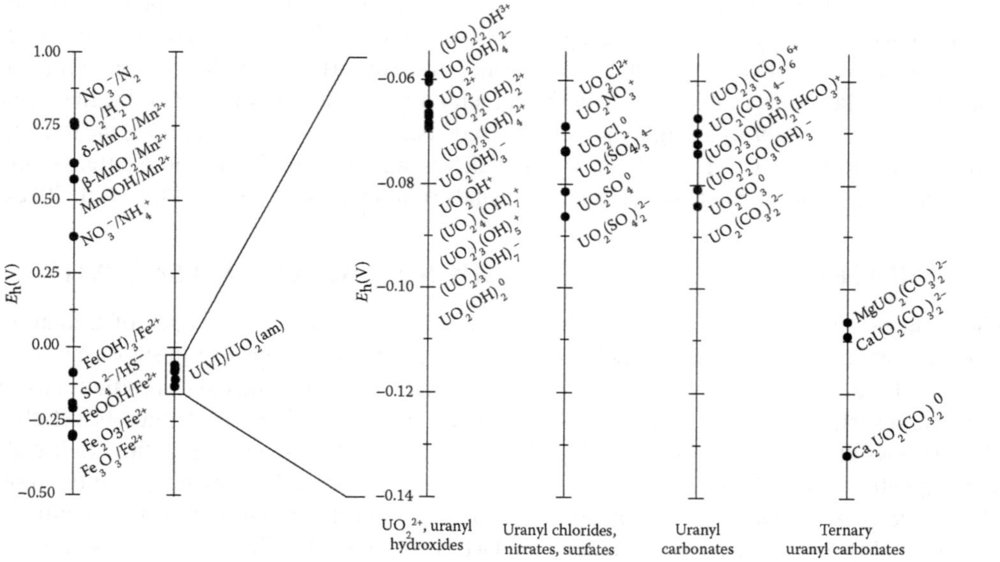

FIGURE 9.5 Redox potentials (E_h) of terminal electron acceptors for microbial respiration at the U.S. DOE IFRC site in Rifle, Colorado. The right side is an expanded view of E_h for individual uranyl species, demonstrating the influence of uranyl speciation on the U(VI)/UO_2(am) redox potential. E_h was calculated with the Nernst equation ($E_h = E_h^0 - [RT/nF]\ln[Q]$, where Q is the reaction quotient) by using groundwater composition at the Rifle site (Campbell et al. 2011; Zachara et al. 2013). Uranyl speciation was calculated as a function of pH by using PHREEQC, updated with the most recent thermodynamic data for solid and aqueous uranyl species (Dong and Brooks 2006; Guillaumont et al. 2003). A concentration of 10 μM was used for N_2 and HS^-.

by molecular dynamic simulations and density functional theory calculations for the thermodynamically favorable reduction of $UO_2(CO_3)_3^{4-}$ by Fe^{2+} (Wander et al. 2006). The combination of these two effects ultimately determines the rate of U(VI) reduction in the presence of carbonates. Carbonates, in turn, do not prevent U(VI) reduction by magnetite, but reduction is not observed in the presence of Ca^{2+} and carbonates as a result of the formation of ternary Ca uranyl carbonate surface complexes (Singer et al. 2012a,b). Although the mechanism of inhibition remains unclear, Ca^{2+} promotes the formation of ternary uranyl carbonate complexes that have less favorable redox potentials than other uranyl carbonate species (Figure 9.5) and may be below the threshold potential for reduction.

Finally, the speciation of U(VI) greatly impacts its bioavailability and, ultimately, determines the kinetics of bioreduction by MRB (Belli et al. 2015; Brooks et al. 2003; Stewart et al. 2011; Ulrich et al. 2011) and its toxicity to microbial communities (Campbell 1995; Di Toro et al. 2001; Morel 1983). For example, conditions that promote the formation of carbonate and ternary carbonate uranyl species (i.e., alkaline pH and high concentrations of carbonates and alkali earth metal cations) decrease the rate of U(VI) bioreduction (Belli et al. 2015; Brooks et al. 2003; Stewart et al. 2011; Ulrich et al. 2011). Simultaneously, UO_2^{2+} and UO_2OH^+ have been identified as the uranyl complexes that are responsible for uranium toxicity to microorganisms (Konopka et al. 2013; Trenfield et al. 2012; VanEngelen et al. 2010). Because the transport of uranium across the outer membrane is likely an active process (Fortin et al. 2007), the adsorption of U(VI) to the outer membrane is an essential first step in the transport of U(VI) into the cell where it may interact with periplasmic reductases, in the case of bioreduction (Marshall et al. 2006; Senko et al. 2007), or inhibit essential enzymes, in the case of toxicity (Pible et al. 2010; VanEngelen et al. 2011). Thus, toxicity may be avoided in alkaline environments where the formation of uranyl carbonate species decreases the concentration of the toxic uranyl species, prevents sufficient cell–uranyl interactions to induce a toxic response, or a combination of both (Carvajal et al. 2012; Konopka et al. 2013; VanEngelen et al. 2010). Recently, it was demonstrated that bioreduction rates increase as the concentration of bioavailable uranyl species increases until a concentration threshold is reached, above which bioreduction rates decrease as a result of uranium toxicity to MRB (Belli et al. 2015). These findings suggest that toxicity and bioavailability are intimately related, and although carbonates, Ca^{2+}, and Mg^{2+} decrease the potential of the U(VI)/U(IV) redox couple (Figure 9.5), they primarily control U(VI) bioreduction kinetics by decreasing the concentration of the "free" hydrated uranyl ion and other noncarbonate uranyl species that appear to represent the most readily reducible and bioavailable fraction of U(VI) (Belli et al. 2015).

After the reduction of U(VI), ligands continue to play a role in the fate of uranium by altering the stability of U(V) and the solubility of U(IV). Pentavalent U(V) has been implicated as a transient intermediate during U(VI) bioreduction (Renshaw et al. 2005), abiotic reduction of U(VI) by aqueous Fe(II) (Wander et al. 2006), magnetite (Ilton et al. 2010), and ferrous mica (Ilton et al. 2005), and uraninite oxidation (Ulrich et al. 2009). Its stability may, therefore, constitute the rate-limiting step during redox transformations of uranium. Carbonates retard the disproportionation of U(V) (Ikeda et al. 2007; Morris 2002), possibly by limiting electron transfer between the two U(V)-carbonate species because of their similar charge and bulky carbonate ligands. Upon disproportionation of U(V) to U(VI) and U(IV), the precipitation of insoluble reduced U(IV) species is expected to occur. In the presence of strongly complexing organic acids and soil humics, however, organic U(IV) complexes may form and prevent its precipitation (Burgos et al. 2007; Francis and Dodge 2008; Ganesh et al. 1997; Haas and Northup 2004; Robinson et al. 1998; Sheng et al. 2011; Suzuki et al. 2010). Naturally occurring organic acids and synthetic chelating agents promote the mobilization of uranium by decreasing the bioavailability of uranium to MRB and SRB and by increasing the solubility of U(IV) (Francis and Dodge 2008; Ganesh et al. 1997; Haas and Northup 2004; Suzuki et al. 2010). Thus, contrary to the classic view of U(IV) as displaying low solubility and low mobility, uranium may remain mobile even under highly reducing conditions in the form of organic complexes.

Despite the importance of the speciation of uranium on its biogeochemical transformations, the sheer number of possible uranyl species present in solution and the lack of analytical techniques that are able to quantify individual species limit the ability to identify specific mechanisms that are responsible for observations in the field or the laboratory. As a result, most investigations calculate the speciation of uranium from total dissolved uranium measurements using geochemical models. Over the past couple of decades, however, laser-induced spectroscopy (Amayri et al. 2005; Bernhard et al. 2001; Collins et al. 2011; Geipel 2006; Geipel et al. 2008; Götz et al. 2010; Gunther et al. 2007) and synchrotron-based spectroscopy (Allen et al. 1995; Bargar et al. 2000; Kelly et al. 2007; Singh et al. 2012), despite providing bulk average measurements and lacking the sensitivity and selectivity to quantify individual species, have proved invaluable at identifying the coordination and stoichiometry of complex uranium species in both solid phase and solution. Voltammetry with mercury amalgam microelectrodes has also proved to be a powerful analytical technique for investigating the transformations of environmentally relevant redox-active species in sediments (Luther et al. 2008; Taillefert et al. 2000), and it has been used to quantify individual uranyl complexes in laboratory studies (Djogic and Branica 1995b; Maya 1982; Morris 2002) as well as total dissolved uranium in complex environmental matrices (Djogic and Branica 1995a; Djogic et al. 1986, 2001; Newton and Van Den Berg 1987; Van Den Berg and Nimmo 1987). Recently, voltammetry has been used to study the influence of uranyl speciation on the kinetics of U(VI) bioreduction and to provide the first direct quantification of individual aqueous uranyl complexes in real time during bioreduction incubations with MRB (Belli et al. 2011). In a carbonate-dominated system, three distinct voltammetric peaks were identified using a hanging mercury drop electrode (HMDE) (Figure 9.6a) and assigned to UO_2^{2+}, $UO_2(CO_3)_2^{2-}$, and $UO_2(CO_3)_3^{4-}$ by comparing the peak surface area to the calculated equilibrium uranyl speciation as predicted by thermodynamics (Figure 9.6b). Mercury amalgam microelectrodes were used to measure uranyl speciation in incubations with *S. putrefaciens* strain 200 when U(VI) was provided as

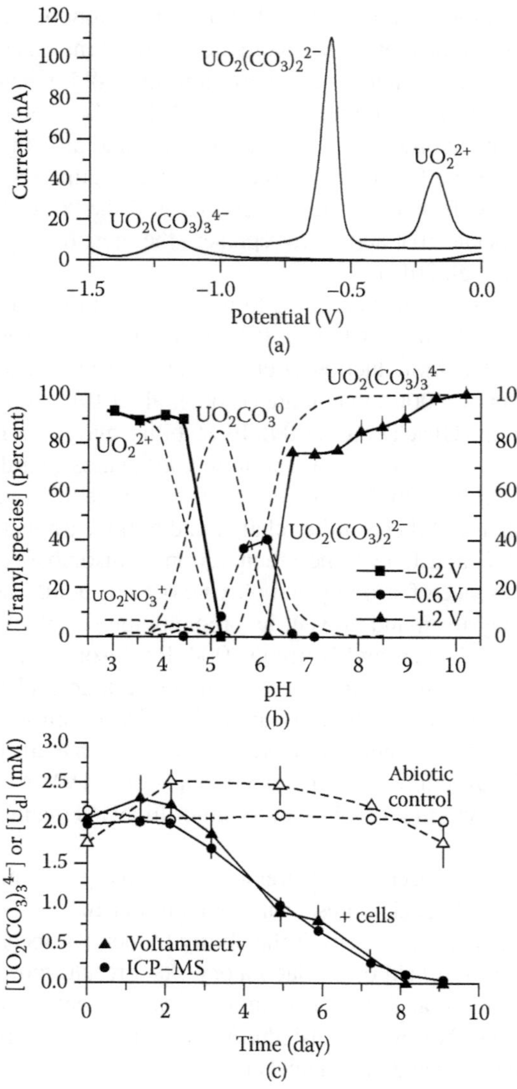

FIGURE 9.6 Quantification of electrochemically active uranyl species by voltammetry on a hanging mercury drop electrode (HMDE) and measurements of uranyl speciation during U(VI) bioreduction incubations with *Shewanella putrefaciens* strain 200. (a) Voltammograms of 100 μM uranyl acetate in 10 mM $NaHCO_3$ show three distinct peaks at pH 2 (-0.2 V), pH 6 (-0.6 V), and pH 9 (-1.2 V). (b) The three peaks were identified as UO_2^{2+}, $UO_2(CO_3)_2^{2-}$, and $UO_2(CO_3)_3^{4-}$ by comparing normalized peak areas (symbols and solid lines) to the abundance of uranyl species at each pH (dashed lines) that were calculated using PHREEQC, updated with the most recent thermodynamic data for solid and aqueous uranyl species (Dong and Brooks 2006; Guillaumont et al. 2003). (c) Uranyl speciation during U(VI) bioreduction incubations with strain 200 was measured *in situ* over time in batch reactors by using mercury amalgam microelectrodes. Under the experimental conditions (2 mM uranyl acetate, 33 mM $NaHCO_3$, 15 mM lactate, pH 8.2, M1 growth medium [Myers and Nealson 1988]), more than 99% of U(VI) is present as $UO_2(CO_3)_3^{4-}$ as confirmed by $UO_2(CO_3)_3^{4-}$ measurements by voltammetry (triangles), total dissolved uranium (U_d) measurements by ICP–MS (circles), and speciation calculations, assuming that equilibrium was reached at each time step. Open symbols represent abiotic controls, and standard deviations represent analytical error.

a TEA under conditions where $UO_2(CO_3)_3^{4-}$ accounted for greater than 99% of U(VI). Voltammetric measurements of $UO_2(CO_3)_3^{4-}$ were in excellent agreement with total dissolved uranium measurements by inductively coupled plasma–mass spectrometry (ICP–MS) (Figure 9.6c), validating the ability of voltammetry to quantify individual uranyl complexes in a dynamic system. Efforts are currently underway to apply this technique to more complex systems and to improve the ability to quantify bioavailable uranyl species (i.e., the "free" hydrated uranyl ion and uranyl hydroxide species) that are the least abundant fraction of U(VI) under most environmentally relevant conditions.

9.7 ANTHROPOGENIC INFLUENCES ON URANIUM CYCLING IN SEDIMENTS

Humans have dramatically perturbed the global uranium cycle by mining and processing uranium ore to produce nuclear weapons and fuel for nuclear power plants. In addition, vast quantities of uranium-contaminated groundwater and soils have been generated from years of poor waste disposal practices and aging of nuclear waste storage containers (Riley et al. 1992). Much of the current understanding of uranium biogeochemistry stems from the need to predict the fate of uranium in contaminated environments, prevent the mobilization of uranium into pristine aquifers, and devise remediation strategies that utilize the natural processes that control uranium mobility in the environment. Initially, natural attenuation of uranium-contaminated soils was expected to play a significant role in decreasing uranium concentrations below the maximum contaminant level (MCL) of 0.126 μmol/L that was tolerated by the U.S. Environmental Protection Agency (EPA) on a time scale of tens of years. The persistence of uranium plumes exceeding regulatory limits, however, highlighted the inability to successfully predict the fate of uranium in subsurface environments. These findings also demonstrated the need for a process-based understanding of uranium biogeochemistry to develop site-specific remediation strategies that exploit the natural biogeochemical and hydrological processes controlling uranium mobility.

9.7.1 Physicochemical Remediation Strategies

Physical remediation strategies to address uranium contaminations are primarily focused on the containment of uranium by erecting impenetrable barriers to restrict surface water infiltration or to impede the subsurface transport of the contaminant plume. Although physical remediation strategies are often the most economic approach, they merely contain the contaminant and do not actually mitigate its toxicity to the environment. Indeed, original efforts to contain the uranium plume at the Y-12 Facility in Oak Ridge, Tennessee, were unsuccessful in preventing uranium migration after soil capping and the construction of a parking lot above the capped waste ponds (Wu et al. 2006a).

Chemical remediation of uranium contaminations involves an alteration of the mobility of uranium by promoting changes in its solubility, either through complexation with ligands or via reduction/oxidation. Soil washing and pump-and-treat strategies are performed *ex situ* by which soil or groundwater is removed from the aquifer, treated above ground, and replaced/reinjected into the subsurface (Kerr 1990; Mackay and Cherry 1989). During treatment, uranium may be removed from the soil matrix through mechanical separation and from groundwater by flocculation, precipitation, adsorption, ion-exchange resins, or reverse osmosis (Dawson and Gilman 2001; Nimmons 2007). Because soil washing and pump-and-treat technologies are cost prohibitive on large scales and require the handling of toxic contaminants, *in situ* remediation strategies have been explored (Jardine et al. 2004; Riley et al. 1992). *In situ* chemical strategies include flushing the soil with a suitable chelating agent by an injection into the subsurface via a gallery of wells upstream of the contaminant plume to increase the mobility of uranium and to allow for extraction via groundwater pumping (Fox et al. 2012; Francis et al. 1998; Shiel et al. 2013). Carbonate and citrate are effective ligands for these applications, as they promote the mobility of uranium by forming stable aqueous complexes that are less prone to adsorption (Fox et al. 2012; Francis et al. 1998; Shiel et al. 2013) and, in the case of citrate, may dissolve potential sorbents (e.g., iron oxides) and release contaminants (Francis et al. 1998; Kantar and Honeyman 2006).

Soil flushing, however, requires high permeability soils and may promote the unwanted mobility of co-contaminants, as a variety of ligands nonspecifically bind to metals.

Conversely, other *in situ* chemical remediation strategies seek to immobilize uranium by promoting adsorption onto solid phases or via the formation of insoluble uranium minerals. Various minerals have been investigated for their ability to remove U(VI) from contaminated environments. The most important are zeolites (Krestou et al. 2004), zero-valent iron (Cantrell et al. 1995), calcium carbonates (Meece and Benninger 1993; Reeder et al. 2000, 2001), and hydroxyapatites (Fuller et al. 2002, 2003; Jerden and Sinha 2003; Krestou et al. 2004). Hydroxyapatites or fluoroapatites precipitate U(VI) as uranium phosphate minerals at a pH less than 6 (Arey et al. 1999; Krestou et al. 2004; Ohnuki et al. 2004; Seaman et al. 2003), whereas concentrations of uranium have to be increased significantly before precipitation of chernikovite ($H_2(UO_2)_2(PO_4)_2$) is observed either at circumneutral or above circumneutral pH (Fuller et al. 2002). Uranyl phosphate minerals (e.g., autunite/*meta*-autunite group minerals) are highly insoluble and stable in mildly acidic and neutral pH over long time scales (Jerden and Sinha 2003; Kanematsu et al. 2014) and are, therefore, appealing candidates for long-term uranium remediation efforts. Although uranyl phosphates are stable in oxidizing environments, the presence of carbonates prevents their precipitation, apparently because carbonates compete with phosphate for uranium (Fuller et al. 2002). The pH plays a significant role, as it controls the concentration of carbonates in natural waters and simultaneously affects the precipitation of uranium phosphate compounds. Indeed, the complex speciation of U(VI) in carbonate-dominated waters in the presence of hydroxyapatite results in the efficient removal of U(VI) by adsorption in the pH range of 6–7, but above circumneutral pH, the pH_{zpc} of hydroxyapatite is exceeded, and U(VI) is in the form of the less negatively charged $(UO_2)_2(CO_3)(OH)_3^-$, which tends to desorb from the solid phase (Wellman et al. 2008). As a result, uranium phosphate phases such as chernikovite or autunites may precipitate preferentially in these conditions, provided that phosphate is present in significant concentrations in the groundwater compared to carbonates (Jerden and Sinha 2003).

As inorganic phosphorus is scarce in most soils, *in situ* remediation strategies have included the injection of inorganic phosphates, organophosphates, such as phytate, or even polyphosphates to provide a phosphorus source that would slowly precipitate U(VI) from groundwater (Nash et al. 1998; Seaman et al. 2003; Wellman et al. 2006b). However, tests performed with Hanford soils at pH 8 revealed that the precipitation of calcium phosphate and calcium phytate rapidly decreases the permeability of these soils, probably precluding their use as a remediation strategy (Wellman et al. 2006b). Tripolyphosphate (TPP), in turn, does not decrease permeability of these soils, suggesting that it could be transported far away from its source and used as a slow source of ΣPO_4^{3-} to immobilize U(VI) (Wellman et al. 2006b). In fact, TPP is rather efficient in precipitating uranium after a small period of remobilization in both water-saturated and -unsaturated conditions (Wellman et al. 2007), though the composition of the precipitate was not obtained in these experiments. It is proposed that uranium is first desorbed from carbonate minerals, which is known to occur in Hanford soils, followed by reprecipitation with TPP. If autunite minerals were the primary uranium phosphate precipitates formed in these experiments, this strategy may be efficient in immobilizing uranium, as their solubility in alkaline pH conditions is generally lower than other forms of mineral precipitates (Wellman et al. 2006a, 2007). Interestingly, the interaction of U(VI) with polyphosphate has been studied, though only in a 1:1 molar ratio between U(VI) and polyphosphate (Vazquez et al. 2007). In these conditions, the chemical precipitation of U(VI) is never complete, but it removes about 80% of uranium at a pH less than 5 compared with 40% at a higher pH, indicating that the chemical hydrolysis of polyphosphate is not very efficient. Indeed, monodentate or bidentate complexes are proposed to be formed between uranium and polyphosphate depending on the pH (Vazquez et al. 2007).

As an alternative to the injection of a reactant, chemical remediation can be achieved using permeable reaction barriers (PRB)—trenches filled with porous, reactive material arranged downstream and perpendicular to groundwater flow such that the contaminant is chemically treated as it flows through the barrier (Dawson and Gilman 2001). Permeable reaction barriers are suitable for shallow surface contaminations, do not require maintenance, and avoid the handling of contaminants that are required in *ex situ* remediation strategies. Permeable reaction barriers constructed from zero-valent iron successfully decrease effluent uranium concentrations via reductive precipitation within the barrier coupled to the oxidation of Fe(0) to Fe(II) (Biermann et al. 2006; Gu et al. 1998; Morrison et al. 2001). Successful PRBs made of apatite have also been used to treat Zn-, Pb-, and Cd-contaminated waters (Conca and Wright 2006) and have shown promise for the treatment of uranium during laboratory studies (Biermann et al. 2006; Simon et al. 2004, 2008). In this treatment, uranium uptake by hydroxyapatite occurs via adsorption at low surface loading and precipitation of uranyl phosphate minerals at higher surface loading (Fuller et al. 2002). Long-term contaminant immobilization by PRBs is limited by the finite amount of surface sites for contaminant adsorption, in the case of hydroxyapatite PRBs, or decreasing permeability due to the buildup of corrosion products with time (i.e., iron oxides), in the case of Fe(0) PRBs (Biermann et al. 2006).

9.7.2 Bioremediation Strategies

In situ bioremediation is an attractive alternative to *in situ* chemical approaches, as it preserves subsurface hydrology, may be applied over a larger scale compared with PRBs, and may be more selective to specific contaminants than soil flushing. The bioremediation of uranium relies on subsurface microbial communities that transform uranium into a less mobile form by either reductive or nonreductive precipitation. Bioreduction is the most studied uranium bioremediation strategy, and it relies on the stimulation of subsurface MRB that catalyze the reduction of soluble U(VI) into sparingly soluble U(IV) minerals to decrease groundwater uranium concentrations. Because subsurface microbial communities are typically carbon limited, an organic carbon substrate that acts as both an electron donor and a carbon source is injected into the subsurface upstream of the contaminant plume via a gallery of injection wells to stimulate microorganisms (Anderson et al. 2003; Istok et al. 2004; Senko et al. 2002; Watson et al. 2013; Williams et al. 2011; Wu et al. 2006a,b, 2007). A variety of carbon substrates have been successfully injected into contaminated subsurfaces to decrease uranium concentrations during large-scale field studies, including acetate (Anderson et al. 2003; Istok et al. 2004; Williams et al. 2011), glucose (Istok et al. 2004), emulsified vegetable oil (EVO) (Watson et al. 2013), and ethanol (Istok et al. 2004).

Although bioreduction is seemingly straightforward, stimulation of diverse microbial communities initiates a complex network of enzymatic and abiotic reactions that ultimately affects the fate of uranium. First, the geochemistry of the sites has to be considered before adopting bioremediation strategies. For example, as uranium-contaminated areas are often associated with nitrate contaminations and low pHs as a result of the use of nitric acid to extract uranium from ore, nitrate reduction via denitrification or dissimilatory nitrate reduction to ammonia (DNRA) is readily stimulated after the addition of a carbon substrate (Istok et al. 2004; Watson et al. 2013; Wu et al. 2006b). During this phase, uranium concentrations remain elevated, as denitrifying bacteria utilize nitrate over U(VI) as a TEA and produce denitrification intermediates that readily oxidize U(IV) (Figure 9.3d). After complete removal of nitrate via denitrification (Senko et al. 2002) or pretreatment of groundwater (Wu et al. 2006a), the growth and activity of MRB are enhanced and associated with the production of Fe(II), Mn(II), and the removal of U(VI) via bioreduction (Anderson et al. 2003; Istok et al. 2004; Watson et al. 2013). Second, after the injection of an organic carbon substrate into the subsurface, microbial diversity declines and favors the growth of organisms that are capable of utilizing the chosen carbon source and byproducts of its mineralization (Barlett et al. 2012; Gihring et al. 2011). Although a variety of organic carbon sources stimulate U(VI) bioreduction, the microbial community response to these substrates during field biostimulation experiments is variable and has to be carefully considered (Barlett et al. 2012). For example, acetate is a common organic carbon substrate that is chosen for stimulating bioreduction, as it promotes the growth of *Geobacter* species, one of the few MRB that are capable of U(VI) bioreduction (Anderson et al. 2003; Williams et al. 2011). Uranium immobilization may be short-lived, however, as SRB that are unable to use U(VI) as a TEA (e.g., Desulfobacteraceae) may eventually dominate the microbial community and outcompete uranium-reducing MRB for acetate (Williams et al. 2011). This decrease in uranium bioreduction rates may result in a rebound of U(VI) concentrations either from advection of U(VI) from background water or via U(VI) desorption due to increased alkalinity that is associated with the elevated consumption of the electron donor during sulfate-reducing conditions (Anderson et al. 2003; Williams et al. 2011). It has been suggested that the continuous supply of Fe(III) to the subsurface along with acetate could prevent the decrease in *Geobacter* abundance and sustain U(VI) bioreduction (Zhuang et al. 2012), though the implementation of this strategy would be challenging, as iron oxides may decrease the hydraulic conductivity of the aquifer, and soluble Fe(III) complexes (Stewart et al. 2013) or freshly formed Fe(III) oxides (Ginder-Vogel et al. 2006, 2010) may oxidize U(IV) and promote U(VI) remobilization. Alternatively, the use of more complex, "slow release" electron donors (i.e., EVO, ethanol) is effective at stimulating bioreduction in the field and may prolong uranium immobilization by stimulating uranium-reducing SRB such as *Desulfovibrio* spp. (Cardenas et al. 2008; Hwang et al. 2009). These findings suggest that the byproducts of microbial respiration stimulated by complex organic substrates promote microbial diversity, which may be more advantageous for uranium immobilization than promoting a specific group of bacteria for a short period (Barlett et al. 2012). Third, in addition to direct enzymatic reduction of U(VI), microorganisms indirectly affect U(VI) reduction via the production of reduced metabolites, which either adsorb onto minerals and make them better abiotic reductants of U(VI) (i.e., Fe(II)) (Liger et al. 1999; Wehrli et al. 1989) or form iron sulfide minerals (i.e., Fe(II) and ΣH_2S), which abiotically reduce U(VI) and immobilize uranium for long periods after biostimulation (Bargar et al. 2013; Williams et al. 2011). In contrast, microorganisms remineralize organic carbon during respiration, which increases alkalinity and promotes desorption of U(VI), dissolution of nonuraninite U(IV), and, ultimately, remobilization of uranium (Anderson et al. 2003; Williams et al. 2011; Wu et al. 2006b).

Finally, the success of bioreduction depends on maintaining reducing conditions to prevent the oxidative dissolution of U(IV) solids, which is challenging over long time scales given the large number of U(IV) oxidants (Figure 9.3). Indeed, the return of aerobic and nitrate-reducing conditions promotes oxidative dissolution of U(IV) species by dissolved oxygen (Campbell et al. 2011; Watson et al. 2013; Wu et al. 2007) and denitrification intermediates (Istok et al. 2004; Watson et al. 2013). Mackinawite and amorphous FeS formed during biostimulation under iron- and sulfate-reducing conditions may protect the oxidation of U(IV) by preferentially reducing dissolved oxygen during short-term oxygenation of the bioreduced zone (Bi and Hayes 2014; Bi et al. 2013), but they are much less effective at preventing U(IV) oxidation by denitrification intermediates (Moon et al. 2009). For this reason, the long-term success of bioreduction is difficult in naturally oxidizing environments without continuous injections of an electron donor to maintain reducing conditions. Furthermore, long-term biostimulation may lead to elevated concentrations of carbonates that lower the potential of the U(VI)/U(IV) redox couple and promote the oxidative dissolution of uranium by amorphous iron and manganese oxides (Ginder-Vogel et al. 2006, 2010), even under highly reducing conditions when uranium-reducing organisms are present and active (Wan et al. 2005, 2008). Finally, uraninite is fairly resistant to ligand-promoted dissolution, whereas mononuclear U(IV) compounds may represent a more significant fraction of U(IV) in the field than previously considered (Bargar et al. 2013; Campbell et al. 2012; Kelly et al. 2008; Latta et al. 2012a; Wang et al. 2013a) and are highly susceptible to dissolution at elevated concentrations of carbonates (Alessi et al. 2012; Cerrato et al. 2013). Non-uraninite U(IV) that is associated with organic matter colloids has recently been demonstrated to be more mobile than originally considered, thus challenging the long-standing paradigm of low uranium mobility in reducing environments (Wang et al. 2013a). The formation of non-uraninite U(IV) products may, therefore, compromise the long-term stability of bioreduction, and more work is needed to understand the mechanism of their formation and their thermodynamic properties.

Although uranium bioreduction may be feasible in naturally reduced aquifers, alternative bioremediation strategies need to be considered in oxidizing environments. Biomineralization of uranium is a promising alternative to bioreduction, and it revolves around promoting the slow release of inorganic phosphate into the environment to immobilize uranium as insoluble uranyl phosphate minerals. To stimulate biomineralization, an organic phosphate substrate is injected into the subsurface where microbial communities catalyze the hydrolysis and release of inorganic orthophosphate from the organic substrate that subsequently precipitates with U(VI) as uranyl phosphate minerals. Unlike bioreduction, uranium biomineralization successfully immobilizes uranium in both oxic (Beazley et al. 2007, 2011; Martinez et al. 2007; Shelobolina et al. 2009) and anoxic (Beazley et al. 2009; Salome et al. 2013; Shelobolina et al. 2009) conditions and under acidic pH (Beazley et al. 2007, 2009, 2011; Martinez et al. 2007; Salome et al. 2013). The microbial production of phosphate ensures widespread phosphate distribution (Beazley et al. 2011) and avoids the rapid precipitation of phosphate minerals at the site of injection that is associated with the direct injection of inorganic phosphate into the subsurface (Wellman et al. 2006b). Large-scale field studies have yet to be conducted to assess the feasibility of uranium biomineralization. In practice, however, bacterial isolates (Beazley et al. 2007, 2009; Martinez et al. 2007) and natural microbial populations (Beazley et al. 2011; Salome et al. 2013, 2016) from the Oak Ridge IFRC successfully hydrolyze glycerol-2-phosphate (G2P), an exogenous organophosphate substrate, and immobilize uranium via the precipitation of chernikovite in laboratory experiments that are conducted under field conditions.

9.8 CONCLUSIONS AND OUTLOOK

In addition to its use in nuclear weapons, uranium is an element of extreme economic and societal importance, as it may represent an attractive alternative to fossil fuel–based electricity generation. Simultaneously, uranium is a toxic contaminant whose poor waste disposal practices at nuclear facilities have made it one of the most important targets of cleanup activities over the past several decades; if nuclear power plants proliferate, uranium will also represent one of the primary environmental concerns of the future. As a result, a mechanistic understanding of its biogeochemical transformations is essential for identifying uranium reserves for industrial uses and for predicting the transport of this contaminant in subsurface environments.

The speciation of uranium in aquatic systems is extremely complex as a result of its existence in the form of multiple redox states and its rapid transformation between them. In addition, its interaction with dissolved organic and inorganic ligands and solid surfaces, along with its high stability under a variety of mineral forms, makes characterization of the biogeochemical transformations of uranium extremely difficult. In this chapter, the current understanding on the biogeochemical processes that regulate the transformations of uranium in aquatic systems has been reviewed (Figure 9.7) based on the most recent findings from the literature. In addition, novel concepts and novel biogeochemical pathways that may have been underappreciated have been highlighted.

Although the reduction of uranium in soils and sediments is influenced by MRB, it appears to be more significantly regulated by reactions with byproducts of the respiration of natural organic matter, likely because uranium is highly toxic and the number of microorganisms that are able to gain energy from the reduction of U(VI) as a TEA appears limited. The reduction of U(VI) by dissolved sulfides is slow, whereas its reduction by iron sulfide minerals is much more efficient. Similarly, the homogenous reduction of U(VI) by Fe(II) is even slower, whereas its heterogeneous reduction by Fe(II)-bearing minerals is much more efficient. Finally, the remineralization of organic matter by microbial communities increases the alkalinity, which changes the speciation of uranyl and increases uranium mobility. Similarly, the resolubilization of U(IV) is mediated directly with chemical oxidants and indirectly via intermediate products of the respiration of natural organic matter. Dissolved oxygen, manganese oxides, Mn(III) compounds, and freshly precipitated iron oxides efficiently oxidize U(IV) products and may be responsible for the resolubilization of uranium in a variety of soils and sediments (Figure 9.7). Simultaneously, denitrification intermediates and electron shuttling between iron and these intermediates may also be largely involved in the oxidative resolubilization of uranium.

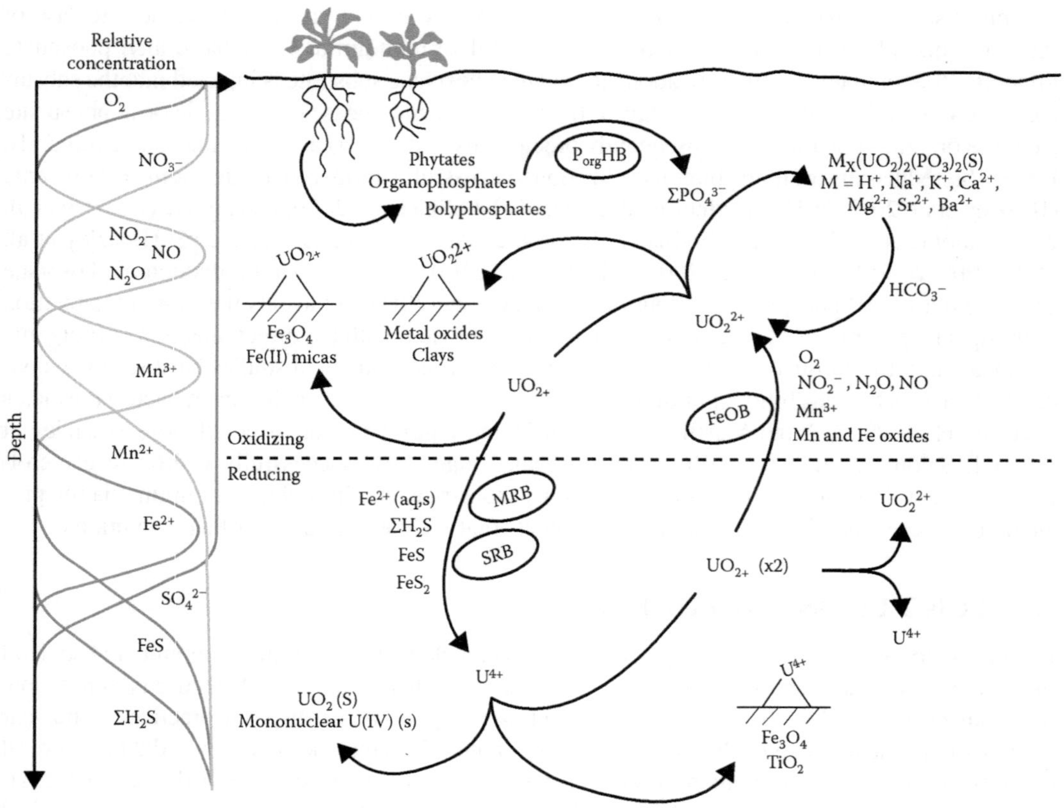

FIGURE 9.7 Schematic of the geochemical processes controlling the mobility of uranium in the subsurface. The left side shows the relative concentrations of redox-active species along with depth in an idealized redox-stratified aquifer. Organophosphate hydrolyzing bacteria ($P_{org}HB$) liberate orthophosphate (ΣPO_4^{3-}) from phytogenic organophosphates, which precipitate with oxidized UO_2^{2+} as uranyl phosphate minerals at high P:DIC ratios and may dissolve at lower P:DIC ratios. In addition, UO_2^{2+} may be immobilized via adsorption/incorporation into metal oxides and clays. UO_2^{2+} can be reduced abiotically by a number of inorganic reductants or microbially by select species of metal-reducing bacteria (MRB) and sulfate-reducing bacteria (SRB). UO_2^{2+} reduction likely proceeds through a pentavalent UO_2^+ intermediate that may undergo disproportionation, thereby regenerating UO_2^{2+} and producing U^{4+}, or it may be stabilized on certain mineral surfaces as a result of heterogeneous reduction. U^{4+} is highly insoluble and may precipitate as U(IV) minerals or noncrystalline U(IV) complexes in the absence of strongly complexing ligands; however, it may be stabilized on mineral surfaces as a result of heterogeneous reduction. U^{4+} can be oxidized abiotically by common oxidants or microbially by select iron-oxidizing bacteria (FeOB).

Although the mobility of uranium is largely dependent on its redox state, adsorption to mineral surfaces is a sink of uranium in oxidizing environments provided mineral surfaces are not saturated. A significant body of work has demonstrated that ligand complexation affects the adsorption of U(VI) species to solid minerals as a result of the net charges of the uranium complex, the surface charge of the minerals, the steric hindrance generated by the ligands, and the eventual formation of ternary complexes. Currently underappreciated, however, is the importance of uranium phosphate minerals in controlling the mobility of uranium in oxidizing environments. Given the low concentrations of dissolved orthophosphates in aquatic systems, these precipitates have not been considered to play a significant role in the biogeochemical cycle of uranium. Uranium phosphate precipitates, however, are extremely stable over a wide range of pH, and they are only destabilized in the presence of relatively high concentrations of carbonates. More importantly, the recent findings that microorganisms that are able to hydrolyze organophosphates and polyphosphates appear

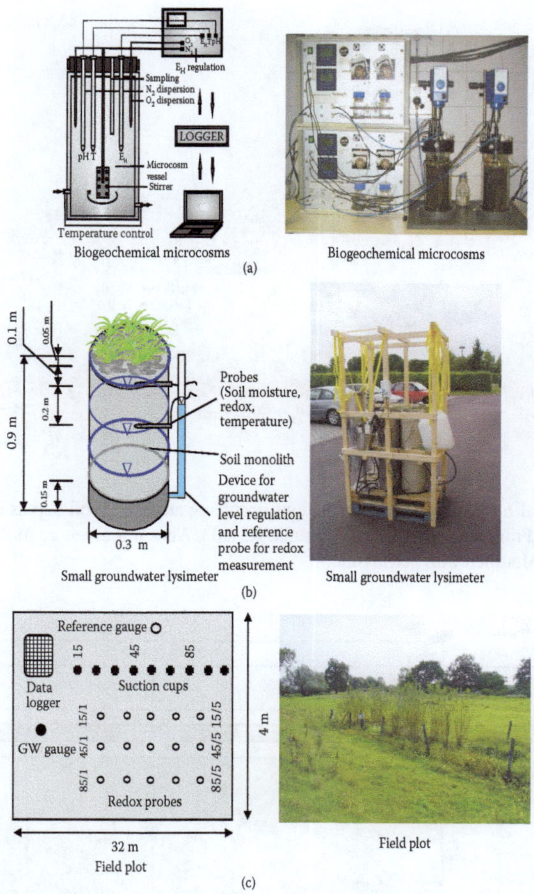

FIGURE 1.2 Experimental setups. (a) Biogeochemical microcosm, (b) groundwater lysimeter, and (c) field plot. (Reprinted from *Ecol Eng.*, 36, Rupp, H., Rinklebe, J., Bolze, S., and Meissner, R., A scale depended approach to study pollution control processes in wetland soils using three different techniques, 1439–1447, Copyright 2010, with permission from Elsevier.)

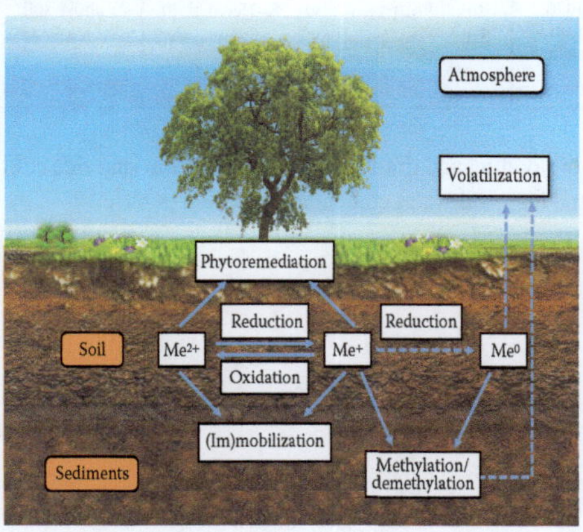

FIGURE 3.1 Dynamics of heavy metal(loid) redox transformation in soils and sediments.

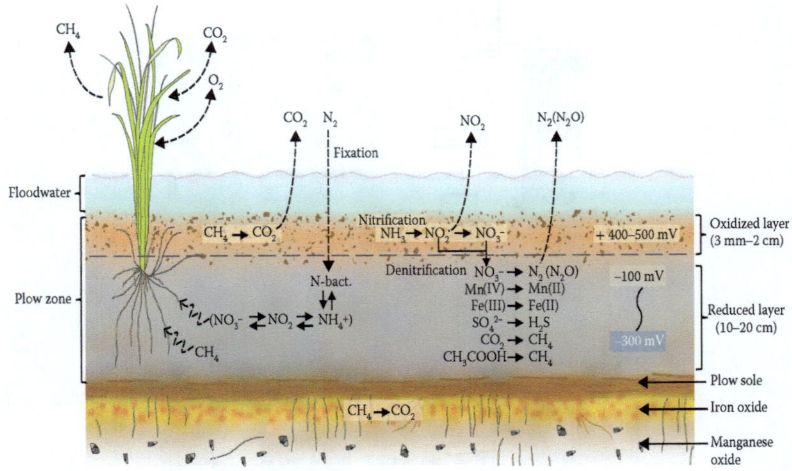

FIGURE 5.1 Typical E_h values and redox reactions occurring at various soil depths in a paddy soil under flooded conditions. (From Wakatsuki, T. In: Kyuma, K. (ed.), *New Soil Science*, Asakura Publishing, Co., Tokyo, Japan, 1997. Modified with permission.)

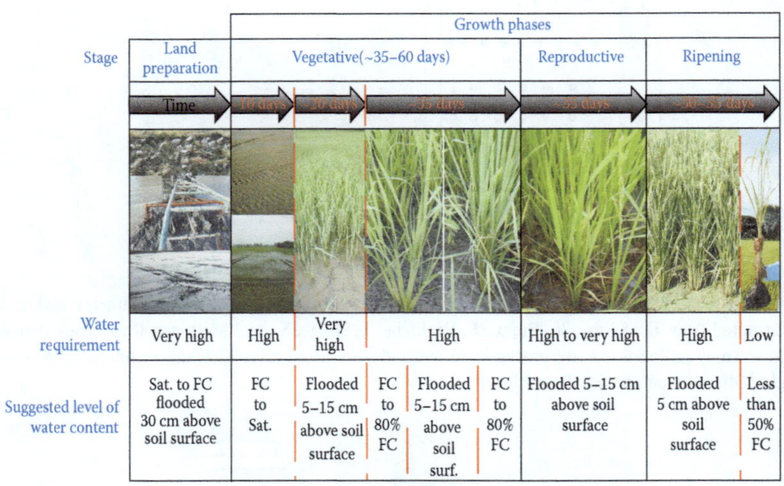

FIGURE 5.5 Examples of flooding and draining cycles used in lowland rice cultivation, with FC: soil water content at field capacity, and Sat.: soil water content at saturation point.

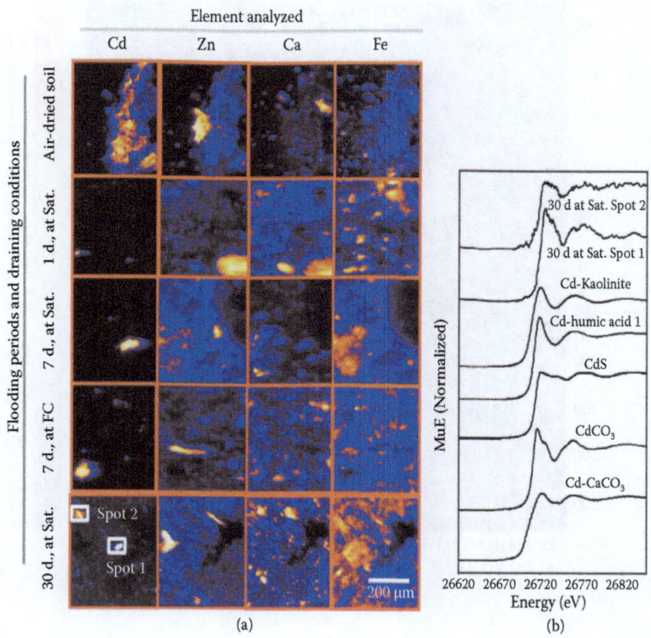

FIGURE 5.12 (a) Elemental distributions of cadmium (Cd), zinc (Zn), calcium (Ca), and iron (Fe) in an air-dried paddy soil flooded for different flooding periods and draining conditions by using μ-XRF mapping; (b) μ-XANES spectra, at the Cd K-edge, of $Cd-CdCO_3$, $CdCO_3$, CdS, Cd-humic acid 1, and Cd-kaolinite standards, and μ-XANES spectra taken at spots 1 and 2. (Reprinted with permission from Khaokaew, S. et al., *Environ. Sci. Technol.*, 45, 4249–4255, 2011. Copyright 2011 American Chemical Society.)

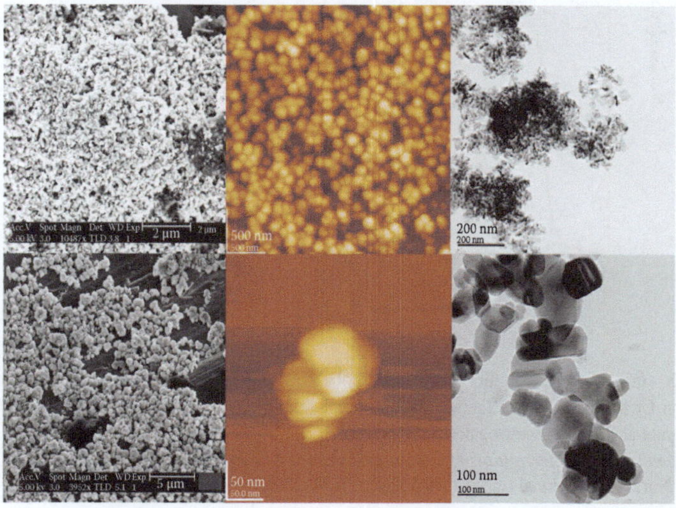

FIGURE 6.8 ZnO (1st row) and TiO_2 (second row) nanoparticles suspended in distilled water, allowed to dry, and imaged in order from left to right by SEM, AFM, and TEM, respectively. (Reprinted from Tiede, K. et al., *Food Addit. Contam.*, 25, 795–821, 2008. With permission from Taylor & Francis Ltd.)

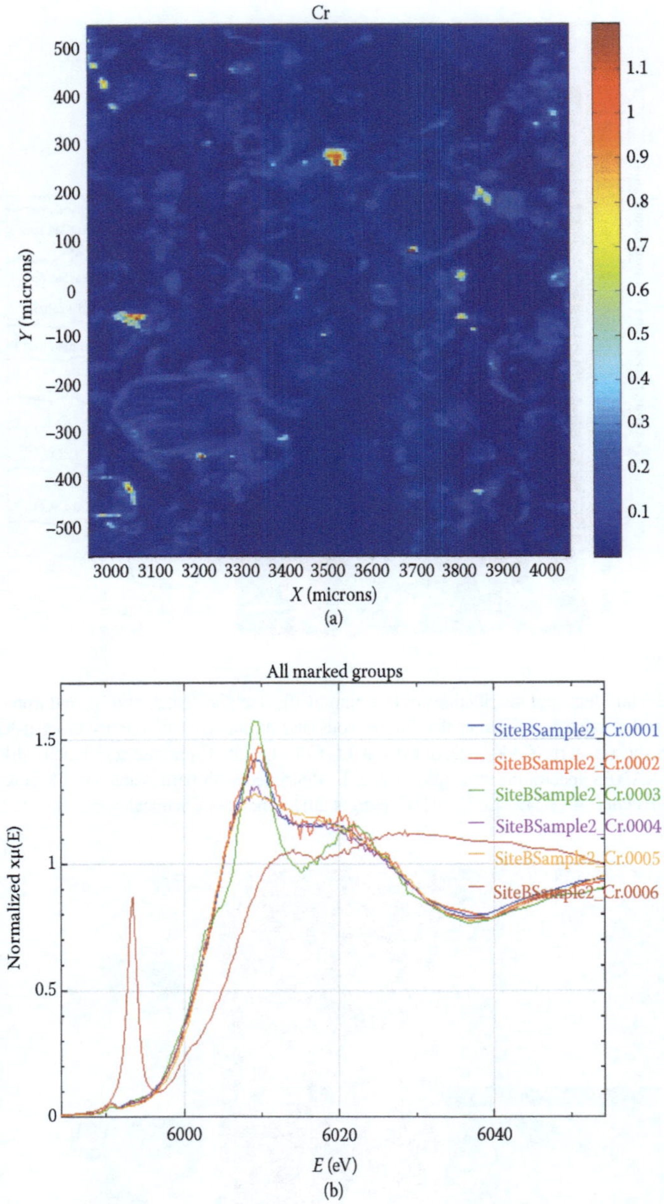

FIGURE 8.3 (a) and (b) Chromium x-ray microprobe mapping and XANES measurements taken in high concentration Cr spots located in the sediments matrix (Hanford sediment). XANES results show varying Cr(VI) contents at different interrogated sites (indicated by the pre-edge peak intensity). (From Dresel, P.E. et al., *Geochemical Characterization of Chromate Contamination in the 100 Area Vadose Zone at the Hanford Site*, Pacific Northwest National Laboratory, PNNL-17674, 2008.)

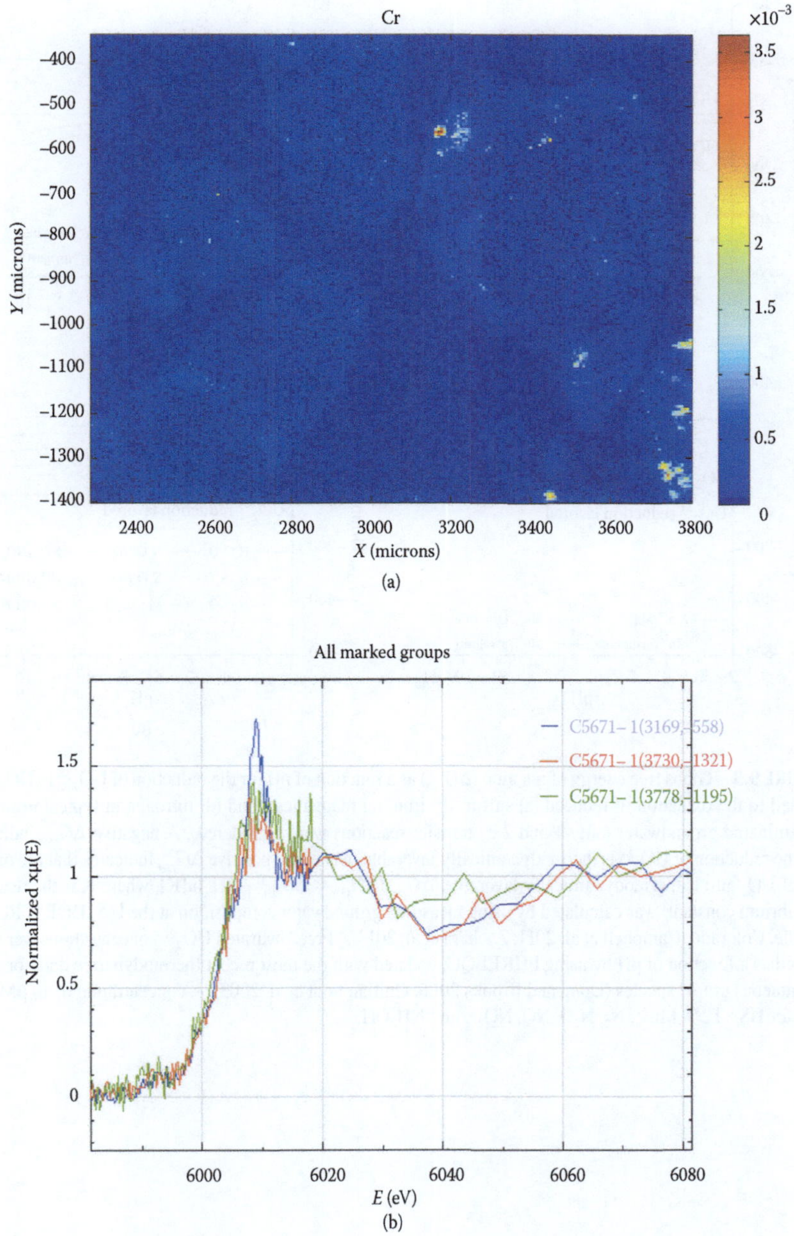

FIGURE 8.4 Chromium x-ray microprobe mapping and XANES measurements taken in high concentration Cr hot spots. The XANES analyses show the presence of insoluble Cr(III). (From Dresel, P.E. et al., *Geochemical Characterization of Chromate Contamination in the 100 Area Vadose Zone at the Hanford Site*, Pacific Northwest National Laboratory, PNNL-17674, 2008.)

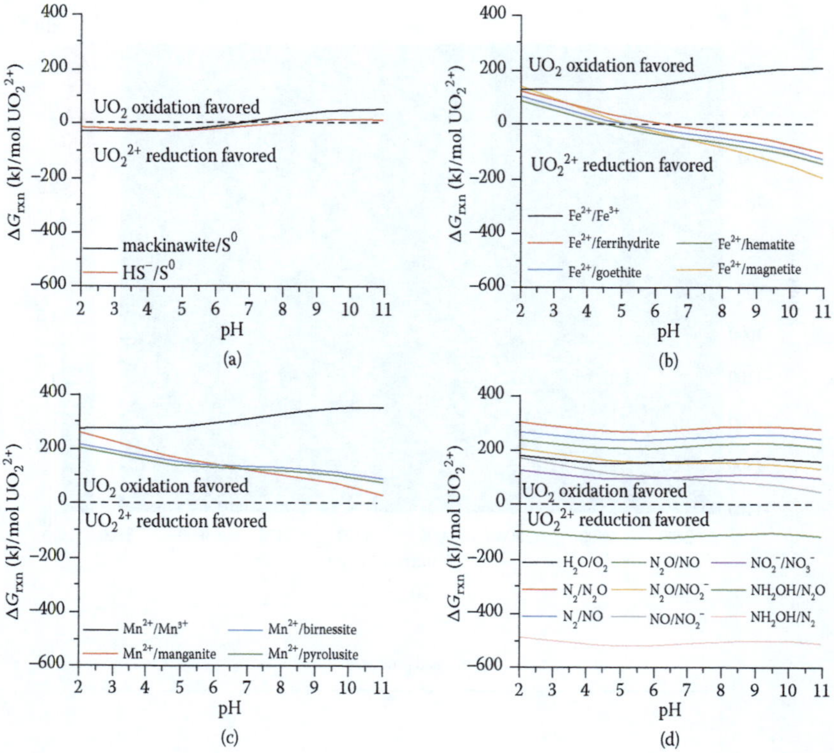

FIGURE 9.3 Gibbs free energy of reaction (ΔG_{rxn}) as a function of pH for the reduction of UO_2^{2+} to UO_2(am) coupled to the oxidation of reduced (a) sulfur, (b) iron, (c) manganese, and (d) nitrogen in typical uranium-contaminated groundwater (only 1 and 2 e^- transfer reactions were considered). A negative ΔG_{rxn} indicates that the reduction of UO_2^{2+} is thermodynamically favorable, whereas a positive ΔG_{rxn} indicates that the oxidation of UO_2(am) is thermodynamically favorable. ΔG_{rxn} ($\Delta G_{rxn} = \Delta G_{rxn}^0 + RT \ln[K]$, where K is the reaction equilibrium constant) was calculated by using measured groundwater composition at the U.S. DOE IFRC site in Rifle, Colorado (Campbell et al. 2011; Zachara et al. 2013). "Free" hydrated UO_2^{2+} concentrations were calculated as a function of pH by using PHREEQC; updated with the most recent thermodynamic data for solid and aqueous uranyl species (Dong and Brooks 2006; Guillaumont et al. 2003). A concentration of 10 μM was used for HS^-, Fe^{3+}, Mn^{3+}, N_2, N_2O, NO, NO_2^-, and NH_2OH.

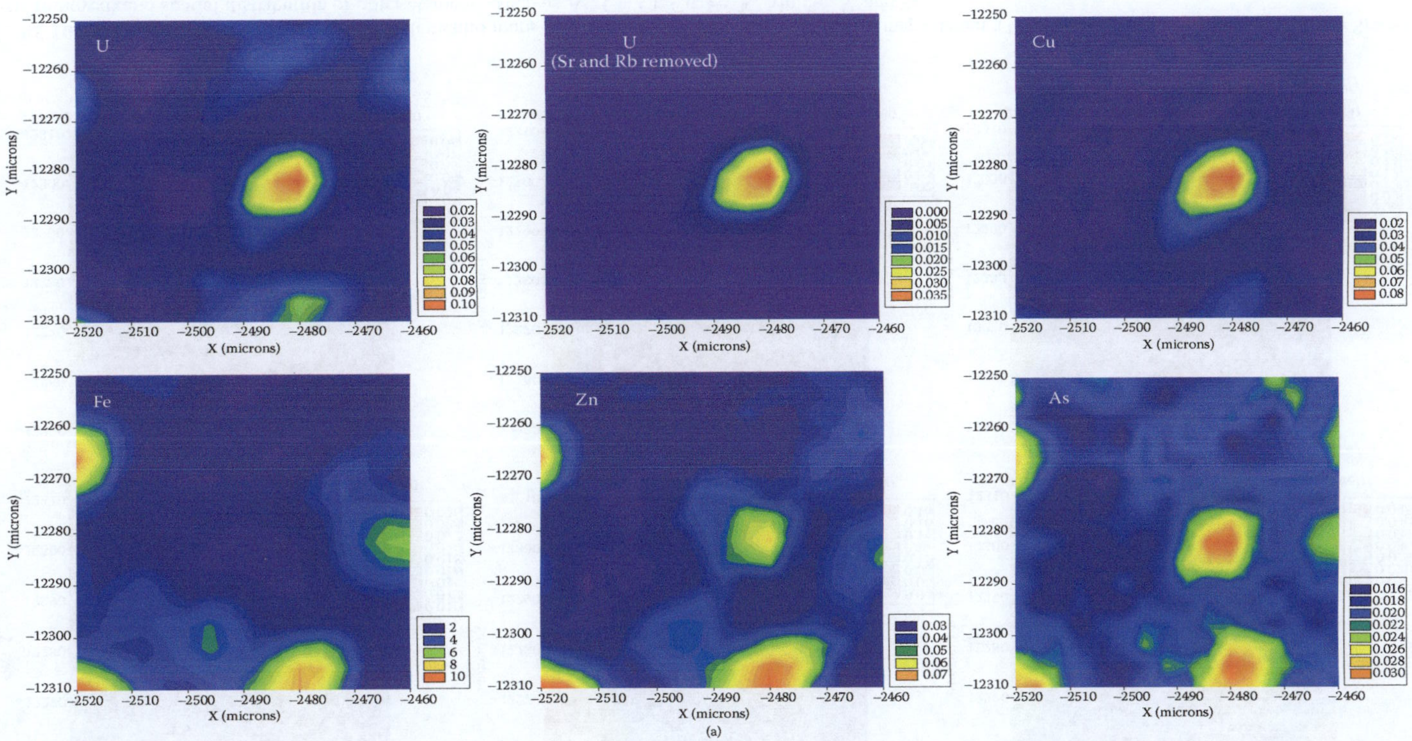

FIGURE 10.13 (a) and (b) μXRF elemental maps for sediment from CD-06 16 ft. These maps depict a U hotspot (shown without and with the Sr and Rb signals removed) and spatial distribution of other elements, such as As, Ca, Cr, Cu, Fe, K, Mn, Se, V, and Zn. *(Continued)*

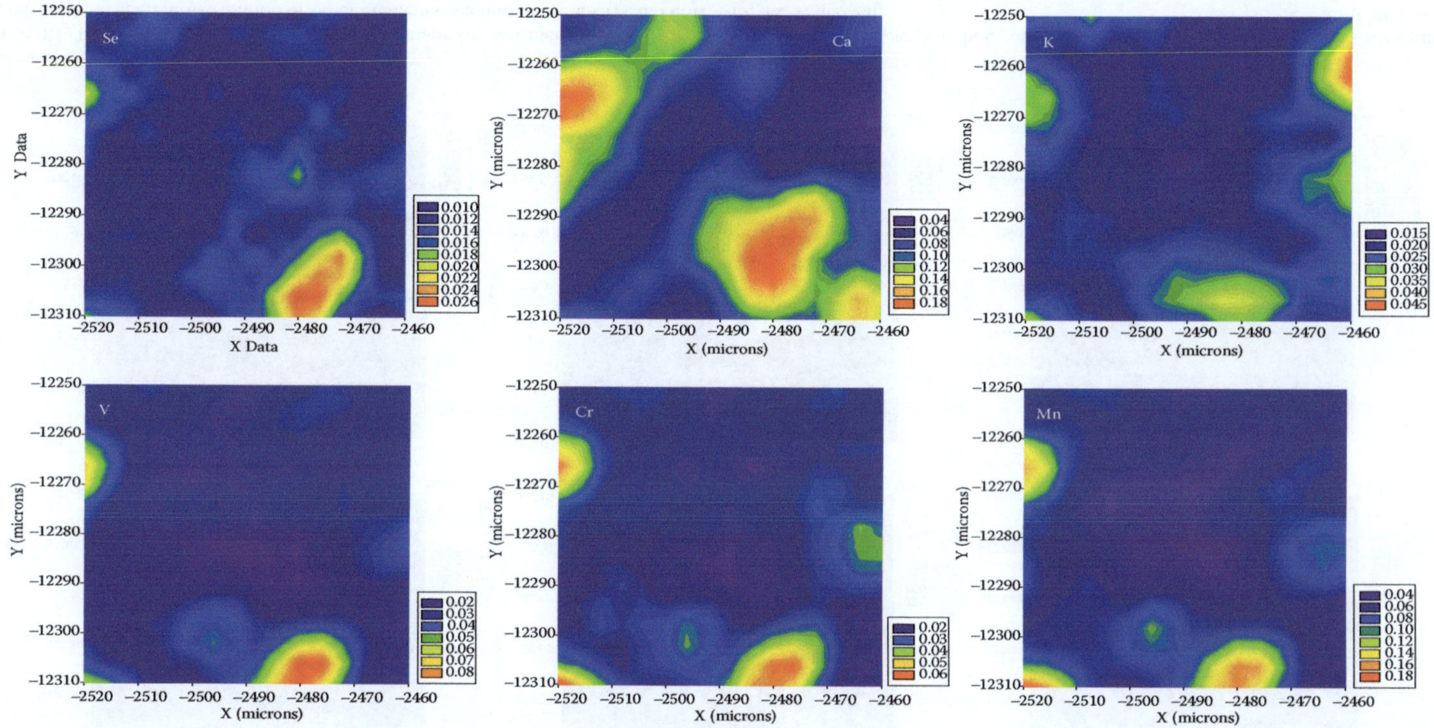

(b)

FIGURE 10.13 (*Continued*) (a) and (b) μXRF elemental maps for sediment from CD-06 16 ft. These maps depict a U hotspot (shown without and with the Sr and Rb signals removed) and spatial distribution of other elements, such as As, Ca, Cr, Cu, Fe, K, Mn, Se, V, and Zn.

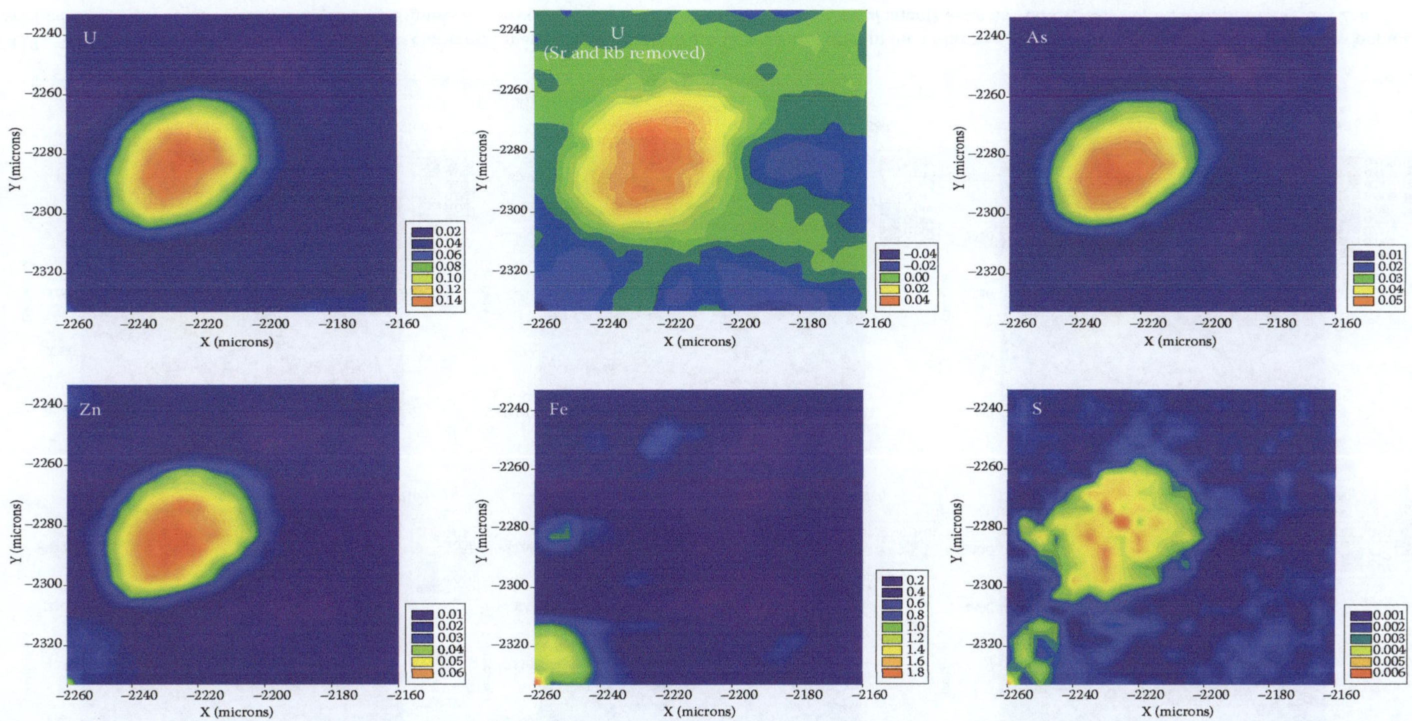

(a)

FIGURE 10.14 (a) and (b) μXRF elemental maps for sediment from CD-04 13 ft (from the column experiment spiked with U). These maps depict a U hotspot (shown without and with the Sr and Rb signals removed) and the spatial distribution of other elements such as As, Cr, Cu, Fe, K, Mn, S, Se, V, and Zn. (*Continued*)

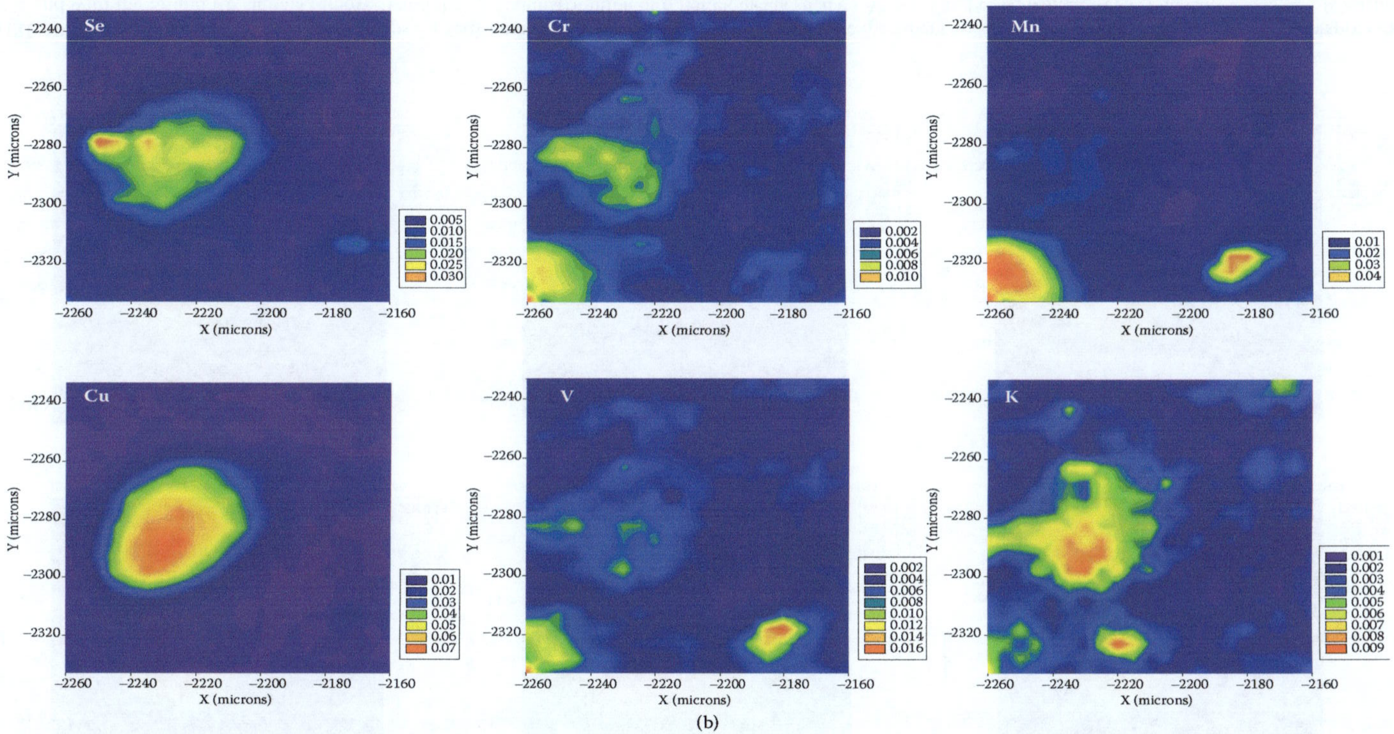

FIGURE 10.14 (*Continued*) (a) and (b) μXRF elemental maps for sediment from CD-04 13 ft (from the column experiment spiked with U). These maps depict a U hotspot (shown without and with the Sr and Rb signals removed) and the spatial distribution of other elements such as As, Cr, Cu, Fe, K, Mn, S, Se, V, and Zn.

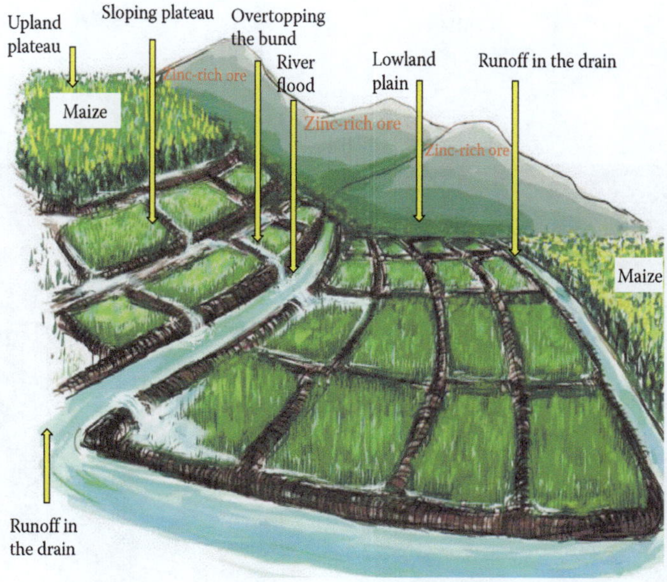

FIGURE 14.1 Schematic sketch of the location of a rice paddy field (land use and irrigation) in Mae Sot, Tak province, Thailand.

FIGURE 14.2 Rice farming in Mae Sot, Tak province, Thailand. (a) Rice field plow up with a driller operated manually. (b) Leveling of puddled rice field by hand (with a hoe). (c) Broadcasting rice seeds manually in the prepared field. (d) Germinated rice seedlings and tillers. (e) Rice and maize fields.

FIGURE 14.3 Rice paddy fields near Mae Sot, Tak province (a) and (b) water runoff through drain; (c) and (d) flooded rice fields.

FIGURE 16.2 (a-b) *Phoomdi*—floating island; (c-d) *Phoomdi* removal and dumping on the bank of Loktak.

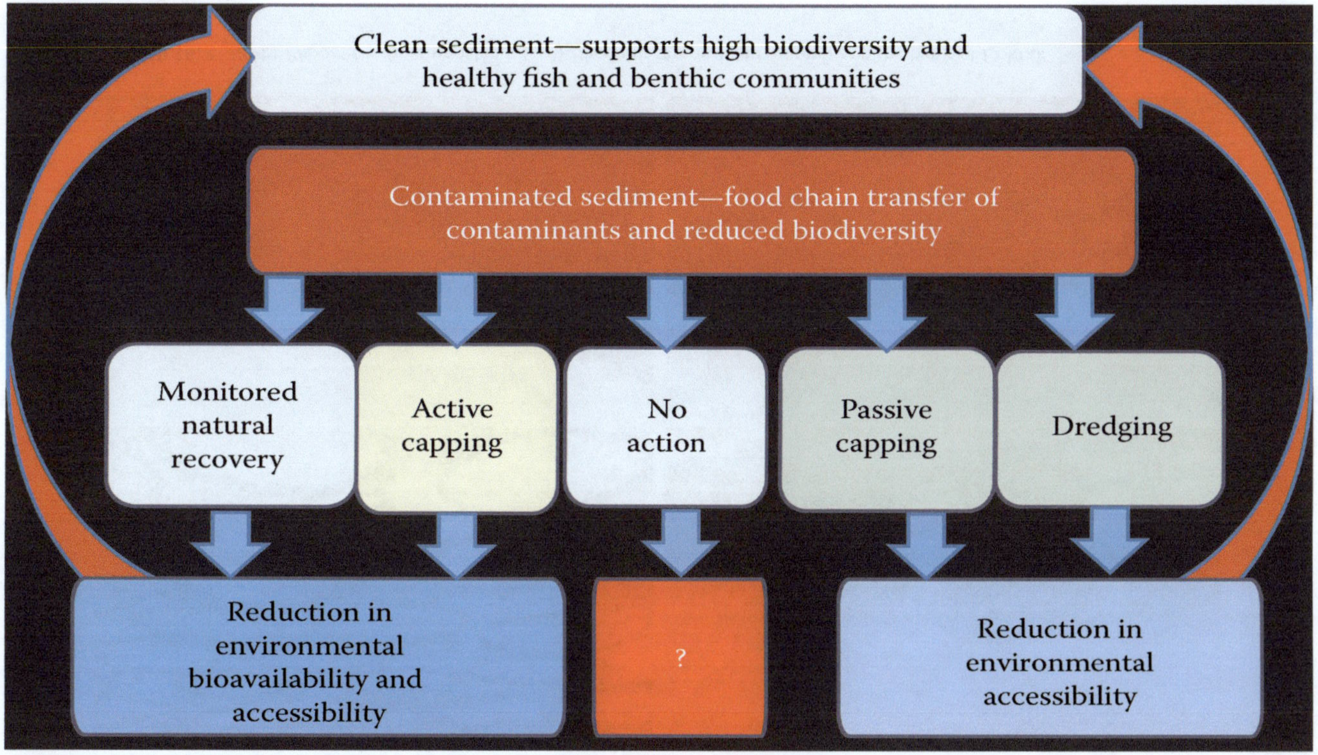

FIGURE 17.1 Current remedial options for contaminated sediments.

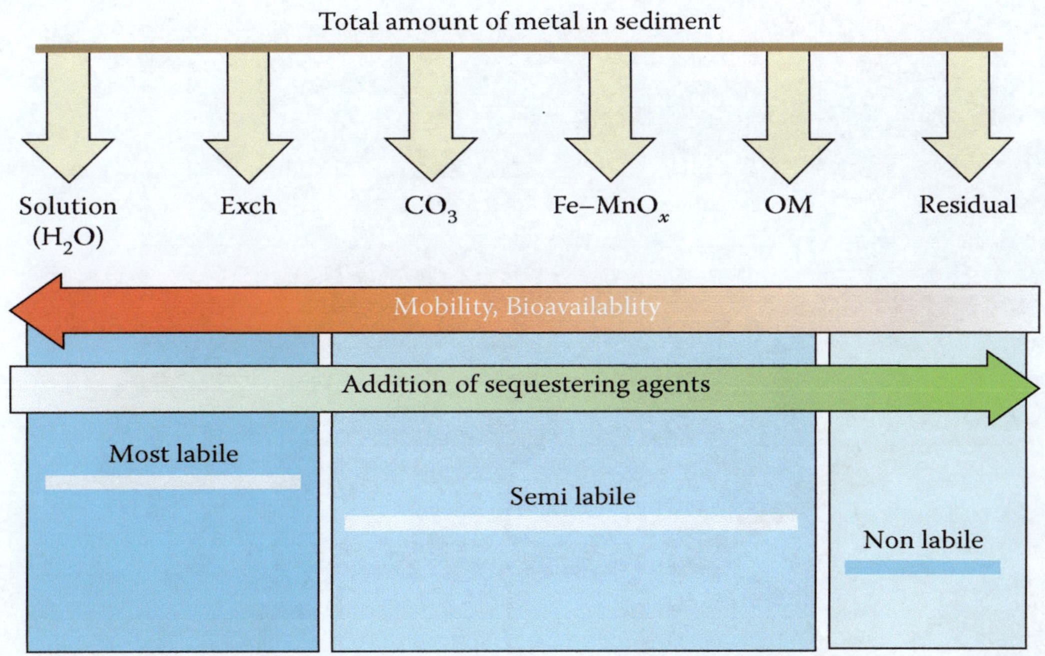

FIGURE 17.3 Reduction of metal mobility/bioavailability with addition of sequestering agents.

FIGURE 18.2 Deployment of active media and sand mix by clamshell bucket.

widespread in uranium-contaminated soils and sediments, and the natural availability of these substrates as plant exudates or cell lysis products in these environments suggest that the biomineralization of uranium phosphate may represent a significant immobilization process of uranium in its oxidized form, possibly as a detoxification mechanism for microorganisms. Simultaneously, in contrast to what was previously believed, new findings demonstrate that U(V) may be stabilized on mineral surfaces and incorporated into mineral structures whereas U(IV) may form mononuclear precipitates that are more labile than uraninite or even remain in solution in the presence of strong ligands or organic colloids. These new findings not only suggest that one-electron transfer reactions with uranium may be more prevalent than previously recognized, but they also suggest that future studies should investigate the exact geochemical conditions resulting in the formation of soluble, colloidal, and particulate U(IV) species.

The aqueous speciation of U(VI) also affects the redox transformations of uranium. Although the reactivity of U(VI) is mostly predictable based on thermodynamic calculations that account for the complexation of U(VI) in solution, the chemical mechanisms that determine the reactivity of uranium remain speculative. In addition, equilibrium is almost never achieved in aquatic systems, and kinetic effects have to be considered. Recent developments in analytical capabilities indicate that the speciation of U(VI) strongly affects its reduction by MRB, and that the bioavailable fraction of uranium, in addition to representing the smallest fraction of U(VI) in solution, is also toxic in high concentrations. These findings explain the effect of carbonates and major divalent cations that are present in natural waters on the rates of U(VI) reduction by MRB, and they indicate the importance of rate laws that account for the changes in aqueous uranium speciation.

Finally, the knowledge gained on the biogeochemical transformations of uranium in aquatic systems has been used to develop alternative *in situ* remediation strategies. Although immobilization of uranium has been demonstrated by adsorption and precipitation onto a variety of solid minerals, recent research has focused on developing bioremediation strategies to immobilize uranium in the solid phase. Interestingly, most bioremediation efforts have investigated the possible use of uranium bioreduction, despite the fact that this process is only efficient in a narrow pH range, it is outcompeted by other respiration reactions, and the U(IV) byproduct may be destabilized by aerobic and anaerobic oxidants. In turn, few efforts have focused on the biomineralization of uranium phosphate minerals as an alternative bioremediation strategy, despite the fact that a variety of microorganisms are able to hydrolyze organophosphate or polyphosphate compounds, that sources of these compounds exist in natural environments, and that this process is capable of immobilizing U(VI) phosphate minerals in both aerobic and anaerobic conditions over a wide pH range. Although the destabilization of uranium phosphate minerals by carbonates may restrain usage of this process in circumneutral pH environments, uranium-contaminated environments are usually characterized by low pH conditions, as acids are often used to extract uranium from ore and are found along with uranium in waste repositories. This process may thus represent an appealing and complementary remediation strategy to uranium bioreduction that is more efficient at circumneutral pH and in anaerobic environments.

REFERENCES

Alessi, D. S., B. Uster, H. Veeramani, E. I. Suvorova, J. S. Lezama-Pacheco, J. E. Stubbs, J. R. Bargar, and R. Bernier-Latmani. 2012. Quantitative separation of monomeric U(IV) from UO2 in products of U(VI) reduction. *Environmental Science and Technology* 46 (11):6150–6157.

Allen, P. G., J. J. Bucher, D. L. Clark, N. M. Edelstein, S. A. Ekberg, J. W. Gohdes, E. A. Hudson, N. Kaltsoyannis, W. W. Lukens, M. P. Neu, P. D. Palmer, T. Reich, D. K. Shuh, C. D. Tait, and B. D. Zwick. 1995. Multinuclear NMR, Raman, EXAFS, and x-ray diffraction studies of uranyl carbonate complexes in near-neutral aqueous solution. X-ray structure of [C(NH2)3]6[(UO2)3(CO3)6]-6.5H2O. *Inorganic Chemistry* 34 (19):4797–4807.

Amayri, S., T. Reich, T. Arnold, G. Geipel, and G. Bernhard. 2005. Spectroscopic characterization of alkaline earth uranyl carbonates. *Journal of Solid State Chemistry* 178 (2):567–577.

Anderson, R. T., H. A. Vrionis, I. Ortiz-Bernad, C. T. Resch, P. E. Long, R. Dayvault, K. Karp, S. Marutzky, D. R. Metzler, A. Peacock, D. C. White, M. Lowe, and D. R. Lovley. 2003. Stimulating the in situ activity of Geobacter species to remove uranium from the groundwater of a uranium-contaminated aquifer. *Applied and Environmental Microbiology* 69 (10):5884–5891.

Arai, Y., M. K. Marcus, N. Tamura, J. A. Davis, and J. M. Zachara. 2007. Spectroscopic evidence for uranium bearing precipitates in vadose zone sediments at the Hanford 300-area site. *Environmental Science and Technology* 41 (13):4633–4639.

Arey, J. S., J. C. Seaman, and P. M. Bertsch. 1999. Immobilization of uranium in contaminated sediments by hydroxyapatite addition. *Environmental Science and Technology* 33 (2):337–342.

Bargar, J. R., R. Reitmeyer, J. J. Lenhart, and J. A. Davis. 2000. Characterization of U(VI)-carbonato ternary complexes on hematite: EXAFS and electrophoretic mobility measurements. *Geochimica Et Cosmochimica Acta* 64 (16):2737–2749.

Bargar, J. R., K. H. Williams, K. M. Campbell, P. E. Long, J. E. Stubbs, E. I. Suvorova, J. S. Lezama-Pacheco, D. S. Alessi, M. Stylo, S. M. Webb, J. A. Davis, D. E. Giammar, L. Y. Blue, and R. Bernier-Latmani. 2013. Uranium redox transition pathways in acetate-amended sediments. *Proceedings of the National Academy of Sciences of the United States of America* 110 (12):4506–4511.

Barlett, M., H. S. Moon, A. A. Peacock, D. B. Hedrick, K. H. Williams, P. E. Long, D. Lovley, and P. R. Jaffe. 2012. Uranium reduction and microbial community development in response to stimulation with different electron donors. *Biodegradation* 23 (4):535–546.

Beazley, M. J., R. J. Martinez, P. A. Sobecky, S. M. Webb, and M. Taillefert. 2007. Uranium biomineralization as a result of bacterial phosphatase activity: Insights from bacterial isolates from a contaminated subsurface. *Environmental Science and Technology* 41 (16):5701–5707.

Beazley, M. J., R. J. Martinez, P. A. Sobecky, S. M. Webb, and M. Taillefert. 2009. Nonreductive biomineralization of uranium(VI) phosphate via microbial phosphatase activity in anaerobic conditions. *Geomicrobiology Journal* 26 (7):431–441.

Beazley, M. J., R. J. Martinez, S. M. Webb, P. A. Sobecky, and M. Taillefert. 2011. The effect of pH and natural microbial phosphatase activity on the speciation of uranium in subsurface soils. *Geochimica Et Cosmochimica Acta* 75 (19):5648–5663.

Behrends, T., and P. Van Cappellen. 2005. Competition between enzymatic and abiotic reduction of uranium(VI) under iron reducing conditions. *Chemical Geology* 220 (3–4):315–327.

Beller, H. R. 2005. Anaerobic, nitrate-dependent oxidation of U(IV) oxide minerals by the chemolithoautotrophic bacterium Thiobacillus denitrificans. *Applied and Environmental Microbiology* 71 (4):2170–2174.

Beller, H. R., T. C. Legler, F. Bourguet, T. E. Letain, S. R. Kane, and M. A. Coleman. 2009. Identification of c-type cytochromes involved in anaerobic, bacterial U(IV) oxidation. *Biodegradation* 20 (1):45–53.

Belli, K. M., T. J. DiChristina, P. Van Cappellen, and M. Taillefert. 2015. Speciation of uranium controls its kinetics of reduction by metal-reducing bacteria. *Geochimica Et Cosmochimica Acta* 157:109–124.

Belli, K. M., L. Pastor, N. Szeinbaum, P. Van Cappellen, T. J. DiChristina, and M. Taillefert. 2011. Influence of uranyl speciation on uranium reduction by metal-reducing bacteria. *American Chemical Society National Meeting* 242:1.

Bencheikh-Latmani, R., and J. O. Leckie. 2003. Association of uranyl with the cell wall of Pseudomonas fluorescens inhibits metabolism. *Geochimica Et Cosmochimica Acta* 67 (21):4057–4066.

Bernhard, G., G. Geipel, T. Reich, V. Brendler, S. Amayri, and H. Nitsche. 2001. Uranyl(VI) carbonate complex formation: Validation of the Ca2UO2(CO3)(3)(aq.) species. *Radiochimica Acta* 89 (8):511–518.

Bernier-Latmani, R., H. Veeramani, E. D. Vecchia, P. Junier, J. S. Lezama-Pacheco, E. I. Suvorova, J. O. Sharp, N. S. Wigginton, and J. R. Bargar. 2010. Non-uraninite products of microbial U(VI) reduction. *Environmental Science and Technology* 44 (24):9456–9462.

Bi, Y. Q., and K. F. Hayes. 2014. Nano-FeS inhibits UO2 reoxidation under varied oxic conditions. *Environmental Science and Technology* 48 (1):632–640.

Bi, Y. Q., S. P. Hyun, R. K. Kukkadapu, and K. F. Hayes. 2013. Oxidative dissolution of UO2 in a simulated groundwater containing synthetic nanocrystalline mackinawite. *Geochimica Et Cosmochimica Acta* 102:175–190.

Biermann, V., F. G. Simon, M. Csovari, J. Csicsak, G. Folding, and G. Simoncsics. 2006. Long-term performance of reactive materials in PRBs for uranium remediation. In: B. J. Merkel and A. HascheBerger (eds.), *Uranium in the Environment: Mining Impact and Consequences*. Berlin, Germany: Springer-Verlag.

Blanchard, R. L., G. G. Eadie, J. M. Hans, and R. F. Kaufmann. 1983. In: U.S. Environmental Protection Agency (eds.), *Potential Health and Environmental Hazards of Uranium Mine Wastes*. U.S. Environmental Protection Agency, Office of Radiation Programs Washington, D.C.

Boyanov, M. I., K. E. Fletcher, M. J. Kwon, X. Rui, E. J. O'Loughlin, F. E. Loffler, and K. M. Kemner. 2011. Solution and microbial controls on the formation of reduced U(IV) species. *Environmental Science and Technology* 45 (19):8336–8344.

Brennecka, G. A., L. E. Wasylenki, J. R. Bargar, S. Weyer, and A. D. Anbar. 2011. Uranium isotope fractionation during adsorption to Mn-oxyhydroxides. *Environmental Science and Technology* 45 (4):1370–1375.

Brooks, S. C., J. K. Fredrickson, S. L. Carroll, D. W. Kennedy, J. M. Zachara, A. E. Plymale, S. D. Kelly, K. M. Kemner, and S. Fendorf. 2003. Inhihition of bacterial U(VI) reduction by calcium. *Environmental Science and Technology* 37 (9):1850–1858.

Brown, M. R. W., and A. Kornberg. 2009. The long and short of it—Polyphosphate, PPK, and bacterial survival. *Trends in Biochemical Science* 33 (6):284–290.

Bruno, J., J. De Pablo, L. Duro, and E. Figuerola. 1995. Experimental study and modeling of U(VI)-Fe(OH)3 surface precipitation/coprecipitation equilibria. *Geochimica et Cosmochimica Acta* 59 (20):4113–4123.

Burgos, W. D., J. T. McDonough, J. M. Senko, G. X. Zhang, A. C. Dohnalkova, S. D. Kelly, Y. Gorby, and K. M. Kemner. 2008. Characterization of uraninite nanoparticles produced by Shewanella oneidensis MR-1. *Geochimica Et Cosmochimica Acta* 72 (20):4901–4915.

Burgos, W. D., J. M. Senko, B. A. Dempsey, E. E. Roden, J. J. Stone, K. M. Kemner, and S. D. Kelly. 2007. Soil humic acid decreases biological uranium(VI) reduction by *Shewanella putrefaciens* CN32. *Environmental Engineering Science* 24 (6):755–761.

Burns, P. C., and R. J. Finch. 1999. Wyartite: Crystallographic evidence for the first pentavalent-uranium mineral. *American Mineralogist* 84 (9):1456–1460.

Butterfield, C. N., A. V. Soldatova, S. W. Lee, T. G. Spiro, and B. M. Tebo. 2013. Mn(II, III) oxidation and MnO2 mineralization by an expressed bacterial multicopper oxidase. *Proceedings of the National Academy of Sciences of the United States of America* 110 (29):11731–11735.

Cai, C. F., H. L. Dong, H. T. Li, X. J. Xiao, G. X. Ou, and C. M. Zhang. 2007. Mineralogical and geochemical evidence for coupled bacterial uranium mineralization and hydrocarbon oxidation in the Shashagetai deposit, NW China. *Chemical Geology* 236 (1–2):167–179.

Campbell, P. G. C. 1995. *In Metal Speciation and Bioavailability in Aquatic Systems*. 1st ed. New York: John Wiley.

Campbell, K. M., R. K. Kukkadapu, N. P. Qafoku, A. D. Peacock, E. Lesher, K. H. Williams, J. R. Bargar, M. J. Wilkins, L. Figueroa, J. Ranville, J. A. Davis, and P. E. Long. 2012. Geochemical, mineralogical and microbiological characteristics of sediment from a naturally reduced zone in a uranium-contaminated aquifer. *Applied Geochemistry* 27 (8):1499–1511.

Campbell, K. M., H. Veeramani, K. U. Urich, L. Y. Blue, D. E. Giammar, R. Bernier-Latmani, J. E. Stubbs, E. Suvorova, S. Yabusaki, J. S. Lezama-Pacheco, A. Mehta, P. E. Long, and J. R. Bargar. 2011. Oxidative dissolution of biogenic uraninite in groundwater at Old Rifle, CO. *Environmental Science and Technology* 45 (20):8748–8754.

Cantrell, K. J., D. I. Kaplan, and T. W. Wiestma. 1995. Zero-valent iron for the in situ remediation of selected metals in groundwater. *Journal of Hazardous Materials* 42:201–212.

Cardenas, E., W. M. Wu, M. B. Leigh, J. Carley, S. Carroll, T. Gentry, J. Luo, D. Watson, B. Gu, M. Ginder-Vogel, P. K. Kitanidis, P. M. Jardine, J. Zhou, C. S. Criddle, T. L. Marsh, and J. A. Tiedje. 2008. Microbial communities in contaminated sediments, associated with bioremediation of uranium to submicromolar levels. *Applied and Environmental Microbiology* 74 (12):3718–3729.

Carvajal, D. A., Y. P. Katsenovich, and L. E. Lagos. 2012. The effects of aqueous bicarbonate and calcium ions on uranium biosorption by Arthrobacter G975 strain. *Chemical Geology* 330:51–59.

Catalano, J. G., and G. E. Brown Jr. 2005. Uranyl adsorption onto montmorillonite: Evaluation of binding sites and carbonate complexation. *Geochimica et Cosmochimica Acta* 69 (12):2995–3005.

Cerrato, J. M., M. N. Ashner, D. S. Alessi, J. S. Lezama-Pacheco, R. Bernier-Latmani, J. R. Bargar, and D. E. Giammar. 2013. Relative reactivity of biogenic and chemogenic uraninite and biogenic noncrystalline U(IV). *Environmental Science and Technology* 47 (17):9756–9763.

Cerrato, J. M., C. J. Barrows, L. Y. Blue, J. S. Lezama-Pacheco, J. R. Bargar, and D. E. Giammar. 2012. Effect of Ca2+ and Zn2+ on UO2 dissolution rates. *Environmental Science and Technology* 46 (5):2731–2737.

Chakraborty, S., F. Favre, D. Banerjee, A. C. Scheinost, M. Mullet, J. J. Ehrhardt, J. Brendle, L. Vidal, and L. Charlet. 2010. U(VI) sorption and reduction by Fe(II) sorbed on montmorillonite. *Environmental Science and Technology* 44 (10):3779–3785.

Cheng, C. W., and B. L. Lim. 2006. Beta-propeller phytases in the aquatic environment. *Archives of Microbiology* 185 (1):1–13.

Cheng, T., M. O. Barnett, E. E. Roden, and J. L. Zhuang. 2004. Effects of phosphate on uranium(VI) adsorption to goethite-coated sand. *Environmental Science and Technology* 38 (22):6059–6065.

Chinni, S., C. R. Anderson, K. U. Ulrich, D. E. Giammar, and B. M. Tebo. 2008. Indirect UO(2) oxidation by Mn(II)-oxidizing spores of Bacillus sp strain SG-1 and the effect of U and Mn concentrations. *Environmental Science and Technology* 42 (23):8709–8714.

Chisholm-Brause, C. J., J. M. Berg, R. A. Matzner, and D. E. Morris. 2001. Uranium(VI) sorption complexes on montmorillonite as a function of solution chemistry. *Journal of Colloid and Interface Science* 233:38–49.

Choudhary, S., and P. Sar. 2011. Uranium biomineralization by a metal resistant Pseudomonas aeruginosa strain isolated from contaminated mine waste. *Journal of Hazardous Materials* 186 (1):336–343.

Ciavatta, L., D. Ferri, I. Grenthe, F. Salvatore, and K. Spahiu. 1983. Studies on metal carbonate equilibria. 4. Reduction of the tris(carbonato)dioxouranate(VI) ion, UO2(CO3)34-, in hydrogen carbonate solutions. *Inorganic Chemistry* 22 (14):2088–2092.

Clark, D. L., D. E. Hobart, and M. P. Neu. 1995. Actinide carbonate complexes and their importance in actinide environmental chemistry. *Chemical Reviews* 95 (1):25–48.

Collins, R. N., T. Saito, N. Aoyagi, T. E. Payne, T. Kimura, and T. D. Waite. 2011. Applications of time-resolved laser fluorescence spectroscopy to the environmental biogeochemistry of actinides. *Journal of Environmental Quality* 40 (3):731–741.

Cologgi, D. L., S. Lampa-Pastirk, A. M. Speers, S. D. Kelly, and G. Reguera. 2011. Extracellular reduction of uranium via Geobacter conductive pili as a protective cellular mechanism. *Proceedings of the National Academy of Sciences of the United States of America* 108 (37):15248–15252.

Conca, J. L., and J. Wright. 2006. An apatite II permeable reactive barrier to remediate groundwater containing Zn, Pb and Cd (vol 21, pg 1288, 2006). *Applied Geochemistry* 21 (12):2187–2200.

Condomines, M., P. J. Gauthier, and G. Sigmarsson. 2003. Timescales of magma chamber processes and dating of young volcanic rocks. *Uranium-Series Geochemistry* 52:125–174.

Costello, A. J. R., T. Glonek, and T. C. Myers. 1976. P-31 nuclear magnetic resonance-pH titrations of myo-inositol hexaphosphate. *Carbohydrate Research* 46 (2):159–171.

Cotton, F. A., G. Wilkinson, C. A. Murillo, and M. Bochmann. 1999. *Advanced Inorganic Chemistry*. New York: John Wiley *and* Sons, Inc.

Crane, R. A., M. Dickinson, I. C. Popescu, and T. B. Scott. 2011. Magnetite and zero-valent iron nanoparticles for the remediation of uranium contaminated environmental water. *Water Research* 45 (9):2931–2942.

Dawson, G. W., and J. Gilman. 2001. Land reclamation technology—Expanding the geotechnical engineering envelope. *Proceedings of the Institution of Civil Engineers-Geotechnical Engineering* 149 (1):49–61.

Degroot, C. J., and H. L. Golterman. 1993. On the presence of organic phosphate in some Camargue sediments: Evidence for the importance of phytate. *Hydrobiologia* 252 (1):117–126.

Di Toro, D. M., H. E. Allen, H. L. Bergman, J. S. Meyer, P. R. Paquin, and R. C. Santore. 2001. Biotic ligand model of the acute toxicity of metals. 1. Technical basis. *Environmental Toxicology and Chemistry* 20 (10):2383–2396.

Djogic, R., and M. Branica. 1995a. Direct determination of dissolved uranyl(VI) in sea-water by cathodic stripping voltammetry. *The Analyst* 120 (7):1989.

Djogic, R., and M. Branica. 1995b. Square-wave cathodic stripping voltammetry of hydrolyzed uranyl species. *Analytica Chimica Acta* 305:159–164.

Djogic, R., I. Pizeta, and M. Branica. 2001. Electrochemical determination of dissolved uranium in Krka river estuary. *Water Research* 35 (8):1915–1920.

Djogic, R., L. Sipos, and M. Branica. 1986. Characterization of uranium(VI) in seawater. *Limnology and Oceanography* 31 (5):1122–1131.

Docrat, T. I., J. F. W. Mosselmans, J. M. Charnock, M. W. Whiteley, D. Collison, F. R. Livens, C. Jones, and M. J. Edmiston. 1999. X-ray absorption spectroscopy of tricarbonatodioxouranate(V), UO2(CO3)(3) (5-), in aqueous solution. *Inorganic Chemistry* 38 (8):1879–1882.

Dong, W., and S. C. Brooks. 2006. Determination of the formation constants of ternary complexes of uranyl and carbonate with alkaline earth metals (Mg2+, Ca2+, Sr2+, and Ba2+) using anion exchange method. *Environmental Science and Technology* 40 (15):4689–4695.

Drot, R., and E. Simoni. 1999. Uranium(VI) and Europium(III) speciation at the phosphate compounds—Solution interface. *Langmuir* 15:4820–4827.

Du, X., B. Boonchayaanant, W. M. Wu, S. Fendorf, J. Bargar, and C. S. Criddle. 2011. Reduction of uranium(VI) by soluble iron(II) conforms with thermodynamic predictions. *Environmental Science and Technology* 45 (11):4718–4725.

Duff, M. C., J. U. Coughlin, and D. B. Hunter. 2002. Uranium co-precipitation with iron oxide minerals. *Geochimica Et Cosmochimica Acta* 66 (20):3533–3547.

Dunham-Cheatham, S., X. Rui, B. Bunker, N. Menguy, R. Hellmann, and J. Fein. 2011. The effects of non-metabolizing bacterial cells on the precipitation of U, Pb and Ca phosphates. *Geochimica Et Cosmochimica Acta* 75 (10):2828–2847.

Duro, L., S. El Aamrani, M. Rovira, J. de Pablo, and J. Bruno. 2008. Study of the interaction between U(VI) and the anoxic corrosion products of carbon steel. *Applied Geochemistry* 23 (5):1094–1100.

Edwards, R. L., C. D. Gallup, and H. Cheng. 2003. Uranium-series dating of marine and lacustrine carbonates. *Uranium-Series Geochemistry* 52:363–405.

Ekstrom, A. 1974. Kinetics and mechanism of disproportionation of uranium(V). *Inorganic Chemistry* 13 (9):2237–2241.

Fein, J. B., C. J. Daughney, N. Yee, and T. A. Davis. 1997. A chemical equilibrium model for metal adsorption onto bacterial surfaces. *Geochimica Et Cosmochimica Acta* 61 (16):3319–3328.

Finch, R., and T. Murakami. 1999. Systematics and paragenesis of uranium minerals. *Reviews in Mineralogy* 38:91–179.

Finneran, K. T., R. T. Anderson, K. P. Nevin, and D. R. Lovley. 2002a. Potential for bioremediation of uranium-contaminated aquifers with microbial U(VI) reduction. *Soil and Sediment Contamination* 11 (3):339–357.

Finneran, K. T., M. E. Housewright, and D. R. Lovley. 2002b. Multiple influences of nitrate on uranium solubility during bioremediation of uranium-contaminated subsurface sediments. *Environmental Microbiology* 4 (9):510–516.

Fletcher, K. E., M. I. Boyanov, S. H. Thomas, Q. Z. Wu, K. M. Kemner, and F. E. Loffler. 2010. U(VI) reduction to mononuclear U(IV) by desulfitobacterium species. *Environmental Science and Technology* 44 (12):4705–4709.

Foerstendorf, H., K. Heim, and A. Rossberg. 2012. The complexation of uranium(VI) and atmospherically derived CO2 at the ferrihydrite-water interface probed by time-resolved vibrational spectroscopy. *Journal of Colloid and Interface Science* 377:299–306.

Fortin, C., F. H. Denison, and J. Garnier-Laplace. 2007. Metal-phytoplankton interactions: Modeling the effect of competing ions (H+, Ca2+, and Mg2+) on uranium uptake. *Environmental Toxicology and Chemistry* 26 (2):242–248.

Fortin, C., L. Dutel, and J. Garnier-Laplace. 2004. Uranium complexation and uptake by a green alga in relation to chemical speciation: The importance of the free uranyl ion. *Environmental Toxicology and Chemistry* 23 (4):974–981.

Fowle, D. A., Jeremy B. Fein, and A. M. Martin. 2000. Experimental study of uranyl adsorption onto Bacillus subtilis. *Environmental Science and Technology* 34:3737–3741.

Fox, P. M., J. A. Davis, M. B. Hay, M. E. Conrad, K. M. Campbell, K. H. Williams, and P. E. Long. 2012. Rate-limited U(VI) desorption during a small-scale tracer test in a heterogeneous uranium-contaminated aquifer. *Water Resources Research* 48:W05512.

Francis, A. J., and C. J. Dodge. 2008. Bioreduction of uranium(VI) complexed with citric acid by Clostridia affects its structure and solubility. *Environmental Science and Technology* 42 (22):8277–8282.

Francis, C. W., M. E. Timpson, S. Y. Lee, M. P. Elless, and J. H. Wilson. 1998. The use of carbonate lixiviants to remove uranium from uranium-contaminated soils. *Journal of Radioanalytical and Nuclear Chemistry* 228 (1–2):15–20.

Francis, C. W., M. E. Timpson, and J. H. Wilson. 1999. Bench- and pilot-scale studies relating to the removal of uranium from uranium-contaminated soils using carbonate and citrate lixiviants. *Journal of Hazardous Materials* 66 (1–2):67–87.

Fredrickson, J. K., J. M. Zachara, D. W. Kennedy, C. X. Liu, M. C. Duff, D. B. Hunter, and A. Dohnalkova. 2002. Influence of Mn oxides on the reduction of uranium(VI) by the metal-reducing bacterium *Shewanella putrefaciens*. *Geochimica Et Cosmochimica Acta* 66 (18):3247–3262.

Fuller, C. C., J. R. Bargar, and J. A. Davis. 2003. Molecular-scale characterization of uranium sorption by bone apatite materials for a permeable reactive barrier demonstration. *Environmental Science and Technology* 37:4642–4649.

Fuller, C. C., J. R. Bargar, J. A. Davis, and M. J. Piana. 2002. Mechanisms of uranium interactions with hydroxyapatite: Implications for groundwater remediation. *Environmental Science and Technology* 36 (2):158–165.

Gabriel, U., J. P. Gaudet, L. Spadini, and L. Charlet. 1998. Reactive transport of uranyl in a goethite column: An experimental and modelling study. *Chemical Geology* 151 (1–4):107–128.

Galindo, C., M. Del Nero, R. Barillon, E. Halter, and B. Made. 2010. Mechanisms of uranyl and phosphate (co) sorption: Complexation and precipitation at a-Al2O3 surfaces. *Journal of Colloid and Interface Science* 347:282–289.

Ganesh, R., K. G. Robinson, L. L. Chu, D. Kucsmas, and G. D. Reed. 1999. Reductive precipitation of uranium by Desulfovibrio desulfuricans: Evaluation of cocontaminant effects and selective removal. *Water Research* 33 (16):3447–3458.

Ganesh, R., K. G. Robinson, G. D. Reed, and G. S. Sayler. 1997. Reduction of hexavalent uranium from organic complexes by sulfate- and iron-reducing bacteria. *Applied and Environmental Microbiology* 63 (11):4385–4391.

Geipel, G. 2006. Some aspects of actinide speciation by laser-induced spectroscopy. *Coordination Chemistry Reviews* 250 (7–8):844–854.

Geipel, G., S. Amayri, and G. Bernhard. 2008. Mixed complexes of alkaline earth uranyl carbonates: A laser-induced time-resolved fluorescence spectroscopic study. *Spectrochimica Acta Part A: Molecular and Biomolecular Spectroscopy* 71 (1):53–58.

Geissler, A., M. L. Merroun, G. Geipel, H. Reuther, and S. Selenska-Pobell. 2009. Biogeochemical changes induced in uranium mining waste pile samples by uranyl nitrate treatments under anaerobic conditions. *Geobiology* 7:282–294.

Ghafar, M., A. Abdul-Hadi, and O. Alhassanieh. 2002. Distribution of some elements in a solid-aqueous system: Mineral phosphate in contact with groundwater. *Journal of Radioanalytical and Nuclear Chemistry* 254:159–163.

Gihring, T. M., G. X. Zhang, C. C. Brandt, S. C. Brooks, J. H. Campbell, S. Carroll, C. S. Criddle, S. J. Green, P. Jardine, J. E. Kostka, K. Lowe, T. L. Mehlhorn, W. Overholt, D. B. Watson, Z. M. Yang, W. M. Wu, and C. W. Schadt. 2011. A limited microbial consortium is responsible for extended bioreduction of uranium in a contaminated aquifer. *Applied and Environmental Microbiology* 77 (17):5955–5965.

Ginder-Vogel, M., C. S. Criddle, and S. Fendorf. 2006. Thermodynamic constraints on the oxidation of biogenic UO2 by Fe(III) (hydr) oxides. *Environmental Science and Technology* 40 (11):3544–3550.

Ginder-Vogel, M., B. Stewart, and S. Fendorf. 2010. Kinetic and mechanistic constraints on the oxidation of biogenic uraninite by ferrihydrite. *Environmental Science and Technology* 44 (1):163–169.

Götz, C., G. Geipel, and G. Bernhard. 2010. The influence of the temperature on the carbonate complexation of uranium(VI): A spectroscopic study. *Journal of Radioanalytical and Nuclear Chemistry* 287 (3):961–969.

Gu, B., L. Liang, M. J. Dickey, X. Yin, and S. Dai. 1998. Reductive precipitation of uranium(VI) by zero-valent iron. *Environmental Science and Technology* 32 (21):3366–3373.

Gu, Baohua, S., C. Brooks, Y. Roh, and P. M. Jardine. 2003. Geochemical reactions and dynamics during titration of a contaminated groundwater with high uranium, aluminum, and calcium. *Geochimica et Cosmochimica Acta* 67 (15):2749–2761.

Gudavalli, R. K. P., Y. P. Katsenovich, D. M. Wellman, M. Idarraga, L. E. Lagos, and B. Tansel. 2013a. Comparison of the kinetic rate law parameters for the dissolution of natural and synthetic autunite in the presence of aqueous bicarbonate ions. *Chemical Geology* 351:299–309.

Gudavalli, R., Y. Katsenovich, D. Wellman, L. Lagos, and B. Tansel. 2013b. Quantification of kinetic rate law parameters of uranium release from sodium autunite as a function of aqueous bicarbonate concentrations. *Environmental Chemistry* 10 (6):475–485.

Guillaumont, R., T. Fanghänel, J. Fuger, I. Grenthe, V. Neck, D. Palmer, and M. Rand. 2003. *Update on the Chemical Thermodynamics of Uranium, Neptunium, Plutonium, Americium, and Technetium. Chemical Thermodynamics*. Vol. 5. Amsterdam, The Netherlands: Elsevier.

Gunther, A., G. Geipel, and G. Bernhard. 2007. Complex formation of uranium(VI) with the amino acids L-glycine and L-cysteine: A fluorescence emission and UV-Vis absorption study. *Polyhedron* 26 (1):59–65.

Guo, Z., C. Yan, J. Xu, and W. Wu. 2009. Sorption of U(VI) and phosphate on γ-alumina: Binary and ternary sorption systems. *Colloids and Surfaces A: Physicochemical and Engineering Aspects* 336:123–129.

Haas, J. R., T. J. Dichristina, and R. Wade. 2001. Thermodynamics of U(VI) sorption onto *Shewanella putrefaciens*. *Chemical Geology* 180 (1–4):33–54.

Haas, J. R., and A. Northup. 2004. Effects of aqueous complexation on reductive precipitation of uranium by *Shewanella putrefaciens*. *Geochemical Transactions* 5 (3):41.

Hansel, C. M., S. G. Benner, J. Neiss, A. Dohnalkova, R. K. Kukkadapu, and S. Fendorf. 2003. Secondary mineralization pathways induced by dissimilatory iron reduction of ferrihydrite under advective flow. *Geochimica Et Cosmochimica Acta* 67 (16):2977–2992.

Hirose, A., and J. K Fawell. 2012. *Uranium in Drinking Water, Background Document for Preparation of WHO Guidelines for Drinking-Water Quality*. Geneva, Switzerland: World Health Organization.

Hsi, C. K. D., and D. Langmuir. 1985. Adsorption of uranyl onto ferric oxyhydroxides—Application of the surface complexation site-binding model *Geochimica Et Cosmochimica Acta* 49 (9):1931–1941.

Hu, M. Z. C., J. M. Norman, B. D. Faison, and M. E. Reeves. 1996. Biosorption of uranium by Pseudomonas aeruginosa strain CSU: Characterization and comparison studies. *Biotechnology and Bioengineering* 51:237–247.

Hua, B., and B. Deng. 2008. Reductive immobilization of uranium(VI) by amorphous iron sulfide. *Environmental Science and Technology* 42 (23):8703–8708.

Hua, B., H. Xu, J. Terry, and B. Deng. 2006. Kinetics of uranium(VI) reduction by hydrogen sulfide in anoxic aqueous systems. *Environmental Science and Technology* 40 (15):4666–4671.

Huber, F., D. Schild, T. Vitova, J. Rothe, R. Kirsch, and T. Schafer. 2012. U(VI) removal kinetics in presence of synthetic magnetite nanoparticles. *Geochimica Et Cosmochimica Acta* 96:154–173.

Hwang, C. C., W. M. Wu, T. J. Gentry, J. Carley, G. A. Corbin, S. L. Carroll, D. B. Watson, P. M. Jardine, J. Z. Zhou, C. S. Criddle, and M. W. Fields. 2009. Bacterial community succession during in situ uranium bioremediation: Spatial similarities along controlled flow paths. *Isme Journal* 3 (1):47–64.

Hyun, S. P., J. A. Davis, K. Sun, and K. F. Hayes. 2012. Uranium(VI) reduction by iron(II) monosulfide mackinawite. *Environmental Science and Technology* 46 (6):3369–3376.

Ikeda, A., C. Hennig, S. Tsushima, K. Takao, Y. Ikeda, A. C. Scheinost, and G. Bernhard. 2007. Comparative study of uranyl(VI) and -(V) carbonato complexes in an aqueous solution. *Inorganic Chemistry* 46 (10):4212–4219.

Ilton, E. S., J. F. Boily, E. C. Buck, F. N. Skomurski, K. M. Rosso, C. L. Cahill, J. R. Bargar, and A. R. Felmy. 2010. Influence of dynamical conditions on the reduction of U-VI at the magnetite-solution interface. *Environmental Science and Technology* 44 (1):170–176.

Ilton, E. S., A. Haiduc, C. L. Cahill, and A. R. Felmy. 2005. Mica surfaces stabilize pentavalent uranium. *Inorganic Chemistry* 44 (9):2986–2988.

Ilton, E. S., J. S. L. Pacheco, J. R. Bargar, Z. Shi, J. Liu, L. Kovarik, M. H. Engelhard, and A. R. Felmy. 2012. Reduction of U(VI) incorporated in the structure of hematite. *Environmental Science and Technology* 46 (17):9428–9436.

Irving, G. C. J., and D. J. Cosgrove. 1971. Inositol phosphate phosphatases of microbiological origin. Some properties of a partially purified bacterial (Pseudomonas sp.) phytase. *Australian Journal of Biological Sciences* 24 (3):547.

Istok, J. D., J. M. Senko, L. R. Krumholz, D. Watson, M. A. Bogle, A. Peacock, Y. J. Chang, and D. C. White. 2004. In situ bioreduction of technetium and uranium in a nitrate-contaminated aquifer. *Environmental Science and Technology* 38 (2):468–475.

Ithurbide, A., S. Peulon, F. Miserque, C. Beaucaire, and A. Chausse. 2009. Interaction between uranium(VI) and siderite (FeCO3) surfaces in carbonate solutions. *Radiochimica Acta* 97 (3):177–180.

Jang, J. H., B. A. Dempsey, and W. D. Burgos. 2008. Reduction of U(VI) by Fe(II) in the presence of hydrous ferric oxide and hematite: Effects of solid transformation, surface coverage, and humic acid. *Water Research* 42 (8–9):2269–2277.

Jardine, P. M., D. B. Watson, D. A. Blake, L. P. Beard, S. C. Brooks, J. M. Carley, C. S. Criddle, W. E. Doll, M. W. Fields, S. E. Fendorf, G. G. Geesey, M. Ginder-Vogel, S. S. Hubbard, J. D. Istok, S. Kelly, K. M. Kemner, A. D. Peacock, B. P. Spalding, D. C. White, A. Wolf, W. Wu, and J. Zhou. 2004. Techniques for assessing the performance of in situ bioreduction and immobilization of metals and radionuclides in contaminated subsurface environments. Lawrence Berkeley National Laboratory, http://www.escholarship.org/uc/item/1d61h80f

Jerden, J. L., and A. K. Sinha. 2003. Phosphate based immobilization of uranium in an oxidizing bedrock aquifer. *Applied Geochemistry* 18 (6):823–843.

Johnson, B. B., E. Quill, and M. J. Angove. 2012. An investigation of the mode of sorption of inositol hexaphosphate to goethite. *Journal of Colloid and Interface Science* 367:436–442.

Jroundi, F., L. Mohamed Merroun, J. Maria Arias, A. Rossberg, and S. Selenska-Pobell. 2009. Spectroscopic and microscopic characterization of uranium biomineralization in Myxococcus xanthus. *Geomicrobiology Journal* 24 (5):441–449.

Jung, H. B., M. I. Boyanov, H. Konishi, Y. Sun, B. Mishra, K. M. Kemner, E. E. Roden, and H. Xu. 2012. Redox behavior of uranium at the nanoporous aluminum oxide water interface: Implications for uranium remediation. *Environmental Science and Technology* 46:7301–7309.

Kanematsu, M., N. Perdrial, W. Um, J. Chorover, and P. A. O'Day. 2014. Influence of phosphate and silica on U(VI) precipitation from acidic and neutralized wastewaters. *Environmental Science and Technology* 48 (11):6097–6106.

Kantar, C., and B. D. Honeyman. 2006. Citric acid enhanced remediation of soils contaminated with uranium by soil flushing and soil washing. *Journal of Environmental Engineering-Asce* 132 (2):247–255.

Katsenovich, Y. P., D. A. Carvajal, D. M. Wellman, and L. E. Lagos. 2012. Enhanced U(VI) release from autunite mineral by aerobic Arthrobacter sp in the presence of aqueous bicarbonate. *Chemical Geology* 308:1–9.

Keasling, J. D., S. J. Van Dien, P. Trelstad, N. Renninger, and K. McMahon. 2000. Application of polyphosphate metabolism to environmental and biotechnological problems. *Biochemistry (Moscow)* 65 (3):324–331.

Kelly, S. D., K. M. Kemner, and S. C. Brooks. 2007. X-ray absorption spectroscopy identifies calcium-uranyl-carbonate complexes at environmental concentrations. *Geochimica Et Cosmochimica Acta* 71 (4):821–834.

Kelly, S. D., K. M. Kemner, J. Carley, C. Criddle, P. M. Jardine, T. L. Marsh, D. Phillips, D. Watson, and W. M. Wu. 2008. Speciation of uranium in sediments before and after in situ biostimulation. *Environmental Science and Technology* 42 (5):1558–1564.

Kelly, S. D., K. M. Kemner, J. B. Fein, D. A. Fowle, M. I. Boyanov, B. A. Bunker, and N. Yee. 2002. X-ray absorption fine structure determination of pH-dependent U-bacterial cell wall interactions. *Geochimica et Cosmochimica Acta* 66 (22):3855–3871.

Kerr, R. S. 1990. *Basics of Pump-and-Tret Ground-Water Remediation Technology*. U.S. Environmental Protection Agency: Ada, OK.

Konopka, A., A. E. Plymale, D. A. Carvajal, X. J. Lin, and J. P. McKinley. 2013. Environmental controls on the activity of aquifer microbial communities in the 300 area of the Hanford Site. *Microbial Ecology* 66 (4):889–896.

Krestou, A., A. Xenidis, and D. Panias. 2004. Mechanism of aqueous uranium(VI) uptake by hydroxyapatite. *Minerals Engineering* 17 (3):373–381.

Kulkarni, S., A. Ballal, and S. K. Apte. 2013. Bioprecipitation of uranium from alkaline waste solutions using recombinant Deinococcus radiodurans. *Journal of Hazardous Materials* 262:853–861.

Lack, J. G., S. K. Chaudhuri, S. D. Kelly, K. M. Kemner, S. M. O'Connor, and J. D. Coates. 2002. Immobilization of radionuclides and heavy metals through anaerobic bio-oxidation of Fe(II). *Applied and Environmental Microbiology* 68 (6):2704–2710.

Langmuir, D. 1978. Uranium solution-mineral equilibria at low-temperatures with applications to sedimentary ore-deposits *Geochimica et Cosmochimica Acta* 42 (6):547–569.

Langmuir, D. 1997. *Aqueous Environmental Geochemistry*. Upper Saddle River, NJ: Prentice Hall, Inc.

Latta, D. E., M. I. Boyanov, K. M. Kemner, E. J. O'Loughlin, and M. M. Scherer. 2012a. Abiotic reduction of uranium by Fe(II) in soil. *Applied Geochemistry* 27 (8):1512–1524.

Latta, D. E., C. A. Gorski, M. I. Boyanov, E. J. O'Loughlin, K. M. Kemner, and M. M. Scherer. 2012b. Influence of magnetite stoichiometry on U-VI reduction. *Environmental Science and Technology* 46 (2):778–786.

Latta, D. E., B. Mishra, R. E. Cook, K. M. Kemner, and M. I. Boyanov. 2014. Stable U(IV) complexes form at high-affinity mineral surface sites. *Environmental Science and Technology* 48 (3):1683–1691.

Latta, D. E., C. I. Pearce, K. M. Rosso, K. M. Kemner, and M. I. Boyanov. 2013. Reaction of U-VI with titanium-substituted magnetite: Influence of Ti on U-IV speciation. *Environmental Science and Technology* 47 (9):4121–4130.

Lee, S. Y., M. H. Baik, and J. W. Choi. 2010. Biogenic formation and growth of uraninite (UO2). *Environmental Science and Technology* 44 (22):8409–8414.

Lenhart, J. J., and B. D. Honeyman. 1999. Uranium(VI) sorption to hematite in the presence of humic acid. *Geochimica et Cosmochimica Acta* 63 (19/20):2891–2901.

Liger, E., L. Charlet, and P. Van Cappellen. 1999. Surface catalysis of uranium(VI) reduction by iron(II). *Geochimica et Cosmochimica Acta* 63 (19–20):2939–2955.

Lim, B. L., P. Yeung, C. Cheng, and J. E. Hill. 2007. Distribution and diversity of phytate-mineralizing bacteria. *Isme Journal* 1 (4):321–330.

Lin, H., N. H. Szeinbaum, T. J. DiChristina, and M. Taillefert. 2012. Microbial Mn(IV) reduction requires an initial one-electron reductive solubilization step. *Geochimica et Cosmochimica Acta* 99:179–192.

Liu, C. X., J. M. Zachara, J. K. Fredrickson, D. W. Kennedy, and A. Dohnalkova. 2002. Modeling the inhibition of the bacterial reduction of U(VI) by beta-MnO2(S)(g). *Environmental Science and Technology* 36 (7):1452–1459.

Lovley, D. R., E. J. P. Phillips, Y. A. Gorby, and E. R. Landa. 1991. Microbial reduction of uranium. *Nature* 350 (6317):413–416.

Luther, G. W., III, B. T. Glazer, S. Ma, R. E. Trouwborst, T. S. Moore, E. Metzger, C. Kraiya, T. J. Waite, G. Druschel, B. Sundby, M. Taillefert, D. B. Nuzzio, T. M. Shank, B. L. Lewis, and P. J. Brendel. 2008. Use of voltammetric solid-state (micro)electrodes for studying biogeochemical processes: Laboratory measurements to real time measurements with an in situ electrochemical analyzer (ISEA). *Marine Chemistry* 108 (3–4):221–235.

Macaskie, L. E., J. D. Blackmore, and R. M. Empson. 1988. Phosphatase overproduction and enhanced uranium accumulation by a stable mutant of a Citrobacter sp. isolated by a novel method. *FEMS Microbiology Ecology* 55:157–162.

Macaskie, L. E., K. M. Bonthrone, P. Yong, and D. T. Goddard. 2000. Enzymically mediated bioprecipitation of uranium by a Citrobacter sp.: A concerted role for exocellular lipopolysaccharide and associated phosphatase in biomineral formation. *Microbiology-UK* 146:1855–1867.

Macaskie, L. E., R. M. Empson, A. K. Cheetham, C. P. Grey, and A. J. Skarnulis. 1992. Uranium bioaccumulation by a Citrobacter sp. as a result of enzymatically mediated growth of polycrystalline HUO2PO4. *Science* 257:782–784.

Mackay, D. M., and J. A. Cherry. 1989. Groundwater contamination: Pump-and-treat remediation. *Environmental Science and Technology* 23 (6):630–636.

Markich, S. J., P. L. Brown, and R. A. Jeffree. 1996. The use of geochemical speciation modelling to predict the impact of uranium to freshwater biota. *Radiochimica Acta* 74:321–326.

Markich, S. J., P. L. Brown, R. A. Jeffree, and R. P. Lim. 2000. Valve movement responses of Velesunio angasi (Bivalvia: Hyriidae) to manganese and uranium: An exception to the free ion activity model. *Aquatic Toxicology* 51 (2):155–175.

Markovarga, G., and L. Gorton. 1990. Postcolumn derivatization in liquid-chromatography using immobilized enzyme reactors and amperometric detection. *Analytica Chimica Acta* 234 (1):13–29.

Marshall, M. J., A. S. Beliaev, A. C. Dohnalkova, D. W. Kennedy, L. Shi, Z. M. Wang, M. I. Boyanov, B. Lai, K. M. Kemner, J. S. McLean, S. B. Reed, D. E. Culley, V. L. Bailey, C. J. Simonson, D. A. Saffarini, M. F. Romine, J. M. Zachara, and J. K. Fredrickson. 2006. c-Type cytochrome-dependent formation of U(IV) nanoparticles by Shewanella oneidensis. *Plos Biology* 4 (8):1324–1333.

Martinez, R. J., M. J. Beazley, M. Taillefert, A. K. Arakaki, J. Skolnick, and P. A. Sobecky. 2007. Aerobic uranium (VI) bioprecipitation by metal-resistant bacteria isolated from radionuclide- and metal-contaminated subsurface soils. *Environmental Microbiology* 9 (12):3122–3133.

Martinez, R. J., Y. L. Wang, M. A. Raimondo, J. M. Coombs, T. Barkay, and P. A. Sobecky. 2006. Horizontal gene transfer of P-IB-type ATPases among bacteria isolated from radionuclide- and metal-contaminated subsurface soils. *Applied and Environmental Microbiology* 72 (5):3111–3118.

Mason, C. F. V., W. R. J. R. Turney, B. M. Thomson, N. Lu, P. A. Longmire, and C. J. Chrisholm-Brause. 1997. Carbonate leaching of uranium from contaminated soils. *Environmental Science and Technology* 31 (10):2707–2711.

Maya, L. 1982. Detection of hydroxo and carbonato species of dioxouranium(VI) in aqueous media by differential pulse polarography. *Inorganica Chimica Acta-Letters* 65 (1):L13–L16.

McKinley, J. P., J. M. Zachara, S. C. Smith, and G. D. Turnert. 1995. The influence of uranyl hydrolysis and multiple site-binding reactions on adsorption of U(VI) to montmorillonite. *Clays and Clay Minerals* 43:586–598.

Meece, D. E., and L. K. Benninger. 1993. The coprecipitation of Pu and other radionuclides with CaCO3. *Geochimica et Cosmochimica Acta* 57:1447–1458.

Meek, J. L., and F. Nicoletti. 1986. Detection of inositol triphosphate and other organic phosphates by high-performance liquid chromatography using an enzyme-loaded post-column reactor. *Journal of Chromatography* 351 (2):303–311.

Merroun, Mohamed L., M. Nedelkova, J. J. Ojeda, T. Reitz, F. Margarita, J. M. Arias, M. Romero-Gonzalez, and S. Selenska-Pobell. 2011. Bio-precipitation of uranium by two bacterial isolates recovered from extreme environments as estimated by potentiometric titration, TEM, and x-ray absorption spectroscopic analyses. *Journal of Hazardous Materials* 197:1–10.

Min, M. Z., H. F. Xu, J. Chen, and M. Fayek. 2005. Evidence of uranium biomineralization in sandstone-hosted roll-front uranium deposits, northwestern China. *Ore Geology Reviews* 26 (3–4):198–206.

Missana, T., U. Maffiotte, and M. Garcia-Gutierrez. 2003. Surface reactions kinetics between nanocrystalline magnetite and uranyl. *Journal of Colloid and Interface Science* 261 (1):154–160.

Moon, H. S., J. Komlos, and P. R. Jaffe. 2007. Uranium reoxidation in previously bioreduced sediment by dissolved oxygen and nitrate. *Environmental Science and Technology* 41 (13):4587–4592.

Moon, H. S., J. Komlos, and P. R. Jaffe. 2009. Biogenic U(IV) oxidation by dissolved oxygen and nitrate in sediment after prolonged U(VI)/Fe(III)/SO42- reduction. *J Contam Hydrol* 105 (1–2):18–27.

Moore, W. S., and T. J. Shaw. 2008. Fluxes and behavior of radium isotopes, barium, and uranium in seven Southeastern US rivers and estuaries. *Marine Chemistry* 108 (3–4):236–254.

Morel, F. M. M. 1983. *Principles of Aquatic Chemistry*. 1st ed. New York: John Wiley.

Morris, D. E. 2002. Redox energetics and kinetics of uranyl coordination complexes in aqueous solution. *Inorganic Chemistry* 41 (13):3542–3547.

Morris, D. E., P. G. Allen, J. M. Berg, C. J. Chisholm-Brause, R. J. Donohoe, N. J. Hess, J. A. Musgrave, and C. D. Tait. 1996. Speciation of uranium in Fernald soils by molecular spectroscopic methods: Characterization of untreated soils. *Environmental Science and Technology* 30:2322–2331.

Morrison, S. J., D. R. Metzler, and C. E. Carpenter. 2001. Uranium precipitation in a permeable reactive barrier by progressive irreversible dissolution of zerovalent iron. *Environmental Science and Technology* 35 (2):385–390.

Mougel, V., B. Biswas, J. Pecaut, and M. Mazzanti. 2010. New insights into the acid mediated disproportionation of pentavalent uranyl. *Chemical Communications* 46 (45):8648–8650.

Moyes, L. N., R. H. Parkman, D. J. Charnock, F. T. Livens, C. R. Hughes, and A. Braithwaite. 2000. Uranium uptake from aqueous solution by interaction with goethite, lepidocrocite, muscovite, and mackinawite: An x-ray absorption spectroscopy study. *Environmental Science and Technology* 34:1062–1068.

Mukherjee, A., G. H. Wheaton, P. H. Blum, and R. M. Kelly. 2012. Uranium extremophily is an adaptive, rather than intrinsic, feature for extremely thermoacidophilic Metallosphaera species. *Proceedings of the National Academy of Sciences of the United States of America* 109 (41):16702–16707.

Mullaney, E. J., and A. H. J. Ullah. 2007. Attributes, catalytic mechanisms and applications. In: B. L. Turner, A. E. Richardson, and E. J. Mullaney (eds.), *Inositol Phosphates: Linking Agriculture and the Environment*. Cambridge, MA: CAB International.

Myers, C. R., and K. H. Nealson. 1988. Bacterial manganese reduction and growth with manganese oxides as the sole electron-acceptor. *Science* 240 (4857):1319–1321.

Nair, S., and B. J. Merkel. 2011. Impact of alkaline earth metals on aqueous speciation of uranium(VI) and sorption on quartz. *Aquatic Geochemistry* 17 (3):209–219.

Nash, K. L., M. P. Jensen, and M. A. Schmidt. 1998. Actinide immobilization in the subsurface environment by in-situ treatment with a hydrolytically unstable organophosphorus complexant: Uranyl uptake by calcium phytate. *Journal of Alloys and Compounds* 271–273:257–261.

Newton, M. P., and C. M. G. Van Den Berg. 1987. Determination of nickel, cobalt, copper, and uranium in water by cathodic stripping chronopotentiometry with continuous flow. *Analytica Chimica Acta* 199:59–76.

Nilgiriwala, K. S., A. Alahari, A. R. Sambasiva, and S. K. Apte. 2008. Cloning and overexpression of alkaline phosphatase PhoK from Sphingomonas sp. strain BSAR-1 for bioprecipitation of uranium from alkaline solutions. *Applied and Environmental Microbiology* 74 (17):5516–5523.

Nimmons, M. J. 2007. Evaluation and Screening of Remedial Technologies for Uranium at the 300-FF-5 Operable Unit, Hanford Site, Washington. Richland, WA: U.S Department of Energy.

O'Loughlin, E. J., S. D. Kelly, R. E. Cook, R. Csencsits, and K. M. Kemner. 2003. Reduction of uranium(VI) by mixed iron(II/iron(III) hydroxide (green rust): Formation of UO(2) manoparticies. *Environmental Science and Technology* 37 (4):721–727.

O'Loughlin, E. J., S. D. Kelly, and K. M. Kemner. 2010. XAFS investigation of the interactions of U-VI with secondary mineralization products from the bioreduction of Fe-III oxides. *Environmental Science and Technology* 44 (5):1656–1661.

Oh, B. C., W. C. Choi, S. Park, Y. O. Kim, and T. K. Oh. 2004. Biochemical properties and substrate specificities of alkaline and histidine acid phytases. *Applied Microbiology and Biotechnology* 63 (4):362–372.

Ohnuki, T., N. Kozai, M. Samadfam, R. Yasuda, S. Yamamoto, K. Narumi, H. Naramoto, and T. Murakami. 2004. The formation of autunite (Ca(UO2)2(PO4)2.nH2O) within the leached layer of dissolving apatite: Incorporation mechanism or uranium by apatite. *Chemical Geology* 211:1–14.

Parthasarathy, R., and F. Eisenberg. 1991. Biochemistry, stereochemistry, and nomenclature of the inositol phosphates. *ACS Symposium Series* 463:1–19.

Pasilis, S. P., and J. E. Pemberton. 2003. Speciation and coordination chemistry of uranyl(VI)-citrate complexes in aqueous solution. *Inorganic Chemistry* 42 (21):6793–6800.

Payne, T. E., J. A. Davis, and T. D. Waite. 1996. Uranium adsorption onto ferrihydrite—Effects of phosphate and humic acid. *Radiochimica Acta* 74:239–243.

Petrie, L., N. N. North, S. L. Dollhopf, D. L. Balkwill, and J. E. Kostka. 2003. Enumeration and characterization of iron(III)-reducing microbial communities from acidic subsurface sediments contaminated with uranium (VI). *Applied and Environmental Microbiology* 69:7467–7479.

Pible, O., C. Vidaud, S. Plantevin, J. L. Pellequer, and E. Quemeneur. 2010. Predicting the disruption by UO22+ of a protein-ligand interaction. *Protein Science* 19 (11):2219–2230.

Plant, J. A., P. R. Simpson, B. Smith, and B. F. Windley. 1999. Uranium ore deposits-products of the radioactive earth. *Reviews in Mineralogy* 38:255–319.

Plathe, K. L., S. W. Lee, B. M. Tebo, J. R. Bargar, and R. Bernier-Latmani. 2013. Impact of microbial Mn oxidation on the remobilization of bioreduced U(IV). *Environmental Science and Technology* 47 (8):3606–3613.

Prikryl, J. D., A. Jain, D. R. Turner, and R. T. Pabalan. 2001. Uranium(VI) sorption behavior on silicate mineral mixtures. *Journal of Contaminant Hydrology* 47 (2–4):241–253.

Qafoku, N. P., R. K. Kukkadapu, J. P. McKinley, B. W. Arey, S. D. Kelly, C. M. Wang, C. T. Resch, and P. E. Long. 2009. Uranium in framboidal pyrite from a naturally bioreduced alluvial sediment. *Environmental Science and Technology* 43 (22):8528–8534.

Rai, D., A. R. Felmy, and J. L. Ryan. 1990. Uranium(IV) hydrolysis constants and solubility product of UO2. xH2O(am). *Inorganic Chemistry* 29 (2):260–264.

Raicevic, S., J. V. Wright, V. Veljkovic, and J. L. Conca. 2006. Theoretical stability assessment of uranyl phosphates and apatites: Selection of amendments for in situ remediation of uranium. *Science of the Total Environment* 355:13–24.

Rakovan, J., R. J. Reeder, E. J. Elzinga, D. J. Cherniak, C. D. Tait, and D. E. Morris. 2002. Structural characterization of U(VI) in apatite by x-ray absorption spectroscopy. *Environmental Science and Technology* 36:3114–3117.

Ray, A. E., J. R. Bargar, V. Sivaswamy, A. C. Dohnalkova, Y. Fujita, B. M. Peyton, and T. S. Magnuson. 2011. Evidence for multiple modes of uranium immobilization by an anaerobic bacterium. *Geochimica et Cosmochimica Acta* 75 (10):2684–2695.

Reeder, R. J., M. Nugent, G. M. Lamble, C. D. Tait, and D. E. Morris. 2000. Uranyl incorporation in calcite and aragonite: XAFS and luminescence studies. *Environmental Science and Technology* 34:638–644.

Reeder, R. J., M. Nugent, C. D. Tait, D. E. Morris, S. M. Heald, K. M. Beck, W. P. Hess, and A. Lanzirotti. 2001. Coprecipitation of uranium(VI) with calcite: XAFS, micro-XAS, and luminescence characterization. *Geochimica Et Cosmochimica Acta* 65 (20):3491–3503.

Renninger, N., R. Knopp, H. Nitsche, D. Clark, and J. Keasling. 2004. Uranyl precipitation by Pseudomonas aeruginosa via controlled polyphosphate metabolism. *Applied and Environmental Microbioloy* 70:7404–7412.

Renock, D., M. Mueller, K. Yuan, R. C. Ewing, and U. Becker. 2013. The energetics and kinetics of uranyl reduction on pyrite, hematite, and magnetite surfaces: A powder microelectrode study. *Geochimica et Cosmochimica Acta* 118:56–71.

Renshaw, J. C., L. J. C. Butchins, F. R. Livens, I. May, J. M. Charnock, and J. R. Lloyd. 2005. Bioreduction of uranium: Environmental implications of a pentavalent intermediate. *Environmental Science and Technology* 39 (15):5657–5660.

Reynolds, R. L., M. B. Goldhaber, and D. J. Carpenter. 1982. Biogenic and nonbiogenic ore-forming processes in the south Texas uranium district; evidence from the Panna Maria deposit. *Economic Geology* 77 (3):541–556.

Rickard, D., and G. W. Luther. 1997. Kinetics of pyrite formation by the H2S oxidation of iron(II) monosulfide in aqueous solutions between 25 and 125 degrees C: The mechanism. *Geochimica et Cosmochimica Acta* 61 (1):135–147.

Rigden, D. J. 2008. The histidine phosphatase superfamily: Structure and function. *Biochemical Journal* 409:333–348.

Rihs, S., C. Gaillard, T. Reich, and S. J. Kohler. 2014. Uranyl sorption onto birnessite: A surface complexation modeling and EXAFS study. *Chemical Geology* 353:59–70.

Riley, R. G, J. M. Zachara, and F. J. Wobber. 1992. Chemical contaminants on DOE lands and selection of contaminant mixtures for subsurface science research. U.S. Department of Energy, Wasington, D.C.

Robinson, K. G., R. Ganesh, and G. D. Reed. 1998. Impact of organic ligands on uranium removal during anaerobic biological treatment. *Water Science and Technology* 37 (8):73–80.

Roh, Y., S. R. Lee, S-K. Choi, M. P. Elless, and S. Y. Lee. 2000. Physicochemical and mineralogical characterization of uranium-contaminated soils. *Soil and Sediment Contamination* 9:463–496.

Rossolini, G. M., S. Schippa, M. L. Riccio, F. Berlutti, L. E. Macaskie, and M. C. Thaller. 1998. Bacterial nonspecific acid phosphohydrolases: Physiology, evolution and use as tools in microbial biotechnology. *Cellular and Molecular Life Sciences* 54 (8):833–850.

Salome, K. R., M. J. Beazley, S. M. Webb, Sobecky, P. A., and M. Taillefert. 2016. Biomineralization of uranium promoted by microbially-mediated phytate hydrolysis in contaminated soils. *Submitted.*

Salome, K. R., S. J. Green, M. J. Beazley, S. M. Webb, J. E. Kostka, and M. Taillefert. 2013. The role of anaerobic respiration in the immobilization of uranium through biomineralization of phosphate minerals. *Geochimica et Cosmochimica Acta* 106:344–363.

Sanford, R. A., Q. Wu, Y. Sung, S. H. Thomas, B. K. Amos, E. K. Prince, and F. E. Loffler. 2007. Hexavalent uranium supports growth of Anaeromyxobacter dehalogenans and Geobacter spp. with lower than predicted biomass yields. *Environmental Microbiology* 9 (11):2885–2893.

Sani, R. K., B. M. Peyton, J. E. Amonette, and G. G. Geesey. 2004. Reduction of uranium(VI) under sulfate-reducing conditions in the presence of Fe(III)-(hydr)oxides. *Geochimica et Cosmochimica Acta* 68 (12):2639–2648.

Santos, I. R., W. C. Burnett, S. Misra, I. Suryaputra, J. P. Chanton, T. Dittmar, R. N. Peterson, and P. W. Swarzenski. 2011. Uranium and barium cycling in a salt wedge subterranean estuary: The influence of tidal pumping. *Chemical Geology* 287 (1–2):114–123.

Sato, T., T. Murakami, N. Yanase, H. Isobe, T. E. Payne, and P. L. Airey. 1997. Iron nodules scavenging uranium from groundwater. *Environmental Science and Technology* 31:2854–2858.

Schofield, E. J., H. Veeramani, J. O. Sharp, E. Suvorova, R. Bernier-Latmani, A. Mehta, J. Stahlman, S. M. Webb, D. L. Clark, S. D. Conradson, E. S. Ilton, and J. R. Bargar. 2008. Structure of biogenic uraninite produced by Shewanella oneidensis strain MR-1. *Environmental Science and Technology* 42 (21):7898–7904.

Seaman, J. C., J. M. Hutchinson, B. P. Jackson, and V. M. Vulava. 2003. In situ of treatment of metals in contaminated soils with phytate. *Journal of Environmental Quality* 32:153–161.

Senko, J. M., J. D. Istok, J. M. Suflita, and L. R. Krumholz. 2002. In-situ evidence for uranium immobilization and remobilization. *Environmental Science and Technology* 36 (7):1491–1496.

Senko, J. M., S. D. Kelly, A. C. Dohnalkova, J. T. McDonough, K. M. Kemner, and W. D. Burgos. 2007. The effect of U(VI) bioreduction kinetics on subsequent reoxidation of biogenic U(IV). *Geochimica et Cosmochimica Acta* 71 (19):4644–4654.

Senko, J. M., Y. Mohamed, T. A. Dewers, and L. R. Krumholz. 2005. Role for Fe(III) minerals in nitrate-dependent microbial U(IV) oxidation. *Environmental Science and Technology* 39 (8):2529–2536.

Shan, Y., I. D. McKelvie, and B. T. Hart. 1994. Determination of alkaline phosphatase-hydrolyzable phosphorus in natural water systems by enzymatic flow injection. *Limnology and Oceanography* 39 (8):1993–2000.

Shelobolina, E. S., H. Konishi, H. F. Xu, and E. E. Roden. 2009. U(VI) Sequestration in hydroxyapatite produced by microbial glycerol 3-phosphate metabolism. *Applied and Environmental Microbiology* 75 (18):5773–5778.

Sheng, L., J. Szymanowski, and J. B. Fein. 2011. The effects of uranium speciation on the rate of U(VI) reduction by Shewanella oneidensis MR-1. *Geochimica et Cosmochimica Acta* 75 (12):3558–3567.

Shiel, A. E., P. G. Laubach, T. M. Johnson, C. C. Lundstrom, P. E. Long, and K. H. Williams. 2013. No measurable changes in U-238/U-235 due to desorption-adsorption of U(VI) from groundwater at the Rifle, Colorado, integrated field research challenge site. *Environmental Science and Technology* 47 (6):2535–2541.

Shin, S., N. C. Ha, B. C. Oh, T. K. Oh, and B. H. Oh. 2001. Enzyme mechanism and catalytic property of beta propeller phytase. *Structure* 9 (9):851–858.

Simon, F. G., V. Biermann, and B. Peplinski. 2008. Uranium removal from groundwater using hydroxyapatite. *Applied Geochemistry* 23 (8):2137–2145.

Simon, F. G., V. Biermann, C. Segebade, and M. Hedrich. 2004. Behaviour of uranium in hydroxyapatite-bearing permeable reactive barriers: Investigation using U-237 as a radioindicator. *Science of the Total Environment* 326 (1–3):249–256.

Singer, D. M., S. M. Chatman, E. S. Ilton, K. M. Rosso, J. F. Banfield, and G. A. Waychunas. 2012a. Identification of simultaneous U(VI) sorption complexes and U(IV) nanoprecipitates on the magnetite (111) surface. *Environmental Science and Technology* 46 (7):3811–3820.

Singer, D. M., S. M. Chatman, E. S. Ilton, K. M. Rosso, J. F. Banfield, and G. A. Waychunas. 2012b. U(VI) sorption and reduction kinetics on the magnetite (111) surface. *Environmental Science and Technology* 46 (7):3821–3830.

Singer, D. M., F. Farges, and G. E. Brown. 2009. Biogenic nanoparticulate UO2: Synthesis, characterization, and factors affecting surface reactivity. *Geochimica et Cosmochimica Acta* 73 (12):3593–3611.

Singh, A., J. G. Catalano, K. U. Ulrich, and D. E. Giammar. 2012. Molecular-scale structure of uranium(VI) immobilized with goethite and phosphate. *Environmental Science and Technology* 46 (12):6594–6603.

Sivaswamy, V., M. I. Boyanov, B. M. Peyton, S. Viamajala, R. Gerlach, W. A. Apel, R. K. Sani, A. Dohnalkova, K. M. Kemner, and T. Borch. 2011. Multiple mechanisms of uranium immobilization by Cellulomonas sp. strain ES6. *Biotechnology and Bioengineering* 108 (2):264–76.

Smeaton, C. M., C. G. Weisener, P. C. Burns, B. J. Fryer, and D. A. Fowle. 2008. Bacterially enhanced dissolution of meta-autunite. *American Mineralogist* 93 (11–12):1858–1864.

Sorensen, J., and L. Thorling. 1991. Stimulation by lepidocrocite (gamma-FeOOH) of Fe(II)-dependent nitrite reduction. *Geochimica et Cosmochimica Acta* 55 (5):1289–1294.

Sowder, A. G., S. B. Clark, and R. A. Fjeld. 2001. The impact of mineralogy in the U(VI)-Ca-PO4 system on the environmental availability of uranium. *Journal of Radioanalytical and Nuclear Chemistry* 248 (3):517–524.

Spycher, N. F., M. Issarangkun, B. D. Stewart, S. S. Sengor, E. Belding, T. R. Ginn, B. M. Peyton, and R. K. Sani. 2011. Biogenic uraninite precipitation and its reoxidation by iron(III) (hydr)oxides: A reaction modeling approach. *Geochimica et Cosmochimica Acta* 75 (16):4426–4440.

Steele, H., and R. J. Taylor. 2007. A theoretical study of the inner-sphere disproportionation reaction mechanism of the pentavalent actinyl ions. *Inorganic Chemistry* 46 (16):6311–6318.

Stewart, B. D., C. Girardot, N. Spycher, R. K. Sani, and B. M. Peyton. 2013. Influence of chelating agents on biogenic uraninite reoxidation by Fe(III) (Hydr)oxides. *Environmental Science and Technology* 47 (1):364–371.

Stewart, B. D., M. A. Mayes, and S. Fendorf. 2010. Impact of uranyl-calcium-carbonato complexes on uranium(VI) adsorption to synthetic and natural sediments. *Environmental Science and Technology* 44 (3):928–934.

Stewart, B. D., P. S. Nico, and S. Fendorf. 2009. Stability of uranium incorporated into Fe (Hydr)oxides under fluctuating redox conditions. *Environmental Science and Technology* 43 (13):4922–4927.

Stewart, B. D., R. T. Amos, P. S. Nico, and S. Fendorf. 2011. Influence of uranyl speciation and iron oxides on uranium biogeochemical redox reactions. *Geomicrobiology Journal* 28 (5–6):444–456.

Strahan, D. 2001. Uranium in glass, glazes and enamels: History, identification and handling. *Studies in Conservation* 46 (3):181–195.

Suzuki, Y., and J. F. Banfield. 2004. Resistance to, and accumulation of, uranium by bacteria from a uranium-contaminated site. *Geomicrobiology Journal* 21:113–121.

Suzuki, Y., S. D. Kelly, K. M. Kemner, and J. F. Banfield. 2005. Direct microbial reduction and subsequent preservation of uranium in natural near-surface sediment. *Applied and Environmental Microbiology* 71 (4):1790–1797.

Suzuki, Y., K. Tanaka, N. Kozai, and T. Ohnuki. 2010. Effects of citrate, NTA, and EDTA on the reduction of U(VI) by *Shewanella putrefaciens*. *Geomicrobiology Journal* 27 (3):245–250.

Sylvester, E. R., E. A. Hudson, and P. G. Allen. 2000. The structure of uranium(VI) sorption complexes on silica, alumina, and montmorillonite. *Geochimica et Cosmochimica Acta* 64 (14):2431–2438.

Taillefert, M., G. W. Luther, and D. B. Nuzzio. 2000. The application of electrochemical tools for in situ measurements in aquatic systems. *Electroanalysis* 12 (6):401–412.

Tao, Z. Y., T. W. Chu, J. Z. Du, X. X. Dai, and Y. J. Gu. 2000. Effect of fulvic acids on sorption of U(VI), Zn, Yb, I, and Se(IV) onto oxides of aluminum, iron, and silicon. *Applied Geochemistry* 15:145–151.

Tapia-Rodriguez, A., A. Luna-Velasco, J. A. Field, and R. Sierra-Alvarez. 2012. Toxicity of uranium to microbial communities in anaerobic biofilms. *Water Air and Soil Pollution* 223 (7):3859–3868.

Tebo, B. M., J. R. Bargar, B. G. Clement, G. J. Dick, K. J. Murray, D. Parker, R. Verity, and S. M. Webb. 2004. Biogenic manganese oxides: Properties and mechanisms of formation. *Annual Review of Earth and Planetary Sciences* 32:287–328.

Tebo, B. M., and A. Y. Obraztsova. 1998. Sulfate-reducing bacterium grows with Cr(VI), U(VI), Mn(IV), and Fe(III) as electron acceptors. *Fems Microbiology Letters* 162 (1):193–198.

Tran, D., J. C. Massabuau, and J. Garnier-Laplace. 2004. Effect of carbon dioxide on uranium bioaccumulation in the freshwater clam Corbicula fluminea. *Environmental Toxicology and Chemistry* 23 (3):739–747.

Trenfield, M. A., J. C. Ng, B. Noller, S. J. Markich, and R. A. van Dam. 2011. Dissolved organic carbon reduces uranium bioavailability and toxicity. 2. Uranium VI speciation and toxicity to three tropical freshwater organisms. *Environmental Science and Technology* 45 (7):3082–3089.

Trenfield, M. A., J. C. Ng, B. Noller, S. J. Markich, and R. A. van Dam. 2012. Dissolved organic carbon reduces uranium toxicity to the unicellular eukaryote Euglena gracilis. *Ecotoxicology* 21 (4):1013–1023.

Tribovillard, N., T. J. Algeo, T. Lyons, and A. Riboulleau. 2006. Trace metals as paleoredox and paleoproductivity proxies: An update. *Chemical Geology* 232 (1–2):12–32.

Turner, B. L., I. D. McKelvie, and P. M. Haygarth. 2002a. Characterisation of water-extractable soil organic phosphorus by phosphatase hydrolysis. *Soil Biology and Biochemistry* 34 (1):27–35.

Turner, B. L., M. J. Paphazy, P. M. Haygarth, and I. D. McKelvie. 2002b. Inositol phosphates in the environment. *Philosophical Transactions of the Royal Society of London Series B-Biological Sciences* 357 (1420):449–469.

Ulrich, K. U., E. S. Ilton, H. Veeramani, J. O. Sharp, R. Bernier-Latmani, E. J. Schofield, J. R. Bargar, and D. E. Giammar. 2009. Comparative dissolution kinetics of biogenic and chemogenic uraninite under oxidizing conditions in the presence of carbonate. *Geochimica et Cosmochimica Acta* 73 (20):6065–6083.

Ulrich, K. U., A. Singh, E. J. Schofield, J. R. Bargar, H. Veeramani, J. O. Sharp, R. Bernier-Latmani, and D. E. Giammar. 2008. Dissolution of biogenic and synthetic UO(2) under varied reducing conditions. *Environmental Science and Technology* 42 (15):5600–5606.

Ulrich, K.-U., H. Veeramani, R. Bernier-Latmani, and D. E. Giammar. 2011. Speciation-dependent kinetics of uranium(VI) bioreduction. *Geomicrobiology Journal* 28 (5–6):396–409.

Van Cleemput, O., and L. Baert. 1983. Nitrite stability influenced by iron compounds. *Soil Biology and Biochemistry* 15 (2):137–140.

Van Den Berg, C. M. G., and M. Nimmo. 1987. Direct determination of uranium in water by cathodic stripping voltammetry. *Analytical Chemistry* 59 (6):924–928.

VanEngelen, M. R., E. K. Field, R. Gerlach, B. D. Lee, W. A. Apel, and B. M. Peyton. 2010. UO22+ speciation determines uranium toxicity and bioaccumulation in an environmental Pseudomonas sp. isolate. *Environmental Toxicology and Chemistry* 29 (4):763–769.

VanEngelen, M. R., R. K. Szilagyi, R. Gerlach, B. D. Lee, W. A. Apel, and B. M. Peyton. 2011. Uranium exerts acute toxicity by binding to Pyrroloquinoline quinone cofactor. *Environmental Science and Technology* 45 (3):937–942.

Vazquez, G. J., C. J. Dodge, and A. J. Francis. 2007. Interactions of uranium with polyphosphate. *Chemosphere* 70:263–269.

Veeramani, H., A. C. Scheinost, N. Monsegue, N. P. Qafoku, R. Kukkadapu, M. Newville, A. Lanzirotti, A. Pruden, M. Murayama, and M. F. Hochella. 2013. Abiotic reductive immobilization of U(VI) by biogenic mackinawite. *Environmental Science and Technology* 47 (5):2361–2369.

Veeramani, H., D. S. Alessi, E. I. Suvorova, J. S. Lezama-Pacheco, J. E. Stubbs, J. O. Sharp, U. Dippon, A. Kappler, J. R. Bargar, and R. Bernier-Latmani. 2011. Products of abiotic U(VI) reduction by biogenic magnetite and vivianite. *Geochimica et Cosmochimica Acta* 75 (9):2512–2528.

Wade, R., and T. J. DiChristina. 2000. Isolation of U(VI) reduction-deficient mutants of *Shewanella putrefaciens*. *Fems Microbiology Letters* 184 (2):143–148.

Waite, T. D., J. A. Davis, T. E. Payne, G. A. Waychunas, and N. Xu. 1994. Uranium(VI) adsorption to ferrihydrite: Application of a surface complexation model. *Geochimica et Cosmochimica Acta* 58 (24):5465–5478.

Wan, J. M., T. K. Tokunaga, E. Brodie, Z. M. Wang, Z. P. Zheng, D. Herman, T. C. Hazen, M. K. Firestone, and S. R. Sutton. 2005. Reoxidation of bioreduced uranium under reducing conditions. *Environmental Science and Technology* 39 (16):6162–6169.

Wan, J. M., T. K. Tokunaga, Y. M. Kim, E. Brodie, R. Daly, T. C. Hazen, and M. K. Firestone. 2008. Effects of organic carbon supply rates on uranium mobility in a previously bioreduced contaminated sediment. *Environmental Science and Technology* 42 (20):7573–7579.

Wander, M. C. F., S. Kerisit, K. M. Rosso, and M. A. A. Schoonen. 2006. Kinetics of triscarbonato uranyl reduction by aqueous ferrous iron: A theoretical study. *Journal of Physical Chemistry A* 110 (31):9691–9701.

Wander, M. C. F., and K. L. Shuford. 2012. A theoretical study of the qualitative reaction mechanism for the homogeneous disproportionation of pentavalent uranyl ions. *Geochimica et Cosmochimica Acta* 84:177–185.

Wang, Y. H., M. Frutschi, E. Suvorova, V. Phrommavanh, M. Descostes, A. A. A. Osman, G. Geipel, and R. Bernier-Latmani. 2013a. Mobile uranium(IV)-bearing colloids in a mining-impacted wetland. *Nature Communications* 4: 2942.

Wang, Z., S.-W. Lee, J. G. Catalano, J. S. Lezama-Pacheco, J. R. Bargar, B. M. Tebo, and D. E. Giammar. 2013b. Adsorption of uranium(VI) to manganese oxides: X-ray absorption spectroscopy and surface complexation modeling. *Environmental Science and Technology* 47:850–858.

Wang, Z. M., S. W. Lee, P. Kapoor, B. M. Tebo, and D. E. Giammar. 2013c. Uraninite oxidation and dissolution induced by manganese oxide: A redox reaction between two insoluble minerals. *Geochimica et Cosmochimica Acta* 100:24–40.

Wang, Z. M., W. Xiong, B. M. Tebo, and D. E. Giammar. 2014. Oxidative UO2 dissolution induced by soluble Mn(III). *Environmental Science and Technology* 48 (1):289–298.

Watson, D. B., W. M. Wu, T. Mehlhorn, G. P. Tang, J. Earles, K. Lowe, T. M. Gihring, G. X. Zhang, J. Phillips, M. I. Boyanov, B. P. Spalding, C. Schadt, K. M. Kemner, C. S. Criddle, P. M. Jardine, and S. C. Brooks. 2013. In situ bioremediation of uranium with emulsified vegetable oil as the electron donor. *Environmental Science and Technology* 47 (12):6440–6448.

Webb, S. M., G. J. Dick, J. R. Bargar, and B. M. Tebo. 2005. Evidence for the presence of Mn(III) intermediates in the bacterial oxidation of Mn(II). *Proceedings of the National Academy of Sciences of the United States of America* 102 (15):5558–5563.

Webb, S. M., C. C. Fuller, B. M. Tebo, and J. R. Bargar. 2006. Determination of uranyl incorporation into biogenic manganese oxides using X-ray absorption spectroscopy and scattering. *Environmental Science and Technology* 40 (3):771–777.

Weber, K. A., J. C. Thrash, J. I. Van Trump, L. A. Achenbach, and J. D. Coates. 2011. Environmental and taxonomic bacterial diversity of anaerobic uranium(IV) bio-oxidation. *Applied and Environmental Microbiology* 77 (13):4693–4696.

Wehrli, B., B. Sulzberger, and W. Stumm. 1989. Redox processes catalyzed by hydrous oxide surfaces. *Chemical Geology* 78 (3–4):167–179.

Wellman, D. M., J. P. Icenhower, A. P. Gamerdinger, and S. W. Forrester. 2006a. Effects of pH, temperature, and aqueous material on the dissolution kinetics of meta-autunite minerals, (Na,Ca)2-1[(UO2)(PO4)2]2.3H2O. *American Mineralogist* 91:143–158.

Wellman, D. M., J. P. Icenhower, and A. T. Owen. 2006b. Comparative analysis of soluble phosphate amendments for the remediation of heavy metal contaminants: Effect on sediment hydraulic conductivity. *Environmental Chemistry* 3 (3):219–224.

Wellman, D. M., J. N. Glovack, K. Parker, and E. L. Richards. 2008. Sequestration and retention of uranium(VI) in the presence of hydroxylapatite under dynamic geochemical conditions. *Environmental Chemistry* 5:40–50.

Wellman, D. M., E. M. Pierce, and M. M. Valenta. 2007. Efficacy of soluble sodium tripolyphosphate amendments for the in-situ immobilisation of uranium. *Environmental Chemistry* 4:293–300.

Wersin, P., M. F. Hochella, P. Persson, G. Redden, J. O. Leckie, and D. W. Harris. 1994. Interaction between aqueous uranium(VI) and sulfide minerals: Spectroscopic evidence for sorption and reduction. *Geochimica et Cosmochimica Acta* 58 (13):2829–2843.

Williams, K. H., P. E. Long, J. A. Davis, M. J. Wilkins, A. L. N'Guessan, C. I. Steefel, L. Yang, D. Newcomer, F. A. Spane, L. J. Kerkhof, L. McGuinness, R. Dayvault, and D. R. Lovley. 2011. Acetate availability and its influence on sustainable bioremediation of uranium-contaminated groundwater. *Geomicrobiology Journal* 28 (5–6):519–539.

Wodzinski, R. J., and A. H. J. Ullah. 1996. Phytase. In: S. L. Neidleman and A. I. Laskin (eds.), *Advances in Applied Microbiology*. Vol. 42. San Diego, CA: Elsevier Academic Press Inc.

Wu, W. M., J. Carley, M. Fienen, T. Mehlhorn, K. Lowe, J. Nyman, J. Luo, M. E. Gentile, R. Rajan, D. Wagner, R. F. Hickey, B. H. Gu, D. Watson, O. A. Cirpka, P. K. Kitanidis, P. M. Jardine, and C. S. Criddle. 2006a. Pilot-scale in situ bioremediation of uranium in a highly contaminated aquifer. 1. Conditioning of a treatment zone. *Environmental Science and Technology* 40 (12):3978–3985.

Wu, W. M., J. Carley, T. Gentry, M. A. Ginder-Vogel, M. Fienen, T. Mehlhorn, H. Yan, S. Caroll, M. N. Pace, J. Nyman, J. Luo, M. E. Gentile, M. W. Fields, R. F. Hickey, B. H. Gu, D. Watson, O. A. Cirpka, J. Z. Zhou, S. Fendorf, P. K. Kitanidis, P. M. Jardine, and C. S. Criddle. 2006b. Pilot-scale in situ bioremedation of uranium in a highly contaminated aquifer. 2. Reduction of U(VI) and geochemical control of U(VI) bioavailability. *Environmental Science and Technology* 40 (12):3986–3995.

Wu, W. M., J. Carley, J. Luo, M. A. Ginder-Vogel, E. Cardenas, M. B. Leigh, C. C. Hwang, S. D. Kelly, C. M. Ruan, L. Y. Wu, J. Van Nostrand, T. Gentry, K. Lowe, T. Mehlhorn, S. Carroll, W. S. Luo, M. W. Fields, B. H. Gu, D. Watson, K. M. Kemner, T. Marsh, J. Tiedje, J. Z. Zhou, S. Fendorf, P. K. Kitanidis, P. M. Jardine, and C. S. Criddle. 2007. In situ bioreduction of uranium (VI) to submicromolar levels and reoxidation by dissolved oxygen. *Environmental Science and Technology* 41 (16):5716–5723.

Yao, M. Z., Y. H. Zhang, W. L. Lu, M. Q. Hu, W. Wang, and A. H. Liang. 2012. Phytases: Crystal structures, protein engineering and potential biotechnological applications. *Journal of Applied Microbiology* 112 (1):1–14.

Zachara, J. M., P. E. Long, J. Bargar, J. A. Davis, P. Fox, J. K. Fredrickson, M. D. Freshley, A. E. Konopka, C. Liu, J. P. McKinley, M. L. Rockhold, K. H. Williams, and S. B. Yabusaki. 2013. Persistence of uranium groundwater plumes: Contrasting mechanisms at two DOE sites in the groundwater-river interaction zone. *J Contam Hydrol* 147:45–72.

Zeng, H., and D. E. Giammar. 2011. U(VI) reduction by Fe(II) on hematite nanoparticles. *Journal of Nanoparticle Research* 13 (9):3741–3754.

Zhuang, K., E. Ma, D. R. Lovley, and R. Mahadevan. 2012. The design of long-term effective uranium bioremediation strategy using a community metabolic model. *Biotechnology and Bioengineering* 109 (10):2475–2483.

10 Uranium Interaction with Soil Minerals in the Presence of Co-Contaminants

Case Study of Subsurface Sediments at or below the Water Table

Brandy N. Gartman and Nikolla P. Qafoku

CONTENTS

10.1 INTRODUCTION AND OBJECTIVES

Many uranium (U) surface- and subsurface-contaminated sites are present around the world, mainly because of its essential role in the production of plutonium for nuclear weapons and other nuclear energy and research activities (Riley et al. 1992; Todorov and Ilieva 2006). Anthropologic sources of U contamination fall into four categories: (1) U from weapon production, (2) U from nuclear energy activities, (3) U from scientific and other uses, and (4) depleted U in army testing grounds and battlefields. Elevated concentrations of U are also found in some agricultural drainage waters that are associated with phosphate fertilizers (Duff and Amrhein 1996).

The presence of U in different environments and conditions that may promote high solubility and mobility is seriously threatening human health. For this reason, understanding sequential and/or simultaneous chemical, biological, and hydrological processes and reactions that affect and/or control U behavior and govern its fate in contaminated media remains one of the most important issues that the scientific community is currently facing as a part of remedial efforts to restore contaminated sites. In addition, an elucidation of the mechanisms of U interactions with soil minerals is an important step in gaining insights into the nature of U contamination and the controlling factors for U short- and long-term behavior and mobility at contaminated sites. Finally, an evaluation of the risks posed by U contamination requires knowledge about U interaction with soils and sediments and transport in groundwater, typically the most significant pathway of exposure to humans (Bargar et al. 1999). For all these reasons, the focus in this chapter is mainly on processes, reactions, and mechanisms controlling U mobility and transport in contaminated subsurface systems that are either close to or below the water table because of concerns related to contamination of the groundwater.

10.1.1 General Characteristics of a Representative U-Contaminated Site

This chapter covers important aspects of U-contaminated sites and the aqueous hexavalent U [U(VI)] interactions with soil minerals in one representative contaminated subsurface system in the United States, that is, the one at Rifle (Colorado, USA), although there are many U-contaminated sites around the world with similar characteristics. One other important site in the United States is the Hanford site (Washington). There have been many studies conducted at this site, and the reader is referred to the following publications for a description of the U contamination at this site (Lindberg and Peterson 2004; Serne et al. 2002, 2001; Szecsody et al. 2013; Zachara et al. 2005).

10.1.1.1 The Rifle Site

The site we will be focusing our attention on in this chapter is the Department of Energy's (DOE's) Integrated Field Research Challenge (IFRC) site at Rifle, Colorado, which overlies a shallow alluvial aquifer adjacent to the Colorado River. The subsurface of this former milling site is contaminated not only with U but also with V, As, Se, Zn, Cu, Cr, and other co-contaminants. In addition, groundwater monitoring efforts over the past decade show varying patterns of behavior of dissolved metals, with significant fluctuations in aqueous U, V, As, and Se that are associated with seasonal variations in water level and organic carbon addition during remediation activities (Qafoku et al. 2014). Based on a distribution coefficient (K_d) model coupled to a groundwater flow and transport model, the U plume was predicted to attenuate below drinking water standard in approximately 10 years. However, the core of the plume has changed little over more than a decade and, at least in one area, aqueous U concentrations have increased (Qafoku et al. 2014). A key challenge at the site is unraveling the complex biogeochemical and hydrological processes that control contaminant mobility and gaining an understanding of variations in the solubility of such contaminants under changing redox conditions. Systematic macroscopic, microscopic, and spectroscopic studies are required to provide useful insights about U fate and interactions in subsurface sediments (an example of such studies is included later in this chapter). The most abundant minerals of the less than 2 mm size fraction separated from the background sediment at Rifle are quartz (55%), plagioclase (20%), K-feldspars (15%), micas (6%), amphibole (2%), and calcite (2%) (Campbell et al. 2012). The clay fraction (<2 μm) is dominated by illite (59%), smectite (25%), chlorite (8%), and kaolinite (8%) (Campbell et al. 2012). The percentage of chlorite [chlorite has strong affinity for U(VI) (Baik et al. 2004)] is greater in the naturally bioreduced zones (NRZs) of the subsurface (Campbell et al. 2012), which have high abundances of recalcitrant or lignitic organic matter compounds; these zones have a remarkable assortment of potential contaminants' host minerals such as framboid pyrite, Fe sulfides, siderite, and potentially complex sulfides of different metals (Qafoku et al. 2014, 2009).

10.1.2 Chapter Objectives

The objectives of this chapter are to: (1) provide a general overview of the contamination levels of U and other co-contaminants in a representative site in the United States, that is, the Rifle site (CO, USA); (2) review and discuss different aspects of mineral–contaminant interactions that focus on surface complex stability and contaminant oxidation state; (3) present results from a recent systematic macroscopic, microscopic, and spectroscopic study conducted with sediments collected in Rifle, Colorado, as an example of current efforts and the state-of-knowledge of sites contaminated with various contaminants; and (4) offer insightful concluding remarks and future research needs on controlling reactions and processes of U and other contaminants' fate and behavior under hydraulically saturated conditions (conditions that are representative of those found in the field at or below the water table).

In this chapter, we discuss different aspects of U(VI) mobility and interactions with minerals in the absence and presence of co-contaminants. However, even though this chapter covers different aspects of U behavior and interaction with an array of soil minerals that are relevant to many sites worldwide, the discussion focuses mainly on recent findings from research conducted with sediments from Rifle. Important aspects of U interaction with organic matter, biologically mediated reactions and reduction of U(VI) to U(IV), the potential for reoxidation of U(IV) back to U(VI), and modeling techniques that are used to fit experimental data and predict U behavior have been effectively described in many other publications and are not within the scope of this chapter.

10.2 URANIUM INTERACTION WITH MINERALS

The movement of U through contaminated soil and subsoil environments may be retarded because of interactions with the soil minerals, which affect and/or even control U mobility and residence time in contaminated subsurface natural systems. Many studies conducted over the past decades have provided clear and unequivocal evidence for such interactions in various surface and subsurface terrestrial systems that contain a large assortment of minerals and are characterized by a different degree of spatial scale-dependent, chemical, physical, and mineralogical heterogeneities.

10.2.1 Oxidized Conditions

Mineral-U contaminant interactions have been the subject of many studies (see Qafoku and Icenhower 2008 and references therein). For example, contaminant U may interact with (1) organic matter (Bruggeman et al. 2012; Joseph et al. 2011, 2013a, 2013b; Schmeide et al. 2000, 2003; Steudtner et al. 2011), (2) Fe oxides (Bargar et al. 1999; Dodge et al. 2002; Duff and Amrhein 1996; Ho and Miller 1986; Hsi and Langmuir 1985; Jang et al. 2007; Lefevre et al. 2006; Moyes et al. 2000; Reich et al. 1998; Rovira et al. 2007; van Geen et al. 1994; Villalobos and Leckie 2001; Waite et al. 1994; Wazne et al. 2003), (3) phyllosilicates (Catalano and Brown 2005), (4) muscovite (Gomez et al. 2013), (5) titanium oxide (Lefevre et al. 2008), and (6) calcite (Kelly et al. 2003).

Minerals such as Fe oxides, phyllosilicates, and calcite are common at U(VI)-contaminated sites with predominant oxidizing conditions, and they play an important role in determining U(VI) mobility. Carbonate coatings may have a significant influence on U(VI) mobility, because they may cover the surfaces of other reactive phases that would otherwise sorb U(VI), such as iron oxyhydroxides and aluminous phases, and, for this reason, U(VI) sorption is sometimes governed by its interaction with calcite (Dong et al. 2005). In addition to these minerals, other minerals such as quartz (Ilton et al. 2012), micas (Ilton et al. 2006), and chlorite (Singer et al. 2009) are good U sorbents as well. On the other hand, a variety of Fe(II)-bearing minerals such as biotite, chlorite (clinochlore), and magnetite may be present in sediments under predominantly oxidizing conditions, which may serve as electron donors and/or may release Fe(II) into the environment on mineral dissolution.

10.2.2 Reduced Conditions

A representative U(VI)-contaminated site where a significant portion of the subsurface is predominantly reduced is the Rifle site in Colorado (USA). U solubility and subsequent mobility change dramatically with redox status; for example, hexavalent U is relatively mobile, although it may undergo some sorption onto reactive mineral surfaces, whereas tetravalent U [U(IV)] is nearly immobile. Studies have been conducted to investigate U(VI) sorption and reduction in these natural sediments (Fox et al. 2013) and minerals that are abundant in sediments under reduced conditions, such as Fe sulfides, for example, pyrite (Aubriet et al. 2006; Eglizaud et al. 2006; Scott et al. 2007; Wersin et al. 1994), magnetite (Singer et al. 2012), mackinawite (Gallegos et al. 2013), biogenic mackinawite (Veeramani et al. 2013), and biogenic mangnetite and vivianite (Veeramani et al. 2011). Structural, sorbed, and/or aqueous Fe(II) may get involved in redox reactions with U(VI), transforming it into less soluble U(IV) (Fox et al. 2013).

Field sampling and characterization of the shallow, unconsolidated alluvial aquifer sediments at the Rifle IFRC demonstrate that unusually high solid phase-U concentrations are associated with the NRZs (Campbell et al. 2012; Kukkadapu et al. 2010); for example, one of these zones contains up to 50 times more U than the typical alluvial sediment at Rifle (Campbell et al. 2012). The NRZs have high abundances of reduced Fe- and S-bearing minerals (Qafoku et al. 2014). Recent studies have shown that in the presence of reducing conditions, high concentrations of organic matter, and reduced Fe and S, microbial activity may promote biogenic mineral formation, especially within NRZs that are uncommon for oxidized sediments. For example, studies have shown that one NRZ at Rifle contained a unique mineral and U host, that is, framboidal pyrite (Qafoku et al. 2009). Other studies have found that a variety of unique minerals, such as siderite, sulfides of different metals such as Fe sulfides [e.g., greigite, which is an Fe(II)/Fe(III) sulfide that serves as a pyrite precursor], Zn sulfides (e.g., framboidal zinc sulfide), multielement sulfides [e.g., (Fe, Ni, Cu, Zn)S], and cupric selenide (CuSe) that may serve as U and/or other contaminant hosts (sorbents and/or electron donors/acceptors) (Qafoku et al. 2014) are also present in the NRZs. This is definitely an area that requires additional attention and research.

10.2.3 Co-Contaminant Effects

One unique characteristic of some U-contaminated sites is that they contain an assortment of co-contaminants that may affect U mobility in contaminated subsurface environments in different ways. Previous studies have suggested that co-contaminants may compete with U for sorption sites and/or available electrons (Campbell et al. 2012; Qafoku et al. 2014). However, micron-scale interactions and spatial distribution and associations of U and other co-contaminants are currently not well studied and documented in the published literature. The degree of U reduction and the identity of the U-bearing solid phases that may interact with co-contaminants are not well known, further reinforcing the need for detailed characterization of these unique contaminated subsurface systems.

A series of studies conducted over the past few years to investigate U behavior in the Rifle subsurface sediments has demonstrated that these sediments had significant solid-phase concentrations of U and other co-contaminants (Campbell et al. 2012). This suggests that competing sorption reactions and complex temporal variations in dissolved contaminant concentrations occur in response to transient redox conditions, compared with single contaminant systems (Qafoku et al. 2014). The network of simultaneous and/or sequential reactions and processes in the Rifle sediments is extremely complex, making the understanding of contaminant fate and mobility challenging. Studies have suggested that a combination of assorted solid-phase species and an abundance of redox-sensitive contaminants may slow U(IV) reoxidation rates, effectively enhancing the stability of U(IV) sequestered via natural attenuation, impeding rapid U flushing, and turning these unique subsurface environments into sinks and long-term, slow-release sources of U contamination to groundwater (Qafoku et al. 2014). Further research efforts are needed to investigate U sorption (adsorption, reduction, precipitation) in the presence of co-contaminants of different types (i.e., redox sensitive or not).

10.3 RESULTS FROM A MACROSCOPIC, MICROSCOPIC, AND SPECTROSCOPIC STUDY

An example of a macroscopic, microscopic, and spectroscopic study is presented next to illustrate current efforts and the state-of-knowledge in this important research area. This study is a combination of macroscopic experiments (e.g., wet chemical batch extractions and column experiments), bulk characterization analyses [e.g., x-ray powder diffraction (XRD)], micron-scale inspections [e.g., scanning electron microscopy with energy-dispersive x-ray spectroscopy (SEM-EDS), micro x-ray fluorescence (μXRF)], and molecular-scale interrogations [e.g., x-ray absorption near edge structure (XANES)]. The sediments used in this study were from the Plot C of the Rifle Integrated Field Research Center (IFRC), Colorado (USA). The main focus of the research work at this IFRC is to reduce the size of the U groundwater plume below the site and to potentially contaminate the Colorado River. One remediation strategy is to form insoluble U(IV) phases through promoting reducing conditions, similar to those found in NRZ (Long et al. 2008). This technique was implemented at the Rifle site through the injection of acetate as an *in situ* method of bio-reduction for long-term bioremediation.

As a part of this investigation, initially, a series of wet chemical extractions were conducted with Rifle sediments, which were well preserved in Mylar bags and stored in a −80°C freezer. They were dried under anoxic conditions and sieved to separate the less than 2 mm size fraction, which was then used for the wet chemical extractions. Fifteen sediments were chosen across the gallery of wells at Plot C to represent spatial and depth variations. For each extraction, 2.5 g of sediment was used with 25 mL of extraction solution; 2 mL aliquots were removed at one-, two-, seven-, and fourteen-day increments to examine solution chemistry changes with time. The following extraction solutions with increasing extraction strengths were used: (1) double de-ionized (DDI) water, (2) a carbonate/bicarbonate solution (0.0144 M $NaHCO_3$/0.0028 M Na_2CO_3), (3) a 0.5 M HNO_3 solution, and (4) an 8 M HNO_3 solution. In addition, the microwave digestion technique (complete digestion of the sediment matrix) was employed to determine total U and co-contaminant concentrations in these sediments. The extract solutions were analyzed with inductively coupled plasma optical emission spectrometry (ICP-OES) and inductively coupled plasma mass spectrometry (ICP-MS) for common cations and trace metals, respectively. This part of the investigation was conducted with the objective to: (1) evaluate the removable amounts of the contaminants from the sediments exposed to different extracting solutions, (2) determine whether the sediments were contaminated with other co-contaminants, in addition to U, and (3) study time-dependent contaminant releases.

The results from extractions conducted with four sediments that contained the highest U concentration (μmol/kg) showed that the DDI water and carbonate solution extracted only small amounts of U and the concentration changed only slightly as a function of time (360 h of experimental time) (Figure 10.1). The nitric acid solutions extracted greater U amounts from the sediments, indicating that the majority of the U pool was strongly held (i.e., did not desorb with the carbonate solution), and that it was associated with phases that are unstable under acid conditions, suggested by increases in the aqueous concentration of Fe (about 5 mg/g), Al (about 2.5 mg/g), Ca (about 20 mg/g), and Si (about 1.3 mg/g). One should emphasize, however, that at least in three sediments, only about one out of three of the total U present in these sediments (determined with the microwave digestion technique) was extracted with the two nitric acid solutions.

Because As and V were determined to be two important co-contaminants (As and V were released in all extraction experiments as compared with Cr, Mo, and Se, which were only released in select samples or extractions), the following discussion will be focused on these three co-contaminants (i.e., U, As, and V). The amount of As released from four sediment samples that contained the highest U concentrations was small in the DDI and carbonate extractions (Figure 10.2). By comparison, results showed that more As was extracted over time with the nitric acid solutions. Similarly, much more V was extracted with the nitric acid solutions than with the DDI water or the carbonate

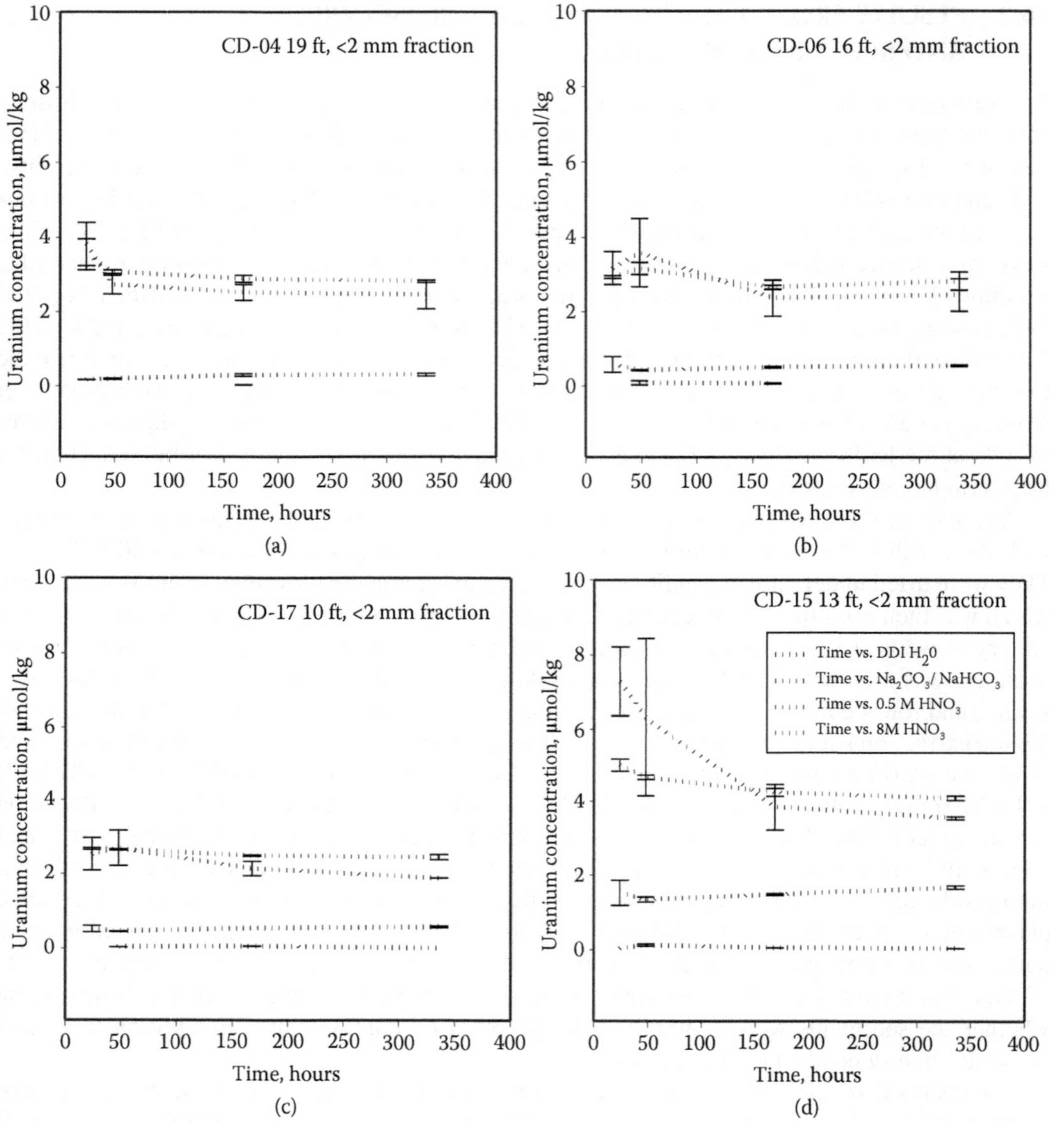

FIGURE 10.1 Results from extractions performed on four Plot C sediments over a 2-week extraction time: U extract concentration. Total U concentrations in the sediments, determined with the microwave digestion, were: (a) CD-04 19 ft below ground surface, U = 11.62 μmol/kg; (b) CD-06 16 ft, U = 12.86 μmol/kg; (c) CD-17 10 ft, U = 11.52 μmol/kg; and (d) CD-15 13 ft, U = 10.07 μmol/kg. For the DDI water extractions, the U concentration was below the detection limit.

solution (Figure 10.3). The sediment samples containing the highest As (i.e., CD-01 13 ft below ground surface) and V (i.e., CD-04, 13 ft) concentrations showed similar U versus As (Figure 10.4), and U versus V (Figure 10.5) releases, suggesting similar sorption mechanisms and interactions with minerals. These three major contaminants exhibited similar behavior during these extraction experiments.

An additional study was conducted with four sediments that were collected in the same well (CD-04) at different depths (i.e., 10, 13, 16, and 19 ft below the ground surface) to characterize the U distribution in the sediment profile. The highest measured U concentrations were found in the CD-04 13 ft sediment for all extraction solutions (Figure 10.6). For the DDI water extract, the

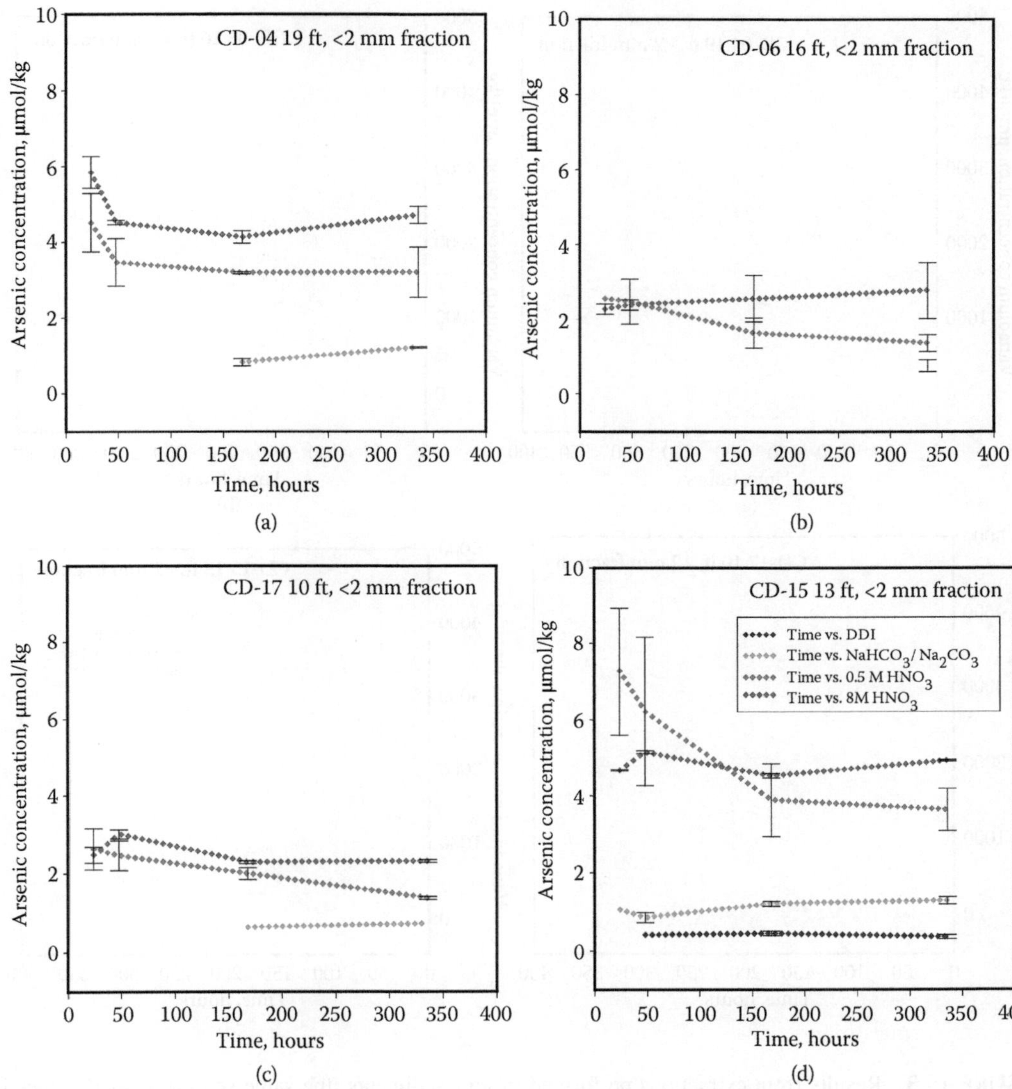

FIGURE 10.2 Results from extractions performed in four sediments (the same sediments as the ones in Figure 10.1): As extract concentration versus extractant type. For some DDI water and bicarbonate–carbonate extractions, As concentration was below the detection limit.

U concentration was below the detection limit for the first two data points, but it was measureable for the third and fourth periods. The CD-04 13 ft sediment also released greater amounts of As (for all extracting solutions) (Figure 10.7).

The release of U and As was further investigated in a column experiment that was conducted with the CD-04 13 ft sediment under controlled atmosphere conditions (95% nitrogen and 4% hydrogen). This was done by packing the less than 2 mm fraction of the CD-04 13 ft sediment in a column (5.7 cm long and with a 2.4 cm inner diameter). Groundwater from Plot C at Rifle, Colorado (with a U concentration of 191 μg/L) was injected through the system with an average flow rate of 0.096 mL/min and alternating flow cycles (i.e., continuous flow and stop flows). Each flow event ran for approximately 12 pore volumes (24 h), whereas intermittent stop-flow events lasted for 24, 72, 264, and 312 h. Before dismantling the column, a pulse of U-rich Rifle

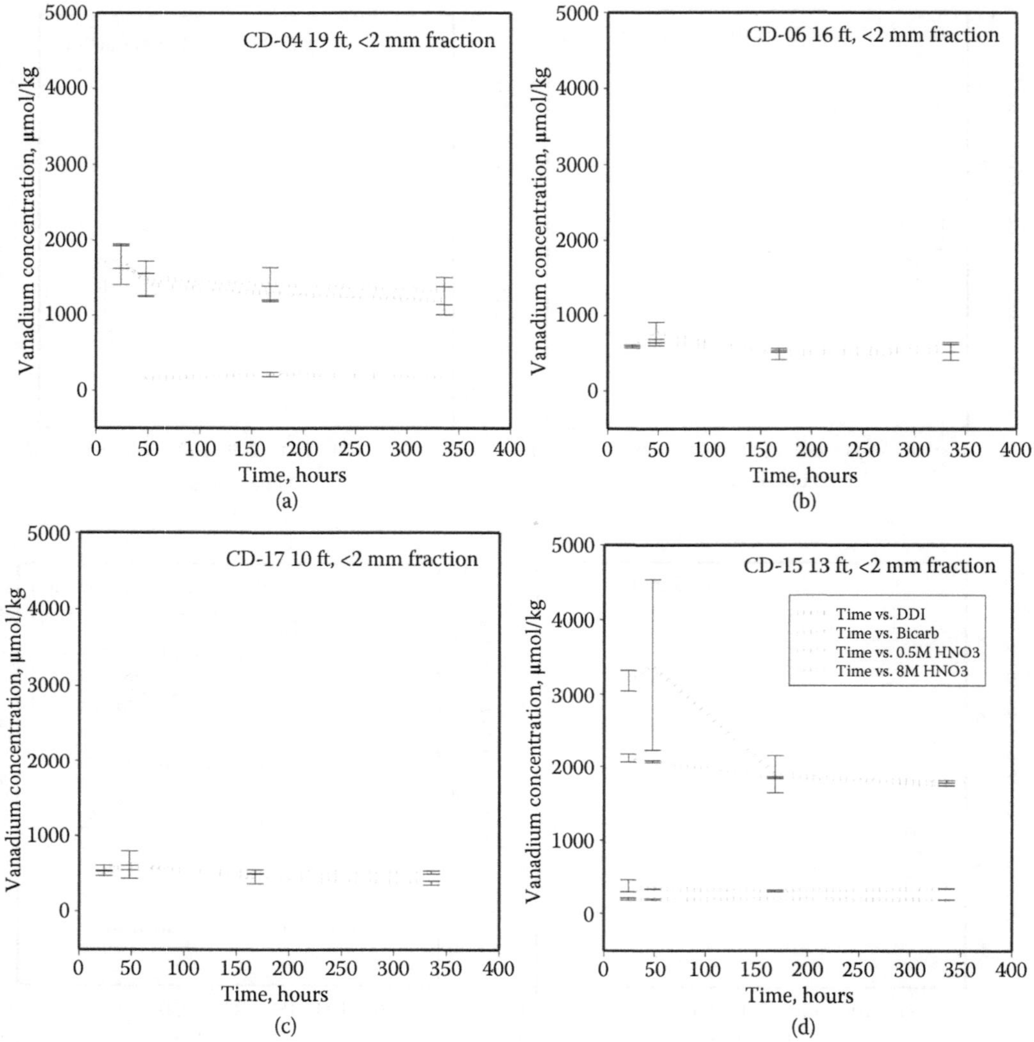

FIGURE 10.3 Results from extractions performed in four sediments (the same sediments as the ones in Figure 10.1): V concentration versus extractant type. For some DDI water and bicarbonate–carbonate extractions, V concentration was below the detection limit.

groundwater (10 ppm in U) was injected into the column. This was done to increase the sorbed U fraction in the sediment, which would facilitate efforts of studying U fate and finding U hotspots while taking μXRF measurements at the Advance Photon Source (APS) at Argonne National Laboratory (USA).

Figures 10.8 through 10.10 show changes in U, As, and Mo versus the pore volumes of groundwater that passed through the column. U release in the column effluents followed an irregular decreasing trend until they reached the U influent concentration (Figure 10.8). Effluent U concentrations were significantly variable during this experiment, especially during the stop-flow events, indicating that more than one process and/or reaction was controlling U desorption into the column effluent. Arsenic effluent concentrations showed an increasing trend during continuous flow, but the effluent concentrations were consistently lower than the influent concentration, and they decreased significantly, especially during the two long stop-flow events, indicating that As was being sorbed in the sediments (Figure 10.9); simultaneously, U was desorbed (Figure 10.8). The effluent

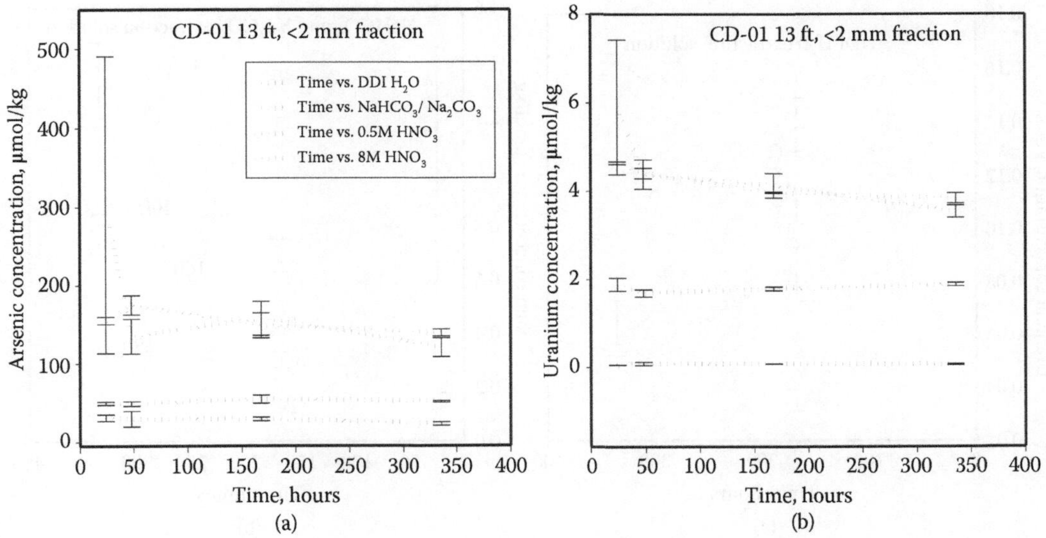

FIGURE 10.4 As and U extract concentrations from the sediment with the highest As concentration. The total U concentration in this sediment that was measured with the microwave digestion technique was 8.20 μmol/kg.

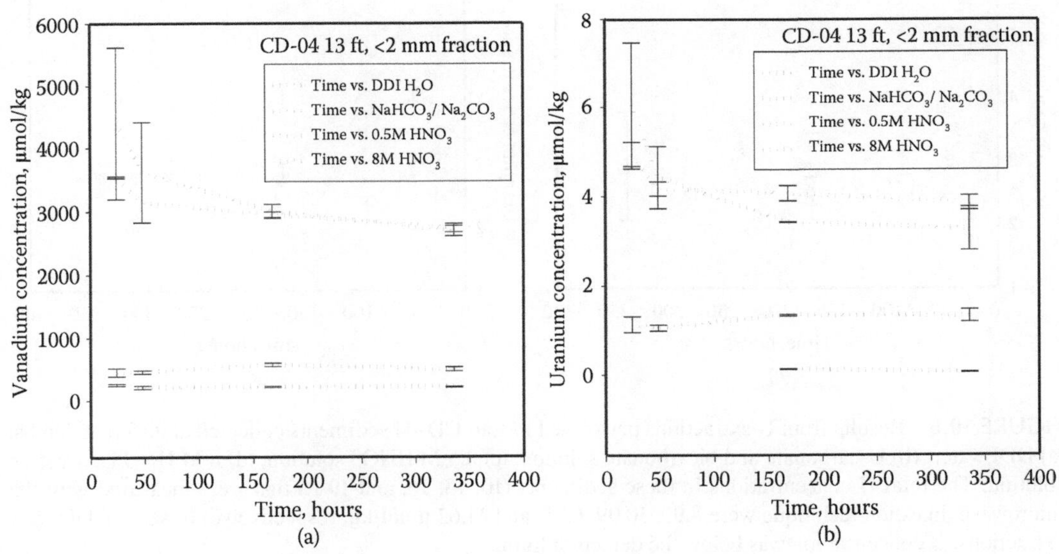

FIGURE 10.5 V and U extract concentrations from the sediment with the highest V concentration. The total U concentration in this sediment that was measured with the microwave digestion technique was 10.09 μmol/kg. For some DDI water extractions, U concentration was below the detection limit.

As concentration never reached the influent concentration for the duration of this experiment, indicating that the As injected into the column with the influent solution was continuously retained by the sediment. Rates of releases of both these contaminants were greater during continuous-flow than during stop-flow events, indicating nonlinear time dependency of the processes or reactions controlling their aqueous concentrations. A much more regular release curve was observed for Mo (Figure 10.10); its effluent concentrations decreased steadily during continuous-flow events and

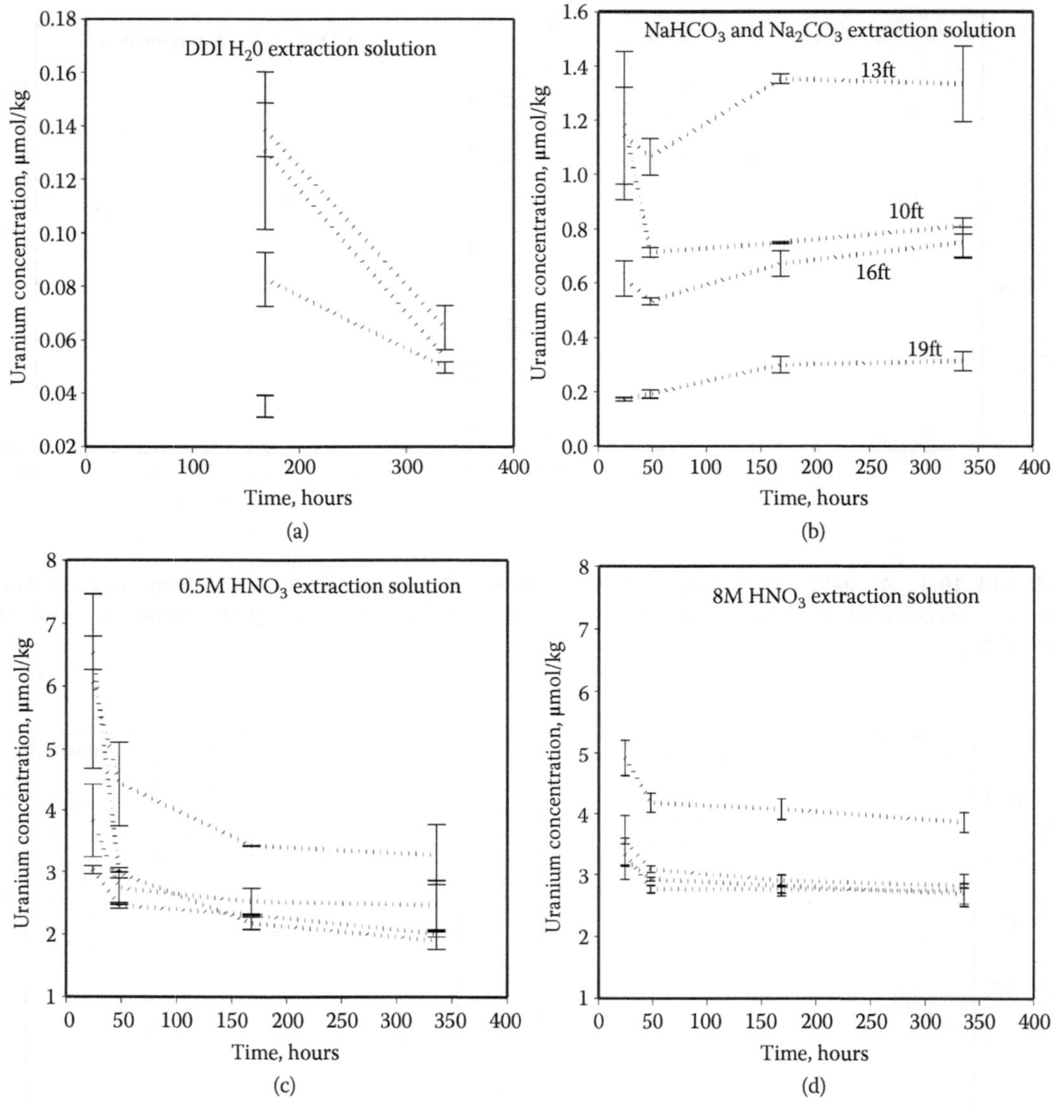

FIGURE 10.6 Results from U extractions performed in four CD-04 sediments collected at different depths. (a) DDI water; (b) U-carbonate and bicarbonate solution; (c) 0.5 M HNO3 solution; (d) 8 M HNO3 extraction solution. The total U concentrations in these sediments (10, 13, 16, and 19 ft) that were measured with the microwave digestion technique were 8.99, 10.09, 6.95, and 11.62 μmol/kg, respectively. For some DDI water extractions, U concentration was below the detection limit.

increased after each stop-flow event, indicating that a time-dependent desorption reaction or dissolution of an Mo-bearing phase was controlling Mo aqueous concentration, reaching equilibrium with the influent concentration at about 15.2 μg/L (Figure 10.10).

Micron-scale inspections of less than 2 mm sized sediment particles were conducted with SEM/EDS to study morphological features of the different phases that are present in Rifle site sediments. The samples chosen for analysis were from Plot C borehole CD-04 10 and 19 ft, and from borehole CD-06 10 and 16 ft; this set included the sediments with the highest U content, that is, CD-06 16 ft and CD-04 13 ft. These samples were examined by using a JEOL 6340f SEM. Results showed the presence of Fe oxide framboid-like structures (Figure 10.11), sulfides, quartz, and silicates

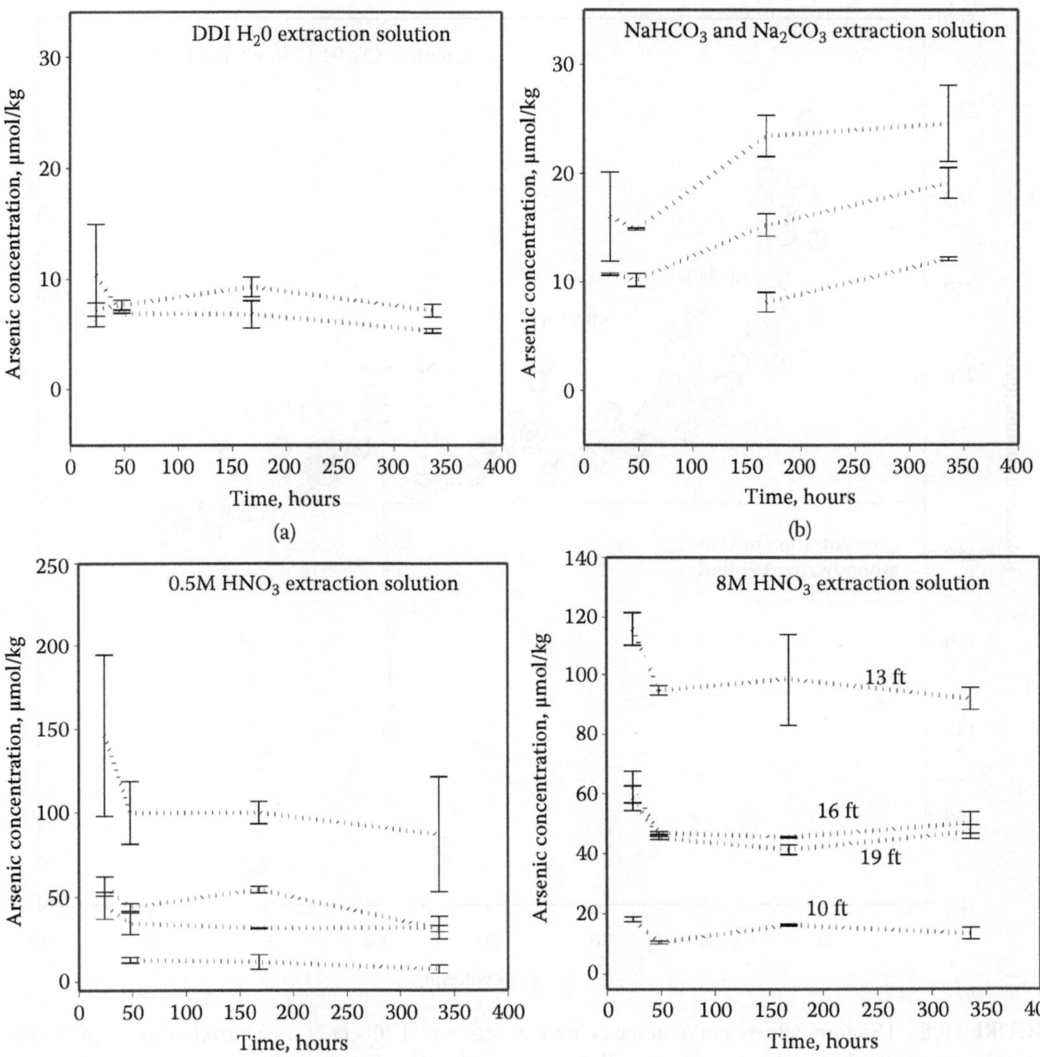

FIGURE 10.7 Results from As extractions performed in four CD-04 sediments collected at different depths. (a) DDI water; (b) U-carbonate and bicarbonate solution; (c) 0.5 M HNO3 solution; (d) 8 M HNO3 extraction solution. For some DDI water and bicarbonate–carbonate extractions, As concentration was below the detection limit.

(Figure 10.12). All these minerals can interact with U via adsorption and/or reduction reactions. The presence of Fe oxide framboids is an indication that although these sediments were collected below the water table (about 10 ft below ground surface at the Rifle site), they were more oxidized than other sediments of the Rifle subsurface where the organic matter is substantially greater and microbial activity has made the system a more reduced one. The presence of other co-contaminants (such as V, Cr, Zn, and Cu) was also confirmed from the EDS measurements. These co-contaminants may affect U behavior in these sediments.

A series of μXRF elemental mapping and XANES analyses were also conducted on these sediments. Thin sections were prepared for analysis at the APS for elemental mapping by using μXRF. The beamline 20-ID-A (Heald et al. 2007) was used. Rh-coated K-B mirrors that operated

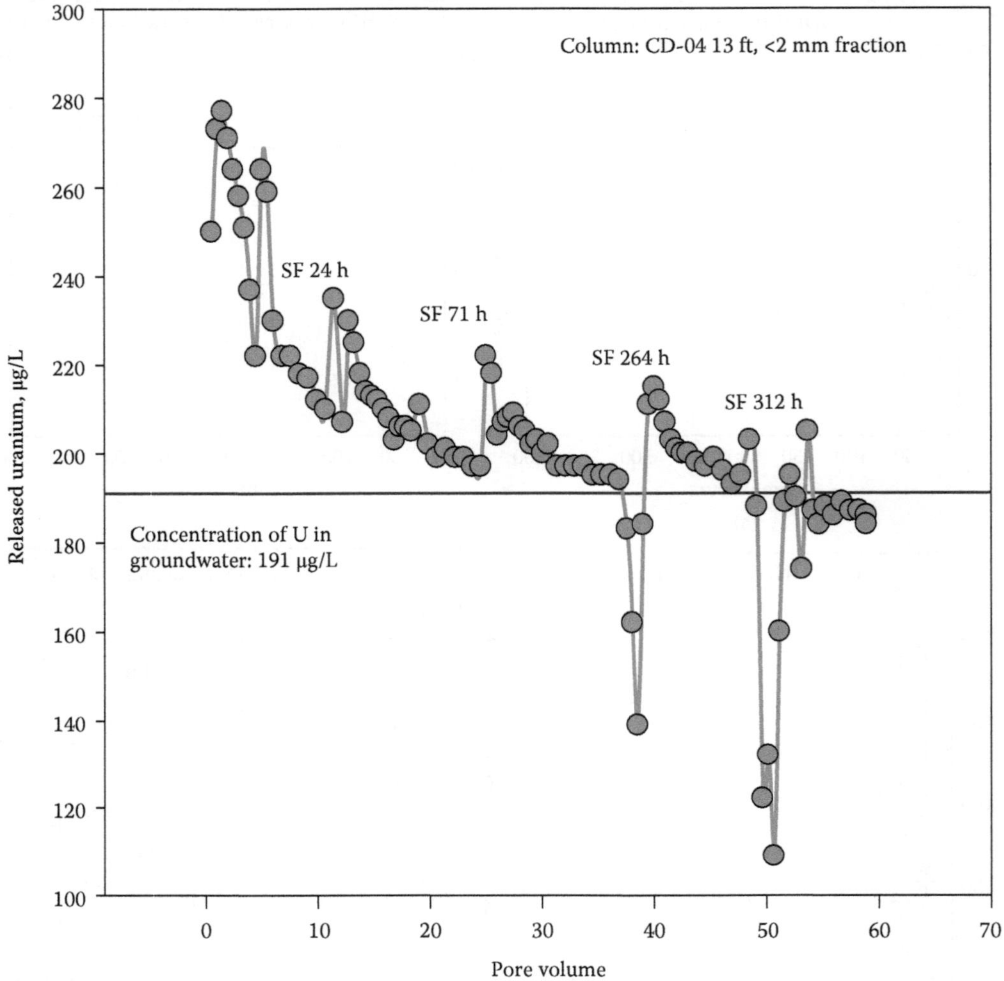

FIGURE 10.8 Uranium release curve in the column experiment. Effluent U concentration was significantly variable, especially during the two long stop-flow events, indicating that more than one process and/or reaction was controlling U release into the column effluent. The 10 ppm U solution was injected at the end of this experiment.

at 3 mrad incident angles were used to focus the beam on 3 μm × 4.8 μm "spots" (CD-06 16 ft, <53 μm fraction) and on 5.6 μm × 4.8 μm (CD-04 13 ft column with U spike, <2 mm fraction). Polished thin sections were placed in the x-ray beam at 45° with respect to the incident x-ray beam direction. The sample fluorescence was measured by using a four-element Vortex detector from SII Nanotechnology. The signals for the various elements were separated by using electronic regions of interest (ROI) around their corresponding peaks. These signals are corrected for dead time and normalized to I_0 measured in an N_2 gas-filled miniature ion chamber located just in front of the sample. For the U peak in the spectrum, there was significant overlap with the peaks from Sr and Rb. The amount of this overlap was determined by scanning through the Sr and Rb edges, and then subtracted from the measured U ROI to obtain an estimated U signal. A U(VI) standard reference sample was measured by using scattered radiation (Cross and Frenkel 1998) using a uranyl nitrate standard. The peak of the first derivative of the uranyl standard was calibrated to 17,171 eV. The XANES spectra from the sediment samples were aligned to the standard that was collected simultaneously with the sample. The sediment spectra were then compared with the

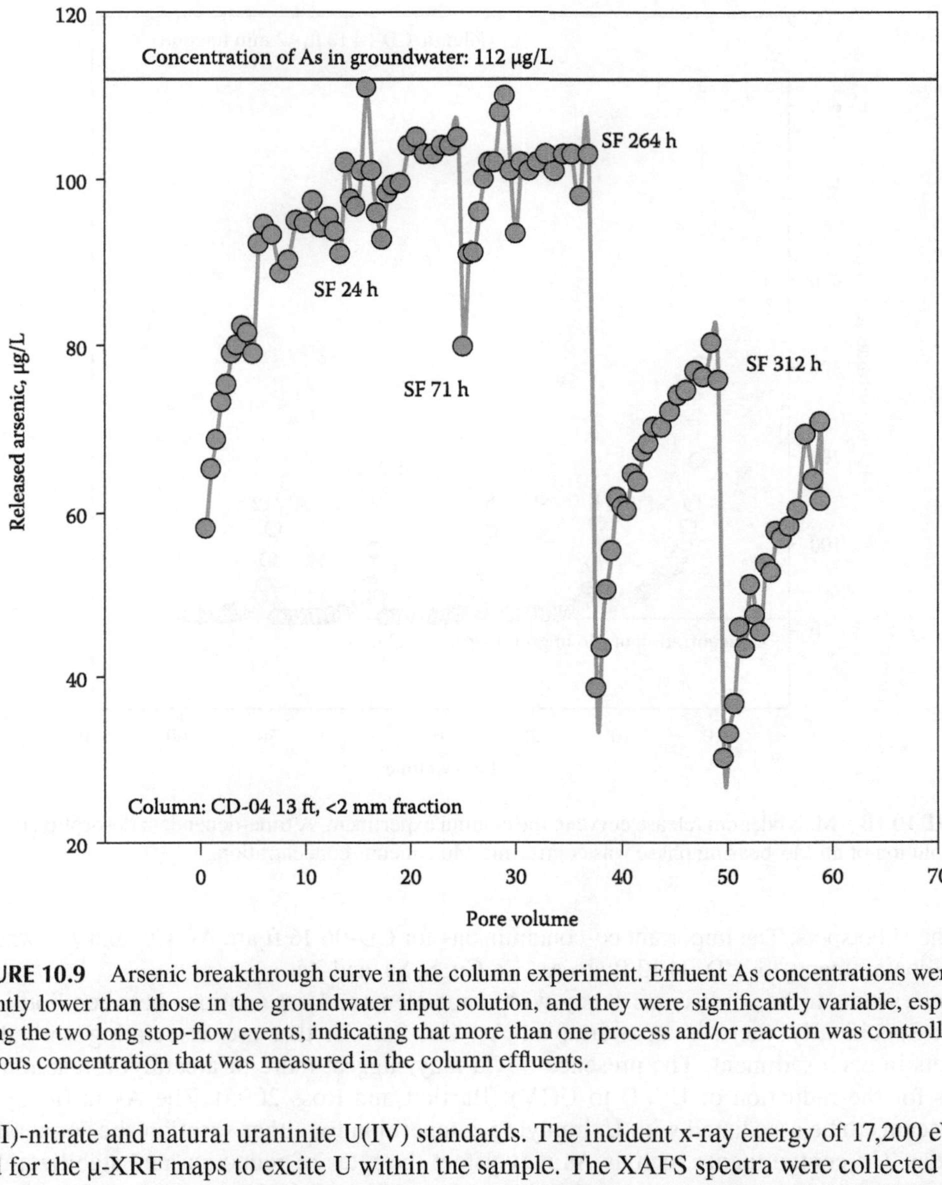

FIGURE 10.9 Arsenic breakthrough curve in the column experiment. Effluent As concentrations were consistently lower than those in the groundwater input solution, and they were significantly variable, especially during the two long stop-flow events, indicating that more than one process and/or reaction was controlling As aqueous concentration that was measured in the column effluents.

U(VI)-nitrate and natural uraninite U(IV) standards. The incident x-ray energy of 17,200 eV was used for the μ-XRF maps to excite U within the sample. The XAFS spectra were collected in the step scanning mode and were processed by using IFEFFIT (Newville 2001) and ATHENA (Ravel and Newville 2005) software.

The μXRF maps and XANES analyses were completed for two different sediments (CD-06 16 ft, <53 μm and CD-04 13 ft, <2 mm fraction, with U spike). The main objective was to interrogate the samples looking for locations of high concentrations of U and determining the U-valence state. For a comprehensive study of our samples, they were first examined in larger areas (1,000 μm × 1,000 μm); then, the areas of higher U concentrations were reexamined with a smaller elemental map (60 μm × 60 μm or 50 μm × 50 μm) (Figures 10.13 and 10.14). The micron-scale spatial distribution of other elements such as As, Ca, Cr, Cu, Fe, K, Mg, Mn, Rb, S, Se, Sr, V, and Zn was investigated as well. In the U hotspots of the CD-06 16 ft sediment sample, XANES analysis determined that the U was present as mainly U(IV) (Figure 10.15); whereas in the CD-04 13 ft sediment, the U was present as mixed U(VI) and U(IV) (Figure 10.16). On closer inspection of elemental maps generated for these U hotspots, there seem to be other redox-sensitive elements that are associated

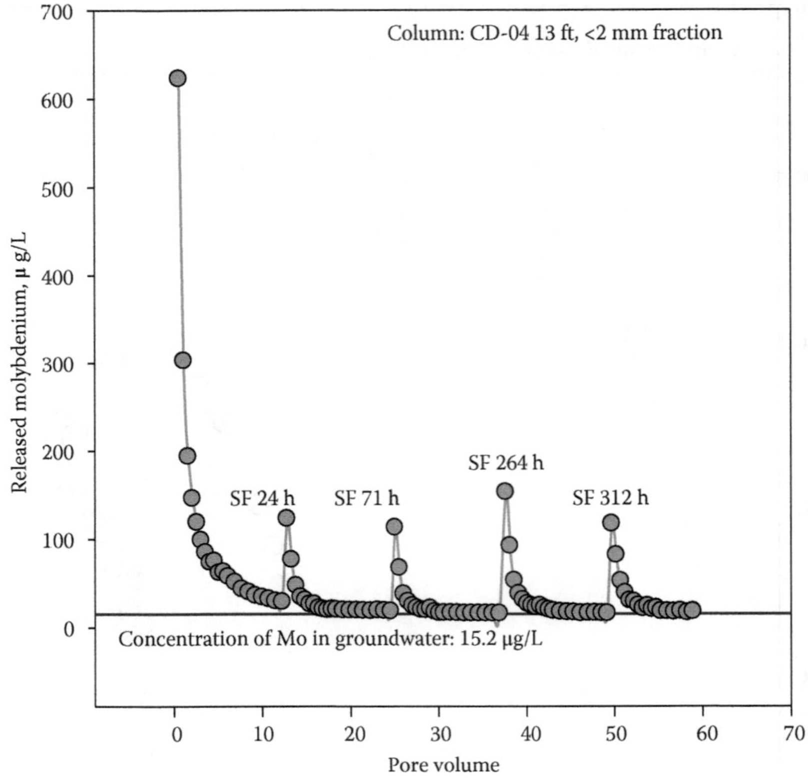

FIGURE 10.10 Molybdenum release curve in the column experiment. A time-dependent desorption reaction or dissolution of an Mo-bearing phase was controlling Mo aqueous concentration.

with the U hotspots. The important co-contaminants for CD-06 16 ft are As, Cu, and Zn; whereas the co-contaminants for CD-04 13 ft are As, Cr, Cu, S, Se, and Zn.

There seems to be some consistency with some elemental associations between both sediments. Specifically, there is an overlap of As, Cu, and Zn that are associating with the U hotspots in each sediment. The presence of Cu may suggest more of a catalytic reaction that allows for the reduction of U(VI) to U(IV) (Bartlett and Ross 2005). The As in the system again seems to be associated with U, implying competition for either sorption sites or available electrons. Co-contaminants such as As may affect the rates of reduction and stabilization of the U present in the system, such as: (1) coating the surface of the U and thus shielding it from oxidizing elements, (2) the co-contaminants could have a stronger affinity for electrons in the system, and/or (3) other components in the system such as natural organic matter, microbes, and pH effects may have a variable influence on the reduction of the U. The results of this work suggest that various processes and interactions are responsible for the fate and mobility of contaminant U(VI) in this subsurface, which is rich in a variety of potential U(VI) sorbents and co-contaminants of different types.

10.4 CONCLUDING REMARKS AND FUTURE RESEARCH NEEDS

In this chapter, we (1) provide an overview of the contamination levels (U and other co-contaminants) in the field sites that we have studied, (2) review mineral–U contaminant interactions in reduced and oxidized environments, and in the presence of co-contaminants, and

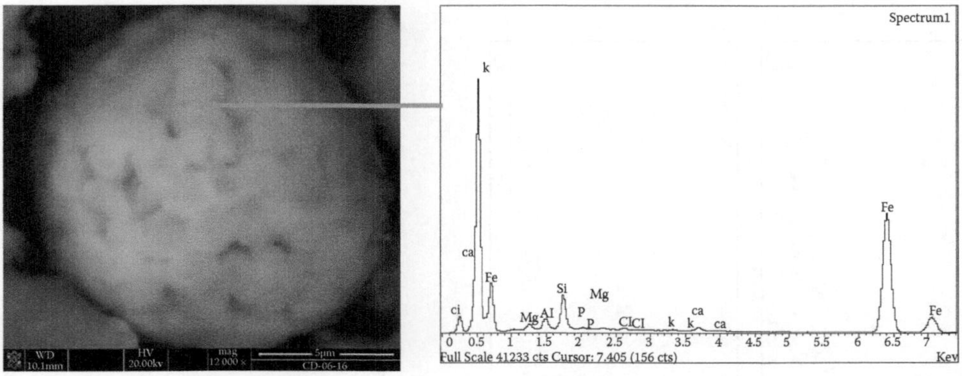

(a)

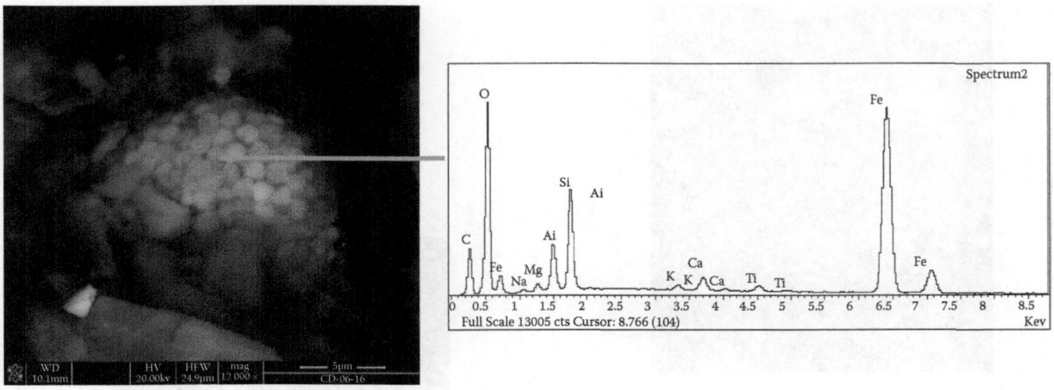

(b)

FIGURE 10.11 These SEM-EDS spectra show the presence of Fe oxide framboid-like structures: (a) sediment CD-04 10 ft and (b) CD-06 16 ft.

(3) present results from a systematic macroscopic, microscopic, and spectroscopic study of sediments from the Rifle (Colorado) site as an example of the current research efforts and the state-of-knowledge in this important research area. The results from this study confirmed that the behavior of contaminant U(VI) may be influenced by the presence of other co-contaminants. High concentration, multicontaminant, and micron-size (ca. 15 and 45 μm) areas with mainly U(IV) and some U(VI) were discovered within the Rifle sediment. The presence of U(IV) not only confirms that the reduction of U(VI) via natural attenuation processes occurred but also demonstrates that the U(IV) is relatively stable toward reoxidation, suggesting that the Rifle sediment may serve as a long-term source of U contamination to groundwater. The redox-sensitive co-contaminants that are present in these sediments, such as As, V, and Se, may occur in the aqueous phase as oxyanions and may compete with the U(VI) anionic species for sorption sites on assorted solid phases and/or available electrons, highlighting (1) the complexity of the multielement contaminated sediments and (2) the fact that the Rifle site may exhibit complex responses to transient redox conditions compared with single contaminant systems. In addition, the valuable information on the spatial correlations among different co-contaminants can be used to explain temporal variations in dissolved contaminants' concentrations in the Rifle subsurface; it could also be used to develop remediation strategies. Finally, the implications and applications from this work are valid for many U-contaminated sites across the world.

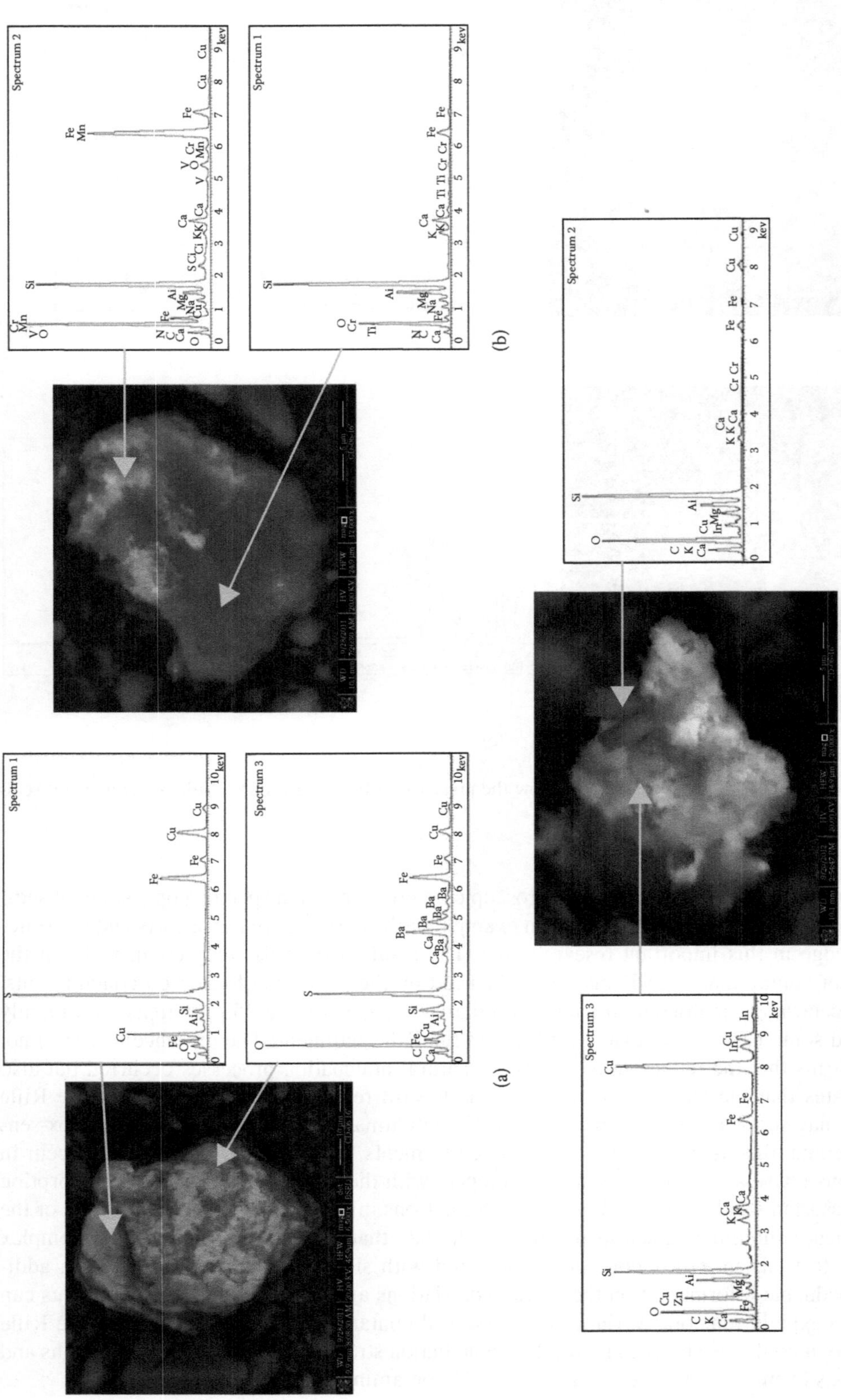

FIGURE 10.12 These SEM micrographs and EDS measurements suggest the presence of sulfides (a) and silicates (b) and (c) in this sediment (CD-06 16 ft).

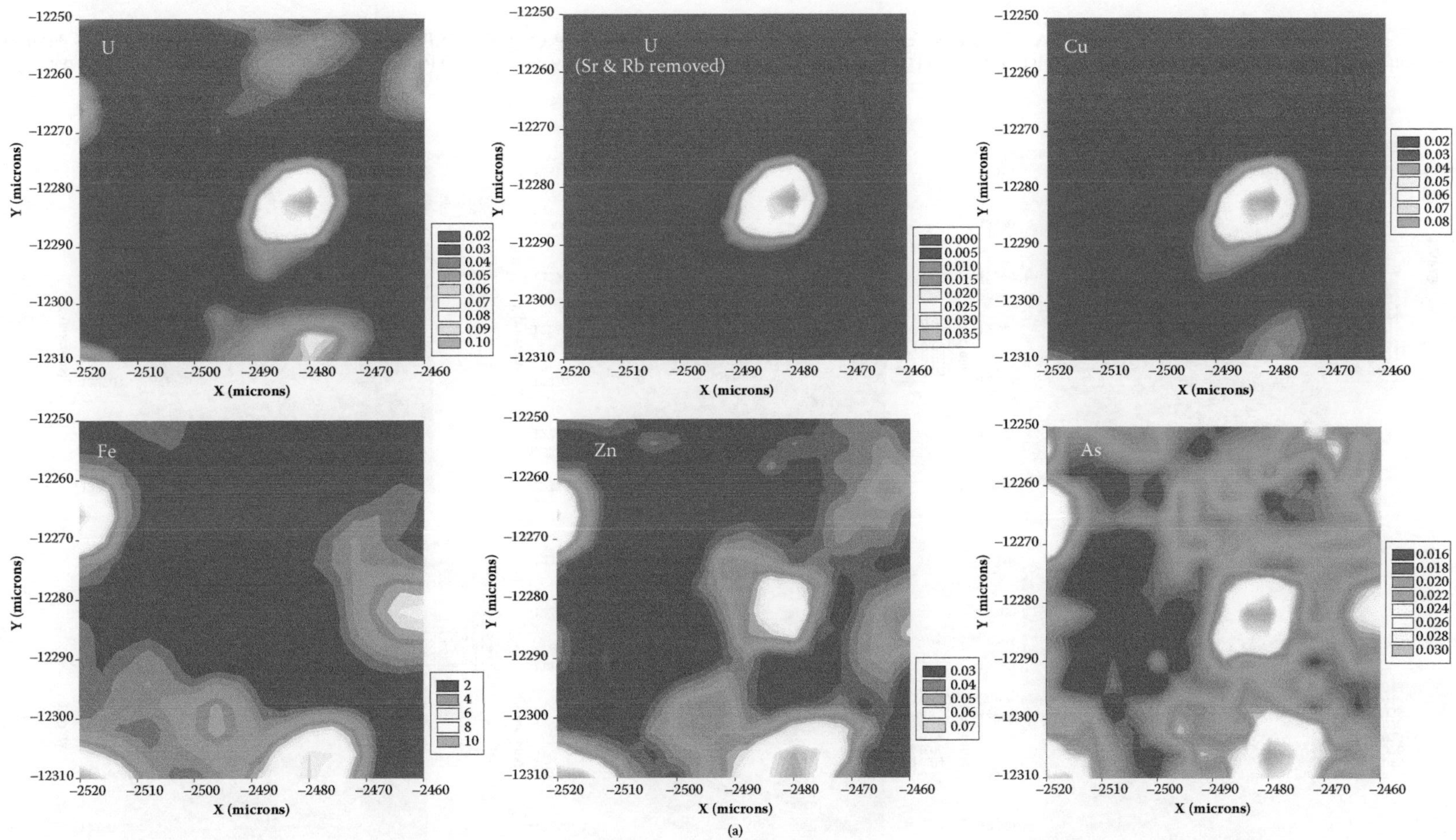

FIGURE 10.13 (See color insert.) (a) and (b) μXRF elemental maps for sediment from CD-06 16 ft. These maps depict a U hotspot (shown without and with the Sr and Rb signals removed) and spatial distribution of other elements, such as As, Ca, Cr, Cu, Fe, K, Mn, Se, V, and Zn. *(Continued)*

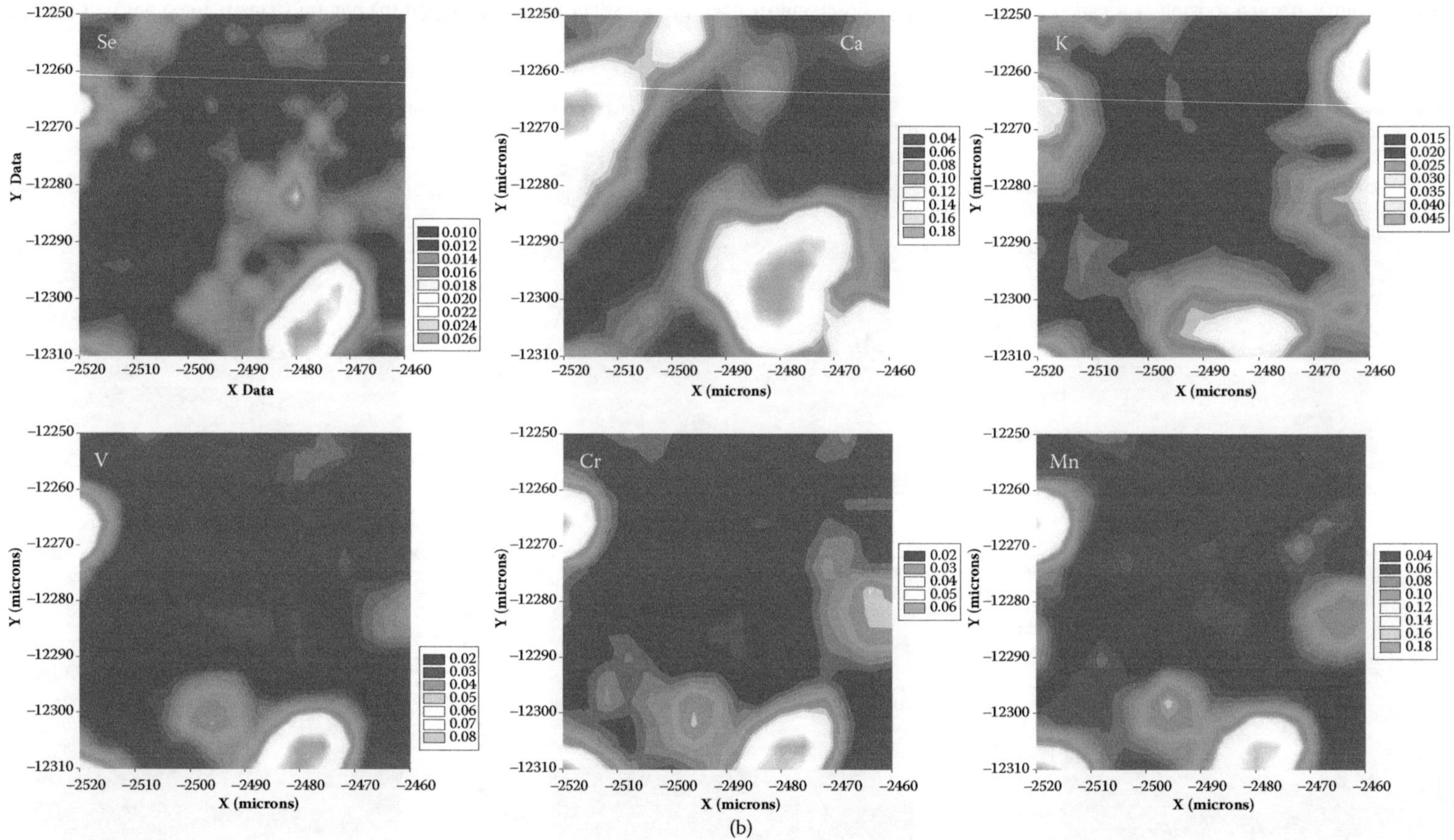

FIGURE 10.13 (*Continued*) **(See color insert.)** (a) and (b) μXRF elemental maps for sediment from CD-06 16 ft. These maps depict a U hotspot (shown without and with the Sr and Rb signals removed) and spatial distribution of other elements, such as As, Ca, Cr, Cu, Fe, K, Mn, Se, V, and Zn.

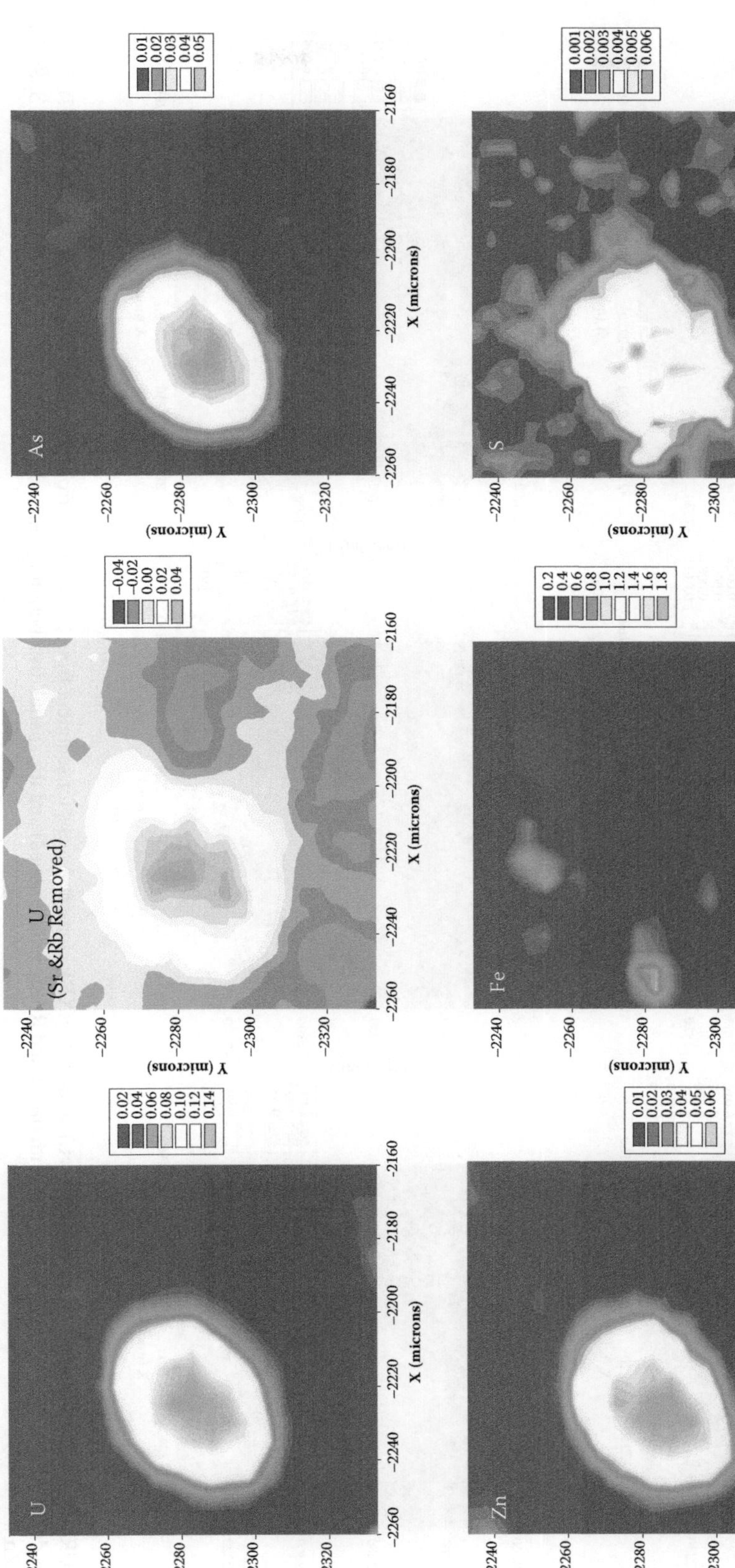

FIGURE 10.14 **(See color insert.)** (a) and (b) μXRF elemental maps for sediment from CD-04 13 ft (from the column experiment spiked with U). These maps depict a U hotspot (shown without and with the Sr and Rb signals removed) and the spatial distribution of other elements such as As, Cr, Cu, Fe, K, Mn, S, Se, V, and Zn. *(Continued)*

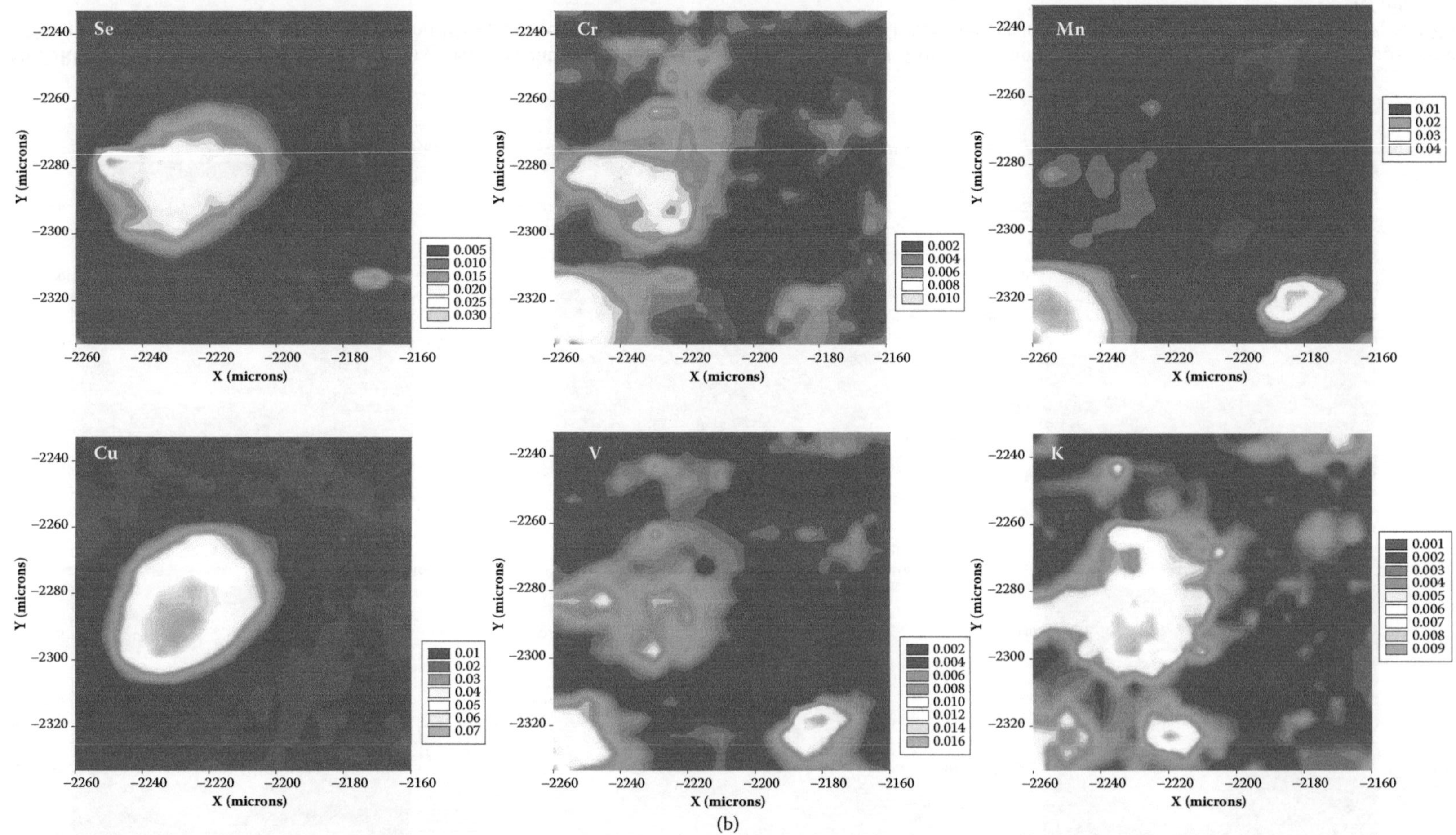

FIGURE 10.14 (*Continued*) **(See color insert.)** (a) and (b) μXRF elemental maps for sediment from CD-04 13 ft (from the column experiment spiked with U). These maps depict a U hotspot (shown without and with the Sr and Rb signals removed) and the spatial distribution of other elements such as As, Cr, Cu, Fe, K, Mn, S, Se, V, and Zn.

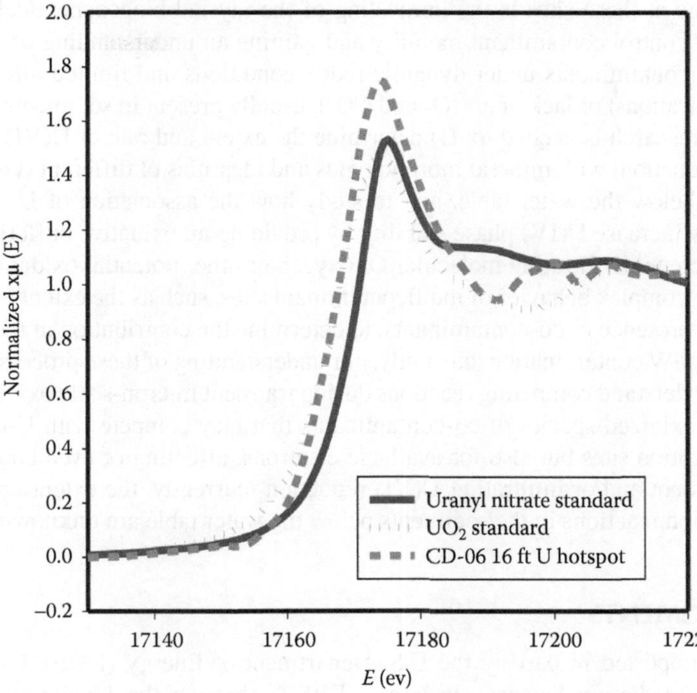

FIGURE 10.15 XANES analysis results collected in sediment CD-06 16 ft. Mostly U(IV) was present in this spot (the coordinates of the U elemental map were $X = -2{,}480$ and $Y = -12{,}282$).

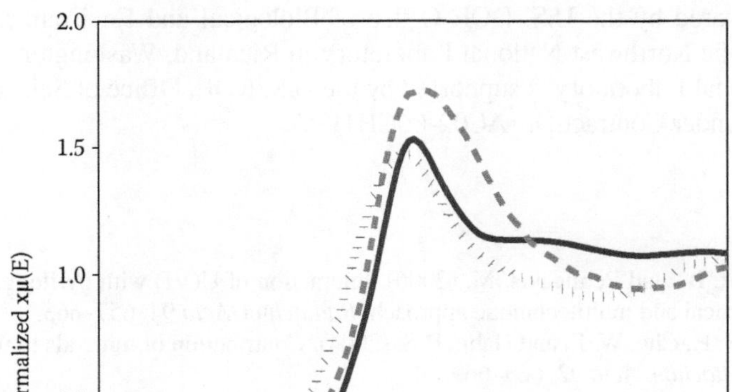

FIGURE 10.16 XANES analysis results collected in sediment CD-04 13 ft that was exposed to a U-rich solution (10 ppm) at the end of the column experiment.

A key challenge at these sites is the unraveling of the myriad biogeochemical and hydrological processes that control contaminant mobility and gaining an understanding of variations in the solubility of such contaminants under dynamic redox conditions and limited amounts (diffusion-controlled concentrations) or lack of air (O_2 and CO_2), usually present in sediments below the water table. Additional research is needed to: (1) determine the extent and rate of U(VI) interaction (i.e., adsorption and reduction) with mineral morphologies and identities of different types that are present in sediments below the water table, and to study how the association of U with the various mineral hosts may increase U(IV) phase stability by providing an oxidative buffering capacity that is capable of preferentially titrating molecular O_2 oxygen or other potential oxidants from solution; (2) understand the complex behavior of multicontaminant sites, such as the extent and rate of U(IV) reoxidation in the presence of co-contaminants, to determine the contribution of the predominantly U(IV) hotspots to GW contamination (currently, our understanding of these processes and reactions is poor); and (3) understand competing reactions during transient micron-scale oxidizing conditions, and the resulting oxidized species of co-contaminants that may compete with U-oxidized species not only for adsorption sites but also for available electrons, affecting or even controlling the rate and extent of resident and/or infiltrating U(VI) reduction (currently, the extent and rates of these competing reduction reactions in the sediments below the water table are unknown).

ACKNOWLEDGMENTS

This effort was supported in part by the U.S. Department of Energy (DOE), Office of Science, Environmental Remediation Sciences Program (ERSP), through the Integrated Field Research Challenge Site (IFRC) at Rifle, Colorado, the United States. We appreciate the support of Drs. Phil Long and Ken Williams (Lawrence Berkeley National Laboratory) and the help of Dr. Steve Heald at APS/ANL. Pacific Northwest National Laboratory is operated for the DOE by Battelle Memorial Institute under Contract DE-AC06-76RLO 1830. The research presented in this chapter was conducted in part in the Environmental Molecular Sciences Laboratory, a national scientific-user facility sponsored by the U.S. DOE Office of Biological and Environmental Research and located at the Pacific Northwest National Laboratory in Richland, Washington. Use of the APS at the Argonne National Laboratory is supported by the U.S. DOE, Office of Science, Office of Basic Energy Sciences, under Contract DE-AC02-06CH11357.

REFERENCES

Aubriet, H., Humbert, B., and Perdicakis, M. (2006). Interaction of U(VI) with pyrite, galena and their mixtures: A theoretical and multitechnique approach. *Radiochim. Acta* 94, 657–663.

Baik, M. H., Hyun, S. P., Cho, W. J., and Hahn, P. S. (2004). Contribution of minerals to the sorption of U(VI) on granite. *Radiochim. Acta* 92, 663–669.

Bargar, J. R., Reitmeyer, R., and Davis, J. A. (1999). Spectroscopic confirmation of uranium(VI)-carbonato adsorption complexes on hematite. *Environ. Sci. Technol.* 33, 2481–2484.

Bartlett, R. J., and Ross, D. S. (2005). Chemistry of redox processes in soils. In Tabatabai, M. and Sparks, D.L. (eds.), *Chemical Processes in Soils*. Madison, WI: Soil Science Society of America, Inc., pp. 461–488. Soil Science Society of America, Inc., Madison, WI.

Bruggeman, C., Maes, N., Christiansen, B. C., Stipp, S. L. S., Breynaert, E., Maes, A., Regenspurg, S., Malstrom, M. E., Liu, X., Grambow, B., and Schafer, T. (2012). Redox-active phases and radionuclide equilibrium valence state in subsurface environments—New insights from 6th EC FP IP FUNMIG. *Appl. Geochem.* 27, 404–413.

Campbell, K. M., Kukkadapu, R. K., Qafoku, N. P., Peacock, A., Lesher, E., Figueroa, L., Williams, K. H., Wilkins, M. J., Resch, C. T., Davis, J. A., and Long, P. E. (2012). Characterizing the extent and role of natural subsurface bioreduction in a uranium-contaminated aquifer. *Appl. Geochem.* 27, 1499–1511.

Catalano, J. G., and Brown, G. E. (2005). Uranyl adsorption onto montmorillonite: Evaluation of binding sites and carbonate complexation. *Geochim. Cosmochim. Acta* 69, 2995–3005.

Cross, J. O., and Frenkel, A. I. (1998). Use of scattered radiation for absolute energy calibration. *Rev. Sci. Instrum.* 70, 38–40.

Dodge, C. J., Francis, A. J., Gillow, J. B., Halada, G. P., Eng, C., and Clayton, C. R. (2002). Association of uranium with iron oxides typically formed on corroding steel surfaces. *Environ. Sci. Technol.* 36, 3504–3511.

Duff, M. C., and Amrhein, C. (1996). Uranium(VI) adsorption on goethite and soil in carbonate solutions. *Soil Sci. Soc. Am. J.* 60, 1393–1400.

Eglizaud, N., Miserque, F., Simoni, E., Schlegel, M., and Descostes, M. (2006). Uranium(VI) interaction with pyrite (FeS2): Chemical and spectroscopic studies. *Radiochim. Acta* 94, 651–656.

Fox, P. M., Davis, J. A., Kukkadapu, R., Singer, D. M., Bargar, J., and Williams, K. H. (2013). Abiotic U(VI) reduction by sorbed Fe(II) on natural sediments. *Geochimica et Cosmochimica Acta* 117, 266–282.

Gallegos, T. J., Fuller, C. C., Webb, S. M., and Betterton, W. (2013). Uranium(VI) interactions with mackinawite in the presence and absence of bicarbonate and oxygen. *Environmental Science and Technology* 47, 7357–7364.

Gomez, S. A. S., Jordan, D. S., Troiano, J. M., and Geiger, F. M. (2013). Uranyl adsorption at the muscovite (mica)/water interface studied by second harmonic generation. *Abstracts of Papers of the American Chemical Society* 245, 1.

Heald, S. M., Cross, J. O., Brewe, D. L., and Gordon, R. A. (2007). The PNC/XOR X-ray microprobe station at APS sector 20. *Nuclear Instruments and Methods in Physics Research Section A: Accelerators Spectrometers Detectors and Associated Equipment* 582, 215–217.

Ho, C. H., and Miller, N. H. (1986). Adsorption of uranyl species from bicarbonate solution onto hematite particles. *J. Colloid Interface Sci.* 110, 165–171.

Hsi, C. D., and Langmuir, D. (1985). Adsorption of uranyl onto ferric oxyhydroxides: Application of the surface complexation site-binding model. *Geochim. Cosmochim. Acta* 49, 1931–1941.

Ilton, E. S., Heald, S. M., Smith, S. C., Elbert, D., and Liu, C. X. (2006). Reduction of uranyl in the interlayer region of low iron micas under anoxic and aerobic conditions. *Environmental Science and Technology* 40, 5003–5009.

Ilton, E. S., Wang, Z. M., Boily, J. F., Qafoku, O., Rosso, K. M., and Smith, S. C. (2012). The effect of pH and time on the extractability and speciation of uranium(VI) sorbed to SiO2. *Environmental Science and Technology* 46, 6604–6611.

Jang, J. H., Dempsey, B. A., and Burgos, W. D. (2007). A model-based evaluation of sorptive reactivities of hydrous ferric oxide and hematite for U(VI). *Environ. Sci. Technol.* 41, 4305–4310.

Joseph, C., Schmeide, K., Sachs, S., Brendler, V., Geipel, G., and Bernhard, G. (2011). Sorption of uranium(VI) onto Opalinus Clay in the absence and presence of humic acid in Opalinus Clay pore water. *Chemical Geology* 284, 240–250.

Joseph, C., Stockmann, M., Schmeide, K., Sachs, S., Brendler, V., and Bernhard, G. (2013a). Sorption of U(VI) onto Opalinus Clay: Effects of pH and humic acid. *Applied Geochemistry* 36, 104–117.

Joseph, C., Van Loon, L. R., Jakob, A., Steudtner, R., Schmeide, K., Sachs, S., and Bernhard, G. (2013b). Diffusion of U(VI) in Opalinus Clay: Influence of temperature and humic acid. *Geochimica et Cosmochimica Acta* 109, 74–89.

Kelly, S. D., Newville, M., Cheng, L., Kemner, K. M., Sutton, S. R., Fenter, P., Sturchio, N. C., and Spotl, C. (2003). Uranyl incorporation in natural calcite. *Environ. Sci. Technol.* 37, 1284–1287.

Kukkadapu, R. K., Qafoku, N. P., Arey, B. W., Resch, C. T., and Long, P. E. (2010). Effect of extent of natural subsurface bioreduction on Fe-mineralogy of subsurface sediments. *J. Phys.: Conf. Ser.* 217.

Lefevre, G., Kneppers, J., and Fedoroff, M. (2008). Sorption of uranyl ions on titanium oxide studied by ATR-IR spectroscopy. *Journal of Colloid and Interface Science* 327, 15–20.

Lefevre, G., Noinville, S., and Fedoroff, M. (2006). Study of uranyl sorption onto hematite by in situ attenuated total reflection-infrared spectroscopy. *J. Coll. Inter. Sci.* 296, 608–613.

Lindberg, J. W., and Peterson, R. E. (2004). 300-FF-5 operable unit, Chapter 1.12, Rep. No. PNNL-14548. Pacific Northwest National Laboratory, Richland, WA.

Long, P. E., Banfield, J., Bush, R., Campbell, K., Chandler, D. P., Davis, J. A., Dayvault, R., Druhan, J., Elifantz, H., Englert, A., Figeroa, L., Hettich, R. L., Holmes, D., Hubbard, S., Icenhower, J., Jaffe, P. R., Kerkhof, L. J., Kukkadapu, R. K., Lesher, E., Lipton, M., Li, L., Lovley, D., Morris, S., Morrison, S., Mouser, P., Murray, C., Newcomer, D., N'Guessan, L., Peacock, A., Qafoku, N. Q., Resch, C. T., Spane, F., Steefel, C., VerBerkmoes, N., Wilkins, M., Williams, K. H., and Yabusaki, S. B. (2008). The integrated field-scale subsurface research challenge site (IFC) at Rifle, Colorado: Preliminary results on microbiological, geochemical and hydrologic processes controlling iron reduction and uranium mobility. In: Long, P.E. (ed.), ERSP P.I. Meeting April 2008, Washington, DC.

Moyes, L. N., Parkman, R. H., Charnock, J. M., Vaughan, D. J., Livens, F. R., Hughes, C. R., and Braithwaite, A. (2000). Uranium uptake from aqueous solution by interaction with goethite, lepidocrocite, muscovite, and mackinawite: An X-ray absorption spectroscopy study. *Environ. Sci. Technol.* 34, 1062–1068.

Newville, M. (2001). IFEFFIT: Interactive EXAFS analysis and FEFF fitting. *J. Synch. Rad.* 8, 322–324.

Qafoku, N. P., Gartman, B. N., Kukkadapu, R. K., Arey, B. W., Williams, K. H., Mouser, P. J., Heald, S. M., Bargar, J. R., Janot, N., Yabusaki, S., and Long, P. E. (2014). Geochemical and mineralogical investigation of uranium in multi-element contaminated, organic-rich subsurface sediment. *Appl. Geochem.* 42, 77–85.

Qafoku, N. P., and Icenhower, J. P. (2008). Interaction of aqueous U(VI) with soil minerals in slightly alkaline natural systems. *Rev. Environ. Sci. Biotechnol.* 7, 355–380. DOI 10.1007/s11157-008-9137-8.

Qafoku, N. P., Kukkadapu, R., McKinley, J. P., Arey, B. W., Kelly, S. D., Resch, C. T., and Long, P. E. (2009). Uranium in framboidal pyrite from a naturally bioreduced alluvial sediment *Environ. Sci. Technol.* 43, 8528–8534.

Ravel, B., and Newville, M. (2005). Athena, artemis, hephaestus: Data analysis for X-ray absorption spectroscopy using IFEFFIT. *J. Synch. Rad.* 12, 537–541.

Reich, T., Moll, H., Arnold, T., Denecke, M. A., Henning, C., Geipel, G., Bernhard, G., Nitsche, H., Allen, P. G., Bucher, J. J., Edelstein, N. M., and Shuh, D. K. (1998). An EXAFS study of uranium(VI) sorption onto silica gel and ferrihydrite. *J. Electron Spectrosc. Relat. Phenom.* 96, 237–243.

Riley, R. G., Zachara, J. M., and Wobber, F. J. (1992). Chemical contaminants on DOE lands and selection of contaminant mixtures for subsurface science research; DOE/ER-0547T, Rep. No. DOE/ER-0547T. U.S. Department of Energy, Office of Energy Research, Washington, DC.

Rovira, M., El Aamrani, S., Duro, L., Gimenez, J., de Pablo, J., and Bruno, J. (2007). Interaction of uranium with in situ anoxically generated magnetite on steel. *J. Hazard. Mater.* 147, 726–731.

Schmeide, K., Pompe, S., Bubner, M., Heise, K. H., Bernhard, G., and Nitsche, H. (2000). Uranium(VI) sorption onto phyllite and selected minerals in the presence of humic acid. *Radiochim. Acta* 88, 723–728.

Schmeide, K., Sachs, S., Bubner, M., Reich, T., Heise, K. H., and Bernhard, G. (2003). Interaction of uranium(VI) with various modified and unmodified natural and synthetic humic substances studied by EXAFS and FTIR spectroscopy. *Inorganica Chimica Acta* 351, 133–140.

Scott, T. B., Tort, O. R., and Allen, G. C. (2007). Aqueous uptake of uranium onto pyrite surfaces; reactivity of fresh versus weathered material. *Geochim. Cosmochim. Acta* 71, 5044–5053.

Serne, J. N., Brown, C. F., Schaef, H. T., Pierce, E. M., Lindberg, M. J., Wang, Z., Gassman, P. L., and Catalano, J. G. (2002). 300 area uranium leach and adsorption project, Rep. No. PNNL-14022. Pacific Northwest National Laboratory, Richland, WA.

Serne, R. J., Schaef, H. T., Bjornstad, B. N., Williams, B. A., Lanigan, D. C., Horton, D. G., Clayton, R. E., LeGore, V. L., O'Hara, M. J., Brown, C. F., Parker, K. E., Kutnyakov, I. V., Serne, J. N., Mitroshkov, A. V., Last, G. V., Smith, S. C., Lindenmeier, C. W., Zachara, J. M., and Burke, D. B. (2001). Characterization of uncontaminated sediments from the Hanford Reservation-RCRA borehole core and composite samples, Rep. No. PNNL-2001-1. Pacific Northwest National Laboratory, Richland, WA.

Singer, D. M., Chatman, S. M., Ilton, E. S., Rosso, K. M., Banfield, J. F., and Waychunas, G. A. (2012). U(VI) sorption and reduction kinetics on the magnetite (111) surface. *Environmental Science and Technology* 46, 3821–3830.

Singer, D. M., Maher, K., and Brown, G. E. (2009). Uranyl-chlorite sorption/desorption: Evaluation of different U(VI) sequestration processes. *Geochimica Et Cosmochimica Acta* 73, 5989–6007.

Steudtner, R., Sachs, S., Schmeide, K., Brendler, V., and Bernhard, G. (2011). Ternary uranium(VI) carbonato humate complex studied by cryo-TRLFS. *Radiochimica Acta* 99, 687–692.

Szecsody, J. E., Truex, M. J., Qafoku, N. P., Wellman, D. M., Resch, T., and Zhong, L. (2013). Influence of acidic and alkaline waste solution properties on uranium migration in subsurface sediments. *J. Contam. Hydrol.* 151, 155–175.

Todorov, P. T., and Ilieva, E. N. (2006). Contamination with uranium from natural and anthropological sources *Rom. Journ. Phys.* 51, 27–34.

van Geen, A., Robertson, A. P., and Leckie, J. O. (1994). Complexation of carbonate species at the goethite surface: Implications for adsorption of metal ions in natural waters. *Geochim. Cosmochim. Acta* 58, 2073–2086.

Veeramani, H., Alessi, D. S., Suvorova, E. I., Lezama-Pacheco, J. S., Stubbs, J. E., Sharp, J. O., Dippon, U., Kappler, A., Bargar, J. R., and Bernier-Latmani, R. (2011). Products of abiotic U(VI) reduction by biogenic magnetite and vivianite. *Geochimica et Cosmochimica Acta* 75, 2512–2528.

Veeramani, H., Scheinost, A. C., Monsegue, N., Qafoku, N. P., Kukkadapu, R., Newville, M., Lanzirotti, A., Pruden, A., Murayama, M., and Hochella, M. F. (2013). Abiotic reductive immobilization of U(VI) by biogenic mackinawite. *Environmental Science and Technology* 47, 2361–2369.

Villalobos, M., and Leckie, J. O. (2001). Surface complexation modeling and FTIR study of carbonate adsorption to goethite. *J. Colloid Interface Sci.* 235, 15–32.

Waite, T. D., Davis, J. A., Payne, T. E., Waychunas, G. A., and Xu, N. (1994). Uranium(VI) adsorption to ferrihydrite: Application of a surface complexation model. *Geochim. Cosmochim. Acta* 58, 5465–5478.

Wazne, M., Korfiatis, G. P., and Meng, X. G. (2003). Carbonate effects on hexavalent uranium adsorption by iron oxyhydroxide. *Environ. Sci. Technol.* 37, 3619–3624.

Wersin, P., Hochella, M. F., Persson, P., Redden, G., Leckie, J. O., and Harris, D. W. (1994). Interactions between aqueous uranium(VI) and sulfide minerals-Spectroscopic evidence for sorption and reduction *Geochim. Cosmochim. Acta* 58, 2829–2843.

Zachara, J. M., Davis, J. A., Liu, C., McKinley, J. P., Qafoku, N. P., Wellman, D. M., and Yabusaki, S. B. (2005). Uranium geochemistry in vadose zone and aquifer sediments from the 300 Area uranium plume, Rep. No. PNNL-15121, Pacific Northwest National Laboratory, Richland, WA.

Section II

Bioavailability

11 Metal Bioavailability in Land-Disposed Dredged Sediments

Filip M. G. Tack

CONTENTS

11.1 INTRODUCTION

11.1.1 Metals in Sediments and Dredging Needs

Bottom sediments from waterbodies, including rivers, lakes, and estuaries, may need to be removed and relocated for several reasons, including construction needs and for maintaining the navigability of waterways. Sediments that are dredged from waterways and harbors are frequently contaminated by organic pollutants and metals. This limits the possibilities for subsequent handling of the excavated materials, because hazardous effects of the contamination from the sediments in their new location may be introduced.

Potential effects of the contaminants are related to their availability to biota. Bioavailability is the extent to which compounds, and metals in particular, that are present in an environment can be taken up by living organisms. Therefore, it is crucial in determining how pollutants may affect the health and functioning of an ecosystem.

11.1.2 Bioavailability

The term *bioavailability* is broadly defined, and its measurement is approached in many different ways (Naidu et al. 2011). The National Research Council (NRC) (2003) defines *bioavailability processes* as the individual physical, chemical, and biological interactions that determine the exposure of organisms to chemicals that are associated with soils or sediments. Figure 11.1 conceptually shows such processes. The contaminant is released from the solid (A) and can reach the physiological membrane (B). Direct contact of the solid phase and the membrane is also possible (C). After uptake across a biological membrane (D), the contaminant is internally transported and may trigger a biological

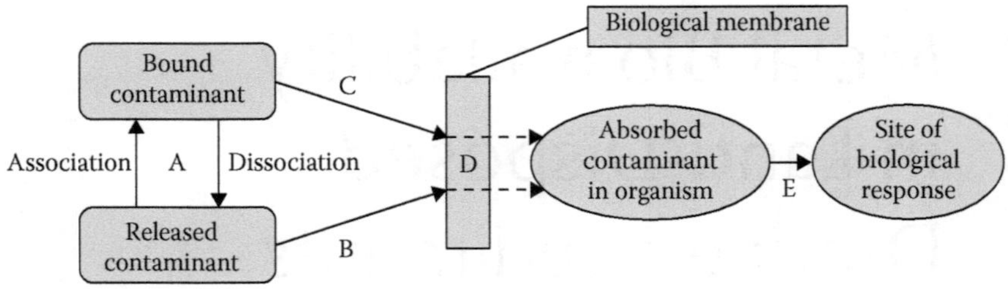

FIGURE 11.1 Schematic of the processes that determine exposure to contamination (Committee on Bioavailability of Contaminants in Soils and Sediments, National Research Council [U.S.], 2003; Ehlers and Luthy, 2003). (Reprinted with permission from the National Academies Press, Copyright 2003, National Academy of Sciences.)

response (E). The processes A through D where soil or sediment plays a role are the "bioavailability processes." The processes can be seen as barriers that must be overcome for a contaminant in soil or sediment to become bioavailable (Ehlers and Luthy 2003). Typically, a few of these barriers will be the slowest and will, thus, control the bioavailability. Semple et al. (2004) provided the following definitions for the terms *bioavailable* and *bioaccessible*. The bioavailable compound is that which is freely available to cross an organism's cellular membrane from the medium that the organism inhabits at a given time (process D on Figure 11.1). Once transfer across the membrane has occurred, storage, transformation, assimilation, or degradation can take place within the organism. The latter processes, shown by E in Figure 11.1, are distinct from the transfer between the medium and the organism. The bioaccessible compound is that which is available to cross an organism's cellular membrane from the environment, provided the organism has access to the chemical. As such, bioaccessibility encompasses what is actually bioavailable now and what is potentially bioavailable, and this would be represented by processes A–D in Figure 11.1 (Semple et al. 2004). Even with a single definition in place, the question remains as to whether the bioavailable portions of contaminants to a specific organism in a given soil or sediment can be measured, and if so, how (Semple et al. 2004). Naidu et al. (2000) defined *bioavailability* as "the fraction of the total contaminant in the interstitial water and soil particles that is available to receptor organisms with the extent of bioavailable fraction varying with time, the nature of soil types, organisms and environmental factors."

11.1.3 Fate of Dredged Materials

Dredging involves the removal of bottom sediments in water bodies, including estuaries, lakes, rivers, and canals, to initiate infrastructural or ecological improvements. The main purpose of dredging is to maintain navigation. "Maintenance dredging" is aimed at maintaining a sufficient depth of water for navigation (Bramley and Rimmer 1988). "Capital dredging" involves dredging activities with the aim of increasing the natural depth of a water body for the first time, for example, in order to allow for trafficking larger ships or in the context of civil engineering work. The additional purposes of dredging are for mining, construction, and the environment. The purpose of "environmental dredging" is to remove contaminated sediments to a place where they could be treated further for the removal and stabilization of contaminants (Yell and Riddell 1995). In Flanders, Belgium, there is currently a need to remove about 2.4 million tons of dry matter from waterways on a yearly basis (Herman and Bogaert 2007). In the north of France, about 200,000 m^3 are dredged yearly (Panfili et al. 2005); however, only a fraction of these amounts is contaminated. It is estimated that about 100–200 million cubic meters of contaminated sediment are produced yearly in Europe (Bortone et al. 2004).

After excavation, dredged materials are transported and relocated. Sediments from marine environments are commonly disposed elsewhere, on an overdepth or at another location where they do not hinder shipping. Formerly, it was common practice to dispose sediments dredged from inland

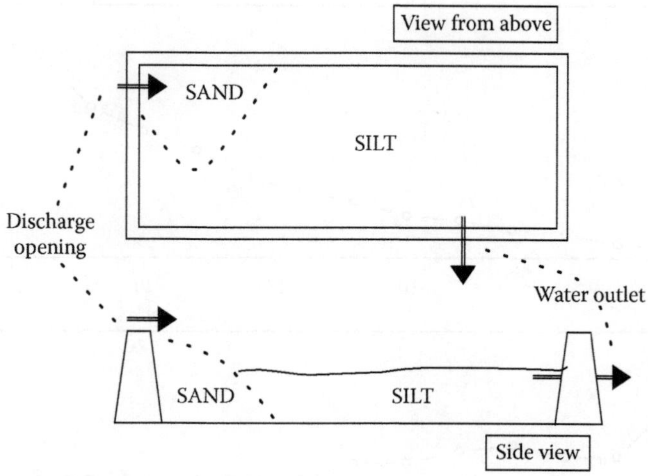

FIGURE 11.2 Structure of a dredged sediment landfill in a horizontal and vertical view with indication of important physical features. (Reprinted from *Sci. Total Environ.*, 290, Vandecasteele, B. et al., Heavy metal contents in surface soils along the Upper Scheldt river (Belgium) affected by historical upland disposal of dredged materials, 1–14, Copyright 2002a, with permission from Elsevier.)

water bodies along the shores and on agricultural land, where they usually contributed to soil fertility. Awareness of potential contamination from dredged sediments caused authorities to stop this practice. In Flanders, between the 1970s and 2000, sediments were disposed in confined disposal facilities (Vandecasteele et al. 2002a; Vervaeke et al. 2001). The dredged materials are hydraulically pumped into the sites (Figure 11.2). Sediments settle down, whereas excess supernatant water flows out of the site though installed outlets. The sediments remain covered by a layer of water and are, therefore, kept in a reduced state during the period that the site is in use. When sediment introduction ceases, the water cover will disappear provided evaporation exceeds input through precipitation. The sediments gradually dry and oxidize. Depending on the disposal conditions, this might take between a few weeks and several years. Wetland plants such as certain willow species may colonize the site. Gradual terrestrialization of the site affects the properties of sediments and causes the ecosystem that develops on it to adapt to the changing conditions (Vandecasteele et al. 2007). The potential bioavailability of metals in these disposal sites is of concern, because it will determine the extent to which trace elements are transferred to the surrounding environment and to the food web.

11.2 CHANGES IN MOBILITY AND BIOAVAILABILITY OF METALS ON LAND DISPOSAL

11.2.1 Redox and pH

Dredged materials that are disposed on land are subject to dewatering and drying (Figure 11.3, Tack et al. 1997). The physicochemical changes that occur may affect both the speciation of metals and the stability of metal-bearing phases in the sediment (Cook and Parker 2003; Singh et al. 1998). During this dynamic transition from reduced sediment to oxic, soil-like material, metal solubility temporarily increases strongly (Stephens et al. 2001) and tends to decreases again as the chemistry of the sediment equilibrates with the oxic conditions (Caille et al. 2003; Cook and Parker 2003), although this might take several years (Piou et al. 2009). Major processes that govern this behavior are the disappearance of sulfide and the formation of hydrated oxides of iron (Fe) and manganese (Mn). In the reduced environment, metal solubility is largely controlled by sulfide, with which they form highly insoluble compounds (Ankley et al. 1996). Mostly because of the oxidation of sulfides, solubility and leaching of most metals increase when strongly reduced sediments are oxidized (Du Laing et al. 2009; Piou et al.

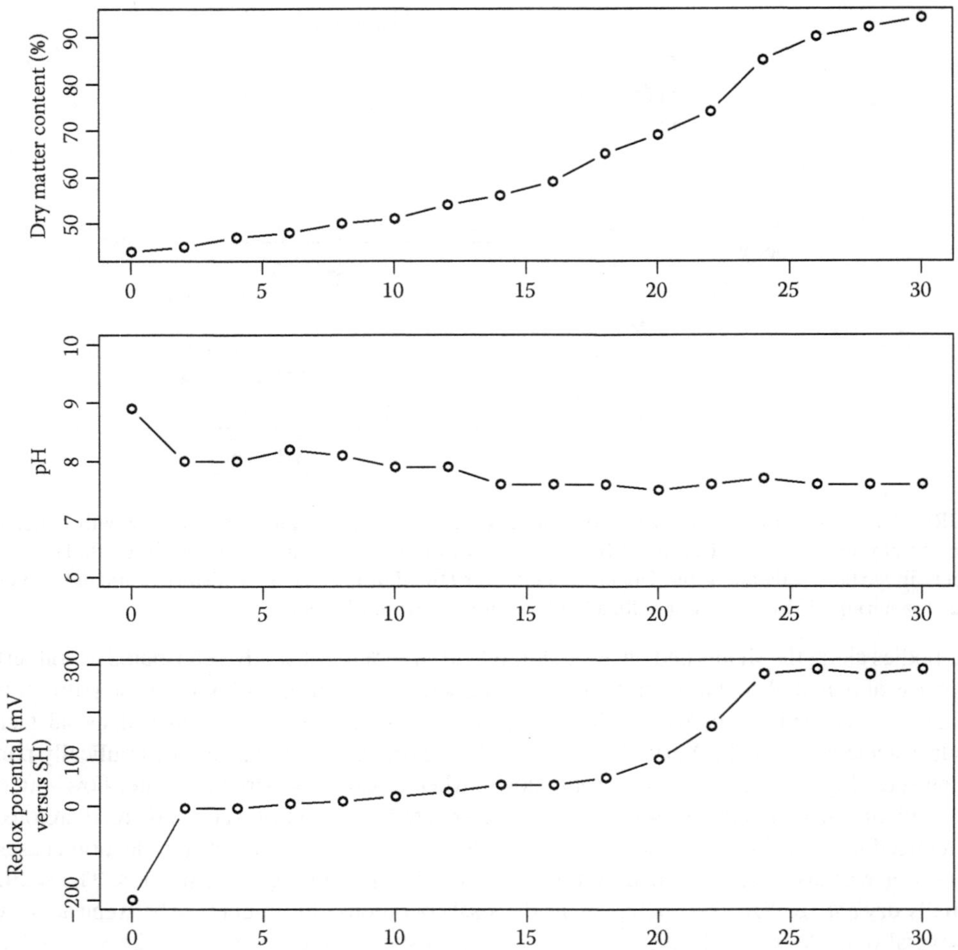

FIGURE 11.3 Evolution in sediment characteristics during drying and oxidation of a sediment. (Data from Tack, F.M. et al., *Talanta*, 44, 2185–2192, 1997.)

2009). Hydrated oxides of Fe and Mn will precipitate in the oxidizing environment. Metals can coprecipitate or adsorb on these oxides (Du Laing et al. 2009). Hence, after a period of enhanced solubility and mobility, metals are again retained more strongly in the oxic sediment, but perhaps less effectively (Gambrell 1994).

Especially during the early years, seasonal effects may be important in maintaining the bioavailability and mobility of metals in an elevated state. In winter conditions in temperate regions, water content of the sediments can remain high for several months, thus limiting oxygen diffusion. Low temperatures limit microbial activity that mediates processes, including acidification, dissolution, and oxidation. Oxidation of sulfides is, therefore, either retarded or temporarily stopped, and amorphous oxides of Fe are subjected to reductive dissolution. Sequential extraction supports the hypothesis that the Fe is complexed by organic matter. Sulfide formation during winter is not excluded, but it is difficult to demonstrate. Even very low concentrations may be effective in decreasing the solubility of metals (Du Laing et al. 2007, 2009; Piou et al. 2009). In summer, microbial activity is intense, favoring processes such as mineralization of organic matter, nitrification, and oxidation of sulfide (Piou et al. 2009). Low water content favors the diffusion of oxygen and the formation of hydrated oxides of Fe and Mn.

Repeated wetting and drying tends, depending on the length of the cycles, to steadily maintain a high leachability of sediments (Du Laing et al. 2007; Hartley and Dickinson 2010). The sediment system continuously shifts between more oxidizing and more reducing conditions, preventing more stable metal-scavenging phases from being established. Management options that are aimed at stabilizing the sediment should focus on stabilizing physicochemical conditions. This involves either limiting the flow of water through the sediment or returning the sediment to anoxic waterlogged conditions (Hartley and Dickinson 2010).

Potential changes in pH on dewatering and oxidation of the sediment may also impact the solubility of metals. Chemically and microbially mediated oxidation processes in sediments tend to be acidifying. The oxidation of sulfide, in particular, generates significant acidity. Carbonates, exchangeable cations, clays, and Al hydroxides are compounds that may buffer against a decrease in pH (Satawathananont et al. 1991). In the absence of carbonates, a quick and strong decrease in pH may be observed, that is, from neutral to values of 3–4 (Gambrell et al. 1991; King et al. 2006). Free carbonates may effectively neutralize such acidity, and they may maintain the pH of the oxidized sediment at a neutral pH. For example, the carbonate content of a sediment was observed to decrease by 2.6% after oxidation (Tack et al. 1996). This amount of carbonate can neutralize 0.52 mol H^+/kg sediment.

11.2.2 Soil Constituents

Metal-binding phases in sediments, that is, sulfides, clays, organic matter, and hydrated oxides of Fe and Mn, reduce metal solubility and hence bioavailability. Complexation by ligands may enhance the solubility of trace metals, but they may, nevertheless, lower the bioavailability (Allen 1993). Sulfides exert major control on metal solubility in reduced sediments. The ratio of acid volatile sulfides (AVS)/simultaneously extracted metal (SEM) has been shown to predict the toxicity of selected metals in reduced metal-contaminated sediments rather well (Ankley 1996; Hansen et al. 1996). For oxic sediments, normalization by organic carbon content and/or hydrated oxides of Fe and Mn have been investigated for their use as indicators of potential toxicity. However, a lack of understanding of the binding constants involved limits the power of this approach (Chapman et al. 1998).

The organic matter content of an upland-disposed dredged material tends to decrease with time (Gambrell 1994; Singh et al. 1998). Decomposition of soil organic matter involves the gradual breakdown and mineralization of nonliving macro-organic matter by soil organisms. This depends directly on the biological activity of the soil and hence is closely related to soil temperature, moisture content, pH, oxygen availability, mineral nutrient status, and stress from toxic chemicals (Lützow et al. 2006). The presence of elevated levels of trace elements may adversely affect the growth and soil organic matter decomposing ability of soil organisms (Gupta et al. 1996).

Hydrous oxides of manganese, iron, and aluminum are important components that control the retention of metals in soils (Kinniburgh et al. 1976). These amorphous and microcrystalline structures have a large sorption capacity for metals. Biological and microbial activity associated with the presence of organic matter contributes to the existence of a periodic reducing environment after rain, irrigation, or flooding. In upland soils, this leads to some reductive dissolution and reprecipitation, which keeps the oxides in a highly reactive state with respect to tracing metal sorption (Cavallaro and McBride 1984). If periods of reduction are more intensive, metal concentrations in the pore water increase significantly, and they may only decrease if the flooding is sufficiently long to allow for the formation of sulfide (Du Laing et al. 2007).

The presence of free $CaCO_3$ generally coincides with a reduced solubility of metals because of the pH effect. If present, they strongly buffer drying sediment against a decrease in pH. The Ca in the soil solution tends to favor the coagulation of colloids and the precipitation of organic matter that otherwise might contribute toward maintaining higher levels of these metals in solution. Furthermore, carbonate/bicarbonate ions will form metal carbonates that are only sparingly soluble (Tack 2010). Sorption on carbonate phases, to some extent, accounts for the low metal solubility, and precipitation of Cd and Cu has been observed in some cases (Martin and Kaplan 1998). Carbonates constitute a significant metal solubility-controlling phase in reduced environments (Guo et al. 1997).

11.2.3 Effects of Vegetation

The growth of phreatophytic trees enhances the removal of water and contributes to the stabilization of dredged sediments. In the short term, the vegetation may lead to an accelerated mobilization of metals, which are associated with the accelerated oxidation of the sediment because of (i) a better aeration of the root zone by oxygen introduced through the transport in roots and (ii) transpiration by the vegetation (Vervaeke et al. 2001). Oxidation of metal sulfides, in particular, is accelerated.

In addition, in the dried substrate, plants tend to increase the mobility of metals compared with unplanted substrates. Effects are markedly apparent by a comparison of the rhizosphere and the bulk substrate (Vervaeke et al. 2004). Combined with the effect of root growth on porosity and soil structure, the leaching of metals may be enhanced (Vervaeke et al. 2004). Other factors that contribute to mobilizing metals are a decrease in pH by root activity, the production of soluble organic compounds by roots and microorganisms that can form soluble organic complexes with the metals, and the accelerated turnover of organic matter because of increased microbial activity (Marseille et al. 2000). These effects and their relative importance differ between plants. In unbuffered sediments, phytostabilization was unsuccessful, because these effects, combined with a decrease in pH, caused significant leaching of metals, a high uptake of metals by the plants, and, accordingly, high plant mortality (King et al. 2006). In calcareous sediments, acidity is effectively neutralized, and plant growth is not hampered. Fluxes of metals depend on species-specific differences in metal uptake and on the quality and turnover of litter (Table 11.1). Metal concentrations in the top layer evolve depending on the accumulation of metals in the leaves and species-specific soil acidification. A long-term study (30 years) (Mertens et al. 2007) revealed that ash and maple had normal tissue concentrations of Cd, Cu, Cr, Pb, and Zn. Poplar and willow take up high amounts of Cd and Zn, which result in increased concentrations of these metals in the upper soil layer through a recycling of the metals in litter (Mertens et al. 2007; Vandecasteele et al. 2009). This effect is even observed on uncontaminated substrates (Vandecasteele et al. 2009). High metal concentrations in the litter on contaminated sites are suspected to retard litter decomposition. This causes low pH values in the litter to be established, which is not expected for these species. Oak does not recycle metal extensively through the litter, but it naturally forms a more acidic litter layer than other trees such as ash and maple. This causes more intense leaching and, accordingly, lower metal concentrations in this soil layer (Mertens et al. 2007).

TABLE 11.1
Metal Stocks (g/ha) in Woody Biomass and Annual Litter Fall of Four Tree Species and Amount of Metals in 0.5 m of Soil

	Ca	Cd	Cr	Cu	Pb	Zn
Wood						
Oak	6534±3340	<54.1	<108	130 ± 22	<108	<541
Ash	16,160 ± 6152	<94.4	208 ± 24	245 ± 78	<189	1322 ± 761
Maple	14,748 ± 4324	<42.6	128 ± 52	94 ± 18	<85	4819 ± 83
Poplar	13,970 ± 2962	168 ± 50	74 ± 15	101 ± 29	<67	4819 ± 809
Litter fall						
Oak	4244 ± 322	<1.4	7.6 ± 1.9	17.4 ± 1.0	12.3 ± 0.2	120 ± 11
Ash	7961 ± 591	<1.6	<3.1	15.8 ± 2.1	13.6 ± 0.4	79 ± 5
Maple	11,776 ± 415	<1.9	6.8 ± 0.9	13.3 ± 1.7	20.5 ± 1.0	60 ± 72
Poplar	6005 ± 166	22.0 ± 3.4	2.6 ± 0.8	17.6 ± 2.5	9.9 ± 2.6	2874 ± 232
Soil	42,000,000	65.650	1,326,000	1,196,000	1,885,000	7,579,000

Source: Reprinted from *Environ. Pollut.*, 149, Mertens, J. et al., Tree species effect on the redistribution of soil metals, 173–181, Copyright 2007, with permission from Elsevier.

11.3 BIOAVAILABILITY TOWARD VEGETATION

Ripened dredged materials are rapidly colonized by vegetation (Lepp and Madejón 2007; Vandecasteele et al. 2002a). They constitute an excellent substrate for vegetation because of their favorable physical properties and chemical fertility (Bramley and Rimmer 1988, p. 19; Darmody and Marlin 2002). Formerly, dredged materials were commonly put on arable land and they contributed to the fertility of the land. Crops and vegetation may uptake metal contaminants in excess of normal concentrations in unpolluted soils (Smilde et al. 1982; Vandecasteele et al. 2002b; van Driel et al. 1995). Physicochemical factors that influence the solubilization and mobilization of metals will partly determine the uptake of metals by vegetation. There is, however, also important control exhibited by the plant. In order to maintain the concentration of essential elements within physiological limits and to minimize the detrimental effects of nonessential metals, plants, similar to all other organisms, have evolved a complex network of homeostatic mechanisms that serve to control the uptake, accumulation, trafficking, and detoxification of metals (Clemens 2001). These plant-specific physiological responses are highly species and even clone specific, and they greatly weaken the relationship between mobile element concentrations as determined by soil solution concentrations and extractable contents, and metal concentrations in plant tissues.

Different trees have a different tendency for the uptake of metals. Tables 11.2 and 11.3 summarize foliar concentrations of Cd, Cu, Pb, and Zn in different trees that are grown on dredged sediment-derived soils. Alder, ash, and maple reveal moderate foliar concentrations, whereas concentrations in leaves of poplar and willow show markedly higher concentrations. Clearly, the choice of tree species is an important tool in controlling metal cycling (Vandecasteele et al. 2008).

Unbuffered dredged materials have been shown to poorly support the development of trees because of the low pH they develop, which results in a strong increase in the bioavailability of trace elements and Al toxicity (King et al. 2006). No issues with the development of plants are observed in calcareous dredged sediments. In these nutrient-rich substrates, the presence of free carbonates, a neutral pH, high organic matter, and clay contents favor a low mobility and bioavailability of metals. Nevertheless, some plants on these substrates exhibit elevated metal concentrations in the above-ground parts (Capilla et al. 2006; Vandecasteele et al. 2002b). Willow (*Salix* sp.) naturally invades dredged sediment landfills. It is the climax vegetation on freshwater tidal marshes and on other sediment-derived substrates that are contaminated with metals (Vandecasteele et al. 2006; Vervaeke et al. 2001). Foliar Cd and Zn concentrations in willow (>6.6 mg Cd kg^{-1} DW and >700 mg Zn kg^{-1} DW) were five to tenfold those observed in reference situations (0.5–2.9 and 128–338 mg kg^{-1} DW for Cd and Zn, respectively) (Vandecasteele et al. 2002b). Foliar concentrations of Cd in *Salix dadyclados* "Loden" were in the order of 10 mg kg^{-1} when grown in the greenhouse on a dredged sediment-derived soil with a Cd content of 8.8 mg kg^{-1}. The same willow clone grown in similar conditions on a pH-neutral sandy soil with a Cd content of 5.5 mg kg^{-1} Cd exhibited 40% lower biomass production and foliar Cd concentrations exceeding 30 mg kg^{-1} dry matter (Meers et al. 2007).

Poplar also exhibits elevated Cd and Zn concentrations when growing on contaminated substrates (Laureysens et al. 2004; Vandecasteele et al. 2003). Leaves of maize grown on contaminated, dredged sediment-derived soils contained high levels of Zn and Cd, which were correlated with total metal concentrations in the soils (Vandecasteele et al. 2006). Other plants, such as ash, alder, maple, and black locust, do not accumulate metal levels in excess to these that are encountered in a not-contaminated environment (Mertens et al. 2004; Vandecasteele et al. 2002b). Metal uptake differs greatly between plants, even between different clones of the same species. This has been extensively shown for willow (Greger and Landberg 1999) and poplar (Laureysens et al. 2004). The uptake of metals in vegetation that develops on dredged sediment-derived soils may be of concern. Appropriate management practices for these lands must aim at minimizing the risk of contaminant dispersal into the environment. Within such management, the choice of species constitutes an important tool.

Metal concentrations in willow biomass compartments on oxidized sediments tend to decrease with stand age. This can be attributed to metal accumulation occurring mostly in actively growing

TABLE 11.2
Reported Foliar Concentrations (mg/kg Dry Matter) of Cd and Cu in Trees Growing on Contaminated, Dredged Sediment-Derived Soils

Element	Plant	Foliar Conc.	Soil Conc.	Soil pH[b]	Ref.
Cd	Alder	0.16	20	6.3 → 3.7	(King et al. 2006)
		<0.23	5.7	7.1	(Mertens et al. 2004)
		<0.35	8.3–15.8	7.2–7.5	(Vandecasteele et al. 2008)
	Ash	0.3	5.7	7.1	(Mertens et al. 2004)
		<0.35	8.3–12.5	7.2–8.1	(Vandecasteele et al. 2008)
	Maple	0.5	5.7	7.1	(Mertens et al. 2004)
		<0.35–0.9	8.3–15.8	7.2–8.1	(Vandecasteele et al. 2008)
	Poplar	2.4	20	6.3 → 3.7	(King et al. 2006)
		13.5	391	5.8	(Lepp and Madejón 2007)
		8	5.7	7.1	(Mertens et al. 2004)
	Willow	2.3	20	6.3 → 3.7	(King et al. 2006)
		25–135	391	5.8	(Lepp and Madejón 2007)
		4.9–8.8	1.5–3.1	7.4–7.5	(Meers et al. 2005)
	Normal ranges[a]	0.05–0.2			(Kabata-Pendias 2010)
		0.1–2.4			(Alloway 1990)
Cu	Alder	16	740	6.3 → 3.7	(King et al. 2006)
		5.8	54.2	7.1	(Mertens et al. 2004)
		11.4–14.8	149–258	7.2–7.5	(Vandecasteele et al. 2008)
	Ash	12.4	54.2	7.1	(Mertens et al. 2004)
		4.3–9.9	109–167	7.2–8.1	(Vandecasteele et al. 2008)
	Maple	5.9	54.2	7.1	(Mertens et al. 2004)
		2.5–9	109–258	7.2–8.1	(Vandecasteele et al. 2008)
	Willow	11.1	740	6.3 → 3.7	(King et al. 2006)
		10.0–11.3	41–71	7.4–7.5	(Meers et al. 2005)
	Normal ranges[a]	5–30			(Kabata-Pendias 2010)
		5–20			(Alloway 1990)

[a] Normal ranges in plants.
[b] An arrow denotes a change from an initial pH to a final pH during the growing period. A dash indicates a measured range.

tissues such as shoots and young leaves. Another factor is that accumulated metals are diluted with increased biomass production (Mertens et al. 2006). Especially in young vegetation, concentrations of Cd and Zn in leaves, wood, and bark strongly increase toward the end of the growing season (Mertens et al. 2006; Vandecasteele et al. 2005).

The hydrological condition of a site is very important in determining the bioavailability of metals to plants. In hydromorphic conditions, the uptake of metals by willow is comparable to that in uncontaminated environments. An upland hydrological regime resulted in elevated Cd and Zn concentrations in leaves of *Salix cinerea* (Vandecasteele et al. 2005). The highest uptake occurs on changing hydrological conditions, when physicochemical conditions of the substrate are strongly being altered by changing redox states during drying or immersion. For example, longer submersion periods in the field caused lower Cd and Zn concentrations in the leaves in the first weeks of the growing season. Emergence then sharply increased foliar concentrations to levels that were comparable with plants that had already emerged at the beginning of the growing season (Vandecasteele et al. 2005). On a site subject to gradual dewatering and terrestrialization, Cd, Zn, and Mn concentrations in leaves increased during subsequent growing seasons (Vandecasteele et al. 2007). Especially for Cd, there was a transfer effect from one growing season to the next. Oxic conditions at the end of the previous growing season appeared to determine, at least partly, the foliar concentrations for *Salix cinerea* during the

TABLE 11.3
Reported Foliar Concentrations (mg/kg Dry Matter) of Pb and Zn in Trees Growing on Contaminated, Dredged Sediment-Derived Soils

Element	Plant	Foliar Conc.	Soil Conc.	Soil pH	Ref.
Pb	Alder	11.1	1445	6.3 → 3.7	(King et al. 2006)
		5	75.2	7.1	(Mertens et al. 2004)
		0	157–280	7.2–7.5	(Vandecasteele et al. 2008)
	Ash	5	75.2	7.1	(Mertens et al. 2004)
		0	141–280	7.2–8.1	(Vandecasteele et al. 2008)
	Maple	4.5	75.2	7.1	(Mertens et al. 2004)
		0	141–233	7.2–8.1	(Vandecasteele et al. 2008)
	Poplar	4	1445	6.3 → 3.7	(King et al. 2006)
		3.3	75.2	7.1	(Mertens et al. 2004)
	Willow	2.1	1445	6.3 →3.7	(King et al. 2006)
		5.0–8.7	59–103	7.4–7.5	(Meers et al. 2005)
	Normal ranges[a]	5–10			(Kabata-Pendias and Pendias 1992)
		0.2–20			(Alloway 1990)
Zn	Alder	551	4285	6.3 → 3.7	(King et al. 2006)
		65	358	7.1	(Mertens et al. 2004)
		214–321	969–1785	7.2–7.5	(Vandecasteele et al. 2008)
	Ash	26	358	7.1	(Mertens et al. 2004)
		28–53	837–1429	7.2–8.1	(Vandecasteele et al. 2008)
	Maple	74	358	7.1	(Mertens et al. 2004)
		64–147	837–1785	7.2–8.1	(Vandecasteele et al. 2008)
	Poplar	1110	4285	6.3 → 3.7	(King et al. 2006)
		804	1306	5.8	(Lepp and Madejón 2007)
		465	358	7.1	(Mertens et al. 2004)
	Willow	1220	4285	6.3 → 3.7	(King et al. 2006)
		317–804	1306	5.8	(Lepp and Madejón 2007)
		273–675	503–945	7.4–7.5	(Meers et al. 2005)
	Normal ranges[a]	27–150			(Kabata-Pendias 2010)
		1–400			(Alloway 1990)

[a] Normal ranges in plants.

next growing season (Vandecasteele et al. 2005). The maintenance of a hydrological regime aiming at wetland creation is a potential management option for reducing bioavailability provided submersion can be maintained throughout the growing season (Vandecasteele et al. 2005).

11.4 BIOAVAILABILITY TOWARD BIOTA

Soil dwelling organisms are directly in contact with the metal contamination in the substrate. Metals taken up by the vegetation may be transferred to the food web when the vegetation is consumed or the metal-enriched plant material is returned to the soil litter layer. Different organisms accumulate markedly higher metal levels in ecosystems that are developed on contaminated, dredged sediment-derived soils compared with unpolluted environments. Leaf beetles on poplars showed up to fivefold higher body Cd concentrations than in reference situations. Zn levels in the leaf beetles were in the normal range, although both Zn and Cd in the poplar leaves were elevated (Vandecasteele et al. 2003).

Earthworms have an important role in the biomagnification of heavy metals in terrestrial ecosystems. Colonization of dredged sediment disposal sites by earthworms is mostly determined by grain size distribution and age of a site, rather than by the pollution status. More sandy areas are colonized more rapidly. It may take more than 40 years until all ecological categories have fully colonized sites that are affected by dredged sediment disposal (Vandecasteele 2004).

Relative to the surrounding environment, earthworm biomass in contaminated heavy clay dredged sediment-derived soils was four times lower than in alluvial soils, whereas in sandy loam dredged sediment-derived soils, it was comparable with these. The intensity of metal transfer to the food web at the more polluted, heavy clay soils thus was partially offset by the lower earthworm biomass (Vandecasteele, 2004). Next to the earthworm biomass, there are differences in susceptibility to predation between different categories of earthworms. Endogeic earthworms remain in the soil, whereas anecic earthworms are at the soil surface only at night. For a good assessment of risks that are related to biomagnification, earthworm tissue concentrations, earthworm biomass, and prevalent species must be taken into consideration (Vandecasteele, 2004).

As soil-dwelling organisms, it is not surprising that these organisms can accumulate high concentrations when living in contaminated substrates. Vandecasteele et al. (2010) measured concentrations (mg/kg dry weight) of 15, 36, 27, and 898 for Cd, Cu, Pb, and Zn, respectively, in earthworms living in a contaminated dredged sediment-derived soil (total contents: 26 mg/kg Cd, 136 mg/kg Cu, 227 mg/kg Pb, and 3045 mg/kg Zn). In an uncontaminated reference soil, metal concentrations were 5.9, 15, 2.9, and 325 mg/kg dry weight for Cd, Cu, Pb, and Zn, respectively. Body concentrations (dry mass including ingested soil) in the order of 90 mg/kg Cd, 1.5 mg/kg Hg, and 200 mg/kg Pb were reported by Beyer and Stafford (1993). These authors estimated that lethal or serious sublethal effects on wildlife might be expected at concentrations of 10 mg/kg Cd, 3 mg/kg Hg, and 670 mg/kg Pb in alkaline surface soils derived from dredged material. Soil amendments believed to be effective in reducing bioavailability and/or mobility in calcareous metal-contaminated soils did not decrease metal uptake by earthworms (Vandecasteele et al. 2010).

Terrestrial gastropods are another very important group of organisms in determining transfer of metals from vegetation or plant litter to carnivores. Shells of *Cepaea nemoralis* exhibited increased concentrations of Cd and Zn, whereas the levels of Cr and Pb remained below the limit of detection, despite elevated contents of these metals in the dredged sediment-derived soils (Jordaens et al. 2006). In their organs, both *C. nemoralis* (especially in the digestive gland) and *Succinea putris* (in the body) tend to accumulate high amounts of metals, and surprisingly rather independent from the contamination level of the substrate. Metal accumulation in organs was correlated with soil properties, including pH, texture, and organic matter, but almost not with soil metal concentrations. As such, in contrast to concentrations in shells, organ concentrations did not differ significantly between a contaminated plot and a corresponding reference plot (Boshoff et al. 2013). The authors suggest that there is substantial accumulation of background metal concentrations and/or that there is an exchange of individuals between contaminated and nearby reference plots. The latter hypothesis, however, does not coincide with the observations on metal concentrations in the shell, which were based on samplings in the same areas. In a study in contaminated floodplains, authors found correlations between total metal concentrations in the tissue of snails and concentrations in soil and stinging nettle (*Urtica dioica*). Correlations with contents in vegetation were markedly stronger, suggesting that metal transfer from polluted leaves to *C. nemoralis* is more important than transfer from the soil (Notten et al. 2005). These and other studies indicate that, in general, total metal concentrations and soil physicochemical properties alone are not sufficient to predict observed accumulation patterns. Morphological (e.g., soft-bodied versus hard-bodied organisms), ecological (diet preferences, microhabitats), or physiological (excretion and storage mechanisms) characteristics are other factors that may also affect the bioavailability of pollutants (Mourier et al. 2011).

Mertens et al. (2001) analyzed small mammals living in disposal sites for dredged sediment. Levels of Cd, but not Zn, in small mammals (wood mouse, bank vole, and common shrew) were elevated compared with background levels. There were no significant differences between sites in Cd or Zn levels in animals, despite differences in soil contaminant levels and concentrations in willow leaves. Risk assessment suggested that the Cd in the soil would cause a limited risk for predators (Mertens et al. 2001). High Cd levels were observed in white-footed mice living in a sediment disposal site in Illinois, United States (Levengood and Heske 2008). However, no impact of the metal exposure on population parameters could be demonstrated. High concentrations in vegetation and small mammals, nevertheless, indicate that ecosystem development should be carefully considered.

11.5 CONCLUSION

Dredging contaminated sediments and disposing them upland involves subjecting the materials to major chemical changes. Metals are most effectively immobilized in reduced conditions where their behavior is controlled by even small amounts of sulfide. The drying and oxidation of the sediment at least temporarily causes a major increase in the solubility and bioavailability of the metals. Metal bioavailability tends to decrease when the sediments are stored in relatively stable upland conditions, that is, where a succession of oxidizing and reducing conditions is avoided.

Sediments disposed on land develop into a soil-like material. It is inherent that metals from these materials will, to some extent, be taken up by crops and organisms. On calcareous materials, the extent to which this occurs may be limited in magnitude, may be limited to specific organisms or plant species, and may not visibly hamper biological functioning. Contaminated sediments not containing carbonates are at risk to be subject to acidification. This results in a prolonged enhanced leaching and renders the substrate toxic for plant growth. There is a need for more data on the uptake and effects of metals in dredged sediment disposal on biota, in particular higher biota such as mammals and birds, in order to better understand potential transfer of the metals into the food web.

REFERENCES

Allen, H.E., 1993. The significance of trace metal speciation for water, sediment and soil quality criteria and standards. *Sci. Total Environ.*, Proceedings of the 2nd European Conference on Ecotoxicology 134, Supplement 1, 23–45. doi:10.1016/S0048-9697(05)80004-X.

Alloway, B.J., 1990. *Heavy Metals in Soils*. Glasgow, Scotland: Blackie and Son.

Ankley, G.T., 1996. Evaluation of metal/acid-volatile sulfide relationships in the prediction of metal bioaccumulation by benthic macroinvertebrates. *Environ. Toxicol. Chem.* 15, 2138–2146. doi:10.1002/etc.5620151209.

Ankley, G.T., Di Toro, D.M., Hansen, D.J., Berry, W.J., 1996. Technical basis and proposal for deriving sediment quality criteria for metals. *Environ. Toxicol. Chem.* 15, 2056–2066.

Beyer, W.N., Stafford, C., 1993. Survey and evaluation of contaminants in earthworms and in soils derived from dredged material at confined disposal facilities in the great-lakes region. *Environ. Monit. Assess.* 24, 151–165.

Bortone, G., Arevalo, E., Deibel, I., Detzner, H.-D., de Propris, L., Elskens, F., Giordano, A., Hakstege, P., Hamer, K., Harmsen, J., Hauge, A., Palumbo, L., van Veen, J., 2004. Synthesis of the SedNet work package 4 outcomes. *J. Soils Sediments* 4, 225–232.

Boshoff, M., Jordaens, K., Backeljau, T., Lettens, S., Tack, F., Vandecasteele, B., De Jonge, M., Bervoets, L., 2013. Organ- and species-specific accumulation of metals in two land snail species (Gastropoda, Pulmonata). *Sci. Total Environ.* 449, 470–481. doi:10.1016/j.scitotenv.2013.02.003.

Bramley, R.G.V., Rimmer, D.L., 1988. Dredged materials—Problems associated with their use on land. *J. Soil Sci.* 39, 469–482.

Caille, N., Tiffreau, C., Leyval, C., Morel, J.L., 2003. Solubility of metals in an anoxic sediment during prolonged aeration. *Sci. Total Environ.* 301, 239–250.

Capilla, X., Schwartz, C., Bedell, J.-P., Sterckeman, T., Perrodin, Y., Morel, J.-L., 2006. Physicochemical and biological characterisation of different dredged sediment deposit sites in France. *Environ. Pollut.* 143, 106–116.

Cavallaro, N., McBride, M.B., 1984. Effect of selective dissolution on charge and surface properties of an acid soil clay. *Clays Clay Miner*. 32, 283–290.

Chapman, P.M., Wang, F., Janssen, C., Persoone, G., Allen, H.E., 1998. Ecotoxicology of metals in aquatic sediments: Binding and release, bioavailability, risk assessment, and remediation. *Can. J. Fish. Aquat.* Sci. 55, 2221–2243.

Clemens, S., 2001. Molecular mechanisms of plant metal tolerance and homeostasis. *Planta* 212, 475–486.

Committee on Bioavailability of Contaminants in Soils and Sediments, National Research Council (U.S.), 2003. Bioavailability of contaminants in soils and sediments processes, tools, and applications. Washington, DC: National Academies Press.

Cook, S.R., Parker, A., 2003. Geochemical changes to dredged canal sediments following land spreading: A review. *Land Contam. Reclam.* 11, 405–410. doi:10.2462/09670513.627.

Darmody, R.G., Marlin, J.C., 2002. Sediments and sediment-derived soils in Illinois: Pedological and agronomic assessment. *Environ. Monit. Assess.* 77, 209–227.

Du Laing, G., Rinklebe, J., Vandecasteele, B., Meers, E., Tack, F., 2009. Trace metal behaviour in estuarine and riverine floodplain soils and sediments: A review. *Sci. Total Environ.* 407, 3972–3985.

Du Laing, G., Vanthuyne, D., Vandecasteele, B., Tack, F., Verloo, M., 2007. Influence of hydrological regime on pore water metal concentrations in a contaminated sediment-derived soil. *Environ. Pollut.* 147, 615–625.

Ehlers, L.J., Luthy, R.G., 2003. Peer reviewed: Contaminant bioavailability in soil and sediment. *Environ. Sci. Technol.* 37, 295A–302A. doi:10.1021/es032524f.

Gambrell, R.P., 1994. Trace and toxic metals in wetlands—A review. *J. Environ. Qual.* 23, 883–891.

Gambrell, R.P., Wiesepape, J.B., Patrick, W.H.J., Duff, M.C., 1991. The effects of pH, redox, and salinity on metal release from a contaminated sediment. *Water Air Soil Pollut.* 57–58, 359–367.

Greger, M., Landberg, T., 1999. Use of willow in phytoextraction. *Int. J. Phytoremediation* 1, 115–123.

Guo, T., DeLaune, R.D., Patrick, J., 1997. The influence of sediment redox chemistry on chemically active forms of arsenic, cadmium, chromium, and zinc in estuarine sediment. *Environ. Int.* 23, 305–316.

Gupta, S.K., Vollmer, M.K., Krebs, R., 1996. The importance of mobile, mobilisable and pseudo total heavy metal fractions in soil for three-level risk assessment and risk management. *Sci. Total Environ.* 178, 11–20.

Hansen, D.J., Berry, W.J., Boothman, W.S., Pesch, C.E., Mahony, J.D., Di Toro, D.M., Robson, D.L., Ankley, G.T., Ma, D., Yan, Q., 1996. Predicting the toxicity of metal-contaminated field sediments using interstitial concentration of metals and acid-volatile sulfide normalizations. *Environ. Toxicol. Chem.* 15, 2080–2094. doi:10.1002/etc.5620151204.

Hartley, W., Dickinson, N.M., 2010. Exposure of an anoxic and contaminated canal sediment: Mobility of metal(loid)s. *Environ. Pollut.* 158, 649–657. doi:10.1016/j.envpol.2009.10.030.

Herman, R., Bogaert, G., 2007. Het Sectoraal Uitvoeringsplan Bagger- en Ruimingsspecie: bijna van kracht. BVDA Nieuwsbr.

Jordaens, K., De Wolf, H., Vandecasteele, B., Blust, R., Backeljau, T., 2006. Associations between shell strength, shell morphology and heavy metals in the land snail *Cepaea nemoralis* (Gastropoda, Helicidae). *Sci. Total Environ.* 363, 285–293.

Kabata-Pendias, A., 2010. *Trace Elements in Soils and Plants*. Boca Raton, FL: CRC Press.

King, R.F., Royle, A., Putwain, P.D., Dickinson, N.M., 2006. Changing contaminant mobility in a dredged canal sediment during a three-year phytoremediation trial. *Environ. Pollut.* 143, 318–326.

Kinniburgh, D.G., Jackson, M.L., Syers, J.K., 1976. Adsorption of alkaline earth, transition, and heavy metal cations by hydrous oxide gels of iron and aluminum. *Soil Sci. Soc. Am. J.* 40, 796–799.

Laureysens, I., Blust, R., De Temmerman, L., Lemmens, C., Ceulemans, R., 2004. Clonal variation in heavy metal accumulation and biomass production in a poplar coppice culture: I. Seasonal variation in leaf, wood and bark concentrations. *Environ. Pollut.* 131, 485–494.

Lepp, N.W., Madejón, P., 2007. Cadmium and zinc in vegetation and litter of a voluntary woodland that has developed on contaminated sediment-derived soil. *J. Environ. Qual.* 36, 1123. doi:10.2134/jeq2006.0218.

Levengood, J.M., Heske, E.J., 2008. Heavy metal exposure, reproductive activity, and demographic patterns in white-footed mice (*Peromyscus leucopus*) inhabiting a contaminated floodplain wetland. *Sci. Total Environ.* 389, 320–328.

Lützow, M.v., Kögel-Knabner, I., Ekschmitt, K., Matzner, E., Guggenberger, G., Marschner, B., Flessa, H., 2006. Stabilization of organic matter in temperate soils: mechanisms and their relevance under different soil conditions—A review. *Eur. J. Soil Sci.* 57, 426–445. doi:10.1111/j.1365-2389.2006.00809.x.

Marseille, F., Tiffreau, C., Laboudigue, A., Lecomte, P., 2000. Impact of vegetation on the mobility and bioavailability of trace elements in a dredged sediment deposit: A greenhouse study. *Agronomie* 20, 547–556.

Martin, H.W., Kaplan, D.I., 1998. Temporal changes in cadmium, thallium, and vanadium mobility in soil and phytoavailability under field conditions. *Water Air Soil Pollut.* 101, 399–410.

Meers, E., Vandecasteele, B., Ruttens, A., Vangronsveld, J., Tack, F., 2007. Potential of five willow species (Salix spp.) for phytoextraction of heavy metals. *Environ. Exp. Bot.* 60, 57–68.

Mertens, J., Luyssaert, S., Verbeeren, S., Vervaeke, P., Lust, N., 2001. Cd and Zn concentrations in small mammals and willow leaves on disposal facilities for dredged material. *Environ. Pollut.* 115, 17–22.

Mertens, J., Van Nevel, L., De Schrijver, A., Piesschaert, F., Oosterbaan, A., Tack, F.M., Verheyen, K., 2007. Tree species effect on the redistribution of soil metals. *Environ. Pollut.* 149, 173–181.

Mertens, J., Vervaeke, P., Meers, E., Tack, F.M.G., 2006. Seasonal changes of metals in Willow (*Salix* sp.) Stands for phytoremediation on dredged sediment. *Environ. Sci. Technol.* 40, 1962–1968.

Mertens, J., Vervaeke, P., Schrijver, A.D., Luyssaert, S., 2004. Metal uptake by young trees from dredged brackish sediment: limitations and possibilities for phytoextraction and phytostabilisation. *Sci. Total Environ.* 326, 209–215.

Mourier, B., Fritsch, C., Dhivert, E., Gimbert, F., Cœurdassier, M., Pauget, B., Vaufleury, A. de, Scheifler, R., 2011. Chemical extractions and predicted free ion activities fail to estimate metal transfer from soil to field land snails. *Chemosphere* 85, 1057–1065. doi:10.1016/j.chemosphere.2011.07.035.

Naidu, R., Megharaj, M., Krishnamurti, G.S.R., Vig, K., Kookana, R.S., 2000. Bioavailability, definition and analytical techniques for assessment and remediation of contaminated (inorganic and organic) soils. In: *Contaminated Site Remediation: From Source Zones to Ecosystems*. Proceedings of the 2000 Contaminated Site Remediation Conference, Melbourne, Australia. DTIC Document, pp. 283–290.

Naidu, R., Semple, K., Megharaj, M., Juhasz, A., Bolan, N., Gupta, S., Clothier, B., Schulin, R., 2011. Bioavailability: Definition, assessment and implications for risk assessment. In: *Chemical Bioavailability in Terrestrial Environments*. London: Elsevier, pp. 39–51.

National Research Council, 2003. Bioavailability of contaminants in soils and sediments: Processes, tools, and applications. Washington, DC: National Academies Press.

Notten, M.J.M., Oosthoek, A.J.P., Rozema, J., Aerts, R., 2005. Heavy metal concentrations in a soil–plant–snail food chain along a terrestrial soil pollution gradient. *Environ. Pollut.* 138, 178–190. doi:10.1016/j.envpol.2005.01.011.

Panfili, F., Manceau, A., Sarret, G., Spadini, L., Kirpichtchikova, T., Bert, V., Laboudigue, A., Marcus, M.A., Ahamdach, N., Libert, M.-F., 2005. The effect of phytostabilization on Zn speciation in a dredged contaminated sediment using scanning electron microscopy, X-ray fluorescence, EXAFS spectroscopy, and principal components analysis. *Geochim. Cosmochim. Acta* 69, 2265–2284.

Piou, S., Bataillard, P., Laboudigue, A., Férard, J.-F., Masfaraud, J.-F., 2009. Changes in the geochemistry and ecotoxicity of a Zn and Cd contaminated dredged sediment over time after land disposal. *Environ. Res.* 109, 712–720.

Satawathananont, S., Patrick, W.H.J., Moore, P.A.J., 1991. Effect of controlled redox conditions on metal solubility in acid sulfate soils. *Plant Soil* 133, 281–290.

Semple, K.T., Doick, K.J., Jones, K.C., Burauel, P., Craven, A., Harms, H., 2004. Peer reviewed: Defining bioavailability and bioaccessibility of contaminated soil and sediment is complicated. *Environ. Sci. Technol.* 38, 228A–231A. doi:10.1021/es040548w.

Singh, S.P., Tack, F.M.G., Verloo, M., 1998. Heavy metal fractionation and extractability in dredged sediment derived surface soils. *Water Air Soil Pollut.* 102, 313–328.

Smilde, K.W., Van Driel, W., Van Luit, B., 1982. Constraints in cropping heavy-metal contaminated fluvial sediments. *Sci. Total Environ.* 25, 225–244. doi:10.1016/0048-9697(82)90016-X.

Stephens, S.R., Alloway, B.J., Parker, A., Carter, J.E., Hodson, M.E., 2001. Changes in the leachability of metals from dredged canal sediments during drying and oxidation. *Environ. Pollut.* 114, 407–413.

Tack, F.M., Callewaert, O.W.J.J., Verloo, M.G., 1996. Metal solubility as a function of pH in a contaminated, dredged sediment affected by oxidation. *Environ. Pollut.* 91, 199–208.

Tack, F.M., Lapauw, F., Verloo, M.G., 1997. Determination and fractionation of sulphur in a contaminated dredged sediment. *Talanta* 44, 2185–2192.

Tack, F.M.G., 2010. Trace elements: general soil chemistry, principles and processes. In: *Trace Elements in Soils*. Chichester, UK: Wiley-Blackwell, pp. 9–37.

Vandecasteele, B., 2004. Earthworm biomass as additional information for risk assessment of heavy metal biomagnification: A case study for dredged sediment-derived soils and polluted floodplain soils. *Environ. Pollut.* 129, 363–375. doi:10.1016/j.envpol.2003.12.007.

Vandecasteele, B., Buysse, C.A., Tack, F.M.G., 2006. Metal uptake in maize, willows and poplars on impoldered and freshwater tidal marshes in the Scheldt estuary. *Soil Use Manag.* 22, 52–61. doi:10.1111/j.1475-2743.2005.00007.x.

Vandecasteele, B., De Vos, B., Tack, F.M.G., 2002a. Heavy metal contents in surface soils along the Upper Scheldt river (Belgium) affected by historical upland disposal of dredged materials. *Sci. Total Environ.* 290, 1–14.

Vandecasteele, B., De Vos, B., Tack, F.M.G., 2002b. Cadmium and zinc uptake by volunteer willow species and elder rooting in polluted dredged sediment disposal sites. *Sci. Total Environ.* 299, 191–205.

Vandecasteele, B., Du Laing, G., Lettens, S., Jordaens, K., Tack, F.M., 2010. Influence of flooding and metal immobilising soil amendments on availability of metals for willows and earthworms in calcareous dredged sediment-derived soils. *Environ. Pollut.* 158, 2181–2188.

Vandecasteele, B., Du Laing, G., Quataert, P., Tack, F.M.G., 2005. Differences in Cd and Zn bioaccumulation for the flood-tolerant *Salix cinerea* rooting in seasonally flooded contaminated sediments. *Sci. Total Environ.* 341, 251–263.

Vandecasteele, B., Lauriks, B., De Vos, B., Tack, F.M.G., 2003. Cd and Zn concentration in hybrid poplar foliage and leaf beetles grown on polluted sediment-derived soils. *Environ. Monit. Assess.* 89, 263–283.

Vandecasteele, B., Quataert, P., Genouw, G., Lettens, S., Tack, F.M.G., 2009. Effects of willow stands on heavy metal concentrations and top soil properties of infrastructure spoil landfills and dredged sediment-derived sites. *Sci. Total Environ.* 407, 5289–5297.

Vandecasteele, B., Quataert, P., Tack, F., 2005. The effect of hydrological regime on the metal bioavailability for the wetland plant species Salix cinerea. *Environ. Pollut.* 135, 303–312. doi:10.1016/j.envpol.2004.09.024.

Vandecasteele, B., Quataert, P., Tack, F.M.G., 2007. Uptake of Cd, Zn and Mn by willow increases during terrestrialisation of initially ponded polluted sediments. *Sci. Total Environ.* 380, 133–143.

Vandecasteele, B., Samyn, J., De Vos, B., Muys, B., 2008. Effect of tree species choice and mineral capping in a woodland phytostabilisation system: A case-study for calcareous dredged sediment landfills with an oxidised topsoil. *Ecol. Eng.* 32, 263–273. doi:10.1016/j.ecoleng.2007.12.002.

Van Driel, W., van Luit, B., Smilde, K.W., Schuurmans, W., 1995. Heavy-metal uptake by crops from polluted river sediments covered by non-polluted topsoil I. Effects of topsoil depth on metal contents. *Plant Soil* 175, 93–104.

Vervaeke, P., Luyssaert, S., Mertens, J., De Vos, B., Speleers, L., Lust, N., 2001. Dredged sediment as a substrate for biomass production of willow trees established using the SALIMAT technique. *Biomass Bioenergy* 21, 81–90.

Vervaeke, P., Tack, F.M.G., Lust, N., Verloo, M., 2004. Short- and longer-term effects of the willow root system on metal extractability in contaminated dredged sediment. *J. Environ. Qual.* 33, 976–983.

Yell, D., Riddell, J., 1995. *Dredging*. London: Thomas Telford.

12 Metal Bioavailability in Sediments and Its Role in Risk Assessment

Michael H. Paller and Anna Sophia Knox

CONTENTS

12.1 INTRODUCTION

Surface water quality has improved in the United States because of better regulation and treatment, but sediments remain as sinks for historical releases of metals and organic contaminants. Sediment contamination is present in more than 70% of U.S. watersheds, with a total estimated surficial volume of more than 1 billion m^3 [U.S. Environmental Protection Agency (USEPA), 1997]. Both point and nonpoint pollution sources contribute to this problem. Metals, including arsenic (As), cadmium (Cd), cobalt (Co), copper (Cu), mercury (Hg), nickel (Ni), lead (Pb), and zinc (Zn), are often found in the sediments of harbors and other areas that are affected by anthropogenic activities. Sediments containing these contaminants act as secondary sources of contamination, posing significant direct and indirect environmental risks through bioaccumulation in aquatic organisms and incorporation into aquatic food webs that may ultimately lead to humans [National Research Council (NRC), 2003]. Episodic physical redistribution of contaminated sediments within dynamic waterways can disperse such environmental risks, potentially affecting biological and water quality conditions and moving contaminated sediment far from the original source (Admiral et al., 2000).

Sediments are essential and valuable resources in river basins and other aquatic environments. They support a large biodiversity, and sediment organisms are ecologically important and an integral part of aquatic food webs. Because sediments are also a sink for contaminants, these organisms are often exposed to higher concentrations of contaminants than organisms that occupy the water column. Metal toxicity and bioaccumulation is poorly predicted by total metal concentrations in sediments, because only the metal fraction that is available for biological uptake has the potential to cause human health or ecological risks. The key to predicting risk is understanding and accurately assessing contaminant bioavailability (USEPA, 2005). However, bioavailability is complex and is influenced by a number of geochemical characteristics of the environment and biological characteristics of the exposed organisms.

12.2 BIOAVAILABILITY OF METALS IN SEDIMENT

Bioavailability can be defined as the extent to which chemicals present in the sediment/soil may be absorbed or metabolized by human or ecological receptors or are available for interactions with biological systems (ISO, 2005). It is a dynamic process that is determined by three factors: environmental bioaccessibility, environmental bioavailability, and toxicological bioavailability (Peijnenburg et al., 1997; Lanno et al., 2004; Harmsen, 2005).

12.2.1 Environmental Accessibility

Metal contaminants can be separated into environmentally accessible (i.e., bioaccessible) fractions and inaccessible fractions. Bioaccessibility in aquatic environments is largely determined by physical and biological factors. Contaminants in deeply buried sediments are usually inaccessible to benthic organisms (Figure 12.1). Infaunal organisms, which burrow into and live within the sediments, such as clams, worms, burrowing crustaceans, and insect larvae, have access to contaminants that are found at the sediment depths they occupy. Epifaunal organisms, which live primarily on the sediment surface, such as many gastropods, insect larvae, crustaceans, and echinoderms, may only be affected by contaminants near the sediment–water column interface.

Although sediments are generally viewed as a metal sink, they may secondarily act as a metal source to organisms of the water column (Figure 12.1). Metals in particulate form can be resuspended by wave action, currents, and propeller washes that scour the bottom; dissolved metals can be released from the sediments by diffusion and advective groundwater flow (Admiral et al., 2000; Beachler and Hill, 2003). Ebullition (bubbling of gasses through the sediment) can increase the rates of both diffusion and particle resuspension (Barabás et al., 2013). Sediment dwelling organisms can release dissolved and particle-bound contaminants into the water column by bioturbation (reworking sediments) and bioirrigation (flushing burrows with water) (Schaller, 2014). Within the water column, particulate metals can be ingested by filter feeders, and dissolved metals (including metals released from resuspended particles) can be adsorbed onto plankton and absorbed through the gills and other membranes of larger aquatic organisms.

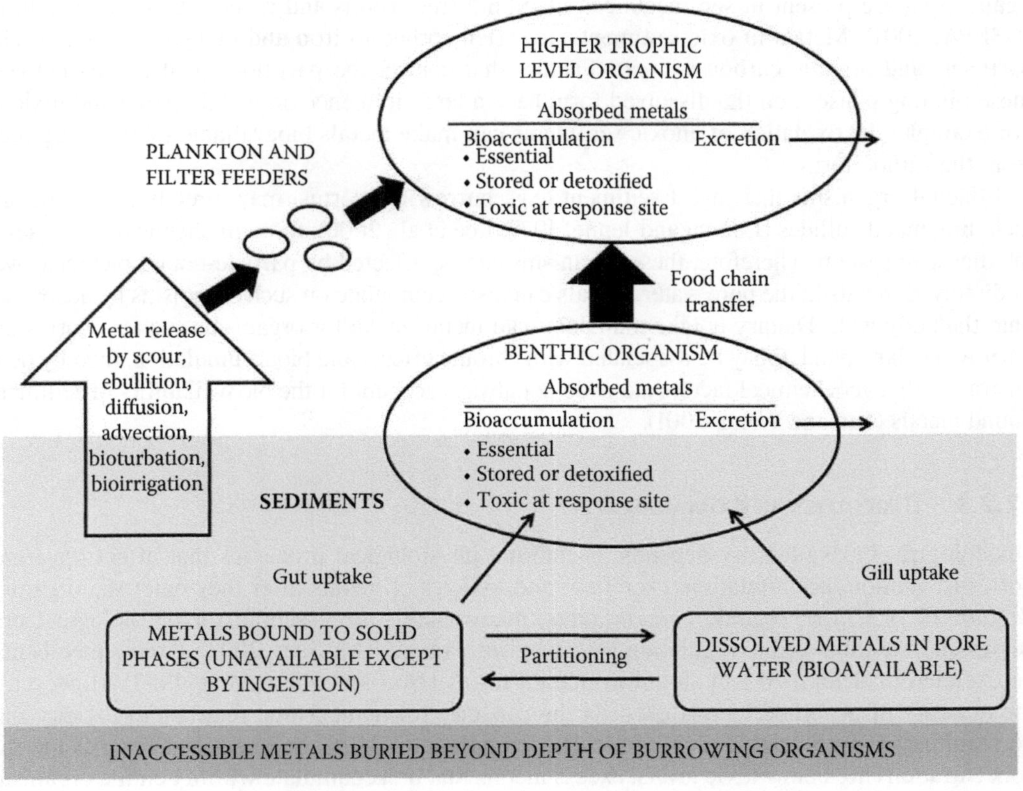

FIGURE 12.1 Processes affecting the bioavailability, uptake, and toxicity of metal contaminants in sediment.

12.2.2 Environmental Bioavailability

The bioaccessible contaminant fraction can be separated into bioavailable and unavailable fractions, with only the former able to be absorbed by organisms. Metals released from the solid phase into interstitial pore water as free metal ions can be absorbed through respiratory surfaces and other membranes (Figure 12.1) (Campbell, 1995; McGeer et al., 2002). The release of contaminants from the solid phase is affected by numerous factors, including pH, redox potential, DOC, inorganic complexes, organic complexes, and sediment particle size (Table 12.1). Anoxic sediments contain iron monosulfides that react with metal cations to form insoluble metal sulfides that are not absorbed through respiratory surfaces and other membranes that are exposed to interstitial water (Di Toro et al., 1992). Organic carbon and other solid- and dissolved-phase

TABLE 12.1
Some Factors Affecting Metal Bioavailability in Sediments

Geochemical Factors	Biological Factors
pH	Feeding behavior
Redox oxidation potential	Habitat (e.g., infauna vs. epifauna)
Iron monosulfides	Life stage
Iron and manganese oxides	Ingestion rate
Dissolved organic carbon (DOC)	Growth rate
Inorganic and organic complexes	Physiology
Sediment particle size	Age

ligands that are present in sediments can also bind free metals and reduce their bioavailability (USEPA, 2001). Metals in oxic sediments are often sorbed to iron and manganese oxides, clay particles, and organic carbon sources. Factors that control the partitioning of metals between these binding phases and the dissolved form have a large influence on metal uptake and toxicity. For example, the oxidation of anoxic sediments can make metals bioavailable by releasing them from the sulfide form.

Infaunal organisms that ingest sediment (e.g., burrowing worms) may directly absorb metals, including metal sulfides (Luoma and Jenne, 1977; Lee et al., 2000), through chemical processes in the digestive system. Therefore, these organisms can be affected by particle-bound metals as well as dissolved metals in the pore water. Metals can also accumulate on such organisms by adsorption onto the body wall. Dietary uptake may dominate metal intake for organisms such as worms and burrowing clams, and it may be associated with chronic effects and bioaccumulation, thereby demonstrating that geochemical factors alone do not always account for the bioavailability of sediment-bound metals (Fan and Wang, 2001).

12.2.3 Toxicological Bioavailability

Toxicological bioavailability depends on internal physiological processes that affect the transport, distribution, accumulation, excretion, and toxicity of metals after they enter the organism (Figure 12.1). Unlike organic contaminants, many metals are essential for physiological processes and are subjected to regulation within optimal ranges (McGeer, 2003). Excess metals may be excreted or detoxified and stored in benign forms (Nott and Nicolaidou, 1993). However, if the quantity of absorbed metals exceeds the capacity for homeostatic regulation, toxicity may be manifested at vulnerable sites of metal accumulation within the organism. Even if internal concentrations are below toxic levels, excess metals that bioaccumulate within benthic organisms pose an ecological risk, because they may be transferred to higher level consumers, including humans.

12.3 ASSESSING BIOAVAILABILITY

There are a number of methods for estimating the bioavailable fraction of metals. In this chapter, bioavailability methods are divided into chemical and biological methods: chemical methods are subdivided into methods for whole sediments and methods for pore water; biological methods are divided into analysis of tissues, laboratory toxicity testing, and benthic community surveys (Table 12.2).

12.3.1 Whole Sediment Methods

12.3.1.1 Total Metals

Metal concentrations in sediments can be determined by digestion with acids (total acid digestion and strong acid digestion) (USEPA, 1994). These methods break down most or all mineral components and liberate metals that may be tightly bound to the sediments. Bulk metal concentrations can be used to predict the absence of toxicity; that is, if results fall below reference values or sediment quality guidelines (SQGs), toxicity should not occur (Wenning et al., 2005). However, they correlate poorly with toxicity, because they include metals that are bound to solid phases that are not bioavailable. Bulk metal concentrations are the most useful for screening sites where metal concentrations are high enough that toxicity is possible but should be followed by other methods to determine whether toxicity actually occurs.

TABLE 12.2
Methods for Assessing Bioavailability of Metals in Sediments

Method	Description	Advantages	Disadvantages
	Chemical Methods—Whole Sediments		
Total sediment metals	Metal concentrations measured by acid digestion in sediment core or grab samples; often compared with sediment quality criteria for screening purposes.	Standard methods available; may be a regulatory requirement.	Bulk metal concentrations are poorly correlated with toxicity.
Simultaneously extracted metal (SEM)/Acid volatile sulfide (AVS)	Molar ratio of SEM (Cd, Cu, Pb, Ni, Ag, and Zn) and AVS liberated from sediment with 1N HCl; estimates free metal concentrations.	Accurately predicts absence of toxicity by accounting for a major binding phase of cationic metals in anoxic sediments.	Predicts absence but not onset of toxicity; does not account for dietary uptake; only for reducing conditions.
(SEM–AVS)/F_{oc}	SEM–AVS normalized for the fraction of organic carbon.	Better accuracy than SEM/AVS, because it accounts for an additional major binding phase of metals in anoxic sediments.	Does not account for dietary uptake; only for reducing conditions.
Biomimetic methods	Determination of available fraction in ingested sediments by extraction with digestive gut fluids—extractions can be performed on bulk sediments or metal fractions that are produced by sequential extractions.	Accounts for the effects of ingested metals that may dominate metal intake for burrowing organisms that ingest sediments.	Accounts only for dietary uptake; not widely used.
Sequential extractions on sediment samples	Quantification of metal fractions in sediments by sequentially leaching with different chemicals of varying strength and reactivity.	Can produce a site-specific desorption K_d corresponding to the bioavailable fraction; can measure binding phases (e.g., organic C, Fe–Mn oxides, carbonates) in aerobic sediments where AVS is oxidized.	There is no metal fraction or extraction technique that consistently identifies the bioavailable fraction.
	Chemical Methods—Pore Water		
Pore-water extraction	Removal of pore water from sediment by centrifugation, suction, etc. followed by chemical analysis; results may be compared with acute or chronic criteria for water.	Interstitial dissolved metal concentrations are correlated with toxicity; results can be refined by adjusting for hardness and other factors that affect toxicity.	Sample processing may affect chemical speciation (e.g., by changing redox conditions); does not account for dietary uptake.

(*Continued*)

TABLE 12.2 *(Continued)*
Methods for Assessing Bioavailability of Metals in Sediments

Method	Description	Advantages	Disadvantages
Diffusion samplers	Diffusive gradients in thin films (DGT) and sediment "peepers" that passively remove metals from pore water; can be used in situ or ex situ.	May mimic biological uptake by benthic organisms, have potential for fine-scale spatial resolution, and can minimize artifacts caused by sediment disturbance.	Insufficient understanding of relationship between analytical results and actual uptake by organisms.
	Biological Methods—Analysis of Tissue Residuals		
Indigenous organisms	Indigenous organisms harvested from the field and analyzed for metal concentrations in tissues.	Direct measure of contaminant availability that integrates all routes of exposure under prevailing environmental conditions for relevant organisms.	Difficult to find sufficient organisms of the same type, especially for impact vs. reference; uncertain exposure scenario for mobile organisms.
Caged organisms (in situ bioassay)	Caged organisms deployed in field and analyzed for metals on retrieval.	Better control over conditions and time of exposure, realistic exposure scenario, can control sample size; can measure multiple endpoints, e.g., growth and mortality.	Non-native organisms may lack ecological relevance; artifacts caused by caging may affect contaminant uptake and other endpoints.
	Biological Methods—Laboratory Toxicity Testing		
Sediment toxicity test	Test organisms exposed to contaminated sediments and compared with reference sediments: acute or chronic.	Standard methods available; may be a regulatory requirement.	Results subject to laboratory artifacts; may be difficult to identify source of toxicity with multiple contaminants.
Pore-water toxicity test	Test organisms exposed to pore water extracted from sediments and compared with reference pore water: acute or chronic.	Toxicity usually related to pore water rather than to sediment concentrations; can be compared with sediment tests to identify routes of exposure.	Results subject to laboratory artifacts; may be difficult to identify source of toxicity with multiple contaminants.
Toxicity identity evaluation (TIE)	A series of tests in which different groups of chemicals are isolated and tested to identify sources of toxicity; can be performed on pore water or bulk sediments.	Can identify particular sources of toxicity in sediments containing multiple contaminants.	Complex and relatively expensive; results subject to laboratory artifacts.
	Biological Methods—Benthic Community Surveys		
Natural substrates	Analysis of types and numbers of benthic macroinvertebrates by protocols that are designed to represent particular or multiple naturally occurring habitats.	Direct, integrative, and in situ measure of the health of benthic communities.	Affected by multiple habitat factors that can be confounded with toxicity; sources of impact may be hard to identify; high inter-replicate variance.

(Continued)

TABLE 12.2 ***(Continued)***
Methods for Assessing Bioavailability of Metals in Sediments

Method	Description	Advantages	Disadvantages
Artificial substrates	Analysis of types and numbers of benthic macroinvertbrates that colonize artificial substrates.	Standardized habitat reduces the effects of benthic habitat variations on bioassessment results.	Organisms that colonize artificial substrates may differ from those naturally present and may experience different exposure scenarios.
Sediment profile imaging (SPI)	Photographic technique for producing profiles of the upper sediment, pore water, and benthic boundary layer of water.	Integrative image of benthic habitat plus sediment organisms and organism traces; can rapidly investigate large areas.	Does not provide detailed information on organism taxonomy and abundance; limited effectiveness in coarse-grained bottoms.

12.3.1.2 Simultaneously Extracted Metal/Acid Volatile Sulfides

Equilibrium partitioning theory assumes that metals that are bound to solids in the sediment are in equilibrium with metals in sediment pore water, which represent the phase that is bioavailable to most benthic organisms. In anoxic sediments, iron monosulfides constitute an important binding phase for divalent metals:

$$M^{2} + FeS(s) \rightarrow MS(s) + Fe^{2+}$$

where M^{2+} = any metal that forms a sulfide that is more insoluble than FeS. Acid-volatile sulfides (AVS), the sulfides that are removed from sediments by extraction with cold one-molar hydrochloric acid, are an important factor affecting the bioavailability of divalent metals (Ag, Cu, Pb, Cd, Zn, and Ni in order of increasing solubility) with a high affinity for sulfide (Di Toro et al., 1990). The AVS present in anaerobic sediments is mostly bound to Fe as solid iron monosulfide (FeS) (Hansen et al., 1996); however, if these metals are present, the iron in FeS is displaced and the metals bind to AVS, removing them from the interstitial water (Di Toro et al., 1992; Hansen et al., 1996). The concentration of bioavailable metals can be estimated from the molar ratio of the metals that are simultaneously extracted with the AVS to the AVS concentration. An SEM to AVS ratio greater than one (or SEM–AVS > 1) indicates that excess metals are present in the sediment relative to AVS. These metals are unbound and can be bioavailable.

12.3.1.3 SEM–AVS/F_{oc}

Research has shown that that toxicity is usually lacking when AVS exceeds SEM; however, an excess of SEM is not always associated with toxicity, because there are other binding phases that can influence the partitioning of metals between the sediment and interstitial water. One of the most important is organic carbon, which can be expressed as a fraction of the sediment mass (F_{oc}) (Mahony et al., 1996). This finding has led to the enhancement of the SEM/AVS method by normalizing for the fraction of organic matter in the sediment: (SEM–AVS)/F_{oc}. The interpretation of this calculation, which improves accuracy by accounting for two binding phases, is greater than 130 μmol/g_{oc} = low risk, 130–3000 μmol/g_{oc} = possible risk, and greater than 3000 μmol/g_{oc} = high risk (Hansen et al., 2005). An example is shown in Table 12.3.

The AVS method is useful in sediments with high levels of sulfate and reducing conditions. Such conditions typically prevail during warm weather in sediments with a high organic matter content. The penetration of oxygen into the sediments as a result of resuspension or diminished bacterial

TABLE 12.3
Example Calculation of SEM–AVS/F_{oc} for Metals in Saltwater Sediments

Analyte	Sediment		Pore Water		
	μg/g	μmol/g	μg/L	FCV	IWBU
SEM Ni	ND[a]	ND	ND (<0.5)	8.2	<0.06
SEM Zn	34.8	0.53	8.0	81	0.10
SEM Cd	11.0	0.10	1.4	8.8	0.16
SEM Pb	56.5	0.27	0.8	8.1	0.10
SEM Cu	20.8	0.33	0.8	3.1	0.26
SEM Ag	ND	ND	ND	ND	–
AVS	25.0	0.80			
TOC	20,000.0				
1.23 (ΣSEM) – 0.80 (AVS) = 0.43 μmol/g SEM				ΣIWBU = <0.68	
(ΣSEM – AVS μmol/g)/F_{oc} (g/g) = 0.43/0.02 = 21.5 μmol/g_{oc}					

[a] ND, detection level not provided.

activity at lower temperature can oxidize metal sulfides and release metals into the water. In addition, bioaccumulation of metals can occur when AVS exceeds SEM through dietary uptake, because metal geochemistry is changed by conditions in the gut of an organism (Lee et al., 2000).

12.3.1.4 Biomimetic Methods

Correlations have been found between metal uptake and the metal fractions extracted by gut juices (Mayer et al., 1996; Chen and Mayer, 1999). This has led to biomimetic approaches that assess the pool of bioavailable metals in sediment by extraction with the digestive gut fluids of sediment-feeding organisms such as the peanut worm *Sipunculus nudus* or lugworm *Arenicola marina* (Yan and Wang, 2002). Extraction with the digestive fluid of *A. marina* has been shown to predict mercury bioaccumulation by the deposit feeding amphipod *Leptocheirus plumulosus* (Lawrence et al., 1999). Gut fluid extractions can be conducted on bulk sediment samples or the metal fractions produced by sequential extractions (Peng et al., 2004). The latter can determine the geochemical metal phase(s) that are likely to be physiologically mobilized in the digestive systems of sediment-feeding invertebrates, and it facilitates linkage between geochemical and biomimetic methods of assessing bioavailability. The results from digestive fluid extractions can be used to calculate a modified desorption distribution coefficient (K_{dg}) based only on fractions showing metal release by the gut. This method is relevant for sediment-feeding infauna that may experience significant exposure through the dietary route.

12.3.1.5 Sequential Extraction

Sequential extractions based on methods developed by Tessier et al. (1979) can identify different metal fractions in sediments by successively extracting samples with chemical reagents of increasing strength and reactivity. The following geochemical fractions are commonly measured: exchangeable, carbonate, amorphous Fe and Mn oxides, crystalline Fe and Mn oxides, organic, sulfides, and residual (Tessier, 1979; Hall, 1996). Sequential extractions can be used to measure metals that are associated with binding phases that are important in aerobic sediments, including organic C, Fe and Mn oxides, and carbonates. A soluble organic fraction can be included as the first step for sediments with a high content of organic matter. However, there is no universally agreed-upon metal fraction or extraction technique that can identify bioavailable metal fractions in all sediments.

Desorption K_d values for oxic environments can be calculated from sequential extraction results:

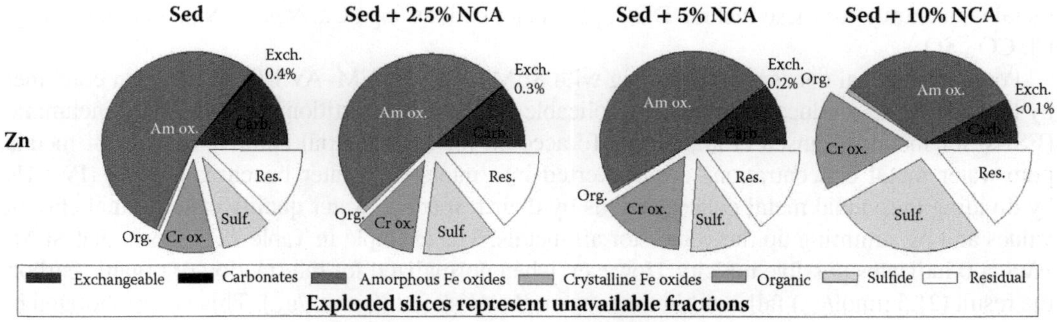

FIGURE 12.2 Changes in sequential extraction fractions of zinc following different levels of treatment with a phosphate amendment (NCA).

$$\text{Desorption } K_d = \frac{C_{\text{Exch}} + C_{\text{AmorphFe}} + C_{\text{Org}}}{C_{\text{aq}}}$$

where C_{Exch} is the metal concentration in exchangeable and/or carbonate sequential fractions (μg/g), C_{AmorphFe} is the metal concentration in the amorphous Fe-oxide sequential extraction fraction (μg/g), C_{Org} is the metal concentration in the organic sequential extraction fraction (μg/g), and C_{aq} is the metal pore-water concentration (μg/L) that is typically derived from pore water that is extracted by using conventional methods.

This site-specific desorption K_d theoretically identifies the metal fraction in sediment that is readily available for entry into interstitial or surface water, that is, the potentially bioavailable fraction (PBF) (Knox et al., 2006). It provides a conservative (high) estimate of the amount of a contaminant that may be released from the sediment. The contaminant pool that is not associated with the PBF can only be extracted from sediments with extremely strong acids and reductants that dissolve minerals. The magnitude of the PBF depends mostly on the sorption mechanism of the metal to the sediment. An example of changes in sequential extraction fractions resulting from the application of a phosphate source to contaminated sediments is shown in Figure 12.2.

12.3.2 Pore-Water Methods

12.3.2.1 Pore-Water Extraction

Dissolved metals in interstitial sediment pore water constitute the main route of contaminant exposure for most benthic organisms. Toxicity can be assessed by comparing metal concentrations in pore water with acute or chronic criteria for aquatic life, because biological effects are correlated with pore-water concentrations (rather than total sediment concentrations) (Di Toro et al., 1991; Hansen et al., 2005). Methods of pore-water collection include centrifugation and filtration, squeezing of core samples, and suction. Regardless of the method used, care must be taken to avoid artifacts such as changes in pore-water redox state or temperature resulting from sample preparation.

In freshwater applications, it is possible to calculate site-specific aquatic life criteria for some metals by using the Biotic Ligand Model (BLM). The BLM, which was developed from the gill surface interaction and free-ion activity models, predicts the lethal accumulation of metals on the gill surface, resulting in mortality of 50% of the exposed individuals (Di Toro et al., 2001; Niyogi and Wood, 2004). It accounts for the effects of cations, organic matter, and pH on metal speciation and absorption on the gill as well as dose–response relationships within the organism. Use of the BLM necessitates the measurement of parameters that are required by the model along with pore-water metal concentrations; these include DOC, pH, major cations (e.g., Ca, Mg, K, Na), and anions (e.g., Cl, CO_3, SO_4).

Pore-water metal concentrations along with SEM/AVS (or SEM–AVS/F_{oc}) have been combined by the USEPA to produce two equally applicable Equilibrium Partitioning Sediment Benchmarks (ESBs) for metals (Hansen et al., 2005). To account for the potential additive toxicity of metals, pore-water metal concentrations are converted into interstitial water benchmark units (IWBUs) by dividing individual metal concentrations by their respective water quality criteria final chronic values and by summing up the values for all metals. The example in Table 12.3 shows that SEM–AVS > 0 indicates possible toxicity. However, when normalized for the fraction of organic carbon, the result (21.5 μmol/g_{oc}) falls within the range of low risk (<130 μmol/g_{oc}). This is corroborated by the interstitial water ESB, which is less than one.

12.3.2.2 Diffusion Samplers

In addition to the methods mentioned earlier, pore-water metals can be evaluated with diffusion samplers that employ a gel, a semipermeable membrane, or a dialysis tube filled with distilled water. Sediment peepers employ a hollow casing with openings covered by a semipermeable membrane that permits the selective diffusion of potentially bioavailable metals over time (Hesslein, 1976). They require knowledge of the rate of flux of metals across the membrane for accurate analysis. Vertically deployed peepers with stacked cells can permit the investigation of vertical gradients in pore-water metal concentrations. Peepers are an effective method for sampling pore waters at specific depths in the sediment and in the overlying water near the sediment–water interface (Hesslein, 1976). This allows for determination of metal concentration profiles in the pore water as well as contaminant release from the sediments near the sediment–water interface.

Diffusive gradient in thin films is another type of diffusion sampler that consists of a housing containing gels that are specific for target compounds (e.g., Chelex or acrylamide gel for metals) (Davison et al., 2000). The DGT probes can accumulate dissolved metals in a controlled fashion that permits the measurement of trace metals and other substances in sediment. It removes metals from water, pore water, and sediments, potentially mimicking the effects of biota while providing quantitative measurement of the mean concentration of metals in pore water at the surface of the device (Davison and Zhang, 1994). The DGT can theoretically measure labile species that correspond to bioavailable metal fractions in sediment, although the relationship between DGT results and uptake by different types of benthos is not well established. The DGT can measure fine vertical profiles of available contaminants by segmenting the gel, which can provide information on contaminant migration rates, identify micro-niche environments, and quantify fluxes of solutes within the sediment environment. Other surrogate sampling technologies include gel probe equilibrium samplers (Campbell et al., 2008) and acid volatile gel probes (Edenborn, 2005).

12.3.3 Analysis of Tissue Residuals

12.3.3.1 Indigenous Organisms

Instead of measuring metals in pore water or using predictive methods to estimate bioavailable metal fractions from sediment concentrations, bioavailability can be determined by collecting benthic organisms from the assessment area and measuring metal concentrations in the whole body or particular tissues, such as edible portions (USEPA, 2000; Beyer and Meador, 2011). Edible portions are appropriate when humans are the receptor of prime interest. This approach directly assesses bioavailability and provides information concerning the food chain transfer of contaminants for risk assessment. It has the advantage of integrating numerous factors that affect exposure, including physical and chemical conditions in the sediments; multiple routes of contaminant intake; changes in contaminant levels over time; and organism-specific factors such as feeding behavior, ingestion, metabolic rates, and biochemical factors that affect metal assimilation. However, obtaining measurements that accurately represent the study area can be difficult because of the spatial heterogeneity that is typical of contaminant distribution and due to problems in ascertaining the territory size of motile organisms relative to the size of the contaminated area. Problems stemming from the movement of organisms among areas with different levels of contamination can be minimized by concentrating on sessile organisms such as mussels, although it is sometimes difficult to obtain sufficient organisms of the same type for statistically valid comparisons between impacted and reference areas, particularly if these areas differ in habitat. In addition, relationships between tissue contaminant levels and adverse effects are variable, particularly for metals that satisfy essential needs or that can be detoxified and stored. In summary, tissue residues provide evidence of bioaccumulation but not necessarily of the route of exposure or the magnitude of adverse effect (ITRC, 2011).

12.3.3.2 Caged Organisms

Some of the problems that are associated with the collection of indigenous sediment organisms can be avoided by translocating organisms from clean reference sites or laboratory cultures to cages within contaminated sites where the organisms can be maintained for the in situ study of contaminant uptake and toxicity. This procedure, termed *active biomonitoring* or *in situ bioassay* (Bervoets et al., 2004), has potential advantages over the collection of indigenous sediment organisms, including the ability to provide adequate sample sizes, definite knowledge of exposure periods, and more control of potentially confounding variables such as life stage. Additional advantages include realistic exposure scenarios, integration of site-specific conditions, the ability to deploy and assess quickly, and the ability to partition exposures of key compartments (e.g., surficial sediment, sediment–water interface, water column). In situ bioassays make it possible to compare organisms exhibiting different feeding behaviors, physiologies, and depths of residence within the sediments such as bivalves, annelids, and amphipods. Endpoints can include tissue residues, growth, survival, and evidence of toxicity such as lesions or biochemical markers. Exposure of test organisms in situ (as opposed to the laboratory) increases realism and reduces the chances of altering metal bioavailability due to sample manipulation (Anderson et al., 2004; Chappie and Burton, 2000).

Active biomonitoring can be conducted with simple equipment such as the small screened cages used by Paller and Knox (2010) to house California blackworms (*Lumbriculus variegates*) for in situ sediment bioassays in a freshwater stream (Figure 12.3). Larger cages can be used for larger organisms such as bivalves. Alternatively, Rosen et al. (2009) used a more complex integrated field exposure system, the SEA Ring, for active biomonitoring at coastal saltwater sites. This system includes multiple chambers that can house a variety of sediment-dwelling organisms for comparative purposes. It can incorporate battery-powered pumps to maintain circulation of surface water through chamber screens, DGT probes, and instrumentation for continuous measurement of water quality inside the bioassay chambers. When conducting in situ bioassays, a subsample of the test organisms is generally analyzed for contaminants before initiation of the experiments to provide a baseline for assessing metal accumulation during the exposure period. Protocols for in situ bioassays in aquatic environments can be found in Chappie and Burton (2000) and Burton et al. (2005).

12.3.4 Laboratory Toxicity Testing

Laboratory bioassays determine the bioavailability of metals by exposing organisms to contaminated sediment under controlled conditions and by measuring effects on endpoints such as survival, growth, behavior, reproduction, and tissue accumulation. Short-term tests (e.g., 10 days) that measure acute toxicity are the most commonly used, but longer tests (e.g., 28 days) can be used to identify sublethal effects (e.g., depressed reproduction or growth) that are associated with chronic exposure to low concentrations. A variety of test organisms can be used, including

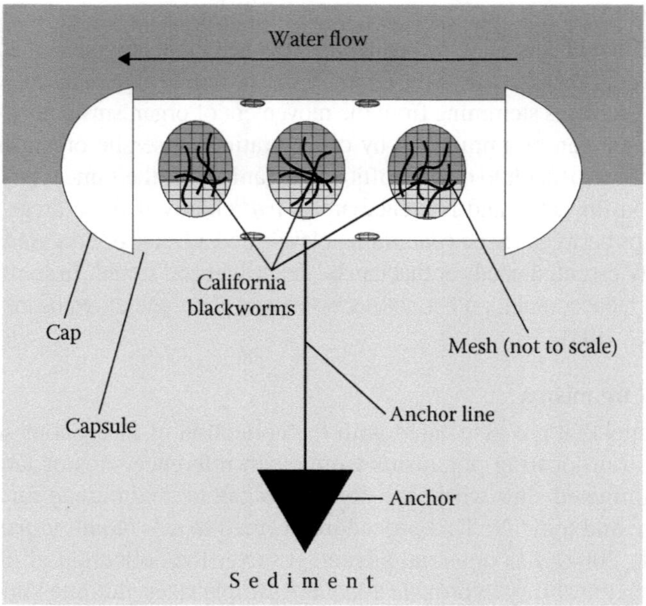

FIGURE 12.3 Screened cage used to hold California blackworms (*Lumbriculus variegates*) for an in situ study of metal bioavailability in a South Carolina (USA) stream.

amphipods, midges, polychaetes, oligochaetes, mayflies, and cladocerans; methods for assessing the toxicity and bioavailability of sediment contaminants are well established (e.g., USEPA, 1994; USEPA, 2000; American Society for Testing and Materials [ASTM], 2008). Laboratory bioassays provide the greatest control over confounding variables that can affect bioavailability but are susceptible to laboratory artifacts that may reduce congruence with actual conditions in the sediment.

12.3.4.1 Sediment Toxicity Tests

Laboratory bioassay methods for metals in sediment include whole-sediment toxicity tests, interstitial water toxicity tests, and sediment TIE methods. Whole sediment tests generally involve the exposure of test organisms such as the amphipod *Hyalella azteca* (freshwater) to potentially contaminated sediments for 10–28 days in static systems in which the overlying water is renewed daily or in flow-through systems (USEPA, 1996). At the end of the exposure period, the response of the test organism is compared with the response of control organisms that are exposed to uncontaminated reference sediment. Sediments that are spiked with different concentrations of metals can be used to establish relationships between chemicals and biological response, but bioavailability in spiked sediment may not be the same as in sediment from the field.

12.3.4.2 Pore-Water Toxicity Tests

Interstitial water tests expose test organisms to water separated from sediment samples based on the supposition that pore-water metals are in equilibrium with solid-phase metals and that the former more accurately represents the bioavailable fraction (Doe et al., 2003). Pore-water tests can be compared with whole sediment tests to identify routes of exposure: A positive whole sediment test and a negative interstitial water test suggest that toxicity occurs through sediment ingestion. Problems with interstitial water tests are that pore-water chemistry may be altered during collection or testing, creating results that are unrepresentative of conditions in the field and difficulties in obtaining sufficient pore water for testing.

12.3.4.3 Toxicity Identity Evaluation

The previously described tests can determine whether a sediment is toxic, but they cannot identify the contaminant that is responsible for toxicity, a significant problem when there are multiple contaminants and it is necessary to identify the key ones for remediation or linkage to specific sources. Toxicity Identity Evaluation (TIE) procedures can identify and quantify the chemicals or chemical classes causing toxicity (USEPA, 2007). The basic TIE approach is to manipulate sediment samples with a series of chemical and physical procedures, each of which eliminates the toxic effects of a class of chemicals. Bioassays are used to assess the changes in toxicity resulting from each procedure and to identify toxic classes of chemicals by observing changes in toxicity from baseline levels after each manipulation. The TIEs can be conducted on interstitial water or whole sediment samples. Interstitial water manipulations include graduated pH for pH-dependent toxicants, aeration for easily oxidized toxicants, reverse-phase chromatography for nonpolar organic toxicants, addition of ethylenediaminetriacetic acid (EDTA) for cationic metals, and addition of zeolite for ammonia. Whole sediment manipulations include addition of *Ulva lactuca* (algae) or zeolite for ammonia; addition of cation-exchange resin and sulfide for cationic metals; and addition of Ambersorb, powdered coconut charcoal, or carbonaceous resin for nonionic organic chemicals (ITRC, 2011). The choice of whether to use interstitial water or whole sediment tests is important, because they do not always agree. Interstitial water tests can be affected by oxidation-related changes in chemistry, absorption of chemicals to test chambers, removal of the dietary exposure route, and differences in the extent of interstitial water exposure by different types of organisms (USEPA, 2007). It can also be difficult to collect sufficient volumes of interstitial water for toxicity testing. The USEPA (2007) provides guidance on when to use whole sediment and interstitial water tests.

12.3.5 Benthic Community Surveys

Benthic communities are a key component of aquatic ecosystems, because they support high biodiversity, are critical in organic matter processing, and constitute the base of many aquatic food chains. Thus, the assessment of benthic communities is widely used in freshwater and marine environments as an integrative and ecologically relevant measure of the health of aquatic ecosystems (Lazorchak et al., 1988). It usually focuses on macroinvertebrates, which consist of benthic organisms that are visible without the aid of a microscope and retained by a 0.5-mm sieve. Benthic organisms are generally sensitive to metals and other contaminants, reflect cumulative exposure, and are more likely to represent site-specific conditions than highly mobile organisms such as fish. However, benthic surveys do not indicate which metals (or other contaminants) cause toxicity and are subject to confounding by habitat and biological interactions that affect benthic communities independent of toxicity.

12.3.5.1 Natural Substrates

Benthic macroinvertebrate surveys often involve the collection, processing, and identification of samples from benthic habitats in potentially impacted sites and from environmentally similar reference sites that are free from disturbance. Such surveys can be restricted to specific benthic habitats within the study area (e.g., riffle habitats in streams) or can include all benthic habitats (e.g., riffles, pools, and runs), which are sampled in proportion to their prevalence (Hilsenhoff, 1987; Lenat, 1988). The analysis of benthic data for assessment purposes often entails the computation of metrics, which are community, population, and organism-level variables that are ecologically important and sensitive to environmental disturbance (Karr and Chu, 1999). Examples include species richness, functional feeding group composition, percent composition of dominant taxonomic groups, pollution tolerance indices, diversity indices, and community similarity indices. In multimetric approaches, multiple metrics are normalized and combined to produce a single number or score that reflects the degree to which the assessment site resembles the range of conditions represented by carefully selected reference sites (Karr et al., 1986). Other commonly used methods of analysis include multivariate statistical methods (e.g., cluster analysis, nonmetric multidimensional scaling, canonical correspondence analysis) that summarize data matrices and graphically depict the results and predictive models that relate the observed taxonomic composition to that expected in the absence of human disturbance (e.g., Observed/Expected [O/E]) (McCune et al., 2002; Hawkins et al., 2000).

12.3.5.2 Artificial Substrates

Although providing a direct and ecologically relevant measure of the effects of sediment contamination, the assessment of benthic communities has several limitations. Benthic communities tend to be spatially heterogeneous because of small-scale differences in ecologically significant habitat features such as sediment particle size, current velocity, and amounts of organic matter. They also vary temporally as a result of seasonal changes and episodic events such as floods. These differences cause metrics that are calculated from benthic communities to have large spatiotemporal variances that result in low statistical power unless sample sizes are large. A method that attempts to overcome this problem is the use of artificial substrates such as concrete blocks, multiple-plate hardboard samplers, and rock-filled baskets that can be deployed at sample sites where they are colonized by benthic organisms (Taylor and Kovats, 1995). These devices offer substrates with uniform characteristics, thereby eliminating the confounding effects of substrate variability among sampling locations. However, the organisms that colonize artificial substrates may not be representative of the naturally occurring benthic fauna, particularly the benthic infauna, and the exposure scenarios may be different from those in the sediments (Rosenberg and Resh, 1982).

12.3.5.3 Sediment Profile Imaging

The large sample sizes often required in benthic surveys necessitate significant labor and time requirements for collection, processing, and identification. Sediment profile imaging (SPI) is a newer approach that permits rapid surveys of benthic communities with less labor (Keegan and Rhoads, 2001). The SPI photographs the water column/seabed interface to a sediment depth of 15–20 cm. It employs an apparatus resembling a prism or an upside-down periscope that is driven into the sediment. A downward-facing camera photographs the cross-section of sediment that is in contact with the faceplate of the prism, producing vertical, cross-sectional profiles of undisturbed sediment. The SPI can record the presence of epibenthic and infaunal organisms, indirect indicators of organism abundance such as the number and length of burrows and the presence of feeding voids and fecal pellets, and characteristics of the sediment environment such as grain size and redox discontinuity depth (transition between oxygenated surface sediments and deeper sediments with little or no oxygen) that determine suitability for benthic organisms (Rhoads and Germano, 1982). Image analysis software can help with the assessment of organism abundance and higher level taxonomic identity. The SPI permits rapid, cost-effective surveys, but it does not generate the detailed taxonomic and quantitative data provided by conventional methods.

12.4 THE USE OF BIOAVAILABILITY IN RISK ASSESSMENT

The previously discussed bioavailability tools are used to better estimate the risks posed by sediment contaminants to ecological and human receptors, identify the factors primarily responsible for the risks, and discern the best methods of risk mitigation. However, all bioavailability tools have limitations, and there is no definitive bioavailability measurement method that is appropriate in all situations. This is because bioavailability is complex and strongly affected by numerous site-specific factors such as the physical and chemical heterogeneity of sediments and organism-specific differences in feeding behavior, physiology, depth in the sediments, and sensitivity to contaminants. Some disadvantages of individual methods are summarized in Table 12.2. For these reasons, it is often desirable to use independent but complementary methods to assess the degree of impact, correctly identify the source of toxicity, and better understand environmental factors that may be affecting bioavailability. Different approaches provide more insight into the causes and extent of sediment contamination than individual methods, provide independent verification when conclusions agree, and suggest the need for additional information when methods do not agree.

The sediment quality triad (SQT) is a method of integrating multiple lines of inquiry concerning sediment contamination. It combines complementary information on sediment chemistry, toxicity testing, and benthic community condition collected from the same location to better understand the effects of contamination and to mitigate the limitations of individual methods (Chapman et al., 1987; Chapman, 1996). The SQT results can be arranged in a contingency table or plotted on a triaxial graph to facilitate interpretation (MacDonald and Ingersoll, 2002). In the latter case, data can be normalized so that the surface area of the graph becomes a representation of impact. An example of the use of bioavailability information in an SQT type of approach for generating a weight-of-evidence conclusion is shown in Figure 12.4. In this case, an initial screening is conducted by comparing bulk sediment chemistry to SQGs. No impact can be assumed if the results of this conservative method are negative. Positive results indicate the need for chemical tests that determine the degree of contaminant bioavailablity (e.g., SEM-AVS or interstitial pore-water tests). However, the limitations of these methods may necessitate additional studies, including toxicity testing and benthic community surveys, especially if the contaminants involve ecologically important habitats or have the potential for transfer to important higher level consumers.

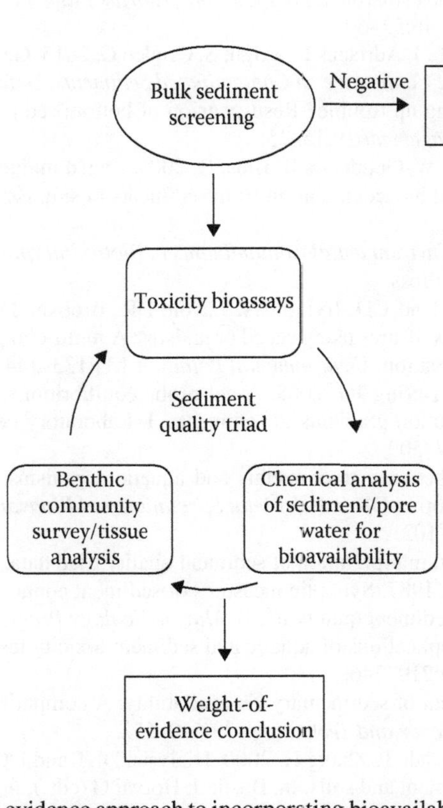

FIGURE 12.4 Weight-of-evidence approach to incorporating bioavailability information into the assessment of metal toxicity in sediments.

12.5 SUMMARY

- Total metal concentrations are seldom related to toxicity: the bioavailable fraction is the key factor.
- Methods for estimating the toxic bioavailable fraction have been developed for anoxic sediments. However, they are better at predicting no toxicity than the onset of toxicity.
- In situ bioavailability measurement technologies (e.g., DGT) are promising but require additional refinement and validation.
- Currently available chemical methods for sediments do not predict bioaccumulation, food chain transfer, or effects on higher trophic-level organisms.
- Toxicity testing and community surveys are important, because bioavailability is affected by numerous site-specific and organism-specific factors that might not be reflected in chemical measurements.
- Bioavailability can be effectively incorporated in risk assessments as part of a weight-of-evidence approach.

REFERENCES

Admiral DM, Garcia MH, Rodriguez JF. 2000. Entrainment response of bed sediment to time varying flows. *Water Resources Research* 36(1):335–348.

American Society for Testing and Materials (ASTM). 2008. *Standard Test Method for Measuring the Toxicity of Sediment-Associated Contaminants with Estuarine and Marine Invertebrates* (ASTM E1367–03e1). West Conshohocken, PA: American Society for Testing and Materials.

Anderson BS, Hunt JW, Phillips BM, Nicely PA, Tjeerdema RS, Martin M. 2004. A comparison of *in situ* and laboratory toxicity tests with the estuarine amphipod *Eohaustorius estuarius*. *Archives of Environmental Contamination and Toxicology* 46:52–60.

Barabás N, Redder T, DePinto J, Wolfe J, Adriaens P, Wright S, Gruden C. 2013. Gas ebullition in sediments, an overview. *Seventh International Conference on Contaminated Sediments*. Dallas, TX, February 4–7, 2013.

Beachler MM, Hill DF. 2003. Stirring up trouble? Resuspension of bottom sediments by recreational watercraft. *Lake and Reservoir Management* 19:15–25.

Bervoets L, Meregalli G, DeCooman W, Goddeeris B, Blust R. 2004. Caged midge larvae (*Chironomus riparius*) for the assessment of metal bioaccumulation from sediments in situ. *Environmental Toxicology and Chemistry* 23:443–454.

Beyer WN, Meador JP (eds.). 2011. *Environmental Contaminants in Biota: Interpreting Tissue Concentrations*, 2nd ed. Boca Raton, FL: CRC Press.

Burton GA Jr., Greenberg MS, Rowland CD, Irvine CA, Lavoie DR, Brooker JA, Moore L, Raymer DFN, McWilliam RA. 2005. In situ exposures using caged organisms: A multi-compartment approach to detect aquatic toxicity and bioaccumulation. *Environmental Pollution* 134:133–144.

Campbell KM, Root R, O'Day PA, Hering JG. 2008. A gel probe equilibrium sampler for measuring arsenic porewater profiles and sorption gradients in sediments: I. Laboratory development. *Environmental Science and Technology* 42:497–503.

Campbell PGC. 1995. Interactions between trace metals and aquatic organisms: A critique of the free-ion activity model. In: Tessier A, Turner DR (eds.). *Metal Speciation and Bioavailability in Aquatic Systems*. Chichester, UK: Wiley, pp. 45–102.

Chapman PM. 1996. Presentation and interpretation of sediment quality triad data. *Ecotoxicology* 5:327–339.

Chapman PM, Dexter RN, Long ER. 1987. Synoptic measures of sediment contamination, toxicity and infaunal community structure (the sediment quality triad). *Marine Ecology Progress Series* 37:75–96.

Chappie DJ, Burton GA Jr. 2000. Applications of aquatic and sediment toxicity testing in situ. *Journal of Soil and Sediment Contamination* 9:219–246.

Chen Z, Mayer LM. 1999. Assessment of sedimentary Cu availability: A comparison of biomimetic and AVS approaches. *Environmental Science and Technology* 33:650–652.

Davison W, Fones G, Harper M, Teasdale P, Zhang H. 2000. Dialysis, DET and DGT: In situ diffusional techniques for studying water, sediment and soils. In: Buffle J, Horvai G (eds.), *In Situ Monitoring of Aquatic Systems: Chemical Analysis and Speciation*. New York: Wiley. pp. 495–569.

Davison W, Zhang H. 1994. In situ speciation measurements of trace components in natural waters using thin-film gels. *Nature* 367:546–548.

Di Toro DM, Allen HE, Bergman HL, Meyer JS, Paquin PR, Santore RC. 2001. Biotic ligand model of the acute toxicity of metals. I. Technical basis. *Environmental Toxicology and Chemistry* 20:2383–2396.

Di Toro DM, Mahony JD, Hansen DJ, Scott KJ, Carlson AR, Ankley GT. 1992. Acid volatile sulfide predicts the acute toxicity of cadmium and nickel in sediments. *Environmental Science and Technology* 26:96–101.

Di Toro DM, Mahony JJ, Hansen DJ, Scott KJ, Hicks MB, Mayr SM, Redmond MS. 1990. Toxicity of cadmium in sediments: The role of acid volatile sulfide. *Environmental Toxicology and Chemistry* 9:1487–1502.

Di Toro DM, Zarba CS, Hansen DJ, Berry WJ, Swartz RC, Cowan CE, Pavlou SP, Allen HE, Thomas NA, Paquin PR. 1991. Technical basis for establishing sediment quality criteria for nonionic organic chemicals using equilibrium partitioning. *Environmental Toxicology and Chemistry* 10:1541–1583.

Doe KJ, Burton GA, Ho K. 2003. Porewater toxicity testing: an overview. In: Carr RS, Nipper M, (eds.), *Porewater Toxcity Testing: Biological, Chemical, and Ecological Considerations*, chap. 6. Pensacola, FL: SETAC Press, pp. 125–142.

Edenborn HM. 2005. Rapid detection of bioavailable heavy metals in sediment porewaters using acid-volatile sulfide gel probes. *Environmental Geology* 47:660–669.

Fan W, Wang W-X. 2001. Sediment geochemical controls on Cd, Cr, and Zn assimilation by the clam *Ruditapes philippinarum. Environmental Toxicology and Chemistry* 20:2309–2317.

Hall GEM, Vaive JE, Beer R, Hoashi M. 1996. Selective leaches revisited, with emphasis on the amorphous Fe oxyhydroxide phase extraction. *Journal of Geochemical Exploration* 56:59–78.

Hansen DJ, Berry WJ, Mahony JD, Boothman WS. 1996. Predicting the toxicity of metal-contaminated field sediments using interstitial concentration of metals and acid-volatile normalizations. *Environmental Toxicology and Chemistry*. 15:2080–2094.

Hansen DJ, DiToro DM, Berry WJ, Boothman WS, Burgess RM, Ankley GT, Mount DR, McGrath JA, DeRosa LD, Bell HE, Reiley MC, Zarba CS. 2005. *Procedures for the Derivation of Equilibrium Partitioning Sediment Benchmarks (ESBs) for the Protection of Benthic Organisms: Metal Mixtures (Cadmium, Copper, Lead, Nickel, Silver and Zinc)* (EPA-600-R-02-011). Washington, DC: U.S. Environmental Protection Agency, Office of Research and Development.

Harmsen J, Rulkens W, Eijsackers H. 2005. Bioavailability: Concept for understanding or tool for predicting. *Land Contamination and Reclamation* 13:161–171.

Hawkins CP, Norris RH, Hogue JN, Feminella JW. 2000. Development and evaluation of predictive models for measuring the biological integrity of streams. *Ecological Applications* 10:1456–1477.

Hesslein RH. 1976. An in situ sampler for close interval pore water studies. *Limnology and Oceanography* 21:912–914.

Hilsenhoff WL. 1987. An improved biotic index of organic stream pollution. *Great Lakes Entomologist* 20:31–39.

Interstate Technology and Regulatory Council (ITRC). 2011. *Incorporating Bioavailability Considerations into the Evaluation of Contaminated Sediment Sites* (CS-1). Washington, DC: Interstate Technology and Regulatory Council, Contaminated Sediments Team. www.itrcweb.org.

ISO. 2005. *11074: Soil quality-Vocabulary*. Geneva, Switzerland: ISO.

Karr JR, Chu EW. 1999. *Restoring Life in Running Waters: Better Biological Modeling*. Washington, DC: Island Press.

Karr JR, Fausch KD, Angermeier PL, Yant PR, Schlosser IJ. 1986. *Assessing Biological Integrity in Running Waters: A Method and its Rationale* (Special Publication 5). Champaign, IL: Illinois Natural History Survey.

Keegan BF, Rhoads DC. 2001. Sediment profile imagery as a benthic monitoring tool: Introduction to a 'long-term' case history evaluation (Galway Bay, west coast of Ireland). *Organism-Sediment Symposium*. Columbia: University of South Carolina Press, pp. 43–62.

Knox AS, Paller M, Nelson E, Specht W, Gladden J. 2006. Contaminant assessment and their distribution and stability in constructed wetland sediments. *Journal of Environmental Quality* 35:1948–1959.

Lanno R, Wells J, Conder J, Bradham K, Basta N. 2004. The bioavailability of chemicals in soil for earthworms. *Ecotoxicology and Environmental Safety* 57:39–47.

Lawrence AL, McAloon KM, Mason R, Mayer M. 1999. Intestinal solubilization of particle-associated organic and inorganic mercury as a measure of bioavailability to benthic invertebrates. *Environmental Science and Technology* 33:1871–1876.

Lazorchak JM, Klemm DJ, Peck DV (eds.). 1998. *Environmental Monitoring and Assessment Program—Surface Waters: Field Operations and Methods for Measuring the Ecological Condition of Wadeable Streams* (EPA/620/R-94/004F). Durham, NC: U.S. Environmental Protection Agency, National Exposure Research Laboratory, Research Triangle Park.

Lee B-G, Lee J-S, Luoma SN, Choi HJ, Koh C-H. 2000. Influence of acid volatile sulfide and metal concentrations on metal bioavailability to marine invertebrates in contaminated sediments. *Environmental Science and Technology* 34:4517–4523.

Lenat DR. 1988. Water quality assessment of streams using a qualitative collection method for benthic macroinvertebrates. *Journal North American Benthological Society* 7:222–233.

Luoma SN, Jenne EA. 1977. The availability of sediment bound cobalt, silver, and zinc to a deposit-feeding clam. In: Wildung RE, Drucker E (eds.). *Biological Implications of Metals in the Environment* (CONF-750929). Springfield, VA: NTIS.

MacDonald DD, Ingersoll CG. 2002. *A Guidance Manual to Support the Assessment of Contaminated Sediments in Freshwater Ecosystems. An Ecosystem-Based Framework for Assessing and Managing Contaminated Sediments*, vol. 1. (EPA-905-B02-001-A). Chicago, IL: U.S. Environmental Protection Agency, Great Lakes National Program Office, 149 p.

Mahony JD, Di Toro DM, Gonzalez AM, Curto M, Dilg M. 1996. Partitioning of metals to sediment organic carbon. *Environmental Toxicology and Chemistry* 15:2187–2197.

Mayer LM, Chen Z, Findlay RH, Fang J, Sampson S, Self RFL, Jumars PA, Quetel C, Donard OFX. 1996. Bioavailability of sedimentary contaminants subject to deposit-feeder digestion. *Environmental Science and Technology* 30:2641–2645.

McCune B, Grace JB, Urban DL. 2002. *Analysis of Ecological Communities*. Gleneden Beach, OR: MJM Software Design.

McGeer JC, Brix KV, Skeaff JM, DeForest DK, Brigham SI, Adams WJ, Green A. 2003. Inverse relationship between bioconcentration factor and exposure concentration for metals: Implications for hazard assessment of metals in the aquatic environment. *Environmental Toxicology and Chemistry* 22:1017–1037.

McGeer JC, Szebedinszky C, McDonald DG, Wood CM. 2002. The role of DOC in moderating the bioavailability and toxicity of copper to rainbow trout during chronic waterborne exposure. *Comparative Biochemistry and Physiology* 133C:147–160.

National Research Council (NRC). 2003. *Bioavailability of Contaminants in Soils and Sediments*. Washington, DC: National Research Council, The National Academies Press.

Niyogi S, Wood CM. 2004. Biotic ligand model, a flexible tool for developing site-specific water quality guidelines for metals. *Environmental Science and Technology* 38:6177–6192.

Nott JA, Nicolaidou A. 1993. Bioreduction of zinc and manganese along a molluscan food chain. *Comparative Biochemistry and Physiology* 104A:235–238.

Office of Water, USEPA. 2005. Contaminated Sediment Remediation Guidance for Hazardous Waste Sites. EPA-540-R-05-012. USEPA, Office of Solid Waste and Emergency Response, Washington, DC.

Paller MH, Knox AS. 2010. Amendments for the remediation of contaminated sediments: Evaluation of potential environmental impacts. *Science of the Total Environment* 408:4894–4900.

Peijnenburg W, Posthuma L, Eijsackers H, Allen H. 1997. A conceptual framework for implementation of bioavailability of metals for environmental management purposes. *Ecotoxicology and Environmental Safety* 37:163–172.

Peng S, Wang W, Li X, Yen Y. 2004. Metal partitioning in river sediments measured by sequential extraction and biomimetic approaches. *Chemosphere* 57:839–851.

Rhoads DC, Germano JD. 1982. Characterization of organism-sediment relations using sediment profile imaging: An efficient method of remote ecological monitoring of the seafloor (Remots (tm) System). *Marine Ecology Progress Series* 8:115–128.

Rosen G, Chadwick DB, Greenberg MS, Burton GA Jr. 2009. Development of a novel in situ based monitoring approach for contaminated sediment assessment. *Oral presentation, Fifth International Conference on Remediation of Contaminated Sediments*. Jacksonville, FL, February 2–5, 2009.

Rosenberg DM, Resh VH. 1982. The use of artificial substrates in the study of freshwater benthic macroinvertebrates. In: Cairns J, Jr. (ed.). *Artificial Substrates*. Ann Arbor, MI: Ann Arbor Science Publishers, Inc., pp. 175–235.

Schaller J. 2014. Bioturbation/bioirrigation by Chironomus plumosus as main factor controlling elemental remobilization from aquatic sediments? *Chemosphere* 107:336–343.

Taylor BR, Kovats Z. 1995. *Review of Artificial Substrates for Benthos Sample Collection*. Ottawa, ON: Prepared for Canada Centre for Mineral and Energy Technology (CANMET).

Tessier A, Campbell PGC, Bisson M. 1979. Sequential extraction procedure for the speciation of particulate trace metals. *Analytical Chemistry* 51:844–850.

U.S. Environmental Protection Agency. 1994. *Methods for Measuring the Toxicity and Bioaccumulation of Sediment-Associated Contaminants with Freshwater Invertebrates* (EPA 600/R-94/024). Washington DC: USEPA, Office of Research and Development.

U.S. Environmental Protection Agency. 1996. *Ecological Effects Test Guidelines, OPPTS 850.1735, Whole Sediment Acute Toxicity Invertebrates, Freshwater* (EPA 712–C–96–354). Washington, DC: USEPA, Prevention, Pesticides,and Toxic Substances (7101).

U.S. Environmental Protection Agency. 1997. *The Incidence and Severity of Sediment Contamination in Surface Waters of the United States* (EPA–823–R–97–006, 007, and 008). Washington, DC: USEPA.

U.S. Environmental Protection Agency. 2000. *Bioaccumulation Testing and Interpretation for the Purpose of Sediment Quality Assessment: Status and Needs* (EPA/823/R-00/001). Washington, DC: USEPA.

U.S. Environmental Protection Agency. 2001. *The Incidence and Severity of Sediment Contamination in Surface Waters of the United States* (EPA/823/R-01/01). National sediment survey. Washington, DC: USEPA.

U.S. Environmental Protection Agency. 2007. *Sediment Toxicity Identification Evaluation (TIE): Phases I, II, and III Guidance Document* (EPA/600/R-07/080). Washington, DC: Office of Research and Development, USEPA.

Wenning RJ, Batley GE, Ingersoll CG, Moore DW (eds.). 2005. *Use of Sediment Quality Guidelines and Related Tools for the Assessment of Contaminated Sediments*. Pensacola, FL: SETAC Press.

Yan Q, Wang W. 2002. Metal exposure and bioavailability to a marine deposit-feeding sipuncula, *Sipunculus nudus. Environmental Science and Technology* 36:40–47.

13 Potential Mobility, Bioavailability, and Plant Uptake of Toxic Elements in Temporary Flooded Soils

Sabry M. Shaheen, Christos D. Tsadilas, Yong Sik Ok, and Jörg Rinklebe

CONTENTS

13.1 INTRODUCTION

Contamination of flooded soils with potentially toxic elements (PTEs) is a serious issue confronting the cultivation and management of these soils. Large areas of flooded soils in the world are polluted with PTEs, including arsenic (As), cadmium (Cd), chromium (Cr), cobalt (Co), copper (Cu), mercury (Hg), nickel (Ni), lead (Pb), selenium (Se), and zinc (Zn) (e.g., Rinklebe et al., 2007, 2010; Rennert and Rinklebe, 2010; Ok et al., 2011a,b; Frohne et al., 2014, 2015; Shaheen and Rinklebe, 2014; Rinklebe and Shaheen, 2014; Shaheen et al., 2014a,b,c; Shaheen et al., 2015a). Elevated levels of PTEs in flooded soils may increase the mobilization and leaching of the elements that negatively impact agricultural ecosystems (Lee et al., 2013; Rinklebe and Shaheen, 2014; Rinklebe et al., 2016a,b; Shaheen et al., 2016).

Redox processes play an important role in soil organic matter (SOM) decomposition, nutrient availability, biogeochemical cycling of PTEs, and ecological functions of flooded ecosystems (Reddy and DeLaune, 2008; Rinklebe and Du Laing, 2011; DeLaune and Seo, 2011; Shaheen et al., 2014a,b,c; Rinklebe et al., 2016a,b; Shaheen et al., 2016). Flooded soils are affected by pronounced seasonal variations in the surrounding temperature. These affect the biogeochemical processes that control the fate of PTEs in the soils, including their potential release into surface water and groundwater during periodic flooding. The potential mobility of PTEs in soils depends on their total concentration, speciation, environmental factors, distribution coefficient, and soil properties (DeLaune and Seo, 2011; Shaheen et al., 2013; Frohne et al., 2014; Rinklebe and Shaheen, 2014; Shaheen and Rinklebe, 2014). Moreover, the mobilization and bioavailability of PTEs in flooded soils are strongly affected by the periodic soil flooding that triggers major variations (Shaheen et al., 2014a,b). The solubility of PTEs under flooding conditions is controlled by redox potential (E_H), pH, and carriers of metals such as dissolved organic carbon (DOC), Fe, Mn, and SO_4^{2-} (Du Laing et al., 2009; Rupp et al., 2010; Rinklebe and Du Laing, 2011; Shaheen et al., 2014a,b,c; Rinklebe et al., 2016a,b; Shaheen et al., 2016). The PTEs can be redistributed into different geochemical fractions, released from the soil to soil solution, and transferred to the ecosystem and food chain, thereby posing a hazard to human and environmental health (Du Laing et al., 2009; Shaheen et al., 2014a,b, 2015a; Schulz-Zunkel et al., 2015; Shaheen et al., 2016).

The distribution of PTEs in the flooded soils, particularly of their mobile fractions, controls the magnitude of their uptake and accumulation in the vegetation (e.g., Baum et al., 2006). This fact can be used to remediate floodplain soils with plants producing large quantities of biomass belonging to the natural floodplain vegetation such as willows and poplars as well as oil-producing crops such as rapeseed (Marston et al., 1995; Vysloužilova et al., 2003; Rockwood et al., 2004; Baum et al., 2006; Unterbrunner et al., 2007; Zummer et al., 2009; Shaheen and Rinklebe, 2015a,b; Shaheen et al., 2015b,c; Rinklebe and Shaheen, 2015). However, for designing a successful remediation strategy of the floodplain soils, the spatial variation in concentrations, geochemical fractionation, mobilization, and bioavailability of PTEs must be known (Shaheen and Rinklebe, 2015a,b; Shaheen et al., 2015b,c; Rinklebe and Shaheen, 2015). Spatiotemporal alteration of stability or mobilization of PTEs in the soil can be described by using sequential extractions (Dold, 2003; Bacon and Davidson, 2008; Rinklebe and Shaheen, 2014).

The main threat of soil PTEs to animals and humans is associated to their introduction in the food chain through contaminated agricultural soils. The PTEs are accumulated in plants that are grown in flooded soils, that is, grasslands and rice paddy, and can be translocated into the edible parts (Chaney et al., 2007; Ok et al., 2011a,b; Shaheen and Rinklebe, 2015a). Because the solubility and phytotoxicity of many PTEs in soils depends on their oxidation state, studies on the uptake of these elements by plants in temporary flooded soils are essential to understand the soil–plant–interactions of PTEs. There is a need for detailed knowledge about the behavior of PTEs under alternating inundation regimes in temporary flooded soils; this might serve as a precondition for the development of innovative technologies to manage contaminated flooded areas. Therefore, in this chapter, the research results about the potential mobility, bioavailability, and uptake of PTEs in temporary flooded soils to evaluate their pollution status in these soils are reviewed.

13.2 FRACTIONATION AND POTENTIAL MOBILITY OF TOXIC ELEMENTS IN TEMPORARY FLOODED SOILS

The total content of PTEs has been used to assess the risk of soil pollution by comparing it with background or guideline values of PTEs and by creating a pollution index such as enrichment factor, geo-accumulation index, contamination factor, and complex quality index (Bhuiyan et al., 2010). However, the use of the total concentration of PTEs in the soil for environmental risk assessment is insufficient, because their potential mobility depends on their binding forms (Pueyo et al., 2008; Zimmer et al., 2011; Rinklebe and Shaheen, 2014; Shaheen and Rinklebe, 2014; Shaheen et al., 2015a). For example, the water soluble + exchangeable and carbonate (easily mobilizable) forms are considered the most mobile and bioavailable; the reducible bound to Fe/Mn oxides and the oxidizable bound to SOM may be potentially bioavailable; and the residue bound to the soil matrix is not phytoavailable (Delgado et al., 2011; Shaheen et al., 2015a). Therefore, determination of the geochemical fractions of PTEs is necessary to assess their mobilization and pollution status in soils and it is a key issue in many environmental studies (Shaheen and Rinklebe, 2014; Rinklebe and Shaheen, 2014; Shaheen et al., 2015a).

Zimmer et al. (2011) studied the fractions of Cd, Cu, Ni, and Zn in surface soils from different length scales in the floodplain of Schoenberg meadow, Germany. About 44%–63% of the total Cd was found to be in the mobile fraction. The mobile fraction of Cu was equivalent to only 4% of the total concentration since the largest amount (64%–68%) of Cu was distributed in the organic fraction. Similar to Cu, only up to 18% of the total Ni was observed in the mobile fraction, but up to 64% of total Ni was found in the residual fraction. In addition, Antić-Mladenović et al. (2011) found that about 94% of total Ni was found in the residual fraction. Similarly, the highest concentration of Zn was observed in the residual fraction and only about 21% of total Zn was observed in the mobile fraction. In the study by Zimmer et al. (2011), the metal portions present in the fractions were similar to those in fractionated sediments from along the River Elbe, Germany (Van der Veen, 2006). In both investigations, the total Cu, Ni, and Zn concentrations were dominated by less available fractions, and large portions were assigned to fractions that were associated to Mn- and Fe-sorbed or to the residual fraction.

Miao and Yang (2012) studied the speciation and availability of Cu, Ni, Pb, and Zn fractions in mine drainage-contaminated paddy soil. The residual fraction was the predominant fraction for Cu, Ni, Pb, and Zn among the tested soils, whereas the exchangeable fraction was the lowest among all fractions. The mobilization coefficients of metals decreased in the order of Cu > Ni > Zn > Pb, but the bioavailability coefficients of metals decreased in the order of Pb > Cu > Zn / Ni (Miao and Yang, 2012).

Jalali and Hemati (2013) fractionated Cd, Cu, Fe, Mn, Ni, Pb, and Zn in 28 surface paddy soils of Isfahan Province in central Iran. Iron, Mn, Ni, Pb, and Zn existed in these paddy soils mainly in Fe–Mn oxides (53.6%, 65.2%, 40.4%, 40.8%, and 53.3%, respectively), whereas Cu and Cd occurred essentially as residual mineral phase (41.4%) and carbonate (36.1%), respectively. The mobile and bioavailable fractions of Cd, Cu, Fe, Mn, Ni, Pb, and Zn in paddy soils averaged 48.8%, 20.8%, 0.79%, 29.2%, 28.5%, 41.1%, and 24.8%, respectively, suggesting that the mobilization and bioavailability of the seven metals probably decline in the following order: Cd > Pb > Mn $\geq$ Ni > Zn > Cu $\gg$ Fe.

Rinklebe (2004) described and determined the properties of seven floodplain soil profiles in detail. Those samples were used by Shaheen and Rinklebe (2014), who determined the geochemical fractions and vertical distribution of Cr, Cu, and Zn in those soil profiles in relation to flooding conditions and relevant soil properties. These soil profiles represent two different soil groups (Mollic Fluvisols and Eutric Gleysols) that vary significantly in flooding duration and soil properties. The metals were fractionated sequentially to seven fractions as follows: F1: soluble + exchangeable, F2: easily mobilizable, F3: bound to Mn oxides, F4: bound to SOM (might include sulfides), F5: bound to low crystalline (amorphous) Fe oxides, F6: bound by crystalline Fe oxides, and F7: residual fraction. The total metal concentrations exceeded the international trigger action values (BBodSchV, 1999; Kabata-Pendias, 2011). The residual fraction was dominant for Cr, the organic bound for Cu, and crystalline Fe oxides for Zn. The potential mobile fraction (PMF = ΣF1–F6) ranged from 38.4% to 71.4%, 63.9% to 85.1%, and 51.5% to 83.3% of the total Cr, Cu, and Zn, respectively. However, the mobile fraction (MF = ΣF1–F2) ranged from 0.96% to 1.84%, 2.1% to 4.1%, and 9.1% to 28.7% of the total concentrations of Cr, Cu, and Zn, respectively (Figure 13.1).

Rinklebe and Shaheen (2014) assessed the potential mobility of Cd, Ni, and Pb in the soil profiles as described in detail by Rinklebe (2004) and Shaheen and Rinklebe (2014). They found that the MF was dominant for Cd; Pb was mainly bound in F4/F5, whereas Ni was bound in F7/F6. The MF ranged from 37% to 59% for Cd, 3% to 14% for Ni, and 1.4% to 12% for Pb. The PMF ranged from 90% to 97%, 44% to 61%, and 83% to 92% of the (pseudo) total Cd, Ni, and Pb, respectively. The results from both studies indicated that the order of PMF was as follows: Cd > Pb > Cu > Zn > Cr > Ni; however, the order of MF was as follows: Cd >Zn> Ni > Pb> Cu > Cr. The Gleysols had a higher metal mobility (MF) compared with Fluvisols due to their longer flooding duration; however, the Fluvisols had a higher potential metal mobility (PMF) compared with Gleysols (Figure 13.1). Their potential metal mobility was very high and thus, their transfer into the groundwater, grassland, and the food chain can be assumed to be harmful for the ecosystem and the environment. Since Cd in the soil is highly mobile and easily accumulated by plants through the root system, the concentration of Cd in the paddy soils could be a concern to human health.

In the study by Shaheen et al. (2015a), three different floodplain soils close to the rivers Nile (Egypt), Elbe (Germany), and Pinions (Greece) were used to link soil development and properties to geochemical fractions and the mobility of Cd, Cu, Ni, Pb, and Zn. The German soil showed the highest total concentration of the studied elements except for Ni, in which the Greek soil had the highest amount. A significant amount (55%–94%) of the studied elements was present in the German soil in the PMF (ΣF1–F4 [Figure 13.2]), whereas the amount of this fraction ranged between 9% and 39% in the Greek soil and between 9% and 34% in the Egyptian soil. In the German soil, most of the potential mobile Ni, Pb, and Zn were associated with the Fe–Mn oxide fraction; Cd was distributed in the soluble plus exchangeable fraction; and Cu was distributed in the organic fraction (Figure 13.2). In the Egyptian and Greek soils, the Fe–Mn oxide fraction was an abundant pool for Cu, Ni, Pb, and Zn whereas Cd had the highest amount in both the soluble + exchangeable and the carbonate fractions (Figure 13.2).

The geochemical fractions and the availability of Co, Ni, Se, and V in two paddy rice soils that developed on the lacustrine and fluvial sediments was studied by Shaheen et al. (2014c). They found that a significant amount (70%–92%) of the studied elements in both soils was present in the residual fraction (F7). Therefore, the potential mobility of a large fraction of the metals might be very low in the studied soils. Among the potentially mobile (nonresidual; ΣF1–F6) fractions, the fractions that bound to Fe–Mn oxide were the most abundant pool for the studied elements (especially V).

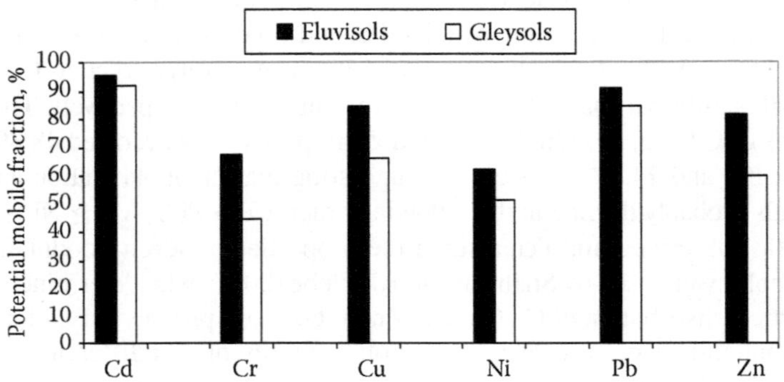

FIGURE 13.1 Distribution of Cd, Cr, Cu, Ni, Pb, and Zn among the mobile and potential mobile fractions (average values as % of total) in soil profiles represent Mollic Fluvisols and Eutric Gleysols along the Central Elbe River, Germany (values calculated from results published in Rinklebe, and Shaheen, 2014 and Shaheen and Rinklebe, 2014).

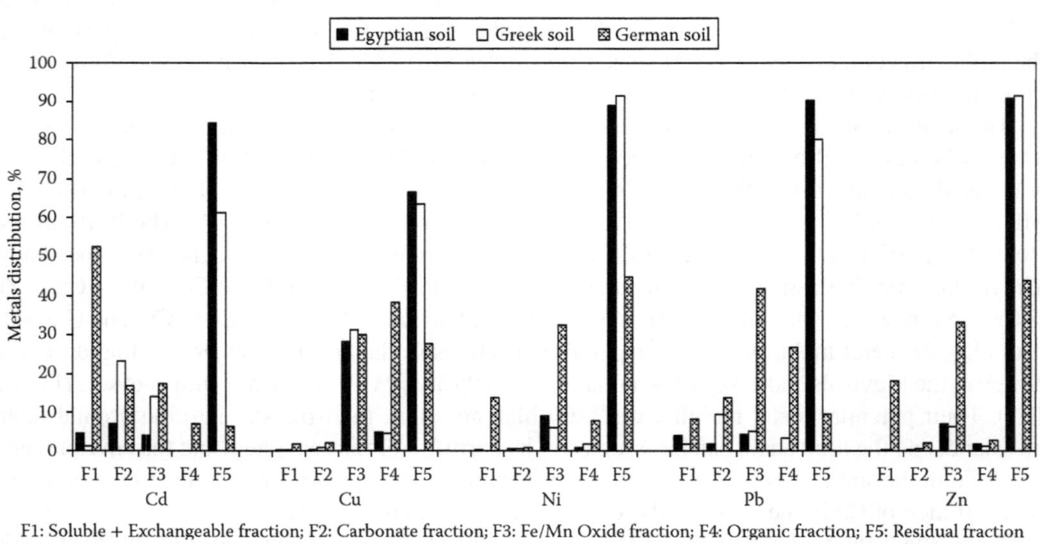

FIGURE 13.2 Chemical fractions of Cd, Cu, Ni, Pb and Zn (% of ΣF1-F5) in floodplain soils from Egypt, Germany and Greece (values calculated from results published in Shaheen et al., 2015a).

The dominance of the residual and Fe–Mn oxide fractions of the four elements might indicate that the major source of these elements in the soils might be of geologic origin.

Kosolsaksakul et al. (2014) studied the geochemical associations and availability of Cd in rice paddy in northwestern Thailand. They found that exchangeable Cd accounted for the majority of Cd in the paddy field soil samples (67%–84%), followed by the reducible or Fe/Mn oxide fraction (15%–27%) (Figure 13.3). The predominance of Cd in the mobile fraction is considered to be indicative of its high lability and bioavailablity in soils (Filgueiras et al., 2002).

The release dynamics and mobilization of PTEs under flooded conditions is mainly controlled by soil E_H and pH, DOC, Fe/Mn oxides, and sulfate (Rupp et al., 2010; Frohne et al., 2011, 2014, 2015).

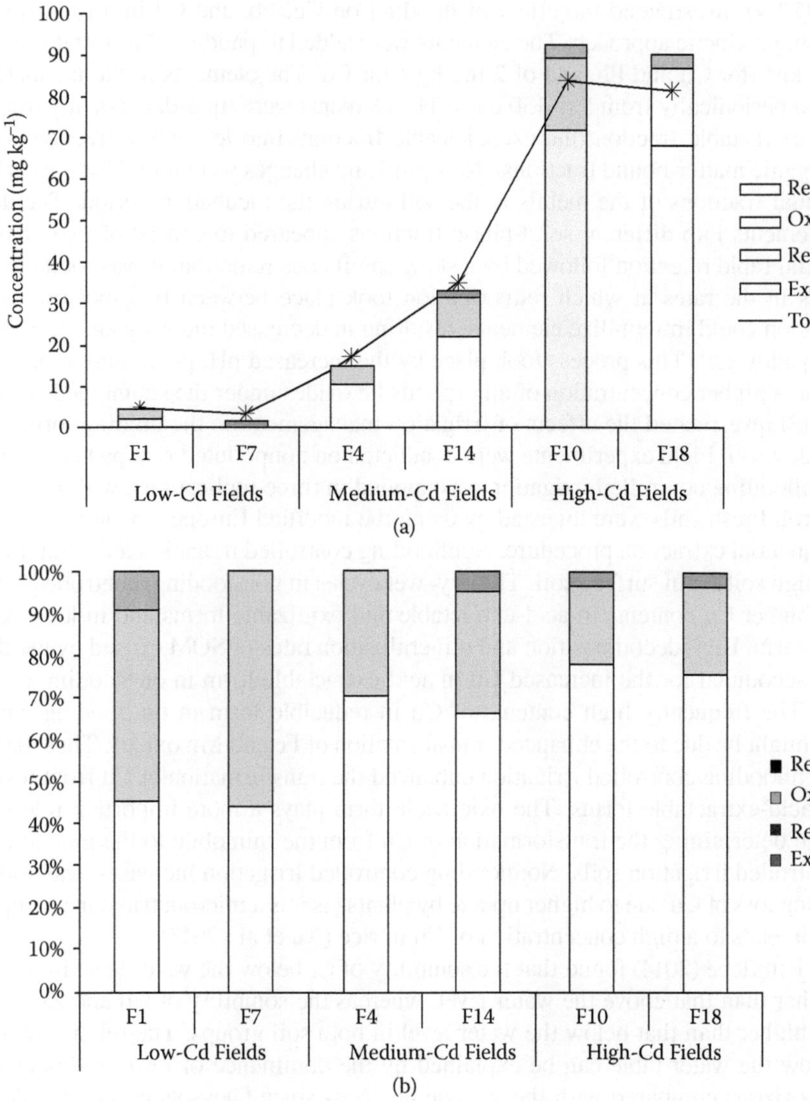

FIGURE 13.3 Cd associations in soils determined by the BCR sequential extraction scheme: (a) total (HF/HNO3) Cd concentration (line) and Cd concentrations in BCR fractions and aquaregia-digested residual phases (stacked bars); and (b) percentage distribution amongst BCR fractions and aquaregia-digested residual phase. (Reprinted from *Environmental Pollution*, 187, Kosolsaksakul, P. et al., Geochemical associations and availability of cadmium (Cd) in a paddy field system, northwestern Thailand, 153–161, Copyright 2014, with permission from Elsevier.)

With flooding conditions, the fluctuation of the water table, and the linked changes of soil redox chemistry, the transformation and mobility of PTEs might be changed. In this respect, DeLaune and Seo (2011) reported that low pH and E_H in sediment–water systems tend to favor the formation of soluble species of many metals; whereas in oxidized, nonacidic systems, slightly soluble or insoluble forms tend to be predominant. Many investigations (e.g., Yu et al., 2007; Shaheen et al., 2014a,b; Rinklebe et al., 2016b) indicated that the solubility of Fe and Mn was higher under reducing than oxidizing conditions. An increase in Fe and Mn solubility under reducing conditions might be due to the decrease of soil pH (e.g., He et al., 2010; DeLaune and Seo, 2011; Frohne et al., 2011, 2014). However, Rinklebe et al. (2016a) found that the concentrations of soluble Al, As, Cd, Cu, Fe, Mn, Ni, and Zn increased under high E_H due to the associated decrease of pH.

Wang et al. (2014) investigated the effect of flooding on Cu, Pb, and Cd in a typical Chinese paddy soil by using a kinetic approach. The elements were added to paddy soil as nitrate salts at the rates of 100 mg kg^{-1} for Cu and Pb, and of 2 mg kg^{-1} for Cd. The elements in the incubated soils were fractionated periodically from 5 to 150 days. The elements were time-dependently transferred from the easily extractable fraction (the exchangeable fraction) into less labile fractions (Fe–Mn oxide and the organic matter-bound fractions). No significant changes were found for the carbonate-bound and residual fractions of the metals in the soil during the incubation period. The distribution of added elements into different solid-phase fractions appeared to consist of two phases that involved the initial rapid retention followed by a slow, continuous retention. It was found that there were differences in the rates at which redistribution took place between two moisture regimes. Flooding incubation could immobilize elements, resulting in decreased mobility and availability of the elements in paddy soil. This process took place by the increased pH, precipitation of the PTEs with sulfides, and a higher concentration of amorphous Fe oxides under the submerged condition.

Xu et al. (2013) investigated the effects of irrigation management on the binding forms and the fate of Cu in paddy soil. Field experiments were conducted on nonpolluted rice paddy in Kunshan, East China. Nonflooding controlled irrigation was applied in three replications, with flooding irrigation as a control. Fresh soils were digested by using the modified European Community Bureau of Reference sequential extraction procedure. Nonflooding controlled irrigation led to multiple dry–wet cycles and high soil E_H in surface soil. The dry–wet cycles in nonflooding, controlled irrigation soil resulted in higher Cu contents in acid-extractable and oxidizable forms and in lower Cu contents in residual form. High decomposition and mineralization rates of SOM caused by the dry–wet cycles partially accounted for the increased Cu in acid-extractable form in on-flooding controlled irrigation soils. The frequently high contents of Cu in reducible form in on-flooding controlled irrigation fields might be due to the enhanced transformation of Fe and Mn oxides. Thus, one might conclude that nonflooding controlled irrigation enhanced the transformation of Cu from residual to oxidizable and acid-extractable forms. The oxidizable form plays a more important role than the reducible form in determining the transformation of Cu from the immobile to the mobile forms in nonflooding controlled irrigation soils. Nonflooding controlled irrigation increases availability and decreases leaching loss of Cu due to higher uptake by plants (as it is a micronutrient in a nonpolluted paddy soil), but it leads to a high concentration of Cu in rice (Xu et al., 2013).

Shaheen and Rinklebe (2014) found that the solubility of Cr below the water level in the studied profiles was higher than that above the water level, whereas the solubility of Cu and Zn above the water level was higher than that below the water level in both soil groups. The relative increase in Cr mobility below the water table can be explained by the dominance of lower redox conditions in the deepest horizons compared with the surface horizons since Gleysols can be flooded up to nine months per year whereas Fluvisols are flooded one to three months per year. For instance, Rupp et al. (2010) reported that Cr concentrations in soil solution were high when E_H was low in a lysimeter study. On the other hand, the mobile fraction of Cu was higher above the mean water table than below it. Prevailing oxidizing conditions above the water table facilitate the mobility of Cu in the studied soils, which may be attributed to the dissolution of sulfides and the resulting release of metals (Frohne et al., 2011). In addition, Schulz-Zunkel et al. (2013) reported that water-soluble

Cu concentrations increased with increased redox potentials in floodplain soils in the Central Elbe River. Elements that are bound to sulfides can be released during the oxidation of the soil due to the oxidation of sulfides to sulfates (DuLaing et al., 2009; Rinklebe and DuLaing, 2011). This might contribute to the increasing solubility of Cu with an increase in E_H.

In the study by Rinklebe and Shaheen (2014), the solubility of Cd in the surface horizons was higher than that in the deepest horizons, which means that the mobile Cd was higher under aerobic conditions than under anaerobic conditions. Therefore, oxidizing conditions facilitate the mobility of Cd in the Gleysols, which may also be attributed to the dissolution of sulfides under oxidizing conditions and the resulting release of Cd (Weber et al., 2009; Frohne et al., 2011, 2014). In addition, the role of soil organic carbon (SOC) should be considered a possible explanation for the relatively high concentrations of Cd at aerobic conditions in the surface layers. Organic matter has a high capacity to complex and to adsorb cations due to the presence of many negatively charged groups. For example, Shaheen et al. (2014b) suggested that the abundance of SOC in the soils involved in their study might contribute toward the formation of mobile metal–dissolved organo-clay (OC) complexes under oxidizing conditions, which prevented Cd from co-precipitating with or adsorbing to Fe (hydr)oxides.

Shaheen et al. (2014a,b) assessed the temporal dynamics of water-soluble As, Cd, Co, Cr, Cu, Mo, Ni, V, and Zn in a contaminated floodplain soil located at the Elbe River (Germany) under different flood-dry cycles by using a specific groundwater lysimeter technique. The impact of major controlling factors such as E_H, soil pH, DOC, Fe, Mn, and SO_4^{2-} on the dynamics of these elements was also quantified. The results indicated that the dynamics of these elements largely depends on the different flood-dry cycles. Shaheen et al. (2014a) indicated that the interactions between As, Cr, Mo, and V and their carriers were stronger during the long flood-dry cycles than during the short cycles. Long-term flooding increased the concentrations of the four elements in soil solution because of the direct metal reduction, the release of metal bound to reductively dissolved Fe- and Mn-oxides and OM, and/or the indirect changes in pH caused by changes in E_H. In particular, their findings suggest that a release of As, Cr, and V even under aerobic/alkaline conditions might occur. Shaheen et al. (2014b) found that reducing conditions under long-term flooding caused a decrease in E_H and pH, as well as an increase in DOC, Co, Fe, Mn, and Ni as compared with the oxidizing conditions under the long drying term. However, in short-term flood-dry cycles, Cd, Cu, Zn, and SO_4^{2-} showed the highest concentrations under high E_H. Therefore, the studies by Shaheen et al. (2014a,b) suggest that a release of As, Cd, Co, Cr, Cu, Mo, Ni, V, and Zn under different flood-dry cycles might create potential environmental risks in using metal-enriched soils in temporary flooded agricultural systems. In addition, the results confirm the role of the metal carriers and common ions in the dynamics and mobility of PTEs in the flooded soils. In this respect, de Livera et al. (2011) investigated the factors influencing Cd solubility relative to Fe and Zn during the pre-harvest drainage of paddy soils, in which soil oxidation is accompanied by the grain-filling stage of rice growth. This was simulated in temperature-controlled "reaction cell" experiments by first excluding oxygen to incubate soil suspensions anaerobically and then inducing aerobic conditions. In the treatments with no sulfur (S) addition, the ratios of Cd:Fe and Cd:Zn in solution increased during the aerobic phase, the Cd concentrations were unaffected, and the Fe and Zn concentrations decreased. However, in the treatments that included sulfur (as sulfate), up to 34% of S was precipitated as sulfide minerals during the anaerobic phase and the Cd:Fe and Cd:Zn ratios in solution during the aerobic phase were lower than those for treatments without S addition. When S was added, Cd solubility decreased whereas Fe and Zn were unaffected. When the soil was spiked with Zn, the Cd:Zn ratio was lower in solution during the aerobic phase, due to higher Zn concentrations. Decreased Cd:Fe and Cd:Zn ratios during the grain-filling stage could potentially limit Cd enrichment in paddy rice grain due to competitive ion effects for root uptake.

Previous studies (e.g., Neal et al., 1987; Haudin et al., 2007; Reddy and DeLaune, 2008; He et al., 2010; DeLaune and Seo, 2011) reported that a high Se solubility could be expected in soils with high E_H and pH. However, Shaheen et al. (2014c) found an opposite trend where the water-soluble Se increased under low E_H and pH. The explanation for this high Se solubility under low E_H and pH is that under reducing conditions, Se may be released from the solid phases into solution by decreases in soil pH that occur in these alkaline soils due to reducing conditions. In addition, the high Se solubility under low E_H and pH might be due to the direct reduction of Se (6), the decomposition of SOM, or the reductive dissolution of Fe and Mn oxides (Shaheen et al., 2014c).

The pronounced seasonal variations in soil and water temperature in floodplains, especially in temperate regions, affects the rates and interplay of microbial and abiotic geochemical processes that control the fate of PTEs in contaminated floodplain soils (e.g., Rinklebe et al., 2010). Hofacker et al. (2013) investigated how temperature affects chalcophile trace metal contaminants (Cu, Cd, Pb), especially flooding of a riparian soil contaminated by past mining activities. Their results demonstrated that temperature controlled trace metal dynamics during soil flooding via its influence on microbial reduction of terminal electron acceptors. Even at low temperatures, soil flooding may trigger the release of chalcophile metals from contaminated floodplain soils by sorbent reduction, competitive sorption, and formation of nanoparticulate metal-bearing colloids (Hofacker et al., 2013).

Huang et al. (2013) examined the changes in the speciation of Cd in contaminated rice paddy soils with prolonged submergence. Three Changhua soils from central Taiwan and three Taoyuan red soils from northern Taiwan with different levels of heavy metal contamination were selected for the study. The Cd, Fe, Mn, soil pH, and E_H in soil solutions were determined as a function of submerging time. During submergence, the Fe and Mn concentration increased, whereas the SO_4^{2-} concentration decreased. The concentrations of Cd immediately increased in the soil solution after a short submerging time and then subsequently decreased. The sequential extraction showed that the exchangeable fraction decreased and the oxide-bound fraction increased after submergence. The reduction of sulfate to sulfide may have subsequently resulted in the formation of CdS precipitate, which could be attributed to the decrease of Cd concentrations in soil solution after prolonged submergence (Ok et al., 2011b).

Khaokaew et al. (2011) determined Cd speciation and release kinetics in a Cd–Zn contaminated alkaline paddy soil. Linear least-squares fitting (LLSF) of bulk x-ray absorption fine structure (XAFS) spectra of the soils showed that at least 50% of Cd was bound to humic acid. Cadmium carbonates were found to be the major species at most flooding periods, whereas a small amount of cadmium sulfide was found after the soils were flooded for longer periods. Under all flooding and draining conditions, at least 14 mg kg^{-1} Cd was desorbed from the soil after a 2-h desorption experiment. The results obtained by micro x-ray fluorescence (μ-XRF) spectroscopy showed that Cd was less associated with Zn than with Ca in most soil samples (Khaokaew et al., 2011). A recent investigation using XAFS spectroscopy determined the presence of CdS and Cd sorbed with multiple soil colloids in an anthropogenically contaminated paddy soil (Hashimoto and Yamaguchi, 2013).

In this study, linear combination fitting on Cd K-edge x-ray absorption near edge structure (XANES) spectra of soil found that (i) at least 55% of Cd was present in association with soil colloids, including kaolinite, ferrihydrite, and humus, rather than precipitates with hydroxide, carbonate, and sulfate; and (ii) CdS occurred more in the subsurface than in surface soil profiles and more in high-S soil (45%) than in low-S soil (32%) in the presence of ZVI. Sulfur K-edge XANES analysis indicated the reduction of intermediate S species with an increasing depth in the soil microcosm. Overall, ZVI enhanced CdS formations and reduced Cd in the exchangeable fraction and in soil solution, and such effects appeared notably more in high-S than in low-S soils (Hashimoto and Yamaguchi, 2013).

13.3 BIOAVAILABILITY AND UPTAKE OF TOXIC ELEMENTS BY PLANTS IN TEMPORARY FLOODED SOILS

Contamination of agricultural soils by PTEs threatens food safety and disturbs the ecosystem, occasionally leading to plant toxicity and chronic and acute diseases for animals or humans, thereby possibly inducing secondary disaster (Ok et al., 2011a,b; Choppala et al., 2014; Shaheen and Rinklebe, 2015a). Flooding/drying conditions result in changes in oxygen demand for both soil and plant roots (Pezeshki, 2001), affect the nutrient uptake, and may cause production or accumulation of phytotoxic compounds (e.g., methane, sulfides, and reduced Mn and Fe), thus contributing to root injury, growth reduction, and mortality (Pezeshki, 2001; DeLaune and Seo, 2011). Pezeshki et al. (1997) and Pezeshki and DeLaune (1998) reported large differences in plant tolerance and effects on plant physiology and growth that reduce soil E_H conditions. Thus, flooding conditions and the relevant E_H could affect the availability of essential elements and contaminants to plants. Several plant species showed lower Cd, Cu, and Zn uptake in flooded soils under reducing conditions (Gambrell and Patrick, 1989; Gambrell, 1994; Shaheen and Rinklebe, 2015a). Flooding and, subsequently, the reduction of contaminated soils, thus, may result in lower environmental bioavailability.

13.3.1 Wetland Plants

Recently, the interest in the use of constructed wetlands for the removal of PTEs from contaminated soils, sediments, and waters has considerably increased (Guittonny-Philippe et al., 2014; Gill et al., 2014). Constructed wetlands are used for the cleanup of effluents and drainage waters (Vangronsveld et al., 2009). They offer a cost-effective and technically feasible technology and have proved to be both effective and successful in phytoremediation of heavy metal pollution and of various water quality issues (Williams, 2002; Olguin and Sanchez-Galvan, 2010; Rai, 2008, 2012; Ali et al., 2013; Shaheen and Rinklebe, 2015a). In constructed wetlands, substrate interactions remove most metals from contaminated water, with plants serving as a "polishing system" (Matagi et al., 1998). The permanent or temporarily anoxic soils that characterize wetlands help create conditions for the immobilization of PTEs in the highly reduced sulfite or metallic form (Gambrell, 1994; Du Laing et al., 2009), whereas plants play an important role in metal removal via filtration, adsorption, and cation exchange, and through plant-induced chemical changes in the rhizosphere (Carbonell et al., 1998; Wright and Otte, 1999). In addition, techniques of phytoremediation include phytoextraction (or phyto-accumulation), phyto-filtration, phyto-stabilization, phyto-volatilization, and phyto-degradation (Lasat, 2002; Alkorta et al., 2004; Anjum et al., 2012; Ali et al., 2013; Shaheen and Rinklebe, 2015a). There is evidence that wetland plants can accumulate PTEs in their tissues, such as duckweed (*Lemna minor*) (Zayed et al., 1998), water hyacinth (*Eichhornia crassipes*) (Vesk et al., 1999), salix (Stoltz and Greger, 2002), cattail (*Typha latifolia*), and common reed (*Phragmites australis*) (Ye et al., 2001). Cattail and common reed have been successfully used for the phytoremediation of Pb/Zn mine tailings under waterlogged conditions (Ye et al., 1997a,b).

Above- and below-ground parts of four common wetland species growing on an old submerged mine tailing, that is, *Salix* (mixed tissue from *Salix phylicifolia* and *Sclerotinia borealis*), *Carex rostrata*, *Eriophorum angustifolium*, and *Phragmites australis,* were sampled and analyzed for As, Cd, Cu, Pb, and Zn by Stoltz and Greger (2002). Differences in uptake and translocation properties of the plant species were observed among both field-grown plants and plants grown in hydroponics. These differences were probably due to processes in the soil–root interface. Most species were found to have restricted translocation of elements to the shoot, that is, they were root accumulators, and only the shoot concentrations of *Salix* for Cd and Zn and of *E. angustifolium* for Pb might be toxic to grazing animals. Thus, the plant establishment on submerged tailings can be considered an effective method for stabilizing the elements via vegetation (Stoltz and Greger, 2002).

The concentrations of Cd, Cu, Pb, and Zn that were accumulated by 12 emergent-rooted wetland plant species, including different populations of *Leersia hexandra, Juncus effusus,* and *Equisetum ramosisti,* were determined by Deng et al. (2004) under field conditions in China. The metal accumulation by wetland plants differed among species, populations, and tissues. Populations grown in substrata with elevated metals contained significantly higher metals in plants. Metals were mostly distributed in root tissues, suggesting that an exclusion strategy for metal tolerance exists within them. Some species/populations could accumulate relatively high metal concentrations (far above the toxic concentration to plants) in their shoots, thus indicating that internal detoxification metal tolerance mechanism(s) are also included. Plants have developed certain mechanisms for solubilizing heavy metals in soil. Plant roots secrete metal-mobilizing substances in the rhizosphere,

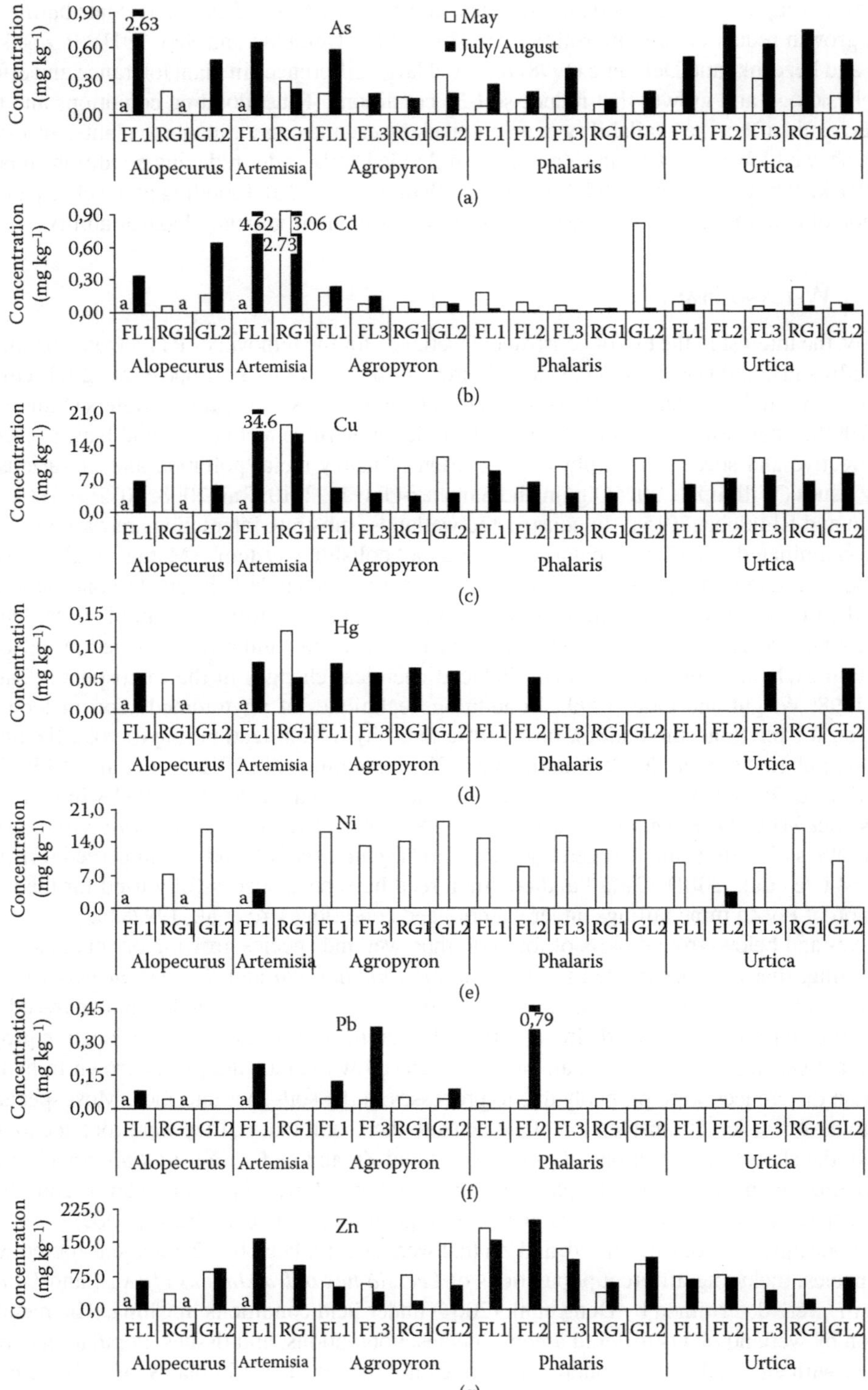

FIGURE 13.4 Element concentrations in selected plant species in May and in July/August. [a]Species could not be sampled at that time. (Reprinted from *Environmental Pollution*, 145, Overesch, M. et al., Metals and arsenic in soils and corresponding vegetation at central Elbe River floodplains (Germany), 800–812, Copyright 2007, with permission from Elsevier.)

and these are called *phytosiderophores* (Lone et al., 2008). The secretion of H^+ ions by roots can acidify the rhizosphere and increase metal dissolution. H^+ ions can displace heavy metal cations that are adsorbed to soil particles (Alford et al., 2010). Root exudates can lower the rhizosphere soil pH generally by one or two units compared with that in the bulk soil. Lower soil pH increases the concentration of PTEs in solution by promoting their desorption (Thangavel and Subbhuraam, 2004). Furthermore, the rhizospheric microorganisms (mainly bacteria and mycorrhizal fungi) may significantly increase the bioavailability of heavy metals in the soil (Vamerali et al., 2010; Sheoran et al., 2011). Interactions of microbial siderophores can increase labile metal pools and uptake by roots (Mench et al., 2009).

The factors affecting metal accumulation by wetland plants include metal concentrations, pH, and nutrient status in substrata. Mostly, concentrations of Pb and Cu in plants are significantly positively related to their total and/or diethylene triamine pentaacetic acid (DTPA)-extractable fractions in substrata whereas they are negatively related to soils N and P, respectively. Thus, a potential use of these wetland plants for phytoremediation might be taken into consideration (Zimmer et al., 2009).

Overesch et al. (2007) quantified metal and As contamination in soils and selected plant species at highly polluted grassland sites at the Elbe River floodplains, and they evaluated the potential for seasonal differences and the associated risks of elevated pollutant uptake by livestock via grazing or feeding hay (Figure 13.4).

The different plant species revealed considerable variations in concentration depending on element and sampling date. By far, the highest concentrations (mg kg^{-1}) of Cd (4.62), Cu (34.6), and Hg (0.13) were measured in *Artemisia vulgaris. Alopecurus pratensis* showed maximum values of As (2.63 mg kg^{-1}), and *Phalaris arundinacea* showed maximum values of Ni (18.6 mg kg^{-1}), Pb (4.15 mg kg^{-1}), and Zn (199.0 mg kg^{-1}). Among five common floodplain plant species, *Artemisia vulgaris* showed the highest concentrations of Cd, Cu, and Hg; *Alopecurus pratensis,* of As; and *Phalaris arundinacea,* of Ni, Pb, and Zn. Surprisingly, relationships between metal concentrations in plants and phytoavailable stocks in soil were weak. Arsenic and Hg uptake seems to be enhanced in long-term submerged soils. The enrichment of Cd and Hg seems to be linked to a special plant community composition.

The bioavailability of Cd, Mn, and Zn and their uptake by *Salix cinerea* on a contaminated dredged sediment landfill with a variable duration of submersion were evaluated by Vandecasteele et al. (2005). The results showed that longer submersion periods in the field caused lower Cd and Zn concentrations in the leaves during the first weeks of the growing season (Figure 13.5).

Zinc availability was the lowest when the soil was submerged, but metal transfer from stems and twigs to leaves may mask the lower availability of Cd in submerged soils. Especially for Cd, a transfer effect from one growing season to the next season was observed. Oxic conditions at the end of the previous growing season seem to determine, at least partly, the foliar concentrations for *S. cinerea*.

Submersion and waterlogging in the first weeks of the growing season resulted in normal foliar Zn concentrations, but emergence then sharply increased foliar concentrations to levels that were comparable with the plots that had already emerged at the beginning of the growing season. Hydrological regime aiming at wetland creation is a potential management option for reducing the bioavailability of metals and thus for establishing a safe management of wetlands that are polluted with metals as long as submersion can be maintained until the end of the growing season.

13.3.2 Rice Soils

Wetland rice ecosystems have spatially and temporally a unique aerobic and anaerobic environment that affects the transformation, mobility, and bioavailability of PTEs in soils, and therefore their uptake by rice plants (Shaheen et al., 2014c; Rinklebe et al., 2016b). Rice is predominantly grown in soils under highly reduced conditions with significantly increased metal protoplasmic fractions

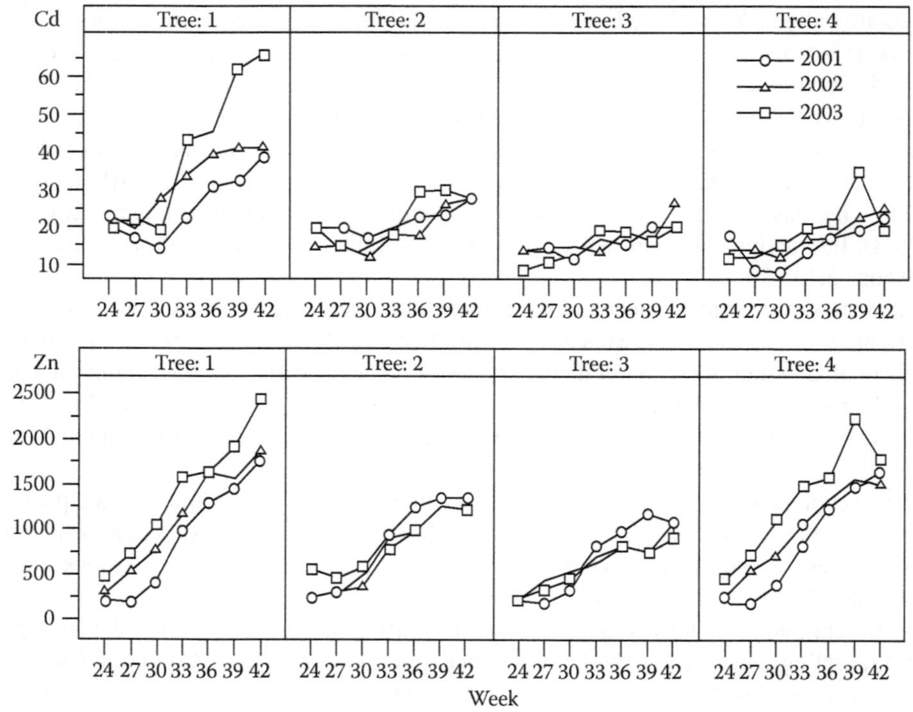

FIGURE 13.5 Foliar uptake patterns for Cd and Zn (mg kg^{-1} DW) during the growing season (weeks 24–42) for four *S. cinereatrees* (trees 1, 2, 3 and 4) on a dredged sediment-derived soil (plot 3) in three consecutive years. (Reproduced from Vandecasteele, B. et al., *Sci. Total Environ.*, 341(1–3), 251–263, 2005. With permission from the publisher.)

of the roots. One possible reason is that the self-adjusting mechanisms of plants play an important role in sequestering the metal in their roots, whereas only small amounts of metal are translocated to the above-ground parts of rice plants (Ye et al., 2012). Rice (*Oryza sativa* L.) has been considered a particular crop that has high potential for Cd uptake and is characterized by the depletion of zinc/iron in its grain (Chaney et al., 2004; Gong and Pan, 2006; Reeves and Chaney, 2008). Consequently, Cd translocation and accumulation in the grain and aerial parts of rice could jeopardize food safety in metal-contaminated rice fields (Kashiwagi et al., 2009). Cadmium contamination in croplands has been a serious concern because of its high health risk through soil–food chain transfer (Choppala et al., 2014). A sudden emergence of Cd-tainted rice from the South China market urged countermeasures to prevent Cd uptake and accumulation in rice grains from Cd-contaminated rice paddies. The Cd content of rice grain grown in metal-contaminated paddy soils near abandoned metal mines in South Korea was found to exceed safety guidelines (0.2 mg Cd kg^{-1}) set by the Korea Food and Drug Administration (KFDA) (Ok et al., 2011a,b). Especially in Korea, the dangerous levels of Cd and Pb in the rice paddy soils near abandoned mining areas have been reported (Ok et al., 2011c; Yang et al., 2006, 2007, 2008). Alrawiq et al. (2014) evaluated the accumulation and translocation of Cd, Cr, Cu, Pb, Zn, Mn, and Fe in soil and in paddy crops that have been irrigated with recycled and nonrecycled water, and they investigated the absorption of these elements by the crops. The average concentrations of studied elements (except Cd) were higher in the soil irrigated with recycled water than in the soil irrigated with nonrecycled water. Meanwhile, in various parts of the rice plant (roots, stem, leaves, straw, and grains), the concentrations of elements (except Cu and Mn) were higher in those plants that had been irrigated with recycled water compared with those that had been irrigated with nonrecycled water. The highest concentration of metals in this study was in the roots rather than in the parts of the rice plant that were above the ground (stem, leaves, straw, and grains), except for Mn, which was present in significant amounts in the leaves. The bioaccumulation factor was higher for Cd compared with other metals in the area irrigated with recycled water, whereas it was higher for Cu compared with other metals in the area irrigated with nonrecycled water. The order for the transfer factors from the soil to the roots for the metals in the areas that were irrigated with recycled and nonrecycled water were Cd > Cu> Zn> Pb> Mn> Fe> Cr and Cu> Mn> Cd> Zn> Pb> Fe> Cr, respectively. From the roots to the straw, the order was Mn> Cd> Zn> Cu> Cr> Pb> Fe for the area irrigated with recycled water, and it was Mn> Zn> Cd> Cr> Cu> Pb> Fe for the area irrigated with nonrecycled water. From the straw to the grains, the order was Zn> Pb> Cu> Cr> Cd> Fe> Mn and Cd> Cu> Fe> Pb> Zn> Cr> Mn for the areas irrigated with recycled and nonrecycled water, respectively. However, these concentrations of metals in the soil and rice grains were still below the maximum levels specified in the guidelines of the Codex Alimentarius Commission and the World Health Organization (WHO) (Alrawiq et al., 2014).

Juen et al. (2014) compared the concentrations of As, Be, Cd, Co, Cr, and Pb that accumulated in various parts (grains, stems, and roots) of rice (*Oryza sativa*) in Malaysia. The mean concentrations (mg kg^{-1}) of metals in grain samples were 0.06 for As, 0.0038 for Be, 0.01 for Cd, 0.14 for Co, and 0.21 for Pb whereas the Cr concentration in all samples was below the inductively coupled plasma–mass spectrometry (ICP–MS) detection limit. From the calculated translocation ratio, metal concentrations were greater in the roots than in grains of rice.

To reveal the impact of nonflooding controlled irrigation on the bioavailability and bioaccumulation of Cu, Pb, Cd, and Cr in rice fields, metal concentrations in different organs of rice plants growing under both flooding irrigation and nonflooding controlled irrigation were measured by Liao et al. (2013). Metal concentrations in roots were always the highest among the plant organs. Compared with flooding irrigation of rice, nonflooding controlled irrigation resulted in higher metal concentrations, bioaccumulation factors, and metal uptake of Cd, Cu, and Pb. This might be ascribed to the higher solubility and bioavailability of metals and to the higher rice root absorbent ability under drying–wetting conditions. However, for Cr, nonflooding controlled irrigation resulted in lower Cr concentration and uptakes in rice roots than did flooding irrigation. This indicated that the Cr bound to the Fe and Mn oxides that were more stable under nonflooding controlled irrigation condition, and they may play an important role in determining the bioavailability of Cr in paddy soil. However, the organic fraction of metals that were more likely released may play an important role in determining the bioavailability of Cu, Cd, and Pb in paddy soils. When the soil is less or not polluted with metals, nonflooding controlled irrigation can help in improving the availability of Cu as a micronutrient and in reducing soil metal accumulation by drawing more metals out of the soil by plant uptakes. If the soil is metal polluted, nonflooding controlled irrigation might result in a higher risk of pollution of food by metals in the short term. However, long-term use of nonflooding controlled irrigation will result in less accumulation of metals in the soil and will finally result in reduced crop metal uptakes. Lei et al. (2011) studied the uptake of As, Cd, and Pb by rice (*Oryza sativa* L.) grown in contaminated soils. The distribution of As, Cd, and Pb in rice plants followed the following order: root > shoot > husk > whole grain. About 30.1%–88.1% of As, 11.2%–43.5% of Cd, and 14.0%–33.9% of Pb were accumulated in iron plaque on root surfaces. The conclusions of this study were that high concentrations of As, Cd, and Pb were observed in paddy soils from mining- and smelting-impacted areas in Hunan province, indicating that those paddy soils are seriously affected by combined heavy metal contamination. In particular, Cd is the dominant contaminant followed by As and Pb in these paddy soils from many locations. The distributions of As, Cd, and Pb in rice tissue were found to be in the following order: root > shoot > husk > whole grain.

A hydroponics experiment was conducted by Cui et al. (2008) to investigate the effect of Cu on Cd, Ca, Fe, and Zn uptake by several rice genotypes. The experiment was carried out as a 2×2×4 factorial with four rice genotypes and two levels of Cu and Cd in nutrient solution. Plants were grown in a growth chamber with a controlled environment. The results showed a significant difference between the biomass of different rice genotypes ($p < .001$). The Cd and Cu concentration in the solution had no significant effect on the biomass. The addition of Cu significantly decreased Cd uptake by the shoots and roots of rice ($p < .001$). The Cd concentration did not significantly influence Ca uptake by plants, whereas the Cu concentration did ($p = .034$). There was a significant influence of Cd on Fe uptake by the shoots and roots ($p < .001$, $p = .003$, respectively). The uptake by Zn decreased significantly as the addition of Cd and Cu increased in the shoots. Cui et al. (2008) concluded that Cu had a significant influence on Cd uptake and discussed the possible mechanisms.

13.4 INFLUENCE OF SOIL AMENDMENTS ON THE PHYTOAVAILABILITY AND UPTAKE OF TOXIC ELEMENTS BY PLANTS IN TEMPORARY FLOODED SOILS

The effect of the addition of municipal solid waste compost (MSWC) on Co and Ni contents of submerged rice paddies was studied by Bhattacharyya et al. (2008) (Figure 13.6).

The metal content in rice straw was higher than in rice grain. The DTPA-extractable Co and Ni and their concentrations in straw and grain under cow dung manure (CDM) treatment were significantly higher than those under the MSWC treatment due to the lesser amount of total and bioavailable metal content in CDM compared with MSWC (Figure 13.6).

A greenhouse pot experiment was conducted by Ok et al. (2011a) to assess the effects of rapeseed residue that was applied as a green manure alone or in combination with mineral N fertilizer on Cd and Pb speciation in the contaminated paddy soil and their availability to the rice plant (*Oryza sativa L.*). Specifically, the following four treatments were evaluated: 100% mineral N fertilizer (N100) as a control, 70% mineral N fertilizer + rapeseed residue (N70 + R), 30% mineral N fertilizer + rapeseed residue (N30 + R), and rapeseed residue alone (R) (Figure 13.7).

Sequential extraction revealed that the addition of rapeseed residue decreased the easily accessible fraction of Cd by 5%–14% and that of Pb by 30%–39% through the transformation into less

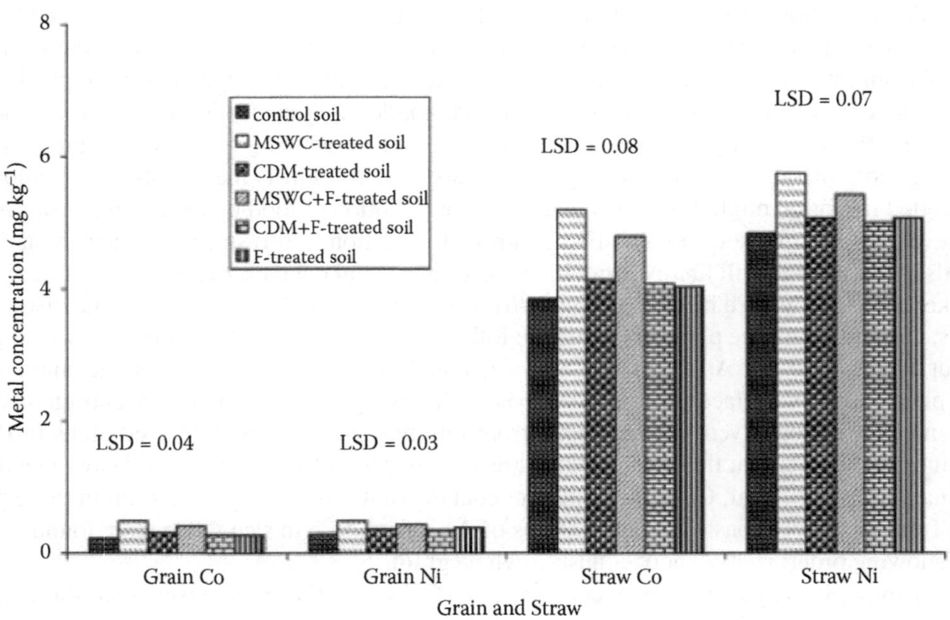

FIGURE 13.6 Co and Ni concentration in grain and straw of rice grown under different treatments (means of 3 years). (Reproduced from Bhattacharyya, P. et al., *Ecotoxicol. Environ. Safety*, 69, 506–512, 2008.)

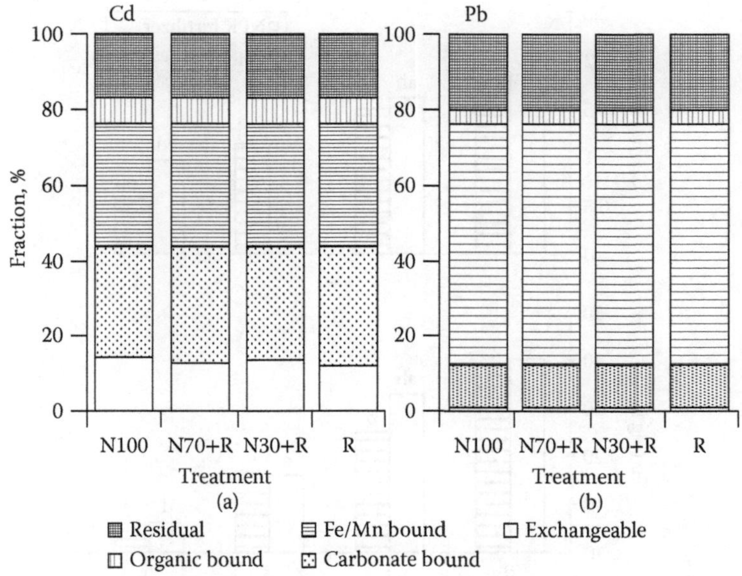

FIGURE 13.7 Effect of the rapeseed residue alone or in combination with mineral N fertilizer (N100: 100% mineral N fertilizer, N70 + R: 70% mineral N fertilizer + rapeseed residue, N30 + R: 30% mineral N fertilizer + rapeseed residue, R: rapeseed residue alone. (Ok, Y.S. et al., *Chemosphere*, 85, 677–682, 2011a.)

accessible fractions (Ok et al., 2011a). The addition of rapeseed residue led to the redistribution of Cd and Pb forms in the studied contaminated paddy soil, resulting in the transformation of elements into more stable fractions. The incorporation of rapeseed residue into the metal-contaminated rice paddy soils reduced metal availability to the rice plant and decreased the metal bioavailability and uptake. A significant decrease in the bioavailability of Pb was observed in response to the addition of rapeseed residue. Thus, the addition of organic materials such as a rapeseed residue could alleviate heavy metal toxicity toward plants by redistributing them to less available fractions (Ok et al., 2011a). Lee et al. (2013) evaluated the effects of eggshell waste in combination with the conventional NPK fertilizer or the rapeseed residue on the solubility and mobility of Cd and Pb in the rice paddy soil (Figure 13.8).

They found that an application of lime-based eggshells increased soil pH, thereby decreasing the extractability of Cd and Pb in rice paddy soils, which possibly might be due to the transformation of Cd and Pb into a more stable form in soils. Cadmium and Pb extractability was tested by using the following two methods: (a) the toxicity characteristics leaching procedure (TCLP) and (b) 0.1 M HCl extraction. The concentrations of TCLP–Cd and Pb were reduced by up to 67.9% and 93.2%, respectively, by the addition of 5% eggshell compared with the control. For the 0.1 M HCl extraction method, the concentrations of 0.1 M HCl–Cd in soils treated with NPK fertilizer and rapeseed residue were significantly reduced by up to 34.01% and 46.1%, respectively, by the addition of 5% eggshell compared with the control (Figure 13.8). Thus, they concluded that combined application of eggshell waste and rapeseed residue could be considered a cost-effective and beneficial method for remediating the soil contaminated with heavy metals.

Further laboratory and greenhouse experiments were conducted by Ok et al. (2011b) to assess the effects of amending contaminated rice paddy soils with zerovalent iron, lime, humus, compost, and combinations of these compounds on Cd solubility and uptake by rice (Figure 13.9).

In all treatments, drastic reductions in bioavailable Cd concentrations in soil solution were observed for the treated soils. Compost showed a 90% decrease in available Cd followed by 87% for zerovalent iron, 85% for lime, and 54% for humus. The results of sequential extractions revealed that bioavailable fractions, such as exchangeable or adsorbed forms, decreased after treatment when

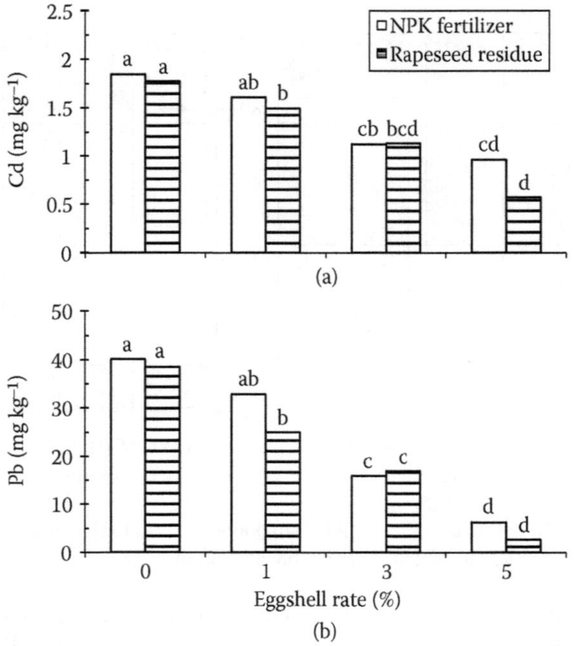

FIGURE 13.8 Toxicity characteristic leaching procedure (TCLP)–extracted concentrations of (a) Cd and (b) Pb in soils treated with NPK fertilizer or rapeseed residue in combination with the eggshell additions at rates of 0%, 1%, 3%, and 5%. The same letters above mean bars indicate no significantly difference as determined by Tukey's studentized range test ($p < .05$; $n = 5$). (With kind permission from Springer Science+Business Media: *Environmental Science Pollution Research*, Heavy metal immobilization in soil near abandoned mines using eggshell waste and rapeseed residue, 20, 2013, 1719-1726, Lee, S.S. et al.)

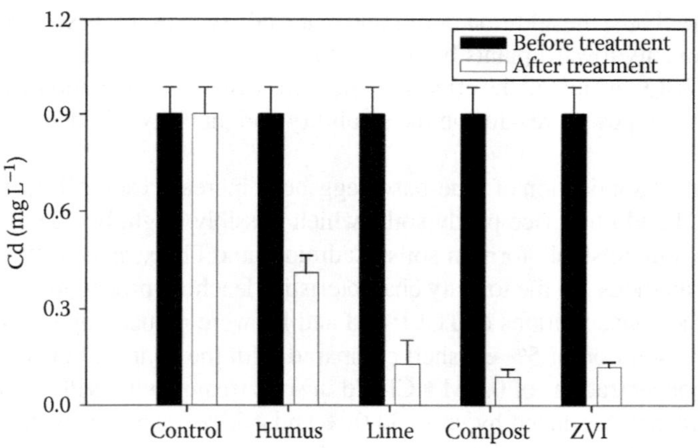

FIGURE 13.9 Changes in bioavailable Cd concentrations in soil solutions treated with humus (1000 mg kg^{-i} as C), lime (3%), compost (2%) and zerovalent iron (ZVI; 5%). (With kind permission from Springer Science+Business Media: *Environmental Geochemistry and Health*, Ameliorants to immobilize Cd in rice paddy soils contaminated by abandoned metal mines in Korea, 33, 2011b, 23–30, Ok, Y.S. et al.)

compared with the control (Figure 13.10) (Ok et al. 2011b). However, forms of nonbioavailable Cd, such as organic, carbonate, or sulfide/residual forms, increased. Treatment with zerovalent iron decreased the amount of exchangeable and adsorbed Cd by 98% and 54%, respectively (Figure 13.10) (Ok et al. 2011b).

The relative immobilization efficiencies generally decreased as follows: zerovalent iron > compost > lime > humus. These results demonstrate that each of these ameliorants can effectively immobilize Cd in contaminated soils. Sequential extraction analysis revealed that treatment with the ameliorants induced a 50%–90% decrease in the bioavailable Cd fractions when compared with the untreated control soil. In addition, the changes in Cd concentrations in paddy soil solutions treated with ameliorants were estimated in a greenhouse experiment (Figure 13.11). The Cd concentrations in soil solutions prior to treatment ranged from 0.15 to 0.38 mg L^{-1}. However, after treatment, the Cd concentration decreased from not-detected levels to 0.18 mg L^{-1} (Figure 13.11). Generally, decreases in Cd in soil solutions resulted from the formation of precipitates such as $Cd(OH)_2$ or the reduction of Cd^{2+} to Cd^0 (Lindsay, 1979). Compared with the control, Cd uptake by rice was decreased in response to treatment with zerovalent iron + humus (69%), lime (65%), zerovalent iron + compost (61%), compost (46%), zerovalent iron (42%), and humus (14%) (Ok et al., 2011b).

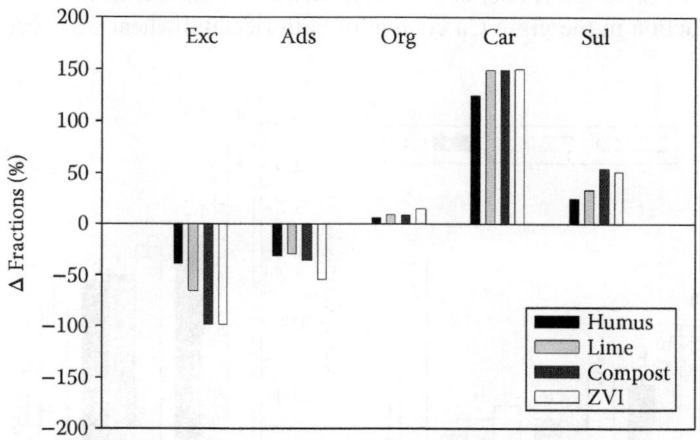

FIGURE 13.10 Changes in chemical forms (%) of Cd in contaminated paddy soils after each treatment when compared with the nontreated control as determined by sequential extraction. Exc, exchangeable; Ads, adsorbed, Org, organic; Car, carbonate; Sul, sulfide/residual. (With kind permission from Springer Science+Business Media: *Environmental Geochemistry and Health*, Ameliorants to immobilize Cd in rice paddy soils contaminated by abandoned metal mines in Korea, 33, 2011b, 23–30, Ok, Y.S. et al.)

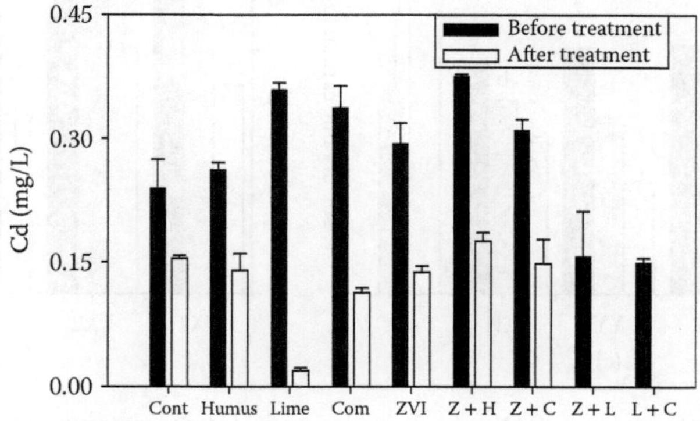

FIGURE 13.11 Changes in Cd concentrations in soil solutions treated with ameliorants in greenhouse experiments. Cont, control; Z+H, ZVI + humus; Z+C, ZVI + compost; Z+L, ZVI + lime; L+C, lime + compost. (With kind permission from Springer Science+Business Media: *Environmental Geochemistry and Health*, Ameliorants to immobilize Cd in rice paddy soils contaminated by abandoned metal mines in Korea, 33, 2011b, 23–30, Ok, Y.S. et al.)

A cross-site field experiment with biochar soil amendment (BSA) at rates from 20 to 40 t ha^{-1} in metal-polluted rice fields was conducted by Bian et al. (2013) across South China during 2010–2011 (Figure 13.12). Samples of both topsoil and rice grains under BSA treatment were collected after rice harvest, and soil-extractable Cd pool and rice grain Cd level were analyzed. Across the sites, BSA treatment greatly reduced (by 20%–90%) rice grain Cd content, and it enabled a safe Cd level (<0.4 mg kg^{-1}) of rice grain from all these Cd-contaminated rice fields to be attained by using a 40 t ha^{-1} biochar (BI) application except in one site where soil had a Cd content more than 20 mg kg^{-1}. This could be explained by a reduction in the extractable Cd pool in the BI-treated soil, which was closely correlated to the rise in soil pH with BSA treatment. This study demonstrated a promising role of BSA in preventing dangerous Cd accumulation by rice grain in contaminated rice paddies (Figure 13.12).

The large surface area with functional groups and the generally high pH of BI could be reactive to immobilize heavy metal cations in soils (Beesley et al., 2011). An increasing effect of BI on metal mobility in soil was observed (Fellet et al., 2011; Méndez et al., 2012). In a previous field experiment, a great reduction in the grain Cd content of both rice and wheat was observed with BSA at

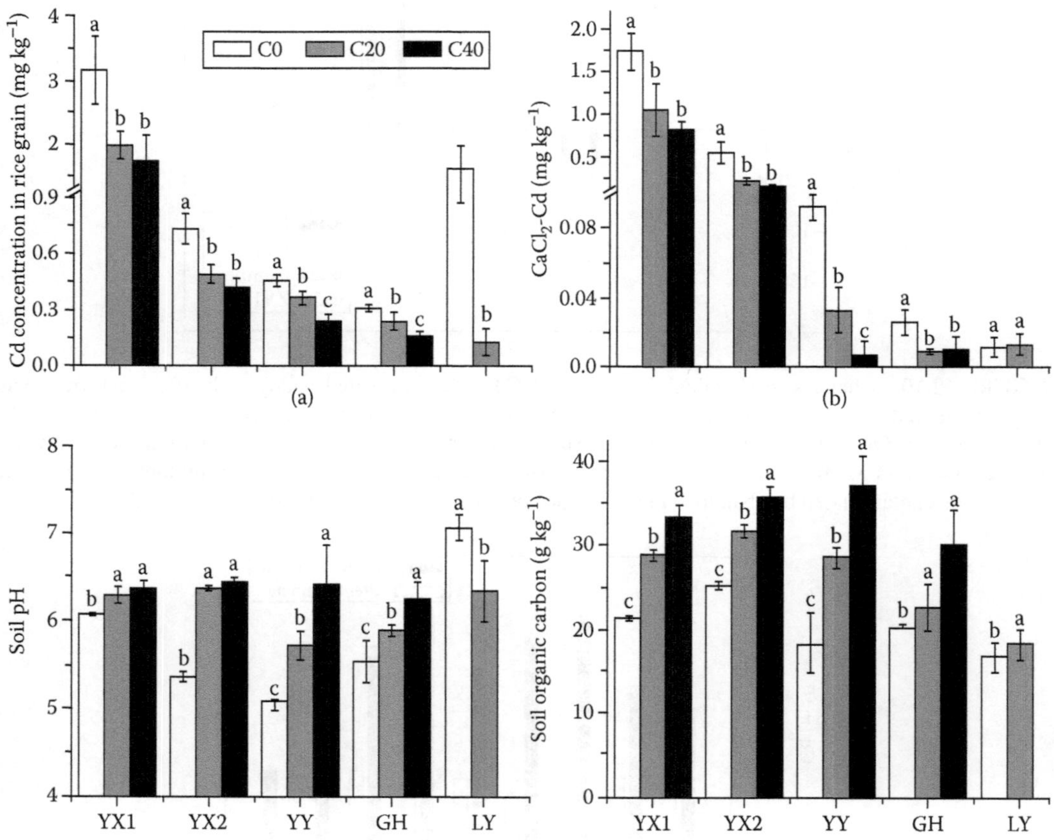

FIGURE 13.12 Changes (mean ± S.D., n = 3) in rice grain Cd concentration (A) and $CaCl_2$-extractable Cd (B), soil pH (C), and SOC (D) with biochar amendment across sites: YX1 and YX2 (Yixing, Jiangsu), YY (Yueyang, Hunan), GH (Guanghan, Sichuan), and LY (Longyan, Fujian). The different lowercase letters indicate a significant difference between the treatments in each site (p < .05). (Reprinted from *Ecological Engineering*, 58, Bian, R. et al., Biochar soil amendment as a solution to prevent Cd-tainted rice from China: Results from a cross-site field experiment, 378–383, Copyright 2013, with permission from Elsevier.)

20 and 40 t ha^{-1} in a heavily Cd-contaminated paddy from eastern China (Cui et al., 2011, 2012). A number of studies have reported the effect of BI amendment on reducing metal mobility in soils, and they have attributed it to the role of functional groups in BI particles in the reactivity for the metal cations (Park et al., 2011; Beesley et al., 2011). The finding of an enhanced immobilization of soil Cd with BSA seemed similar to other lab/pot studies or single-site studies (Cui et al., 2011, 2012; Zhang et al., 2012; Beesley et al., 2010). The results of these studies could serve as proof that BI in soil could act as an engineer within an ecosystem that is able to reduce soil Cd mobility, thus preventing rice Cd uptake by raising soil pH.

Ping et al. (2008) investigated the effects of seven amendments on the growth of rice and the uptake of PTEs from a paddy soil that was contaminated by Cu and Cd. The best results were seen in the application of limestone that increased grain yield 12.5- to 16.5-fold, and in that which decreased Cu and Cd concentrations in grain by 23.0%–50.4%. Application of calcium magnesium phosphate, calcium silicate, pig manure, and peat also increased the grain yield 0.3- to 15.3-fold, and it effectively decreased the Cu and Cd concentrations in grain. Cadmium concentration in grain was slightly reduced in the treatments of Chinese milk vetch and zinc sulfate. Concentrations of Cu and Cd in grain and straw were dependent on the available Cu and Cd in the soils, and soil-available Cu and Cd were significantly affected by the soil pH. The concentrations of available Cu and Cd in the soils decreased consistently with increasing rates of amendment application, suggesting that these amendments reduced the bioavailability of Cu and Cd. The decrease in Cu and Cd concentrations in grain and straw of rice was related to the change in Cu and Cd fractions in the soil. Castaldi et al. (2005) found a significant decrease in plant tissue concentrations of Cd after the addition of compost and lime; they also found that compost and lime decreased the Cd fraction extracted with H_2O and $Ca(NO_3)_2$, and it increased the residual fraction of Cd. Narwal and Singh (1998) also reported that pig manure and peat decreased the concentration of DTPA-extractable Cd (available) in the soil, but it increased Cd in the residual fraction. Therefore, both soil pH and available heavy metal are important factors in predicting the metal concentrations in plants (Ping et al., 2008).

The studies by Shaheen and Rinklebe (2015b), Shaheen et al. (2015b), and Rinklebe and Shaheen (2015) examined the influence of various emerging amendments [e.g., nano-hydroxyapatite (HA), BI, chitosan (CH), and OC] and several low-cost amendments [e.g., activated carbon (AC), bentonite (BE), cement kiln dust (CBD), fly ash (FA), limestone (LS), sugar beet factory lime (SBFL), and zeolite (Z)] on the (im)mobilization of Cd and Pb (Shaheen and Rinklebe, 2015b), Ni and Zn (Shaheen et al., 2015b), and Cu (Rinklebe and Shaheen, 2015) and their uptake by rapeseed (*Brassica napus*) in a contaminated, temporary flooded riparian grassland soil. Application of the amendments significantly increased the biomass production of rapeseed compared with the control (except for OC, HA, and Z). Water-soluble Ni and Zn decreased significantly after the addition of the amendments (except OC). The SBFL, CBD, LS, BE, AC, and BI were the most effective, resulting in a 58%–99% and a 56%–96% decrease of water-soluble Ni and Zn, respectively. The addition of SBFL, CBD, and LS leads to the highest decreasing rate of plant tissue concentrations of Ni (56%–68%) and Zn (40%–49%) (Shaheen et al., 2015b).

The amendments (except OC) decreased soluble + exchangeable Cd (4%–60%) compared with the control. Although the CBD, SBFL, and LS showed the highest decreasing rate of soluble + exchangeable Cd, they increased Cd in plants. The Z, BE, AC, BI, and CH decreased Cd in plants (22%–36%). The amendments (except OC and Z) decreased soluble + exchangeable Pb by 6%–87% and decreased Pb in the plant by 35%–99% compared with the control. The SBFL, CBD, and LS showed the highest decreasing rate of water-soluble Pb, and HA showed the highest decreasing rate of Pb uptake (Shaheen and Rinklebe, 2015b).

Application of the amendments changed the distribution of Cu among geochemical fractions: alkaline materials (i.e., CKD, SBFL, and LS) lead to increased carbonate-bound fractions, and the acid rhizosphere zone might cause a release of this Cu. Thus, the mobilization of Cu and the uptake of Cu by rapeseed were increased compared with the control (except for OC) under the prevailing conditions (Rinklebe and Shaheen, 2015).

13.5 CONCLUSIONS

Temporary flooded soils around the world are often polluted by PTEs. Wetlands are characterized by a highly dynamic and variable hydrological regime, which, in turn, has considerable impacts on the mobilization and bioavailability of metals. Studies on factors controlling the dynamics of PTEs in soils is challenging, because the obtained results should elucidate the underlying processes and encourage an explanation of their fate in the environment. Knowledge about the fate of PTEs in the ecosystem is required to answer both scientific and practical questions regarding the protection of groundwater and plants, sustainable management of soils, or to explain the pathways of environmental harmful substances. The dynamics of redox-sensitive processes is of large importance for flooded soils, as the location of the oxic–anoxic interface is subject to change due to fluctuating water table levels. The dynamics and release of PTEs in temporary flooded soils is determined by a complex of effects such as metal concentrations, E_H, pH, adsorption/desorption processes, the presence of Fe–Mn oxides, SOM, and total sulfur (Du Laing et al., 2009; DeLaune and Seo, 2011).

In many investigations, a significant concentration of the total Cd in flooded soils was observed in the mobile fraction, whereas the total Cu, Co, Cr, Ni, V, and Zn concentrations were dominated by less available fractions and large portions were assigned to fractions that were interpreted as Mn- and Fe-sorbed metals or to the residual fraction. This suggests that the potential mobility and bioavailability of many PTEs probably decline in the following order: Cd > Pb > Zn > Cu > Cr > Ni, thus suggesting the greater mobility of Cd, which means that Cd in these soils is easily accumulated by plants through the root system. Thus, the concentration of Cd in the paddy soils could be a concern to human health (Zimmer et al., 2011; Jalali and Hemati, 2013; Rinklebe and Shaheen, 2014; Shaheen and Rinklebe, 2014; Shaheen et al., 2015a).

The dynamics of PTEs depends on the inundation periods and the reducing–oxidizing conditions. Therefore, the release of PTEs under different flood-dry cycles might be increased, which might create potential environmental risks in using metal-enriched soils in temporary flooded agricultural systems (Du Laing et al., 2009; Rupp et al., 2010; DeLaune and Seo, 2011; Frohne et al., 2011, 2014, 2015; Shaheen et al., 2014a,b,c; Rinklebe et al., 2016a,b). In particular, numerous findings suggest that the release of PTEs from temporally flooded soils should be considered due to increased mobility and the potential environmental risks in using metal-enriched soils in flooded agricultural systems (Shaheen et al., 2014a,b,c; Rinklebe and Shaheen, 2014).

Flooding conditions affect the availability of nutrients and PTEs to plants. Several plant species showed lower Cd, Cu, and Zn uptake in reduced soils (Gambrell and Patrick, 1989; Gambrell, 1994). As this evidence suggests that flooding and, subsequently, reduction of contaminated soils may result in lower environmental bioavailability, it may constitute a valid management option for polluted soils. This means that permanent or temporarily anoxic soils that characterize wetlands help in creating conditions for the immobilization of PTEs in the highly reduced sulfite or metallic form (Gambrell, 1994). However, other types of plants play an important role in metal removal via filtration, adsorption, and cation exchange, and through plant-induced chemical changes in the rhizosphere (Carbonell et al., 1998; Wright and Otte, 1999). There is evidence that some of the wetland plants can accumulate PTEs in their tissues (Zayed et al., 1998; Vesk et al., 1999; Stoltz and Greger, 2002; Ye et al., 1997a,b, 2001; Shaheen and Rinklebe, 2015a). This means that these type of plants might be successfully used for the phytoremediation of PTE-contaminated temporary waterlogged conditions (Stoltz and Greger, 2002; Overesch et al., 2007; Zimmer et al., 2009; Shaheen and Rinklebe, 2015a).

Rice had been considered a particular crop that had high PTEs, especially Cd uptake, with the depletion of Zn/Fe in its grain (Chaney et al., 2004; Gong and Pan, 2006; Reeves and

Chaney, 2008). Consequently, Cd translocation and accumulation in the grain and aerial plant parts of rice could jeopardize food safety in regions with metal-contaminated rice fields (Kashiwagi et al., 2009; Choppala et al., 2014). Thus, the development of adequate remediation approaches of those contaminated sites leading to a reduction in the release of metals under reducing conditions in floodplain ecosystems should be considered a challenge for the near future, thus aiming at minimizing the potential risk to both humans and the environment. Various organic and inorganic soil amendments were used to reduce the solubility and the phytoavailability of PTEs by rice plants (Ok et al., 2011a,b; Bian et al., 2013; Lee et al., 2013). This is important since contaminated flooded areas are extended over the entire world. Therefore, certain trials should to be done by law to conduct an appropriate risk assessment and to implement practical actions to eliminate (or reduce) these environmental problems in floodplain ecosystems.

ACKNOWLEDGMENTS

The authors are grateful to the Egyptian Science and Technology Development Fund (STDF–STF; Project ID: 5333) and the German Academic Exchange Foundation (Deutscher Akademischer Austauschdienst, DAAD) (DAAD–WAP program; Code number A/14/05113) for the financial support of the postdoctoral scholarship of Dr. Shaheen at the University of Wuppertal, Germany. The authors are also thankful to the publishers for their permission to reuse figures, tables, and/or data from the mentioned publications.

REFERENCES

Alford, E.R., Pilon-Smits, E.A.H., Paschke, M.W., 2010. Metallophytes—A view from the rhizosphere. *Plant Soil* 337, 33–50.

Ali, H., Khan, E., Sajad, M.A., 2013. Phytoremediation of heavy metals—Concepts and applications: A review. *Chemosphere* 91, 869–881.

Alkorta, I., Hernandez-Allica, J., Becerril, J., Amezaga, I., Albizu, I., Garbisu, C., 2004. Recent findings on the phytoremediation of soils contaminated with environmentally toxic heavy metals and metalloids such as zinc, cadmium, lead, and arsenic. *Reviews of Environmental Science in Biotechnology* 3, 71–90.

Alrawiq, N., Khairiah, J., Talib, M.L., Ismail, B.S., Anizan, I., 2014. Accumulation and translocation of heavy metals in soil and paddy plant samples collected from rice fields irrigated with recycled and non-recycled water in MADA Kedah, Malaysia. *International Journal of ChemTech Research* 6, 2347–2356.

Anjum, N.A., Pereira, M.E., Ahmad, I., Duarte, A.C., Umar, S., Khan, N.A., 2012. *Phytotechnologies: Remediation of Environmental Contaminants*. Boca Raton, London, New York: Taylor & Francis Group, LLC.

Antić-Mladenović, S., Rinklebe, J., Frohne, T., Stärk, H.-U., Wennrich, R., Tomić, Z., Ličina, V., 2011. Impact of controlled redox conditions on nickel in a serpentine soil. *Journal of Soils and Sediments* 11, 406–415.

Bacon, J.R., Davidson, C.M., 2008. Is there a future for sequential chemical extraction? *The Analyst* 133, 25–46.

Baum, C., Hrynkiewicz, K., Leinweber, P., Meisner, R., 2006. Heavy-metal mobilization and uptake by mycorrhizal and nonmycorrhizal willows (*Salix dasyclados*). *Journal of Plant Nutrition in Soil Science* 169, 516–522.

BBodSchV, 1999. Bundes-Bodenschutz- und Altlastenverordnung (BBodSchV) vom 12. Juli 1999. Bundesgesetzblatt I 1999, 1554 [Federal Soil Protection and Contaminated Sites Ordinance dated 12 July 1999].

Beesley, L., Marmiroli, M., 2011. The immobilisation and retention of soluble arsenic, cadmium and zinc by biochar. *Environmental Pollution* 159, 474–480.

Bhattacharyya, P., Chakrabarti, K., Chakraborty, A., Tripathy, S., Kim, K., Powell, M.A., 2008. Cobalt and nickel uptake by rice and accumulation in soil amended with municipal solid waste compost. *Ecotoxicology and Environmental Safety* 69, 506–512.

Bhuiyan, M.A.H., Parvez, L., Islam, M., Dampare, S.B., Suzuki, S., 2010. Heavy metal pollution of coal mine-affected agricultural soils in the northern part of Bangladesh. *Journal of Hazardous Materials* 173, 384–392.

Bian, R., Chen, D., Liu, X., Cui, L., Li, L., Pan, G., Xie, D., Zheng, J., Zhang, X., Zheng, J., Chang, A., 2013. Biochar soil amendment as a solution to prevent Cd-tainted rice from China: Results from a cross-site field experiment. *Ecological Engineering* 58, 378–383.

Carbonell, A.A., Aarabi, M.A., DeLaune, R.D., Gambrell, R.P., Patrick, W.H., Jr., 1998. Arsenic in wetland vegetation: Availability, phytotoxicity, uptake and effects on plant growth and nutrition. *Science of the Total Environment* 217, 189–199.

Castaldi, P., Santona, L., Melis, P. 2005. Heavy metal immobilisation by chemical amendments in a polluted soil and influence on white lupin growth. *Chemosphere* 60, 365–371.

Chaney, R.L., Angle, J.S., Broadhurst, C.L., Peters, C.A., Tappero, R.V., Sparks, D.L., 2007. Improved understanding of hyperaccumulation yields commercial phytoextraction and phytomining technologies. *Journal of Environmental Quality* 36, 1429–1443.

Chaney, R.L., Reeves, P.G., Ryan, J.A., Simmons, R.W., Welch, R.M., Angle, J.S., 2004. An improved understanding of soil Cd risk to humans and low cost methods to phytoextract Cd from contaminated soils to prevent soil Cd risks. *Biometals* 17, 549–553.

Choppala, G., Saifullah, Bolan, N., Bibi, S., Iqbal, M., Rengel, Z., Kunhikrishnan, K., Ashwath, N., Ok, Y.S., 2014. Cellular mechanisms in higher plants governing tolerance to cadmium toxicity. *Critical Reviews in Plant Sciences* 33, 374–391.

Cui, L., Li, L., Zhang, A., Pan, G., Bao, D., Chang, A., 2011. Biochar amendment greatly reduces rice Cd uptake in a contaminated paddy soil: A two-year field experiment. *Bioresources* 6, 2605–2618.

Cui, L., Pan, G., Li, L., Yan, J., Zhang, A., Bian, R., Chang, A., 2012. The reduction of wheat Cd uptake in contaminated soil via biochar amendment: A two-year field experiment, *Bioresources*, 7, 5666–5676.

Cui, Y., Zhang, X., Yongguan Zhu, Y., 2008. Does copper reduce cadmium uptake by different rice genotypes? *Journal of Environmental Sciences* 20 (3), 332–338.

de Livera, J., McLaughlin, M.J., Hettiarachchi, G.M., Kirby, J.K., Beak, D.J., 2011. Cadmium solubility in paddy soils: Effects of soil oxidation, metal sulfides and competitive ions. *Science of the Total Environment* 409, 1489–1497.

DeLaune, R.D., Seo, D.C., 2011. Heavy metals transformation in wetlands. In: H.M. Selim (ed.). *Dynamics and Bioavilablity of Heavy Metals in the Rootzone*. Boca Raton, London, New York: Taylor & Francis Group, LLC, p. 219–244.

Delgado, J., Barba-Brioso, C., Nieto, J.M., Boski, T., 2011. Speciation and ecological risk of toxic elements in estuarine sediments affected by multiple anthropogenic contributions (Guadiana saltmarshes, SW Iberian Peninsula): I. Surficial sediments. *The Science of the Total Environment* 409, 3666–3679.

Deng, H., Ye, Z.H., Wong, M.H., 2004. Accumulation of lead, zinc, copper and cadmium by 12 wetland plant species thriving in metal-contaminated sites in China. *Environmental Pollution* 132, 29–40.

Dold, B., 2003. Speciation of the most soluble phases in a sequential extraction procedure adapted for geochemical studies of copper sulfide mine waste. *Journal of Geochemistry Explorer* 80, 55–68.

Du Laing, G., Rinklebe, J., Vandecasteele, B., Meers, E., Tack, F.M.G., 2009. Trace metal behaviour in estuarine and riverine floodplain soils and sediments: A review. *Science of the Total Environment* 407, 3972–3985.

Stoltz, E., Greger, M., 2002. Accumulation properties of As, Cd, Cu, Pb and Zn by four wetland plant species growing on submerged mine tailings. *Environmental and Experimental Botany* 47, 271–280.

Fellet, G., Marchiol, L., Delle Vedove, G., Peressotti, A., 2011. Application of biochar on mine tailings: Effects and perspectives for land reclamation. *Chemosphere*, 83, 1262–1297.

Filgueiras, A.V., Lavilla, I., Bendicho, C., 2002. Chemical sequential extraction for metal partitioning in environmental solid samples. *Journal of Environmental Monitoring* 4, 823–857.

Frohne, T., Diaz-Bone, R.A., Du Laing, G., Rinklebe, J., 2015. Impact of systematic change of redox potential on the leaching of Ba, Cr, Sr, and V from a riverine soil into water. *Journal of Soils and Sediments* 15, 623–633.

Frohne, T., Rinklebe, J., Diaz-Bone, R.D., 2014. Contamination of floodplain soils along the Wupper River, Germany, with As, Co, Cu, Ni, Sb, and Zn and the impact of pre-definite redox variations on the mobility of these elements. *Soil and Sediment Contamination* 23, 779–799.

Frohne, T., Rinklebe, J., Diaz-Bone, R.A., Du Laing, G., 2011. Controlled variation of redox conditions in a floodplain soil: Impact on metal mobilization and biomethylation of arsenic and antimony. *Geoderma* 160, 414–424.

Gambrell, R.P., 1994. Trace and toxic metals in wetlands: A review. *Journal of Environmental Quality* 23, 883–891.

Gambrell, R.P., Patrick Jr., W.H., 1989. Cu, Zn, and Cd availability in a sludge-amended soil under controlled pH and redox potential conditions. In: B. Bar-Yosef, N.J. Barrow, J. Goldshmid (eds.). *Inorganic Contaminants in the Vadose Zone: Ecological Studies*. Vol. 74. Berlin, Germany: Springer–Verlag, pp. 89–106.

Gill, L.W., Ring, P., Higgins, N.M.P., Johnston, P.M., 2014. Accumulation of heavy metals in a constructed wetland treating road runoff. *Ecological Engineering* 70, 133–139.

Gong, W.Q., Pan, G.X., 2006. Issues of grain Cd uptake and the potential health risk of rice production sector of China. *Science and Technology Review* 24, 43–48 (in Chinese with English abstract).

Guittonny-Philippe, A., Masotti, V., Höhener, P., Boudenne, J., Viglione, J., Laffont-Schwob, I., 2014. Constructed wetlands to reduce metal pollution from industrial catchments in aquatic Mediterranean ecosystems: A review to overcome obstacles and suggest potential solutions. *Environment International* 64, 1–16.

Hashimoto, Y., Yamaguchi, N., 2013. Chemical speciation of cadmium and sulfur K-Edge XANES spectroscopy in flooded paddy soils amended with zerovalent iron. *Soil Science Society of America Journal* 77, 1189–1198.

Haudin, C.S.P., Renault, V., Hallaire, E., Leclerc-Cessac, S., Staunton, S., 2007. Effect of aeration on mobility of selenium in columns of aggregated soil as influenced by straw amendment and tomato plant growth. *Geoderma* 141, 98–110.

He, Z.L., Shentu, J., Yang, X.E., 2010. Manganese and Selenium. In: P.S. Hooda (ed.). *Trace Elements in Soils*. 1st ed. Chichester, UK: John Wiley & Sons Ltd, pp. 481–497.

Hofacker, A.F., Voegelin, A., Kaegi, R., Weber, F., Kretzschmar, R., 2013. Temperature-dependent formation of metallic copper and metal sulfide nanoparticles during flooding of a contaminated soil. *Geochimica et Cosmochimica Acta* 103, 316–332.

Huang, J.-H., Wang, S.-L., Lin, J.-H., Chen, Y.-M., Wang, M.-K., 2013. Dynamics of cadmium concentration in contaminated rice paddy soils with submerging time. *Paddy and Water Environment* 11, 483–491.

Jalali, M., Hemati, N., 2013. Chemical fractionation of seven heavy metals (Cd, Cu, Fe, Mn, Ni, Pb, and Zn) in selected paddy soils of Iran (Article). *Paddy and Water Environment* 11, 299–309.

Juen, L.L., Aris, A.Z., Ying, L.W., Haris, H., 2014. Bioconcentration and translocation efficiency of metals in paddy (*Oryza sativa*): A case study from Alor Setar, Kedah, Malaysia. *Sains Malaysiana* 43, 521–528.

Kabata-Pendias, A., 2011. *Trace Elements in Soils and Plants*. 4th ed. Boca Raton, FL: CRC Press.

Kashiwagi, T., Shindoh, K., Hirotsu, N., Ishimaru, K., 2009. Evidence for separate translocation pathways in determining cadmium accumulation in grain and aerial plant parts in rice. *BMC Plant Biology* 9 (1), 8.

Khaokaew, S., Chaney, R.L., Landrot, G., Ginder-Vogel, M., Sparks, D.L., 2011. Speciation and release kinetics of cadmium in an alkaline paddy soil under various flooding periods and draining conditions. *Environmental Science and Technology* 45, 4249–4255.

Kosolsaksakul, P., Farmer, J.G., Oliver, I.W., Graham, M.C., 2014. Geochemical associations and availability of cadmium (Cd) in a paddy field system, northwestern Thailand. *Environmental Pollution* 187, 153–161.

Lasat, M.M., 2002. Phytoextraction of toxic metals: A review of biological mechanisms. *Journal of Environmental Quality* 31, 109–120.

Lee, S.S., Lim, J.E., Abd El-Azeem, S.A.M., Choi, B., Oh, S-E., Moon, D.H., Ok, Y.S., 2013. Heavy metal immobilization in soil near abandoned mines using eggshell waste and rapeseed residue. *Environmental Science Pollution Research* 20, 1719–1726.

Lei, M., Tie, B., Williams, P.N., Zheng, Y., Huang, Y., 2011. Arsenic, cadmium, and lead pollution and uptake by rice (*Oryza sativa* L.) grown in greenhouse. *Journal of Soils and Sediments* 11, 115–123.

Liao, L., Xu, J., Peng, S., Qiao, Z., Gao, X., 2013. Uptake and bioaccumulation of heavy metals in rice plants as affect by water saving irrigation. *Advance Journal of Food Science and Technology* 5, 1244–1248.

Lindsay, W.L., 1979. *Chemical Equilibria in Soils*. New York: John Wiley & Sons.

Lone, M.I., He, Z., Stoffella, P.J., Yang, X., 2008. Phytoremediation of heavy metal polluted soils and water: Progresses and perspectives. *Journal of Zhejiang University. Science B* 9, 210–220.

Marston, R.A., Girel, J., Pautou, G., Piegay, H., Bravard, J.-P., Arneson, C., 1995. Channel metamorphosis, floodplain disturbance, and vegetation development: Ain River, France. *Geomorphology* 13, 121–131.

Matagi, S.V., Swai, D., Muganbe, R., 1998. A review of heavy metal removal mechanisms in wetlands. *African Journal of Tropical Hydrobiology and Fisheries* 8, 23–35.

Mench, M., Schwitzguebel, J.-P., Schroeder, P., Bert, V., Gawronski, S., Gupta, S., 2009. Assessment of successful experiments and limitations of phytotechnologies: Contaminant uptake, detoxification and sequestration, and consequences for food safety. *Environmental Science Pollution Research* 16, 876–900.

Méndez, A., Gömez, A., Paz-Ferreiro, J., Gascö, G., 2012. Effects of biochar from sewage sludge pyrolysis on Mediterranean agricultural soils. *Chemosphere* 89, 1354–1359.

Miao, D., Yang, W., 2012. Speciation distributions and availability of Zn, Pb, Cu and Ni in a mine drainage contaminated paddy soils (Conference Paper). *Advanced Materials Research* 347–353, 826–831.

Narwal R.P., Singh B.R., 1998. Effect of organic materials on partitioning, extractability and plant uptake of metals in an alum shale soil. *Water, Air and Soil Pollution* 103, 405–421.

Neal, R.H., Sposito, G., Holtzclaw, K.M., Traina, S.J., 1987. Selenite adsorption on alluvial soils: II. Solution composition effects. *Soil Science Society of America Journal* 51, 1165–1169.

Ok, Y.S., Kim, S.C., Kim, D.K., Skousen, J.G., Lee, J.S., Cheong, Y.W., Kim, S.J., Yang, J.E., 2011b. Ameliorants to immobilize Cd in rice paddy soils contaminated by abandoned metal mines in Korea. *Environmental Geochemistry and Health* 33, 23–30.

Ok, Y.S., Lim, J.E., Moon, D.H., 2011c. Stabilization of Pb and Cd contaminated soils and soil quality improvements using waste oyster shells. *Environmental Geochemistry and Health* 33, 83–91.

Ok, Y.S., Usman, A.R.A., Lee, S.S., Abd El-Azeem, S.A.M., Choi, B.S., Hashimoto, Y., Yang, J.E., 2011a. Effect of rapeseed residue on cadmium and lead availability and uptake by rice plants in heavy metal contaminated paddy soil. *Chemosphere* 85, 677–682.

Olguin, E.J., Sanchez-Galvan, G., 2010. Aquatic phytoremediation: Novel insights in tropical and subtropical regions. *Pure and Applied Chemistry* 82, 27–38.

Overesch, M., Rinklebe, J., Broll, G., Neue, H.-U., 2007. Metals and arsenic in soils and corresponding vegetation at central Elbe River floodplains (Germany). *Environmental Pollution* 145, 800–812.

Park, J.H., Choppala, G.K., Bolan, N.S., Chung, J.W., Chuasavathi, T., 2011. Biochar reduces the bioavailability and phytotoxicity of heavy metals. *Plant Soil* 348, 439–451.

Pezeshki, S.R., 2001. Wetland plant responses to soil flooding. *Environmental and Experimental Botany* 46, 299–312.

Pezeshki, S.R., DeLaune, R.D., 1998. Responses of seedlings of selected woody species to soil oxidation-reduction conditions. *Environmental and Experimental Botany* 40, 123–133.

Pezeshki, S.R., DeLaune, R.D., Meeder, J.F., 1997. Carbon assimilation and biomass partitioning in Avicennia germinans and Rhizophora mangle seedlings in response to soil redox conditions. *Environmental and Experimental Botany* 37, 161–171.

Ping, L., Xingxiang, W., Taolin, Z., Dongmei, Z., Yuanqiu, H., 2008. Effects of several amendments on rice growth and uptake of copper and cadmium from a contaminated soil. *Journal of Environmental Sciences* 20, 449–455.

Pueyo, M., Mateu, J., Rigol, A., Vidal, M., López-Sánchez, J.F., Rauret, G., 2008. Use of the modified BCR three-step sequential extraction procedure for the study of trace element dynamics in contaminated soils. *Environmental Pollution* 152, 330–341.

Rai, P.K., 2008. Phytoremediation of Hg and Cd from industrial effluents using an aquatic free floating macrophyte Azolla pinnata. *International Journal of Phytoremediation* 10, 430–439.

Rai, P.K., 2012. An eco-sustainable green approach for heavy metals management: Two case studies of developing industrial region. *Environmental Monitoring and Assessment* 184, 421–448.

Reddy, K.R., DeLaune, R.D., 2008. *Biogeochemistry of Wetlands: Science and Applications*. Boca Raton, London, New York: Taylor & Francis Group, LLC.

Reeves, P.G., Chaney, R.L., 2008. Bioavailability as an issue in risk assessment and management of food cadmium: A review. *The Science of the Total Environment* 398, 13–19.

Rennert, T., Rinklebe, J., 2010. Release of Ni and Zn from contaminated floodplain soils under saturated flow conditions. *Water, Air, and Soil Pollution* 205, 93–105.

Rinklebe, J., 2004. Differenzierung von Auenböden der Mittleren Elbe und Quantifizierung des Einflusses von deren Bodenkennwerten auf die mikrobielle Biomasse und die Bodenenzymaktivitäten von b-Glucosidase, Protease und alkalischer Phosphatase. Dissertation. Institut für Bodenkunde und Pflanzenernährung der Landwirtschaftlichen Fakultät der Martin-Luther-Universität Halle-Wittenberg. PhD Thesis. 113 pp. and Appendix.

Rinklebe, J., Du Laing, G., 2011. Factors affecting the dynamics of trace metals in frequently flooded soils. In: H.M. Selim (ed.). *Dynamics and Bioavilablity of Heavy Metals in the Rootzone*. New York: Taylor & Francis Group, pp. 245–270.

Rinklebe, J., During, A., Overesch, M., Du Laing, G., Wennrich, R., Stärk, H.-J., Mothes, S., 2010. Dynamics of mercury fluxes and their controlling factors in large Hg-polluted floodplain areas. *Environmental Pollution* 158, 308–318.

Rinklebe, J., Franke, C., Neue H.U., 2007. Aggregation of floodplain soils as an instrument for predicting concentrations of nutrients and pollutants. *Geoderma* 141, 210–223.

Rinklebe, J., Shaheen, S.M., 2014. Assessing the mobilization of cadmium, lead, and nickel using a seven-step sequential extraction technique in contaminated floodplain soil orofiles along the central Elbe river, Germany. *Water Air, and Soil Pollution* 225, 2039. doi:10.1007/s11270-014-2039-1

Rinklebe, J., Shaheen, S.M., 2015. Miscellaneous additives can enhance plant uptake and affect geochemical fractions of copper in a heavily polluted riparian grassland soil. *Ecotoxicology and Environmental Safety* 119, 58–65.

Rinklebe, J., Shaheen S.M., Frohne, T., 2016a. Amendment of biochar reduces the release of toxic elements under dynamic redox conditions in a contaminated floodplain soil. *Chemosphere.* doi:10.1016/j.chemosphere.2015.03.067

Rinklebe, J., Shaheen S.M., Yu, K., 2016b. Release of As, Ba, Cd, Cu, Pb, and Sr under pre-definite redox conditions in different rice paddy soils originating from the U.S.A. and Asia. *Geoderma.* doi:doi.org/10.1016/j.geoderma.2015.10.011

Rockwood, D.L., Naidu, C.V., Carter, D.R., Rahmani, M., Spriggs, T.A., Lin, C., Alker, G.R., Isebrands, J.G., Segrest, S.A., 2004. Short-rotation woody crops and phytoremediation: Opportunities for agroforestry? *Agroforestry Systems* 61, 51–63.

Rupp, H., Rinklebe, J., Bolze, S., Meisner, R., 2010. A scale-dependent approach to study pollution control processes in wetland soils using three different techniques. *Ecological Engineering* 36, 1439–1447.

Schulz-Zunkel, C., Krueger, F., Rupp, H., Meissner, R., Gruber, B., Gerisch, M., Bork, H., 2013. Spatial and seasonal distribution of trace metals in floodplain soils. A case study with the middle Elbe river, Germany. *Geoderma* 211–212, 128–137.

Schulz-Zunkel, C., Rinklebe, J., Bork, H.-R., 2015. Trace element release patterns from three floodplain soils under simulated oxidized-reduced cycles. *Ecological Engineering* 83, 485–495.

Shaheen, S.M., Rinklebe, J., 2014. Geochemical fractions of chromium, copper, and zinc and their vertical distribution in soil profiles along the Central Elbe River, Germany. *Geoderma* 228–229, 142–159.

Shaheen, S.M., Rinklebe, J., 2015a. Phytoextraction of potentially toxic elements from a contaminated floodplain soil using Indian mustered, rapeseed, and sun flower. *Environmental Geochemistry and Health* 37 (6), 953–967.

Shaheen, S.M., Rinklebe, J., 2015b. Impact of emerging and low cost alternative amendments on the (im)mobilization and phytoavailability of Cd and Pb in a contaminated floodplain soil. *Ecological Engineering* 74, 319–326.

Shaheen S.M., Rinklebe, J., Frohne, T., White, J., DeLaune, R., 2014c. Biogeochemical factors governing Co, Ni, Se, and V dynamics in periodically flooded Egyptian north Nile delta rice soils. *Soil Science Society of America Journal* 78, 1065–1078.

Shaheen S.M., Rinklebe, J., Frohne, T., White, J., DeLaune, R., 2016. Redox effects on mobility of arsenic, cadmium, cobalt and vanadium in Wax Lake deltaic soils. *Chemosphere.* doi:doi.org.10.1016/j.chemosphere.2015.12.043

Shaheen, S.M., Rinklebe, J., Rupp, H., Meissner, R., 2014a. Lysimeter trials to assess the impact of different flood-dry-cycles the dynamics of pore water concentrations of As, Cr, Mo, and V in a contaminated floodplain soil. *Geoderma* 228–229, 5–13.

Shaheen, S.M., Rinklebe, J., Rupp, H., Meissner, R., 2014b. Temporal dynamics of soluble Cd, Co, Cu, Ni, and Zn and their controlling factor in a contaminated floodplain soil using undisturbed groundwater lysimeter. *Environmental Pollution* 191, 223–231.

Shaheen, S.M., Rinklebe, J., Selim, H.M., 2015b. Impact of various amendments on the bioavailability and immobilization of Ni and Zn in a contaminated floodplain soil. *International Journal of Environmental Science and Technology* 12 (9), 2765–2776.

Shaheen, S.M., Rinklebe, J., Tsadilas, C.D., 2015a. Fractionation and mobilization of toxic elements in floodplain soils from Egypt, Germany and Greece: A comparison study. *Eurasian Soil Sciences* 48 (12), 1317–1328.

Shaheen, S.M., Tsadilas, C.D., Rinklebe, J., 2013. A review of the distribution coefficient of trace elements in soils: Influence of sorption system, element characteristics, and soil colloidal properties. *Advances in Colloid Interface Science* 201–202, 43–56.

Shaheen, S.M., Tsadilas, C.D., Rinklebe, J., 2015c. Immobilization of soil copper using organic and inorganic amendments. *Journal of Plant Nutrition and Soil Sciences* 178 (1), 112–117.

Sheoran, V., Sheoran, A., Poonia, P., 2011. Role of hyperaccumulators in phytoextraction of metals from contaminated mining sites: A review. *Critical Reviews of Environmental Science and Technology* 41, 168–214.

Thangavel, P., Subbhuraam, C., 2004. Phytoextraction: role of hyperaccumulators in metal contaminated soils. *Proceedings of the Indian National Science Academy. Part B* 70, 109–130.

Unterbrunner, R., Puschenreiter, M., Sommer, P., Wieshammer, G., Tlustoš, P., Zupan, M., Wenzel, W.W., 2007. Heavy metal accumulation in trees growing on contaminated sites in Central Europe. *Environmental Pollution* 148, 107–114.

Vamerali, T., Bandiera, M., Mosca, G., 2010. Field crops for phytoremediation of metal-contaminated land. A review. *Environmental Chemistry Letters* 8, 1–17.

Vandecasteele, B., Du Laing, G., Quataert, P., Tack, F.M.G., 2005. Differences in Cd and Zn bioaccumulation for the flood-tolerant Salix cinerea rooting in seasonally flooded contaminated sediments. *Science of the Total Environment* 341, 251–263

Van der Veen, A., Ahlers, C., Zachmann, D.W., Friese, K., 2006. Spatial distribution and bonding forms of heavy metals in sediment along the middle course of the river Elbe (km 287...390). *Acta Hydrochimica et Hydrobiologica* 34, 214–222.

Vangronsveld, J., Herzig, R., Weyens, N., Boulet, J., Adriaensen, K., Ruttens, A., Thewys, T., Vassilev, A., Meers, E., Nehnevajova, E., Van der Lelie, D., Mench, M., 2009. Phytoremediation of contaminated soils and groundwater: Lessons from the field. *Environmental Science and Pollution Research* 16, 765–794.

Vesk, P.A., Nockold, C.E., Allaway, W.G., 1999. Metal localization in water hyacinth roots from an urban wetland. *Plant, Cell and Environment* 22, 149–159.

Vysloužilova, M., Tlustoš, P., Szakova, J., 2003. Cadmium and zinc phytoextraction potential of seven clones of *Salix* spp. planted on heavy metal contaminated soils. *PSE* 49, 542–547.

Wang, F., Qiu, L., Zheng, S.A., 2014. Kinetic approach for assessing the effect of flooding on the redistribution of heavy metals in paddy soil. *Fresenius Environmental Bulletin* 23, 113–121.

Weber, F.-A., Voeglin, A., Kretzschmar, R., 2009. Multi-metal contaminant dynamics in temporarily flooded soil under sulfate limitation. *Geochimica et Cosmochimica Acta* 73, 5513–5527.

Williams, J.B., 2002. Phytoremediation in wetland ecosystems: Progress, problems, and potential. *Critical Reviews of Plant Science* 21, 607–635.

Wright, D.J., Otte, M.L., 1999. Wetland plant effects on the biogeochemistry of metals beyond the rhizosphere. *Biology and Environment: Proceedings of the Royal Irish Academy* 99B, 3–10.

Xu, J., Peng, S., Qiao, Z., Yang, S., Gao, X., 2014. Binding forms and availability of Cd and Cr in paddy soil under non-flooding controlled irrigation. *Paddy and Water Environment* 12, 213–222

Yang, J.E., Kim, H.J., Ok, Y.S., Lee, J.Y., Park, J., 2007. Treatment of abandoned coal mine discharged waters using lime wastes. *Geoscience Journal* 11, 111–114.

Yang, J.E., Ok, Y.S., Kim, W.I., Lee, J.S., 2008. Heavy metal pollution, risk assessment and remediation in paddy soil environment: Research experiences and perspectives in Korea. In: M.L. Sanchez (ed.). *Ecological Research Progress*. New York: Nova Science Publishers, pp. 341–369.

Yang, J.E., Skousen, J.G., Ok, Y.S., Yoo, K.R., Kim, H.J., 2006. Reclamation of abandoned coal mine wastes using lime cake by-products in Korea. *Mine Water and the Environment* 25, 227–232.

Ye, X.X., Sun, B., Yin, Y.L., 2012. Variation of As concentration between soil types and rice genotypes and the selection of cultivars for reducing As in the diet. *Chemosphere* 87, 384–389.

Ye, Z.H., Baker, A.J.M., Wong, M.H., Willis, A.J., 1997a. Zinc, lead and cadmium tolerance, uptake and accumulation by Typha latifolia. *New Phytologist* 136, 469–480.

Ye, Z.H., Baker, A.J.M., Wong, M.H., Willis, A.J., 1997b. Zinc, lead and cadmium tolerance, uptake and accumulation by the common reed, Phragmites australis (Cav.) Trin. Ex Steudel. *Annals of Botany* 80, 363–370.

Ye, Z.H., Whiting, S.N., Lin, Z.Q., Lytle, C.M., Qian, J.H., Terry, N., 2001. Removal and distribution of iron, manganese, cobalt and nickel within a Pennsylvania constructed wetland treating coal combustion by-product leachate. *Journal of Environmental Quality* 30, 1464–1473.

Zayed, A., Growthaman, S., Terry, N., 1998. Phytoaccumulation of trace elements by wetland plants: I. Duckweed. *Journal of Environmental Quality* 27, 715–721.

Zhang, Z., Solaiman, Z.M., Meney, K., Murphy, D.V., Rengel, Z., 2012. Biochars immobilize soil cadmium, but do not improve growth of emergent wetland species Juncus subsecundus in cadmium-contaminated soil. *Journal of Soils and Sediments* 13,140–151.

Zimmer, D., Baum, C., Leinweber, P., Katarzyna, H., Meissner, R., 2009. Associated bacteria increase the phytoextraction of Cd and Zn from a metal contaminated soil by mycorrhizal Willows. *International Journal of Phytoremediation* 11, 200–213.

Zimmer, D., Kiersch, K., Baum, C., Meissner, R., Muller, R., Jand, G., Leinweber, P., 2011. Scale-dependent variability of As and heavy metals in a River Elbe Floodplain. *Clean—Soil, Air, Water* 39, 328–337.

14 Fate of Trace Elements in Rice Paddies

Abin Sebastian, Natthawoot Panitlertumpai, Woranan Nakbanpote, and Majeti Narasimha Vara Prasad

CONTENTS

14.1 INTRODUCTION

Rice occupies a position of overwhelming importance in the global food system. More than a third of the world's population, predominantly in Asia, depends on rice as a primary dietary staple. More than 90% of the rice in the world is produced and consumed in the Asia–Pacific Region. Increase of 51 million rice consumers are predicted annually (Papademetriou 2000). Rice cultivation is unique in nature because of management and irrigation practices. It is well known that the development of paddy soils is dependent on field management practices such as submergence, puddling, and addition of nutrients in the form of manure and fertilizers (Kögel-Knabner et al. 2010). This kind of cultivation practice leads to spatiotemporal variations in soil processes that affect trace element (TE) mobility.

Rice grains from Asian countries are reported to have TEs such as Cd above the international standard (Simmons et al. 2005; Bian et al. 2014; Zhou et al. 2014). For example, 91 polished rice samples collected from markets in China contain about 10% of Cd, Zn, and Se concentrations in samples that are higher than those of ATSNFR NY5115–2008 (Zhen et al. 2008). Therefore, toxic TE contamination in paddy soil has become a major topic of wide concern, because it is a primary source of toxic metal contamination of rice. Many of the toxic TEs are transition elements with incompletely filled *d* orbitals and a density of 5.0 g/cm^3. The *d* orbital provides toxic TE cations with the ability to form complex compounds that may or may not be redox active. Rice plants growing near mines are prone to toxic TE pollution due to the mismanagement of waste water release. Rice grain harvest from mining areas is often contaminated with TEs such as Cd (Sebastian and Prasad 2014a). Hence, an exploration of the mobility of TEs in paddy soil deserves attention. This not only helps prevent the accumulation of TEs in rice grain but also helps formulate sustainable agricultural practices that aid in bringing down the amount of plant-available TEs in the paddy soil.

14.1.1 TEs as Contaminants in Paddy Soils

A TE in biochemistry focuses on a dietary mineral that is needed in very minute quantities for the proper growth, development, and physiology of the organism (Kabata-Pendias and Pendias 2001). However, when present in excess essential functions, they may give rise to toxic manifestation even when intakes are only moderately in excess of the natural intake (Friberg and Nordberg 1986). Many TEs are essential or beneficial micronutrients for microorganisms, plants, and animals at low concentrations (e.g., Cu, Cr, Se, Zn, Fe, Mn, Mo, Ni, and Co), but they become toxic to organisms at high concentrations (Nedelkoska and Doran 2000; Munzuroglu and Geckil 2002). In addition, some TEs are also nonessential and even low concentrations in the environment can cause toxicity to both plants and animals, such as, Cd, Pb, and Hg (Kabata-Pendias and Pendias 2001). Any metal (or metalloid) species may be considered a "contaminant" if it occurs where it is unwanted or in a form or concentration that causes a detrimental human or environmental effect. TE contamination of paddy soils became severe from both natural and anthropogenic activities, such as mining or industrial activities and improper use of metal-enriched materials in agriculture, including chemical fertilizers and pesticides, industrial effluents, sewage sludge, and wastewater irrigation (Bolan et al. 2014; Ramadan and Al-Ashkar 2007).

14.1.2 Soil Processes Regulate TE Availability

The persistence of TE contaminants in the soil is much longer than in other compartments of the biosphere, and it appears to be virtually permanent (Kabata-Pendias and Pendias 2001). The first half-life of TEs for the soil varies greatly for Zn, Cu, Ni, Pb, and Se (1000–3000 years); Cd (13–1100 years); and Hg (500–1000 years) (Iimura et al. 1977; Bowen 1979). However, the residence time of TEs in soil depends on the mobility fraction of TEs. Kabata-Pendias and Pendias (2001) reported that the rate of leaching of the TEs in the soil of tropical rain forests is much shorter than the soil of temperate climate and the soil in lysimetric condition. In general, the fate of TEs in paddy soils depends on the soil process, such as adsorption and desorption characteristics (Krishnamurti et al. 1999). The adsorption and desorption of TEs are governed by soil properties, including pH, organic matter content, cation-exchange capacity (CEC), oxidation reduction status (E_h), the contents of clay minerals, calcium carbonate, and metal oxides (Kabata-Pendias and Pendias 2001; Alloway 1995; Kashem and Singh 2001; Antoniadis et al. 2008). For example, with low pH, the TE availability and dissolution in soil solution were observed for Cd, Pb, and Zn (Sukreeyapongse et al. 2002; Bang and Hesterberg 2004). In addition, toxic TE concentrations in surface soil are likely to increase, on a global scale, with growing industrial and agricultural activity (Kabata-Pendias and Pendias 2001). Clay hydroxides assist TE retention in paddy soils. The adsorption of TEs onto clay acts as a barrier to cation exchange. This accounts for the relatively lower mobility of nutrient ions in alfisol having more clay content that assists the mobilization of toxic TEs such as Cd compared with vertisol (Sebastian and Prasad 2014d). Soil that is contaminated with TEs can produce apparently normal crops (rice, soybean, wheat, maize, and vegetables), which may be unsafe for human or animal consumption (Arunakumara et al. 2013). The accumulation effect strongly depends on the physiological properties of the crop, the mobility of the TEs, and the availability of TEs in soils but not entirely on the total TE concentrations in the soils (Liu et al. 2005).

TEs undergo temporal cycling in paddy soils (Semple et al. 2004). Most often, these elements are adsorbed to compounds, especially oxides of Fe or Mn or Al in the soil. This process is one of the major immobilizing processes of TEs in submerged paddy soils. For example, the precipitation and coprecipitation reactions mainly limit the dissolution of Cd in paddy soils, as the occurrence of free Cd ions is not controlled by chemical solubility equilibrium with the mineral phase. Dissolved soil organic matter often forms a complex with TEs, which makes them soluble in the soil. The increase in the rhizospheric zone also enhances the solubility of ions such as Cd because of the root-mediated process. Metalloids such as As also contaminate paddy soils. Arsenic contamination is often associated with the application of As-containing pesticides. Arsenic occurs as the anionic form $H_2AsSO_4^-$ in soil solution. This causes the absorption of As by hydrous ions and other metal oxides, especially in acidic soil. The potential immobilizing agents of As include sulfate salts of Zn, Fe, and Al.

Biological activities in the soil also play a significant role in the mobilization of toxic TEs, especially siderophore-producing and plant-growth-promoting bacteria (Rajkumar et al. 2012). Rice plants growing in submerged conditions favor radical oxygen loss (diffusion of oxygen to the rhizosphere from aerenchyma tissue in roots) in the rhizosphere that forms an oxidative layer around roots. This prevents toxicity due to both toxic TEs such as Cd^{2+} and nutrient elements such as Fe^{2+}. However, radical oxygen loss enhances the mobility of TEs in soils. This is due to a decrease of soil pH during oxidation reactions in the rhizosphere. The metabolic fate and role of TEs in rice plants related with many processes operate in plants, for instance, in the form of uptake (absorption), transport and translocation within a plant, enzymatic processes, distribution of heavy metal in parts of plants, deficiency and toxicity, and competition and interaction of TEs (Kabata-Pendias and Pendias 2001; Greger 2004). The mechanisms of uptake of TEs involve processes of adsorption/sequestration by numerous reactive groups and constituents in apoplast and transport crossing the plasma membrane by chelating agents or other carriers in the symplast (Greger 2004; Peer et al. 2006). Roots and associated microorganisms in the rhizosphere are known to produce various amino-acid (e.g., aspartic, glutamic, prolinic) and organic compounds (e.g., malic acid, maleic acid, citric acid, oxalic acid, succinic acid, acetic acid, and formic acid), which are important in the chelation of TE cations from firmly fixed species in soils; influence nutrient availability; and assist in nutrient uptake by plants (Wenzel et al. 2004). Dong et al. (2007) reported that plant roots excrete some organic compounds to the rhizosphere under TE stress, and the rhizosphere controls the entrance of metals, such as Cd, to the plant. The transport and translocation of TEs within tissues and organs involves cation exchange, transpiration of water via leaves, and immobilization at the xylem (Kabata-Pendias and Pendias 2001; Greger 2004). The distribution and accumulation patterns of TEs vary considerably for each element, kind of plant, and growth season. The functions and forms of the elements in plants involve their incorporation into structural materials and their ability to relate to organelles or their parts (e.g., mitochondria, chloroplast, enzyme systems) (Cobbett and Goldbrough 2002; Prasad 2004; Peer et al. 2006).

14.1.3 Irrigation and Temporal Changes in TE Solubility

Rice cultivation is either wetland or dry land culture. However, only 10%–15% of the area planted to rice in South and Southeast Asia uses dry land rice culture. In Asia, most rice is grown under wetland or flooding rice field conditions, because dry land rice generally has a shorter growing season, low levels of productivity, and lower resistance to drought (Barker et al. 1985). In wetland rice culture, rice is directly seeded or transplanted in the paddy fields after flood establishment through an enclosure with an earth levee or dyke to contain the water. Water is supplied by natural rainfall, floodwater, runoff from higher ground, or irrigation, and the fields typically remain flooded throughout much of the growing season, depending on rainfall or water availability (Barker et al. 1985). In a rice paddy field located near a zinc mine in Mae Sot, Tak province, Thailand (Figure 14.1), farmers grow rice in the way of wetland culture (Figure 14.2). Ploughing enhance plant available TEs pool in the rice field (Figure 14.2a). The runoff and irrigation through drained and flooded rice fields in that area (Figures 14.1 and 14.3) are the reason behind zinc and cadmium contaminating the rice fields (Simmons et al. 2005).

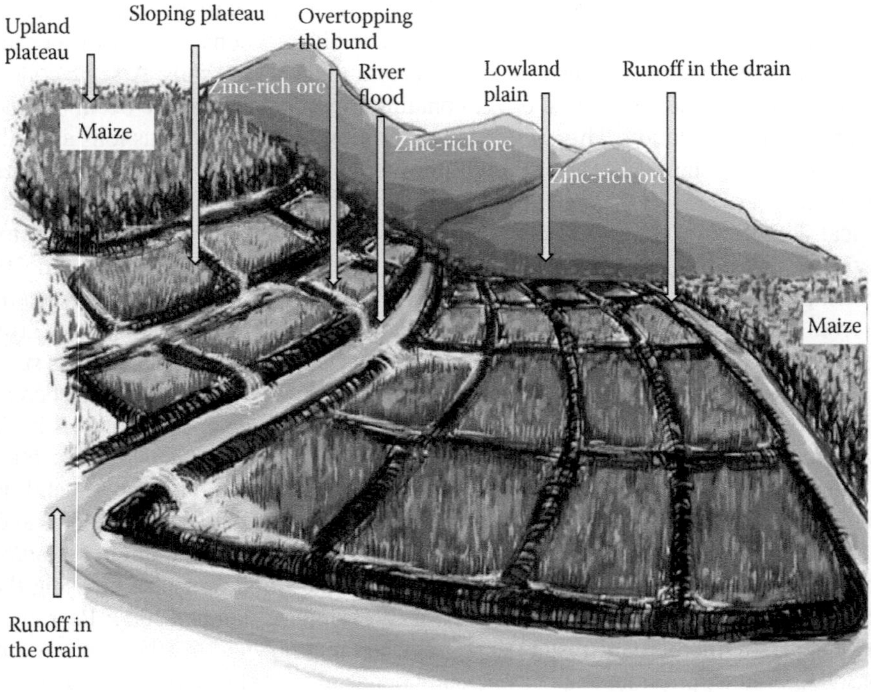

FIGURE 14.1 **(See color insert.)** Schematic sketch of the location of a rice paddy field (land use and irrigation) in Mae Sot, Tak province, Thailand.

FIGURE 14.2 **(See color insert.)** Rice farming in Mae Sot, Tak province, Thailand. (a) Rice field plow up with a driller operated manually. (b) Leveling of puddled rice field by hand (with a hoe). (c) Broadcasting rice seeds manually in the prepared field. (d) Germinated rice seedlings and tillers. (e) Rice and maize fields.

FIGURE 14.3 **(See color insert.)** Rice paddy fields near Mae Sot, Tak province (a) and (b) water runoff through drain; (c) and (d) flooded rice fields.

Flooding and draining cycles modify the physiochemical properties of the paddy soil (e.g., pH, E_h, and soil organic matter) and cause dynamics in the solubility of TEs. Figure 14.4 shows a schematic representation of various mechanisms that are used for the mobilization and immobilization of metals in rice paddies during the water-drained period and the flooded period. The drained period provides aerobic conditions to soil microbes (Figure 14.4a). Aerobic microorganisms can metabolite organic carbon to CO_2, and they can produce acid (H^+) to solute metals ions (Me^{n+}) and TEs in the soil. Root exudates such as organic acids and phenolic compounds also affect TEs and Me^{n+} solubility. Microbes adsorb and desorb TEs and Me^{n+}, which results in both a decrease and an increase in the amount of plant-available TEs. Ion exchange between H^+ and Me^{n+} can occur on a clay surface. Under flooding conditions (Figure 14.4b), anaerobic microbes can proliferate, generating energy by the catabolism of organic compounds and reducing electron acceptors other than O_2, such as Fe^{3+}, and sulfate (SO_4^{2-}). Some of the products in these processes are phytotoxic. For example, Fe^{2+} and Me^{n+} are particularly soluble under reducing conditions and they are taken up by plants in toxic amounts. Hydrogen sulfide (H_2S) produced during hypoxia inhibits the activity of cytochrome *c* oxidase in mitochondria, blocks energy production, and negatively affects a range of metal-containing enzymes (Atwell et al. 1999; Lamers et al. 2013). Reduced forms of nitrogen also become dominant during flooding conditions when denitrification of nitrate (NO_3^-) to gaseous N_2 occurs, and ammonium (NH_4^+) often accumulates after the mineralization of organic nitrogen. Denitrification causes the loss of nitrogen in the soil. After prolonged waterlogging, the gas methane (CH_4) is produced as a consequence of the reduction of organic compounds. Me^{n+} can react with H_2S to become as immobile as metal sulfide (MeS).

Microbiological processes in the paddy soil induce pH alteration, siderophore production, and modification of the physical environment (e.g., changing the redox potential); they release organic compound association through mineralization (Tate 2000). In addition, flooding and draining cycles, rice root dehydrogenase activity might be ascribed to increasing the solubility and bioavailability of metals (e.g., Ca, K, Mg, Cd, Cu, and Pb) (Liao et al. 2013). Therefore, the accumulation of TEs can be explained by the following hypotheses: (1) with a weak ability to selectively extract essential metal from the soil, rice will take up nonessential (Si, Co) and even highly toxic (Hg, Pb, Cd); (2) after flooding, water will increase the solubility and bioavailability of TEs; and

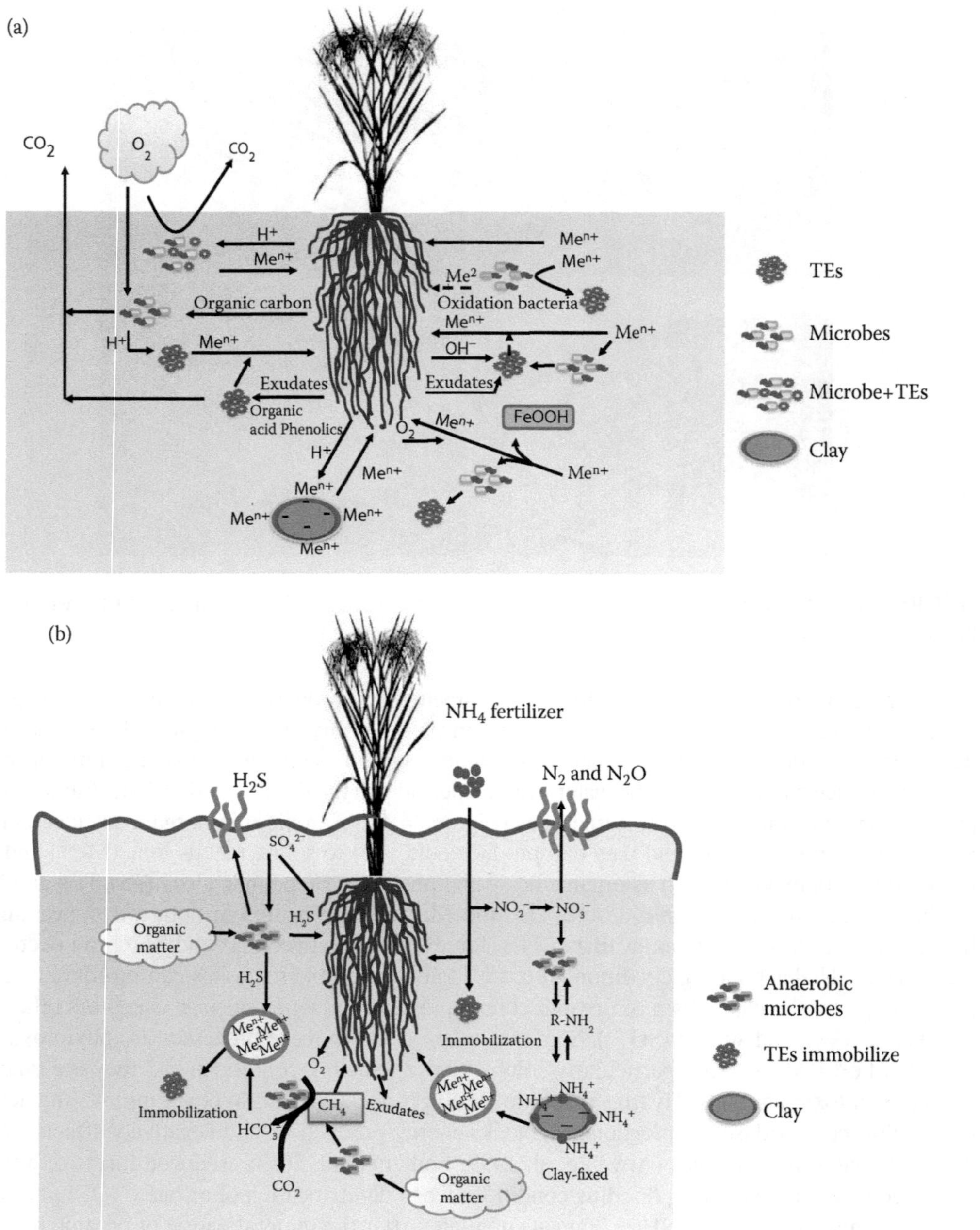

FIGURE 14.4 Schematic representation of various mechanisms for mobilization and immobilization of metals in rice paddies during water drained period (a) and flooded period (b).

(3) the biological process in the rhizosphere, such as root exudates, can associate with the mineralization process. Many scientists have proposed processes to decrease the accumulation of toxic TEs in rice grain. The processes can be separated into four groups: (1) soil amendments by organic matter, inorganic matter, char, mixture, and so on; (2) bioaugmentation of bacteria to promote plant growth and/or to reduce the phytoavailability of toxic TEs; (3) cultivar selection of rice species; and (4) cultivation management (Sebastian and Prasad 2014a; Prasad et al. 2014). Examples of the processes are shown in Table 14.1.

TABLE 14.1
Examples of Processes to Decrease the Accumulation of Heavy Metals in Rice Growing in Heavy Metal-Contaminated Soil

Method	Experiment	Metals	Country/ Source of Metal	Results	References
Soil Amendments					
Biochar	Field	Cd, Pb	China, Yixing municipality, Jiangsu province	↓ [Cd, Pb] in rice plant tissues —Al, Fe, and P on biochar particles might bond with Cd and Pb.	Bian et al. (2014)
Biochars from rice residue	Greenhouse	Cd, Zn, Pb, As	China, Hunan province	↓ [Cd, Zn, Pb], ↑ [As] transport to rice shoots —Soil pore water influenced solubility of Cd, Zn, Pb, and As.	Zheng et al. (2012)
Limestone + sepiolite and hydroxyhistidine + zeolite	Field	Pb, Cd, Cu, Zn	China, Shuikoushan Mine Zone	↓ [Pb, Cd, Cu, Zn] in brown rice by 10.6%–31.8%, 16.7%–25.5%, 11.5%–22.1%, and 11.7%–16.3% —Amendments inhibited uptake and accumulation of Pb, Cd, Cu, and Zn in rice plants.	Zhou et al. (2014)
Ameliorants (zero-valent iron, lime, humus, compost, and combinations)	Greenhouse	Cd	Korea, Chung-buk province	↓ bioavailable Cd fractions 50%–90% ↓ [Cd] uptake by rice —Ameliorants did not influence rice yield.	Ok et al. (2011a)
Silicon-rich amendments	Pot and field	Cd, Cu, Zn, Pb	China, Guangdong province	↓ phytoavailability of Cd, Cu, Zn, and Pb by at least 60% ↑soil pH from 4.0 to 5.0–6.4 —The mobile metals were mainly deposited as their silicates, phosphates, and hydroxides.	Gu et al. (2011)
Eggshell waste	Pot	Cd, Pb	Korea, Seoseong mine, Seosan-si, Chungnam province	↓ [Cd, Pb] uptake in rice —Eggshell waste can be used as an alternative to $CaCO_3$ for the immobilization of heavy metals in soils.	Ok et al. (2011b)
Organic fertilizer	Pot	Cd	Thailand, Mae Sot District, Tak province	↓ phytoavailability of Cd ↓ bioavailability by fertilizer amendment, a concentration of organic fertilizer in the level ranging between 1000 and 2000 kg/rai	Sampanpanish and Pongpaladisai (2011)

(Continued)

TABLE 14.1 (*Continued*)
Examples of Processes to Decrease the Accumulation of Heavy Metals in Rice Growing in Heavy Metal-Contaminated Soil

Method	Experiment	Metals	Country/ Source of Metal	Results	References
Planting oilseed rape + compost	Pot	Cd, Pb	China, Jiaxing, Zhejiang province	↓ 46%–80% and 17%–86% (Cd, Pb) grains of the rice rotated with oilseed rape ↓ bioavailability and uptake by compost amendment	Wu et al. (2011)
Agricultural residues (coir pith, coir pith + NaOH, and corncob)	Pot	Cd	Thailand, Mae Sot District, Tak province	↓ [Cd] uptake in rice —Coir pith and corncob reduce Cd uptake.	Siswanto et al. (2013)
Bioaugmentation					
Bioaugmentation of bacteria	Filter papers	Cd	Thailand, Mae Sot District, Tak province	↓ [Cd] accumulation in rice ↑ Cd tolerance, ↑ growth seedlings and numbers of fibrous roots —Isolate KKU2500-3; transform $CdCl_2$ into nontoxic, insoluble CdS.	Siripornadulsil and Siripornadulsil (2013)
Cultivar Selection					
Selection of rice species	Field	Hg	China (eastern Guizhou province, Hunan province, and Guangdong province)	—Cultivar tended toward stability in Hg accumulation across sites. —Some cultivars accumulated low [Hg]. —Cultivar selection ↓ [Hg] in seeds of rice grown in Hg-contaminated regions.	Li et al. (2013)
Select rice cultivation	Pot	Cd	China, Shenyang Ecological Experimental Station of the Chinese Academy of Science	—Four cultivars of rice: Shendao 5, Tianfu 1, Fuhe 90, and Yanfeng 47 showed Cd-exclusive characteristic and better foreground application.	Zhan et al. (2013)
Select rice cultivation	Pot	Pb	China	—[Pb] in ears and grains were in the order *Hybrid Indica* > *Indica* > *Japonica*. —Translocation factors were in the order *Hybrid Indica* > *Indica* > *Japonica*.	Liu et al. (2013)

TABLE 14.1 (*Continued*)
Examples of Processes to Decrease the Accumulation of Heavy Metals in Rice Growing in Heavy Metal-Contaminated Soil

Method	Experiment	Metals	Country/ Source of Metal	Results	References
Select rice cultivation	Pot	Cd	Thailand, Mae Sot District, Tak province	—[Cd] accumulated in seeds and branches were in the order of Chainart 1< Supanburee 1 < Phitsanulok 2 < Khao Dawk Mali 105. —Root system and transpiration rate involved Cd accumulation in shoot parts.	Boonjun (2005)
Cultivation Management					
Rice straw ash amendment with flooding condition	HDPE box	Cu	Taiwan, Changhua Country	↓ [Cu] uptake in rice —Rice straw ash enhances the changes in pH and redox potential of the flooded soils, and it results in immobilization of Cu in the soils.	Huang et al. (2011)
Flooding and nonflooding conditions with addition of organic matter	Pot	Cd, Ni, Zn	Bangladesh and Norway	—Flooding condition ↓ [Cd, Ni, Zn] in rice grown —Organic matter ↓ [Ni], no reduction was seen in [Cd, Zn]. —[Cd, Zn] was 82% and 55% higher than that of [Ni] in the plant.	Kashem and Singh (2001)
Flooding and draining conditions	Field	Cd, Zn	Thailand, Mae Sot District, Tak province	—At least 14 mg/kg of Cd and 22% of Zn were desorbed from the soil. —At least 50% of Cd was bound to humic acid. —Zn is immobilized in the phase of Zn-layered double hydroxides (Zn/Mg–hydrotalcite like and ZnAl–LDH) and Zn–phyllosilicates (Znkerolite).	Khaokaew et al. (2011, 2012)

(Continued)

TABLE 14.1 (*Continued*)
Examples of Processes to Decrease the Accumulation of Heavy Metals in Rice Growing in Heavy Metal-Contaminated Soil

Method	Experiment	Metals	Country/ Source of Metal	Results	References
Flooding and aerobic conditions	Pot	As	England, Rothamsted farm	—Flooding of soil led to a rapid mobilization of As, mainly as arsenite, in the soil solution. —[As] in soil by flooding > aerobic conditions. —As accumulation by flooded rice and growing rice aerobically can dramatically decrease the As transfer from the soil to grains.	Xu et al. (2008)
Growing rice after hyperaccumulation	Pot	As	India (West Bengal and Bangladesh) and China (Chenzhou and Qiyang)	—Rice grown after *P. vittatahad* significantly lower [As] in straw and grain, being 17%–82% and 22–58%	Ye et al. (2011)
Irrigation and soil washing	Field	Cd	Thailand, Mae Sot District, Tak province	↓[Cd] rice grains	Sukreeyapongse et al. (2010)

14.1.4 Field Management Influences TE Utilization

Field management in paddy soils mainly focuses on fertilization and crop rotation. Nutrient availability stands among the key factors that limits the grain yield of rice plants. Hence, the fertilization of rice paddies is essential for meeting grain demand. The major nutrient that limits grain yield is N. This makes N fertilization inevitable in rice paddies. Phosphorous and potassium are also reported to constrain rice plant growth (Mutert and Fairhurst 2002). Hence, the application of fertilizers in paddy soil is often a mixture of NPK. The chemical form of fertilizers and the ability of fertilizers to participate in chemical reactions play an important role in the mobility of toxic TEs. The acidifying effect of many of the mineral nitrogen fertilizers often enhances the mobility of toxic TEs in paddy soils. On the other hand, N fertilizers with buffering activity help retain soil pH without causing soil acidification. The nitrate in the soil solution reacts with cations of TEs, which increase soil acidity by depleting the cation pool (Su and Puls 2004). The ability among nitrogen fertilizers to cause soil acidification is in the decreasing order, where ammonium sulfate is followed by ammonium phosphate, ammonium nitrate, and urea. It is noteworthy that the fertilization with an ammonium-based fertilizer reduces Cd translocation by assisting photosynthetic-dependent processes of rice plants grown in vertisol, even though the treatment shows a tendency to enhance Cd accumulation in roots (Sebastian and Prasad 2014b). Thus, ammonium fertilizers help in reducing the root-to-grain allocation of toxic TEs.

Rock phosphates, which are ingredients of the NPK fertilizers, are sources of toxic TEs in rice paddies (Sandalio et al. 2001). This is because of the higher level of metal in parent rock material being used for the manufacture of fertilizers. On the other hand, phosphates in the paddy soil react with toxic TEs such as Cd and precipitate the metal. This makes Cd unavailable for plant uptake. Diphosphate is better than trisuperphosphate in reducing plant-available toxic TEs, because it causes efficient sorption of metal onto the soil (McLaughlin et al. 1995). It is noticeable that the application of phosphate has a disadvantage in nutrient-rich soil with regard to reducing Cd uptake by plants, because P decreases Zn uptake, which enhances Cd uptake into plants. It is also reported that the presence of essential nutrients such as Ca, Zn, Mn, Fe, and Cu inhibits Cd uptake in plants (Sarwar et al. 2010). This points toward the potential of the application of micronutrients in the field to decrease Cd accumulation in rice.

Biodynamic farming is often practiced in rice cultivation, especially in developing countries such as India. The application of manure such as cow dung not only increases nutrients in the field but also helps improve biological and physiochemical processes in the soil that mobilize TEs. Another important benefit of this kind of agricultural practice is the elevation of organic matter contents in the field. An increase in organic matter causes the adsorption of toxic TEs, resulting in a decrease in the availability of these metals for plant uptake (Brams and Anthony 1983). However, the degradation of organic matter has a reverse effect. It is noticeable that dissolved organic matter often increases the chance of reducing events. Thus, the prevalence of reducing the state of the soil after flooding through the incorporation of organic matter increases the chance of the binding of toxic TEs to organic matter and reduces the plant-available toxic TEs in paddy soils (Sauve et al. 2003).

Crop rotation with legume also has a significant role in determining plant-available heavy metals in paddy fields. It is well known that the soil surrounding the legume rhizosphere gets acidified through the nitrogen fixation process and favors the plant uptake of toxic TEs (Rao et al. 2002). So, the growth of legume in the field helps mobilize toxic TEs. Another field management practice that is characteristic of paddy fields is the puddling. The puddled layer and the plow pan form a reducing soil horizon due to the segregation of the Fe and Mn matrix, which could precipitate the soluble fraction of toxic TEs (So and Ringrose-Voase 2000). The addition of soil amendments is considered a practical solution against TE contamination in rice fields. Amendments cause a partitioning of TEs between solid and liquid phases of the soil. The addition of organic amendments enhanced the toxic TE-binding properties of the soil by approximately 30 times more than mineral soils (Sauve et al. 2000). Organic amendments are characterized by the presence of lignin, cellulose, tannins, and carbonates that increase the natural capability of the soil to retain toxic TEs. These amendments often form a coating over particulate matter that acts as a metal-binding agent. The addition of clay materials that are characterized by high metal adsorption capacity due to high specific surface area, high CEC, Brönsted and Lewis acidities, and flexible surface charge against pH in the field also enhance the immobilization of TE (Sebastian and Prasad 2014a).

14.1.5 Root Architecture and TE Mobilization

Rice roots play a critical role in the survival of adverse environmental conditions such as heavy metal stress. Being monocot plants, rice plants are characterized by the presence of adventitious roots. This kind of root often acts as a stress avoidance mechanism during nutrient deficiency and anoxia. In a typical rice root development program, the radical gives rise to a seminal root that is short-lived and replaced by a number of equal-sized adventitious roots that are known as *crown roots*. These roots arise from nodes of the stem, and each crown root (primary root) gives rise to large and short lateral roots (secondary roots). These roots are the most important component of the rice root system that assists in water and nutrient acquisition. The entire rice root system is often referred to as the *fibrous root system*. The development of adventitious roots is controlled by genes such as *RAL1*, which inhibit radical elongation, and *crl1*, which initiate crown roots (Wu and Cheng 2014). The extensive fibrous root system clings to the soil particles and prevents paddy soil erosion. Thus, the morphological development of roots of rice plants plays a crucial role in the outflow of TEs from the paddy soil through soil erosion.

Rice varieties often show differences in root density. The differences between root densities of deep-rooted and shallow-rooted varieties are commonly found in the soil layers that are deeper than 30 cm below the soil surface. It is noticeable that rice plants with few and early tillers tend to have a deep root system. The variation in the depth of the root system leads to a difference in tolerance to water deficit and nutrient stress among rice varieties. Rice plants with a higher root biomass tend to accumulate more TEs. Thus, agricultural practices that enhance root biomass also enhance a higher rate of mobilization of TEs from the soil to plants. This phenomenon has been exemplified in the case of toxic TE accumulation too. It is well known that vermicompost enhances soil properties such as water-holding capacity and soil porosity. Both these properties help in accelerating root growth during compost application in the field. It has been reported that vermicompost stimulated root growth, leading to more Cd accumulation in rice plants (Sebastian and Prasad 2013). On the other hand, sufficient availability of nutrients in the rhizosphere that retards the lengthening of roots reduced Cd accumulation in rice plants. The application of synthetic fertilizers also often reduces the lengthening of roots in search of nutrients. Studies with rice revealed that ammonium fertilizer reduces root growth and helps in reducing Cd accumulation in rice varieties (Sebastian and Prasad 2014b). It is documented that rice varieties with a deep root system tend to accumulate more Cd than those with a shallow root system in the study mentioned earlier. Thus, genetic manipulation that brings changes in the depth and density of rice roots is highly promising for the mobilization of TEs in paddy soils. It is reported that the depth of roots increases with the expression of the *DRO1* gene (Uga et al. 2013). The *DRO*1 gene is negatively regulated by auxin and assists cell elongation in the root tip. This results in asymmetric root growth and downward bending of the root in response to gravity. Hormones such as auxins, cytokinin, and jasmonic acid also take part in the development of rice roots. External application of auxin and cytokinin has been found to induce a gene called *WOX11* that is involved in crown root development and cell division (Wu and Cheng 2014). This indicates that hormone treatment that stimulates root growth can play a role in the mobilization of TE through a root-mediated process. The development of aerenchyma and related radical oxygen-loss–mediated mobility of TEs are also matters of concern in the cycling of TEs in paddy soils.

14.1.5.1 Role of Root Metabolism

Rice roots metabolically respond to metals. Tricarboxylic acids (TCAs) such as citrate, oxalate, and malate are reported to be secreted into the rhizosphere as a tolerance mechanism against toxic metals such as Al (Kochian et al. 2004). It is the presence of the anionic carboxylic group that enables these acids to bind with cations such as Al^{2+} and Cd^{2+}. The binding of Cd with these organic acids prevents the involvement of Cd in redox reactions in the cell. Apart from this organic acid, bound Cd is reported to be transported into vacuoles in the cells (Tong et al. 2004). It is well known that respiration in root cells of plants increase during TEs stress. This indicates that the triggering of cellular metabolism in favor of organic acid synthesis helps in chelating toxic TEs. The enhancement of photosynthesis helps channel sugars for respiratory metabolism, and hence enhances organic acid synthesis. It has also been reported that the increase of organic acid content in the roots enhance Cd rhizocomplexation in rice seedlings (Sebastian and Prasad 2014c). Being the precursor pathway of carbon fixation, photosynthesis ensures the supply of carbon compounds for the TCA cycle. It is the TCA cycle that modulates the cellular flux of organic acids such as citrate and malate. Hence, the respiratory peak that is observed in plants during heavy metal exposure helps alleviate the metal-induced changes in plant metabolism by enhancing metal chelation (Keunen et al. 2013).

Rice root metabolism is also characteristic with respect to Fe uptake. Fe uptake into the rice plant is through either Type I or Type II (Kobayashi and Nishikawa 2012). During Type I strategy, rice roots acidify the rhizosphere with the help of H^+-ATPase and secrete phenolics into the rhizosphere. This makes Fe^{3+} compounds soluble in soil solution and enables them to form a complex with phenolics. Ferric chelate reductase–oxidase (FRO) present in the root plasma membrane further converts Fe^{3+} to Fe^{2+}, and this leads to the release of Fe^{2+} from phenolic complexes. Free Fe^{2+} ions are taken up by plants through iron-regulated transporter 1 (IRT1) transporters into the root plasma membrane.

Being cytotoxic, Fe^{2+} ions are converted into Fe^{3+} ions with the help of ferroxidase (FOX) enzyme in the cytoplasm. Strategy II plants secrete muigenic acid derivatives called *phytosiderophores*. These compounds chelate Fe^{3+} in the soil and are able to cross the plasma membrane of roots with the help of YS1/YSL transporters. Once inside the cell, the Fe–phytosiderophore complex is broken up and Fe^{3+} is released into the cytoplasm. Fe^{3+} in the cytoplasm is trafficked into various plant parts. It is clear that both types of Fe uptake take place with the help of metal chelators. The wide range of substrate specificity of these metal-chelating ligands promotes the solubility of TEs other than Fe. Apart from this, metals other than Fe that can bind metal chelators with a high affinity cause a disturbance in the mobility of TEs in paddy soils. The activity of alcohol dehydrogenase enzymes is another noticeable metabolic alteration in rice roots that is commonly seen in flooded rice paddies (Xie and Wu 1989). These enzymes aid in the regeneration of nicotinamide adenine dinucleotide (NADH) during anoxia. It is reported that this enzyme takes part in the prevention of cytosolic acidosis due to lactate dehydrogenase activity during anoxia. Alcohol dehydrogenase enzyme contains Zn as a cofactor, and the formation of sparingly soluble ZnS in the flooded paddy field reduces the activity of this enzyme. This causes a reduction in plant growth and the mobilization of TEs from the soil to plants.

14.2 OUTLOOK

TEs in the mobile phase of paddy soils require periodic monitoring. Rice paddies near mines pose the risk of toxic TE contamination, and hence sustainable agricultural practices that reduce accumulation of these metals need to be practiced (Sebastian and Prasad 2014a). Irrigation with the help of various natural and artificial water flow regulators helps to not only reduce water requirement but also modulate TE solubility. The changes in the soil physiochemical characteristics in relation to the mobility of TEs during the application of soil amendments in paddy fields must be characterized. Field management practices such as crop rotation help in reducing the risk of metal accumulation in edible crops such as rice. With the root being the mining agent of TEs in soil, a better understanding of rice root architecture and the mobilization of TEs through the root-mediated biochemical process will help in regulating the mobility of TEs though rhizosphere-mediated processes.

ACKNOWLEDGMENTS

We would like to acknowledge the receipt of financial support under the auspices of India–Thailand bilateral scientific cooperation reference DST/INT/THAI/P-02/2012 dated January 31, 2013.

REFERENCES

Alloway, B.J. (1995) Soil processes and the behavior of heavy metals. In: Alloway, B.J. (ed.), *Heavy Metals in Soils*, Second edition. New York: Blackie, pp. 11–37.

Antoniadis, V., Robinson, J.S. and Alloway, B.J. (2008) Effects of short-term pH fluctuations on cadmium, nickel, lead, and zinc availability to ryegrass in a sewage sludge-amended field, *Chemosphere*. 71, 759–764.

Arunakumara, K., Walpola, B.C. and Yoon, M.H. (2013) Agricultural methods for toxicity alleviation in metal contaminated soils: A review, *Korean Journal of Soil Science and Fertilizer*. 46, 73–80.

Atwell, B.J., Kriedemann, P.E., Turnbull, C.G.N., Eamus, D., Bieleski, R.L. and Farquhar, G. (1999) *Plants in Action*, Melbourne, Australia: Macmillan Education Australia Pty Ltd.

Bang, J. and Hesterberg, D. (2004) Dissolution of trace element contaminants from two coastal plain soils as affected by pH, *Journal of Environmental Quality*. 33, 891–901.

Barker, R., Herdt, R.W. and Rose, B. (1985) *The Rice Economy of Asia*, Washington, DC: Resources for the Future, Inc.

Bian, R., Joseph, S., Cui, L., Pan, G., Li, L., Liu, X., Zhang, A., Rutlidge, H., Wong, S., Chia, C., Marjo, C., Gong, B., Munroe, P. and Donne, S. (2014) A three-year experiment confirms continuous immobilization of cadmium and lead in contaminated paddy field with biochar amendment, *Journal of Hazardous Materials*. 272, 121–128.

Bolan, N., Kunhikrishnan, A., Thangarajan, R., Kumpien, J., Park, J., Makino, T., Kirkham, M.B. and Scheckel, K. (2014) Remediation of heavy metal(loid)s contaminated soils—To mobilize or to immobilize? *Journal of Hazardous Materials*. 266, 141–166.

Boonjun, P. (2005) Cadmium accumulation in rice: Chainat 1, Supanburee 1, Pitsunulok 2 and KhaoDawkMali 105 varieties planted in soil sample collected from paddy field in TambonPra Thad Pha-Dang, Mae Sod District, Tak province. Master of Science Thesis. Naresuan University (in Thai).

Bowen, H.J.M. (1979) *Environmental Chemistry of the Elements*, New York: Academic Press.

Brams E. and Anthony W. (1983) Cadmium and lead through an agricultural food chain, *Science of the Total Environment.* 28, 295–306.

Cobbett, C. and Goldsbrough, P. (2002) Phytochelatins and metallothionens: Roles in heavy metal detoxification and homeostasis, *Annual Review of Plant Biology*, 53, 159–182.

Dong, J., Mao, W.H., Zhang, G.P., Wu, F.B. and Cai, Y. (2007) Root excretion and plant tolerance to cadmium toxicity—A review, *Plant Soil Environment*. 53, 193–200.

Friberg, L. and Nordberg, G.F. (1986) Introduction, In: *Handbook on the Toxicology of Metals*, Amsterdam, the Netherlands: Elsevier Science.

Greger, M. (2004) Metal availability, uptake, transport and accumulation in plants. In: Prasad, M.N.V. (ed.), *Heavy Metal Stress in Plants from Biomolecules to Ecosystems*, First edition. Hyderabad, India: Springer-Verlag Berlin Heidelberg, pp. 1–27.

Gu, H.H., Gu, H., Tian, T., Zhan, S.S., Deng, T.H.B., Chaney, R.L., Wang, S.Z., Tang, Y.T., Morel, J.L. and Qiu, R.L. (2011) Mitigation effects of silicon rich amendments on heavy metal accumulation in rice (*Oryza sativa* L.) planted on multi-metal contaminated acidic soil, *Chemosphere*. 83, 1234–1240.

Huang, J.H., Hsu, S.H. and Wang, S.L. (2011) Effects of rice straw ash amendment on Cu solubility and distribution in flooded rice paddy soils, *Journal of Hazardous Materials*. 186, 1801–1807.

Iimura, K., Ito, H., Chino, M., Morishita, T. and Hirata, H. (1977) Behavior of contaminant heavy metals in soil-plant system, *Proceedings of the International Seminar on Soil Environment and Fertility Management in Intensive Agriculture (SEFMIA)*, Japan, pp. 357–368.

Kabata-Pendias, A. and Pendias, H. (2001) *Trace Element in Soils and Plants*, Third edition, Boca Raton, FL: CRC Press.

Kashem, M.A. and Singh, B.R. (2001) Metal availability in contaminated soils: II. Uptake of Cd, Ni and Zn in rice plants grown under flooded culture with organic matter addition, *Nutrient Cycling in Agroecosystems*. 61, 257–266.

Keunen, E., Peshev, D., Vangronsveld, J., Van den Ende, W. and Cuypers, A. (2013) Alternative respiration as a primary defense during cadmium-induced mitochondrial oxidative challenge in *Arabidopsis thaliana*, *Environmental and Experimental Botany.* 91, 63–73.

Khaokaew, S., Chaney, R.L., Landrot, G., Ginder-Vogel, M. and Sparks, D.L. (2011) Speciation and release kinetics of cadmium in an alkaline paddy soil under various flooding periods and draining conditions, *Environmental Science and Technology*. 45, 4249–4255.

Khaokaew, S., Landrot, G., Chaney, R.L., Pandya, K. and Sparks, D.L. (2012) Speciation and release kinetics of zinc in contaminated paddy soils, *Environmental Science and Technology*. 46, 3957–3963.

Kobayashi, T. and Nishizawa, N.K. (2012) Iron uptake, translocation, and regulation in higher plants, *Annual Review of Plant Biology.* 63, 131–152.

Kochian, L.V., Hoekenga, A.O. and Pineros, A.M. (2004) How do crop plants tolerate acid soils? Mechanism of aluminium tolerance and phosphorous efficiency, *Annual Review of Plant Biology.* 55, 459–493.

Kögel-Knabner, I., Amelung, W., Cao, Z., Fiedler, S., Frenzel, P., Jahn, R., Kalbitz, K., Kölbl, A. and Schloter, M. (2010) Biogeochemistry of paddy soils, *Geoderma.* 157 (1–2), 1–14.

Krishnamurti, G.S.R., Huang, P.M. and Kozak, L.M. (1999) Sorption and desorption kinetics of cadmium from soils: Influence of phosphate, *Soil Science*. 164, 888–898.

Lamers, L.P., Govers, L.L., Janssen, I.C., Geurts, J.J., Van der Welle, M.E., Van Katwijk, M.M., Van der Heide, T., Roelofs, J.G. and Smolders, A.J. (2013) Sulfide as a soil phytotoxin—A review, *Frontiers in Plant Science*. 4, 1–14.

Li, B., Shi, J.B., Wang, X., Meng, M., Huang, L., Qi, X.L., He, B. and Ye, Z.H. (2013) Variations and constancy of mercury and methylmercury accumulation in rice grown at contaminated paddy field sites in three Provinces of China, *Environmental Pollution*. 181, 91–97.

Liao, L., Xu, J., Peng, S., Qiao, Z. and Gao, X. (2013) Uptake and bioaccumulation of heavy metals in rice plants as affect by water saving irrigation, *Advance Journal of Food Science and Technology*. 5, 1244–1248.

Liu, H., Probst, A. and Liao, B. (2005) Metal contamination of soils and crops affected by the Chenzhou lead/zinc mine spill (Hunan, China), *Science of the Total Environment*. 339, 153–166.

Liu, J., Ma, X., Wang, M., Sun, X. (2013) Genotypic differences among rice cultivars in lead accumulation and translocation and the relation with grain Pb levels, *Ecotoxicololgy and Environmental Safety*. 90, 35–40.

McLaughlin, M.J., Maier, N.A., Freeman, K., Tiller, K.G., Williams, C.M.J. and Smart, M.K. (1995) Effect of potassic and phosphatic fertilizer type, fertilizer Cd concentration and zinc rate on cadmium uptake by potatoes, *Fertility Research*. 40, 63–70.

Munzuroglu, O. and Geckil, H. (2002) Effects of metals on seed germination, root elongation, and coleoptile and hypocotyl growth in *Triticum aestivum* and *Cucumis sativus*, *Archives of Environmental Contamination and Toxicology*. 43, 203–213.

Mutert, E. and Fairhurst, T.H. (2002) Developments in rice production in Southeast Asia, *Better Crops International*. 15 (Suppl.), 12–17.

Nedelkoska, T.V. and Doran, P.M. (2000) Characteristics of metal uptake by plants species with potential for phytoremediation and phytomining, *Minerals Engineering*. 13, 549–561.

Ok, Y.K., Kim, S.C., Kim, D.K., Skousen, J.G., Lee, J.S., Cheong, Y.W., Kim, S.J. and Yang, J.E. (2011a) Ameliorants to immobilize Cd in rice paddy soils contaminated by abandoned metal mines in Korea, *Environmental Geochemistry and Health*. 33, 23–30.

Ok, Y.K., Lee, S.S., Jeon, W.T., Oh, S.E., Usman, A.R.A. and Moon, D.H. (2011b) Application of eggshell waste for the immobilization of cadmium and lead in a contaminated soil, *Environmental Geochemistry and Health*. 33, 31–39.

Papademetriou, M.K. (2000) Rice production in the Asia-Pacific region: Issues and perspective. In: Papademetrious, M.K., Dent, F.J., Herath, E.M. (eds.), *Bridging the Rice Yield Gap in the Asia-Pacific Region*> Bangkok, Thailand: Food and Agriculture Organization of the United Nations Regional Office for Asia and the Pacific, pp. 4–26.

Peer, W.A., Baxter, I.R., Richards, E.L., Freeman, J.L. and Murphy, A.S. (2006) Phytoremediation and hyperaccumulator plants. In: *Molecular Biology of Metal Homeostasis and Detoxification*. Berlin/Heidelberg: Springer, pp. 299–340.

Prasad, M.N.V. (2004) Metallothioneins, metal binding complexes and metal sequestration in plants. In: Prasad, M.N.V. (ed.), *Heavy Metal Stress in Plants from Biomolecules to Ecosystems*, First edition. Hyderabad, India: Springer-Verlag Berlin Heidelberg, pp. 47–83.

Prasad, M.N.V., Nakbanpote, W., Sebastian, A., Panitlertumpai, N. and Phadermrod, C. (2014) Phytomanagement of Padaeng zinc mine waste, Mae Sot district, Tak province, Thailand. In: Hakeem, K.R, Sabir, M., Ozturk, M., Murmut, A. (eds.), *Soil Remediation and Plants: Prospects and Challendes*. New York: Academic Press, Elsevier.

Rajkumar, M., Sandhya, S., Prasad, M.N.V. and Freitas, H. (2012) Perspectives of plant-associated microbes in heavy metal phytoremediation, *Biotechnology Advances*. 30(6), 1562–1574.

Ramadan, M.A.E and Al-Ashkar, E. A. (2007) The effect of different fertilizers on the heavy metals in soil and tomato plant, *Australian Journal of Basic and Applied Sciences*. 1, 300–306.

Rao T.P., Yano, K., Iijima, M. and Yamauchi, A. (2002) Regulation of rhizosphere acidification by photosynthetic activity in cowpea (*Vigna unguiculata* L. Walp.) seedlings, *Annals of Botany*. 89 (2), 213–220.

Sampanpanish, P. and Pongpaladisai, P. (2011) Effect of organic fertilizer on cadmium uptake by rice grown in contaminated soil, *International Conference on Environmental and Agriculture Engineering*. 15, 103–109.

Sandalio, L.M., Dalruzo, H.C., Gomez, M., Romero-Puetras, M.C. and del-Rio, L.A. (2001) Cadmium-induced changes in the growth and oxidative metabolism of pea plants, *Journal of Experimental Botany*. 52 (364), 2115–2126. doi:10.1093/jexbot/52.364.2115.

Sarwar, N., Saifullah, Malhi, S.S., Zia, M.H., Naeem, A., Bibi, S. and Farid, G. (2010) Role of mineral nutrition in minimizing cadmium accumulation by plants, *Journal of Science of Food Agriculture*. 90, 925–937.

Sauve, S., Hendershot, W. and Allen, H.E. (2000) Solid-solution partitioning of metals in contam-inated soils: Dependence on pH, total metal burden, organic matter, *Environmental Science and Technology* 34, 1125–1131.

Sauve, S., Manna, S., Turmel, M.C., Roy, A.G. and Courchesne, F. (2003) Solid-solution partitioning of Cd, Cu, Ni, Pb, and Zn in the organic horizons of a forest soil, *Environmental Science and Technology*. 37, 5191–5196.

Sebastian, A. and Prasad, M. N. V. (2013) Cadmium accumulation retard activity of functional components of photo assimilation and growth of rice cultivars amended with vermicompost, *International Journal of Phytoremediation*. 15, 965–978.

Sebastian, A. and Prasad, M.N.V. (2014a) Cadmium minimization in rice. A review, *Agronomy for Sustainable Development*. 34, 155–173.

Sebastian, A. and Prasad, M.N.V. (2014b) Photosynthesis-mediated decrease in cadmium translocation protects shoot growth of *Oryza sativa* seedlings up on ammonium phosphate—Sulfur fertilization, *Environmental Science and Pollution Research.* 21, 986–997.

Sebastian, A. and Prasad, M.N.V. (2014c) Red and blue lights induced oxidative stress tolerance promote cadmium rhizocomplexation in *Oryza sativa*, *Journal of Photochemistry and Photobiology B: Biology* 137, 135–143. doi:10.1016/j.jphotobiol. 2013 12 011.

Sebastian, A. and Prasad, M.N.V. (2014d) Vertisol prevents cadmium accumulation in rice: Analysis by ecophysiological toxicity markers, *Chemosphere.* 108, 85–92.

Semple, K.T., Doick, K.J., Jones, K.C., Burauel, P., Craven, A. and Harms, H. (2004) Defining bioavailability and bioaccessibility of contaminated soil and sediment is complicated, *Environmental Science and Technology.* 38, 228–231.

Simmons, R.W, Pongsakul, P., Saiyasitpanich, D. and Klinphoklap, S. (2005) Elevated levels of Cd and zinc in paddy soils and elevated levels of Cd in rice grain downstream of a zinc mineralized area in Thailand: Implications for public health, *Environmental Geochemistry and Health.* 27, 501–511.

Siripornadulsil, S. and Siripornadulsil, W. (2013) Cadmium-tolerant bacteria reduce the uptake of cadmium in rice: Potential for microbial bioremediation, *Ecotoxicology and Environmental Safety.* 94, 94–103.

Siswanto, D., Suksabye, P. and Thiravetyan, P. (2013) Reduction of cadmium uptake of rice plants using soil amendments in high cadmium contaminated soil: A pot experiment, *Journal of Tropical Life Science.* 3, 132–137.

So, H.B. and Ringrose-Voase, A.J. (2000) Management of clay soils for rainfed lowland rice-based cropping systems: An overview, *Soil and Tillage Research.* 56, 3–14.

Su, C. and Puls, R.W. (2004) Nitrate reduction by zerovalent iron: Effects of formate, oxalate, citrate, chloride, sulfate, borate, and phosphate, *Environmental Science and Technology.* 38, 2715–2720.

Sukreeyapongse, O., Holm, P.E., Strobel, B.W., Panichsakpatana, S., Magid, J. and Hansen, H.C.B. (2002) pH-dependent release of cadmium, copper, and lead from natural and sludge-amended soils, *Journal of Environmental Quality.* 31, 1901–1909.

Sukreeyapongse, O., Srisawat, L., Chomsiri, O., Notesir, N. (2010) *Soil management for reduce Cd concentration in rice grains*, 19th World Congress of Soil Science, Soil Solutions for a Changing World. Brisbane, Australia, 51–56.

Tate, R.L. (2000) Sulfur, phosphorus, and mineral cycles. In: *Soil Microbiology*, Second edition. Canada: John Wiley & Sons, Inc.

Tong, Y.P., Kneer, R. and Zhu, Y.G. (2004) Vacuolar compartmentalization: A second-generation approach to engineering plants for phytoremediation, *Trends in Plant Science.* 9, 7–9.

Uga, Y., Sugimoto, K., Ogawa, S., Rane, J., Ishitani, M., Hara, N., Kitomi, Y., Inukai, Y., Ono, K., Kanno, N., Inoue, H., Takehisa, H., Motoyama, R., Nagamura, Y., Wu, J., Matsumoto, T., Takai, T., Okuno, K. and Yano, M. (2013) Control of root system architecture by DEEPER ROOTING 1 increases rice yield under drought conditions, *Nature Genetics.* 45, 1097–1102.

Wenzel, W., Lombi, E. and Adriano, D.C. (2004) Root and rhizosphere processes in metal hyperaccumulation and phytoremediation technology. In: Prasad, M.N.V. (ed.) *Heavy Metal Stress in Plants*, First edition. India: Springer-Verlag Berlin Heidelberg, pp. 313–344.

Wu, W. and Cheng, S. (2014) Root genetic research, an opportunity and challenge to rice improvement, *Field Crops Research.* doi:http://dx.doi.org/10.1016/j.fcr.2014.04.013.

Xie, Y. and Wu, R. (1989) Rice alcohol dehydrogenase genes: Anaerobic induction, organ specific expression and characterization of cDNA clones, *Plant Molecular Biology.* 13, 53–68.

Xu, X.Y., Mcgrath, S.P., Meharg, A.A. and Zhao, F.J. (2008) Growing rice aerobically markedly decreases arsenic accumulation, *Environmental Science and Technology.* 42, 5574–5579.

Ye, W.L., Khan, M.A., McGrath, S.P. and Zhao, F.J. (2011) Phytoremediation of arsenic contaminated paddy soils with *Pteris vittata* markedly reduces arsenic uptake by rice, *Environmental Pollution.* 159, 3739–3743.

Zhan, J., Wei, S., Niu, R., Li, Y., Wang, S. and Zhu, J. (2013) Identification of rice cultivar with exclusive characteristic to Cd using a field-polluted soil and its foreground application, *Environmental Science Pollution Research.* 20, 2645–2650.

Zhen, Y.H., Cheng, V.J., Pan, G.X. and Li, L.Q. (2008) Cd, Zn and Se content of the polished rice samples from some Chinese open markets and their relevance to food safety, *Journal of Safety and Environmental.* 8, 119–122.

Zheng, R.L., Cai, C., Liang, J.H., Huang, Q., Chen, Z., Huang, Y.Z., Arp, H.P.H. and Sun, G.X. (2012) The effects of biochars from rice residue on the formation of iron plaque and the accumulation of Cd, Zn, Pb, As in rice (*Oryza sativa* L.) seedling, *Chemosphere.* 89, 856–862.

Zhou, H., Zhou, X., Zeng, M., Liao, B.H., Liu, L., Yang, W.T., Wu, Y.M., Qiu, Q.Y. and Wang, Y.J. (2014) Effects of combined amendments on heavy metal accumulation in rice (*Oryza sativa* L.) planted on contaminated paddy soil, *Ecotoxicology and Environmental Safety.* 101, 226–232.

15 Reduction Induced Immobilization of Chromium and Its Bioavailability in Soils and Sediments

Girish Choppala, Anitha Kunhikrishnan, Balaji Seshadri, Shiv Shankar, Richard Bush, and Nanthi Bolan

CONTENTS

15.1 INTRODUCTION

Chromium (Cr) enters terrestrial and aquatic environments through the indiscriminate disposal of wastes from industries, including electroplating, leather tanning, timber treatment, pulp production, and petroleum refining (Zhitkovich, 2011). Chromium is widely used as trivalent Cr [Cr(III)] in the tannery industry and as hexavalent Cr [Cr(VI)] in the timber treatment industry (Barnhart, 1997). Chromium(VI) is highly toxic and carcinogenic even when present in very low concentrations in water (Owlad et al., 2009), and it represents the greatest concern and focus for remediation. The maximum recommended concentration of Cr(total) in groundwater and wastewater is 2 mg L^{-1}, whereas for Cr(VI), it is only 0.05 mg L^{-1} (Park et al., 2004).

Cr(III) is strongly retained on soil particles, whereas Cr(VI) is weakly adsorbed and is readily available for plant uptake and leaching to groundwater (Leita et al., 2009). Chromium(VI) can be reduced to Cr(III) in environments provided there is a ready source of electrons such as carbon and reduced iron [Fe(II)] available and under acidic and waterlogged conditions, the reduction of Cr(VI) is enhanced (Chen et al., 2010; Hsu et al., 2009). The reduction of Cr(VI) to Cr(III), and subsequent immobilization of Cr(III) through adsorption and precipitation reactions, underpins the most common methods for treating Cr(VI)-contaminated soils, sediments, and industrial effluents (Wielinga et al., 2001).

Carbon-based organic amendments such as manures, biosolids, biochars, and black carbon (BC) can be considered excellent electron donors and provide the energy source for soil microorganisms that are involved in the reduction of metal(loid)s, such as Cr [Cr(VI) to Cr(III)] and arsenic {arsenate [As(V) to arsenite [As(III)]}, and nonmetals, such as nitrogen (N) [dentrification of nitrate (NO_3^-) to gaseous N (NO, N_2O, and N_2)] (Jardine et al., 1999; Park et al., 2011). The reduction of Cr(VI) to Cr(III) in soils is a proton (H^+) consumption or hydroxyl (OH^-) release reaction, resulting in an increase in soil pH (Park et al., 2006). A rise in soil pH causes an increase in negative surface charge, which enhances the immobilization of Cr(III) though adsorption and precipitation as $Cr(OH)_3$ (Adriano et al., 2004). Thus, redox reactions play an important role in the interconversion of Cr(VI) and Cr(III), controlling the bioavailability of Cr.

This chapter provides an overview of the sources and reactions of Cr in relation to bioavailability and remediation. It goes on to describe a case study that examines the effect of two carbon matrices [chicken manure biochar (CMB) and BC] on the concomitant reduction of Cr(VI) to Cr(III), and subsequent Cr(III) immobilization.

15.2 SOURCES OF CHROMIUM

Chromium enters into the environment by natural and anthropogenic pathways and it is present in various environmental matrices, including surface and groundwater, soil and sediments.

15.2.1 Natural Sources

Chromium ranks 21st among the elements in crustal abundance at about 100 mg kg^{-1} (Barnhart, 1997). The average Cr concentration in the continental crust is 125 mg kg^{-1}, with a common range of 80–200 mg kg^{-1} [National Academy of Sciences (NAS), 1974]. Most Cr in the atmosphere is from windblown dust, forest fires, meteoric dust, sea salt spray, and volcanic eruptions, and it ranges from 0.015 to 0.03 μg m^{-3} (Nriagu and Nieboer, 1988).

The soil Cr content is largely dependent on parent material; concentrations usually range from 10 to 150 mg kg^{-1}, with an average of 40 mg kg^{-1} (Bertine and Goldberg, 1971). Chromium levels average from 20 to 35 mg kg^{-1} in granitic igneous rock, limestone, and sandstones, and they are up to 220 mg kg^{-1} in basaltic igneous rock and 1800 mg kg^{-1} in ultramafic rock (Bowen, 1979). Another oxide mineral, chromite ($FeCr_2O_4$), contains nearly 40% chrome (Cr_2O_3). Soil minerals such as Fe(III) (oxy) hydroxides and clay particles serve to sequester Cr(III) that is released from Cr-bearing minerals. In addition, Cr(III) forms inner-sphere complexes and precipitates at a pH greater than 4 (Oze et al., 2004). However, Cr(VI) is not strongly adsorbed, thereby resulting in high solubility and bioavailability.

15.2.2 Anthropogenic Sources

Hazardous Cr [particularly Cr(VI)] is primarily derived from industries. There are several sources, such as direct infiltration of leachate from landfill, disposal of solid wastes, sewage sludge, leachate from mining wastes, seepage from industrial lagoons, spills and leaks from industrial metal processing or wood-preserving facilities, and other industrial operations.

Nearly 80% of the Cr produced from mines is used in metallurgical applications, and the total annual input of Cr into soils has been estimated to be between 4.35×10^5 and 1.18×10^6 t (Nriagu and Pacyna, 1988). Nearly 170,000 t of Cr waste are discharged into the environment annually around the globe because of industrial and manufacturing activities (Gadd and White, 1993). Chromium production began in the 1950s, and by 2000 the worldwide cumulative production was estimated at 105.4 t (Su et al., 2005). In the United States alone, nearly 50,000 t of Cr(VI) have been generated through industrial wastewaters (Bokare and Choi, 2010). In New Zealand, approximately 6400 t of tannery and 1600 t of timber treatment effluents are generated annually. These effluents are considered the major sources of Cr contamination into aquatic and terrestrial environments

(Carey et al., 1996). In 2007, in Australia, the lack of a take-back scheme for computers led to around 6 t of Cr(VI) being dumped in landfills. Around 157,290 hectares of land in Australia are used for wine cultivation, and approximately 3400 t of Cr(VI) are used to treat the poles in Australian vineyards [Australian Bureau of Statistics (ABS), 2009].

Chromium is released into the environment from effluents of metal plating, anodizing, ink manufacture, dyes, pigments, glass, ceramics, glues, tanning, wood preservatives, textiles, and corrosion inhibitors in cooling water (Shanker et al., 2005). The largest direct application of Cr to the land surface is through the disposal of trapped and bottom fly ash from coal combustion (Nriagu and Pacyna, 1988). Other important anthropogenic sources of Cr in the environment include fuel combustion, cement production, and sewage sludge incineration/deposition.

15.3 BIOGEOCHEMISTRY OF CHROMIUM

The dominant Cr species in soil and sediments, including Cr(III) and Cr(VI), undergo a number of reactions, including sorption, precipitation, and redox reactions. Chromium in the natural environment is influenced mainly by redox reactions, which alter its speciation and chemical properties.

15.3.1 Sorption and Precipitation Reactions

Chromium(III) is adsorbed strongly onto negative charges on soil particles. However, Cr(VI) is adsorbed onto positive charges, which are limited in most soils under optimum pH conditions. Iron, manganese (Mn) oxides, and clay minerals strongly sorb Cr(III) and this sorption is rapid; about 90% of Cr(III) is sorbed by clay minerals and Fe oxides within 24 h (Richard and Bourg, 1991). Moreover, Cr(III) sorption increases with rising soil pH, as the clay surface becomes more negative. Sorption also increases with soil organic matter content (Paya Perez et al., 1988). However, Cr(III) adsorption decreases when other inorganic cations or dissolved organic ligands are present (Avudainayagam et al., 2001). An understanding of these various reactions and controls on the retention of Cr(VI) can be increased either by increasing the positive charge on the soil particles or by reducing Cr(VI) to Cr(III). Anions of Cr, $HCrO_4^-$, and CrO_4^{2-} species are sorbed by Mn, aluminum (Al), Fe oxides and hydroxides (positively charged surfaces), and clay minerals (Rai et al., 1987).

A broad range of practical experiments involving the addition of organic amendments, such as plant bark, increases the sorption of Cr(VI) (Bolan and Thiagarajan, 2001). Cellulose in the bark absorbs Cr(VI) strongly and forms Cr(VI)–cellulose complexes (Miretzky and Cirelli, 2010). Low-cost adsorbents such as agricultural byproducts (sugar beet pulp, coir pith, rice straw, saw dust), as well as activated carbon of chars and bio-sorbents (algae, fungi, bacteria, chitosan in fungal cell wall) can be used for the sorption of Cr(VI) and Cr(III) (Mohan and Pittman, 2006; Zhou and Richard, 2010). Competing anions have a detrimental effect on Cr(VI) sorption, depending on dissolved concentrations of the competing anion and CrO_4^{2-}, their relative affinities for the solid surface, and surface site concentration (Rai et al., 1987).

Chromium(III) species do not occur naturally and are less stable in the environment. The precipitation of Cr(III) as mixed Fe–Cr hydroxide (Cr, Fe)$(OH)_3$ enhances the precipitation of Cr(III) in waters at neutral pH levels (Sass and Rai, 1987). The removal of Cr(III) from industrial effluent can be achieved by using alkaline materials such as lime and magnesium oxide to precipitate Cr(III) as chromic hydroxide [$Cr(OH)_3$]. Precipitation is reported to be the most effective at a pH between 8.5 and 9.5, due to the low solubility of $Cr(OH)_3$ in that range (Patterson, 1985). The reduction of Cr(VI) to Cr(III), and the subsequent hydroxide precipitation of Cr(III) ion, is the most common method that is used for treating Cr(VI)-contaminated industrial effluents (Loyaux-Lawniczak et al., 2001; James, 2001).

15.3.2 Redox Reactions

Oxidation–reduction (redox) reactions involve the transfer of electrons. The redox transformation of Cr(III) to Cr(VI) requires another redox couple (oxidizing/reducing agent) that accepts or donates the necessary electrons. In natural waters, sediments, and soils, the significant redox couples are H_2O/O_2 (aq); Mn(II)/Mn(VI); NO_2/NO_3; Fe(II)/Fe(III); S^{2-}/SO_4^{2-}; and CH_4/CO_2 (Buerge and Hug, 1997).

In oxidizing or moderately anoxic environments, Cr exists as Cr(VI). Many oxidizing agents are known to oxidize Cr(III) to Cr(VI), but only a few of them are found in the environment in sufficient concentrations to initiate this reaction. Water, MnO_2, O_3, H_2O_2, and PbO_2 are good sources of oxygen for the oxidation of Cr(III) (Landrot et al., 2009). Dissolved oxygen can also oxidize Cr(III) to Cr(VI), but the rate of oxidation is relatively slow (Rai et al., 1987).

In the presence of electron donors, Cr(VI) can be reduced to Cr(III), which is less toxic. Iron(II), reduced sulfur, and organic matter are the chief sources of electrons for Cr(VI) reduction (Kotaś and Stasicka, 2000). Both chemical (abiotic) and biological (biotic) processes can reduce Cr(VI) to Cr(III). Although abiotic reduction occurs at a much faster rate, the biotic reduction process is more common in soils (Wielinga et al., 2001).

Ferrous iron [Fe(II)] plays a dominant role in the reduction of Cr(VI) to Cr(III) in water and soil. The discharge of industrial waste generates dissolved Fe(II) ions in environmental waters. The weathering of Fe(II)-containing minerals releases Fe(II) into soils. In natural anaerobic systems at redox levels less than +100 mV and in natural to alkaline conditions, Fe(II) controls the reduction of Cr(VI) (Pettine et al., 1998). The rate of reduction of Cr(VI) depends on the dissolution and oxygenation of Fe(II) in soils (Fendorf and Li, 1996).

Reduced sulfur, such as S^{2-} and $S_2O_3^{2-}$ species, is significant in the reduction of Cr(VI) to Cr(III) in soils. Most sulfides are not soluble, but some dissolved S^{2-} species from industrial waste discharges and the decomposition of organic matter may reduce Cr(VI) (Equation 15.1). The reduction of Cr(VI) in the presence of gaseous H_2S is dependent on both the relative humidity of the gaseous H_2S stream and the particle size on which Cr(VI) is retained in the contaminated soil (Hua and Deng, 2003). However, Cr(VI) salts such as $PbCrO_4$ and $BaCrO_4$ are not reduced due to the low dissolution of these salts.

$$2CrO_4^{2-} + 3H_2S + 4H^+ \longrightarrow 2Cr(OH)_3 + 3S_{(S)} + 2H_2O \quad (15.1)$$

Natural organic matter comprises humic acid and fulvic acids, which promote the reduction of Cr(VI) in soils (Equation 15.2) (Wittbrodt and Palmer, 1995). The hydroquinone group in the dissolved organic matter is the main source of electrons for Cr(VI) reduction in soils (Elovitz and Fish, 1995). Eckert et al. (1990) observed that the rate of reduction decreased with an increase in pH. The half-life for Cr(VI) was 3 days, at very low pH when reduced with humic acids; whereas in the pH range of 4–7, reduction took several days (Park et al., 2011). Incorporation of organic matter into soil increases the dissolved organic carbon, which both stimulates microbes as a source of energy and serves as a pool of electrons with acidic and basic functional groups (Bolan et al., 2011).

$$2Cr_2O_7^{2-} + 3CH_2O\text{ (organic matter)} + 16H^+ \rightarrow 4Cr^{3+} + 3CO_2 + 11H_2O \quad (15.2)$$

15.3.3 Aquatic Environments

The behavior of Cr in the aquatic environment is governed by its oxidation state. Chromium exists in different oxidation states; however, in the aquatic environment, only Cr(III) and Cr(VI) species are prevalent. Chromium(VI) can be transported great distances in groundwater owing, in part, to its high solubility, but it may be transformed by reduction to Cr(III) and precipitated as $Cr(OH)_3$ (Tokunaga et al., 2001). Chromium(III) is less soluble and not easily transported into groundwater; however, it can be oxidized by Mn oxides in sediments (Bartlett and James, 1979).

The mobility of Cr species is mainly controlled by adsorption in the aquatic environment. Solution pH is critical for adsorption since it influences the surface properties and ionic form of the Cr in the aqueous media (Garg et al., 2007). Under neutral and slightly acidic pH conditions, Cr(III) adsorbs on sediments (Richard and Bourg, 1991). Since Cr(VI) exists as anionic species, adsorption is limited on negatively charged sediment colloids (Fendorf, 1995). Therefore, an increase in positive charge on sediments increases the adsorption of Cr(VI).

15.4 CONCOMITANT REDUCTION AND IMMOBILIZATION

This section presents a case study that examines the effect of two carbon matrices on the concomitant reduction of Cr(VI) to Cr(III) in a spiked and a naturally contaminated soil, and the subsequent immobilization of Cr.

An uncontaminated, calcic red, and sandy loam mineral soil that was low in organic matter and a naturally occurring Cr-contaminated soil collected from a long-term tannery waste-contaminated site were used in this study. Two carbon amendments (CMB and BC) were used to examine their efficiency in reducing Cr(VI) to Cr(III). The uncontaminated soil was mixed with two levels of Cr(VI) (0 and 600 mg Cr kg^{-1} soil) of the form $K_2Cr_2O_7$ and incubated at field capacity. The spiked and naturally contaminated soil samples were then mixed with two levels of carbon amendment (0 and 50 g organic kg^{-1} soil) and incubated at field capacity for 4 weeks. During incubation, the soil subsamples were taken at various intervals and extracted with 1 M KH_2PO_4 at a 1:10 soil:solution ratio for 16 h. The soil extracts were analyzed for total Cr and Cr(VI). Chromium(VI) reduction was estimated from the decrease in Cr(VI) concentration in the soil solution. The rate of Cr(VI) reduction was described by using Equation 15.3.

$$Y = Y_m(1 - \text{Exp}^{-rx}) \tag{15.3}$$

where Y is the amount of Cr(VI) reduced (mg kg^{-1}), Y_m is the amount of maximum reduction (mg k^{-1}), r is the rate constant (per day), and x is the incubation period (days).

A leaching experiment was conducted by using the spiked and naturally contaminated soil samples that were incubated for 4 weeks with carbon amendments. The leachate samples were analyzed for Cr(VI) and Cr(III) species. A greenhouse plant growth experiment was used to investigate the effect of carbon amendments on the plant uptake of Cr(VI) from the spiked and naturally contaminated soil samples. Sunflower (*Helianthus annuus* L.) was used as a test plant due to its ability to tolerate high levels of heavy metals in soils. The plants were harvested eight weeks after sowing and were analyzed for Cr content.

15.4.1 REDUCTION OF CR(VI)

The amount of maximum reduction (Y_m) and rate constant (r) parameters of the equation describing the data (Equation 15.3) are presented in Table 15.1. Both the Y_m and the r values increased with the addition of carbon amendments. Black carbon achieved greater Cr(VI) reduction than did CMB. The BC addition achieved greater reduction in the calcic red sandy loam (75.4%) than in the tannery waste soil (62.9%). This was in contrast to the CMB additions, which achieved 55.4% and 38.2% Cr(VI) reduction, respectively (Figure 15.1). The reduction rates of Cr(VI) in soils treated with carbon amendments, relative to the unamended soils, were calculated from the r values (Table 15.1). These values indicated that BC and CMB additions caused 1.2- to 2.3- and 2.5- to 4.0-fold increases in the rate of Cr(VI) reduction in the soils. This was consistent with the findings of Hsu et al. (2009), Bolan et al. (2003), and Losi et al. (1994), where a significant increase in Cr(VI) reduction rates occurred in the presence of various carbon amendments such as manures, biosolids, and biochar.

Various processes could be involved in the observed increase in Cr(VI) reduction in the presence of carbon amendments. Black carbon and biochar comprise high-surface areas with several surface functional groups. These functional groups serve as electron donors for Cr(VI) reduction. The higher reducing capacity of BC may be due to the protonation of the material during acid treatment to remove silica. Furthermore, acid treatment increases the acidic functional groups (Chen and Wu,

TABLE 15.1
Parameters of the Equation Describing the Rate of Reduction of Cr(VI) in the Unamended and Amended Soils

Treatment	Y_m (Maximum Reduction; mg Cr kg^{-1})	r (Rate factor)	Regression Coefficient (R^2)	Relative Rate of Cr(VI) Reduction[a]
Spiked soil	135.6	0.201	0.985	1.00
Spiked soil + black carbon	520.5	0.410	0.867	2.04
Spiked soil + biochar	437.7	0.321	0.781	1.60
Tannery soil	67.70	0.305	0.812	1.00
Tannery soil + black carbon	380.6	0.551	0.875	1.81
Tannery soil + biochar	198.7	0.487	0.756	1.60

[a] Relative rate of Cr(VI) reduction = (r [rate factor])/(r [rate factor] for soil alone).

Note: $Y = Y_m$ (1-Exp^{-rx}), where Y = amount of Cr(VI) reduced (mg kg-1); Y_m = maximum amount of Cr(VI) reduction (mg kg^{-1}); r = rate constant; and x = incubation period (days).

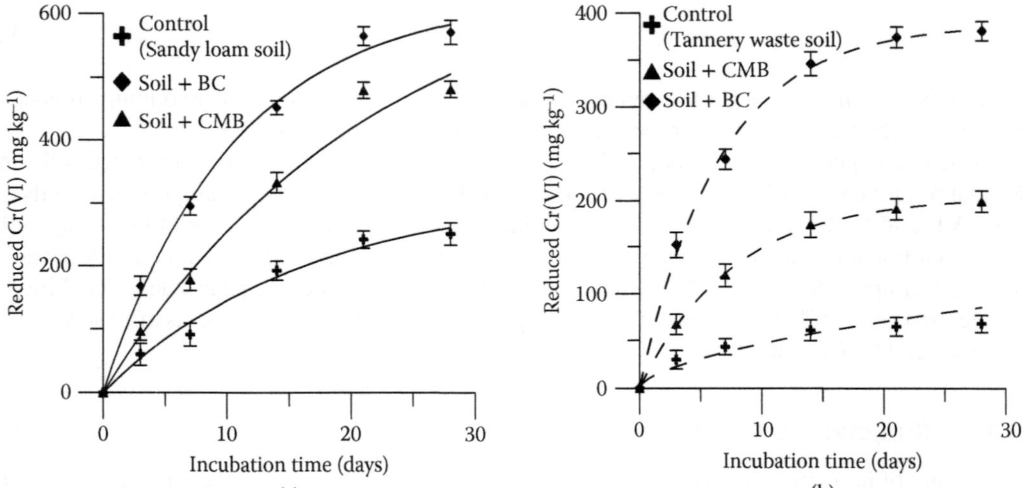

FIGURE 15.1 Amounts of Cr(VI) reduced during the incubation for calcic red (a) sandy loam and (b) tannery waste soil (fitted curves are shown as lines).

2004) that are critical in the reduction of Cr(VI). The reduction of Cr(VI) to Cr(III), being a H^+ consumption (or $^-OH-$ release) reaction (Equation 15.2), increased as pH dropped (Lan et al., 2007), which may be one of the reasons for higher Cr(VI) reduction in the presence of protonated BC.

The increase in pH due to H^+ consumption in the Cr(VI) reduction reaction is likely to result in the immobilization of Cr(III) through adsorption (Equation 15.4) and precipitation (Equation 15.5) reactions.

$$Cr(III) + soil \rightarrow Cr^{3+} + H^+ \quad \text{(Adsorption – proton release reaction)} \tag{15.4}$$

$$Cr(III) + H_2O \rightarrow Cr(OH)_3 + H^+ \quad \text{(Precipitation – proton release reaction)} \tag{15.5}$$

The increase in Cr(VI) reduction in the presence of carbon materials may also result from an increase in microbial activity. The addition of biochar increases the microbial activity of the soil, as measured by increased respiration (Choppala et al., 2012; Steinbeiss et al., 2009). Although Cr(VI) reduction can occur through both chemical (abiotic) and biological (biotic) processes, the latter is considered the dominant process, especially in soils that are low in ferrous iron. The carbon, nutrients, and stimulation of microorganisms are considered important factors in enhancing the reduction of Cr(VI) to Cr(III) (Tokunaga et al., 2003; Park et al., 2008).

15.4.2 Leaching of Chromium

The cumulative leachate concentration of Cr(VI) in the two tested soils was higher than that of Cr(III). The amount of Cr(VI) leached from the calcic red sandy loam was 334.7 mg kg^{-1}. However, only 123.1 mg kg^{-1} of Cr(III) was leached from this spiked soil. The amounts of Cr(VI) and Cr(III) that were leached from tannery waste-contaminated soils were higher than those from spiked soil (Figure 15.2). The leaching experiment confirmed the high mobility and low-sorption capacity of Cr(VI), where it leached rapidly from contaminated soils (Weng et al., 2002). The leaching of total Cr decreased with the addition of carbon amendments, which is ascribed to the increase in Cr(VI) reduction and the subsequent Cr(III) retention. The amount of leachable Cr(VI) decreased by 87.2% in calcic red sandy loam with the addition of 50 g BC kg^{-1} soil. In a column experiment, Boni and Sbaffoni (2009) confirmed that Cr(VI) removal efficiency was greater than 99% in soils mixed with organic amendments, and they attributed this to adsorption to the organic-based matrix and the reduction to Cr(III).

In the calcic red sandy loam soil, Cr(III) leaching decreased by 56.1% and 61.3% in BC and CMB-amended soils, respectively. However, the amount of leached Cr(III) slightly increased in the tannery waste soil (7.7% for both amendments). This was attributed to the reduction of Cr(VI) species. Adsorption of Cr(III) by soils resulting from both nonspecific cation exchange and specific adsorption processes at low-surface coverage has been reported (Fendorf et al., 1994). The sorption of metal(loid)s such as Pb, Cu, and Cd by BC and CMB occurs through cation-exchange reactions (Park et al., 2011).

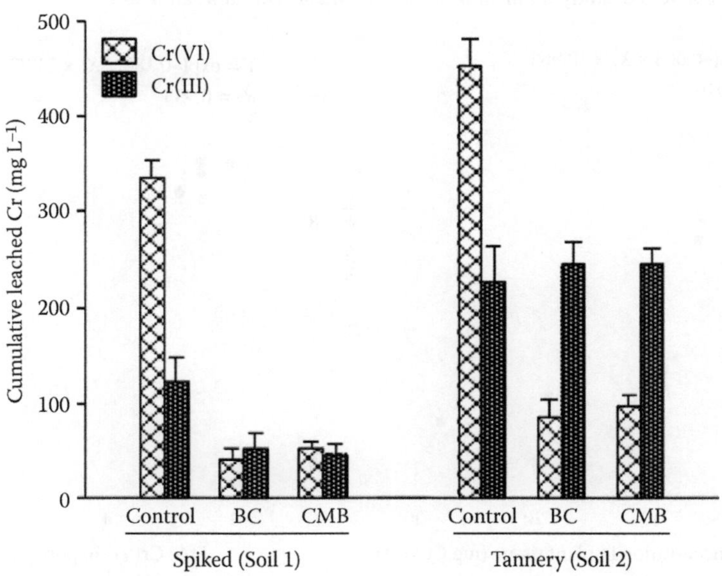

FIGURE 15.2 Effect of carbon amendments on leaching of Cr(III) and Cr(VI) species in soils (soil 1: calcic red sandy loam and soil 2: tannery waste-contaminated soil; BC, black carbon; CMB, chicken manure biochar).

15.5 BIOAVAILABILITY OF CHROMIUM

Plant growth decreased with Cr(VI) addition. In the calcic red sandy loam soil spiked with 250 mg Cr(VI) kg^{-1} soil, the addition of 50 g BC and CMB kg^{-1} soil increased the dry matter yield by 63% and 65%, respectively. In the tannery waste-contaminated soils, the dry matter yield was increased by 83% and 81% with the addition of 50 g BC and CMB kg^{-1} soil, respectively.

The addition of Cr(VI) to the soils retarded plant growth, and the Cr concentration in plant biomass was strongly influenced by the concentration of Cr(VI) in the soil solution. In the calcic red sandy loam and tannery waste-contaminated soils, the Cr levels in plant biomass grown in unamended soils were 24.63 and 27.85 mg kg^{-1}, respectively. The relatively higher concentration of

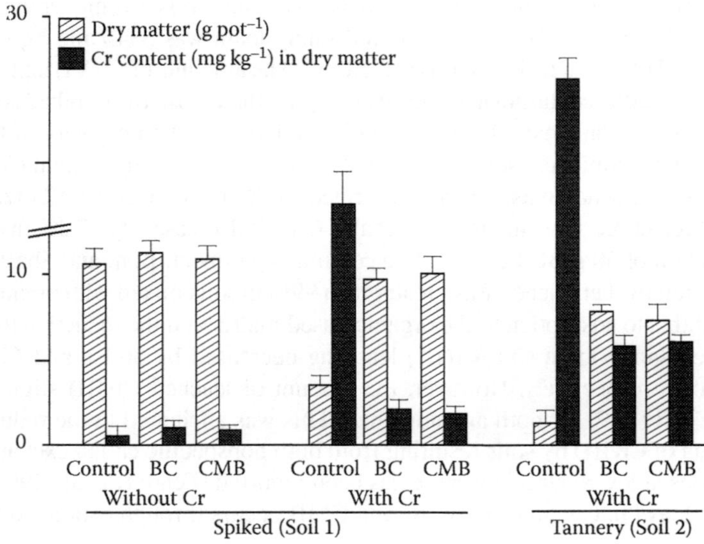

FIGURE 15.3 Effect of carbon amendments on dry matter yield (g pot^{-1}) and Cr concentration in plant tissue (mg kg^{-1}) (soil 1: calcic red sandy loam; soil 2: tannery waste-contaminated soil).

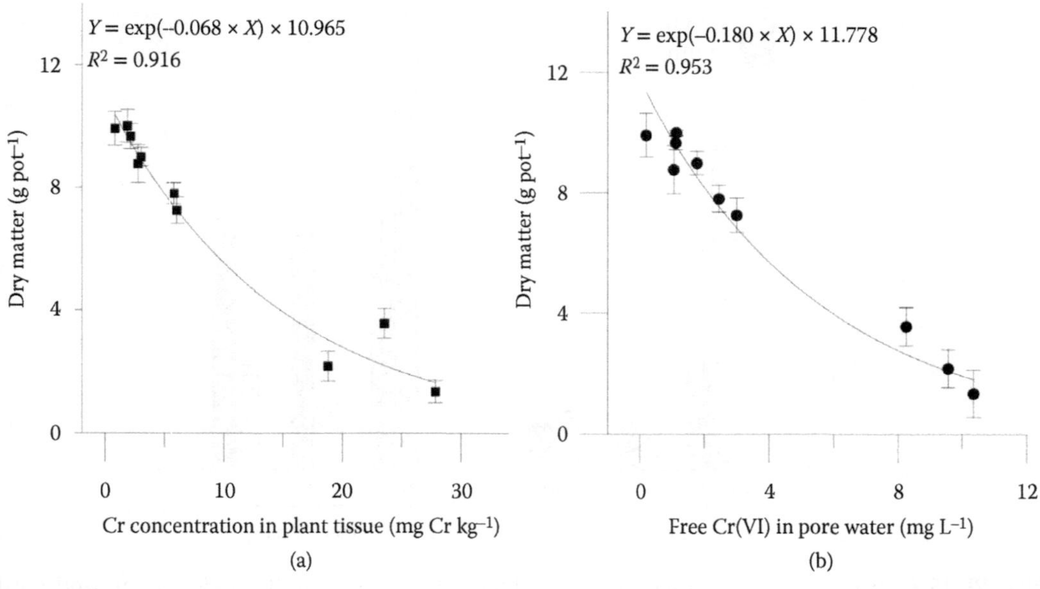

FIGURE 15.4 Relationships between dry matter yield and total (a) Cr concentration in plant tissue and (b) Cr(VI) concentration in soil solution.

Cr in the tannery waste soil compared with that in the spiked soil may have been due to the high levels of Cr from tannery sludge and the in situ oxidation of Cr(VI). Sorption sites on the tannery waste soil that were preoccupied by Cr(III) from tannery waste may have led to low-Cr(III) retention by this soil, thereby increasing the plant uptake of Cr. However, with the addition of 50 g BC and CMB kg^{-1} soil, the Cr concentration in biomass decreased by 92% and 90%, respectively, in the calcic red sandy loam soil. This trend also occurred in the tannery waste soil, where tissue Cr concentration was decreased by 79% and 77% for BC and CMB amendments, respectively.

The addition of carbon amendments is known to be very good at decreasing the phytotoxic effect of Cr(VI) on plant growth. The effect was more pronounced in the spiked soil than in the naturally contaminated soil, which may be attributed to the difference in the bioavailability of Cr between these two soil groups. In the spiked soil, Cr(VI) was more readily available for microbial reduction than in naturally contaminated soils, where the bioavailability of Cr(VI) was considerably lower. The increase of Cr(VI) levels in the soil enhanced the uptake of Cr in the plants. The dry matter yield decreased with an increase in concentrations of Cr in the plant tissue and Cr(VI) in soil solution (Figures 15.3 and 15.4). Han et al. (2004) observed that the number of palisade and spongy parenchyma cells decreased in plant shoots that were grown in Cr(VI)-treated soils. The addition of carbon amendments was very effective in reducing Cr concentration in plant tissue. The metal(loid) transfer coefficient, as defined by the ratio between plant tissue concentration and soil concentration, decreased with the addition of carbon compounds. This indicates that the addition of carbon compounds decreased the bioavailability of Cr by reducing the readily available Cr(VI) to less available Cr(III) and by enhancing the subsequent immobilization of this species.

15.6 CONCLUSIONS AND OUTLOOK

The case study found that the addition of carbon amendments to Cr(VI)-contaminated mineral soil enhanced the reduction of Cr(VI) to Cr(III) and the subsequent immobilization of Cr(III), thereby reducing the bioavailability of Cr for plant uptake. These carbon amendments provided a source of electrons, thereby facilitating the reduction of Cr(VI) to Cr(III) in soils (Figure 15.5). The Cr(VI) reduction reaction resulted in an increase in soil pH, thereby increasing the pH-induced

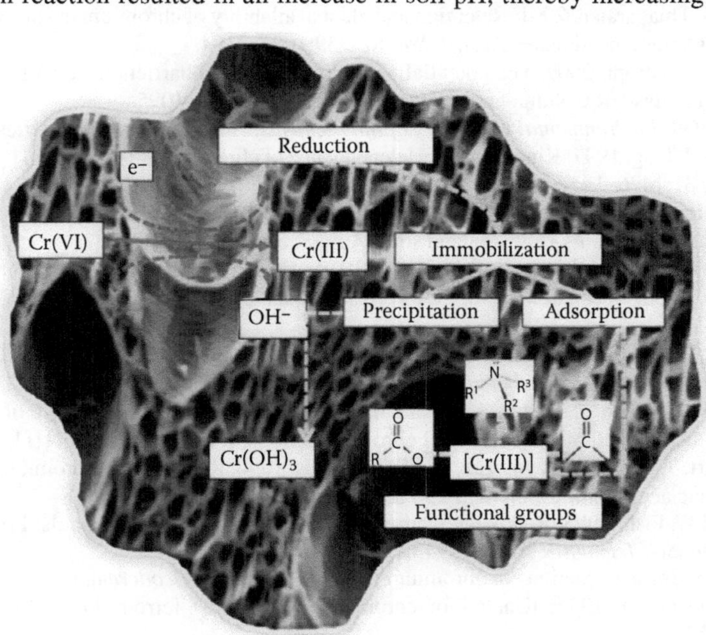

FIGURE 15.5 Reduction induced immobilization of Cr(VI) in black carbon-amended soils. (From Bolan, N.S. et al., *Rev. Environ. Contam. Toxicol.*, 225, 1–56, 2013.)

negative charge on variable charge surfaces in both soil and carbon amendments. The increase in surface negative charge facilitated the adsorption of Cr(III), thereby reducing its mobility and bioavailability. Since bioavailability is the key factor for remediation technologies, the reduction of Cr(VI) to Cr(III) that is followed by chemical or biological immobilization of Cr(III) may be a preferred option (Frankenberger and Losi, 1995; James, 1996). A major inherent problem with immobilization techniques, in general, is that although the metal(loid)s become less bioavailable, the total soil contaminant concentration remains unchanged. The immobilized metal(loid) may become bioavailable with time through natural weathering processes or advanced decomposition of soil carbon.

REFERENCES

Adriano, D. C., W. W. Wenzel, J. Vangronsveld, and N. S. Bolan. 2004. Role of assisted natural remediation in environmental cleanup. *Geoderma* 122:121–142.

Australian Bureau of Statistics (ABS). 2009. *1301.0–Year Book*, Australia.

Avudainayagam, S., R. Naidu, R. Kookana, A. M. Alston, S. McClure, and L. Smith. 2001. Effects of electrolyte composition on chromium desorption in soils contaminated by tannery waste. *Aust. J. Soil Res.* 39:1077–1090.

Barnhart, J. 1997. Occurrences, uses, and properties of chromium. *Regul. Toxicol. Pharm.* 26:S3–S7.

Bartlett, R., and B. James. 1979. Behavior of chromium in soils: III. Oxidation. *J. Environ. Qual.* 8:31–35.

Bertine, K., and E. D. Goldberg. 1971. Fossil fuel combustion and the major sedimentary cycle. *Science* 173:233–235.

Bokare, A. D., and W. Choi. 2010. Chromate-induced activation of hydrogen peroxide for oxidative degradation of aqueous organic pollutants. *Environ. Sci. Technol.* 44:7232–7237.

Bolan, N. S., D. C. Adriano, A. Kunhikrishnan, T. James, R. McDowell, and N. Senesi. 2011. Dissolved organic matter: Biogeochemistry, dynamics, and environmental significance in soils. *Adv. Agron.* 110:1–75.

Bolan, N. S., D. C. Adriano, R. Natesan, and B. J. Koo. 2003. Effects of organic amendments on the reduction and phytoavailability of chromate in mineral soil. *J. Environ. Qual.* 32:120–128.

Bolan, N. S., G. Choppala, A. Kunhikrishnan, J. Park, and R. Naidu. 2013. Microbial transformation of trace elements in soils in relation to bioavailability and remediation. *Rev. Environ. Contam. Toxicol.* 225:1–56.

Bolan, N. S., and S. Thiagarajan. 2001. Retention and plant availability of chromium in soils as affected by lime and organic matter amendments. *Aust. J. Soil Res.* 39:1091–1104.

Boni, M. R., and S. Sbaffoni. 2009. The potential of compost-based biobarriers for Cr(VI) removal from contaminated groundwater: Column test. *J. Hazard. Mater.* 166:1087–1095.

Bowen, H. J. M. 1979. *Environmental Chemistry of the Elements*. New York: Academic Press.

Buerge, I. J., and S. J. Hug. 1997. Kinetics and pH dependence of chromium(VI) reduction by iron(II). *Environ. Sci. Technol.* 31:1426–1432.

Carey, P., R. K. McLaren, K. Cameron, and J. Sedcole. 1996. Leaching of copper, chromium, and arsenic through some free-draining New Zealand soils. *Aust. J. Soil Res.* 34:583–597.

Chen, C. P., K. W. Juang, T. H. Lin, and D. Y. Lee. 2010. Assessing the phytotoxicity of chromium in Cr(VI)-spiked soils by Cr speciation using XANES and resin extractable Cr(III) and Cr(VI). *Plant Soil.* 334:299–309.

Chen, J. P., and S. Wu. 2004. Acid/base-treated activated carbons: Characterization of functional groups and metal adsorptive properties. *Langmuir.* 20:2233–2242.

Choppala, G. K., N. S. Bolan, Z. Chen, M. Megharaj, and R. Naidu. 2012. The influence of biochar and black carbon on reduction and bioavailability of chromate in Soils. *J. Environ. Qual.* 41:1175–1184.

Eckert, J., J. Stewart, T. Waite, R. Szymczak, and K. Williams. 1990. Reduction of chromium (VI) at sub-μg l^{-1} levels by fulvic acid. *Anal. Chim. Acta* 236:357–362.

Elovitz, M. S., and W. Fish. 1995. Redox interactions of Cr(VI) and substituted phenols: Products and mechanism. *Environ. Sci. Technol.* 29:1933–1943.

Fendorf, S. E. 1995. Surface reactions of chromium in soils and waters. *Geoderma.* 67:55–71.

Fendorf, S. E., and G. Li. 1996. Kinetics of chromate reduction by ferrous iron. *Environ. Sci. Technol.* 30:1614–1617.

Fendorf, S. E., G. M. Lamble, M. G. Stapleton, M. J. Kelley, and D. L. Sparks. 1994. Mechanisms of chromium(III) sorption on silica. 1. Chromium(III) surface structure derived by extended x-ray absorption fine structure spectroscopy. *Environ. Sci. Technol.* 28:284–289.

Frankenberger Jr, W., and M. Losi. 1995. Applications of bioremediation in the cleanup of heavy metals and metalloids. In: H. D. Skipper, R. F. Turco (eds.), *Bioremediation: Science and Applications*. SSSA Special Publ. 43, Madison, WI: SSSA, 173–210.

Gadd, G., and C. White. 1993. Microbial treatment of metal pollution—a working biotechnology. *Trends Biotechnol.* 11:353–359.

Garg, U. K., M. Kaur, V. Garg, and D. Sud. 2007. Removal of hexavalent chromium from aqueous solution by agricultural waste biomass. *J. Hazard. Mater.* 140:60–68.

Han, F. X., B. B. Sridhar, D. L. Monts, and Y. Su. 2004. Phytoavailability and toxicity of trivalent and hexavalent chromium to *Brassica juncea*. *New Phytol.* 162:489–499.

Hsu, N. H., S. L. Wang, Y. C. Lin, G. D. Sheng, and J. F. Lee. 2009. Reduction of Cr(VI) by crop-residue-derived black carbon. *Environ. Sci. Technol.* 43:8801–8806.

Hua, B., and B. Deng. 2003. Influences of water vapor on Cr(VI) reduction by gaseous hydrogen sulfide. *Environ. Sci. Technol.* 37:4771–4777.

James, B. R. 1996. The challenge of remediating chromium-contaminated soil. *Environ. Sci. Technol.* 30:248–251.

James, B. R. 2001. Remediation-by-reduction strategies for chromate-contaminated soils. *Environ. Geochem. Health.* 23:175–179.

Jardine, P. M., S. E. Fendorf, M. A. Mayes, I. L. Larsen, S. C. Brooks, and W. B. Bailey. 1999. Fate and transport of hexavalent chromium in undisturbed heterogeneous soil. *Environ. Sci. Technol.* 33: 2939–2944.

Kotaś, J., and Z. Stasicka. 2000. Chromium occurrence in the environment and methods of its speciation. *Environ. Pollut.* 107: 263–283.

Lan, Y., B. Deng, C. Kim, and E. C. Thornton. 2007. Influence of soil minerals on chromium(VI) reduction by sulfide under anoxic conditions. *Geochem. Trans.* 8:4.

Landrot, G., M. Ginder-Vogel, and D. L. Sparks. 2009. Kinetics of chromium (III) oxidation by manganese (IV) oxides using quick scanning x-ray absorption fine structure spectroscopy (Q-XAFS). *Environ. Sci. Technol.* 44:143–149.

Leita, L., A. Margon, A. Pastrello, I. Arcon, M. Contin, D. Mosetti. 2009. Soil humic acids may favour the persistence of hexavalent chromium in soil. *Environ. Pollut.* 157:1862–1866.

Losi, M. E., C. Amrhein, and W. T. Frankenberger Jr. 1994. Factors affecting chemical and biological reduction of hexavalent chromium in soil. *Environ. Toxicol. Chem.* 13:1727–1735.

Loyaux-Lawniczak, S., P. Lecomte, and Ehrhardt, J. J. 2001. Behavior of hexavalent chromium in a polluted groundwater: Redox processes and immobilization in soils. *Environ. Sci. Technol.* 35:1350–1357.

Miretzky, P., and A. F. Cirelli. 2010. Cr(VI) and Cr(III) removal from aqueous solution by raw and modified lignocellulosic materials: A review. *J. Hazard. Mater.* 180:1–19.

Mohan, D., and C. U. Pittman Jr. 2006. Activated carbons and low cost adsorbents for remediation of tri- and hexavalent chromium from water. *J. Hazard. Mater.* 137:762–811.

National Academy of Sciences (NAS). 1974. *Chromium*. Washington, DC: National Academy of Sciences.

Nriagu, J. O., and E. Nieboer. 1988. *Chromium in the Natural and Human Environments*. New York: Wiley Interscience.

Nriagu, J. O., and J. M. Pacyna. 1988. Quantitative assessment of worldwide contamination of air, water and soils by trace metals. *Nature* 333:134–139.

Owlad, M., M. K. Aroua, W. A. W. Daud, and S. Baroutian. 2009. Removal of hexavalent chromium-contaminated water and wastewater: A review. *Water Air Soil Pollut.* 200:59–77.

Oze, C., S. Fendorf, D. K. Bird, and R. G. Coleman. 2004. Chromium geochemistry in serpentinized ultramafic rocks and serpentine soils from the Franciscan complex of California. *Am. J. Sci.* 304:67–101.

Park, D., C. K. Ahn, Y. M. Kim, Y. S. Yun, and J. M. Park. 2008. Enhanced abiotic reduction of Cr(VI) in a soil slurry system by natural biomaterial addition. *J. Hazard. Mater.* 160:422–427.

Park, D., Y. S. Yun, D. S. Lee, S. R. Lim, and J. M. Park. 2006. Column study on Cr(VI)-reduction using the brown seaweed *Ecklonia* biomass. *J. Hazard. Mater.* 137:1377–1384.

Park, D., Y. S. Yun, and J. M. Park. 2004. Reduction of hexavalent chromium with the brown seaweed *Ecklonia* biomass. *Environ. Sci. Technol.* 38:4860–4864.

Park, J. H., D. Lamb, P. Panneerselvam, G. Choppala, N. Bolan, and J. W. Chung. 2011. Role of organic amendments on enhanced bioremediation of heavy metal (loid) contaminated soils. *J. Hazard. Mater.* 185:549–574.

Patterson, J. W. 1985. *Industrial Wastewater Treatment Technology*. Stoneham, MA: Butterworth Publishers.

Paya Perez, A., L. Gotz, S. Kephalopoulos, and G. Bignoli. 1988. Sorption of chromium species on soil. In: M. Astruc and J. N. Lester (eds.) *Heavy Metals in the Hydrocycle*. London: Selper, 59–66.

Pettine, M., L. D'Ottone, L. Campanella, F. J. Millero, and R. Passino. 1998. The reduction of chromium(VI) by iron(II) in aqueous solutions. *Geochim. Cosmochim. Acta* 62: 1509–1519.

Rai, D., B. M. Sass, and D. A. Moore. 1987. Chromium(III) hydrolysis constants and solubility of chromium(III) hydroxide. *Inorg. Chem.* 26:345–349.

Richard, F. C., and A. Bourg. 1991. Aqueous geochemistry of chromium: A review. *Water Res.* 25:807–816.

Sass, B. M., and D. Rai. 1987. Solubility of amorphous chromium(III)-iron(III) hydroxide solid solutions. *Inorg. Chem.* 26:2228–2232.

Shanker, A. K., C. Cervantes, H. Loza-Tavera, and S. Avudainayagam. 2005. Chromium toxicity in plants. *Environ. Int.* 31:739–753.

Steinbeiss, S., G. Gleixner, and M. Antonietti. 2009. Effect of biochar amendment on soil carbon balance and soil microbial activity. *Soil Biol. Biochem.* 41:1301–1310.

Su, Y., F. X. Han, B. Sridhar, D. L. Monts. 2005. Phytotoxicity and phytoaccumulation of trivalent and hexavalent chromium in brake fern. *Environ. Toxicol. Chem.* 24:2019–2026.

Tokunaga, S. R., M. K. Firestone, K. R. Olson, J. Wan, T. K. Sutton, A. Lanzirotti, T. C. Hazen, and D. J. Herman. 2003. In situ reduction of chromium(VI) in heavily contaminated soils through organic carbon amendment. *J. Environ. Qual.* 32:1641–1649.

Tokunaga, T. K., J. Wan, M. K. Firestone, T. C. Hazen, E. Schwartz, S. R. Sutton, and M. Newville. 2001. Chromium diffusion and reduction in soil aggregates. *Environ. Sci. Technol.* 35:3169–3174.

Weng, C. H., C. P. Huang, and P. F. Sanders. 2002. Transport of Cr(VI) in soils contaminated with chromite ore processing residue (COPR). *Pract. Period. Hazard. Toxic Radioact. Waste Manage.* 6:6–13.

Wielinga, B., M. M. Mizuba, C. M. Hansel, and S. Fendorf. 2001. Iron promoted reduction of chromate by dissimilatory iron-reducing bacteria. *Environ. Sci. Technol.* 35:522–527.

Wittbrodt, P. R., and C. D. Palmer. 1995. Reduction of Cr(VI) in the presence of excess soil fulvic acid. *Environ. Sci. Technol.* 29:255–263.

Zhitkovich, A. 2011. Chromium in drinking water: Sources, metabolism and cancer risks. *Chem. Res. Toxicol.* 24:1617–1629.

Zhou, Y. F., and J. H. Richard. 2010. Sorption of heavy metals by inorganic and organic components of solid wastes: Significance to use of wastes as low-cost adsorbents and immobilizing agents. *Crit. Rev. Environ. Sci. Technol.* 40:909–977.

16 Metal Bioavailability in *Phoomdi*—Compost from Loktak Lake (Ramsar Site), Manipur, North-East India

Maibam Dhanaraj Meitei and Majeti Narasimha Vara Prasad

CONTENTS

16.1 INTRODUCTION

Manipur, a state in north-east India in the eastern Himalayas, has a rich and diverse heritage of wetlands (Jain et al., 2011). The wetlands of Manipur cover 524.5 km^2 and account for 2.4% of the total geographical area (Singh and Singh, 1994). Loktak (93°46′ E to 93°55′ E, and 24°22′ N to 24°42′ N), a Ramsar site covering an area of 287 km^2, is the largest natural freshwater wetland in the State of Manipur (Figure 16.1) and north-east India (Trishal and Manihar, 2002). Currently, Loktak is marked by nutrient enrichment, heavy metal pollution, loss of biodiversity, a high rate of siltation and garbage dumping (Meitei and Prasad, 2015), and an alarming proliferation of floating *phoomdi* covering more than 50% of the wetland (Trishal and Manihar, 2002). The unmanaged *phoomdi* accelerates the ecological succession of Loktak from open water to marshes and swamps. *Phoomdi* is a heterogenous mass of soil, vegetation, and organic matter in different stages of decay that floats in the wetland (Meitei and Prasad, 2013). Hence, to tackle the problem with *phoomdi*, the Loktak Development Authority (LDA) has been relentlessly searching for ways to clean *phoomdi* from Loktak (LDA, 2011). Recent reports suggest the conversion of *phoomdi* biomass into compost and its use in agriculture (Singh et al., 2014; Singh and Kalamdhad, 2014). In this chapter, the speciation and fate of various heavy metals from *phoomdi* compost are discussed, with an emphasis on the possible risks of heavy metal bioavailability from *phoomdi*.

16.2 *PHOOMDI* AS ACCUMULATOR OF HEAVY METALS

Phoomdi is a habitat of a large variety of aquatic, semiaquatic, and terrestrial plants, including 128 species (Trishal and Manihar, 2002; Singh and Singh, 1994). The dominant plants are several species of grass (Poaceae) and sedge (Cyperaceae). The reeds (*Phragmites karka* Cav. Trin. Ex Steud) alone constitute about half the vegetation, particularly in the Keibul Lamjao National Park (Singh and Singh, 1994). A fully formed *phoomdi* is about 1 to 2 m thick with one-fifth of it above water

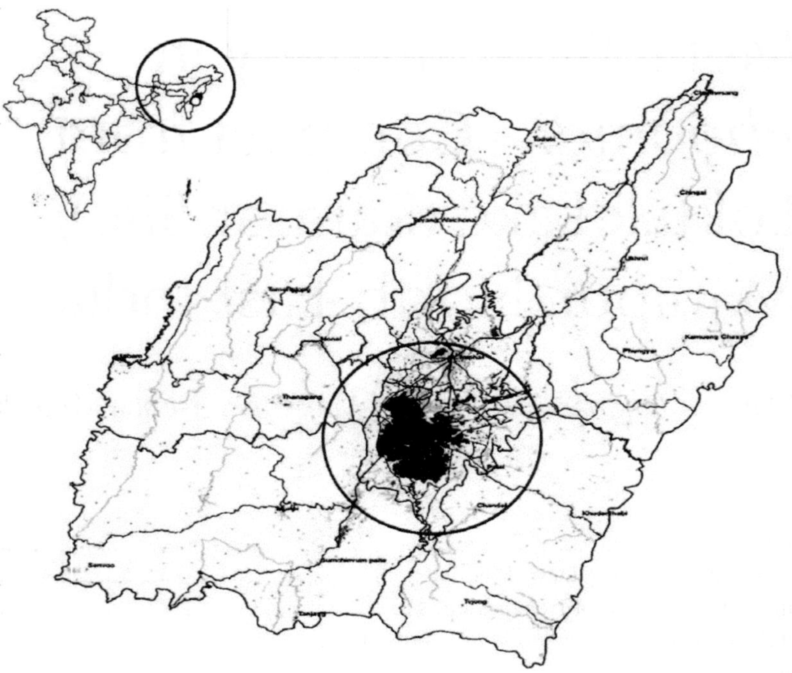

FIGURE 16.1 Geographic location of Loktak Lake, Manipur, north-east India.

and four-fifths of it submerged (Singh and Khundrakpam, 2011). A survey conducted in 2002 estimated that 134.6 km^2 or 46% of the wetland was covered by *phoomdi* (Manipur Remote Sensing Application Center, 2005). Wetland plants are reported to have a comprehensive capacity for the accumulation of heavy metals (Klink et al., 2013; Lafabrie et al., 2013) (Table 16.1). Because of their fibrous root systems with a large contact area, they have the ability to concentrate higher amounts of metals than the surrounding water (Zhang et al., 2009; Arora et al., 2008).

Heavy metals severely reduce sediment and water quality, affecting organisms in aquatic ecosystems due to their persistence, toxic nature, and ability to bioaccumulate in the food chain (Xiong et al., 2013). Small-scale industrial effluents; municipal sewage disposal; agricultural runoff; and the use of fertilizers, herbicides, and pesticides have increased pollution of the Manipur River basin wetlands, including Loktak (Meitei and Prasad, 2015). Nearly 0.28 million people living within the Nambul River catchment generate, on a daily basis, an estimated 72.2 million tons of solid waste and 31,027 m^3 of sewage. Similarly, 4.9 million tons of solid waste and 2,121 m^3 of sewage from the Nambol River drain into Loktak on a daily basis (Trishal and Manihar, 2002). Singh et al. (2013) reported that the discharge of various heavy metals by the Nambul River into Loktak plays a significant role in the pollution of the wetland.

Phoomdi includes plants that accumulate various heavy metals (Table 16.1) (Meitei and Prasad, 2015; Trishal and Manihar, 2002; Singh and Singh, 1994). Singh et al. (2013) compared the concentrations of metals between Loktak and the Nambul River, a main tributary. Metal concentrations were found to be higher in the Nambul River than in Loktak, that is, 1966.7/468.4 (Mn), 1486.5/646.2 (Fe), 244.8/45 (Zn), 5.5/3.7 (Cr), 2/1.4 (Ni), and 3/0.8 (Hg) ppb, respectively. These results signify the possible role of sediments and *phoomdi* that act as sinks and accumulators (The metal accumulation capability of *phoomdi* plants has not been studied in detail so far.). In addition, management plans for Loktak have failed to include the ecosystem services of *phoomdi* and the implications of *phoomdi* removal to the freshwater ecosystem (Figure 16.2a through d). Viewing this scenario, Singh and Kalamdhad (2014) and Singh et al. (2014) reported the use of *phoomdi* biomass as compost to protect the wetland from unwanted growth of plants and for possible application in agricultural fields.

TABLE 16.1
***Phoomdi*—A Unique Biosystem Comprising Plants That Accumulate Various Metals**

Species	Accumulation	Reference
Alisma plantago-aquatica	Removal efficiency of 50, 92, 74.1, 41, 89, and 48.3% Pb, Cd, Fe, Ni, Cr, and Cu from industrial wastewater in constructed wetlands	Khan et al. (2009)
Alternanthera sp.	Metal concentration factor of Cu (1051), Cr (722), Fe (1156), Mn (6395), Cd (23000), and Pb (555) was observed	Rai et al. (1995)
Arundo donax	Accumulation and biomonitoring of Al, As, Cd, Cr, Cu, Hg, Mn, Ni, Pb, and Zn pollution in aquatic ecosystems	Bonanno (2013)
Azolla pinnata	Accumulation of Cu, Cr, Pb, and Ni in the roots and fronds	Pandey (2012)
Ceratophyllum demersum	Uptake of Cd, Cu, Fe, Mo, Ni, and Zn up to 0.37–151, 14–84, 16.1–430, 0.18–6.43, 15.8–1430, and 267–853 μg g^{-1} dw from oligotrophic lake water	Andresen et al. (2013)
	Bioaccumulation of Cd, Co, Cr, Cu, Ni, Pb, Zn, and Fe up to 0.37, 10, 4.4, 3.7, 17, 0.9, 47, and 1250 μg g^{-1} dw from hypertrophic lake water	Hassan et al. (2010)
	Uptake of Cu, Fe, Ni, Zn, and Mn up to 43.5, 1706.2, 16.6, 372.6, and 2977.3 mg kg^{-1} dw from water of Iset River	Borisova et al. (2014)
	Removal efficiency of 50, 92, 74.1, 41, 89, and 48.3% Pb, Cd, Fe, Ni, Cr, and Cu from industrial wastewater in constructed wetlands	Khan et al. (2009)
Eichhornia crassipes	Accumulation of Fe up to 1905–6707 mg kg^{-1} dw from constructed wetlands	Jayaweera et al. (2008)
	Bioconcentration of Cr in the root tissues up to 70 μg g^{-1} dw from river water of Cachoeira	Mangabeira et al. (2004)
	Phytoextraction of Cr, Cd, Cu, Ni, Pb, and Zn from Morichal Largo River	Schorin et al. (1991)
	Uptake of 0.57, 0.27, 0.16, 0.36, and 0.48 mg kg^{-1} dw of Fe, Zn, Ni, Cd, and Cr from coal mine effluent	Mishra et al. (2008)
	Phytoextraction of Zn, Mn, Cu, Pb, Cd, Hg, and Ni up to 70–301, 269–1708, 15–18, 8.7–72, 0.95–2, 1.3–3.6, and 4.2–24 ppm from wastewater of Bangkok city	Chunk et al. (2012)
Euryale ferox	Bioconcentration of heavy metals in the order Pb>Cr>Cu>Cd from wastewater	Rai et al. (2002)
Ipomoea aquatica	Bioaccumulation of Cu, Mn, Fe, Zn, Pb, and Cd from Beung Boraphet reservoir	Dummee et al. (2012)
Lemna sp.	Cr bioaccumulation up to 0.12–4.42 mg kg^{-1} dw from wastewater by using a pilot system with continuous flow	Uysal (2013)
	Phytoaccumulation of B upto 639–2711 mg kg^{-1} dw from wastewater	Bocuk et al. (2013)
	Removal of 94, 80, 36, 33, and 27% Cr, Zn, Pb, Cd, and Cu from textile wastewater	Sekomo et al. (2012)
	Uptake of 0.28, 0.26, 0.17, 0.30, and 0.30 g kg^{-1} dw Fe, Zn, Ni, Cd, and Cr from coal mine effluent	Mishra et al. (2008)
Nelumbo nucifera	Uptake of Fe, Zn, Mn, and Cu up to 116–378, 22–46, 12–69, and 5.2–16.8 mg kg^{-1} dw from wastewater	Arora et al. (2008)
	Bioaccumulation of Cd, As, Pb, and Hg in the roots •	Xiong et al. (2013)
Neptunia oleracea	Accumulation of Cu and Zn up to 0.61 and 5.4 mg kg^{-1} dw from wastewater of Hanoi city	Khai et al. (2007)
Oenanthe javanica	Cu and Zn uptake of 0.69–1.10 and 2.44–3.0 mg kg^{-1} dw from wastewater of Hanoi city	Khai et al. (2007)

(Continued)

TABLE 16.1 (*Continued*)
***Phoomdi*—A Unique Biosystem Comprising Plants That Accumulate Various Metals**

Species	Accumulation	Reference
Phragmites sp.	Higher accumulation of Fe, Mn, and Zn compared with Hg and Cd from municipal wastewater in constructed wetlands	Vymazal et al. (2009)
	Bioaccumulation of 0.68–1.1, 0.4–7, 2.3–15, 1.1–5.2, 28–475.8, 0.5–9.1, 9.9–16.5, and 10–104.1 mg kg^{-1} dw Cd, Cr, Cu, Hg, Mn, Ni, Pb, and Zn from river water of Meridionale	Bonanno and Giudice (2010)
	Bioconcentration of Hg, Zn, As, Cu, and Cr from eutrophic water of Hengshuihu wetland	Zhang et al. (2009)
Pistia stratiotes	Accumulation of As up to 0.42–0.57 mg g^{-1} dw from coal mine effluent	Mishra et al., 2008
Polygonum sp.	Removal of 50, 92, 74.1, 41, 89, and 48.3 % Pb, Cd, Fe, Ni, Cr, and Cu from industrial wastewater in constructed wetlands	Khan et al. (2009)
Salvinia sp.	Zn, Cu, Ni, and Cr removal to the extent of 84.8, 73.8, 56.8, and 41.4 % from wastewater	Dhir and Srivastava (2011)
Typha latifolia	Accumulation of 215–340, 40–55, 30–50, 0.05–3, 10–24, 860–990, 0.21–0.44, and 21–44 mg kg^{-1} dw Zn, Ni, Cu, Pb, Co, Mn, Cd, and Cr from secondary effluents	Sasmaz et al. (2008)
	Accumulation of 113–8431, 226–1943, 20.8–373, 2.2–8.6, 0.04–7.3, 2–12.1, 2.1–27.8, 0.08–2.6, and 2.8–35.7 mg kg^{-1} dw Fe, Mn, Zn, Cu, Cd, Pb, Ni, Co, and Cr from wastewater	Klink et al. (2013)

FIGURE 16.2 (**See color insert.**) (a–b) *Phoomdi*—floating island; (c–d) *Phoomdi* removal and dumping on the bank of Loktak.

16.3 COMPOSTING AND SPECIATION OF HEAVY METALS

Composting and vermicomposting of biomass obtained from aquatic weeds are the best known approaches for biological stabilization of green waste by transforming them into a safer and more stabilized material, compost, which can be used as a soil conditioner in agricultural practices (Singh and Kalamdhad, 2014). *Composting* can be defined as the biological decomposition of organic matter to form stable humus-like end products under controlled aerobic conditions (Farrell and Jones, 2009). During composting, the organic components undergo several transformations and produce various metabolites that may inhibit or stimulate the growth of the plants before they mature into compost, which is biologically safe and stable and contains newly formed organic and inorganic humic-like substances (Singh and Kalamdhad, 2013a). The quality of the compost depends on various factors such as the composting facility design, feedstock source and proportions, procedures used, and length of maturation (Hargreaves et al., 2008). Heavy metals do not degrade and tend to concentrate during composting due to microbial degradation of the organic matters and the loss of carbon and water, posing a threat if the compost is to be applied in agricultural systems for growing food crops (Singh et al., 2014). Table 16.2 summarizes the total heavy metal content in compost from different sources reported in the literature.

Singh and Kalamdhad (2012) employed the Tessier sequential extraction method to investigate the change in heavy metal speciation during water hyacinth composting. Their results showed that metal concentrations increased due to weight loss after organic matter decomposition, release of CO_2 and water, and mineralization processes. The total metal content followed the order Fe>Pb>Mn>Ni>Cr>Zn>Cd>Cu. The exchangeable and residual fractions of Zn in trials 1 (water hyacinth) and 2, 3, 4, and 5 (water hyacinth, cattle manure, and sawdust) decreased; however, the carbonate, reducible, and oxidizable fractions of Zn increased during composting. The heavy metal concentrations after 30 days of composting ranged from 5 to 107.7 (Zn), 1.5 to 29.3 (Cu), 145.7 to 281.6 (Mn), 0.08 to 12.5 (Fe), 3.5 to 29.6 (Cr), 3.4 to 260 (Ni), 0.8 to 59 (Cd), and 79 to 1077 (Pb) mg kg^{-1} dw for trial 1 compared with 4.3 to 143 (Zn), 1.2 to 72.3 (Cu), 52 to 570.5 (Mn), 0.02 to 6.2 (Fe), 5.6 to 233.8 (Cr), 3.4 to 343.5 (Ni), 0.7 to 72.3 (Cd), and 1.8 to 1375 (Pb) mg kg^{-1} dw for trials 2, 3, 4, and 5. The trial 1 compost showed a decrease in Zn and Mn exchangeable (18 to 5 mg kg^{-1} dw Zn and 273.6 to 145.7 mg kg^{-1} dw for Mn) and residual fractions (11 to 107.7 mg kg^{-1} dw Zn and 400 to 182 mg kg^{-1} dw for Mn) and an increase in the carbonate (9.7 to 13.7 mg kg^{-1} dw Zn and 81.5 to 226.4 mg kg^{-1} dw Mn), reducible (14 to 19.7 mg kg^{-1} dw Zn and 111.8 to 278.2 mg kg^{-1} dw Mn), and oxidizable fractions (21.4 to 35.4 mg kg^{-1} dw Zn and 124.7 to 281.6 mg kg^{-1} dw Mn). Bioavailability factors (BFs) for Zn in trials 1, 2, and 5 increased from 0.36, 0.42, and 0.5 (initial) to 0.4, 0.44, and 0.53 (final); in trials 3 and 4, during composting, they decreased from 0.53 and 0.75 (initial) to 0.4 and 0.48 (final). The availability of Zn is influenced by total Zn content, pH, organic matter content, and availability of adsorption sites; phytotoxicity could be caused by these higher Zn BFs. For Cu in trial 1, the oxidizable fraction decreased and the carbonate-bound, reducible, and residual fractions increased. The BFs in trials 1 and 5 decreased from 0.44 and 0.59 at the beginning to 0.4 and 0.35, respectively, at the end due to organic matter loss and the conversion of organic matter into humic substances. Lead occurred predominantly in the residual fraction in all the raw materials. In trial 5, the exchangeable and carbonate-bound fractions of Pb decreased whereas the residual fractions increased. Intense microbial degradation of the organic matter with the release of Pb from the compost was observed during the thermophilic phase of composting. For Cd, the exchangeable and carbonate-bound fractions increased whereas the residual fractions decreased. The exchangeable fractions indicated the possible availability of Cd for plant uptake and its bioavailability in the environment. Chunk et al. (2012) investigated the speciation of heavy metals during the composting of water hyacinth grown in the wastewaters of Beung Makkasa,

TABLE 16.2
Heavy Metal Concentrations in *Phoomdi*—Compost and Compost Derived from Various Sources (mg kg^{-1} dry mass)

	Concentration (mg kg^{-1})								
Species / Type	**Cd**	**Pb**	**Ni**	**Cr**	**Cu**	**Mn**	**Zn**	**Fe**	**References**
Phoomdi only[1]	66.5	667.5	143.0	182.0	40.0	692.0	221.0	15665	Singh et al. (2014)
Phoomdi + cattle manure + rice husk[2]	59.0	852.5	222.0	117.0	56.0	596.0	230.8	13023	Singh et al. (2014)
Water hyacinth only[3]	59.0	1077.0	260.0	115.3	29.3	281.6	107.7	12.5	Singh and Kalamdhad (2012)
Water hyacinth + cattle manure + sawdust[4]	59.0	1115.0	231.5	138.0	42.7	570.5	153.3	5.7	Singh and Kalamdhad (2012)
Water hyacinth only	0.85	81.5	–	93.2	–	–	–	–	Chunk et al. (2013)
Water hyacinth + cattle manure + sawdust[5]	83.8	235.8	279.0	1537	103.3	1105	297.8	13.3	Singh and Kalamdhad (2013a)
Water hyacinth + cattle manure + sawdust + lime[6]	71.4	107.3	311.7	182.5	109.0	671.3	183.4	23.9	Singh and Kalamdhad (2013a)
Water hyacinth only[7]	57.5	1037.5	246.0	54.5	54.1	1928	177.1	43.0	Singh and Kalamdhad, 2013b
Water hyacinth + cattle manure + sawdust[8]	56.2	1115.0	231.5	138.0	42.7	570.5	153.3	5.7	Singh and Kalamdhad (2013b)
MSW - compost[9] (from various sources)	0.3–20.4	3.4–972	6.5–168	7.0–123	10.3–337	322–639	92.8–1825	1.37–14368	Saha et al. (2010); Cherif et al. (2009); Paradelo et al. (2009); Weber et al. (2007); Castaldi et al. (2005); Lasaridi et al. (2006); Soumare et al. (2003); Chwastowska and Skalmowski (1997); Breslin (1999)

Notes: [1-8] represents average values; [9] represents range values.

Thailand. The Department of Land Development used the Beung Makkasa compost for growing food crops in its agricultural systems. The compost from water hyacinth, however, was found to contain high metal concentrations that ranged from 93.2 to 179 (Cr), 47 to 181.5 (Pb), and 0.83 to 1.1 (Cd) ppm, respectively. It was recommended that Beung Makkasa compost be used for growing floral crops but not food crops.

Singh and Kalamdhad (2013a) investigated the effects of lime addition on metal bioavailability and leachibility during water hyacinth composting with cattle manure and saw dust. The metal concentrations showed significant variations from 297.8 to 183.4 (Zn), 103.3 to 89.4 (Cu), 110.5 to 612.5 (Mn), 13.3 to 24 (Fe), 279 to 342.5 (Ni), 235.8 to 1199 (Pb), 83.8 to 66.2 (Cd), and 1537 to 169 (Cr) mg kg^{-1} dw, respectively. Zinc, copper, cadmium, and chromium concentrations decreased significantly with lime amendment due to greater reduction of organic matter and the formation of organo-metallic complexes. However, concentrations of Mn, Fe, Ni, and Pb increased. The DTPA-extractable metals followed the order MN>Zn>Cu>Fe>Cr>Ni. The leachable fractions of the metals in the compost amended with lime ranged from 56.1 to 36.7 (Zn), 6.8 to 12.1 (Cu), 164 to 210 (Mn), 64.6 to 94.6 (Fe), 1.5 to 1.7 (Ni), 5.6 to 10.7 (Pb), 0.26 to 0.74 (Cr), and 7.1 to 7.6 (Cd) mg kg^{-1} dw, respectively. Likewise, the leachable fractions of Zn, Fe, Ni, Cr, and Cd decreased significantly in the control and lime-amended compost, with reductions of 67.5, 77.8, 49, 53.7, and 85.7%, respectively, in the total metal during composting. The initial pH increase after lime addition might have reduced the solubility of metals, resulting in the reduction of their leachability. Singh and Kalamdhad (2013b) reported the influence of temperature, pH, and organic matter content on the distribution of metal availability and leachability during water hyacinth composting. The results showed high metal concentrations that ranged from 144 to 187.7 (Zn), 50.3 to 54.1 (Cu), 658 to 1928 (Mn), 14.1 to 23.4 (Fe), 54.5 to 76.3 (Cr), 196 to 246 (Ni), 56.2 to 66 (Cd), and 842.5 to 1111 (Pb) mg kg^{-1} dw, respectively. Metal fractionation was influenced by their release through organic matter mineralization or metal solubilization; this occurred by the decrease of pH or metal complexation with the newly formed humic substances that made the metals less soluble and extractable. The study showed a decrease in the fractions of leachable metals compared with the total metal content because of cattle manure application. The order of the leachable metal content was Mn>Fe>Pb>Zn>Ni>Cr>Cu>Cd. The leachable fractions of the metals ranged from 21.3 to 37.2 (Zn), 4.5 to 5.4 (Cu), 175.5 to 777.6 (Mn), 63.3 to 306 (Fe), 3.6 to 10.8 (Cr), 3.6 to 10 (Ni), 0.8 to 2.5 (Cd), and 17 to 50 (Pb) mg kg^{-1} dw, respectively.

Singh et al. (2014) evaluated the availability of heavy metals and nutrients during the agitated pile composting of *phoomdi*, and this was one of the first reports of its kind from Loktak. Metal concentrations ranged from 187.3 to 230.8 (Zn), 40 to 60.3 (Cu), 587.6 to 692.4 (Mn), 13023 to 16563 (Fe), 143 to 222.2 (Ni), 667.5 to 852.5 (Pb), 59 to 68 (Cd), and 117.3 to 232.3 (Cr) mg kg^{-1} dw, respectively. The concentrations increased during composting due to weight loss of the matter that was caused by the decomposition of organic matter, resulting in the release of CO_2 and subsequent mineralization. The metal content of *phoomdi* compost followed the order Fe>Pb>Mn>Zn>Ni>Cr>Cd>Cu. The water-soluble fraction of metals may be readily bioavailable due to the decomposition of organic matter and the formation of complex compounds with newly formed humic substances. It represents the most toxic fraction of the metals to plants. Water-soluble metal fractions decreased in the range of 12.6 to 38.3% for Zn, 12.1 to 56% for Cu, 16.4 to 58.3% for Mn, 6.8 to 28.3% for Fe, 11.7 to 47.7% for Pb, and 31.6 to 46.3% for Cr after composting. The concentrations of DTPA-extractable metals ranged from 24.8 to 66.7% for Zn, 29 to 49% for Cu, 22.1 to 45.4% for Mn, 42.4 to 50% for Fe, 42.4 to 53% for Ni, 32.3 to 50.3% for Pb, 4.2 to 37% for Cd, and 17 to 52.4% for Cr, respectively, during composting. The reduction order was Zn>Ni>Cr>Pb>Cu>Mn>Fe>Cd. The reduction of DTPA-extractable metals could be due to the formation of organo-metallic complexes. Further, it could be explained as the transformation of easily available metal fractions into a more or less soluble form during composting. The concentrations of leachable metal fractions ranged from 39 to 89 (Zn), 12 to 16 (Cu), 214.5 to 371 (Mn), 138 to 347.5 (Fe), 3.3 to 6 (Ni), 18.2 to 42.8 (Pb), 0.81 to 2.4 (Cd), and 34.3 to 57.8 (Cr) mg kg^{-1} dw, respectively. The leachibility of metals was reduced most for Cd and Ni.

Increased pH affected the leachability of metals either through solubility or equilibria or due to the complexation of soluble and surface ligands (Singh and Kalamdhad, 2014). This study showed that the composting of *phoomdi* biomass with the appropriate proportion of cattle manure is the best method for reducing the bioavailability and leachability of metals. The metal concentration in the *phoomdi* compost ranged from 165.3 to 201 (Zn), 562 to 637.2 (Mn), 62.2 to 64.7 (Cd), 37.3 to 41 (Cu), and 10,800 to 16,500 (Fe) mg kg^{-1} dw, respectively. Singh and Kalamdhad (2014) reported that the composting of *phoomdi* could serve as the best alternative for the utilization of this huge floating biomass as a fertilizer in agricultural practices and as a means of protecting the wetland.

16.4 HEAVY METAL BIOAVAILABILITY IN *PHOOMDI* COMPOST

Singh et al. (2014) concluded that the composting of *phoomdi* biomass followed by land application can be an economical option for the disposal of the available green biomass. The transformation of *phoomdi* into compost may help protect the wetland and reduce the application of chemical fertilizers for agricultural practices (Singh and Kalamdhad, 2014). However, *phoomdi* is reported to act as a filter for the purification of the lake water and to accumulate large amounts of hazardous wastes (Trishal and Manihar, 2002). From the literature, it is observed that *phoomdi* species, namely, *Alternanthera* sp., *Alisma plantago aquatica*, *C. demersum*, *E. crassipes*, *N. nucifera*, *Lemna* sp., *Polygonum* sp., *S. polyrhiza*, *Salvinia cucullata*, *Pistia stratiotes,* and *Phragmites* sp. are either accumulators or hyperaccumulators of various metals (Bocuk et al., 2013; Dhir and Srivastava, 2011; Khan et al., 2009; Zhang et al., 2009; Arora et al., 2008; Jayaweera et al., 2008; Rai et al., 1995). Singh et al. (2014) reported that the ability of *phoomdi* to concentrate metals increases in the compost, thereby creating a possible threat or making the compost unfit for applications in agricultural soils without proper long-term field studies on metal bioavailability. Further, the compost reported in the study crossed the threshold limits of various European, American, and Australasian standards for metals that were allowed in compost for agricultural applications (WRAP, 2002). Cadmium, lead, nickel, and chromium contents were significantly higher in the compost due to weight loss in the course of composting. These results presented a possible risk to the health of humans and ecosystems because of metal leaching from *phoomdi* compost. Further, the amount of metals in the compost will differ based on the species selected for composting, as *phoomdi* is composed of nearly 128 plants and different plants will have different accumulation capabilities for various metals.

In addition, *phoomdi* sequesters nearly 50% of the mineral nutrients present in the wetland system and helps in reducing their concentration within water, thereby suppressing algal growth (Trishal and Manihar, 2002). The conversion of *phoomdi* biomass into compost ignores the role of *phoomdi* as a biological sink and the negative consequences that may occur to the wetland ecosystem if harvesting is unsustainable. *Phoomdi* also plays an important role in the socioeconomic and traditional life of the locals (Meitei and Prasad, 2015). The presence of edible, medicinal, fodder, fuel, and house-making materials, and plants that are useful in making handicrafts are reported from *phoomdi* (Meitei and Prasad, 2015; Trishal and Manihar, 2002). During the process of composting, edible plants such as *Oenanthe javanica*, *Centella asiatica*, and *Colocasia esculenta* or medicinal plants such as *Hedychium coronarium* and *Zizania latifolia* will be converted into compost. Further, Loktak represents the largest fishery resource of Manipur, accounting for more than 50% of its fish production. Thirty-nine percent of the harvest from Loktak comes from *athaphoom* fishing, contributing significantly to the socio economy of the locals (Meitei and Prasad, 2015; Trishal and Manihar, 2004). *Athaphoom* or *phoom* fishing represents a traditional practice of fishing in Loktak using *phoomdi.* It involves the preparation of a fishing enclosure made up of *phoomdi* (Meitei and Prasad, 2015).

Thus, the various aspects mentioned earlier need to be considered before conversion of the floating islands—*phoomdi* of Loktak—into compost; otherwise, it may lead to disturbance of the socio economy and ecology of the wetland and the surrounding environment (Figure 16.3).

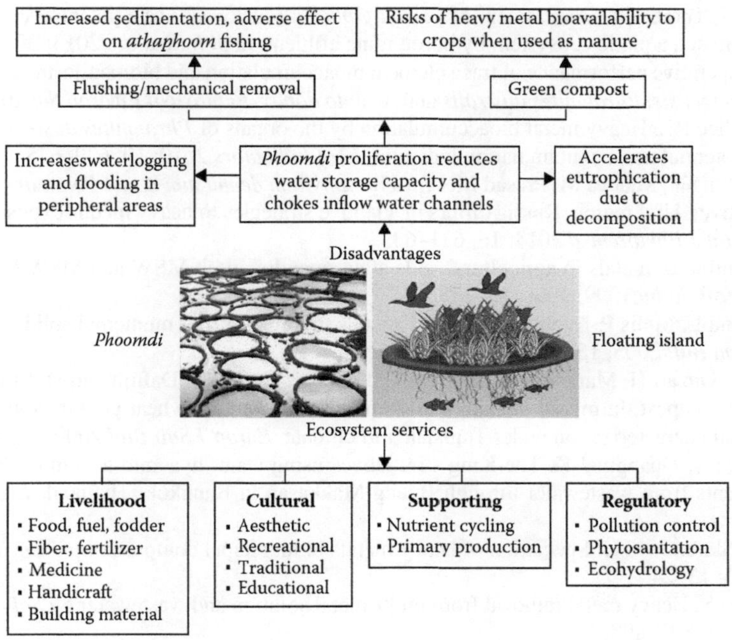

FIGURE 16.3 An approach toward sustainable utilization and conservation of *phoomdi*—floating island of Loktak.

16.5 CONCLUSIONS

Although *phoomdi* compost is proposed to serve as an alternative for artificial fertilizers, there is an inherent risk of metal bioavailability to humans and other organisms if the compost is applied to agricultural areas. The *phoomdi* species showed significant metal accumulation capabilities, indicating potential risks if the plants are converted into compost and used for growing food crops. *Phoomdi* compost conversion requires an in-depth compositional analysis of the unique *phoomdi* biosystem that consists of nearly 128 species. The selection of species such as *E. crassipes*, *Alternanthera* sp., and *Phragmites* sp. is not advisable, as these species are reported to be good accumulators of various metals. Furthermore, long-term field analysis is needed to investigate the bioavailability and leaching of metals into the environment from compost applied to the soil. Therefore, the application of *phoomdi* compost and its use as a means to control *phoomdi* proliferation in Loktak is a promising hypothesis that needs further in-depth scientific study before use in the field.

ACKNOWLEDGMENTS

Maibam Dhanaraj Meitei gratefully acknowledges University of Hyderabad for the scholarship. Majeti Narasimha Vara Prasad gratefully acknowledges the award of the Pitamber Pant National Environment Fellowship by the Ministry of Environment and Forests, GOI, New Delhi (MoEF Ref. No. 17/3/2010-RE Dt 29-2-2012).

REFERENCES

Andresen E, Opitz J, Thomas G, Stark HJ, Dienemann H, Jenemann K, Dickson BC, Kupper H. Effects of Cd and Ni toxicity to *Ceratophyllum demersum* under environmentally relevant conditions in soft and hard water including a German lake. *Aquat Toxicol* 2013; 142–143: 387–402.

Arora M, Kiran B, Rani S, Rani A, Kaur B, Mittal N. Heavy metal accumulation in vegetables irrigated with water from different sources. *Food Chem* 2008; 111: 811–815.

Bocuk H, Yakar A, Turker OC. Assessment of *Lemna gibba* L. (duckweed) as a potential ecological indicator for contaminated aquatic ecosystem by boron mine effluent. *Ecol Indicators* 2013; 29: 538–548.

Bonanno G. Comparative performance of trace element bioaccumulation and biomonitoring in the plant species *Typha domingensis*, *Phragmites australis* and *Arundo donax*. *Ecotoxicol Environ Saf* 2013; 97: 124–130.

Bonanno G, Giudice RL. Heavy metal bioaccumulation by the organs of *Phragmites australis* (common reed) and their potential use a contamination indicator. *Ecol Indicators* 2010; 10: 639–645.

Borisova G, Chukina N, Maleva M, Prasad MNV. *Ceratophyllum demersum* L and *Potamogeton alpinus* Balb. from Iset River, Ural region, Russia differs in adaptive strategies to heavy metals exposure—A comparative study. *Int J Phytoremed* 2014; 16: 621–633.

Breslin VT. Retention of metals in agricultural soils after amending with MSW and MSW-biosolids compost. *Water Air Soil Pollut* 1999; 109: 163–178.

Castaldi P, Santona L, Melis P. Evolution of heavy metals mobility during municipal solid waste composting. *Fres Environ Bull* 2005; 15.

Cherif H, Ayari F, Ouzari H, Marzorati M, Brusetti L, Jedidi N, Hassen A, Daffonchio D. Effects of municipal solid waste compost, farmyard manure and chemical fertilizers on wheat growth, soil composition and soil bacterial characterization under Tunisian arid climate. *Europ J Soil Biol* 2009; 45: 138–145.

Chunk K, Nimpee C, Duangmal K. The King's initiatives using water hyacinth to remove heavy metals and plant nutrients from wastewater through Bueng Makkasan in Bangkok, Thailand. *Ecol Eng* 2012; 39: 40–52.

Chwastowska J, Skalmowki K. Speciation of heavy metals in municipal composts. *Int J Environ Analyt Chem* 1997; 68(1): 13–24.

Dhir B, Srivastava S. Heavy metal removal from multi-metal solution and wastewater by *Salvinia natans*. *Ecol Eng* 2011; 37: 893–898.

Dummee V, Kruatrachue M, Trinachartvanit W, Tanhan P, Pokethitiyook P, Damrongphol P. Bioaccumulation of heavy metals in water, sediments, aquatic plant and histopathological effects on the golden apple snail in Beung Boraphet reservoir, Thailand. *Ecotoxicol Environ Saf* 2012; 86: 204–212.

Farrell M, Jones DL. Critical evaluation of municipal solid waste composting and potential compost markets. *Bioresour Technol* 2009; 100: 4301–4310.

Hargreaves JG, Adl MS, Warman PR. A review of the use of composted municipal solid waste in agriculture. *Agricult Ecosyst Environ* 2008; 123: 1–14.

Hassan S, Schmieder K, Böcker R. Spatial patterns of submerged macrophytes and heavy metals in the hypertrophic, contaminated, shallow reservoir Lake Qattieneh/Syria. *Limnol-Ecol Manage Inland Waters* 2010; 40(1): 54–60.

Jain A, Sundriyal M, Roshnibala S, Kotoky R, Kanjilal PB, Singh HB, Sundriyal RC. Dietary use and conservation concern of edible wetland plants of Indo-Burma region: A case study from North-Eastern India. *J Ethnobiol Ethnomed* 2011; 7: 29.

Jayaweera MW, Kasturiarachchi JC, Kularatne RKA, Wijeyekoon SLJ. Contribution of water hyacinth (*Eichhornia crassipes* (Mart.) Solms) grown under different nutrient conditions to Fe- removal mechanisms in constructed wetlands. *J Environ Manage* 2008; 87: 450–460.

Khai NM, Ha PQ, Öborn I. Nutrient flows in small-scale peri-urban vegetable farming systems in Southeast Asia: A case study in Hanoi. *Agricul Ecosyst Environ* 2007; 122(2): 192–202.

Khan S, Ahmad I, Shah MT, Rehman S, Khaliq A. Use of constructed wetlands for the removal of heavy metals from industrial wastewater. *J Environ Manage* 2009; 90: 3451–3457.

Klink A, Maciol A, Wislocka M, Krwaczyk J. Metal accumulation and distribution in the organs of *Typha latifolia* L. (cattail) and their potential use in bioindication. *Limnologicia* 2013; 43: 164–168.

Lafabrie C, Major KM, Major CS, Cebrian J. Trace metal contamination of the aquatic plant *Hydrilla verticillata* and associated sediment in a Coastal Alabama Creek (Gulf of Mexico—USA). *Marine Pollut Bull* 2013; 68: 147–151.

Lasaridi K, Protopapa I, Kotsou M, Pilidis G, Manios T, Kyriacou A. Quality assessment of composting in the Greek market: The need for standards and quality assurance. *J Environ Manage* 2006; 80: 58–65.

Loktak Development Authority (LDA). Annual report. LDA, Government of Manipur, India; 2011.

Mangabeira PAO, Labejot L, Lamperti A, Almeida AAF, Oliveira AH, Escaig F, Severo MIG, Silva DC, Saloes M, Mielke MS, Lucena ER, Martins MC, Sanatana KB, Gavrilov KL, Galle P, Levi-Setti R. Accumulation of chromium in root tissues of *Eichhornia crassipes* (Mart.) Solms in Cachoeira river—Brazil. *Appl Surf Sci* 2004; 231–232: 497–501.

Manipur Remote Sensing Application Centre. Wetlands of Manipur. Manipur Remote Sensing Application Centre, Imphal, Manipur, 2005.

Meitei MD, Prasad MNV. *Phoomdi*—A unique plant biosystem of Loktak lake, Manipur, North-East India: Traditional and ecological knowledge. *Plant Biosyst* 2015; 149(4): 777–787.

Meitei MD, Prasad MNV. Phytotechnological applications of *phoomdi*, Loktak lake, Manipur, Northeast India. *Curr Sci* 2013; 105(5): 569–570.

Mishra VK, Upadhyaya AR, Pandey SK, Tripathi BD. Heavy metal pollution induced due to coal mining effluent on surrounding aquatic ecosystem and its management through naturally occurring aquatic macrophytes. *Bioresour Technol* 2008; 99: 930–936.

Pandey VC. Phytoremediation of heavy metals from fly ash pond by *Azolla caroliniana*. *Ecotoxicol Environ Saf* 2012; 82: 8–12.

Paradelo K, Moldes AB, Rodriguez M, Barral MT. Relationship between heavy metals and phytotoxicity in composts. *Ciencia Y Tecnologia Alimentaria* 2009; 6(2): 143–151.

Rai UN, Sinha S, Tripathi RD, Chandra P. Wastewater treatability potential of some aquatic macrophytes: Removal of heavy metals. *Ecol Eng* 1995; 5: 5–12.

Rai UN, Tripathi RD, Vajpayee P, Jha V, Ali MB. Bioaccumulation of toxic metals (Cr, Cd, Pb and Cu) by seeds of *Euryale ferox* Salisb. (Makhana). *Chemosphere* 2002; 46(2): 267–272.

Saha JK, Panwar NR, Singh MV. Determination of lead and cadmium concentration limits in agricultural soil and municipal solid waste compost through an approach of zero tolerance to food contamination. *Environ Monit Assess* 2010; 168: 397–406.

Sasmaz A, Obek E, Hasan H. The accumulation of heavy metals in *Typha latifolia* L. grown in a stream carrying secondary effluent. *Ecol Eng* 2008; 33: 278–284.

Schorin H, De Benzo ZA, Bastidas C, Velosa M, Marcano E. The use of water hyacinths to determine trace metal concentrations in the tropical Morichal Largo river, Venezuela. *Appl Geochem* 1991; 6(2): 195–200.

Sekomo CB, Rousseau DPL, Saleh SA, Lens PNL. Heavy metal removal in duckweed and algae ponds as a polishing step for textile wastewater treatment. *Ecol Eng* 2012; 44: 102–110.

Singh AL, Khundrakpam ML. *Phumdi*—Proliferation: A case study of Loktak lake, Manipur. *Water Environ J* 2011; 25: 99–105.

Singh J, Kalamdhad AJ. Effects of lime on bioavailability and leachability of heavy metals during agitated pile composting of water hyacinth. *Bioresour Technol.* 2013a; 138: 148–155.

Singh J, Kalamdhad AJ. Assessment of bioavailability and leachability of heavy metals during rotary drum composting of green waste (water hyacinth). *Ecol Eng* 2013b; 52: 59–69.

Singh J, Kalamdhad AS. Concentration and speciation of heavy metals during water hyacinth composting. *Bioresour Technol* 2012; 124: 169–179.

Singh NKS, Devi CHB, Sudarshan M, Meetei NS, Singh TB, Singh NR. Influence of Nambul river on the quality of freshwater in Loktak lake. *Int J Water Res Environ Eng* 2013; 5(6): 321–327.

Singh TH, Singh RKS. Ramsar sites of India, Loktak lake, WWF-India, New-Delhi, India; 1994.

Singh WR, Kalamdhad AS. Potential for composting of green *phumdi* biomass of Loktak lake. *Ecol Eng* 2014; 67: 119–126.

Singh WR, Kumar PS, Jiwan S, Kalamdhad AS. Evaluation of bioavailability of heavy metals and nutrients during agitated pile composting of green *phumdi*. *Res J Chem Environ* 2014; 18(4): 37–48.

Soumare M, Tack F, Verloo M. Characterisation of Malian and Belgian solid waste composts with respect to fertility and suitability for land application. *Waste Manage* 2003; 23: 517–522.

Trishal CI, Manihar T. Management of *phumdis* in Loktak lake, Manipur and proceedings of a workshop held at Imphal. WISA (Wetlands International South-Asia)-LDA, Imphal, 2002.

Uysal Y. Removal of chromium ions from wastewater by duckweed (*Lemna minor* L.) by using a pilot system with continuous flow. *J Hazard Mater* 2013; 263: 486–492.

Vymazal J, Kropfelova L, Svehla J, Chrastny V, Stichova J. Trace elements in *Phragmites australis* grown in constructed wetlands for treatment of municipal wastewater. *Ecol Eng* 2009; 35: 3030–309.

Weber J, Karczeuska A, Drozd J, Licznar M, Licznar S, Jamroz E, Kocowicz A. Agricultural and ecological aspects of a sandy soil as affected by the application of municipal solid waste composts. *Soil Biol Biochem* 2007; 39: 1294–1302.

WRAP (Waste and Resources Action Programme). Comparison of compost standards within the EU, North America and Australasia. The Waste and Resources Action Programme, Oxford, UK, 2002.

Xiong C, Zhang Y, Xu X, Lu Y, Quyung BO, Ye Z, Li H. Lotus roots accumulate heavy metals independently from soil in main production regions of China. *Sci Horticulturae* 2013; 164: 295–302.

Zhang M, Cui L, Sheng L, Wang Y. Distribution and enrichment of heavy metals among sediments, waterbody and plants in Hengshiuhu wetland of Northern China. *Ecol Eng* 2009; 35: 563–569.

Section III

Remediation

17 Remediation of Metal-Contaminated Sediments

Anna Sophia Knox and Michael H. Paller

CONTENTS

17.1 INTRODUCTION

Water is a highly precious resource. More than half of the world's animal and plant species live in the water. Less than 5% is nonsaline (Yong, 2001), and only 0.2% and 0.3% are found in lakes and rivers, respectively, and are readily available for human use. Therefore, the protection of water and sediment quality is highly important. Rapid industrialization and urbanization leads to the contamination of sediments with heavy metals and creates a pervasive problem worldwide. Major sources of metal pollutants for the aquatic environment include agricultural and urban lands, industrial activities, spills, and accidents. In general, pollutants enter the environment through surface runoff that discharges into rivers, lakes, and groundwater or through point sources from municipalities, industries, or other sources. These contaminants pose a potential risk to the environment and human health, because they can harm aquatic organisms and enter aquatic food chains that lead to humans. Metals that enter the aquatic environment often accumulate in sediments that, subsequently, act as a source for contaminant remobilization. As, Cd, Cu, Hg, Ni, Pb, and Zn are often found in harbor sediments and in other areas that are affected by anthropogenic activities.

17.2 REMEDIATION OF CONTAMINATED SEDIMENTS

Contaminated sediments have reduced biodiversity and pose a significant risk to aquatic environments; however, the ability of contaminated sediments to support diverse benthic communities and healthy fish populations can be restored by remedial technologies (Figure 17.1). Current remedial options for contaminated sediments include no action, monitored natural recovery (MNR), institutional controls (land use restrictions, etc.), removal (dredging), in situ treatment and management, and ex situ treatment and management (Figure 17.1). These methodologies have different modes of action. For example, passive caps, which are composed of inert materials, reduce the environmental accessibility of sediment contaminants by imposing a relatively thick physical barrier between them and benthic organisms. Active caps, which are usually thinner and composed of chemically active

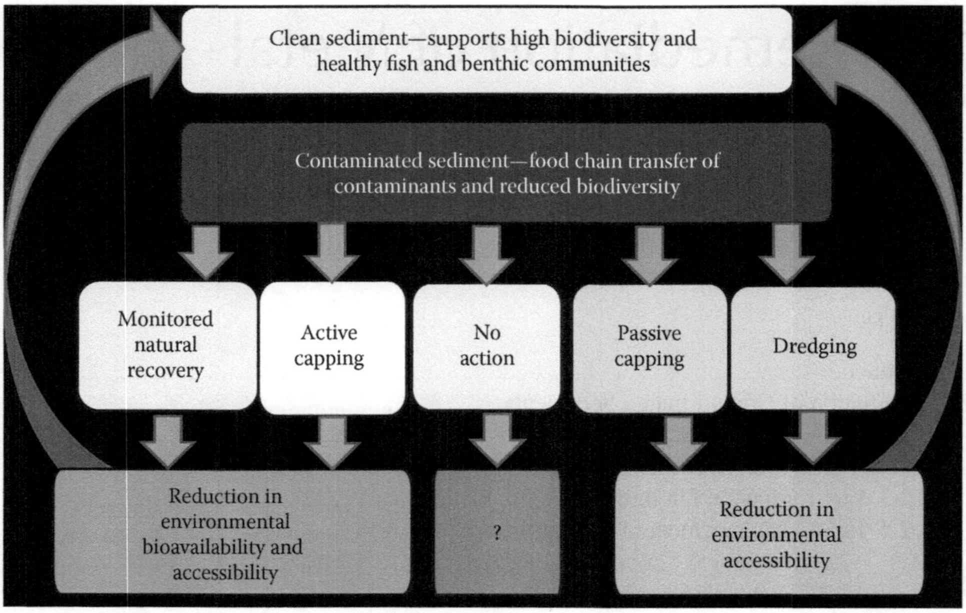

FIGURE 17.1 **(See color insert.)** Current remedial options for contaminated sediments.

materials, may have some influence on environmental accessibility but act primarily by altering the chemistry of contaminants to reduce their bioavailability.

17.2.1 No Action

"No action" is not equivalent to monitored natural recovery or natural attenuation. If no risk or minimal risk is determined from a risk assessment, no feasibility study is needed and a no further action determination is appropriate.

17.2.2 Monitored Natural Recovery

The National Research Council (NRC) defines MNR as a practice that "relies on un-enhanced natural processes to protect human and environmental receptors from unacceptable exposures to contaminants" (NRC, 2000). The definition provided by the United States Environmental Protection Agency (USEPA, 2005) is slightly different. It says that MNR employs the ongoing, naturally occurring processes to contain, destroy, or otherwise "reduce the bioavailability or toxicity of contaminants" in sediments. Monitoring is included to confirm that the risk is reduced as expected.

The MNR relies on natural physical, chemical, and biological processes to achieve remedial objectives. It is viable where the short-term risks posed by the contaminants are acceptable, the sediments are fairly stable, and the overall cost/benefit of MNR is desirable. The weaknesses of natural recovery are risk of changes in the site's natural processes, risk of contaminant dispersion due to natural or anthropogenic events, and risks posed by the current sediment concentrations/distribution until acceptable levels are achieved. Therefore, the successful implementation of MNR depends on the following conditions:

1. Natural recovery processes are transforming, immobilizing, isolating, or removing chemical contaminants in sediments to levels that achieve acceptable risk reduction within an acceptable period.

2. Source control has been achieved or sources are sufficiently minimized such that these natural recovery processes can be effective. This condition is common to all sediment remedies but particularly to MNR, because slow rates of recovery could be outpaced by ongoing releases.

The natural processes of interest for MNR include a variety of mechanisms that, under favorable condition, act without human intervention to reduce the mass, toxicity, mobility, or concentration of contaminants in the sediment bed. The major natural processes are summarized in Table 17.1.

Chemical processes in sediments are very important for metals. Many environmental parameters govern the chemical state of metals in sediments, which affects their mobility, toxicity, and bioavailability. The formation of relatively insoluble metal sulfides under reducing conditions can effectively control the risk posed by metal contaminants if reducing conditions are maintained. The master factors controlling the speciation of metals in sediments include pore water pH and alkalinity, sediment grain size, oxidation–reduction conditions, the concentration of sulfides, and the organic carbon in sediments. In addition, many chemical processes in sedimentary environments are affected by the biological community in the sediments.

Physical processes do not directly change the chemical nature of contaminants. Instead, they bury, mix, dilute, or transfer contaminants to another medium. Physical processes of interest for MNR include sedimentation, erosion, diffusion, dilution, dispersion, advection, and volatilization (Table 17.1). All these processes may reduce contaminant concentrations in surface sediment and, thus, reduce the risk associated with the sediment. Physical processes in sediments can operate at vastly different rates. In general, processes in which contaminants are transported by bulk movement of particles or pore water (e.g., erosion) occur at faster rates than processes in which contaminants are transported by diffusion or volatilization and, therefore, could be more important when evaluating MNR. Some physical processes are continuous, and others are either seasonal or episodic. For example, episodic flooding could have a positive or negative effect on risk over an entire site. Flooding is most likely to cause erosion in some areas, whereas it causes significant deposition in others. Transport and deposition of cleaner sediments may lead to natural burial of contaminated sediments, reduce the bioavailability of contaminants to aquatic plants and animals, and, therefore, reduce toxicity and bioaccumulation. The overlaying cleaner sediments will also reduce the flux of contaminants into the surface water.

Biological processes include biodegradation, a chemical change that is facilitated by microorganisms living in the sediments. Unlike organic contaminants, metals are not destroyed by biodegradation, although biological action can affect metal chemistry. One limitation of biodegradation as a risk-reduction mechanism for organic contaminants is that the greater the molecular weight of the contaminant, the greater the partitioning to sorption sites on sediment particles and the lower the contaminant availability to microorganisms.

A site-specific evaluation is important to determine whether MNR is a successful remedy. There are a variety of empirical and modeling methods to assess sediment and contaminant fate and transport, that is, MNR effectiveness (Magar et al., 2009). The first level can include simple correlations and statistical models. The second level would be a conceptual model with the incorporation of various trends and observations. The next level would include numerical modeling. However, models can be difficult to verify and calibrate, and model predictions can be uncertain; therefore, models should be not be used as a replacement for data collection.

To demonstrate that natural attenuation is taking place, lines of evidence should be established to indicate decreases in contaminant concentrations (USEPA, 2005; Magar et al., 2009). The most important lines of evidence for MNR are: (1) decreases in contaminant concentration in highertrophic-level biota (e.g., fish) over time, (2) decreasing concentrations in the water column, (3) decreasing concentrations in the sediment profile over time, and (4) trends of decrease in surface sediment concentrations and sediment toxicity.

TABLE 17.1
Overview of Natural Recovery Processes

Natural Recovery Processes	Mechanisms	Effectiveness	Examples
Chemical	Change in chemical structure or valence state: • Abiotic chemical reaction • Mineralization • Redox transformation	Achieves risk reduction to the extent that the transformation process eliminates, detoxifies, or reduces the bioavailability of the contaminant	Cr(VI) + Sulfide / TOC / Iron → Cr(III) Hackensack River, Jersey City, NY (Martello et al., 2007) Me^{2+} + Sulfide → MeS Foundry Cove, NY (USEPA, 2005)
	Sequestration via sorption: • Association or bonding with solids	Contaminant sorption reduces solubility and bioavailability	
Physical	Physical containment: • Burial via natural sedimentation • Surface sediment dilution via mixing with clean sediment • Consolidation and cohesion of sediment bed • Natural sediment winnowing and bed armoring	Achieves risk reduction by reducing direct exposure to contaminants in the surface sediment, reducing contaminated sediment suspension, and off-site transport	Burial of mercury-contaminated sediment in Eight-Day Swamp, NJ (Weis et al., 2005)
	Physical mechanisms: • Advection (water flow) • Diffusion (molecular) • Dispersion (tortuosity, mixing)	Effective in achieving risk reduction via reduced exposure and bioavailability	Dispersion of selenium-contaminated sediment in Belews and Hyco Lakes, North Carolina (Finley and Garrett, 2007)
Biological	• Biological transformation	Biological transformation is effective in reducing risk and bioavailability, mostly of organic contaminants	Degradation of polycyclic aromatic hydrocarbons (PAHs) in tidal marsh sediments, Charleston, SC (Boyd et al., 2000)

17.2.3 Environmental Dredging

Environmental dredging is defined as the removal of contaminated sediments from a water body for purposes of sediment remediation. As defined by the USEPA (2005), a hierarchy of objectives for the most contaminated sediment remediation projects can be described in terms of remedial action objectives (RAOs), remediation goals (RGs), and sediment cleanup levels (CULs). However, from an engineering standpoint, active remedies for contaminated sediment sites are formulated and designed to achieve sediment CULs. Environmental dredging involves the following seven major steps (Table 17.2):

1. Initial evaluation
2. Site and sediment assessment
3. Performance standards
4. Selection of dredging equipment
5. Dredging operations
6. Monitoring
7. Disposal and treatment options

Table 17.2 discusses these environmental dredging steps in more detail and presents key considerations for each step.

Environmental dredging is relatively expensive, and it can mobilize contaminants and destroy benthic ecosystems. The National Research Council (2007) summarized the effectiveness of environmental dredging at 26 projects. This study revealed that data from most sites was insufficient to determine whether long-term reduction goals were met and risk-based goals were achieved. This research also suggested that dredging alone is unlikely to be effective in reaching short-term or long-term goals at sites with unfavorable conditions, because increased contaminant resuspension, contaminant release, and residual contamination tend to limit the ability to meet CULs (NRC, 2007).

At present, there are different guidelines for the disposal of dredged sediments, and countries with different guidelines are seeking harmonization in treating dredged materials (Bolam et al., 2006). However, there are only two basic options for the disposal of dredged contaminated sediments: (1) disposal in a secure landfill and (2) treatment of the contaminated sediments and reuse of the treated sediments (Figure 17.2). Dredged contaminated sediments can be treated by the following technologies (Mulligan et al., 2010):

1. Physical remediation technologies—physical separation, sediment washing, flotation, and ultrasonic cleaning
2. Chemical/thermal remediation technologies—oxidation, electrokinetic remediation, solidification/stabilization, vitrification, thermal extraction
3. Biological remediation technologies—slurry reactors, landfarming, composting, bioleaching, bioconversion processes, and phytoremediation

Most ex situ remediation technologies for soils or mineral ores can be used for dredged sediments; however, they may be more expensive due to the environmental properties of sediments. The most promising technologies for metal-contaminated dredged sediments include washing, electrochemical remediation, flotation, ultra-assisted extraction, and immobilization/stabilization.

Sediment washing is a relatively simple and useful ex situ remediation technology that uses wash water or wash solution to separate contaminants from the bulk of the sediment. To enhance the performance of sediment washing, various additives can be employed, for example acids (HNO_3), chelating agents [e.g., ethylenediaminetetraacetic acid (EDTA), diethylene triamine pentaacetic acid (DTPA), and ethylenediamine disuccinate (EDDS)], or surfactants (e.g., rhamnolipid). This technology is the most appropriate for weakly bound metals with exchangeable, hydroxides, carbonates,

TABLE 17.2
Environmental Dredging: Steps and Key Considerations

Steps	Key Considerations
Initial evaluation	• Should be part of a feasibility study or remedial design • Is dredging feasible at the site? • What area and volumes should be considered? • What type of dredging is appropriate?
Site and sediment assessment	• Site access • Presence of debris that could affect dredging • Physical conditions, including sediment characteristics • Sediment geochemistry • Compatibility with disposal requirements
Performance standards	• Should achieve all RAOs and CULs
Selection of dredging equipment	• Volume to be dredged • Site conditions • Physical and chemical characteristics of the sediment • Presence of debris, vegetation, or loose rock • Physical site constraints, e.g., bridges • Distance to the disposal site • Treatment/disposal methods • Availability/cost of equipment • Performance standards
Dredging operations	• Operation and monitoring plan • Resuspension and contaminant release • Residual sediments • Passing the cleanup limits
Monitoring	• Monitoring approaches • Equipment • Data management and interpretation • Requiring a written plan
Disposal and treatment	• Transport • Staging and dewatering • Sediment treatment technologies: biological, chemical, and physical • Sediment disposal: sanitary/hazardous waste landfills, confined disposal facilities, and contained aquatic disposal

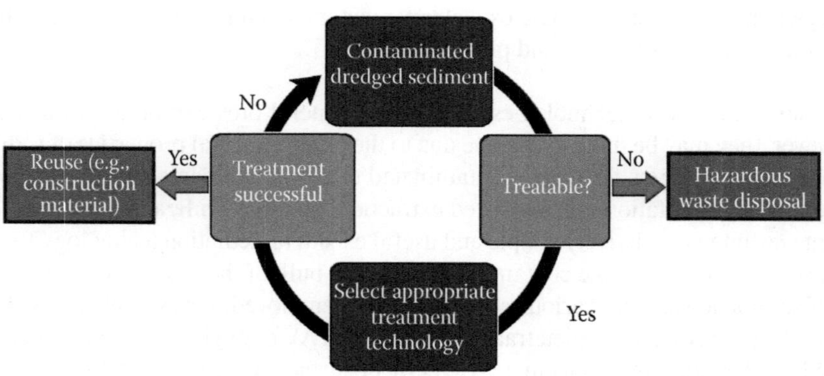

FIGURE 17.2 Strategies for dredged contaminated sediments.

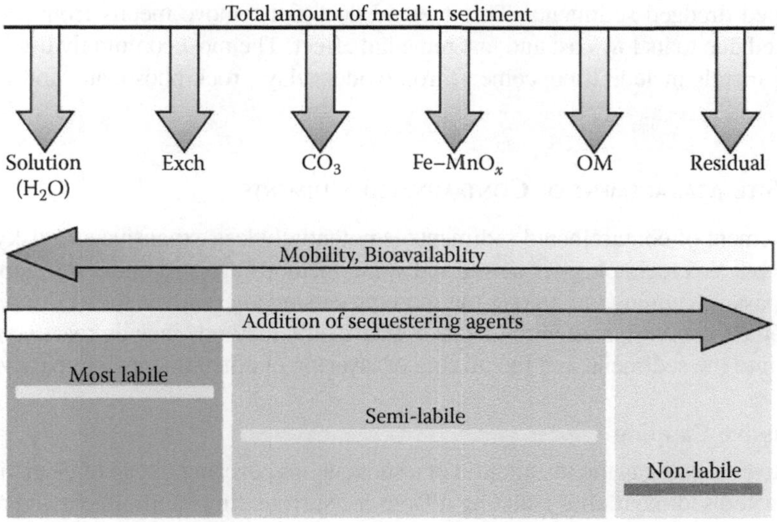

FIGURE 17.3 **(See color insert.)** Reduction of metal mobility/bioavailability with addition of sequestering agents.

and reducible oxide fractions (Figure 17.3). Removal of metals from the residual fraction (i.e., metals in mineral structures) (Figure 17.3) is the most difficult, and it is not affected by the washing process (Catherine et al., 2001; Ortega et al., 2008). Also, because fine grain sediments are difficult to decontaminate through washing, this technology is most applicable to sands and gravels (Peng et al., 2009).

Electrochemical remediation involves applying a low-DC current or a low-potential gradient to electrodes that are inserted into the contaminated sediment, resulting in the migration of charged ions. Positive ions move toward the negatively charged cathode, and negative ions move toward the positively charged anode. For example, under an induced electric potential, the anionic Cr(VI) migrates toward the anode, whereas the cationic Cr(III), Ni(II), and Cd(II) migrate toward the cathode. Once remediation is over, the contaminants that have accumulated at the electrodes are extracted by methods such as electroplating, precipitation/co-precipitation, pumping water near the electrodes, or complexing with ion-exchange resins (Krishna et al., 2001). Electric conductivity is the highest in fine sediment particles; therefore, this method is well suited for fine-grained dredged sediments.

Flotation is a method that uses gas bubble attachment to the dispersed phase (separation of heterophase systems). Flotation is widely used in the mining industry to separate mineral ores (Matis, 1995). This technology is used to remove metal sulfides from anaerobic sediments, especially fine-textured (20–50 μm) sediments. It is expected that various metal ions (e.g. Cd, Ni, Pb, and Zn) would be present as sulfides in dredged anaerobic sediments. The surface of these metal sulfides is hydrophobic by nature and can be selectively separated from suspensions by means of collectorless flotation. For most metals in sediments, up to 80% removal efficiencies can be achieved by flotation (Cauwenberg et al., 1998).

Ultrasonic-assisted extraction is based on ultrasound coupled with vacuum pressure. Ultrasound can cause high-energy acoustic cavitation, that is, the formation, growth, and implosive collapse of bubbles in liquid. Intense heating of the bubbles occurs during cavitational collapse. These localized hot spots have temperatures of roughly 5000°C, pressures of 500 atm, and a lifetime of a few microseconds, the impact of which is sufficient to melt most metals. This method depends on particle size, and the best results are achieved when the ultrasound is applied to treat coarse grains, sand, or silt. Therefore, this method is an effective and economical remediation process for sediments with low-clay content (Hanna et al., 2004).

The immobilization/stabilization method uses sequestering agents to change metal speciation in contaminated dredged sediments. This method does not remove metals from sediment but is commonly used due to its low cost and fast remedial effect. The most commonly used materials for immobilizing metals include lime, cement, iron oxides, clays, rock phosphate, and zeolites (Peng et al., 2009).

17.2.4 In Situ Management of Contaminated Sediments

In situ management of contaminated sediments is potentially less expensive and risky than ex situ management, but this technology is newer and some methods are still under development. Active capping and passive capping are among the more promising alternatives for in situ remediation of sediments that are contaminated with metals. Other in situ methods include reactive mats, mixing amendments into the sediment, and the mixing or layering of amendments into passive caps.

17.2.4.1 Passive Capping

Passive (inactive) capping is the installation of a subaqueous covering or cap of clean, inert material over contaminated sediment, thus isolating it from the surrounding environment and reducing contaminant migration in the water phase. This alternative can be considered an effective approach for remediating contaminated sediments under certain static conditions and is relatively economical. However, the inert materials (sand, gravel, or similar nonreactive materials) used in passive caps do not provide permanent stabilization, because they are subject to leaching and mechanical disturbance, which, in turn, can release toxic contaminants. In addition, the application of passive caps can cause severe disruption and alteration of benthic ecology. Additional factors that can reduce the effectiveness of passive caps include

- Erosion
- High groundwater upwelling rates
- Mobile (low sorption) contaminants
- High rates of gas ebullition

Cap thickness can be increased to reduce some of the issues with passive caps. However, the required thickness can reach unacceptable levels in shallow streams or navigable water bodies.

17.2.4.2 Active Capping

Caps that incorporate chemical sequestering agents may be utilized to avoid some of the problems that are associated with passive caps. Such caps are referred to as *active caps*, because their reactive components alter the chemistry of sediment contaminants in situ. The sequestering agents used in active caps are designed to interact with sediment contaminants to enhance the containment properties of the cap without increasing cap thickness. The contaminants are not removed from the sediment, but their mobility is reduced (Figure 17.3). Sequestering agents reduce metal mobility/bioavailability and increase the nonlabile fraction of metals (Figure 17.3). The design of active sediment caps must consider a wide variety of factors, including chemistry of the contaminants and sequestering agents, burrowing habits of potential receptors, erosive forces acting on the surface of the caps, and geotechnical characteristics of the native sediment (Palermo et al., 1998).

In recent years, there has been considerable focus on capping amendments and active capping (Hawkins et al., 2011; Reible et al., 2006; Knox et al., 2011; Paller and Knox, 2010; Dixon and Knox, 2012). Some of the amendments used in active caps include activated carbon, phosphate materials (e.g., apatite), zeolite, organo-clays, bentonite, and others (Table 17.3). Most of these sequestering agents are effective in fresh and salt water (Knox et al., 2008) and have proven abilities to sequester sediment contaminants in situ (Knox et al., 2012; Kwon et al., 2010).

TABLE 17.3
Sequestrating Agents for Heavy Metals: Possible Application for In Situ Remediation, Including Active Caps

Sequestering Agent	Immobilization Mechanisms	Key Components	Metal Removed	References
Rock phosphate	Shift nonresidual metal to residual fraction	Apatite	Pb, Mn, Co, Cu, Zn, Mg, Ba, U, Th, Cr	Knox et al. (2008)
Zeolite	High cation-exchange capacity	Aluminosilicates (e.g., clinoptilolite)	Cd, Cu, and Zn	Knox et al. (2008)
Iron oxides/ hydroxides	Adsorption	Steel shot; limonite; goethite	Cd, Cu, Zn, and As	Enzo et al. (2002)
Clays	Adsorption	Sepiolite; palygorkite; bentonite	Cd, Cu, and Zn	Gracia et al. (1999) and Knox et al. (2012)
Beringite	Increase pH values; directly adsorb and fix metals	Fe- and Mn-bearing materials; aluminosilicates	Cd, Zn, Pb, Cu, and Ni	Enzo et al. (2002) and Mench et al. (1994)
Biopolymers	Adsorption, fixation	Chitosan	Cd, Cr, Cu, Co, Pb, and Zn	Knox et al. (2008)
Organo-clays	Adsorption	Modified bentonite with quaternary amines	Cd, Cr, Cu, Co, Pb, and Zn As and Hg	Knox et al. (2008), (CETCO Remediation Technologies, 2012)
Activated carbon	Adsorption	Carbon	Anions metals	Reed (2001)
Thiol–Self-Assembled Monolayers on Mesoporous Supports (Th–SAMMS)	Adsorption	Thiol group assembled on mesoporous silica	Hg, Cd, and Pb	Kwon et al. (2010)

17.3 RECONTAMINATION

A challenge to all remedial approaches is the continued influx of contaminants from uncontrolled sources after remediation. Dissolved contaminants may gradually enter remediated sediments until the upper several centimeters of sediments, which constitute the habitat zone for most benthic organisms, are incapable of supporting normal benthic biodiversity. Similarly, the influx of contaminated particulate matter (e.g., contaminated sediments) that settles over remediated sediments can lead to the production of a polluted habitat zone that overlies remediated sediments, thereby negating the benefits accrued by the remedial action.

The severity of the problem posed by the continued influx of contaminants on remediated sediments will be partly determined by the type of remediation that has been undertaken. Influxes of particulate contaminants (e.g., contaminated sediment from offsite sources) on sediments remediated by dredging or by capping with inert materials can be expected to degrade the remediated benthic environment at a rate that is proportional to the rate of contaminated sediment deposition. Remedial effectiveness will be largely negated when the deposited sediments accumulate to a depth that encompasses the habitat zone for most benthic organisms (the upper 10 cm or less of sediment). However, particulate contaminants that are deposited on sediments remediated with chemically active sequestering agents (such as those used in active caps) may be affected by the sequestering agents, resulting in a subsequent reduction in the environmental impacts arising from contaminant influx. Knox et al. (2012) demonstrated that active caps can influence the speciation of metals that are located in sediments beneath the caps, resulting in reduced metal mobility and potential bioavailability. For example, water-extractable Pb concentrations in the sediment beneath an apatite cap were substantially lower than those in the sediment from a control treatment (i.e., uncapped sediment) or those in the sediment beneath a sand cap (Knox et al., 2012). Knox et al. (2012) described the layer of contaminated sediment below the active caps in which contaminants were immobilized as the zone of influence (ZOI), and this was usually several centimeters in depth. The mechanisms of formation of a ZOI may include the mixing of dissolved- or solid-phase cap components with underlying sediment by diffusion, bioturbation, mixing of sediment and cap material by turbulent currents (especially during flood flows), or advective movements of dissolved cap materials into the underlying sediment (e.g., as a result of cap consolidation) (Knox et al., 2012).

17.4 CONCLUSIONS

This chapter provides a brief summary of the state of contaminated sediment management and remediation. Although problems remain, contaminated sediment management is improving. Currently, combinations of technologies are emphasized in the management and remediation of contaminated sediments (Bridges et al., 2012). A logical progression is to start with source control and, perhaps, the dredging of "hot spots." After source control, natural recovery processes should be relied upon where feasible, but if more than MNR is required, additional engineering can be incrementally added, including active capping.

Consideration of the current state-of-the-art suggests that there is a need for new technologies or modifications of existing ones, for example, capping technologies that can sequester sediment contaminants and create a reliable, stable, and long-lasting cap in a range of aquatic environments. Current capping technologies typically produce caps with limited physical stability that are suitable primarily for low energy, depositional aquatic environments. However, depositional environments can become erosive as a result of unpredictable natural events such as floods and storms and anthropogenic actions such as boating and construction activities. Future research should emphasize the development of caps that have the capacity to both stabilize contaminants and resist the physical disturbances that are likely to occur over the required lifetime of the cap. In situ remediation methods such as active caps, reactive mats, mixing amendments in the sediment, and the mixing or layering of amendments in sand caps have the potential to provide relatively long-lasting and cost-effective remediation with less damage to the benthic environment than dredging. These methods should be emphasized in the selection of remedial technologies for contaminated sediments, and more field applications should be seen in the future.

REFERENCES

Bolam, S.G., H.L. Rees, P. Somerfield et al. 2006. Ecological consequences of dredged material disposal in the marine environment: A holistic assessment of activities around the England and Wales coastline. *Marine Pollut. Bull.* 52:415–546.

Boyd, T.J., M.T. Montgomery, B.J. Spargo et al. 2000. Source reduction effect on creosote PAH bioremediation in marsh sediments. Second International Conference on Remediation of Chlorinated and Recalcitrant Compounds, Monterey, CA.

Bridges, T.S., S.C. Nadeau, M.C. McCulloch. 2012. Accelerating progress at contaminated sediment sites: Moving from guidance to practice. *Integr. Environ. Assess. Manage.* 8:331–338.

Catherine, N.M., R.N. Yong, B.F. Gidds. 2001. An evaluation of technologies for the heavy metal remediation of dredged sediments. *J. Hazard. Mater.* 85:145–163.

Cauwenberg, P., F. Verdonckt, A. Maes. 1998. Flotation as a remediation technique for heavily polluted dredged material. 2. Characterization of flotated fractions. *Sci. Total. Environ.* 209:121–131.

CETCO Remediation Technologies. 2012. Company Website. www.cetco.com.

Dixon, K.L., A.S. Knox. 2012. Sequestration of metals in active cap materials: A laboratory and numerical evaluation. *Remediation* . 22(2):81–91.

Enzo, L., F.J. Zhao, G.Y. Zhang et al. 2002. In situ fixation of metals in soils using bauxite residues: Chemical assessment. *Environ. Pollut.* 118:435–443.

Finley, K.A., R. Garrett. 2007. Recovery at Belews and Hyco Lakes: Implications for fish and tissue selenium thresholds. *Integr.Environ. Assess. Manage.* 3(2):297–299.

Gracia, S.A., A. Alastuey, Q. Querol. 1999. Heavy metal adsorption by different minerals: Application to the remediation of polluted soils. *Sci. Total Environ.* 242:179–188.

Hanna, K., P. Pentti, H. Vaino et al. 2004. Ultrasonically aided mineral processing technique for remediation of soil contaminated by heavy metals. *Ultraso. Sonochem.* 11:211–216.

Hawkins, A.L., G.A. Tracey, J.J. Swanko et al. 2011. Reactive Capping Mat Development and Evaluation for Sequestering Contaminants in Sediments. Final Report; ER-1493, DOD SERDP. Technical report. TR-2366-ENV.

Knox, A.S., M.H. Paller, K.L. Dixon et al. 2011. Innovative in-situ remediation of contaminated sediments for simultaneous control of contamination and erosion. Final Report. PART I, SRNL-RP-2010-00480.

Knox, A.S., M.H. Paller, D.D. Reible et al. 2008. Sequestering agents for active caps—Remediation of metals and organics. *Soil Sediment Contam.* 17(5):516–532.

Knox, A.S., M.H. Paller, J. Roberts. 2012. Active capping technology—New approaches for in situ remediation of contaminated sediments. *Remediation.* 22(2):93–117.

Krishna, R.R., C.Y. Xu, C. Supraja. 2001. Assessment of electronic removal of heavy metals from soils by sequential extraction analysis. *J. Hazard. Mater.* 84:279–296.

Kwon, S., J. Thomas, B.E. Reed et al. 2010. Evaluation of sorbent amendments for in situ remediation of metal-contaminated sediments. *Environ. Toxicol. Chem.* 29:1883–1892.

Magar, V.S., D.B. Chadwick, T.S. Bridges et al. 2009. Monitored natural recovery at contaminated sites. Technical guide. ESTCP Project ER-0622.

Martello, L.B., M.T. Sorensen, P.C. Fuchsman et al. 2007. Chromium geochemistry and bioaccumulation in sediments from the lower Hackensack River, New Jersey. *Arch. Environ. Contam. Toxicol.* 53:337–350.

Matis, K.A. 1995. *Flotation Science and Engineering.* New York: Mercel Dekker.

Mench, M., J. Vangronsveld, V. Didier et al. 1994. A mimicked in situ remediation study of metal contaminated soils with emphasis on cadmium and lead. *Environ. Pollut.* 86:279–286.

Mulligan, C., M. Fukue, Y. Sato. 2010. *Sediments Contaminations and Sustainable Remediation.* Boca Raton, FL: Taylor & Francis.

National Research Council (NRC). 2000. *Natural Attenuation for Groundwater Remediation.* Washington, DC: National Academy Press.

National Research Council (NRC). 2007. *Sediment Dredging at Super-Fund Megasites: Assessing the Effectiveness.* Washington, DC: National Academy Press.

Ortega, L.M., R. Lebrun, J.F. Blai et al. 2008. Effectiveness of soil washing, nanofiltration and electrochemical treatment for the recovery of metal ions coming from a contaminated soil. *Water Res.* 42:1943–1952.

Palermo, M.R., S. Maynord, J. Miller et al. 1998. *Guidance for In-Situ Subaqueous Capping of Contaminated Sediments.* Chicago, CA: Great Lakes National Program Office.

Paller, M.H., A.S. Knox. 2010. Amendments for the remediation of contaminated sediments: Evaluation of potential environmental impacts. *Sci. Total Environ.* 408:4894–4900.

Peng, J., Y. Song, P. Yuan et al. 2009. The remediation of heavy metal contaminated sediment. *J. Hazard. Mater.* 161:633–640.

Reed, B. 2001. Removal of heavy metals by activated carbon. In: *Environmental Separation of Heavy Metals: Engineering Processes.* Pittsburgh, PA: Technical Publishing Inc.

Reible, D.D., D. Lampert, W.D. Constant et al. 2006. Active capping demonstration in the Anacostia River, Washington, DC. *Remediation.* 17(1):39–53.

USEPA. 2005. *Contaminated Sediment Remediation Guidance for Hazardous Waste Sites.* EPA-540-R-05-012.

Weis, P., K.R. Barrett, T. Proctor, R.F. Bopp. 2005. Studies of contaminated brackish marsh in the Hackensack Meadowlands of northeastern New Jersey: An assessment of natural recovery. *Mar. Pollut. Bull.* 50:1405–1415.

Yong, R.N. (2001). *Geoenvironmental Engineering: Contaminated Soils, Pollutant Fate and Mitigation.* Boca Raton, FL: CRC Press.

18 Passive and Active Capping for In Situ Remediation of Contaminated Sediments

Jim Olsta

CONTENTS

18.1 INTRODUCTION

This chapter discusses the use of active media for in situ remediation of metal-contaminated sediments. Before choosing active media, sand caps should first be considered. Trace metal mobility can be reduced by (1) complexing with organic matter, (2) chemisorbing or co-precipitating with minerals, or (3) precipitating as insoluble sulfides, carbonates, phosphates, or oxides (McBride 1994). Most of the active media proposed for sediment capping have functional groups consisting of natural minerals or synthetic materials that utilize this chemistry. Langmuir or Freundlich parameters can be inserted into capping models to help estimate the time to breakthrough (Lampert and Reible 2009; Reible and Lampert 2014) if sorption phenomena control the performance of active cap media. In choosing active media, some understanding of the binding mechanism is important, because media can be impacted differently by competing ions, pH, redox conditions, and so on. It is also important to evaluate the biotoxicity of media toward representative organisms.

18.2 PASSIVE CAPS

Sand caps, sometimes referred to as "passive" or "conventional" caps, can be effective in metal-contaminated sediment remediation by enhancing reducing conditions in the underlying sediment

(Reible and Lampert 2014). The presence of excess sulfides in a reducing environment can lead to the formation of divalent metal sulfides that exhibit low solubility and bioavailability.

$$M^{2+} + FeS_{(s)} \rightarrow MS_{(s)} + Fe^{2+}$$

Porewater measurements of acid volatile sulfides (AVS) and simultaneously extracted metals (SEM) provide an indicator of the potential for a sand cap. In anaerobic sediments, divalent cation metals should not cause toxicity to benthic organisms if the ΣSEM–AVS is less than or equal to 0 [U.S. Environmental Protection Agency (EPA) 2005a]. This is because under reducing conditions, AVS can bind divalent metals in proportion to their molar concentration.

The effect of sand caps on mercury (Hg) was observed in laboratory simulation cells (Lui and Reible 2007). Over an approximate 8-month period, the total Hg flux in the overlying water of an exposed uncapped sediment was ~10^{-3} ng m^{-2} s^{-1}; this was compared with a nondetectable Hg concentration for a 10-mm sand cap over the same sediment. Redox potential measurements and oxygen concentrations showed that aerobic conditions extended only a few millimeters into the sand cap, resulting in reducing conditions in the underlying contaminated sediment. Measurements also showed that the proportion of methylmercury (MeHg) in underlying sediments decreased in the presence of a sand cap.

Thin sand caps may not be effective for achieving remedial goals at sites with high transport rates (significant groundwater upwelling or tidal pumping effects), where contamination levels are high, when there is the presence of nonaqueous-phase liquids, or when there is a high rate of gas ebullition. Thicker sand caps may decrease navigation channel depth or flow capacity to unacceptable levels. When these factors are present, then alternative materials that exhibit greater containment effectiveness than sand can be considered an option.

18.3 ACTIVE CAPS

The use of active media, in place of sand, to enhance the containment properties of the cap are often referred to as "active" caps. Direct amendment or capping of sediment with active media can reduce pollutant bioavailability and retard contaminant transport, thus reducing risk to acceptable levels (U.S. EPA 2005b).

18.4 ACTIVE MEDIA

The following is a review of active media for the remediation of metal-contaminated sediment. Active media that met one of the following criteria have been included: (1) have been used in a pilot-scale or full-scale sediment remediation project, (2) were cited by a regulatory agency as a possible amendment, or (3) have been laboratory tested extensively under sediment-specific conditions.

18.4.1 Activated Carbon

Activated carbon has a long history of use in potable water treatment and industrial wastewater treatment of organic compounds. Activated carbon is made from a number of different carbonaceous raw materials (e.g., coal, coconut shells, wood); it is made by thermal decomposition of the raw material and is, subsequently, followed by an activation process that results in a network of pores.

The ability of activated carbon to remove heavy metals has been established by numerous researchers (Reed 2001). The rate and the extent at which inorganic compounds are adsorbed is affected by the type and number of activated carbon surface groups. It is not known what surface functional groups are formed on activated carbon during the activation process. The method of activation determines whether the surface oxides are acidic, basic, or amphoteric. L-type carbons are generally formed at temperatures of 200–500°C during the activation process. H-type carbons are exposed to activation methods that remove indigenous surface oxide groups. Activated carbons develop a surface charge and exhibit acid–base properties when added to water. The zero point of charge, pH_{zpc}, is defined as the pH at which the proton excess on the surface is zero. At solution pH values below the pH_{zpc}, the surface has a net positive charge. At solution pH values above the pH_{zpc}, the surface has a net negative charge. L-type carbons tend to have a low pH_{zpc} and a negative charge over a wide range of pH. However, H-type carbons tend to have a high pH_{zpc} and a positive surface charge.

Metals can be removed via the following processes: chemisorption, filtration, hydrogen bonding, ion exchange, physical adsorption, and surface precipitation. Heavy metals will preferably precipitate onto the carbon surface given the proper chemical conditions. If the heavy metal is associated with other solid particles (e.g., sediment-bound metal), the metal can be removed by filtration.

Heavy metal removal by activated carbon is affected by the following factors: solution pH, metal type and concentration, presence of complexing ligands and other competing adsorbates, ionic strength, temperature, and carbon type. It has been observed that heavy metal removal by activated carbon was inversely proportional to the solubility of the metal. For uncomplexed cationic heavy metals, removal generally increases with an increase in pH. An exception to this rule is Hg(II). Hg removal was the highest in the acidic region, and it decreased with an increase in pH. Removal increases with a decrease in pH when the heavy metal exists as an anion. However, the effect of pH on heavy metals that exist as anions, such as arsenic (As), is not as well understood. In general, the lower the pH at which cationic metals form aqueous hydroxide complexes, the better the metal will be removed. Lead (Pb), copper (Cu), nickel (Ni), and zinc (Zn) form hydroxide complexes in the following order based on an increase in pH:

$$Pb > Cu > Zn > Ni$$

The concentration of the metal ion affects the removal mechanism. If the concentration of a divalent cation is large enough such that $M(OH)_2$ forms, then surface precipitation can occur. If the metal concentration is less than the solubility of the metal, then the primary removal mechanisms are physical/chemical sorption or ion exchange.

The presence of complexing ligands can alter metal removal. A metal ion that has been complexed in solution may adsorb more strongly, weakly, or the same as the uncomplexed metal species. Generally, the adsorption of complexed cationic metals increases at lower pH values and decreases at higher pH values compared with the metal-only ligand-free case. The effectiveness of complexed heavy metal removal by activated carbon should be determined using treatability studies on a case-by-case basis.

Ionic strength is higher in marine and estuarine sediments compared with freshwater sediments. Changes in ionic strength will alter the aqueous chemistry of the metal and the electric double-layer surrounding the carbon surface. As ionic strength increases, the carbon surface is swamped by the electrolyte, and access to the surface is more difficult.

Corapcioglu and Huang (1987) concluded that there is a large variability in the ability of carbons to adsorb metals. It has been proposed that metal removal can be quantitatively predicted based on whether a carbon is L-type or H-type. As discussed earlier, L-type carbon more readily takes on a negative charge that favors cation removal. However, H-type carbon more likely has a positive charge and the removal of anions [e.g., Cr(VI), Se (IV)] is favored.

Most research on activated carbon as in situ active media for heavy metal sediment remediation has focused on Hg and MeHg, a highly toxic form of Hg. Gilmour et al. (2013) tested a 177 μm × 62 μm (80 × 235 mesh) activated carbon as an in situ amendment for Hg and MeHg contamination. The study utilized 2 L sediment/water microcosms with 14-day bioaccumulation assays. The key parameters were porewater concentrations, and bioaccumulation of total Hg and MeHg by oligochaete Lumbriculus variegatus. Activated carbon was tested by using Hg-contaminated sediments from two freshwater and two estuarine sites in four separate microcosm assays. Activated carbon added at 2.4%–7.3% of the dry weight of sediments significantly reduced both MeHg concentrations in porewaters and the bioaccumulation of MeHg by Lumbriculus, relative to unamended controls. The activated carbon also reduced inorganic Hg concentrations, relative to unamended controls, but to a lesser degree than MeHg reductions. Porewater concentrations of inorganic Hg and MeHg were significant predictors of uptake by Lumbriculus for the sediments tested. The activated carbon amendment also resulted in significant reductions in porewater dissolved organic matter (Gilmour et al. 2013).

Activated carbons have also been impregnated with various salts, elements, and oxides in an effort to enhance their performance for both organics and metals. Ghosh et al. (2008) examined several of these impregnated activated carbons. But there are no known pilot-scale or full-scale projects to show how effectively the impregnated carbons will reduce porewater concentrations and the bioavailability of metals and organics simultaneously in various sediment environments.

Activated carbon can be difficult to place through the water column because of its low density and air entrapment in pores. However, granular activated carbon and powdered activated carbon have been deployed successfully at a number of pilot-scale and full-scale sites. In the United States, two sites selected full-scale remedies of activated carbon in a geotextile mat, REACTIVE CORE MAT™. At one site in Indiana, flow was diverted from the river so that the mat could be placed in the dry. At another site in Minnesota, sand was added to the carbon mat so that it could be placed by barge and sink through the water column (U.S. EPA 2013). These sites are beginning to be monitored as a part of the remediation activities conducted by the U.S. EPA Great Lakes Legacy Act.

Pilot studies in the United States evaluated the following methods for placement of activated carbon: (1) broadcasting a slurry of water and activated carbon through a tremie pipe above the sediment surface; (2) injection of activated carbon into sediment through hollow tines; or (3) delivery of activated carbon in a composite pelletized form, such as AQUAGATE™ or SEDIMITE™. AQUAGATE™ is a pellet composed of powdered activated carbon, aggregate stone, and a cellulosic polymer binder. SEDIMITE™ is a pellet containing activated carbon, sand, and a bentonite clay binder. Ghosh et al. (2008) showed that the activated carbon–sand–bentonite pellet was nontoxic to freshwater oligochaetes. Also, similar methods were used in Norway, where activated carbon was mixed with clean clay and deployed between a 30-m and 100-m depth by using a hopper dredger, a pump, and a tremie pipe (Cornelissen et al. 2011).

18.4.2 Apatite

Apatite consists primarily of a matrix of calcium phosphate. Apatite is a granular media that comes in two forms: mineral and biological (animal bones). Apatite phosphate minerals usually refer to hydroxylapatite, fluorapatite, and chlorapatite, named for high concentrations of OH−, F−, and Cl− ions, respectively, in the crystal. The formula of the admixture of the four most common endmembers is written as $Ca_{10}(PO_4)_6(OH,F,Cl)_2$, and the crystal unit cell formulae of the individual minerals are written as $Ca_{10}(PO_4)_6(OH)_2$, $Ca_{10}(PO_4)_6(F)_2$, and $Ca_{10}(PO_4)_6(Cl)_2$. Mineral apatite that is supplied to the environmental market is directly from rock phosphate or a waste byproduct from fertilizer production. APATITE II™ is a patented crushed biological fish bone product that is used for metal remediation applications (Figure 18.1). Biological fish bone is available seasonally around the time that seafood canning production is active. Phosphates react with heavy metals through both surface complexation reactions and precipitation to form metal phosphate minerals (e.g., pyromorphite) and other low-solubility phases that are stable.

Research has been conducted on both the mineral and biological forms of apatite. Tests of fish bone apatite with contaminated soil, groundwater, and wastewater have shown sequestration of Pb, uranium (U), cadmium (Cd), Zn, aluminum (Al), and Cu, and uptake between 5% and 50% of its weight in metals (Wright et al. 2004). Analysis indicated that fish bone apatite works by four general, nonmutually exclusive processes depending on the metal, the concentration of the metal, and the aqueous chemistry.

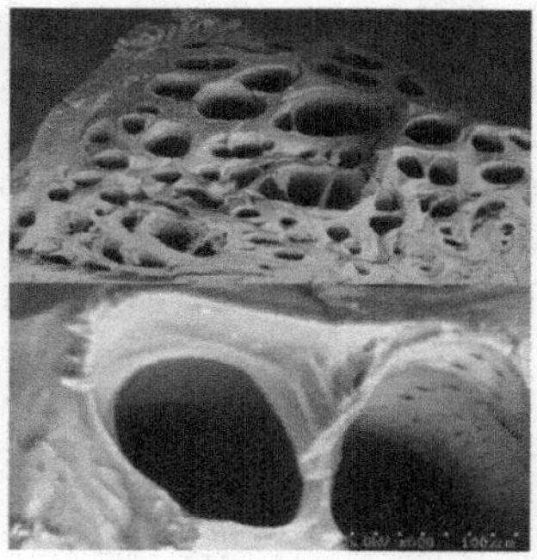

FIGURE 18.1 Magnified photos of fish bone apatite (Apatite II™).

Crannell et al. (2004) ran batch sorption tests on several mineral apatites to determine partition coefficients. North Carolina mineral apatite was shown to adsorb/precipitate Pb, U, Cu, Ni, As, Zn, selenium (Se), Cd, vanadium (V), and chromium (Cr) in the order of decreasing efficiency. The measured K_d for Pb was greater than 500,000 L/kg. The metal sequestration by the mineral apatite was very stable over a wide range of pH and redox potential. Some Florida mineral apatite sources were found to leach phosphorus at a relatively high rate. This is an important characteristic and could be considered a limitation for sediment application of some apatites, because soluble phosphorus can increase eutrophication in water bodies.

Results of another study showed that certain apatites were effective at removing many metals from freshwater (Knox et al. 2011). Sorption studies were performed on As, Cd, Cr, cobalt (Co), Cu, Pb, Ni, Se, U, and Zn in freshwater with both biological apatite and mineral apatite. Partition coefficients, K_d, were calculated. Certain apatites exhibited greater than 80% retention of most metals. Sets of experiments on metal speciation and retention in contaminated sediment showed that apatite can immobilize a broad range of metals under both oxidized and reduced conditions. Sequential extraction studies of the treated sediments were conducted to evaluate the bioavailability and retention of metals. Metal fractions recovered in early extraction steps, which are more likely to be bioavailable, were reduced for several metals, especially Pb, Zn, cobalt (Co), and Cd. Less bioavailable fractions collected in later extraction steps increased. Aqueous extracts and toxicity characteristic leaching procedure (TCLP) extracts from mineral apatite were analyzed for metal concentrations. Aqueous extracts were below ambient water quality criteria. The TCLP extract results were below TCLP metal limits. Biotoxicity tests with freshwater and estuarine organisms showed that mineral apatite was nontoxic (Paller and Knox 2010). In separate bioassay studies, mineral apatite was tested in two forms (incorporated into a geotextile mat, and bulk mixed 5% by weight into sediment) for biotoxicity to marine organisms, and it was found to be nontoxic (Rosen et al. 2011).

A 24 m × 30 m apatite pilot cap was deployed using a barge-mounted crane and a clamshell bucket (Melton 2007). Push cores indicated that the as-deployed layer thickness varied from 10 to 20 cm. The sediment–apatite interface was fairly well defined, with a sediment–apatite mixing band that was 1–2 cm thick. Core samples taken at 540 days did not show any statistically significant migration of metals Pb, Zn, Cu, or Cd into the apatite cap.

18.4.3 Bauxite

Bauxite is an aluminum ore that is relatively inexpensive and readily available in bulk quantities. Bauxite consists mostly of the minerals gibbsite $Al(OH)_3$, boehmite γ-AlO(OH), and diaspore α-AlO(OH), mixed with the two iron oxides goethite and hematite, the clay mineral kaolinite, and small amounts of anatase. Bauxite has surface properties that allow it to sequester both cationic and oxy-anionic forms of heavy metals through a combination of sorption, ion exchange, and precipitation. Bauxite is able to sequester the following metals: Hg, As, Cr, Cd, Pb, and Ni (U.S. EPA 2013).

A bench-scale study was conducted on the effectiveness of treating metals in sediments (Gavaskar et al. 2005). The group evaluated a number of natural minerals, including ores of iron, aluminum, manganese, and titanium, and determined that bauxite had the set of properties that are the most suitable for a sediment cap. Depending on its source and composition, bauxite specific gravity varies between 2.0 and 3.0, which allows it to settle in water. For their treatability studies, a particle size range of 75–150 μm (100–200 mesh) was chosen as a balance between sorption capacity and expected particle settling rate.

In the study conducted by Gavaskar et al. (2005), the following set of tests was conducted to evaluate the use of bauxite for capping contaminated sediments:

- Batch sorption tests to evaluate the removal of Hg, As, Cr, Pb, Cd, and Zn from water solution and to develop adsorption isotherms for each metal
- Desorption tests with spent bauxite from the earlier adsorption tests and fresh deionized water
- Batch sorption tests with varying ionic strength (typical of freshwater), pH, and redox potential to evaluate the stability of the sorption and tests with total organic carbon (TOC) content to evaluate potential interference
- Equilibration tests with bauxite and water in the pH range of 5.0–8.4 to evaluate potential leaching of 16 trace metals from bauxite to water phase, and TCLP testing of eight U.S. EPA Resource Conservation and Recovery Act (RCRA) metals
- Biotoxicity tests with fish (*Pimephales promelas*, 96 h), zooplankton (*Ceriodaphnia dubia*, 48 h), and benthic amphipods (*Leptocheirus plumulosus*, 96 h)

Batch sorption tests at metal equilibrium concentrations of 600–800 mg/l showed that sorption capacities were approximately 2,500 mg/kg of Hg, 2,500 mg/kg of As(III), 2,000 mg/kg of As(V), 2,500 mg/kg of Cr(VI), 3,000 mg/kg of Cu, and 750 mg/kg of Zn.

The only metal that showed signs of partial desorption was Zn. Hg (cation) and As (oxy-anion) were chosen as the representative cation and oxy-anion for testing with variable conditions. Sorption of both metals showed relatively little or no impact from low (5 mg/L), medium (200 mg/L), or high (1,000 mg/L) levels of sulfate. At 100–250 mg/L levels of calcium, the sorption efficiency of both Hg and As was slightly impacted, although more than 90% of the metals remained bound to the bauxite. There was little or no change in the sorption of either metal at low (60 mg/L) or medium (250 mg/L) levels of TOC. The water redox state had little to no impact on Hg sorption. However, As sorption was higher under anaerobic conditions, as As(III), than under aerobic conditions, as As(V). Under these conditions, contaminant sequestration was relatively unaffected by interferences from competing ions and natural organic matter. None of the eight RCRA metals exceeded their respective limits in the TCLP extract. Biotoxicity tests with fish, benthic amphipods, and zooplankton have shown bauxite to be nontoxic.

However, bauxite performance as an active media for sediment remediation still needs to be verified on a pilot- or full-scale basis.

18.4.4 Bentonite

Bentonite is a relatively inexpensive industrial clay mineral that is readily available worldwide. A commercially available product, bentonite primarily comprises montmorillonite, with small quantities of feldspar, quartz, calcite, and other minerals. For industrial purposes, there are two main types of bentonite: calcium bentonite and sodium bentonite.

Montmorillonite has a permanent structural negative charge on its silicate layers that yields a high cation-exchange capacity (CEC) relative to other clays, such as kaolinite and illite. In recent decades, research has been conducted on the heavy metal retention capabilities of bentonite. An early bench-scale sorption study was conducted with Turkish bentonite and solutions containing dissolved Pb, Cd, Cu, and Zn (Bereket et al. 1997). Experimental results showed that bentonite adsorption of Pb(II), Cd(II), and Cu(II) fits a Langmuir isotherm model, whereas Zn(II) fits a Freundlich isotherm model much better.

Two more recent studies have tried to further determine Pb adsorption mechanisms. In one study, three bentonites (one US and two Japanese) with varying calcite content were tested (Nakano and Li 2009). Carbonate content was 0.8% for the US bentonite and 2.4 to 3.7% for the Japanese bentonites. The CEC was closer with 702 mmol kg^{-1} for the US bentonite and 734–826 mmol kg^{-1} for the Japanese bentonites. Results of batch Pb sorption tests showed Pb concentrations of 323 mmol kg^{-1} for the US bentonite and 415–540 mmol kg^{-1} for the Japanese bentonites. After the bentonites were spiked with Pb batch solutions, selective sequential extraction (SSE) was used to investigate the Pb retention mechanism of the bentonites. The SSE indicated that 80% of the Pb was retained through exchangeable and carbonate phases in all bentonite samples. X-ray diffraction compared the mineral composition of the bentonites both before and after Pb sorption tests. By x-ray analysis, cerussite ($PbCO_3$) peaks were identified for the Japanese bentonites that were spiked with Pb. The following reaction between calcite and Pb was theorized:

$$CaCO_3\,(\text{calcite}) + Pb^{+2} \leftrightarrow PbCO_3\,(\text{cerussite}) + Ca^{+2}$$

In another study, a US bentonite was examined for the effect of pH on Pb(II) adsorption (Xu et al. 2008). The adsorption of Pb(II) on bentonite increased with the increasing pH value at a pH less than 6, and it reached a maximum at pH 10. The strong pH-dependent adsorption of Pb(II) on the bentonite suggests surface complexation that may be attributed to the attachment of Pb(II) to clay platelet edge hydroxyl groups. Adsorption isotherms were well fitted to Langmuir, Freundlich, and D–R models.

Although its sorptive properties are of interest, bentonite is most often used in sediment remediation as a low-permeability barrier. Sodium bentonite can also be used as a low-permeability cap to divert upwelling groundwater away from a metal-contaminated sediment area in freshwater waterways. Geosynthetic clay liners, bentonite chips (e.g., BENTOBLOCK™), and bentonite-aggregate pellets (AQUABLOK™) are products that enable easier deployment. The chips and pellets are capable of settling relatively quickly through the water column and of creating a layer with permeabilities in the order of 10^{-8} to 10^{-9} cm/s. However, due to the high ionic strength of saltwater (SW) bentonite, permeability increases several orders of magnitude in seawater environments. Attapulgite may be able to be substituted for bentonite as a low-permeability layer in SW applications. AquaBlok, Ltd. reports an SW permeability that is approximately less than or equal to 10^{-7} cm/s for the SW version of their low-permeability product (AquaBlok 2006).

A bentonite-aggregate pellet pilot cap was placed by using a barge and a clamshell on the Anacostia River in Washington, DC (Reible et al. 2006). The bentonite-aggregate pellet cap, installed as a demonstration of permeability control, reduced porewater advection rates to zero versus a control area and a sand cap, for a certain period after placement. However, inclinometers indicated that gas accumulation led to substantial movement of the low-permeability cap. Periodic displacement of low-permeability caps could potentially reduce long-term containment. Thus, upwelling groundwater flow and/or gas accumulation need to be considered in the design of low-permeability caps.

18.4.5 Nitrate

The addition of nitrate to sediments inhibits sulfate reduction and, therefore, reduces the formation of MeHg in sediments. The concept was developed for Onondaga Lake in upstate New York (Nolan et al. 2013). Historically, in late summer/early fall, a cold, stratified bottom layer formed and the sediment became depleted of oxygen and nitrate. During this period, MeHg was produced by sulfate-reducing bacteria and released into the hypolimnion. Periodic sampling in 2004 on Onondaga Lake revealed a noticeable change in redox conditions and a decrease in MeHg concentrations compared with the previous year. The decrease correlated to a recent change at the publically owned treatment works (POTW) that were being discharged into the lake. The 3.5 m^3/s (80 MGD) POTW had expanded their biologically aerated filter nitrification system to convert ammonia-N in the treated municipal wastewater into nitrate-N.

A full-scale application of additional nitrate throughout the hypolimnion layer took place during the summer and early fall of 2011 and 2012, as deep waters became anoxic and nitrate concentrations dropped. Deployments continued until the lake turned over, typically in mid-to-late October. The addition of nitrate to Onondaga Lake was conducted by applying a diluted calcium nitrate solution (calcium nitrate) to waters that were two to five meters above the lake bottom. Nitrate–nitrogen concentrations in the deep, stratified waters were maintained above 1 milligram per liter on average.

The first two years of the whole-lake nitrate addition pilot test in Onondaga Lake demonstrated the ability of nitrate to inhibit the release of MeHg to overlying, stratified waters from sediments. During 2011 and 2012, MeHg concentrations in deep, stratified waters were lower than those during recent recorded years. The release of MeHg appears to be inhibited by maintaining nitrate concentrations above 0.5–1.0 mgN/l near sediments.

18.4.6 Organophilic Clay

Organophilic clay is a mineral clay that has been converted from being hydrophilic into being organophilic by ion exchange with a polymer or a surfactant, typically a quaternary amine compound. Since the media is organophilic, it may also sorb dissolved organically complexed metals.

The results of an extensive study showed that organophilic clay, which was modified with quaternary amine, was effective at removing many metals from both freshwater and saltwater (Knox et al. 2011). The organophilic clay exhibited greater than 80% retention of most metals. Sets of experiments on metal speciation and retention in contaminated sediment showed that organophilic clay can immobilize a broad range of metals under both oxidized and reduced conditions. Sequential extraction studies of the treated sediments were conducted to evaluate the bioavailability and retention of metals. Metal fractions recovered in early extraction steps, which are more likely to be bioavailable, were reduced for several metals, especially Pb, Zn, Ni, Cr, and Cd. Less bioavailable fractions collected in later extraction steps increased. In the aqueous extracts, metal concentrations were below EPA ambient water quality criteria. The TCLP extract results were below TCLP metal limits. In bioassay studies, organophilic clay was tested in two forms (incorporated into a geotextile mat, and bulk mixed 5% by weight into sediment) for biotoxicity to marine organisms and found to be nontoxic (Rosen et al. 2011).

Sulfur-amended organophilic clay is available on the market (ORGANOCLAY MRM™). This media is primarily utilized on off-shore oil and gas platforms in filtration vessels for the removal of both residual oil and grease and for Hg from produced water prior to discharge to sea. However, when studied in laboratory tests with freshwater and estuarine sediment the sulfur-amended organophilic clay has been shown to increase sulfur compounds and conductivity in the microcosms (Gilmour et al., 2013). The increased conductivity was likely due to free chloride ions in the sulfur-amended organophilic clay. Increased chloride concentrations in freshwater and estuarine sediments can result in Hg(II) forming soluble metal chloride. Also, the sulfur compounds may stimulate sulfate-reducing bacteria in the formation of MeHg as well. Water-washing to rinse residual sulfur compounds and chloride from the sulfur-amended organoclay may reduce these effects.

Organophilic clay has been installed in numerous sediment remediation capping projects by using several methods: in bulk (alone or mixed with sand), in geotextile mats, and in gabion mattresses (EPRI, 2011; U.S. EPA, 2013). But the primary purpose of the organophilic clay in these sediment caps was the sorption of nonaqueous phase liquids (NAPL), particularly coal tar and creosote, and polycyclic aromatic hydrocarbons.

18.4.7 Siderite

Siderite, $FeCO_3$, is found primarily as a mineral in certain natural ore deposits. Since pH is an important factor in metal speciation, and in turn its resulting bioavailability, active media that can provide pH buffering capacity may be used as part of a metal-contaminated sediment remediation strategy.

Porewater pH neutralization and buffering ability of siderite were studied for an alkaline-contaminated sediment. Brine (pH 12) porewater from contaminated lake sediments and siderite concentrates were investigated in a series of kinetic batch tests (Vlassopoulos et al., 2013). The tests showed that pH was neutralized and consistently buffered within weeks to months. Reaction products in the batch experiments were identified as mainly consisting of calcite and iron oxides/oxyhydroxides. Column testing under dynamic flow conditions that were representative of field situations confirmed the effectiveness of siderite for pH neutralization and buffering.

Based on the study mentioned earlier, a full-scale sediment remediation capping project is underway on Onondaga Lake in upstate New York. The source siderite ore contains a minimum specified siderite content of 74%. The minimum siderite ore application rate is 8.65 kg/m^2, which will yield a minimum siderite application rate of 6.40 kg/m^2 in this area.

18.4.8 Thiol-Amended Mesoporous Silica

THIOL-SAMMS® (TS) is a powdered thiol-amended mesoporous silica developed by the U.S. Department of Energy Pacific Northwest National Laboratory. The media has a surface area of ~900 m^2/g with functional groups that have a high affinity for certain cations and anions (Mattigod et al. 2004). A bench-scale treatability test was conducted with TS and a waste condensate stream containing 4.64 mg/L Hg. Adsorption was rapid with ~99% of the dissolved Hg being adsorbed in the first 5 min. The Hg adsorption by the media fits well to the Langmuir isotherm model. The maximum media adsorption was determined to be 625 mg of Hg/g. An additional set of tests showed that variations in pH (4–8) and ionic strength did not significantly affect Hg adsorption with all partition coefficients greater than 10^7 L/kg.

A 14-day bioaccumulation assay study was conducted with TS- and Hg-contaminated sediment from two freshwater and two estuarine sites (Gilmour et al. 2013). The bioaccumulation of total Hg and MeHg by a deposit-feeding oligochaete worm, Lumbriculus variegatus, was measured. The TS added at 4.4%–5% of the dry weight of sediments significantly reduced both MeHg concentrations in porewaters, relative to unamended controls, and the bioaccumulation of MeHg by the oligochaete worm.

Another study examined the ability of TS to adsorb Cd and Pb (Ghosh et al. 2008). In this study, the TS removed greater than 90% of both Cd and Pb over a wide range of pH and the biouptake of Cd was reduced by 98% after the amendment of TS.

Although the media appears quite effective in controlling several metals, the cost of the TS is currently orders of magnitude higher than most other active media. There are no known pilot-scale or full-scale applications for sediment remediation.

18.4.9 Zeolite

Zeolites are porous crystalline aluminosilicates. Both natural minerals and synthetic zeolites are used commercially for their adsorption, ion exchange, molecular sieve, and catalytic properties. Zeolites are used in water treatment for the removal of nitrates and metals such as Pb, Zn, and Cu (Crittenden and Thomas 1998). It should be noted from the earlier subsection on nitrate that nitrates inhibit sulfate reduction and, subsequently, MeHg formation. So if MeHg is a concern, then the removal of nitrate by zeolite amendment may have an effect of increasing MeHg concentrations.

Chabazite, a tectosilicate mineral of the zeolite group, has been shown effective for the treatment of cesium (Cs)- and strontium (Sr)-contaminated wastes (Mimura and Kanno 1985). Chabazite has been used to treat Cs- and Sr-contaminated wastewater at damaged nuclear reactor sites. However, in marine environments, the high ionic strength of SW reduces the Cs adsorption efficiency of chabazite (Ojovan and Lee 2014). In another study, the most abundant and readily available natural zeolite, clinoptilolite, was evaluated (Knox et al. 2011). Clinoptilolite effectively removed Cr, Cu, Pb, and Zn in both freshwater and saltwater. Neither As nor Se was significantly sorbed by clinoptilolite.

There have been no pilot-scale or full-scale sediment applications to test the longevity and efficiency of zeolites as active media (U.S. EPA 2013).

18.4.10 Zero-Valent Iron

Zero-valent iron (ZVI) is an elemental metallic iron available in various sizes from granules to nanoparticles. ZVI is an effective amendment for soil and groundwater remediation. ZVI particles have a reactive surface that can both reduce and immobilize metals (U.S. EPA 2013). Laboratory-scale treatability studies have shown the ability for ZVI to treat As (Kanel et al. 2005), Cr (VI), and Pb in aqueous solutions (Ponder et al. 2000).

There may be limitations with the use of ZVI in sediment applications. Laboratory-scale treatability studies indicated changes to sediment geochemistry and passivation of iron by the formation of a thin layer of iron oxide, as shown in Gardner (2004). In theory, however, insertion of the iron in the reducing zone prior to the occurrence of significant oxidation could prove to be an effective application (Reible and Lampert 2014).

The use of ZVI to treat sediments has been limited to the bench-scale studies mentioned earlier. There have been no pilot-scale or full-scale applications of ZVI for in situ sediment remediation (U.S. EPA 2013). Pilot-scale and field-scale projects are necessary to evaluate the long-term feasibility of ZVI as a sediment capping media.

18.5 DEPLOYMENT

A brief discussion of some of the methods previously used to deploy each active medium has been provided earlier in this chapter. When designing a sediment remediation cap, geotechnical considerations and erosion protection are important. Guidance is available on geotechnical aspects and armor layer design of in situ capping (Palermo et al. 1998).

Active media can be placed in bulk in several ways. Amendments can also be mixed with sand prior to placement. Figure 18.2 shows deployment of a mix of active media and sand by a clamshell bucket. The bucket is opened slowly close to the surface. This allows the media to "rain" down

FIGURE 18.2 **(See color insert.)** Deployment of active media and sand mix by clamshell bucket.

through the water column onto the sediment. Media can also be slurried with water and deployed using a rotating broadcast spreader.

For deep or fast-moving water, media and water can be slurried and pumped down over the sediment via a tremie pipe (Figure 18.3). Since most sediments have relatively low shear strength, bulk cap materials may need to be placed in multiple lifts, which enables the sediment to consolidate and gain strength.

Geosynthetics of sufficient strength and permeability can improve the geotechnical stability of an active media cap as well as ease placement. Active media can be incorporated into a geotextile mat or a geosynthetic gabion. The geotextile mat and geosynthetic gabions also provide a bioturbation barrier, prevent the mixing of amendments with underlying sediments, allow a more uniform application of amendments, and reduce erosion (U.S. EPA 2013). The geotextile mats are covered with a minimum 15-cm conventional sand cap and/or, if needed, an armoring layer. Active media-filled geotextile mats can be deployed from a barge in several ways. Figure 18.4 shows a mat being unrolled from an attachment on an extended boom backhoe. A line and winch fixed on shore are helping guide placement. Geotextile mats can also be sewn side by side and slid off the side of a barge as was done at Harbor Point in upstate New York.

Geogrid gabions have also been used to deploy active media (Figure 18.5). The gabion is lined inside with a geotextile, filled with media and the geotextile heat welded shut. A pilot with 333 m^2 of active media gabion was deployed in 10 m deep water in the Hudson River (EPRI 2011).

Contractors are constantly devising new ways to deploy active media.

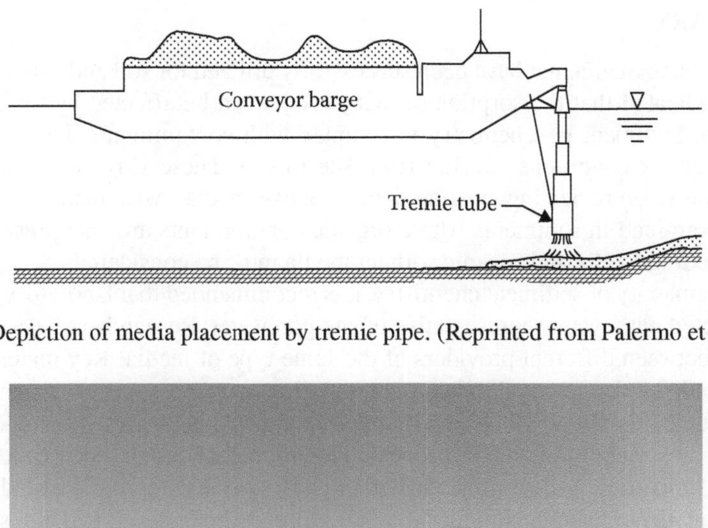

FIGURE 18.3 Depiction of media placement by tremie pipe. (Reprinted from Palermo et al., 1998.)

FIGURE 18.4 Media-filled geotextile mat being unrolled.

FIGURE 18.5 Gabion cages lined with geotextile and filled with media.

18.6 SUMMARY

There are many active media that have been successfully utilized for soil and groundwater remediation. Research indicated that the sorption of many media can be affected by the chemistry of the aqueous solution. Sediment geochemistry is complex with contaminants, ionic strength, organic matter, pH, and redox conditions varying from site to site. These variable conditions should be taken into account when reviewing existing data on active media. Also, metal-binding media have not been fully examined in sediments where organic contaminants are also present. The effect of mixing metal-binding media and organic sorbent media must be considered.

Due to the complexity of sediment chemistry, it is recommended that laboratory studies of active media be conducted with water representative of site porewater. Researchers have found variability in performance between different providers of the same type of media. Key material properties of virgin media should be documented in laboratory studies.

Laboratory studies should be followed by a pilot-scale field trial before moving on to a full-scale one. Spatiotemporal variability should be accounted for. Remedial design specifications should clearly define the media properties, as determined in earlier phases, so that appropriate media are sourced.

There are many deployment techniques that have already been developed and successful placement of active media on pilot-scale and full-scale projects. Geotechnical considerations and erosion protection should be considered. Geosynthetics of sufficient strength and permeability can improve the geotechnical stability of an active media cap.

Finally, future advances in material science and increases in our understanding of sediment geochemistry may result in new, cost-effective active media for use in metal-contaminated sediment remediation.

REFERENCES

AquaBlok, Ltd. 2006. Test Report #10, Bench-scale hydraulic conductivity as a function of product formulation and permeant salinity. http://www.aquablok.com/clientuploads/library/Technical_Product_Detail/10%20Test%20Report%20(color,%202-sides,%203%20pages).pdf. Accessed June 30, 2014.

Bereket, G.Z., A.Z. Aroğuz, and M.Z. Özel. 1997. Removal of Pb(II), Cd(II), Cu(II), and Zn(II) from aqueous solutions by adsorption on bentonite. *J. Colloid Sci.* 187, 338–343.

Corapcioglu, M.O. and C.P. Huang, 1987. The adsorption of heavy metals onto hydrous activated carbon. *Water Res.* 9(9), 1031–1044.

Cornelissen, G., M.E. Krusa, G.D. Breedveld, E. Eek, A.M.P. Oen, H.P.H. Arp, C. Raymond, G. Samuelsson, J.E. Hedman, O. Stokland, and J.S. Gunnarsson. 2011. Remediation of contaminated marine sediment using thin-layer capping with activated carbon—A field experiment in Trondheim Harbor, Norway. *Environ. Sci. Tech.* 45(14), 6110–6116.

Crannell, B.S., T.T. Eighmy, C. Wilson, D.D. Reible, and M. Yin. 2004. Pilot-scale reactive barrier technologies for containment of metal-contaminated sediments and dredged materials phosphate-based heavy metal stabilization and reactive barrier technologies for contaminated sediments and dredged materials. Final NOAA/CICEET Project Report. Durham: UNH.

Crittenden, B. and W.J., Thomas. 1998. *Adsorption Technology* and *Design*. Oxford, UK: Butterworth-Heinemann.

Electric Power Research Institute. 2011. Reactive cap shows potential to remediate coal tar. *EPRI Journal.* 32. http://mydocs.epri.com/docs/CorporateDocuments/EPRI_Journal/2011-Summer/1023458_InTheField.pdf

Gardner, K. 2004. In-situ treatment of PCBs in marine and freshwater sediments using colloidal zero-valent iron. A Final Report Submitted to the NOAA/UNH Cooperative Institute for Coastal and Estuarine Environmental Technology. Durham: UNH.

Gavaskar, A., S. Chattopadhyay, M. Hackworth, V. Lal, B. Sugiyama, and P. Randall. 2005. A reactive cap for contaminated sediments at the navy's dodge pond site. *Technology Innovation News Survey*. www.clu-in.org/products/tins/tinsone.cfm?id=47254881. Accessed June 28, 2014.

Ghosh, U., B.E. Reed, S. Kwon, J. Thomas, T. Bridges, D. Farrar, V. Magar, and L. Levine, 2008. Rational selection of tailored amendment mixtures and composites for in situ remediation of contaminated sediments. SERDP Project # ER 1491.

Gilmour, C.C., G.S. Riedel, G. Riedel, S. Kwon, R. Landis, S.S. Brown, C.A. Menzie, and U. Ghosh. 2013. Activated carbon mitigates mercury and methylmercury bioavailability in contaminated sediments. *Environ. Sci. Technol.* 47, 13001–13010.

Kanel, S. R., B. Manning, L. Charlet, and H. Choi. 2005. Removal of Arsenic (III) from groundwater by nanoscale zero valent iron. *Environ. Sci. Technol* 39, 1291–1298.

Knox, A., M. Paller, K. Dixon, D. Reible, J. Roberts, and I. Petrisor. 2011. Innovative in-situ remediation of contaminated sediments for simultaneous control of contamination and erosion. Final Report. ESTCP Project ER-1501.

Lampert, D. and D.D. Reible. 2009. An analytical modeling approach for evaluation of capping of contaminated sediments. *Soil Sediment Contam.* 18, 470–488.

Lui, J. and D.D. Reible. 2007. Observations of mercury fate and transport beneath a sediment cap. In: E.A. Foote and G.S. Durell (eds.). *Proceedings of Fourth International Conference on Remediation of Contaminated Sediments*. Columbus, OH: Battelle.

Mattigod, S.V., G.E. Fryxell, R. Skaggs, and K.E. Parker. 2004. Novel nanoporous sorbents for removal of mercury. *Prepr. Pap. Am. Chem. Soc. Div. Fuel Chem.* 49(1), 287.

McBride, M.B. 1994. *Environmental Chemistry of Soils.* New York: Oxford University Press.

Melton, J.S. 2007. Review of the apatite reactive barrier in the Anacostia river. In: E.A. Foote and G.S. Durell (eds.). *Proceedings of Fourth International Conference on Remediation of Contaminated Sediments.* Columbus, OH: Battelle.

Mimura, H. and T. Kanno. 1985. Distribution and fixation of cesium and strontium in zeolite A and chabazite. *J. Nuc. Sci. Technol.* 22(4), 284–291.

Nakano, A. and L. Li. 2009. Evaluation of lead retention for Japanese and US bentonites using selective sequential extraction and X-ray analysis. *Proceedings of ASTM Fourth International Symposium on Contaminated Sediments*. Dublin, OH: Trinity College.

Nolan, J., D. Babcock, and D. Matthews. 2013. Effective methylmercury control in Onondoga Lake via nitrate addition. In: A.K. Ballard and E.A. Stern (eds.). *Proceedings of Seventh International Conference on Remediation of Contaminated Sediments*. Columbus: Battelle.

Ojovan, M.I. and W.E. Lee. 2014. *An Introduction to Nuclear Waste Immobilisation.* Oxford, UK: Elsevier.

Palermo, M., S. Maynord, J. Miller, and D. Reible. 1998. *Guidance for In-Situ Subaqueous Capping of Contaminated Sediments*. EPA 905-B96-004. Chicago, CA: Great Lakes National Program Office. http://www.epa.gov/greatlakes/sediment/iscmain/

Paller, M.H. and A.S. Knox, 2010. Amendments for the in situ remediation of contaminated sediments: evaluation of potential environmental impacts. *Sci. Total Environ.* 408, 4894–4900.

Ponder, S.M., J.B. Durab, and T.E. Mallouk, 2000. Remediation of Cr(VI) and Pb(II) aqueous solutions using supported, nanoscale zero-valent iron. *Environ. Sci. Technol.* 34, 2564–2569.

Reed, B.E. 2001. Removal of heavy metals by activated carbon. In: A.K. SenGupta (ed.). *Environmental Separation of Heavy Metals: Engineering Processes*. Boca Raton, FL: CRC Press, pp. 205–225.

Reible, D.D. and D.J. Lampert. 2014. Capping for remediation of contaminated sediments. In: D.D. Reible (ed.). *Processes, Assessment and Remediation of Contaminated Sediments.* New York: Springer, pp. 325–363.

Reible, D.D., D.J. Lampert, D. Constant, R.D. Mulch Jr., and Y. Zhu. 2006. Active capping demonstration in the Anacostia River, Washington, D.C. *Remediat. J.* 17, 39–53.

Rosen, G., J. Leather, J. Kan, and Y.M. Arias-Thode. 2011. Ecotoxicological response of marine organisms to inorganic and organic sediment amendments in laboratory exposure. *Ecotoxicol. Environ. Safety*, 74(7), 1921–1930.

U.S. Environmental Protection Agency. 2005a. *Procedures for the Derivation of Equilibrium Partitioning Sediment Benchmarks (ESBs) for the Protection of Benthic Organisms: Metal Mixtures (Cadmium, Copper, Lead, Nickel, Silver and Zinc)*. EPA-600-R-02-011.

U.S. Environmental Protection Agency. 2005b. *Contaminated Sediment Remediation Guidance for Hazardous Waste Sites*. EPA-540-R-05-012; OSWER 9355.0-85. December.

U.S. Environmental Protection Agency. 2013. *Use of Amendments for In Situ Remediation at Superfund Sediment Sites*. OSWER Directive 9200.2-128FS, April.

Vlassopoulos, D., J. Goin, and M. Carey. 2013. Siderite amendment for in-situ pH control in hyperalkaline environments. *Mineral. Mag.* 77(5), 2421.

Wright, J., J.L. Conca, K.R. Rice, and B. Murphy. 2004. PIMS using Apatite II™: How it works to remediate soil and water. In: R.E. Hinchee and B. Alleman (eds.). *Proceedings of the Conference on Sustainable Range Management*. B4-05. Columbus, OH: Battelle.

Xu, D., X.L. Tan, C.L. Chen, and X.K. Wang. 2008. Adsorption of Pb(II) to MX-80 bentonite: Effect on pH, ionic strength, foreign ions and temperature. *Appl. Clay Sci.* 41, 37–46.

Index

D

E

An environmentally friendly book printed and bound in England by www.printondemand-worldwide.com

PEFC Certified

This product is from sustainably managed forests and controlled sources

www.pefc.org

PEFC/16-33-415

This book is made of chain-of-custody materials; FSC materials for the cover and PEFC materials for the text pages.